Proceedings from the Ninth International Zeolite Conference

Montreal
1992

Volume I

Edited by
R. von Ballmoos, J.B. Higgins and M.M.J. Treacy

Butterworth-Heinemann
Boston London Oxford Singapore Sydney Toronto Wellington

Recognizing the importance of preserving what has been written, it is the policy of Butterworth-Heinemann to have the books it publishes printed on acid-free paper, and we exert our best efforts to that end.

ISBN 0-7506-9488-2
LC 93-071336

Butterworth-Heinemann
80 Montvale Avenue
Stoneham, MA 02180

10 9 8 7 6 5 4 3 2 1

Printed in the United States of America.

Editors' Preface

These Proceedings of the Ninth International Zeolite Conference, held in Montreal, Canada, from July 5th to July 10th, 1992, continue the proud tradition of the previous zeolite conference proceedings. For any student or researcher entering the field of zeolites as well as for the experts who have spent most of their professional carreer in this fascinating area, these conference volumes constitute an invaluable tool. They not only represent a snapshot of the current state-of-the-art, they also serve as continuous sources of information and reference for years to come. Surprisingly, only a few textbooks on zeolites have been published (the out-of-print "Breck" is still an often consulted book). The conference proceedings therefore take on the role of regular textbooks and are regarded as such by zeolite researchers.

The contents of the two-volume Proceedings are compiled from the 150 papers presented at the 9th IZC. The five plenary lectures, given by experts in their respective areas, give an overview of recent progress as well as an outlook into the future of zeolite research. 94 of the full papers were presented in 30 sessions during the meeting. 50 additional full papers were presented as posters in one of the two two-day poster exhibits. We regard the discussion sections at the end of each of the major chapters as one of the most important and informative aspects of these Proceedings. The Program Committee and the Editors agreed early on that the Proceedings should be published after the 9th IZC and that they should contain the discussions of the full papers. The discussion sections add depth to the papers, put the papers in perspective, and give them a "reality check." Often, the real issues in a particular aspect of zeolite science are only brought out in the discussion. We therefore are of the opinion that delaying the publication of the proceedings by including the discussion was well worth while.

We have organized the two volumes into the classical segments of zeolite research. Volume One contains the following sections: synthesis (26 papers), structure (8 papers), characterization (27 papers), and modelling (4 papers). Volume Two consists of papers in adsorption (18 papers), solid state (4 papers), natural zeolites (3 papers), acid site characterization (7 papers), catalysis (42 papers), and application (5 papers). The complete Table of Contents, the full Author Index, and the extensive Subject Index are printed in both volumes.

Many people supported us in our efforts to edit these Proceedings and we owe them many thanks. Foremost, we want to acknowledge the major contributions of our secretaries, Ruth Ann Sitarski (Engelhard), Lillian Cole (Mobil R&D Corp.), and Peggy Bisher and Demorris Capers (NEC). The editors are grateful to their companies for the support of their work, and the permission to carry out such a major task. Mobil R&D Corporation provided ideal meeting facilities for the editors throughout the whole project. The computer graphics of the $AlPO_4$-8 framework on the cover were provided by Biosym. Our collaboration with Butterworth-Heinemann was essential for producing the high quality volumes. Chris Rawlins was responsible for the pre-work and Christine Swanson was in charge of production.

Roland von Ballmoos
John B. Higgins
Micheal M.J. Treacy

Princeton, New Jersey
December, 1992

Preface: Chairman, Executive Committee of the 9th IZC

In the twenty-five years since the 1st International Zeolite Conference (or Molecular Sieve Conference as it was before the name change at the 4th Conference) the science, economic impact and diversity of the subject matter has grown enormously. As this represents a generational time change it is interesting to look for the differences. There are twelve times as many contributions at this conference compared to the 1967 meeting - from 33 to over 380 papers and posters. The 120 participants in London in 1967 compare with about 600 in Montreal. Whereas the earlier participation was largely European and North American, today it is truly global.

In reviewing the research "hot spots" of the late 1960's we note surprising similarities with the present - possibly a reflection of the refractory nature of the science problems and that progress is made by repeated cycles of incremental advances. The idea bank expands slowly, but as new methods, materials, instruments and techniques emerge and develop, ideas are tested and retested until new constructions are possible. Faujasites and L are still of major interest, as is faujasite resynthesis, modelling and computational chemistry, synthesis mechanisms, phosphate effects and separations. Whereas catalysis was one amongst many equals in 1967, at least in a research sense, it is now the dominant interest. Other great changes include the massive expansion of molecular sieve chemical compositions and the new diversity of characterization methods and data handling techniques. (In 1967 the standard was the desk top adding machine and mainframe computers were new, expensive and of restricted availability.) From scientific curiosities, zeolitic materials have grown into a major branch of inorganic chemistry having an economic impact of great proportions. The 1967 domination of the field by a small group of research centers (Mobil, Union Carbide, Imperial College and the Russian Institutes) and manufacturers has given way to numerous excellent research centers and producers in most of the industrial and industrializing world. Research, invention and commercialization in molecular sieves is now a global, highly competitive activity. Detergent additives have become the largest tonnage zeolite application and zeolites have penetrated many catalyst and separations processes. Auto NOx catalysts seem to be just around the corner, as do catalysts for a wide range of chemical specialties.

The 9th IZC was a new experiment for the IZA in that we organized a meeting in a remote mode, which may be a useful precedent for future meetings. The organization was from the USA with the site organizer 4000 miles from Montreal, and most of the rest of the Organizing Committee 400 miles away in the New York-Philadelphia area. Using only the services of a well organized Conference Hotel and modern methods of communication and transportation, few problems were encountered and those were readily solved with few site visits. Particular thanks go to the members of the Organizing Committee who have toiled much in the service of the 9th International Zeolite Conference, and to their companies, without whose support they could not have effectively participated. Particular thanks go to the Proceedings editors who have continued to toil long after the Conference to bring you these volumes, and to the Treasurer who will be the last one off the hook when he closes the accounts and passes them to the IZA.

Particular thanks go to several people who gave their time or expertise to the success of the meeting. They include Dr. Howard Donahoe of the Nova Scotia Department of Mines and Energy for his excellent organization of the wonderful Field Trip; Phillipa O'Docherty at Concordia University for her assistance in organizing the Zeolite School; Jesse Vezza and Eric Baumgarten at Exxon for producing the Program and Circulars; Joyce Stoneking for handling registration lists and numerous enquiries from participants; and Jeannine von Ballmoos, Peggy Bisher, Anne Vaughan and Gail Smith for running the 9th IZC Registration and Conference Offices at the Queen Elizabeth Hotel, Montreal.

D. E. W. Vaughan

Executive Committee of the 9th IZC

Program Committee of the 9th IZC

Program Committee Chairs:

E. M. Flanigen
UOP Research, Tarrytown
New York, USA

D. H. Olson
Mobil Research and Development Corporation
Princeton, New Jersey, USA

Members:

C. G. Coe
Air Products and Chemicals, Inc.,
Allentown, Pennsylvania, USA

R. Gorte
University of Pennsylvania
Philadelphia, Pennsylvania, USA

J. B. Higgins
Mobil Research and Development Corporation
Princeton, New Jersey, USA

S. B. Rice
Exxon Production & Research Co.,
Houston, Texas, USA

M. M. J. Treacy
NEC Research Institute, Inc.,
Princeton, New Jersey, USA

R. von Ballmoos
Engelhard Corporation,
Iselin, New Jersey, USA

International Advisory Board to the 9th IZC

Chairpersons:

E.M. Flanigen	UOP Research, Tarrytown, NY
D.H. Olson	Mobil R&D Corporation, Princeton, NJ

Members:

Alberti, A.	Universita di Modena, Italy
Barthomeuf, D.	University of Paris, France
Bein, T.	University of New Mexico, USA
Bell, A.T.	University of California, Berkeley, USA
Bibby, D.M.	DSIR Chemistry Division, New Zealand
Chang, C.D.	Mobil R&D Corporation, Princeton, NJ
Chon, H.	Korea Advanced Institute of Science
Corma, A.	Inst. Technologia Quimica CSIC-UPV, Spain
Derouane, E.G.	University of Namur, Belgium
Dwyer, J.	University of Manchester, UK
Galabova, I.M.	Chemical Technological Institute, Bulgaria
Guth, J.L.	END de Chimie de Mulhouse, France
Hay, R.L.	University of California, Berkeley, USA
Inui, T.	Kyoto University, Japan
Ione, K.	Institute of Catalysis, Novosibirsk, Russia
Kaliaguine, S.	Laval University, Quebec, Canada
Kallo, D.	Hungarian Academy of Sciences, Hungary
Karge, H.	Fritz-Haber Institut der MPG, Germany
Kärger, J.	Leipzig Universität, Germany
Koizumi, H.	Ryokoku University, Japan
Kutz, N.A.	Amoco Chemicals, Naperville, IL, USA
Ma, E.	Worcester Polytechnic Inst., MA, USA
Martens, J.	Katholieke Universiteit Leuven, Belgium
McCusker, L.	Institut für Kristallographie, ETH, Switzerland
Mumpton, F.A.	State University College at Brockport, NY, USA
Newsam, J.M.	Biosym Technologies, San Diego, CA, USA
Ozin, G.	University of Toronto, Canada
Price, G.	Louisiana State University, USA
Ratnasamy, P.	National Chemical Laboratory, India
Rees, L.V.C.	Imperial College, London University, UK
Ruthven, D.M.	University of New Brunswick, Canada
Smith, J.V.	University of Chicago, IL, USA
Stucky, G.D.	University of California, Santa Barbara, USA
Suib, S.	University of Connecticut, Storrs, CT, USA
Subotic, B.	Rugier Boskovic Institute, Yugoslavia
Szostak, R.	Georgia Institute of Technology, USA
Thompson, R.	Worcester Polytechnic Inst., MA, USA
Tominaga, H.	University of Tokyo, Japan
van Bekkum, H.	Delft University of Technology, The Netherlands
van Santen, R.	Eindhoven University of Technology, The Netherlands
Weitkamp, J.	Universität Stuttgart, Germany
Whyte, T.E.	PQ Corporation, Conshohocken, PA, USA
Wilson, S.	UOP Technical Center, Tarrytown, NY, USA
Xu, R.	Jilin University, China
Yang, R.	State University of New York, Amherst, NY, USA

Financial Support

The Executive Committee gratefully acknowledges the receipt of financial support from:

AIR PRODUCTS AND CHEMICALS, INC.
ALLIED-SIGNAL INC.
AMOCO CORPORATION
BP AMERICA, INC.
BUTTERWORTH – HEINEMANN
CHEVRON RESEARCH & TECHNOLOGY COMPANY
CMP – CATIVCO INTERNATIONAL
DOW USA
ENGELHARD CORPORATION
ETHYL CORPORATION
EXXON RESEARCH & ENGINEERING COMPANY, USA
GRACE DAVISON CHEMICAL DIVISION
IFP/PROCATALYSE
MOBIL RESEARCH & DEVELOPMENT CORPORATION
NEC RESEARCH INSTITUTE, INC.
SHELL OIL COMPANY
TEXACO INC.
THE AMERICAN CHEMICAL SOCIETY – PETROLEUM RESEARCH FUND
THE PQ CORPORATION
UNION CARBIDE CHEMICALS AND PLASTICS COMPANY, INC.
UOP

CONTENTS

VOLUME I

Keynote Lecture

Plenary Lectures

Synthesis

Structure

Characterization

Modelling

VOLUME II

Adsorption

Solid State

Natural Zeolites

Acid Site Characterization

Catalysis

Applications

Keynote Lecture

OBSERVATIONS AND DESCRIPTIONS:

On an Unknown Mineral-Species Called Zeolites

by

Axel Fredrick Cronstedt

Translated from the Swedish by:

Dr. John. L. Schlenker* and Dr. Günter H. Kühl**

*Central Research Laboratory Mobil Research and Development Corporation

P. O. Box 1025 Princeton, New Jersey 08540

**Paulsboro Research Laboratory Mobil Research and Development Corporation

P. O. Box 480 Paulsboro, New Jersey 08066

INTRODUCTION

The following manuscript with the original title: "Om en obekant bärg art som kallas Zeolites" is a translation of the paper in which Axel Fredrick Cronstedt (1722 - 1765) reports the discovery of that subclass of the tectosilicate minerals which we now call zeolites.(1) The term zeolite, derived from the Greek words ζειν and λιθοσ ("boiling stone"), was coined by Cronstedt for use in this paper and appears for the first time in this manuscript. The entire body of zeolite literature, which now comprises thousands of papers, books, and patents (According to Breck, circa 7,000 papers and 2,000 United States patents dealing with zeolite science and technology had appeared by the year 1972.) had its beginning with the appearance of this paper in the year 1756.(2) This paper is cited in many monographs and reviews of the zeolite literature; however, it is to be doubted that most of those referencing said paper have ever attempted to read it. Although Breck and others have maintained that the particular zeolite species discovered by Cronstedt (see page 187) was stilbite, ($Na_2Ca_4(AlO_2)_{10}(SiO_2)_{26}\cdot 28H_2O$), this is by no means apparent from the Cronstedt text.(2,3)

An "Attempted translation of the original old-Swedish Paper of Cronstedt" was circulated in manuscript form by Dr. Göran Sumelius (Finland) at the Eighth International Zeolite Conference in Amsterdam and has since been published.(4) We believe the present translation represents a significant improvement over that of Dr. Sumelius who was in the difficult position of translating from one foreign tongue into another. The translation is difficult because of the changes in the Swedish language since the publication in 1756; words have disappeared and spelling reforms have altered the orthography. In addition, Cronstedt's geological terminology appears heavily influenced by the technical German vocabulary of his day and by Latin. Geological and chemical concepts regarding the origin and nature of mineral species have changed radically. In this connection, it should be noted that the Svappavari Copper Mine of Torneå, Lappland, specified as the source of one of Cronstedt's specimens, is now in Finland.

According to Dr. Solfrid Söderlund, Associate Curator of the Department of Royal Castles' Collection of the Swedish National Art Museum, the reputed portrait of Cronstedt accompanying the Sumelius translation is actually that of "his cousin Carl Johan Cronstedt. The portrait in

From the Proceedings of the 9th International Zeolite Conference, Montreal 1992, Eds. R. von Ballmoos et al.,

question belongs to the Gripsholm Collection (Grh 1861)...." Dr. Söderlund further states that there is no known likeness of Cronstedt apart from a "medallion made in 1870 [105 years subsequent to his death] as an exterior decoration of a building in Stockholm."(5) This medallion is reproduced by Weeks and Leicester with the notation that it was cast by Kjellberg and that the source of the likeness is unknown.(6)

The authors solicit possible improvements and criticisms of their translation from those more familiar with eighteenth century mineralogical and analytical terminology and with the Swedish of the same period. The biography of Cronstedt accompanying this translation appeared in The Resonator, a newsletter of the American Chemical Society, South [New] Jersey Section, and is reproduced with permission.

REFERENCES

1. Axel Fr. Cronstedt (1756) Rön och beskrifning om en obekant bärg art, som kallas Zeolites, Svenska Vetenskaps Akademiens Handlingar Stockholm, 17, pages 120 - 123.
2. Donald W. Breck (1984) Zeolite Molecular Sieves: Structure, Chemistry, and Use, Robert E. Krieger Publishing Company, Malabar, Florida, page 172.
3. F. A. Mumpton (1981) Reviews in Mineralogy, Volume 4, Mineralogy and Geology of Natural Zeolites, Mineralogical Society of America, BookCrafters, Inc., Chelsea, Michigan 48118, pages 2 and 15.
4. G. Sumelius (M. L. Occelli and H. E. Robson (editors)) (1992) Synthesis of Microporous Materials Volume I, Molecular Sieves, van Nordstrand Reinhold, 115 Fifth Avenue, New York, New York 10003, pages 1 - 5.
5. Private communication (attached) dated 28 February 1992 from Dr. Solfrid Söderlund, Associate Curator of the Department of Royal Castles' Collection (Slottssamlingarna) of the Swedish National Art Museum (Statens konstmuseer), Stockholm, Sweden.
6. M. E. Weeks and H. M. Leicester (1968) Discovery of the Elements, 7th Edition, The Journal of Chemical Education, Easton, Pennsylvania 18042, page 527.

OBSERVATIONS AND DESCRIPTIONS
On an Unknown Mineral-Species called Zeolites

by

Axel Fr. Cronstedt

Among the minerals I have collected and whose properties I have sought to investigate, the species described below exhibits so strange and peculiar a behavior in a blowpipe-flame that it is impossible to identify it with any known family; it can only be classified provisionally on the basis of the primary elements [oxides] of stones: lime-, silica-, clay-, and talc-earth.

I acquired my samples from two localities, namely, through a Mr. ADLERHEIM of the Svappavari Copper Mine at Torneå, Lappland and from a Mr. SCHINDEL, a mine owner from Iceland, but neither in sufficient amount to permit comparison with other mineral varieties by testing in a crucible.

Nevertheless, the following have been established with certainty:

1. The stone from Svappavari has a light yellow colour, while that from Iceland is white- in part translucent and in part opaque.

2. The two specimens differ somewhat in texture and in the morphology of their constituent particles. The Svappavari material is composed of spherulites and wavy fragments which consist of radiating pyramids whose apices terminate in a center. The Icelandic specimen is a mixture of a massive opaque chalk-like material and tangled concentric wedges. The radiating component seems to be a druse or the initial stage of a crystallization analogous to spar from limestone, and rock-crystal from quartz, garnet and schorl from their respective rocks, and none of these exhibits a regular habit because the space necessary for the formation of crystalline forms was lacking.

3. The hardness corresponds to that of normal spar or massive limestone earth and there is neither triboluminescence on striking with steel nor is there effervescence with acid.

4. In the blowpipe-flame they emit gas and puff up almost like borax. This behavior is more pronounced in the Svappavari specimen where the previously described pyramids separate and divide into a multitude of fine threads which, though still cohering, are initially transformed into a white spongy mass which subsequently fuses with a phosphorescent glow to a white glass which, on stronger heating, becomes clear, that is, colourless, when the air bubbles, which seem to contribute to the opacity, disappear.

6. When mixed with borax or fusible microcosmic salt(1) and heated in a flame, they dissolve slowly without effervescence.

7. Sal soda(2) exhibits a strong affinity for both materials and dissolves them vigorously. When placed on charcoal, the Svappavari, but not the Icelandic specimen, can be driven to a pure glass. Because the Svappavari material is found together with copper-lazurite, one sees in this case a trace of copper impurity which initially colours the glass an opaque reddish-brown. The green colour of the flame also indicates copper; however, this copper cannot be regarded as anything but a nonessential contaminant.

Judging from their behavior, these materials cannot be any known type of spar. Their visual appearance and hardness combined with the failure to effervesce in fusible salt and the easy fusibility with soda salt are characteristics which, based on my limited experience, contrast strongly with those of minerals having lime as their base.

Asbestos-like substances behave quite differently, and radiating schorl, which is still frequently confused with the former, fuses readily without addition of a fluxing agent, as does the entire schorl family generally, though not under similar conditions. Nevertheless, the behavior is quite reminiscent of the schorls so that the question arises as to whether these owe their easy fusibility to the incorporated metals or to the earth [oxide] which is their ultimate constituent. Their characteristics differ greatly from those of the silica earths proper, and it may be maintained with greater justification that these may be termed vitrescents, provided any class of stone or earth may be thusly designated.

A more copious supply of a pure material, which is not presently available, would enable us to ascertain whether these substances are of utility in any handicraft.

In order to avoid an awkward nomenclature which could give rise to various inconveniences and with the aim of precluding an inappropriate designation suggestive of some characteristic shared with other species, I have boldly, but I trust not too presumptuously, designated these minerals Zeolites.

submitted February 21.

1. $NaNH_4HPO_4 \cdot 4H_2O$.
2. Na_2CO_3.

[TRANSCRIPTION OF ORIGINAL TEXT]

RÖN och BESKRIFNING

Om en obekant bärg art som kallas Zeolites

Af

AXEL FR. CRONSTEDT

Ibland bärg-arter som jag samlat och til egenskaper sökt känna, har den arten, som nedanföre beskrifves, i elden vist et så besynnerligt och särskildt förhållande, at den icke kan föras under något af de bekanta slag, och allenast gissningsvis under någon Class, efter de utrönte grundämnen i stenar: Kalk- Kisel- Ler- och Talc-Jord.

Jag har fått honom ifrån tvänne ställen, nemligen genom Herr ADLERHEIM, från Svappavari Koppargrufva i Torneå Lappmark, och ifrån Herr Bärgjunkern SCHINDEL från Island; men icke til den myckenhet, at prof i degel mot andra bärgarter derpå kunnat göras.

Imedlertid är följande med säkerhet utrönt:

1. Färgen är i stenen från Svappavari ljusgul, och i den från Island, hvit, dels halfigenomskinlig, dels opac.

2. Texturen och Particlarnas figur är något skiljaktig i bägge profstenarne. Svappavari arten faller i klot och vågige drummer, ihopsatte af stråliga pyramider, som sluta sina spetsar i et centro: den Isländska är dels tät af particlar som krita, då den och är opac, dels fallande i förvirrade concentrique kilar. Den stråliga lärer vara en drus eller början til en Chrystallisation af den täta, likasom spaten af kalkstenen, Bärg-Crystallen af qvartsen, granaten och skörlen af deras särskilda stenslag, och visa alle desse icke någon regulier figur, där de icke haft fritt utrymme til anskjutning.

3. Hårdheten är lika med den, som vanlig spat eller tät kalksten äger, och gifver således icke eld för stålet: effervescerar icke eller med syrliga Spiritus.

4. I elden för blås-rör gäser och pöfer up nästan såsom Borax, hvilket Phenomene Svappavari arten bättre visar, då de omtalte pyramider skilja sig från hvarannan och dela sig i många fina trådar, som dock hålla ihop och förvandla sig först i en hvit Spongieuse massa, samt sedan smälta med et Phosphorescerande sken til et hvit glas, hvilket i starkare hetta drifvit, blifver klart, ofärgadt, sedan luftblåsorne försvunnit, som tyckas bidraga til opaciteten.

6. Af borax och sal fusibili microcosmica löses den up i elden utan gäsning, fast långsamt.

7. Sal soudæ har därtil stark attraction samt löser honom up med häftighet; och kan man på

kolet drifva Svappavari arten; men icke den Isländska, til rent glas. Som den förra finnes jämte Koppar-lasur, ser man vid detta tilfället spor af Kopparens medfölje, i det glaset först blir rödbrunt opact. Så visar sig ock Kopparen i den gröna lågan: icke destomindre kan man icke anse det för annat än accidentelt.

Efter et sådant förhållande at dömma, kann denna icka föras til de bekanta spat-arter, dit ögnamåttet och hårdheten synas vilja bringa honom, hälst som han ej gäser med sale fusibili, och smälter lätt med souda saltet, tvärt emot egenskaperne hos de stenar som hafva kalk til grund-ämne, efter min ringa förfarenhet.

Asbester hafva långt ifrån icke dylikt förhållande; och strålskörlen, som med den förra nog almänt confunderas, smälter väl, som hela skörlslägtet, tämeligen lätt för sig sjelf, men icke med dylika omständigheter. Imedlertid kommer denna närmast överens med skörlen, och står ännu åter at röna, om desse alle hafva sin lättsmällthet af inblandade metaller, eller af den jord, som är deras grund-ämne, hvilken då är vida skild från den stränga Kisel-jorden, och snarare än denna må kalas Vitrescens, om det namnet kan behållas för någon Clas af sten eller jord.

En myckenhet of ren art, som hos oss icke ännu är till fångs, skulle gifva lägenhet til försök at nyttja honom i någon handaslögd.

At förekomma vidlyftiga epitheta, som medföra hvarjehanda olägenheter, och namn, som innebära någon egenskap, gemensam med flera arter, tager man sig dristighet at kalla denna utan afsigt, Zeolites.

Inkommit d. 21 Febr.

THE LIFE AND TIMES OF AXEL FREDRICK CRONSTEDT*

The Swedish chemist and mineralogist Axel Fredrick Cronstedt, born December 23, 1722 at Ströppsta in the Parish of Turinge in Dalarna/Sweden, was the first to describe a natural zeolite. He also discovered nickel and popularized the use of the blow-pipe for the chemical analysis of ores.

Swedish history had been quite turbulent during the century preceeding Cronstedt's birth, and it is necessary to take the reader back about eighty years in order to gain an appreciation of the situation in Sweden at this time.

As a result of the thirty years war (1618 - 1648) and a war against Denmark (1657 - 60), Sweden had acquired the Danish Province of Skåne and the Norwegian provinces of Jämtland, Härjedalen and Bohuslän. At this time, the Baltic coast as far south as Riga was Swedish and Courland belonged to Poland. Russia had no access to the Baltic Sea. In Germany, Hither-Pomerania, the City of Wismar in Mecklenburg, and the bishoprics of Bremen and Verden were Swedish. Holstein-Gottorp was linked to Sweden by a dynastic alliance.

At the age of fifteen, Charles XII acceded to the Swedish throne in 1697. Russia, Denmark and Poland, natural enemies of Sweden, concluded a treaty in 1699. War started in 1700 with an overwhelming Swedish victory at Narva. Charles then turned toward Poland (1702 - 06), giving Peter the Great time to found St. Petersburg, build a Baltic fleet and reorganize the Russian army. When Charles began his Russian offensive the Russians used, for the first time, a scorched earth strategy diverting the Swedish armies from Moscow to the Ukraine where the Swedes suffered a crushing defeat at Poltava in 1709.

Peter occupied all the Swedish annexations on the Baltic coast and even Finland. Hanover occupied Bremen and Verden, and Denmark took Holstein-Gottorp. Charles governed Sweden from Turkish controlled Moldavia during this catastrophe, trying in vain to persuade the Turks to attack Russia. He returned in 1715, but was killed during the siege of the Norwegian fortress of Fredricksten on November 30, 1718. His death ended the "Age of Greatness." By the peace of Nystad (1721) Sweden formally resigned the Baltic provinces, part of Karelia and the city of Vyborg to Russia; Hither-Pomerania was ceeded to Prussia.

Charles XII had no successor. In 1718 his sister Ulrike Eleonora was elected queen. She abdicated in 1720 in favor of her husband, Frederick of Hesse (ruled 1720 - 51). This period saw a transition from Carolingian absolutism to a parliamentary form of government brought about by the complete failure of the policy of "greatness." Although the king was accorded a double vote in the Diet, he had no right to make decisions. In the Diet, decisions were made by the "Secret Committee" from which the peasants were excluded. The public sessions were reserved for speeches and debates. The members of the Diet were mainly state servants. Thus, the "Age of Freedom" (until 1722) was also an age of bureaucracy.

A dual party system evolved in this period, the parties were known as the "Night-caps" and the "Hats." The former were in power until 1738 leading a most careful foreign policy so as not to provoke Russia. In 1738 power passed to the "Hats," who made treaties with France in order to obtain support against Russia. War with Russia (1741 - 43) led to temporary Russian occupation of Finland and further loss of Finnish provinces west of St. Petersburg. A very expensive participation (1757 - 62) in the Seven Years War against Prussia led to inflation and financial collapse, and the regime of the "Hats" came to an end in 1756.

Axel Fredrick Cronstedt bore, as his middle name, the name of the ruling Swedish king, Frederick of Hesse. He spent his youth in Siggesta in Värmdö parish where he was taught at home by a private tutor. He began his studies at Upsala in 1738 or '39 without registering as a student. He was interested in mathematics and natural science, and one of his professors was the astronomer Anders Celsius, inventor of the Celsius thermometer. For financial reasons he was unable to continue his studies at Upsala, and changed over to the mining school. However, his studies were interrupted in the summer of 1742 when his father, General Cronstedt, required his assistance for the inspection of the fortifications of Hörningholm, Nyköping and Norrköping. In the following summer, when the peasants revolted and marched toward Stockholm and an attack of the Russian fleet was expected, his father was appointed commander-in-chief of Stockholm and the fortifications off Stockholm. Axel was given six weeks time to join his father in Vaxholm. In the fall, when war with Denmark was feared, the elder Cronstedt had to inspect the souther border fortifications, and, since no funds could be located to cover travel expenses for the regular fortifications secretary, Axel had to take his place. On his three month trip he acquired his first mine, the long closed Gåsebäcks anthracite mine near Hälsingborg.

As a mineralogist, Cronstedt made contributions of such significance as to fully justify the honorary title "Reformer of Mineralogy" bestowed on him in 1780 by the famous mineralogist and geologist Abraham Gottlob Werner of Freiberg/Saxony. Cronstedt's "An Attempt to Systematize Mineralogy and the Realm of Minerals" published in 1758, signaled a turning point in the history of mineralogy. Based on his chemical tests, Cronstedt concluded that the old classification into earth and rock ores was unfounded. He pointed out, for the first time, the difference between simple minerals on the one hand and rocks consisting of a composite of different minerals on the other. Cronstedt subsequently excluded the latter from the mineral system proper.

In his chemical studies, Cronstedt made extensive use of the blow-pipe. In a paper entitled: "Observations and Experiments on an Ore from Los (Hälsingland) Cobalt Mines," published in 1751, he reported the discovery of a new element which he named nickel. On February 21, 1756, he submitted a paper: "Observations and Descriptions on an Unknown Mineral-Species Called Zeolites," in which he described the properties of two mineral specimens and proposed the name "zeolites" for them.

Toward the end of his life Cronstedt settled near the Nisshytte smelter in Dalarne Province where he established a famous garden from whence source the cultivation of potatoes and other crops was introduced to the Dalarne region. Cronstedt died on August 19, 1765 at Nissshytte.

---

* G. H. Kühl (1987) The Resonator, American Chemical Society, South [New] Jersey Section, P. O. Box 67, Gibbstown, New Jersey 08027. Reprinted with permission.

Plenary Lectures

HOST/GUEST CHEMISTRY AND CATALYSIS IN ZEOLITES

JENS WEITKAMP

Institute of Chemical Technology I, University of Stuttgart, Pfaffenwaldring 55, D-W-7000 Stuttgart 80, Germany

ABSTRACT

With the advent of a broad assortment of synthetic zeolites with different pore widths and architecture of the intracrystalline voids, their use as hosts for the selective chemical conversion of guest molecules has become particularly attractive. Such chemical reactions can be conducted stoichiometrically or in a catalytic cycle. The progress in these fields is reviewed. Within host/guest chemistry, particular attention is payed to ship-in-the-bottle syntheses, photochemistry with guests in zeolite hosts, reptation strategies and pseudo high pressure cycloadditions. Modern trends in the field of shape selective catalysis are discussed. Guests deliberately introduced into zeolite hosts may act as catalysts, and this principle is mainly used for base catalysis and enantioselective catalysis. The role of zeolites as solid solvents is more and more recognized, e. g. in Wacker-Hoechst type catalysis. The similarities and dissimilarities between zeolite chemistry and modern supramolecular chemistry are briefly addressed.

INTRODUCTION AND SCOPE

Host/guest interactions have been known in chemistry for a long time. There are countless examples for guests accommodated inside the cavities of a host, such as (i) small hydrocarbons in ice (gas hydrates [1]), (ii) aromatics in Hofmann [2] or Werner [3] type compounds, (iii) n-alkanes or their linear derivatives in urea [4], (iv) neopentane in cadmium cyanide [5], (v) metals or noble gases in buckminsterfullerene [6-9], (vi) aromatics or dyes in cyclodextrins [10], (vii) cations or small molecules in crown ethers [11], or (viii) aromatics in calixarenes [12,13]. The guest may either be fully encapsulated (clathrates or inclusion compounds) or have free access to and egress from the cavities of the host. The binding energies in host/guest complexes are usually much lower than those of covalent chemical bonds. Necessarily, the host possesses cavities or channels with molecular dimensions and with concave surfaces as opposed to the convex surface of the guest.

In modern chemistry, two completely different classes of host systems experience a particularly rapid development, viz. organic molecular container compounds [14] and zeolites which are inorganic crystalline macromolecules. The term *supramolecular chemistry* has been coined for the selective synthesis of and host/guest interactions in hollow organic molecules [15-19]. While the similarities between modern supramolecular chemistry and zeolite chemistry and catalysis are

From the Proceedings of the 9th International Zeolite Conference, Montreal 1992, Eds. R. von Ballmoos et al.,

evident, principle dissimilarities might stem from the much lower molecular weight of organic container compounds. For instance, organic host/guest complexes can exist in solution or, perhaps, even in the gas phase. In spite of such principle differences, we believe that it is worthwhile for the zeolite community to watch the fascinating results of modern host/guest chemistry in hollow molecules.

It is the aim of the present article to review the role of zeolites as hosts inside which guest molecules can undergo selective chemical reactions. Host/guest interactions without chemical transformations, such as encapsulation of gases, will not be covered. Nor does this review, for reasons of space limitations, address the use of zeolites for the manufacture of advanced materials, such as conducting polymer chains, novel semiconductors, chemical sensors, non-linear optical materials etc. For these developments, the reader is referred to various recent articles [20-22].

Chemical reactions inside zeolites can, of course, proceed stoichiometrically or in a catalytic cycle, and sometimes, a clear-cut distinction is difficult. The catalytically active sites within the channels or cavities may be directly associated with the framework constituents (as, e. g., a Brönsted acid site in the H^+-form of a zeolite) or reside at guests deliberately introduced into the pores. A unique feature in zeolite catalysis is shape selectivity which can occur if the molecules, intermediates or transition states involved in a catalytic reaction are of the same size as the pore width. In this article, we will discuss selected developments in host/guest chemistry, shape selective catalysis and host/guest catalysis inside the pores of molecular sieve materials.

HOST/GUEST CHEMISTRY INSIDE ZEOLITES

Ship-in-the-bottle synthesis strategies

In the synthesis of bulky molecules from small reactants, the product may find itself incarcerated inside the zeolite cavities, in perfect analogy to the preparation of a ship-in-the-bottle (Figure 1). The entrapped guest may be a catalytically active species, such as a monomolecular metal complex or a cluster from metal atoms or metal carbonyls. In particular, encapsulating transition metal chelates inside the cavities of zeolites continues to be an attractive approach for heterogenizing selective catalysts which are traditionally used in a homogeneous liquid phase. Alternatively, the bulky guest could be a valuable product which is difficult to manufacture by conventional synthesis procedures, but which does form in high yields in the steric environment inside the host cavity. Liberation of such products requires destruction of the zeolite, e.g. with HF, hence such synthesis strategies will usually be uneconomic unless the product is of very high value and/or the zeolite host is very cheap.

Romanovsky et al. [23-26] were probably the first authors who described the ship-in-the-bottle synthesis of transition metal phthalocyanines in faujasite. Their technique (cf. Figure 2), i.e.,

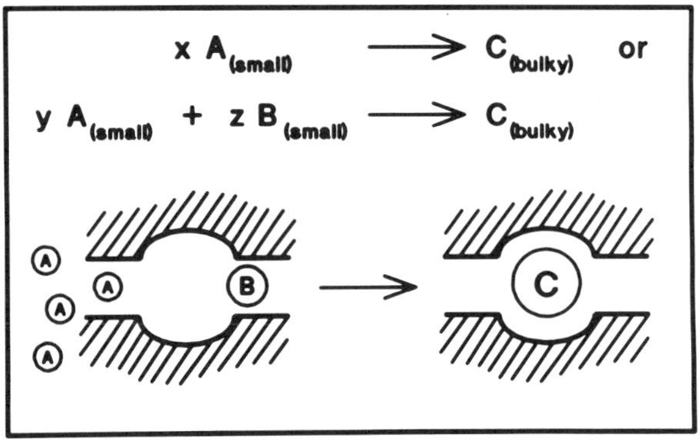

Figure 1. Schematic representation of ship-in-the-bottle syntheses.

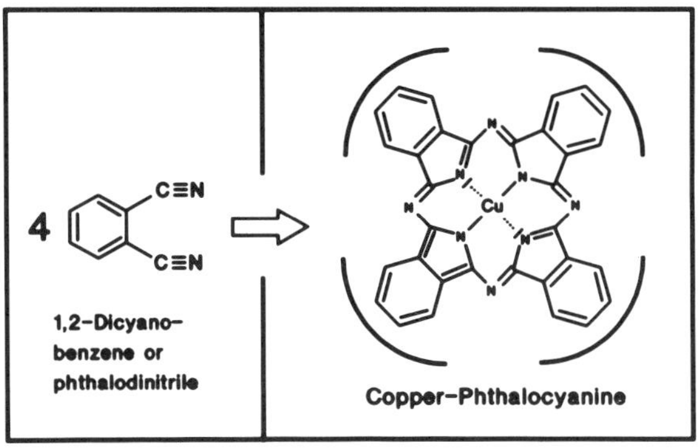

Figure 2. True ship-in-the-bottle synthesis of copper-phthalocyanine.

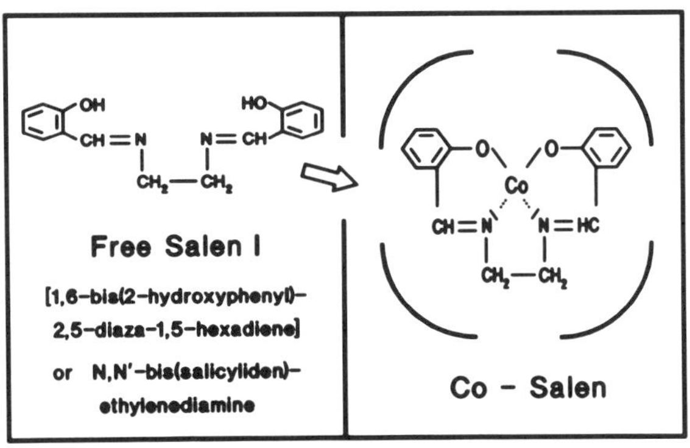

Figure 3. Ship-in-the-bottle synthesis of cobalt-salen via the flexible ligand method.

introduction of the metal via ion exchange or preadsorption of a labile metal complex, e.g., a carbonyl or a metallocene, followed by reaction with gaseous 1,2-dicyanobenzene at 250 to 350 °C has now been adopted by other groups to prepare phthalocyanine complexes of iron, cobalt, nickel and copper in the supercages of zeolite Y [27-30] and, more recently, in the channels of the 18-membered ring aluminophosphate VPI-5 [31]. An alternative method for encapsulating bulky complexes in zeolite cavities is hydrothermal synthesis in their presence [32]. Phthalocyanine complexes in faujasite are currently scrutinized as catalysts for shape selective and regioselective oxidations of hydrocarbons [33,34].

A slightly modified technique referred to as the *flexible ligand method* has been applied for the preparation of salen complexes (Figure 3) of manganese, cobalt, palladium and rhodium in faujasite. Free salen, i.e., the complete ligand system, is flexible enough to enter through the 12-membered ring window, but once the chelate with its square planar coordination is formed, it is entrapped in the cage [35]. Such salen complexes in faujasites have been shown to act as oxygen carriers mimicking hemoglobin [36], as regioselective oxidation catalysts [37-39] or as catalysts for selective hydrogenations [40].

Ship-in-the-bottle syntheses of organic products have been reported by *Anderson et al.* The conversion of 1-hexene in SAPO-34, which has the chabazite structure, under mild conditions (240 °C), led to adamantane and its alkyl derivatives, presumably via tetrahydrocyclopentadiene as intermediate [41]. The adamantanes are, of course, too bulky to escape through the 8-membered ring windows. They were liberated by dissolving the molecular sieve host in HF, followed by extraction in CH_2Cl_2. It was one of the authors' conclusions that the chabazite cage of SAPO-34 is an ideal environment for the formation of sphere-like adamantanes. In a similar study with strongly acidic HZSM-5, it was found [42] that azobenzene with small amounts of oxygen is mainly converted to phenazine, whereas in the absence of host/guest interactions, e.g. in $AlCl_3$, benzo(c)cinnoline is the predominating product (cf. Figure 4). Again, most of the phenazine product was retained inside the ZSM-5 pores. It was concluded that trans-azobenzene enters the pores and is protonated at the N=N-bond which enables homolytic cleavage of this bond. The resulting N-radicals, being guests in the ZSM-5 channels, are severely constrained in their mutual orientation, and this favors recombination towards dihydrophenazine. Oxygen now acts as hydrogen acceptor so that phenazine is ultimately formed.

Host/guest photochemistry in zeolites

This is a rapidly expanding field of modern host/guest chemistry, and the state of the art has recently been reviewed by *Ramamurthy* [43]. So far, the majority of investigations were dealing with either photochlorinations or photolysis of carbonyl compounds, and the zeolites employed as hosts

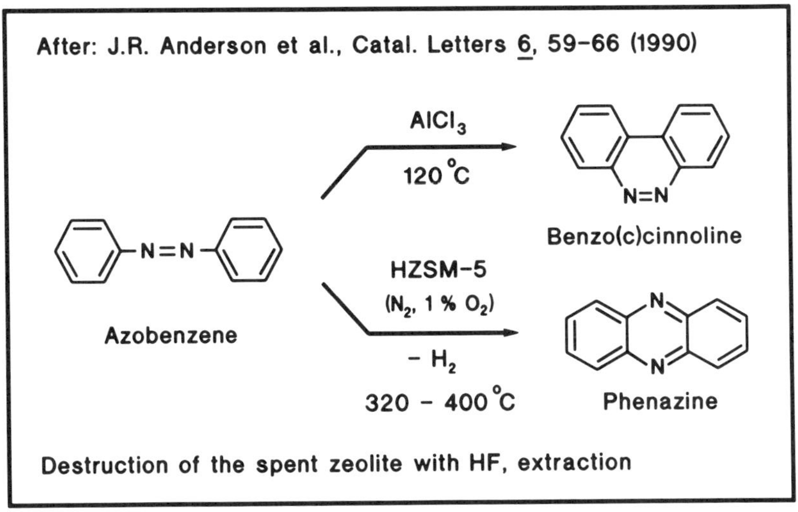

Figure 4. Selective synthesis of phenazine in HZSM-5 after [42].

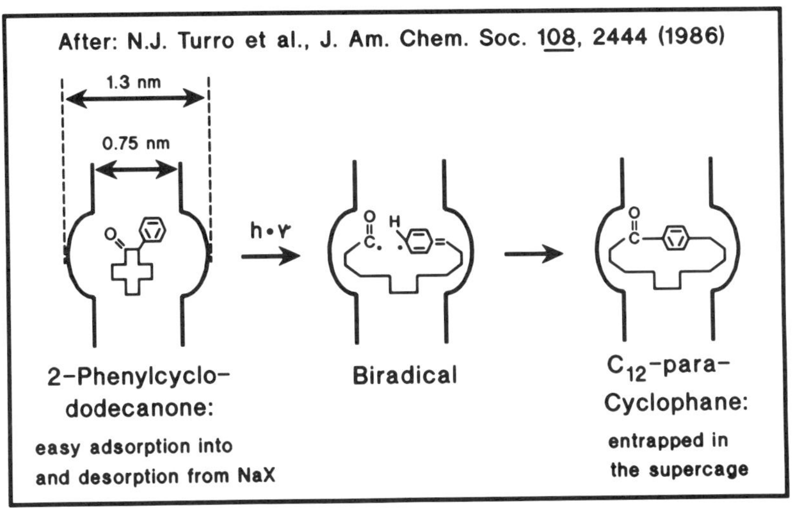

Figure 5. A photochemical ship-in-the-bottle synthesis after [44].

were mainly faujasites and pentasils.

Ship-in-the-bottle effects have been observed here as well. *Turro et al.* found [44,45] that 2-phenylcyclododecanone has access to and can escape from the supercages of NaX zeolite. Upon photolysis, a biradical forms which in the absence of host/guest effects would be expected to stabilize mainly into open chain enals. Within the supercage host, however, the guest prefers to stabilize into a para-cyclophane, probably via the biradical shown in Figure 5. The bulky C_{12}-para-cyclophane is incarcerated in the supercage. It can be recovered in good yields after destruction of the zeolite.

When the ring size of the starting phenylcycloalkanone is larger, the reactant is no longer able to enter the faujasite cage but adsorbs onto the external surface. Upon irradiation of, e.g., 2-phenylcycloheptadecanone, an open-chain biradical forms (Figure 6) and this is believed to slither into a supercage where it recombines and backbites itself to give the para-cyclophane which is, of course, again entrapped [44]. The essential step is the reptation of the biradical intermediate into a supercage, and in its generalized form, this principle can be referred to as a *reptation synthesis strategy* [45].

Norrish type I and type II reactions of ketones inside various zeolites have been extensively studied [46-50]. It has been suggested [48] that with suitable reactants, the Norrish type photochemical reactions can be employed as probes of the interior of zeolites. It has been found, for example, with alkanophenones as starting materials, that the selectivity of the Norrish type II reaction is strongly influenced by the zeolite host. Pertinent results from ref. [47] are summarized in Figure 7. The biradical intermediate can stabilize into two directions, viz. by cyclization (C) to a cyclobutanol or by ß-scission (elimination, E) to acetophenone and the corresponding 1-alkene. In the large-pore zeolite X, essentially the same selectivity is observed as in a liquid solvent like benzene. In pentasils, by contrast, cyclization is completely suppressed and this was interpreted in terms of a restricted rotational motion of the central σ bond in the biradical. In the language of the zeolite community, this is then an example for restricted transition state selectivity. Very interestingly, the cyclization towards phenylbutanol is favored in zeolite Na-beta, especially if the alkyl chain R is very long. It is no doubt worthwhile to look after this effect in more detail in future studies.

Reptation strategies in the thermo-catalytic degradation of waste polymers

Polyethylene and polypropylene are rich in hydrogen and this could make new processes attractive which convert such plastic wastes into valuable products. Various routes for upgrading plastic wastes have recently been envisaged, one of them being the thermo-catalytic degradation using zeolites [51-54]. The original macromolecules are too bulky to enter the pores of any zeolite. They are thermally degraded under mild conditions, and this could happen at the external surface of

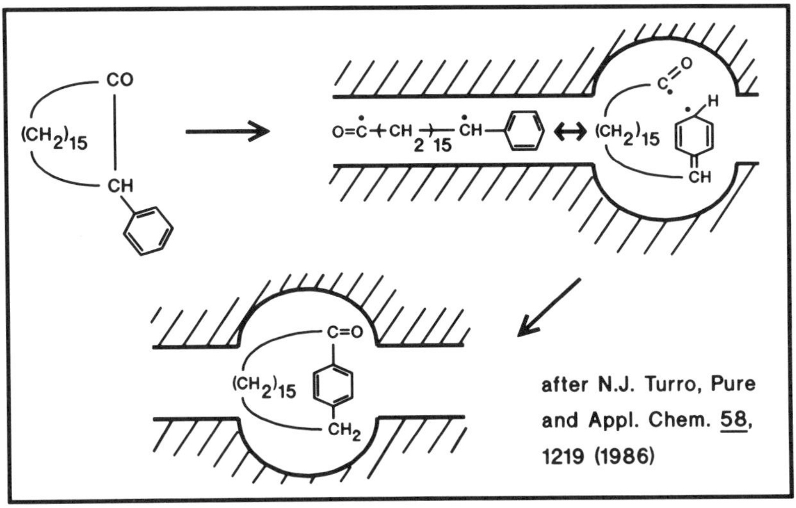

Figure 6. A photochemical reptation strategy after [45].

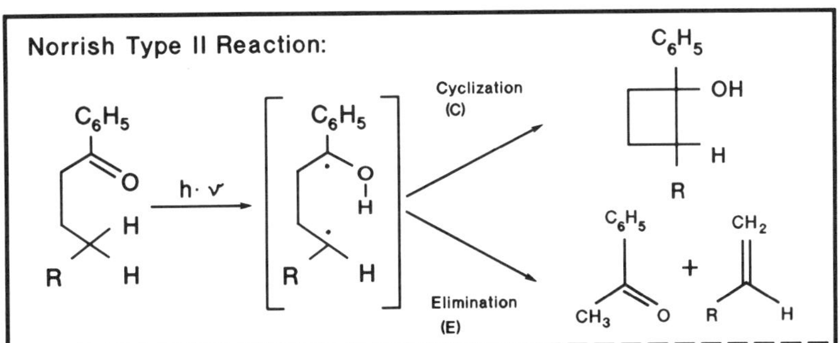

E/C in:	Benzene	NaX	CsX	NaZSM-5	NaZSM-11	Na-Beta
R= C_4H_9	1.9	1.5	1.9	>> 100	>> 100	0.62
R= $C_{14}H_{29}$	2.7	2.7	6.8	>> 100	>> 100	0.48

After V. Ramamurthy et al., J. Chem. Soc., Chem. Commun. 1213 (1989)

Figure 7. Selectivities in the Norrish type II reaction of alkanophenones in various zeolites after [47].

the zeolite crystals. Once the fragments are small and mobile enough, they will slither inside the pore system, and from then on the chemistry of degradation can be influenced catalytically, such as to direct the overall degradation process towards maximum selectivity for valuable products.

Pseudo high pressure Diels-Alder cycloadditions

So far, only relatively few studies are known on Diels-Alder reactions in zeolites [55-58]. It has been generally found that cycloaddition rates are considerably enhanced compared to a homogeneous gas or liquid phase, if the diene and dienophile react inside a zeolite with sufficiently spacious cages or channel intersections, such as faujasite, ZSM-20 or beta [55]. Even substrates which otherwise react very sluggishly enter into the Diels-Alder reaction if it is carried out in appropriate zeolites as hosts. Examples are the cycloadditions of cyclopentadiene and methylacrylate or the dimerization of furan [56,57]. Another marked effect of the zeolite is to increase the selectivity for the endo-adduct. It has been argued [56,58] that using a zeolite host has essentially the same effects as applying high pressures. This has been explained, in part, by a surface concentration of the reactants under Diels-Alder conditions. In addition, one could envisage steric effects, such as an easier formation of the sphere-like transition states under the influence of the concave wall of the cage.

Certain guests were discovered to assist in Diels-Alder cycloadditions. For example, addition of $ZnBr_2$ to CeY zeolite enabled a considerable reduction in the reaction time and led to 100 % selectivity for the endo-adduct in the reaction of cyclopentadiene with methylacrylate [57]. Likewise, Cu^+ inside the cages of some zeolites (introduced as Cu^{2+} followed by reduction with CO and a concommitant formation of Brönsted acid sites) has an enhancing effect on both the rate and the endo-selectivity [55,56]. Transfer of one electron from the dienophile to Cu^+ followed by reaction of the radical with the diene and back-donation of one electron to the adduct radical has been envisaged as the function of the transition metal guest.

It is evident from these few examples that very interesting host/guest interactions must occur during Diels-Alder cycloadditions in zeolites, and a more systematic understanding of these effects is desirable.

SHAPE SELECTIVE CATALYSIS IN ZEOLITES

Trends in conventional (primary) shape selecitivity

Shape selective catalysis is by now a well established field in zeolite science and technology.

Both its fundamentals [59-63] and its industrial applications [64] have been reviewed earlier. The usual classification into mass transfer effects (reactant and product shape selectivity) and intrinsic chemical effects (restricted transition state shape selectivity) has been very helpful for a rough understanding of the numerous reported examples.

In the more recent literature we perceive three major trends concerning shape selective catalysis: (i) While up to the 80's, the vast majority of shape selectivity effects was related to the conversion of aliphatic compounds or mononuclear aromatics in 10-membered ring zeolites, typically ZSM-5, more and more attempts are being made to exploit shape selective catalysis for the manufacture of bulkier molecules, e. g., with the naphthalene or biphenyl skeleton. This requires, of course, catalysts with somewhat more spacious pore systems, hence the increased interest in the puckered 12-membered ring zeolites. (ii) Shape selective catalysis progressively spreads into the manufacture of organic intermediates containing heteroatoms, especially oxygen or nitrogen. (iii) In some instances, attempts are being made to rationalize the origin of observed shape selectivity effects by independent information from, e. g., molecular graphics or molecular dynamics studies. Just a few examples for these trends will be enumerated in the subsequent paragraphs.

Lee et al. [65] investigated the acid catalyzed alkylation of biphenyl with propene in the liquid phase on various zeolite catalysts. The objective was to produce 4,4'-diisopropylbiphenyl (Figure 8), a starting material for valuable liquid crystals, as selectively as possible. Zeolites HY, HL and H-offretite gave very poor results, and so did H-mordenite with a n_{Si}/n_{Al}-ratio of 5. This latter zeolite, however, upon massive dealumination with HNO_3 up to n_{Si}/n_{Al} = 1300 enabled astonishingly high yields (> 70 %) of the desired para-para-isomer. In the authors' interpretation, the effect of dealumination is threefold: Firstly, it creates mesopores which are essential for an efficient mass transport of the bulky molecules; secondly, it leads to completely isolated Brönsted sites, and this was postulated to strongly favor alkylation in the desired para-positions; thirdly, it was argued that the dilution of acid sites slows down the undesired coke formation. Very recently, the acid catalyzed alkylation of biphenyl with propene has been proposed as a test reaction for probing the pore width of large pore zeolites and pillared clays [66].

Another example is isomerization of 1-methylnaphthalene into 2-methylnaphthalene. In the absence of shape selectivity, this acid catalyzed reaction is inevitably accompanied by undesired side reactions, especially by transalkylation to naphthalene and dimethylnaphthalenes, cracking and coking, and this was indeed so for zeolite HY [67] at the typical reaction temperature of 300 °C (at a much lower temperature of 180 °C, transalkylation was the only side reaction; interestingly, it exhibited an induction period, as reported earlier by *Karge et al.* [68] for transalkylation of ethylbenzene in acid faujasites). HZSM-5, on the other hand, was found to have much too narrow pores, there were severe mass transfer limitations for the bulky alkylnaphthalenes resulting in low conversions even at 400 °C. 12-Membered ring zeolites with intermediate pore width, especially

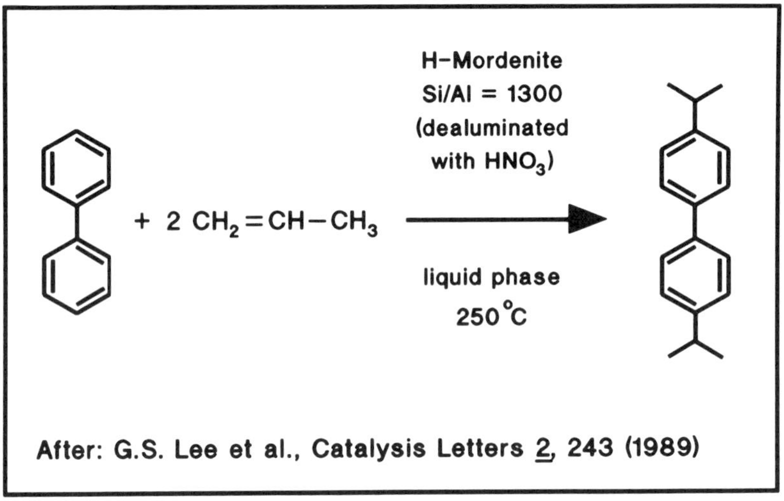

Figure 8. Shape selective alkylation of biphenyl in dealuminated mordenite after [65].

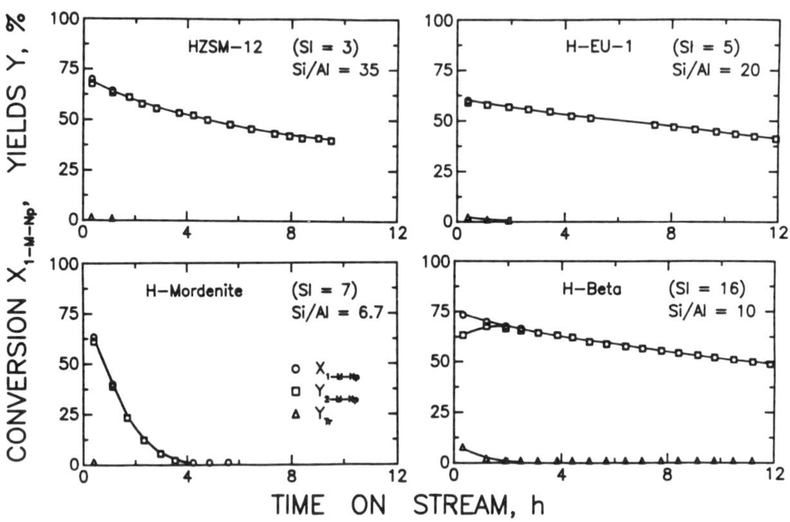

Figure 9. Shape selective isomerization of 1-methylnaphthalene in 12-membered ring zeolites at 300 °C after [67]. SI is the Spaciousness Index after [70].

HZSM-12, H-EU-1 and H-beta gave excellent results, as demonstrated in Figure 9: The yield for the desired 2-methylnaphthalene was essentially identical with the conversion, i. e., the selectivity was 100 % (except for the beginning of the experiments, when the external surface of the zeolites gave, presumably, some catalytic contribution). This is a typical example for restricted transition state shape selectivity with perfect suppression of the bimolecular transalkylation reaction which necessarily goes via a very bulky intermediate. Initially, the equilibrium conversion was attained (ca. 70 %). With HZSM-12, H-EU-1 and H-beta, deactivation was moderate, i. e., coke formation was retarded as shown in a separate study by in-situ IR experiments for HZSM-12 [69]. With H-mordenite only, there was a rapid deactivation due to coking. This rapid coking cannot be attributed to the unidimensional channel system of mordenite, otherwise one would expect the same rapid deactivation for HZSM-12 which has unidimensional pores as well. It is more likely that the coking propensity of mordenite stems from the higher aluminium content. Also given in Figure 9 are the Spaciousness Indices (SI) which characterize the effective pore width [70].

There has been much effort to selectively produce 2,6-dialkylnaphthalenes which can be oxidized into the very valuable 2,6-dicarboxylic acid. It has been reported [71], for instance, that mildly dealuminated H-mordenite (n_{Si}/n_{Al} = 13) enables the shape selective diisopropylation of naphthalene with propene or 2-propanol in the 2,6-positions with more than 50 % selectivity at ca. 70 % conversion at 240 °C in the liquid phase. Earlier, *Fraenkel et al.* had demonstrated that naphthalene or 2-methylnaphthalene can be selectively alkylated with methanol on HZSM-5 to give surprisingly high contents of the 2,6-isomer in the dimethylnaphthalene fraction [72,73]. These authors claimed that this shape selective alkylation occurs in so-called *"half-cavities" at the external surface*. This model, however, has to be discarded, as shown later [74]. Rather, all findings in the methylation of 2-methylnaphthalene at 400 °C in pentasils can be accounted for by conventional product shape selectivity inside the pores [74].

Even in the very open and spacious pore system of faujasite, shape selectivity effects were observed, viz. during the competitive hydrogenation of cyclohexene and cyclododecene in Rh/NaY [75], hydrocracking of hydropyrenes in NiW/ultrastable Y [76] and catalytic cracking of cis-decalin/trans-decalin mixtures in ultrastable Y [77].

Titanium silicalite-1(TS-1) discovered in the early 80's at Eniricerche S.p.A. [78] is of ever increasing attractiveness as a catalyst for selective oxidation reactions using hydrogen peroxide [79,80]. It is, moreover, active in the ammoximation of cyclohexanone with ammonia and hydrogen peroxide or molecular oxygen [81,82]. Much work has recently been directed towards an elucidation of the active sites in TS-1 and an understanding of the mechanisms involved during oxidations with H_2O_2 [83-85]. The shape selective properties of TS-1 in the synthesis of oxygen containing compounds from hydrocarbons were studied by Tatsumi et al. [86,87]. Their results are summarized in Figure 10. Both in the oxifunctionalization of alkanes and in the epoxidation of alkenes linear

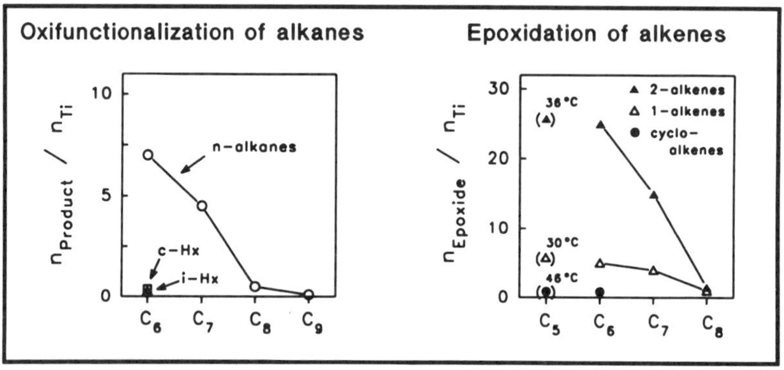

Figure 10. Reactant shape selectivity in oxidations with hydrogen peroxide on titanium silicalite-1 after [86,87].

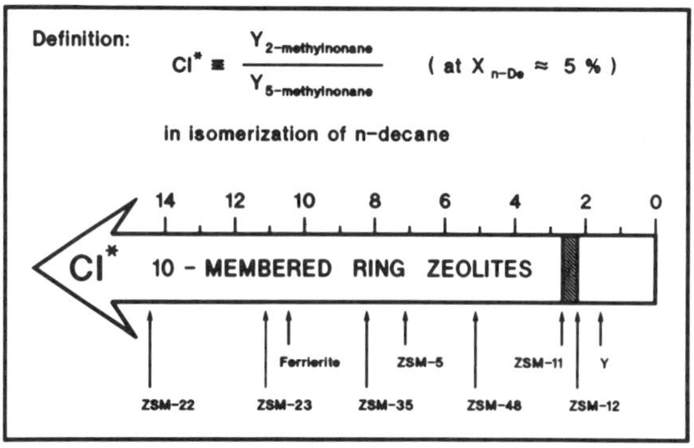

Figure 11. The refined or modified Constraint Index after [90].

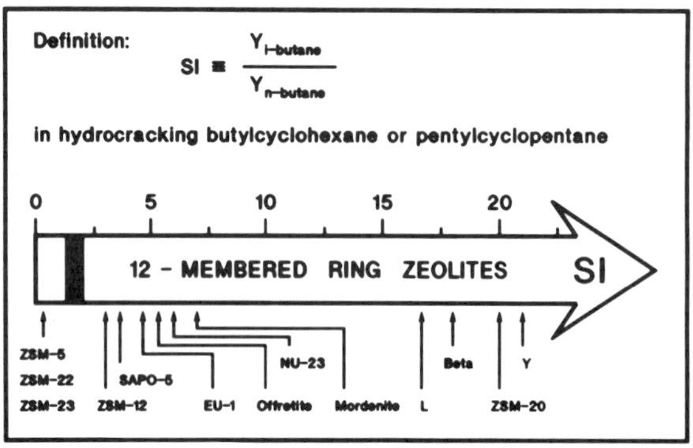

Figure 12. The Spaciousness Index after [91].

hydrocarbons reacted much faster than branched or cyclic ones and for linear homologues the rate decreased with increasing chain length. All these findings were rationalized in terms of diffusional resistances inside the pentasil pores.

Years ago, several shape selective reactions were identified which furnish information on the effective pore width or even the pore architecture of the zeolite catalyst [88,89]. In the meantime, two of these reactions, viz. isomerization of n-decane [90] and hydrocracking of butylcyclohexane [91] turned out to be particularly useful for probing the effective pore width of 10- and 12-membered ring zeolites, respectively. Both test reactions lead to quantitative results concerning the effective pore width expressed as the refined or modified Constraint Index CI^* and the Spaciousness Index SI, respectively. The exact definitions and pertinent values of these indices for common zeolites are given in Figures 11 and 12.

In both test reactions, the zeolite catalyst is used in a bifunctional form, i. e., the Brönsted acid form is loaded with small amounts of a hydrogenation/dehydrogenation metal, typically platinum or palladium. While the mechanisms of hydrocarbon conversion on such catalysts have been elucidated in considerable detail, two essential questions remained open: (i) The refined Constraint Index is based on the discovery that the rate of formation of 2-methylnonane (as compared to its positional isomers) from n-decane at low conversions is relatively high on medium pore zeolites while it is low on large pore zeolites (Figure 13). It has been shown that the latter result can easily be rationalized in terms of modern carbocation chemistry [93]. By contrast, the precise reasons for the catalytic behavior of 10-membered ring zeolites remained obscure. (ii) The Spaciousness Index was repeatedly found to be slighthly below 1 for all 10-membered ring zeolites (cf. Figure 12) including ZSM-5 and ZSM-11. For ZSM-12, SI is significantly larger, viz. around 3. This is difficult to understand as long as one assumes that in the pentasils, hydrocracking of butylcyclohexane occurs at the channel intersections: Structure models indicate that these intersections probably offer more space to reacting guests than the strongly puckered unidimensional channels in ZSM-12.

Helpful arguments are now coming from molecular graphics and molecular dynamics studies. In a recent investigation, *Martens et al.* [94] were led to the conclusion that n-decane isomerization on Pt/HZSM-22 is an example for pore mouth catalysis which could explain the very high CI^* value of ZSM-22. Molecular graphics simulations had suggested this pore mouth catalysis model.

June et al. [95] conducted molecular dynamics studies for n-butane and n-hexane at 300 K inside silicalite-1. The simulations showed, among other results, that the n-alkanes reside with 95 % or greater probability in the channel segments rather than in the intersection region. If one speculates that the same holds for molecules like butylcyclohexane at hydrocracking temperatures (ca. 500 K), at least for the alkyl side chain, this could explain the much lower SI values for the pentasils as compared to, e. g., ZSM-12.

Even if there is much speculation in both examples, they are meant to indicate that challenging ideas and new vistas might emerge from novel techniques and this could, in favorable cases, contribute to a rational understanding of shape selective catalysis.

Shape selectivity due to tip-on adsorption of reactant molecules

A new type of shape selectivity was proposed by *Gallezot et al.* [96] who studied the selective hydrogenation of cinnamaldehyde in the liquid phase. In the first step (cf. Figure 14), either 3-phenylpropanal or the desired cinnamyl alcohol forms, depending on whether hydrogen adds to the carbon-carbon double bond or to the carbonyl group. Upon further hydrogenation, 3-phenylpropanol results from both intermediates. Based on *Gallezot's* earlier experience [97,98], two samples of Pt/NaY zeolite were prepared via ion exchange with $Pt(NH_3)_4^{2+}$ followed by controlled thermal decomposition of the ammine complex, such as to arrive at Pt loadings of 11 and 14 wt.-% and metal particles with average diameters of 1 to 2 nm and ca. 5 nm, respectively, *inside* the zeolite pores (the existence of platinum particles inside faujasite with a diameter much larger than the dimensions of the supercage is by now well established [99,100]). The catalytic data summarized in Figure 14 show that the zeolite catalyst with 1 to 2 nm particles gave considerably better selectivities for cinnamaldehyde than the non-zeolitic reference catalyst; still much better results were obtained with the zeolite containing 5 nm platinum aggregates.

All this was interpreted by a novel type of shape selectivity due to enforced tip-on adsorption of cinnamaldehyde inside the faujasite pores: The 1 to 2 nm platinum clusters most probably reside in supercages, and a cinnamaldehyde molecule can only adsorb onto such a cluster through the 12-membered ring window, in an end-on mode, i. e., either via the carbonyl group or the phenyl ring. Under the mild reaction conditions, no hydrogenation of the phenyl ring occurs, so reduction of the carbonyl group is the only possible reaction coming from such an adsorption. More specifically, the tip-on adsorption prevents the carbon-carbon double bond from being hydrogenated. This is only true, however, if all platinum aggregates are big enough to match perfectly the hollow space of their host. A smaller cluster of, say, six or less metal atoms would leave enough space for a cinnamaldehyde molecule to enter the supercage and adsorb laterally through the carbon-carbon double bond, and this would result in the formation of 3-phenylpropanal. If the average particle size is 5 nm, the number of such very small clusters can be expected to be negligible, hence the excellent selectivities for cinnamyl alcohol. In a more recent investigation with both cinnamaldehyde and 3-methylcrotonaldehyde [101], the above model was essentially confirmed and slightly modified by taking into account electronic effects of the kind that adsorbed 3-phenylpropanol increases the charge density on the metal, thus lowering the rate of carbon-carbon double bond activation.

There is a large number of patents and scientific papers dealing with the aromatization of alkanes, especially n-hexane or n-heptane, on Pt in zeolite KL or BaL. Excellent selectvities for aromatization can be attained on such catalysts which makes them attractive for naphtha reforming. Several models have been advanced to account for the outstanding capability of platinum clusters residing in KL for aromatizing n-hexane. One of these models envisages a similar end-on, i. e., terminal adsorption of n-hexane on platinum in KL [102]. Such a tip-on adsorption of n-hexane is expected to result in either aromatization, via 1,6-ring closure, hydrogenolysis into n-pentane and methane or skeletal isomerization via 1,5-coordination. Nonterminal adsorption of n-hexane cannot lead to aromatization, hence one would expect a catalyst which gives high benzene selectivities to favor hydrogenolysis to n-pentane over the one to n-butane (terminal versus non-terminal hydrogenolysis). The latter selectivity can be quantitatively expressed by the Terminal Cracking Index [103], which is simply the yield ratio of n-pentane and n-butane. Indeed, a clear correlation between the aromatization selectivity and the Terminal Cracking Index could be established [102] for a large number of non-acidic Pt/SiO_2 and Pt/KL catalysts (Figure 15), and this is consistent with the terminal or tip-on adsorption model.

Secondary and inverse shape selectivity

All shape selectivity effects discussed so far result directly from the spatial constraints experienced by guest molecules, reaction intermediates or transition states inside the micropores of the zeolite host. For these effects the generic term *primary* (or conventional) *shape selectivity* has been introduced [104]. When there are mixtures of reactants or products, competitive adsorption is usually superimposed to shape selectivity, i. e., molecules with higher molecular weight and/or higher polarity are preferentially adsorbed at the catalytic sites. The net effect is still primary shape selectivity.

Besides, there are findings which cannot be accounted for in terms of primary shape selectivity. For example, *Namba et al.* found [105] that cracking of n-octane in HZSM-5 at 400 °C is retarded by added 2,2-dimethylbutane or 2,3-dimethylbutane, but not by 3-methylpentane. This cannot be understood in terms of simple competitive adsorption since 2,2- and 2,3-dimethylbutane are neither larger nor more polar than n-octane. Rather, the presence of these doubly branched alkanes slows down the diffusion of n-octane inside ZSM-5. Another example has been reported by *Santilli* and *Zones* [104]. They studied cracking of pure n-hexane, pure n-hexadecane and their 50:50 mixture in an acid form of SSZ-16, a small pore zeolite of unknown structure, at 330 °C. Under similar reaction conditions (residence time, partial pressure), the conversion with pure n-hexane was 50 % and with pure n-hexadecane 74 %. Surprisingly, when the n-hexane/n-hexadecane mixture was used, the n-hexane conversion was 62 % whereas the n-hexadecane conversion dropped to virtually zero. The results of both studies had in common that, in a zeolite with sufficiently narrow pores, one compound affected the reactivity of another one in a manner which is not observed on catalysts with

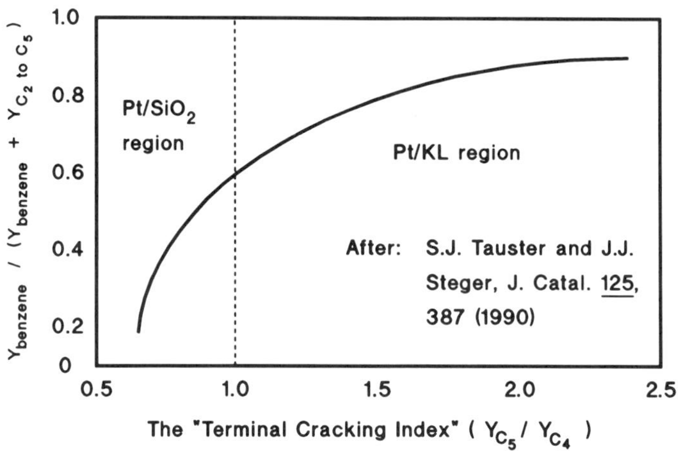

Figure 15. For a large number of Pt/SiO_2 and Pt/KL catalysts, there was a clear correlation between the aromatization selectivity and the Terminal Cracking Index during conversion of n-hexane after [102].

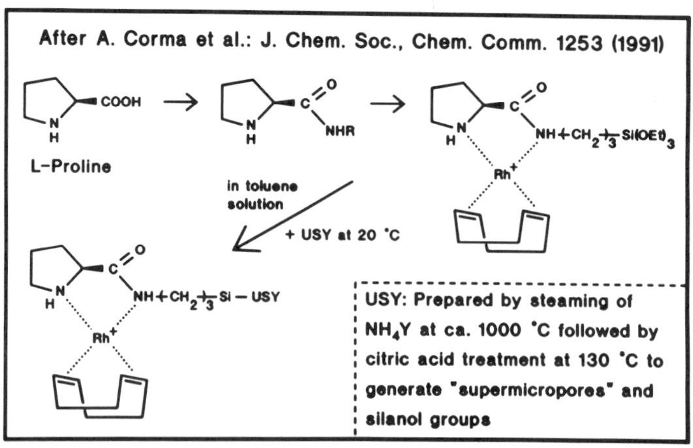

Figure 16. Anchoring of a chiral rhodium complex in ultrastable zeolite Y after [136].

sufficiently large pores. For such effects, Santilli and Zones coined the term *secondary shape selectivity*.

Inverse shape selectivity is an even younger concept introduced by *van Nordstrand et al.* [106]. It is based on the finding that hydrocracking of n-hexadecane on Pd/SAPO-5 and Pd/Al-SSZ-24, the aluminosilicate with the structure of $AlPO_4$-5 (AFI), at 60 to 80 % conversion gives significantly more 2,3-dimethylbutane in the hexane fraction than on both Pd/HZSM-5 and Pd/HY. The lack of the doubly branched hexane isomer in the product formed in ZSM-5 was convincingly accounted for by conventional product or transition state shape selectivity in agreement with earlier hydrocracking studies on Pt/HZSM-5 [92]; from the fact, however, that the two AFI based catalysts gave significantly more 2,3-dimethylbutane in the hexane fraction than Pd/HY, van Nordstrand et al. invoked the novel type of shape selectivity, favoring the formation of products with *larger* molecular dimensions. Unfortunately, there is some ambiguity in the authors' interpretation since the degree of secondary cracking over the various catalysts was not reported. It has been shown for hydrocracking of n-dodecane on faujasites [107] that considerable amounts of 2,3-dimethylbutane can occur in the hexane fraction even at moderate conversions, if there is secondary cracking on account of the acid and hydrogenation functions of the catalyst being out of balance. The above concept of inverse shape selectivity should not be confused with the so-called *reverse molecular-size selectivity*, proposed 25 years ago by *Venuto* and *Hamilton* [108], which would be classified in today's terminology as regular product shape selectivity.

The precise reasons for the occurrence of secondary and inverse shape selectivity are, at this time, not yet clear. It remains to be seen whether additional examples for these effects will be reported and whether these relatively young concepts will find a broader acceptance.

Test reactions for probing the location of noble metals

Much effort has been undertaken in the 80's to identify suitable test reactions for probing the effective pore width of molecular sieve catalysts [88,89]. In almost all these shape selective reactions, the zeolite is used in either a Brönsted acid or a bifunctional form comprising acid and hydrogenation activity. Examples for such reactions are (i) competitive cracking of n-hexane and 3-methylpentane [109], (ii) isomerization of m-xylene [110], (iii) disproportionation of ethylbenzene [111] or propylbenzene [112], (iv) isomerization and hydrocracking of n-decane [90] or n-heptadecane [113] and (v) hydrocracking of butylcyclohexane [70, 91].

Less attention has been paid in the past to using shape selective catalysis for probing the location of noble metals. *Dessau* employed the competitive hydrogenation of hexene-1 and 4,4-dimethylhexene-1 at 100 to 150 °C to discriminate between platinum inside the pores and at the external surface of ZSM-5 [114, 115]. The method has recently been improved by hydrogenating hexene-1 in admixture with an even bulkier olefin, viz. 2,4,4-trimethylpentene-1 [116]. Based on this test, octene-1/2,4,4-trimethylpentene-1 hydrogenation has been successfully used to trace the location of palladium introduced into ZSM-5 via solid state ion exchange [117].

CATALYSIS ON GUESTS IN ZEOLITE HOSTS

Base catalysis in zeolites

As pointed out earlier [118], base catalysis in zeolites is an underdeveloped field. So far, no examples for shape selective base catalysis in zeolites have been reported. The ultimate goal in this area is perhaps a zeolite analogue with anion exchange capacity which, in its OH^--form, could act as a true Brönsted base [118]. As long as such microporous materials are not available, research focusses on the accommodation of basic guests in conventional molecular sieves as hosts. Today, we perceive essentially four approaches to incorporate basic guests into zeolite hosts: (i) Exchange of alkali metal cations into the zeolite [119]. IR investigations with probe molecules such as pyrrole seem to indicate that, in such materials, framework oxygens adjacent to the alkali cations act as Lewis base sites [120]. (ii) Reaction of alkali metal vapors with alkali cation exchanged zeolites which results in two kinds of guests: *Ionic* clusters like Na_4^{3+}, Na_6^{5+}, K_3^{2+} or K_4^{3+} and alkali *metal* clusters. The alkali metal vapors are usually introduced by the method of *Jacobs et al.* [121, 122], i.e. impregnation of the dehydrated zeolite with aqueous alkali azide solution followed by controlled thermal decomposition of the azide [123, 124]. Alternatively, the zeolite can be directly exposed to metal vapors [124] or contacted with sodium solutions in liquid ammonia [125]. Rhodium carbonyl clusters as guests in zeolite Y pretreated with the NaN_3 method showed interesting non-Schulz-Flory-selectivities in the hydrogenation of carbon monoxide [126]. (iii) Basic oxide guests like cesium oxide have been generated by impregnation with the metal salt of an organic acid followed by thermal decomposition of the salt [127-130]. (iv) Finally, it has been suggested that NaX zeolite in which silicon has been substituted by germanium, acts as a basic catalyst [131, 132].

Stereoselective catalysis on chiral guests

So far, only very few examples for enantioselective catalysis in zeolites have been described. In much the same manner as in base catalysis (vide supra), the ultimate goal is a zeolite host with chiral channels or cavities. Polymorph A of zeolite beta could indeed represent such a structure which forms a pair of enantiomorphs [133], but all endeavors to synthesize this chiral host were, up till now, unsuccessful. Again, the alternative approach is accommodation of chiral guests inside the pores, and such guests can either act themselves as catalytically active sites or induce chirality on an adjacent site.

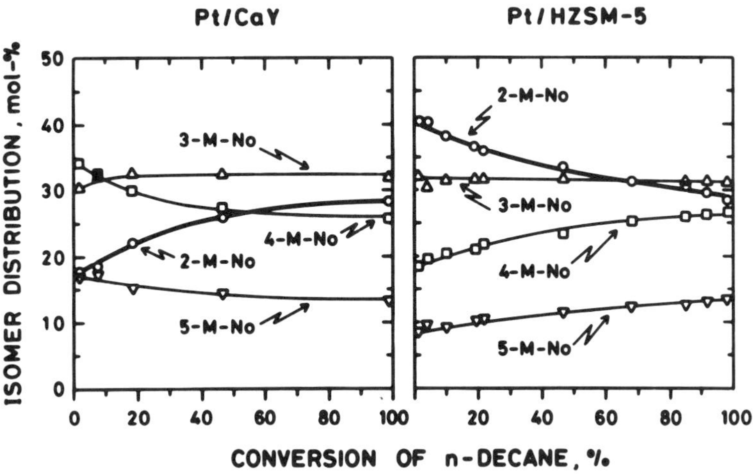

Figure 13. The relative rate of formation of 2-methylnonane from n-decane is the key information for the refined Constraint Index (data after [92]).

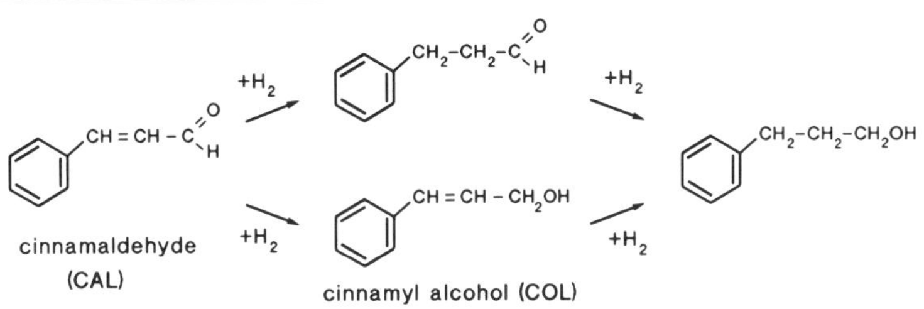

Catalyst	X_{CAL}[1]:	25 %	50 %
4 wt.-% Pt/Charcoal (1–1.5 nm)	S_{COL}:	33 %	55 %
11 wt.-% Pt/Y (1–2 nm, inside)	S_{COL}:	74 %	82 %
14 wt.-% Pt/Y (5 nm, inside)	S_{COL}:	97 %	97 %

[1]Liquid phase, T = 60°C, p_{H_2} = 40 bar

After P. Gallezot et al., Catal. Letters 5, 169 (1990)

Figure 14. Selective hydrogenation of cinnamaldehyde to cinnamyl alcohol on large platinum clusters inside zeolite NaY after [96].

For example, *Dessau* describes chiral hydrogenating catalysts [134] based on platinum in HZSM-5, whose acid sites were neutralized with an optically active ammonium ion, such as S-(−)-α-methylbenzylamine [118, 134, 135]. It is claimed [134] that the hydrogenation of prochiral substrates, e. g., acetophenone or 2-phenylbutene-2, on Pt/ZSM-5 modified by the chiral guest gives one form of 1-phenylethanol or 2-phenylbutane, respectively, in an enantiomeric excess (ee), however, no quantitative ee-values are reported. More recently, *Corma et al.* [136] gave a procedure for anchoring a chiral rhodium complex inside the mesopores of ultrastable zeolite Y. Their synthesis procedure is outlined in Figure 16. Liquid phase hydrogenation of α-acylaminocinnamates at 65 °C on these catalysts was reported to give the corresponding phenylalanine derivatives in an enantiomeric excess of 94 to 99 %, at 100 % conversion. Interestingly, chiral ship-in-the-bottle complexes incarcerated in the cavities of, say, faujasites, have not yet been examined as enantioselective catalysts.

Zeolites as solid solvents— Wacker-Hoechst type catalysis in faujasites

Heterogenizing the palladium/copper catalyzed Wacker-Hoechst technology for the production of acetaldehyde by direct oxidation of ethene continues to be an important goal of catalysis research. Many years ago, the principle feasibility of Wacker-Hoechst type catalysis on Pd/Cu exchanged faujasites was demonstrated [137, 138]. Only recently, however, a more systematic endeavor was undertaken by *Espeel et al.* [139] to arrive at an in-depth understanding of the chemistry involved in the zeolite catalyzed manufacture of acetaldehyde.

According to Espeel et al. [139], the analogies between the conventional Wacker-Hoechst chemistry and faujasite hosted Pd/Cu catalysis are very striking, and the same electron transfer chain mechanism seems to be operative in both systems, provided that the zeolite is selected and pretreated in the appropriate manner: The faujasite structure is mandatory, Pd/Cu catalysts based on chabazite, mordenite or zeolite L were found to be completely inactive. This points to an interesting role of the sphere-like supercage in faujasite as a host for the active site. Furthermore, the n_{Si}/n_{Al} ratio in the lattice seems to have a strong influence on the catalytic activity with maximum activity occurring at n_{Si}/n_{Al} = 2.5. To arrive at a good catalyst, Cu^{2+} should be first exchanged into NaY, followed by exchange of $[Pd(NH_3)_4]^{2+}$ into CuY. Optimum catalytic performance results if $n_{Cu^{2+}}/n_{Pd^{2+}}$ is equal to 2. Highly active and reasonably stable catalysts are obtained only if pretreatment is carried out under such conditions that an average of two ammonia ligands reside on each Pd^{2+}, probably because such entities can be readily coordinated by ethene. In this state, ethene is prone to nucleophilic attack by water, leading to a hydroxy-palladate adduct. Intramolecular hydrogen transfer and decomposition of the complex finally gives acetaldehyde. As a whole then, Wacker-Hoechst type catalysis in faujasites is an interesting example for the role of zeolites as solid solvents and, at the same time, for intricate chemical transformations of various guests present inside the faujasite supercage as hosts.

Catalysis on miscellaneous guests in zeolite hosts

There are probably numerous other examples for host/guest catalysis in zeolites, and only very few of these effects are currently understood in sufficient detail. Just two more classes of catalytic reactions will be briefly addressed in this paragraph, viz. halogenation of aromatics in zeolites containing alkali cations and activation of molecular hydrogen in zeolites lacking traditional hydrogenation/dehydrogenation catalyst components.

One of the most impressive examples for selective halogenations in zeolite catalysts has been reported by *Botta et al.* [140]. They investigated the chlorination of biphenyl in the liquid phase on zeolite L exchanged with various cations. Pertinent results are given in Figure 17: Under optimum conditions almost 100 % of the biphenyl reactant could be converted into the desired product, viz. 4,4'-dichlorobiphenyl. A systematic variation of the zeolite type, of the cation present in zeolite L and of the nature of the solvent led to a number of interesting conclusions: (i) The striking para-selectivity could *not* be interpreted in terms of conventional product shape selectivity; (ii) only catalysts based on zeolite L gave the high para-selectivities, obviously its particular pore width and channel architecture are required; (iii) the nature of cations present in the zeolite pores was also important, lithium or potassium being the best choice (by contrast, zeolite HL gave very poor results); (iv) finally, the nature of the solvent was found to have a marked influence on the para-selectivity, and the best effect was obtained with dichloromethane. While all these findings are, at this time, difficult to rationalize, they point to the existence of intricate and subtle host/guest interactions during the para-selective chlorination of biphenyl in zeolite L.

Molecular hydrogen has been repeatedly observed to influence the rate and/or selectivity of various reactions carried out on zeolite catalysts, even if there is no traditional hydrogenation/dehydrogenation component, such as Pt, Pd, Ni etc. in the system. At high temperatures in the order of 500 °C, molecular hydrogen may be activated by the Brönsted acid sites of the strength typically occurring in zeolites as pointed out by *Chen* and *Haag* [141]. Particularly pronounced effects of molecular hydrogen on acid catalyzed reactions such as alcohol dehydration [142] or ethylbenzene disproportionation [143] have been observed even at low temperatures around 200 °C when small clusters of silver were present. Iron impurities present in small amounts in many zeolites have also often been invoked to account for an unexpected hydrogenation activity. It has now been confirmed in a very systematic study by *Kanai et al.* [144] who used carefully prepared faujasites, mordenite and ZSM-5 in acid forms that both iron impurities and Brönsted acid sites can indeed act as active sites for hydrogenation of ethene. Typically,

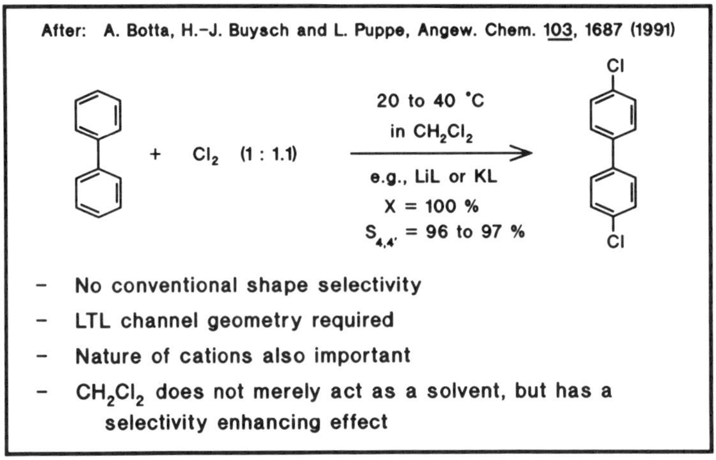

Figure 17. Selective para-para'-chlorination of biphenyl in zeolite L after [140].

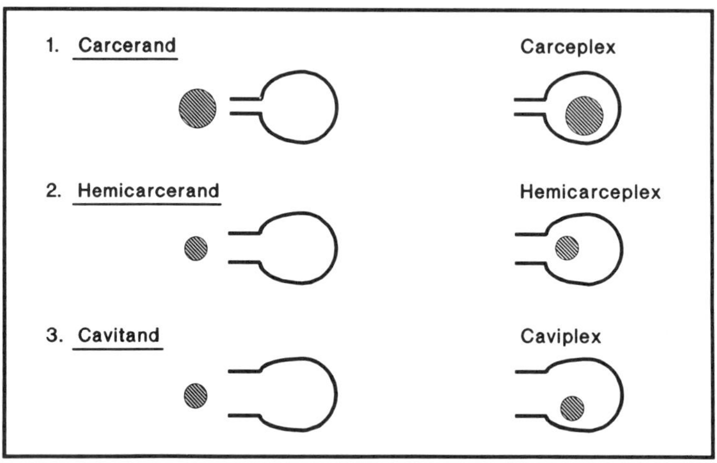

Figure 18. Cram's classification of molecular organic container compounds and the corresponding host/guest complexes after [14].

After D.J. Cram, M.E. Tanner and R. Thomas, Angew. Chem. 103, 1048 (1991) and D.J. Cram, Nature 356, 29 (1992)

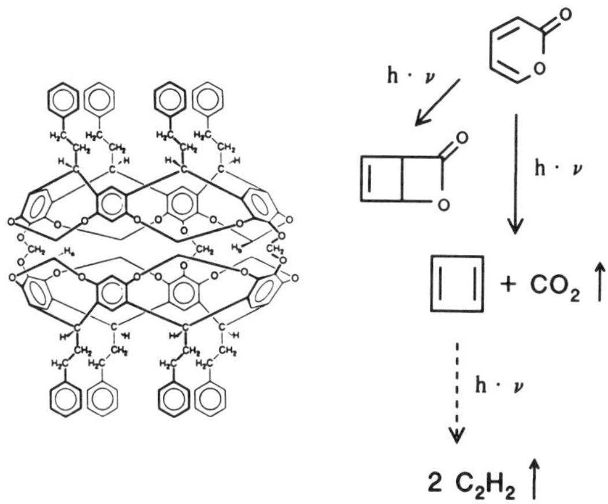

Figure 19. An example for a molecular container compound and a spectacular photochemical reaction performed inside its hollow after [14,19].

Table 1. A comparison of fundamental properties of organic container compounds and zeolite molecular sieves.

	Hollow Molecules	Zeolites
Building units	organic	inorganic
Molecular weight	low or medium	high
Existing in solution	yes	no
Width of portals	tunable	tunable
Shape and width of cavities	tunable	tunable
Properties of internal (concave) surfaces	tunable	tunable
Temperature stability	poor	very good
Cost	high	relatively low

hydrogenation on iron impurities is favored at low hydrogen pressures (< 5 kPa). At higher hydrogen pressures, the main site for ethene hydrogenation depends on the temperature: At about 450 °C, activation of H_2 at the acid sites predominates; at lower temperatures of about 350 °C, olefin hydrogenation is mainly effected by sodium cations. This is an interesting catalytic action of the Na^+ guests, and since many catalytic studies are performed in the temperature range up to 350 °C in the presence of some sodium cations, this type of host/guest catalysis could play a much more important role than hitherto known. Another class of guests which were found to catalyze alkene hydrogenation are ytterbium and europium introduced into zeolite Y by impregnation with the solutions of the metals in liquid ammonia followed by evacuation at 670 °C [145].

HOST/GUEST CHEMISTRY IN ZEOLITES AND SUPRAMOLECULAR CHEMISTRY

Today, the potential of zeolites as hosts inside which guest molecules can be converted stoichiometrically or catalytically, is being recognized by an ever increasing number of chemists. Some of them were referring to zeolites as *reaction vessels with molecular dimensions.* The same holds for organic molecular container compounds, so the question arises as to whether supramolecular chemistry and zeolite chemistry and catalysis are growing together or continue to be separate disciplines of chemistry. *Cram* [14, 19] emphasizes the dissimilarities between organic and zeolitic host/guest complexes. He even goes so far to claim that the internal space of organic container compounds should be looked upon as a new phase of matter [14]. It is attractive, therefore, to look briefly at his classification of host/guest complexes (Figure 18) and a pertinent example for chemical reactions inside molecular organic container compounds (Figure 19).

Depending on the relative size of the guest and the width of the portal leading to the hollow of the organic molecule, the latter is classified as a carcerand, hemicarcerand or cavitand (cf. Figure 18). The respective host/guest complexes are referred to as carceplexes, hemicarceplexes and caviplexes. At this point, the similarities with guests in zeolite hosts become evident: A carceplex immediately reminds a ship-in-the-bottle inside a zeolite cage. In a hemicarceplex, the guest can, in principle, pass in either direction, but with considerable steric hindrance, because its size resembles the width of the portal. This time, the analogue in zeolites is reactant or product shape selectivity due to mass transfer limitations. Finally, a caviplex reminds us a sufficiently small guest inside a spacious cage of a very open zeolite. Figure 19 gives an example for an organic container compound inside which a spectacular reaction could be conducted: α-Pyron was introduced into the hemicarcerand at elevated temperature. The resulting hemicarceplex was irradiated with light of a wavelength typically employed in photochemistry (200 to 250 nm). This resulted in the formation of cyclobutadiene which is otherwise difficult to synthesize, and carbon dioxide. The latter product escaped through the portals while cylobutadiene turned out to be remarkly stable inside the cavity of the molecular container. The whole synthesis immediately reminds host/guest photochemistry in zeolites, and the stabilization of an otherwise labile molecule could as well happen in a zeolite host. We anticipate that many similarities between supramolecular chemistry and host/guest chemistry in zeolites will be disclosed in the near future. In Table 1, some principle properties of both host systems are compared. While hollow organic molecules and molecular sieve zeolites do differ fundamentally in some respect, they have indeed many features in common. We advice the zeolite community to carefully observe the fascinating results which are currently emerging from supramolecular chemistry.

ACKNOWLEDGEMENTS

The author expresses his thanks to the following funding institutions which have, over many years, sponsored his research on zeolite molecular sieves: The German Science Foundation (Deutsche Forschungsgemeinschaft), the Federal Ministery of Research and Technology (Bundesministerium für Forschung und Technologie), Fonds der Chemischen Industrie and Max Buchner-Forschungsstiftung.

REFERENCES

1. Berecz, E. and Balla-Achs, M., *Gas Hydrates*, Studies in Inorganic Chemistry, Vol. 4, Elsevier, Amsterdam, Oxford, New York, 343 pp. (1983).
2. Iwamoto, T., in: *Inclusion Compounds*, Eds. J.L. Atwood, J.E.D. Davies and D.D. McNicol, Vol. 1, Academic Press, London, pp. 29-57 (1984).
3. Lipkowski, J., in: Ref. [2], Vol. 1, pp. 59-103 (1984).
4. Takemoto, K. and Sonoda, N., in: Ref. [2], Vol. 2, pp. 47-67 (1984).
5. Kitazawa, T., Nishikiori, S., Yamagishi, A., Kuroda, R. and Iwamoto, T., J. Chem. Soc., Chem. Commun., 413-415 (1992).
6. Kroto, H.W., Allaf, A.W., and Balm, S.P., Chem. Rev. **91**, 1213-1235 (1991).
7. Kroto, H.W., Angew. Chem. **104**, 113-133 (1992).
8. Schwarz, H., Angew. Chem. **104**, 301-305 (1992).
9. Fischer, J.E., Heiney, P.A. and Smith III, A.B., Acc. Chem. Res. **25**, 112-118 (1992).

10. Saenger, W., in: Ref. [2], Vol. 2, pp. 231-259 (1984).

11. Goldberg, I., in: Ref. [2], Vol. 2, pp. 261-335 (1984).

12. Böhmer, V. and McKervey, M.A., Chemie in unserer Zeit **25**, 195-207 (1991).

13. Andreetti, G.D., Ugozzoli, F., Ungaro, R. and Pochini, A., in: *Inclusion Compounds*, Eds. J.L. Atwood, J.E.D. Davies and D.D. McNicol, Vol. 4, Oxford University Press, Oxford, New York, Tokyo, pp. 64-125 (1991).

14. Cram, D.J., Nature **356**, 29-36 (1992).

15. Lehn, J.M., Pure Appl. Chem. **50**, 871-892 (1978).

16. Lehn, J.M., Science **227**, 849-856 (1985).

17. Vögtle, F. and Weber, E. (Eds.), *Host Guest Complex Chemistry, Macrocycles*, Springer-Verlag, Berlin, Heidelberg, New York, Tokyo, 421 pp. (1985).

18. Seel, C. and Vögtle, F., Angew. Chem. **104**, 542-563 (1992).

19. Cram, D.J., Tanner, M.E. and Thomas, R., Angew. Chem. **103**, 1048-1051 (1991).

20. Ozin, G.A., Kuperman, A. and Stein, A., Angew. Chem. **101**, 373-390 (1989).

21. Schulz-Ekloff, G., in: *Zeolite Chemistry and Catalysis*, Eds. P.A. Jacobs, N.I. Jaeger, L. Kubelkova and B. Wichterlova, Studies in Surface Science and Catalysis, Vol. 69, Elsevier, Amsterdam, Oxford, New York, Tokyo, pp. 65-78 (1991).

22. Rolison, D.R., Chem. Rev. **90**, 867-878 (1990).

23. Shpiro, E.S., Antoshin, G.V., Tkachenko, O.P., Gudkov, S.V., Romanovsky, B.V. and Minachev, Kh.M., in: *Structure and Reactivity of Modified Zeolites*, Eds. P.A. Jacobs, N.I. Jäger, P. Jiru, V.B. Kazansky and G. Schulz-Ekloff, Studies in Surface Science and Catalysis, Vol. 18, Elsevier, Amsterdam, Oxford, New York, Tokyo, pp. 31-39 (1984).

24. Romanovsky, B.V., in: *Proceedings of the 8th International Congress on Catalysis*, Vol. 4, Verlag Chemie, Weinheim, Deerfield Beach, Basel, pp. 657-667 (1984).

25. Romanovsky, B.V., in: *Proceedings of the International Symposium on Zeolite Catalysis*, Siófok, Hungary, May 13-16, 1985, Acta Physica et Chemica Szegediensis, pp. 215-222 (1985).

26. Romanovsky, B.V. and Gabrielov, A.G., in: *New Developments in Selective Oxidation by Heterogeneous Catalysis,* Eds. P. Ruiz and B. Delmon, Studies in Surface Science and Catalysis, Vol. 72, Elsevier, Amsterdam, Oxford, New York, Tokyo, pp. 443-452 (1992).

27. Meyer, G., Wöhrle, D., Mohl, M. and Schulz-Ekloff, G., Zeolites **4**, 30-34 (1984).

28. Herron, N., J. Coord. Chem. **19**, 25-38 (1988).

29. Balkus, Jr., K.J. and Ferraris, J.P., J. Phys. Chem. **94**, 8019-8020 (1990).

30. Mitchell, P.C.H., Chemistry and Industry, pp. 308-311, May 6 (1991).

31. Parton, R.F., Uytterhoeven, L. and Jacobs, P.A., in: *Heterogeneous Catalysis and Fine Chemicals II*, Eds. M. Guisnet et al., Studies in Surface Science and Catalysis, Vol. 59, Elsevier, Amsterdam, Oxford, New York, Tokyo, pp. 395-403 (1991).

32. Balkus, Jr., K.J., Kowalak, S., Ly, K.T. and Hargis, D.C., in: *Zeolite Chemistry and Catalysis*, Eds. P.A. Jacobs, N.I. Jäger, L. Kubelková and B. Wichterlová, Studies in Surface Science and Catalysis, Vol. 69, Elsevier, Amsterdam, Oxford, New York, Tokyo, pp. 93-99 (1991).

33. Herron, N., Stucky, G.D. and Tolman, C.A., J. Chem. Soc., Chem. Commun., 1521-1522 (1986).

34. Huybrechts, D.R.C., Parton, R.F. and Jacobs, P.A., in: *Chemistry of Microporous Crystals*, Eds. T. Inui, S. Namba and T. Tatsumi, Studies in Surface Science and Catalysis, Vol. 60, Kodansha, Tokyo, and Elsevier, Amsterdam, Oxford, New York, Tokyo, pp. 225-254 (1991).

35. Herron, N., Inorg. Chem. **25**, 4714-4717 (1986).

36. Herron, N., Chemtech **19**, 542-548 (1989).

37. Bowers, C. and Dutta, P.K., J. Catal. **122**, 271-279 (1990).

38. Balkus, Jr., K.J., Welch, A.A. and Gnade, B.E., Zeolites **10**, 722-729 (1990).

39. Bedioui, F., de Boysson, E., Devynck, J. and Balkus, Jr., K.J., J. Chem. Soc. Faraday Trans. **87**, 3831-3834 (1991).

40. Kowalak, S., Weiss, R.C. and Balkus, Jr., K.J., J. Chem. Soc., Chem. Commun., 57-58 (1991).

41. Anderson, J.R., Chang, Y.-F. and Western, R.J., J. Catal. **124**, 259-267 (1990).

42. Anderson, J.R., Chang, Y.-F. and Western, R.J., Catal. Letters **6**, 59-66 (1990).

43. Ramamurthy, V., in: *Photochemistry in Organized and Constrained Media*, Ed. V. Ramamurthy, VCH Publishers Inc., New York, Weinheim, Cambridge, pp. 429-493 (1991).

44. Lei, X.-G., Doubleday, Jr., C.E., Zimmt, M.B. and Turro, N.J., J. Am. Chem. Soc. **108**, 2444-2445 (1986).

45. Turro, N.J., Pure Appl. Chem. **58**, 1219-1228 (1986).

46. Corbin, D.R., Eaton, D.F. and Ramamurthy, V., J. Am. Chem. Soc. **110**, 4848-4849 (1988).

47. Ramamurthy, V., Corbin, D.R. and Eaton, D. F., J. Chem. Soc., Chem. Commun., 1213-1214 (1989).

48. Ramamurthy, V., Corbin, D.R. and Eaton, D.F., J. Org. Chem. **55**, 5269-5278 (1990).

49. Ramamurthy, V., Lei, X.-G., Turro, N.J., Thillairaj, J.L. and Scheffer, J.R., Tetrahedron Letters **32**, 7675-7678 (1991).

50. De Keukeleire, D. and He, S.-L., J. Chem. Soc., Chem. Commun., 419-420 (1992).

51. Saito, K., Fukuda, T., Suzuki, S., Sato, H. and Hirota, T., Europ. Patent Appl. 276 081 (Mobil Oil Corp, Research Devel. Corp. Japan, Fuji Tech K.K.), July 27 (1988).

52. Fukuda, T., Saito, K., Suzuki, S., Sato, H. and Hirota, T., US Patent 4 851 601 (Mobil Oil Corp.), July 25 (1989).

53. Ishihara, Y., Nanbu, H., Saido, K., Ikemura, T. and Takesue, T., Bull. Chem. Soc. Japan **64**, 3585-3592 (1991).

54. Mordi, R.C., Fields, R. and Dwyer, J., J. Chem. Soc., Chem. Commun., 374-375 (1992) and 802 (1992).

55. Dessau, R.M., J. Chem. Soc., Chem. Commun., 1167-1168 (1986).

56. Ipaktschi, J., Z. Naturforsch. **41b**, 496-498 (1986).

57. Narayana Murthy, Y.V.S. and Pillai, C.N., Synth. Commun. **21**, 783-791 (1991).

58. Pillai, C.N., Indian J. Technol. **30**, 59-63 (1992).

59. Csicsery, S.M., in: *Zeolite Chemistry and Catalysis*, Ed. J.A. Rabo, ACS Monograph 171, American Chemical Society, Washington, D.C., pp. 680-713 (1976).

60. Weisz, P.B., Pure Appl. Chem. **52**, 2091-2103 (1980).

61. Csicsery, S.M., Zeolites **4**, 202-213 (1984).

62. Weitkamp, J., Ernst, S., Dauns, H. and Gallei, E., Chem.-Ing.-Tech. **58**, 623-632 (1986).

63. Csicsery, S.M., Pure Appl. Chem. **58**, 841-856 (1986).

64. Chen, N.Y., Garwood, W.E. and Dwyer, F.G., *Shape Selective Catalysis in Industrial Applications*, Marcel Dekker, Inc., New York, Basel, 303 pp. (1989).

65. Lee, G.S., Maj, JJ., Rocke, S.C. and Garcés, J.M., Catal. Letters **2**, 243-248 (1989).

66. Butruille, J.-R. and Pinnavaia, T.J., Catal. Letters **12**, 187-192 (1992).

67. Weitkamp, J. and Neuber, M., in: *Chemistry of Microporous Crystals*, Eds. T. Inui, S. Namba and T. Tatsumi, Studies in Surface Science and Catalysis, Vol. 60, Kodansha, Tokyo, and Elsevier, Amsterdam, Oxford, New York, Tokyo, pp. 291-301 (1991).

68. Karge, H.G., Hatada, K., Zhang, Y. and Fiedorow, R., Zeolites **3**, 13-21 (1983).

69. Neuber, M., Karge, H.G. and Weitkamp, J., Catal. Today **3**, 11-22 (1987).

70. Weitkamp, J., Ernst, S. and Chen, C.Y., in: *Proceedings of the 8th International Zeolite Conference*, Eds. P.A. Jacobs and R.A. van Santen, Part B, Elsevier, Amsterdam, Oxford, New York, Tokyo, pp. 1115-1129 (1989).

71. Katayama, A., Toba, M., Takeuchi, G., Mizukami, F., Niwa, S. and Mitamura, S., J. Chem. Soc., Chem. Commun., 39-40 (1991).

72. Fraenkel, D., Cherniavsky, M. and Levy, M., in: *Proceedings of the 8th International Congress on Catalysis*, Vol. 4, Verlag Chemie, Weinheim, Deerfield Beach, Basel, pp. 545-554 (1984).

73. Fraenkel, D., Cherniavsky, M., Ittah, B. and Levy, M., J. Catal. **101**, 273-283 (1986).

74. Neuber, M. and Weitkamp, J., in: *Zeolites for the Nineties, Recent Research Reports Presented at the 8th International Zeolite Conference*, Eds. J.C. Jansen, L. Moscou and M.F.M. Post, Amsterdam, pp. 425-426 (1989).

75. Yamaguchi, I., Joh, T. and Takahashi, S., J. Chem. Soc., Chem. Commun., 1412-1413 (1986).

76. Haynes, Jr., H.W., Parcher, J.F. and Helmer, N.E., Ind. Eng. Chem., Process Des. Devel. **22**, 401-409 (1983).

77. Mostad, H.B., Riis, T.U. and Ellestad, O.H., Appl. Catal. **58**, 105-117 (1990).

78. Perego, G., Bellussi, G., Corno C., Taramasso, M., Buonomo, F. and Esposito, A., in: *Proceedings of the 7th International Zeolite Conference*, Eds. Y. Murakami, A. Iijima and J.W. Ward, Kodansha, Tokyo, and Elsevier, Amsterdam, Oxford, New York, Tokyo, pp. 129-136 (1986).

79. Notari, B., in: *Chemistry of Microporous Crystals*, Eds. T. Inui, S. Namba and T. Tatsumi, Studies in Surface Science and Catalysis, Vol. 60, Kodansha, Tokyo, and Elsevier, Amsterdam, Oxford, New York, Tokyo, pp. 343-352 (1991).

80. Bellussi, G., Carati, A., Clerici, M.G., Maddinelli, G. and Millini, R., J. Catal. **133**, 220-230 (1992).

81. Dreoni, D.P., Pinelli, D., Trifirò, F., Tvaruzkova, Z., Habersberger, H. and Jiru, P., Catal. Letters **11**, 285-294 (1991).

82. Dreoni, D.P., Pinelli, D. and Trifirò, F., J. Molec. Catal. **69**, 171-190 (1991).

83. Notari, B., in: *Innovation in Zeolite Materials Science*, Eds. P.J. Grobet, W.J. Mortier, E.F. Vansant and G. Schulz-Ekloff, Studies in Surface Science and Catalysis, Vol. 37, Elsevier, Amsterdam, Oxford, New York, Tokyo, pp. 413-425 (1988).

84. Huybrechts, D.R.C., Vaesen, I., Li, H.X. and Jacobs, P.A., Catal. Letters **8**, 237-244 (1991).

85. Huybrechts, D.R.C., Buskens, P.L. and Jacobs, P.A., J. Molec. Catal. **71**, 129-147 (1992).

86. Tatsumi, T., Nakamura, M., Negishi, S. and Tominaga, H., J. Chem. Soc., Chem. Commun., 476-477 (1990).

87. Tatsumi, T., Nakamura, M., Yuasa, K. and Tominaga, H., Chem. Letters, 297-298 (1990).

88. Jacobs, P.A. and Martens, J.A., in: *Proceedings of the 7th International Zeolite Conference*, Eds. Y. Murakami, A. Iijima and J.W. Ward, Kodansha, Tokyo, and Elsevier, Amsterdam, Oxford, New York, Tokyo, pp. 23-32 (1986).

89. Weitkamp, J. and Ernst, S., in: *Catalysis 1987*, Ed. J.W. Ward, Studies in Surface Science and Catalysis, Vol. 38, Elsevier, Amsterdam, Oxford, New York, Tokyo, pp. 367-382 (1988).

90. Martens, J.A., Tielen, M., Jacobs, P.A. and Weitkamp, J., Zeolites **4**, 98-107 (1984).

91. Weitkamp, J., Ernst, S. and Kumar, R., Appl. Catal. **27**, 207-210 (1986).

92. Weitkamp, J., Jacobs, P.A. and Martens, J.A., Appl. Catal. **8**, 123-141 (1983).

93. Weitkamp, J., Ind. Eng. Chem., Prod. Res. Devel. **21**, 550-558 (1982).

94. Martens, J.A., Parton, R., Uytterhoeven, L., Jacobs, P.A. and Froment, G.F., Appl. Catal. **76**, 95-116 (1991).

95. June, R.L., Bell, A.T. and Theodoru, D.N., J. Phys. Chem. **96**, 1051-1060 (1992).

96. Gallezot, P., Giroir-Fendler, A. and Richard, D., Catal. Letters **5**, 169-174 (1990).

97. Gallezot, P., Alarcon-Diaz, A., Dalmon, J.-A., Renouprez, A.J. and Imelik, B., J. Catal. **39**, 334-349 (1975).

98. Gallezot, P., Catal. Rev.-Sci. Eng. **20**, 121-154 (1979).

99. Kleine, A., Ryder, P.L., Jaeger, N. and Schulz-Ekloff, G., J. Chem. Soc., Faraday Trans. I **82**, 205-212 (1986).

100. Jaeger, N., Rathousky, J., Schulz-Ekloff, G., Svensson, A. and Zukal, A., in: *Proceedings of the 8th International Zeolite Conference*, Eds. P.A. Jacobs and R.A. van Santen, Part B, Elsevier, Amsterdam, Oxford, New York, Tokyo, pp. 1005-1013 (1989).

101. Blackmond, D.G., Oukaci, R., Blanc, B. and Gallezot, P., J. Catal. **131**, 401-411 (1991).

102. Tauster, S.J. and Steger, J.J., J. Catal. **125**, 387-389 (1990).

103. Tauster, S.J., Steger, J.J., Fung, S.C., Poeppelmeier, K.R., Funk, W.G., Montagna, A.A., Cross, V.R. and Kao, J.-L., Europ. Patent Appl. 145 289, Nov. 9 (1984).

104. Santilli, D.S. and Zones, S.I., Catal. Letters **7**, 383-388 (1990).

105. Namba, S., Sato, K., Fujita K., Kim, J.H. and Yashima, T., in: *Proceedings of the 7th International Zeolite Conference*, Eds. Y. Murakami, A. Iijima and J.W. Ward, Kodansha, Tokyo and Elsevier, Amsterdam, Oxford, New York, Tokyo, pp. 661-668 (1986).

106. Van Nordstrand, R.A., Santilli, D.S. and Zones, S.I., in: *Synthesis of Microporous Materials*, Vol. 1 (Molecular Sieves), Eds. M.L. Occelli and H. Robson, van Nostrand Reinhold, New York, pp. 373-383 (1992).

107. Schulz, H.F. and Weitkamp, J., Ind. Eng. Chem., Prod. Res. Devel. **11**, 46-53 (1972).

108. Venuto, P.B. and Hamilton, L.A., Ind. Eng. Chem., Prod. Res. Devel. **6**, 190-192 (1967).

109. Frilette, V.J., Haag, W.O. and Lago, R.M., J. Catal. **67**, 218-222 (1981).

110. Dewing, J., J. Molec. Catal. **27**, 25-33 (1984).

111. Karge, H.G., Wada, Y., Weitkamp, J., Ernst, S., Girrbach, U. and Beyer, H.K., in: *Catalysis on the Energy Scene*, Eds. S. Kaliaguine and A. Mahay, Studies in Surface Science and Catalysis, Vol. 19, Elsevier, Amsterdam, Oxford, New York, Tokyo, pp. 101-111 (1984).

112. Tsai, T.-C. and Wang, I., J. Catal. **133**, 136-145 (1992).

113. Martens, J.A., Vanbutsele, G.M. and Jacobs, P.A., in: *Proceedings of the 9th International Zeolite Conference*, Montreal, Canada, July 5-10, 1992, accepted for publication.

114. Dessau, R.M., J. Catal. **77**, 304-306 (1982).

115. Dessau, R.M., J. Catal. **89**, 520-526 (1984).

116. Weitkamp, J., Kromminga, T. and Ernst, S., Chem.-Ing.-Tech., accepted for publication.

117. Karge, H.G., Zhang, Y. and Beyer, H.K., in: *Proceedings of the 10th International Congress on Catalysis*, Budapest, Hungary, July 19-24, 1992, accepted for publication.

118. Weitkamp, J., in: *Catalysis and Adsorption by Zeolites*, Eds. G. Öhlmann, H. Pfeifer and R. Fricke, Studies in Surface Science and Catalysis, Vol. 65, Elsevier, Amsterdam, Oxford, New York, Tokyo, pp. 21-46 (1991).

119. Rode, E.J., Gee, P.E., Marquez, L.N., Uemura, T. and Bazargani, M., Catal. Letters **9**, 103-114 (1991).

120. Huang, M. and Kaliaguine, S., J. Chem. Soc. Faraday Trans. **88**, 751-758 (1992).

121. Martens, L.R.M., Grobet, P.J. and Jacobs, P.A., Nature **315**, 568-570 (1985).

122. Martens, L.R.M., Vermeiren, W.J., Huybrechts, D.R., Grobet, P.J. and Jacobs, P.A., in: *Proceedings of the 9th International Congress on Catalysis*, Eds. M.J. Phillips and M. Ternan, Vol. 1, The Chemical Institute of Canada, Ottowa, pp. 420-428 (1988).

123. Xu, B. and Kevan, L., J. Phys. Chem. **96**, 2642-2645 (1992).

124. Xu, B. and Kevan, L., J. Chem. Soc. Faraday Trans. **87**, 2843-2847 (1991).

125. Bordiga, S., Ferrero, A., Giamello, E., Spoto, G. and Zecchina, A., Catal. Letters **8**, 375-378 (1991).

126. Lee, T.J. and Gates, B.C., J. Molec. Catal. **71**, 335-346 (1992).

127. Hathaway, P.E. and Davis, M.E., J. Catal. **116**, 263-278 (1989).

128. Hathaway, P.E. and Davis, M.E., J. Catal. **116**, 279-284 (1989).

129. Hathaway, P.E. and Davis, M.E., J. Catal. **119**, 497-507 (1989).

130. Tsuji, H., Yagi, F. and Hattori, H., Chem. Letters, 1881-1884 (1991).

131. Poncelet, G., Dubru, M.L., Somme, G., Lerot, L., Jacobs, P.A., Tielen, M. and Uytterhoeven, J.B., in: *Proceedings of the International Symposium on Zeolite Catalysis,* Szeged, Hungary, September 11-14, 1978, Acta Physica et Chemica Szegediensis, pp. 273-280 (1978).

132. Corma, A., Martín-Avanda, R.M. and Sánchez, F., J. Catal. **126**, 192-198 (1990).

133. Newsam, J.M., Treacy, M.M.J., Koetsier, W.T. and de Gruyter, C.B., Proc. Royal Soc. London A **420**, 375-405 (1988).

134. Dessau, R.M., US Patent 4 554 262, Mobil Oil Corp., Nov. 19, 1985.

135. Parton, R.F., Jacobs, J.M., Huybrechts, D.R. and Jacobs, P.A., in: *Zeolites as Catalysts, Sorbents and Detergent Builders*, Eds. H.G. Karge and J. Weitkamp, Studies in Surface Science and Catalysis, Vol. 46, Elsevier, Amsterdam, Oxford, New York, Tokyo, pp. 163-192 (1989).

136. Corma, A., Iglesias, M., del Pino, C. and Sánchez, F., J. Chem. Soc., Chem. Commun., 1253-1255 (1991).

137. Arai, H., Yamashiro, T., Kubo, T. and Tominaga, H., Bull. Japan Petrol. Inst. **18**, 39-41 (1976).

138. Elleuch, B., Naccache, C., Ben Taarit, Y. and Wicker, G., in: *Catalysis on the Energy Scene*, Eds. S. Kaliaguine and A. Mahay, Studies in Surface Science and Catalysis, Vol. 19, Elsevier, Amsterdam, Oxford, New York, Tokyo, pp. 139-145 (1984).

139. Espeel, P.H., Tielen, M.C. and Jacobs, P.A., J. Chem. Soc., Chem. Commun., 669-671 (1991).

140. Botta, A., Buysch, H.-J. and Puppe, L., Angew. Chem. **103**, 1687-1689 (1991).

141. Chen, N.Y. and Haag, W.O., in: *Hydrogen Effects in Catalysis — Fundamentals and Practical Applications*, Eds. Z. Paál and P.G. Menon, Marcel Dekker, Inc., New York and Basel, pp. 695-722 (1988).

142. Baba, T., Seo S.G. and Ono, Y., in: *Innovation in Zeolite Materials Science*, Eds. P.J. Grobet, W.J. Mortier, E.F. Vansant and G. Schulz-Ekloff, Studies in Surface Science and Catalysis, Vol. 37, Elsevier, Amsterdam, Oxford, New York, Tokyo, pp. 443-450 (1988).

143. Baba, T. and Ono, Y., Zeolites **7**, 292-294 (1987).

144. Kanai, J., Martens, J.A. and Jacobs, P.A., J. Catal. **133**, 527-543 (1992).

145. Baba, T., Kim, G.J. and Ono, Y., J. Chem. Soc. Faraday Trans. **88**, 891-897 (1992).

CERAMIC MEMBRANES FOR SEPARATIONS AND REACTIONS

A.J. BURGGRAAF, K. KEIZER, R.S.A. de LANGE, Z.A.E.P. VROON, V.T. ZASPALIS

Laboratory of Inorganic Chemistry Materials Science and Catalysis, Department of Chemical Technology, University of Twente, P.O. Box 217, 7500 AE Enschede, The Netherlands

ABSTRACT

An overview is presented of the most important techno-economical and scientific aspects of inorganic membranes. Major attention is given to gas (vapour) separation with meso and especially microporous systems (porediameter < 2 nm) and to reactions in (catalytically active) membrane reactors.

The main types of membrane materials and system architectures and the most important synthesis methods are overviewed. Focus is given to the most important membrane types today, being an asymmetric ceramic membrane produced by slip casting/film formation of colloidal or polymeric solutions followed by some modification treatment. This results in separation layers with a thickness ranging from 30 nm to 30 μm and pore diameters ranging from 0.5 nm to 50 nm or to dense (non porous) layers semi permeable for some gases at high temperature. Possibilities and problems of zeolite membranes are discussed and compared with other types of microporous systems.

High fluxes of H_2 and CO_2 up to 200 $m^3.m^{-2}.bar^{-1}.day^{-1}$ ($2*10^6$ $mol.cm^{-2}.s^{-1}.Pa^{-1}$) combined with high separation factors (α > 100) with respect to hydrocarbon molecules can be obtained with amorphous oxidic micropore systems showing activated transport. Model calculations and extrapolated results indicate the possibility to obtain with zeolite membranes similar or even larger values. Mesoporous systems (pore diameter 20-100 nm) show interesting separation properties due to capillary condensation and surface diffusion. It will be shown that conversion and selectivity of a number of reactions can be increased considerably with the help of membranes and membrane modification methods. Examples are given of partial oxidation and dehydrogenation reactions. Essential to obtain strongly increased large conversions in combination with large selectivities is a proper use of the reaction conditions and a combination of a large value of H (ratio of membrane permeation to reaction rate) in combination with large permselectivities. Finally some model considerations will be given for transport in microporous systems showing the relative importance of adsorptive and diffusion processes. At medium temperature, transport can be considered as a surface mediated diffusion process with an activation energy being a measure for the average pore diameter of a given system.

From the Proceedings of the 9th International Zeolite Conference, Montreal 1992, Eds. R. von Ballmoos et al.,

INTRODUCTION AND OVERVIEW

In the past three decades membrane processes have grown from laboratory-scale devices to well developed industrial installations. This is especially the case for polymeric membranes which have the advantage of a relative simple thin film formation process at relative low cost. However, polymeric membranes have several limitations : high temperature instability, swelling and decomposition in organic solvents, low ability to steam sterilisation and compressibility and creep at high pressures.

These limitations can be overcome by inorganic membranes [1]. After some earlier developments of gas separation membranes in nuclear industry (isotope enrichment) the volume of the research and development increased strongly after 1980 following the introduction of Carbosep membranes [2] and the pioneering work published by Leenaars and Burggraaf [3,4]. Today broad interest is based on the large number of potential applications of considerable economic value. This will be discussed briefly below.

The state of the art is described in a number of overviews [1,2,5,6], a book [7] and the Proceedings of the 2nd International Conference on Inorganic Membranes, Montpellier, France [8].

Inorganic membranes can be produced in a variety of materials, morphologies and configurations and for different types of applications (TABLE I). As will be discussed below the use of asymmetric ceramic membranes in gas (vapour) phase applications (separation, membrane reactions) have a large potential provided the selectivity of the process can be further enhanced. This can be obtained while maintaining a reasonable permeability probably with microporous systems (porediameter < 2 nm) making use of surface diffusion (flow) and of micropore diffusion. Therefore the present paper will focus on these experimental and theoretical aspects of gas transport in membranes, and on synthesis methods and properties of inorganic membranes.

TECHNICAL AND ECONOMICAL ASPECTS AND PROSPECTS.

The main application fields are summarized in Table 1, some figures for present and future sales are given in Table 2. These data are taken from techno economical overviews of Charpin et al. [2], Fain [10] and especially of Crull [9, 8 pp. 279-289].

REALISED AND PROJECTED SALES FOR INORGANIC MEMBRANE MATERIALS (Crull)

Material	1986	1989	1994	1999	growth%
	in million USA dollars				
Ceramics	6	18	75	345	34
Carbon	0	3	9	50	32
Metals and other	5	8	13	25	12
Glass	0	0	1	3	> 100
Other	1	2	4	9	16
Total	12	31	102	432	30

Projection year 2000: $ 1400 * 10^6 (Fain)
Provided: decrease of costs to 1/3 of present
increase of separation factors

TYPE OF APPLICATION	INDUSTRY	REMARKS
Microfiltration, Ultrafiltration and Nanofiltration	Milk, beverage	good resistance to cleaning in alkaline media
	biotechnology and pharmaceutical	steam sterilization
	textile, engineering and paper	high temperature resistance
	oil	high temperature and solvent resistant
Gas separation	petrochemical industry	high temperatures, high selectivity and permeability for H_2, CO_2 or O_2
Membrane reactors (reaction and separation)	petrochemical	high temperature, dehydrogenation, partial oxidation, catalytically active
	environmental	removal of NO_x and SO_x, high temperature, catalytically active

TABLE 1. Application and application possibilities of inorganic membranes.

TABLE 2. Realised and projected sales for inorganic membrane materials (Crull).

From Crulls analysis follows that at present micro and ultrafiltration applications have obtained reasonable industrial interest. For gas separation a rapid growth is foreseen between 1995 and 2000 and spectacular growth after the year 2000. Fain [10] predicts an even more spectacular growth provided the separation factors increase considerably (values > 100) and the production costs decrease to 30-40 % of the present costs.

The projected sales for inorganic membranes are dominated by ceramic membrane systems and are only 5-10 % of the market for polymeric membranes (low temperature application). One of the main reasons for this phenomenon and for the relative slow market penetration is the relatively high production cost of inorganic membranes and especially of membrane modules (about five times that of polymeric ones). This effect can at least partially be overcome by a predicted considerably longer life-time of ceramic membrane systems.

For gas separations and membrane reactors a further increase of the separation factors to 100 or more with reasonable flux values (10-200 m^3/m^2.bar.day or more) is important. This requires very thin separation layers and so asymmetric configurations and microporous (pore-diameter < 2 nm) separation layers.

For membrane reactors an increase of the surface area to volume ratio of the catalytically active surface area is favourable. This suggests that the development of flat stacked modules should be preferred above the tubular modules which at present are developed.

The main conclusion is that inorganic membrane systems and especially asymmetric ceramic composite systems and to a lesser extend carbon based (composite) systems, have good opportunities when use is made of their strong points: high temperature stability (up to 1000 °C), resistance against aggresive conditions (extreme pH, aggressive organic media), in combination with good control of pore size (distribution) and surface characteristics, and a long life-time.

STRUCTURAL ASPECTS AND PREPARATION METHODS.

An overview of the main aspects is given by Burggraaf and coworkers [6,11]. A short summary is given below and new aspects concerning microporous materials are added. Inorganic membranes can subdivided in two main groups i) dense and ii) porous materials. Both groups can again be subdivided in a) unsupported and b) supported membranes. The unsupported systems are relatively thick (1-2 mm) porous systems (plates, tubes) which are mechanically stable. To obtain large flux values the separation layer should be very thin. Therefore thin films are supported by thick macroporous (porediameter d_p relatively large) support materials. Such a combination yields an asymmetric system which combines mechanical stability with low flow resistance. In many cases the system consists of three or more layers as is represented in Figure 1. This type of ceramic membranes is used most frequently (TABLE 2).

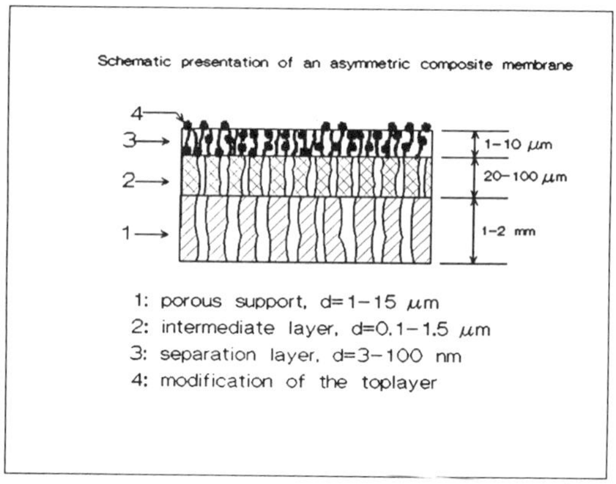

FIGURE 1. Schematic presentation of an asymmetric composite membrane.

The thin (1-10 μm) toplayer can be produced with porosities of 30-50 % and pore diameters (d_p) as small as 3 nm from γ-Al_2O_3, TiO_2, ZrO_2 and mixtures of them on α-Al_2O_3 supports [11].

Interesting is the possibility to synthesize binary composite (two phase) mixtures or sandwich structures of these materials e.g. Al_2O_3-TiO_2, Al_2O_3-SiO_2, SiO_2-TiO_2, ZrO_2-TiO_2 [13] with bifunctional properties.

The toplayer can be modified by a variety of methods principally resulting in 4 types of nanostructures as represented in Figure 2.

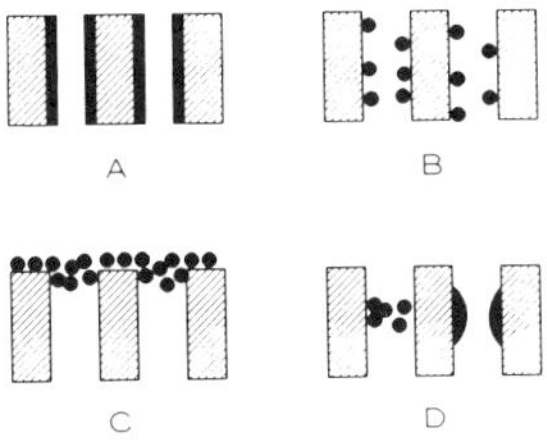

Figure 2. Schematic representation of the nanostructures obtainable in a modified membrane toplayer. A) mono or multilayer deposit. B) micro(nano) particles within pore, partial plugging. C) plugs in the pore entrance, film on top of the membrane. D) plugs or constrictions on certain sites within the pore.

In this way the pore-diameters can be decreased and the chemical nature of the internal surface can be changed e.g. catalytic properties can be introduced [22,23]. Using sol-gel methods, amorphous microporous SiO_2 and SiO_2-TiO_2/ZrO_2 ultra thin layers (thickness 30nm with plugs into the pore of 30 nm thick; fig 2, type c) are produced by Uhlhorn et al. [40] and de Lange et al. [18]. These materials have molecular sieve properties (see further below).

MATERIALS	SUPPORT SYNTHESIS	TOPLAYER(S) SYNTHESIS
composite membranes, oxidic/nonoxidic ceramics	slipcasting, extrusion, sintering, filmcoating techniques	suspension techniques, film coating techniques, sol/gel, polymeric gel, impregnation/adsorption
glass (Vycor)	extrusion, spinodal decomposition, leaching	-
anodized alumina	anodization of Al-foil	-
carbon - large pores - micropores	 pressing,extrusion, pyrolysis hollow fibre pyrolysis and moderate oxidation	 controlled pyrolysis -
carbon composite	extrusion, pyrolysis	stable suspension, film coating technique (ZrO_2)
metal metal composite	extrusion/sintering idem	- stable suspension, coating technique (ZrO_2)

TABLE 3. Main synthesis methods and materials of inorganic membranes.

An overview of main material types and synthesis methods is given in Table 3 and a detailed treatment is given in [7]. Very different pore morphologies can be obtained like i) straight channels with small diameter at one surface and a wider diameter at the other surface (anodised alumina), ii) ("wormlike") interconnected pore systems obtained by phase separation and leaching resulting in porous glass or controlled pyrolysis of a polymeric precursor (carbon) and iii) interconnected pore systems obtained by packing of small, uniform (colloidal or polymeric) particles from a particle dispersion onto a porous substrate in a sol-gel (slip-casting) process. This last process will be discussed into some detail in the next section.

An important aspect is the pore-size distribution (which should be sharp) and the thermal and chemical stability of the toplayer. Oxidic systems in the mesopore range can be made stable up to 1100 °C with d_p = 5-20 nm depending on conditions [20].

In the micropore range thermal stabilities are not yet fully explored but thermal stabilities up to at least 600 °C can be obtained (de Lange/Burggraaf [44]). Chemical stabilities against aggressive conditions (extreme pH, organic media and agressive gases) are good for some oxidic systems like ZrO_2, TiO_2 and carbon. In the last case oxidative conditions should be avoided while in the other cases steam at high temperature and pressure might give problems.

A most important point is to make the membrane free of cracks, pinholes and other defects which will deteriorate the separation (permselective) properties seriously. This depends very critically on a series of technological parameters including support quality, agglomerate structure of the colloidal suspension, additions, and drying process.

MAIN PREPARATION METHODS.

At the moment the most important group of ceramic membranes is produced by slipcasting of a colloidal suspension onto a porous support. In this process a thin gel layer is formed during a certain time by dipping the porous substrate into a liquid medium containing colloidal particles (diameters down to 7 nm) or metal containing polymeric species. During careful drying , capillary forces cause compaction of the gel layer resulting in a regular packing of the elementary (nano) particles. Finally a calcination process transforms the gel material into a layer of oxidic particles and a stable membrane is formed with a sharp pore size distribution. The process is described into detail by Leenaars et al. [3,4], Burggraaf and Keizer [11] and Uhlhorn et al. [22]. With this method the layer thickness is a function of the dipping-time and of other parameters (viscosity, concentration and surface tension). There is a maximum thickness which can be obtained in a single step without cracking which, depending on conditions and composition, ranges from about 2-10 μm. Interesting is that much thicker layers can be obtained by repeating the dipping procedure (multi-dipping process). Zaspalis et al. [12] recently found however that again a maximum exist which for γ-Al_2O_3 amounts to about 24 μm. In this way sandwich structures were formed of up to 10 layers.

This principle of repeated dipping can be used also as a modification procedure to block cracks or pinholes or to obtain ultra thin microporous systems with molecular sieve properties.

Kitao et al. [14] used a multiple dipping procedure (10-20 steps) to obtain a microporous silica membrane. To obtain a good quality it was necessary to apply a Chemical Vapour Deposition treatment to block the remaining defects with SiO_2 plugs [48].

Burggraaf and coworkers [17,18,44] succeeded to synthesize amorphous micropore membranes by modifying γ-Al_2O_3 membranes (pore diameter 3 nm) by controlled hydrolysis of Si, Si-Ti or Si-Zr alkoxide mixtures resulting in metal containing polymeric molecules. A microporous film (SiO_2 or SiO_2-TiO_2 etc.) is formed in a single dipping step, followed by drying and calcination, having in a thickness of 20-30 nm which penetrates („plugs") the pores of the membrane substrate by 20-30 nm. The pore diameters obtained are of the order 0.5-0.7 nm. The crucial point is to start with small and only weakly branched polymeric particles having a fractal dimension smaller than 1.5. (de Lange and Burggraaf et al., to be published).

Gas phase deposition (C.V.D.) is used to block defects in membranes formed in another way or the make very thin dense membranes. Gavalas et al. [15,19] produced dense SiO_2 plugs with an estimated thickness of 100-200 nm by gas phase reaction between O_2 and SiH_4 or $SiCl_4$ in a porous glass substrate . The films are semipermeable for H_2. Lin et al. [21a-c] prepared dense yttria stabilized ZrO_2 films with an estimated thickness of 200-500 nm by gas phase reaction of $ZrCl_4$ and YCl_3 with H_2O in a mesoporous alumina membrane. These layers are semipermeable for oxygen.

Finally (meso) pore systems can be modified by impregnation of the pore system with a liquid solution of a precursor followed by precipitation and decomposition. In this way Uhlhorn et al. [23] produced MgO or Ag particle deposits within γ-Al_2O_3 or titania membranes filling up to 50 % or more of the pore volume.

Polymeric precursors can be attached to the inner walls of the pore system to modify the physicochemical character of the surface [24] and to decrease the pore diameter [25]. In this way the separation properties in liquid and vapour media can be changed. Finally the pores can be filled with a ceramic precursor polymer which is pyrolysed in a controlled way. Shelekhin et al. [27] describe experiments to obtain macroporous plugs of pyrolysed polysilastyrene in the pores of porous glass. Similar experiments were performed by Guizard [28] using heteropolysiloxanes.

Carbon membranes with molecular sieving properties can be obtained in several ways which are briefly reviewed by Burggraaf and Keizer [11] and in [27]. The general principle is to pyrolyse in a controlled way thermo-setting polymers applied on or into a porous support. Interesting experiments have been reported by Koresh and Soffer [29,30] using PAN (polyacrylonitrile) in certain configurations (e.g. hollow fibers). Pyrolysis in inert atmospheres resulted in a dense carbon membrane. Then controlled oxidation treatments created a microporous structure.

The synthesis of zeolite membranes represent a recent development mainly described in patent literature [31-39]. The present authors believe that at present no polycrystalline zeolite membranes with permeation properties mainly determined by intrapore diffusion through the zeolite crystals have been prepared due to remaining pores or cracks in between the zeolite crystals. Useful experiments have however been performed with single crystal membranes of very small size [31] and with mixtures of zeolites and other matrix components [32]. These can give rise to interesting properties also.

The synthesis methods applied to zeolitic membranes follow mainly the combinations of methods described above e.g. dipping followed by after treatment (multi dipping and or C.V.D.) or impregnation methods [34].

New methods for zeolite synthesis are i) conversion of an active gel layer into a zeolite by hydrothermal treatment [34] or impregnation followed by conversion of active material [35]. Lachman [36] converted the surface of the substrate by reacting it with active silica. Geus [37] gave examples of the design possibilities of zeolite membranes following methods mentioned above.

TRANSPORT PROPERTIES

Mesoporous systems

Important properties of porous membranes are their single gas permeability, separation efficiency of gas mixtures and adsorptive properties. An extensive review has been given by Uhlhorn [40], a shortened version of this review is given by Uhlhorn and Burggraaf [41]. The total gas permeability in mesoporous systems consists of three components as expressed by a simplified equation 1 [40,41] ;

$$F_o = F_{o,p} + F_{o,kn} + F_{o,s} = \text{constant(s)}.\epsilon.\mu_k.[c_1.r^2 + c_2.r + c_3.r^{-1}] \qquad (1)$$

with r = pore radius, ϵ =porosity, μ_k = shape factor, c_1, c_2, c_3 are constants which are functions of pressure and of temperature. The first term represents the viscous (Poiseuille flow), the second gives the Knudsen (diffusion) flux while the last term expresses the surface diffusional flux across the internal surface. In small pores and not to high pressure only the last two terms are important and especially in very small pores and not to high temperatures surface diffusion becomes important. Equation 1 does not contain the phenomenon of micropore diffusion in pores with diameters < 2 nm which type of flux occurs in amorphous and zeolite (micropore) membranes.

The separation capability for a binary gas mixture is expressed by the separation factor α, defined by equation 2. It can also be obtained by the quotient of the single gas permeabilities and is then called the permselectivity, which ignores all types of interactions between the components. It is not surprising that in many cases the permselectivity strongly deviates from the phenomenological selectivity defined by equation 2.

$$\alpha_{ideal} = \frac{y}{1-y} * \frac{1-x}{x} \qquad (2)$$

in which x = feed composition and y = permeate concentration. In equation 2 it is assumed that feed and permeate concentration are kept constant. In practice this is not the case. The practical value of α is strongly determined by a number of parameters e.g. by the mean pressure and the mean pressure drop. The highest value is obtained if no back diffusion takes place, that is with the pressure at the low pressure side is kept small with respect to that at the high pressure side. Practical values of α are always smaller then α_{ideal}.

The effect of the porous support on permeability and separation factor can be very important. If the flow resistance of the support is not ignorable then the properties of the toplayer must be calculated from the total system properties [40-42]. An example of the results of such a calculation is given in Figure 3a for a silicalite membrane, synthesised at the University of Twente.

The presence of defects in mesoporous membranes , giving rise to significant leakage, manifest itself as a pressure dependency of the permeability of the toplayer. This can be understood from the fact that Poiseuille flow is a linear function of the pressure, while the Knudsen permeability is independent of pressure. Small leakage flows (with respect to the Knudsen flow) can not be detected in this way but can deteriorate the separation factor considerably. From Figure 3 it can be seen that the alumina supported silicalite zeolite film shows Knudsen flow behaviour. Knudsen permeability depends on the molecular weight (M) and is proportional with $(R.T/M)^{1/2}$ and so the ideal separation factor of component 1 with respect to component 2 is proportional to $(M_2/M_1)^{1/2}$.

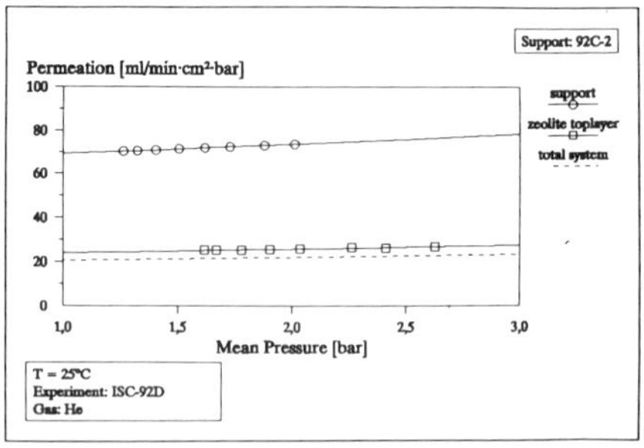

FIGURE 3. Permeation as a function of mean pressure for a silicalite membrane on an α alumina support (Vroon, Burggraaf).

Surface diffusion can contribute considerably to transport in porous system [40,41]. It can be expressed by a two dimensional form of Ficks law (equation 3).

$$F_s = - A.\rho_{app}.D_s(q).dq/dx \quad (3)$$

with F_s = surface flux, A the membrane outer surface, ρ_{app} = apparent density, D(q) the concentration dependent Fickian diffusivity, q the sorbate concentration (mol/m^3) or the occupation degree of the surface and x the distance from the feed side . Several models are presented in literature [40,41] and equation 3 is based on a hopping model of adsorbed species from occupied to non occupied sites across an energy barrier. The activation energy of this process can not exceed approximately 0.5 times the heat of adsorption because in that case desorption occurs. Uhlhorn [16] describes surface diffusion of CO_2 in γ-Al_2O_3 supported membranes around room temperature and of H_2 in γ-Al_2O_3 membranes modified with silver at temperature up to 500 K. In both cases the pore diameter was near 3.5 nm. The contribution of the surface diffusion to the total flux is for CO_2 comparable with the Knudsen flow and for H_2 even considerably larger. At higher temperatures the decrease of the surface concentration with increasing temperature becomes dominant and the surface diffusional flux decreases. However, it can contribute to 40 % of the total flux even at high temperature and with pores as large as 500 nm (Sloot et al. [26])

With moderate temperatures and larger relative pressures $[P/P_o]$ (with P_o the condensation pressure over a free surface) multilayer diffusion and capillary condensation takes places as is given by Uhlhorn et al. [40,43]. In small pores the capillary condensation pressure is a function of the porediameter (Kelvin equation) and blocking of pores by capillary condensation of a condensable compound A in a mixture A-B can occur at P/P_o values smaller then 0.5. If B has only a limited solubility in A very high separation factors can be obtained in combination with high permeabilities. This type of behaviour can be described quantitatively with models [41,45].

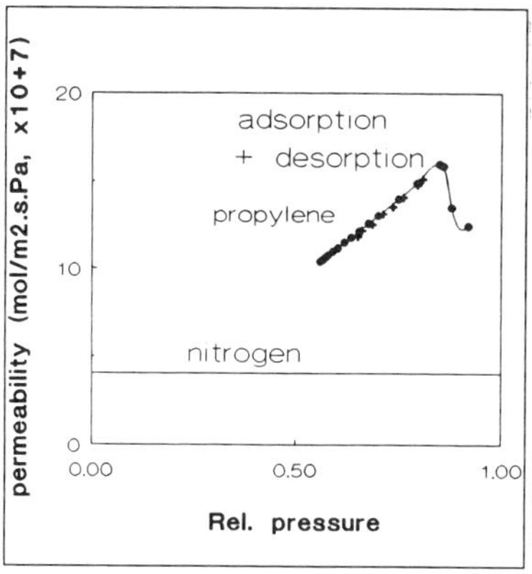

FIGURE 4. Permeability of a magnesia modified supported gamma alumina thin film for nitrogen and propylene at 263 K as a function of the relative pressure of propylene.

System	Pore diam (nm)	T (°C)	Trans. mech.	Sep. fac. α	$F*10^6$ (mole/ m^2.s.Pa)
γ-Al_2O_3 thickness $5*10^{-6}$ m	50	600	Knudsen diffusion	H_2/N_2 : 2	H_2 : 200
	3	200		H_2/N_2 : 2	H_2 : 15
γ-Al_2O_3 modified with MgO thickness $5*10^{-6}$ m	3	- 10	Multilayer adsorption	C_3H_6/N_2: 26	C_3H_6: 30
	2-3	- 10	capillary condens.	C_3H_6/N_2: 84	C_3H_6: 1.6

TABLE 4. Summary of typical permeabilities F and separation factors *α* for mesoporous membranes.

Examples of this behaviour are given by Uhlhorn [43] and an illustration is given in Figure 4 and Table 4 for N_2 and propylene. Multilayer diffusion of propylene causes already at low P/P_0 values a surface flux which is considerably larger then the gas phase flux of N_2. At about P/P_o = 0.75 capillary condensation blocks the pores (diameter 3-4 nm) for N_2 transport and the separation factor *α* increases to 80, propylene being the preferably permeating component in the mixture. Recently Sperry et al. [50] showed that this principle can be used up to high temperatures provided the pressures of the condensable gas is increased sufficiently.

Micropore systems

According to the IUPAC definition, systems with a porediameter smaller than 2 nm are called microporous. In literature there is much confusion because frequently systems with pores having diameters > 1 μm are called incorrectly microporous.

Micopore diffusion and adsorption is a well known phenomenon in adsorbents (carbon, silica-gel) and zeolites. Classical equations like the Kelvin equation break down and there is no generally accepted model to determine porediameters below 1.5 nm or to describe the diffusional and adsorption flux. The problem here is that molecules always move in the potential field of the pore walls, therefore, a liquid meniscus does not form and the properties of the liquid after condensation are different from a bulk liquid.

Nevertheless it is shown e.g. by Asaeda et al. [14,15,46,47] that (in a capillary condensation process at 60-80 °C.) reducing the porediameter of mesoporous membranes into the micropore range increased the separation factor of water -alcohol mixtures to 60 and more.

Permeation and separation of gases in microporous systems membranes is a new field. Results are reported for carbon [29,30], for amorphous SiO_2 and SiO_2-TiO_2 membranes [17,18] and for zeolite membranes [31,33].

Gas mixture	Sep. fact.	$E_{act.\ H2}$ (kJ/mole)	F (10^{-8} mol/m2.s.Pa)	T (°C)	Material	Author (year)
H_2/C_3H_6 H_2/CO_2	160@ 3	7 (11)#	160	200	Sol-Gel SiO_2	Uhlhorn (1990)
H_2/CH_4 H_2/CO_2	200@ 5	11	20-80	150	Sol-Gel SiO_2-TiO_2	de Lange (1991)
H_2/N_2	142 185	8.3	4.5 5.0	200 270	Sol-Gel SiO_2	Kitao (1990)
H_2/N_2	119 227	12.2	1.1 2.1	200 270	Sol-Gel- CVD-SiO_2	Kitao (1991)
H_2/N_2	2000	31 (35)#	1.6	450	CVD-SiO_2 * glass	Gavalas (1989)
H_2/N_2 H_2/N_2	28000 1000	37 27	0.2 0.6	450 450	CVD-SiO_2 CVD-B_2O_3 * glass	Tsapatsis (1991)
H_2/N_2	12	5.3	0.4	200	CVD-SiO_2	Okubo (1989)
H_2/CH_4	56	0.9	11	200	carbon	Koresh/ Soffer
	32 35	-	18 8	200 25	carbon	(1987)

@ : Separation experiment with gas mixture
# : Corrected for Knudsen contribution

TABLE 5. Permeabilities and activation energies of H_2 and separation factors (permselectivity) of H_2 / ... through microporous (and dense *) inorganic membranes.

A summary of the results for H_2 permeabilities and of separation factors of H_2 from other gases in microporous and dense glass membrane is given in Table 5. The dense membranes are obtained in these cases by high temperature Chemical Vapour Deposition.

Recent results from de Lange/ Burggraaf and Keizer are shown in figures 5 and 6 and represent the permeabilities of H_2 and CO_2 in microporous silica membranes respectively as a function of pressure and temperature.

The results show several important things. In the first place the permeation is activated with activation energies depending on type of gas and pore diameter. This is typical for micropore diffusion. The activation energy for the dense glass membranes is considerable larger and only H_2 permeates through them. High permeabilities in combination with high

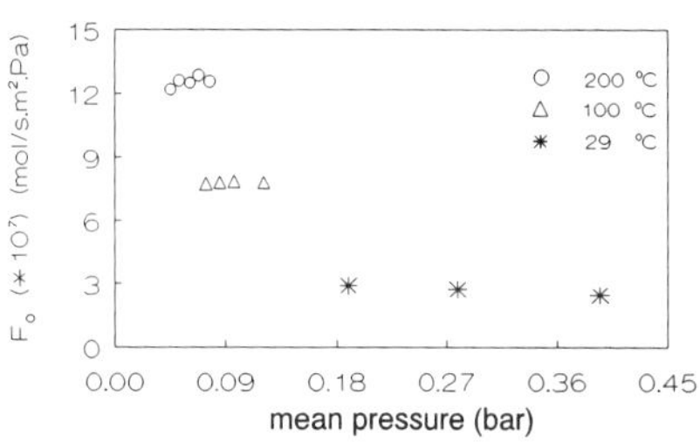

FIGURE 5. H_2 permeability of a supported silica membrane.

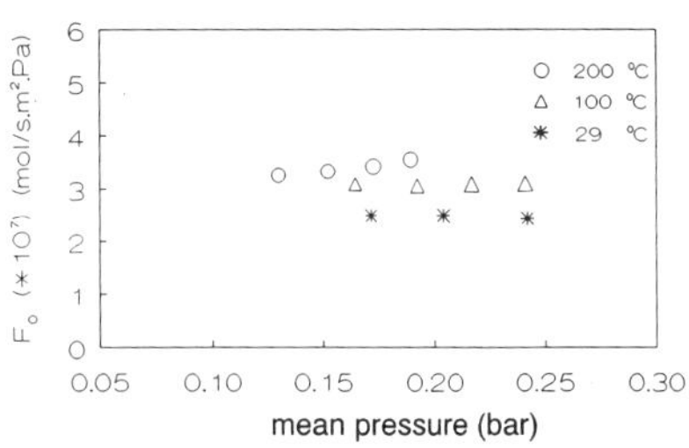

FIGURE 6. CO_2 permeability of a supported silica membrane.

separation factors are obtained by Burggaaf and coworkers for H_2 and CO_2 with respect to other gases which show no detectable or relatively small permeability. (CH_4, higher alkanes). The permeabilities increase with temperature in the temperature range investigated up to 200 °C. The pore diameters of these membranes are not known but considering the molecular diameters of penetrating and non-penetrating molecules it is probably 0.5-0.7 nm. The large permeabilities are probably due to the extremely small thickness of the toplayer (60 nm, see previous section on preparation methods).

Results for zeolite single crystal membranes are reported by Geus et al. [49] and Wernicke and Osterhuber [31]. In both cases crystals with typical dimensions of 50-150 nm are attached across a hole in a support and the permeation flow has been measured with a pressure difference of a about one bar with low pressure (<100 Pa) at the permeate side. For comparison reasons the authors of this review paper have recalculated their results to a certain dimension. Wernicke found with NaX single crystals a maximum in the flux of butane at 40-50 °C. The permeability (F) at this temperature is 1-2 $mol.m^{-2}.s^{-1}.bar^{-1}$ for a membrane thickness of 100 μm. At higher temperature the flux decreases and obviously overrules the effect of the increasing intra crystalline diffusivity.

Geus [49] measured the permeability of CH_4, C_2H_6, C_3H_8 and C_4H_{10} through silicalite single crystals at T = 30 °C and 130 °C. They reported a considerable scatter of the results between different but nominally equal single crystals. The reported permeability values are of the same order of magnitude for all gases. The permeability values for a crystal thickness of 100 μm are $3*10^{-1}$ - $3\text{-}10^{-3}$ $mol/m^{-2}.s^{-1}.bar^{-1}$ at 30 °C and $4*10^{-3}$ $mol.m^{-2}.s^{-1}.bar^{-1}$ at 130 °C (different values for different crystals). It should be noted that all authors assume a linear dependence between flux and membrane thickness. It must be noted that these permeability values are really high (600-4000 $m^3.m^{-2}.bar^{-1}.day^{-1}$). The problem is that this type of measurement is very sensitive to small leakages in the experimental set-up.

Finally Geus et al. [33] also studied the permeability of a zeolite membrane of intergrown, randomly oriented, crystals with a thickness of about 50-80 μm on a (porous) clay support. The membrane was said to be gas tight but no evidence was given how this was measured. The permeability was measured of CH_4, Ne, n-C_4H_{10} and i-C_4H_{10} at 20 °C using the Wicke-Kallenbach method (no absolute pressure gradient, only partial pressure gradients in the carrier gas). Permeation rates were highest for weakly adsorbing molecules (CH_4, Ne) and lowest for the better adsorbing ones and ranged from $6*10^{-4}$ (CH_4) to $1.3*10^{-4}$ $mol.m^{-2}.s^{-1}$. Direct comparison with the values obtained under a pressure gradient is not possible but it can be seen that the results with and without pressure gradient differ by less than one order of magnitude. On the other hand, calculated values using are larger by two to three orders of magnitude compared with the experimental results. It is interesting to note that the calculations indicate a linear dependence between permeation and partial pressure for weakly adsorbing species (CH_4) and saturation at low pressure for the stronger adsorbing ones (butane) with an increasing permeability for decreasing pressure. We believe theoretical considerations indicate that in these cases an increase of the temperature will initially give rise to an increase of the permeability at given pressure as is found also for permeation in amorphous silica (de Lange/Burggraaf [44]).

CATALYTICALLY ACTIVE MEMBRANES AND MEMBRANE REACTORS.

In recent years the application of ceramic inorganic membranes in high temperature reactors attract increasing attention. The field has been reviewed and analyzed by Zaspalis and Burggraaf [7,51], by Armor [53] and Hsieh [52].

In a high temperature reactor the incorporation of membranes can serve three different purposes:

i) the enhancement of the conversion of a chemical reaction i.e. by shifting the equilibrium situation by selective removal of one of the reaction products.

ii) to influence the path of the chemical reaction i.e. to increase the reaction selectivity.

iii) to obtain special effects i.e. to avoid or decrease catalyst de-activation i.e. by de-coking simultaneously with the reaction.

In all cases the membrane divides the reactor into two compartments and mediates the contact between various species or process streams. This is especially the case where undesirable consecutive gasphase reactions should be avoided e.g. in partial oxidation. Note that in mesoporous membranes gas transport takes place by Knudsen diffusion, so direct gas phase reactions within the pore system are avoided (processes related with gas molecule-wall collisions dominate).

Two membrane caterogies can be distinguished on basis of the function of the membrane in a high temperature membrane reactor: a) passive and b) catalytically active membrane.

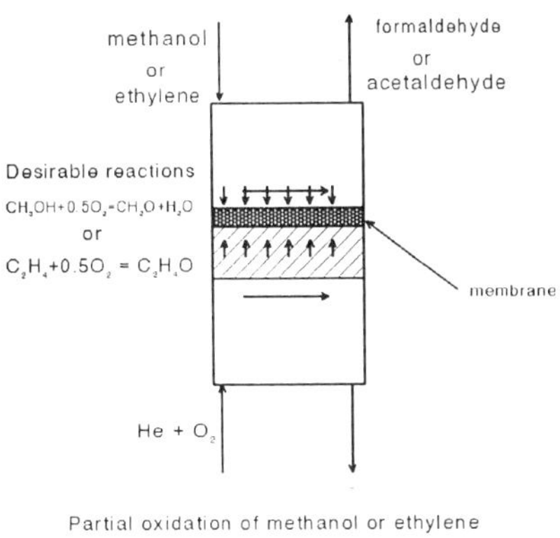

FIGURE 7. Schematic diagram of the passive membrane reactor operation.

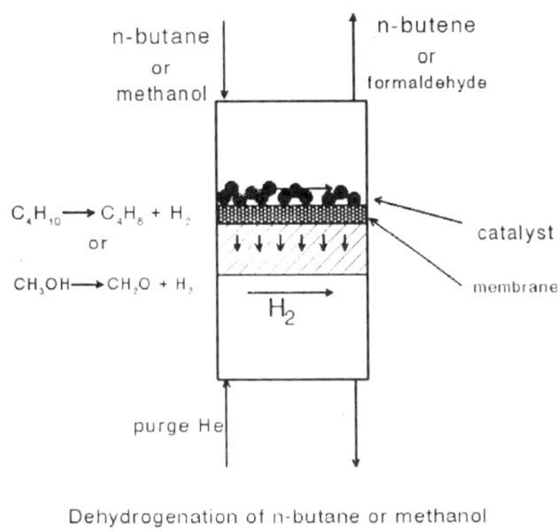

FIGURE 8. Schematic diagram of a catalytically active membrane reactor.

The case of passive membranes is shown in Figure 7. The membrane is used here exclusively for its separation function. In the case shown this is an equilibrium limited dehydrogenation reaction which is considerably shifted to higher conversion by selectively removing H_2 from the equilibrium. The catalyst forms not an integral part of the membrane but it is present either as a packed bed or as a thick layer on top of the membrane.

The case of catalytically active membranes is shown in Figure 8. The catalytic activity is an inherent property of the membrane material or a property added to the membrane by modification with a catalytically active phase within the pores (or as a thin layer on top of it). A quantative analysis by Zaspalis et al. shows that the distribution of the active phase can be important to obtain optimum results [54].

A number of examples of membrane catalysis has been given in a series recent papers [54,55]. A recent result of Zaspalis et al. concerns the selective oxidation of ethylene to acetylene with the configuration given in Figure 8. The catalytically active membrane consists of an γ-Al_2O_3 membrane toplayer (pore diameter 4 nm) modified with a V_2O_5 catalyst. By supplying the oxygen separately (from the support side) instead of together with the ethylene feed, the undesirable total oxidation to CO_2 could be greatly decreased. At 300 °C the selectivity to acetaldehyde was increased from 20 % to 70 % , while the total conversion decreased from 70 % to 25 %. The case of butane dehydrogenation to butene with a passive γ-Al_2O_3 membrane (porediameter 4 nm) is shown in figure 7 and has been discussed fully in [55b]. Here the catalyst is a commercial Pt/SiO_2 placed in the form of small pellets on top of the membrane and H_2 is selectively removed from the support side. At 500 °C the selectivity and conversion are increased from 40 to 60 % and 9 to 15 % respectively compared with the packed bed reactor. In this case the mesoporous membrane has only a low separation factor based on Knudsen diffusion.

The effect of increasing the separation factor has been discussed by Zaspalis and Burggraaf [51] and is quantitatively analyzed for mesoporous membranes by e.g. Mohan and Govind [57] and by Zaspalis et al. [56]. For reactions with relatively low conversions in the equilibrium situation and membranes with relatively low separation factors there is maximum increase in the conversion by a factor of 1.3 to 2.3. An example is the dehydrogenation of cyclohexane for which the maximum obtainable conversion using a membrane reactor is 60-75 % under conditions where the equilibrium conversion is 40 %. Recently Zaspalis [58] made a theoretical analysis of the effect of a drastic increase of the selectivity factor in combination with a large permeability which is possible by the use of thin microporous (amorphous) membranes. The results are summarized in Figure 9 and are compared with the results under the same conditions obtained with dense membranes (infinity high separation factor, low permeability) and mesoporous membranes. The reference is the equilibrium conversion which is about 7 % under the conditions chosen. The important parameters are shown to be the ratio H of the permeation rate with respect to the reaction rate. The permeation rate is also a function of the purging rate. The last parameter is related with the rate at which H_2 coming from the

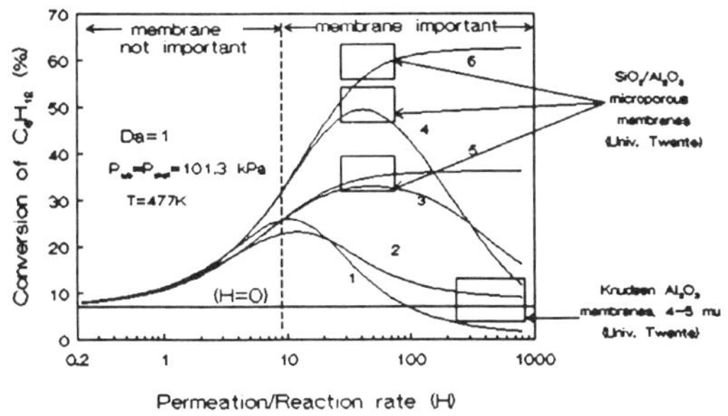

FIGURE 9. Schematic representation of the effect of membranes with high separation factors and permeation rates on the dehydrogenation of cyclohexane at T = 205°C. 1 = mesoporous γAl_2O_3 membrane, Knudsen separation; 3 = Molecular Sieve, amorphous SiO_2 toplayer on γAl_2O_3 mesoporous membrane with separation factor 100; 5 = semilar to 1 but separation factor infinite, 2,4 and 6 are similar to 1,3 and 5 respectively but with purging gas flow strongly increased.

membrane is removed and therefore with the H_2 concentration at the permeate side. The lower this concentration is the less the backdiffusion to the permeate side will be. At low H values the use of membranes has no technical benefit, immaterial whether they are dense, meso- and macroporous. This is the reason that dense materials with low permeabilities are only useful with very slow reactions. Membranes with finite separation factors show an optimum H value above which the loss of reactant due to slow permeation becomes detrimental for the total conversion. This is the reason that γ-Al_2O_3 Knudsen type membranes as discussed before give only a limited improvement for the dehydrogenation of cyclohexane. Increase of the thickness (lower permeability) and selectivity will increase their effectiveness. The cases of curves numbers 3 and 4 are typical for the microporous silica-alumina membranes with typical values of H ≈ 30-80 (marked (□) areas in figure 9) for the reaction chosen and permeabilities of $3*10^{-6}$ $mol.m^{-2}.s^{-1}.Pa^{-1}$ (H_2 selectivity = 100) in the high H value range. For these preferred H values conversion factors of 50-55 % are obtainable which are about 7 times the equilibrium conversion in a packed bed reactor. Increase of the selectivity factor to > 1000 increase the conversion factor further to 65 % (at high H value). The same results can be expected for zeolite membranes provided large H values, and a combination of good permeability and large separation factor can be realized.

Zeolite membranes mentioned in literature [34,35] in conjuncture with catalytic properties can hardly be expected to have good separation factors and can be considered as catalytic active meso- or macroporous zeolite membranes. Nevertheless interesting effects in a number of reactions have been reported.

Suzuki [35] uses ultrathin zeolite films (thickness ≤ 100 Å). According to our investigations with amorphous films this thickness is smaller then the roughness of the substrates and the flow through this type of film is a mixture of Knudsen and micropore diffusion for which no pure molecular sieve action can be expected. Nevertheless, Suzuki reports reasonable to good separation factors could be obtained with NaA type zeolite film for CH_4 with respect to ethane and propane or for n heptane with respect to cyclohexane. This can be understood if preferential absorption phenomena on the zeolite surface dominate the separation mechanism. At 450 °C and 18 bar this type of zeolitic film shows catalytic activity for the conversion of cumene to propylene. Under similar conditions n hexane was converted partially to ethylene and propylene with NaY and CaY type zeolite films. A mixture of these zeolite types converted kerosine into a mixture of C_{10} - C_{15} paraffins (mainly C_{12} and C_{13}) at room temperature and 17 bar. No quantitative data were given however, so an analysis of their results is not possible.

Suzuki [34] reports systems consisting of a porous support (Vycor glass) with deposits of zeolite crystals within the pores of the support. Catalytic activity is reported for this type of systems despite the fact that pure molecular sieve action cannot be expected. An example is the partial conversion of toluene to benzene, para xylene (18.5 wt%) and others with a HZSM5 Vycor composite system at 400 °C and 27 bar.

MODEL CONSIDERATIONS FOR TRANSPORT IN MICROPOROUS MEMBRANES.

Model descriptions for transport in microporous systems are given especially in carbon materials (with slit shaped pores) and zeolites and especially in conjunction with pore diameter determination and diffusivity studies. Useful studies connected with gas permeability are given by Barrer [59] and by Nicholson and Petropoulos [60]. A qualitative picture is discussed by Uhlhorn et al. [44].

Barrer treats the case of zeolite systems with pore diameters of about the same size as the molecular diameter of the diffusing species.

The model is schematically given in Figure 10. Very briefly this model describes that first adsorption from the gas phase will take place at the surface (f_1J and f_2.J) followed by surface diffusion to the pore entrance (f_2J). A parallel, direct flow component is present for which gas molecules directly enter the pore without being adsorbed at the external surface (F_1J). An

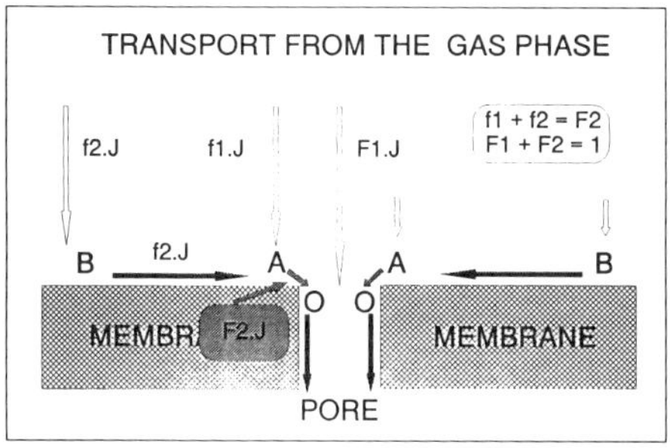

FIGURE 10. Schematic representation of gasphase transport through a microporous membrane (Barrer).

extra energy barrier (E_s) is present to enter the pore. In the micropores activated diffusion takes place with activation energy E_1. For the intracrystalline steady state flow J is derived:

$$J = \frac{s * k}{n-1} \exp(-E_1/RT)[\theta_1 - \theta_2] \tag{4}$$

s is the number of pore entrances per unit surface, n is the number of intracrystalline energy barriers E_1 (so is proportional to the membrane thickness) and Θ_l en Θ_n are the surface concentrations at entrance and exit surface respectively, Equation (4) is equivalent to

$$J_{ideal} = \frac{D_{id} C_{sat} (\theta_{1,eq} - \theta_{2,eq})}{l} \tag{5}$$

with J_{id} is the ideal (maximum flux) obtainable given equilibrium conditions, the surface process is ignorable and C_{sat} given by equilibrium adsorption conditions.

Using equation (4) Barrer estimates the magnitude of J_{id} and of J_{id}/J. For silicalites and small hydrocarbons, molecular flux values at 302 K are calculated ranging from 6 to 900 $cm^3.cm^{-2}.hr^{-1}$ with E_1 values of 3 and 6 $kcal.mol^{-1}$ respectively. The gas phase concentration at the pore entrance (C_{go}) and exit (C_{gl}) differs by a factor of 100 and $K*C_{go} \approx 10$ (K = sorption constant). The zeolite crystal dimension was taken as $5*10^{-8}$ cm. The value of J_{id}

calculated under these conditions is about the same order of magnitude as the ones measured by Geus [49].

Nicholson and Petropolus [60] analyzed the influence of adsorption forces on the flow of gases through slit shaped porous media as a function of pore diameter and temperature. This model is only applicable to pores which are at least twice the diameter of the molecule of interest. The total flow is considered to be a combination of a modified Knudsen flow J'_g and a modified surface flow J'_s. In the surface flow component the molecule is adsorbed on the wall, moves across a certain distance and is desorbed to the gas phase. The potential energy of activated molecules perpendicular to the (plane) surface of movement is expressed by a Lennard Jones (9-3 type) potential. The height of the desorption barrier on wall 1 inside the pore is now affected by the potential field of the opposing wall 2 and the result is a decreasing desorption barrier as the slit becomes narrower. In pores of 0.5 nm diameter (2.5 times the molecular diameter) and with a desorption energy $E^* = 4$ kcal.mol^{-1}, $J'_s = 2.5.J'_g$ at 500 K and even at 1000K $J'_s > J'_g$. Even for relative large pores of 2.2 nm, $J'_s \approx 0.25\ J'_g$ at 500 K. These results indicate that even at relative high temperatures and relative large micro pores adsorption forces and related surface flow phenomena can be important for permeability.

Nicholson et al. found that the flux through the gas phase J'_g within the slit is different from a "free" Knudsen flow J_g. The density concentration distributions of molecules belonging to J'_g increase in density away from the surface at low temperatures just opposite to that belonging to J'_s. At increasing temperature both distributions become more uniform. The total flux J_{tot} is now the sum of J'_g and J'_s with J'_s/J'_g becoming larger at smaller pores, larger adsorption energy and lower temperature.

The essential point is that given a certain membrane material the experimentally measured activation energy in micro pores at intermediate temperature can be considered a measure for pore size (changes) in the micropore range with the activation energy becoming smaller with decreasing pore size. This picture breaks down at pores with sizes almost equal to that of the molecule.

The molecular dynamics study of Yashonath [61] drew similar conclusions for transport in super-cages in zeolites, that is in pores with more or less spherical symmetry. Yashonath studied the transport of Xe through supercages (1.8 nm diameter) of zeolite X and Y with window diameters of 0.74 nm. At low adsorbate concentration and room temperature simulation results show that particles predominantly move along the inner surface of the supercage and that the energy barrier for this surface mediated cage to cage migration is low (0.5 kJ.mol^{-1}) This means that a type of surface flow dominated by site to site hopping, similar to that discussed by Nicholson, prevails. Within the window plane the potential energy function is a function of the distance from the centre. It shows a minimum close to the wall and a maximum in the centre, the difference being relatively small. Particles essentially pass through a channel with a radius of 0.2 nm from the line connecting the window centres.

At high concentration or high temperature transport without surface migration can occur. In this case particles migrate predominantly along a line connecting the window centres staying away from the walls of the supercage. It is shown that there is an attractive well close to the window centre in stead of an energy barrier.

CONCLUSIONS

1. Ceramic membrane systems have a considerable growth potential. Recent estimates indicate an economic value in the year 2000 between $430 and $1400 million depending on technological developments.
2. Asymmetric (supported) membranes with pores in the mesopore range ($2nm < d_p < 100nm$) can be produced in a variety of materials and forms.

3a. Microporous membranes ($d_p \leq 2.0nm$) form a new class of membranes. They can combine high gas (vapour) permeabilities (F) with large separation values (α).
The best combination of F and α values at present is based on silica derived systems with typical values of $F_{H2} = 100\text{-}200 m^3 m^{-2} day^{-1} bar^{-1}$ and $\alpha \geq 100$ for H_2 versus hydrocarbons.

3b. Zeolite membranes are in a very early development. Reliable data combinations of F and α for (polycrystalline) zeolite systems are hardly reported. Estimates based on calculations and model experiments indicate the possibility of practical zeolite membrane systems with $F \geq 100 m^3 m^{-2} day^{-1} bar^{-1}$ and separation factors mainly determined by the zeolite properties.

4. Membrane reactors with catalytically active membranes increase considerably the conversion and selectivity of e.g. dehydrogenation and partial oxidation reactions. Calculations show best conversion results for microporous membranes with large H values (ratio of permeation rate/reaction rate) <u>and</u> large permeability. The mode of operation is very important to obtain large selectivities. The further development of material combinations and reactor types yielding large F and α values is an important research direction. The phenomenological and mechanistic description of transport properties of microporous systems as well as the development of synthesis methods allowing to vary the (effective) pore diameter form intriguing new fields.

REFERENCES

1. Hsieh, H.P., Inorganic Membranes, AIChE Symp.Ser. **84**, pp. 1-18 (1988).
2. Charpin, J., Burggraaf, A.J. and Cot, L., Industrial Ceramics **11**, 83 (1991).
3. Leenaars, A.F.M., Keizer,K. and Burggraaf, A.J., J.of Mat. Sc. **19**, 1077 (1984).
4. Leenaars, A.F.M. and Burggraaf, A.J., J. Colloid and Interface Science **105**, 27 (1985).
5. Burggraaf, A.J. and Keizer, K., in: Proceedings of the 1st International Conference on Inorganic Membranes (ICIM1989), Ed. L. Cot, Montpellier, France, pp. 311-322 (1989).
6. Burggraaf, A.J., Bouwmeester, H.J.M., Boukamp, B.A., Uhlhorn, R.J.R. and Zaspalis, V.T. in: Science of Ceramics Interfaces, Ed. J. Nowotny, Elsevier Science Publ., 1991, pp. 525-568 (1991).
7. Bhave, R. Ed., Inorganic Membranes, Synthesis Characteristics and Applications, v.Nostrand Reinhold, New York, USA (1991).
8. Burggraaf, A.J., Charpin, J. and L. Cot Eds., Inorganic Membranes, Trans Tech. Publications Ltd, Zürich, Switzerland (1991).
9. Crull, A., Key Engineering Materials **6162**, 279 (1991), Trans Tech.Publ., Zürich, Switserland
10. Fain, E., Key Engineering Materials **6162**, 327 (1991), Trans Tech.Publ., Zürich, Switzerland.
11. Burggraaf, A.J. and Keizer, K., see ref. 7 Chapter 2, Synthesis of inorganic membranes, pp. 10-64 (1991).
12. Zaspalis, V.T., Keizer, K. and Burggraaf, A.J.B., Faculty of Chemical Technology, University of Twente, Enschede, The Netherlands, Internal Report (CT92/246/128) (1992)
13. Zaspalis, V.T., van Praag, W., Keizer, K., Ross, J.R.H. and Burggraaf, A.J., J.Mat.Science **27**, 1023 (1992).
14. Kitao, S., Kameda, H. and Asoeda, M., Membrane **15(4)**, pp. 222-227 (1990).
15. Tsapatsis, M., Kim, S., Nam, S.W. and Gavalas, G., I and EC Research **30**, pp. 2152-2159 (1991).
16. Uhlhorn, R.J.R., Keizer, K. and Burggraaf, A.J., J.Membrane Science **46**, pp. 225-241 (1989).
17. Hekkink, J.H.A., de Lange, R.S.A., ten Hoeve, A.A., Blankenvoorde, P.J.A.M., Keizer, K. and Burggraaf, A.J., see ref. 8, p. 375.
18. de Lange, R.S.A., Hekkink, J.H.A., Keizer, K. and Burggraaf, A.J.:
    a) see ref. 8, 77.

b) Key Engineering Materials 61/62,(1991),77.

19. Gavalas, G.R., Megiris, C.E. and Nam, S.W., Chem.Eng.Sci. **44**, 1829 (1989).
20. Lin, Y.S. and Burggraaf, A.J., J.Am.Cer.Soc. **74**, 219 (1991).
21. a) Lin, Y.S., Chemical and electrochemical Vapour Deposition of Zirconia-yttria solid solutions in porous ceramic media, PhD Thesis, University of Twente, Enschede, The Netherlands (1991).
21. b) Lin, Y.S., de Haart, L.G.J. and Burggraaf, A.J., J.Electrochem.Soc. **137**, 3960 (1990).
21. c) Lin, Y.S., Meijerink, J., Brinkman, H.W., de Vries, K.J. and Burggraaf, A.J., see ref.8, p.465.
22. Uhlhorn, R.J.R., Huis in het Veld, M.B.B.J., Keizer, K. and Burggraaf, A.J., J. of Mat.Science, **27**, 527 (1992).
23. Uhlhorn, R.J.R., Zaspalis, V.T., Keizer, K. and Burggraaf, A.J., J.of Mat. Science, **27**, 538 (1992).
24. Guizard, C., Julbe, A., Labort, A. and Cot, L., Key Engineering Materials, **122**, 47 (1991).
25. Abeles, B., Chen, L.F., Johnson, J.W. and Drake, J.M., Israel J.chem. **31** (1991) in press.
26. Sloot, H.J., PhD Thesis, A non permselective membrane reactor for catalytic gasphase reactions, University of Twente, Enschede, The Netherlands (1991).
27. Shelekhin, A.B., Grosgogeat, E.J. and Hwang, S.T.:
    a) Key Engineering Materials, **61/62**, 122, (1991), 1.
    b) see ref 8, 1.
28. Guizard, C., Ajaka, N., Besland, M.P., Larbot, A. and Cot, L.,Key Engineering Materials **61/62**, (1991), 537.
29. Koresh, J. and Soffer, A., J.C.S.Far.Transl, **76**, 2057 (1980) and **82**, 2057 (1986).
30. Koresh, J. and Soffer, A., Separation and Technology, **22**, 972 (1987).
31. Wernicke, D.L. and Osterhuber, E.J., J. Membrane Science, **22**, 137 (1985).
32. Geus, E.R., van Veen, H., Veringa, H., Schoonman, J. and van Beckum, H. in: Inorganic Membranes ICIM$_2$-91, Eds. A.J. Burggraaf, J. Charpin and L. Cot, Trans.Tech.Publ.Ltd., Zürich, Switzerland, p. 461 (1991).
33. Geus, E.R., Bakker, W.J.W., Verheyen, P.J.T., den Exter, M.J., Moulijn, J.A. and van Beckum, in: Proceedings of the 9th International Zeolite Conference, Montreal, Canada (1992).
34. Suzuki, H., Europ. Patent Application, 180200 (1986).

35. Suzuki, H., U.S. Patent 4699892 (1987).
36. Lachman, R., Pohl, U.S. Patent 4800187 (1989).
37. Geus, E.R., Mulder, A., Vischjager, D.J., Schoonman, J. and van Bekkum, H., see ref. 32, p. 57.
38. Sano, T., et al., Zeolites, **11**, 842 (1991).
39. Suzuki, K., Kiyozomi, Y., Sahne, T., Obate, K., Shando, Y., Shin, S., in: Proceedings of the 8th International Zeolite Conference, Eds. J.C. Jansen, L. Moscou, M.F.M. Post, Amsterdam, p. 65 (1989).
40. Uhlhorn, R.J.R., Ceramic Membranes for gas separation, PhD Thesis, University of Twente, Enschede, The Netherlands (1990).
41. Uhlhorn, R.J.R. and Burggraaf, A.J., in: Inorganic Membranes, ed. R. Bhave, see ref. 7, Chapter 6, p. 155.
42. van Vuren, R.J., Bonekamp, B.C., Keizer, K., Uhlhorn, R.J.R., Veringa, H.J. and Burggraaf, A.J., in: in High Tech Ceramics, Ed. P. Vincenzini, Elsevier Science Publ., p. 2235 (1987).
43. Uhlhorn, R.J.R., Keizer, K. and Burggraaf, A.J., J. Membrane Science **66**, 271 (1992).
44. R.S.A. de Lange, J.H.A. Hekkink, K. Keizer and A.J. Burggraaf, accepted for publication in: Better Ceramics Through Chemistry V, eds. W.G. Klemperer et al., (Mater.Res.Soc.Symp.Proc., **271** (1992).
45. Lee, K.H., Hwang, S.J., J. Coll. Interface Science, **110**, 544 (1986).
46. a) Asaeda, M., Du, L.D., Ikeda, R. and Fuji, M., J.Chem.Eng.Japan 19, 84, 238 (1986).
    b) Kitao, S. and Asaeda, M., J.Chem. Eng. Japan **23**, 367 (1990).
47. Kitao, S., Ishizaki, M. and Asaeda, M., see ref. 8, 175.
48. Kitao, S. and Asaeda, M., see ref. 8, 267.
49. Geus, E.R., Jansen, A.E., Jansen, J.C., Schoonman, J. and van Beckum, H., Poster session Symposium Membraantechnologie, Utrecht, The Netherlands (1990).
50. Sperry, D., Falconer, J.L. and Noble, R., J.Membrane Science **60**, 185 (1991).
51. Zaspalis, V.T. and Burggraaf, A.J., in: Inorganic Membranes, see ref. 7, Chapter 7, pp. 77-107.
52. a) Hsieh, H.P., Catal. Rev., Sci.Eng. **33** (1 and 2), pp. 1-70 (1991).
    b) Hsieh, H.P., AIChE Symposium Series **268** (85), 1 (1989).
53. Armor, J.N., Applied Cat., **49**, 1 (1989).
54. Zaspalis, V.T., Catalytically active ceramic membranes, PhD Thesis, University of Twente, Enschede, The Netherlands (1990).
55. Zaspalis, V.T., Keizer, K., van Ommen, J.G., Ross, J.R.H. and Burggraaf, A.J., in:

a) Applied Catalysis **74**, 205 (1991);

b) ibid p. 223;

c) ibid p. 235;

d) ibid p. 249.

56. Zaspalis, V.T., Harold, M., Keizer, K. and Burggraaf, A.J., Intermediate Product Yield Enhancement with a Catalytic Inorganic Membrane, to be published in Chem.Eng.Sci.
57. Mohan, K. and Govind, R., AIChE Journal **34**, 1493 (1988).
58. a) Zaspalis, V.T., Keizer, K. and Burggraaf, A.J., Internal Report CT92/136/128, Faculty of Chemical Technology, University of Twente, Enschede, The Netherlands (1992).
58. b) Zaspalis, V.T., Harold, M.P., Keizer, K. and Burggraaf, A.J., in: Proceedings of the Eurogas'92, Trondheim, Norway (1992).
59. Barrer, R.M., J.Chem.Soc.Faraday Trans. **86**, 1123 (1990).
60. Nicholson, D. and Petropoulos, J.H., J.Colloid and Interface Science **45**, 459 (1973).
61. Yashonath, S., J.Phys.Chem. **95**, 5877 (1991).

RECENT ADVANCES AND PERSPECTIVES IN MOLECULAR SIEVE SYNTHESIS

H. KESSLER

Laboratoire de Matériaux Minéraux (URA-CNRS 428) ENSCMu-UHA,

3, rue Alfred Werner, 68093 MULHOUSE Cedex (FRANCE)

ABSTRACT

A review of the major advances in molecular sieve synthesis based on the literature back to 1985 is given. The new reported framework compositions and topologies of tetrahedral and mixed frameworks are described. Progress made in templating with unusual templates such as crown ethers or with a combination of organic and inorganic species (organics + boric acid or organic + hydrofluoric acid) is discussed, as well as the advances in organic free synthesis. Some specific synthesis routes including conversion of an organic-free zeolite to an organozeolite, nonaqueous synthesis and fast heating of reaction mixtures are reviewed. Finally, the perspectives in the synthesis of molecular sieves are outlined.

INTRODUCTION

In this review, the major advances in molecular sieve synthesis will be described on the basis of the patent literature and the open literature back to 1985. Much progress concerns the composition and framework topology of tetrahedral frameworks. Numerous all-silica forms and metallosilicates of known frameworks were obtained by direct synthesis. The number of known microporous phosphates and arsenates has increased dramatically, in particular numerous beryllium and zinc compounds were reported. A new field was opened with the discovery of microporous germanium and tin sulfides based on tetrahedral sulfide frameworks. There has also been an expansion of the number of mixed frameworks based on tri- and tetracoordination (aluminoborates), tetra- and pentacoordination (zinc phosphate), and tetra- and hexacoordination (phosphomolybdates and titanosilicates). An ultralarge-pore aluminosilicate (pore size larger than 15 Å) with good thermal stability and extremely high adsorption capacity of organic molecules was patented.

Concerning the organic templates, novel nitrogen-containing species were used but also unusual ones such as crown ethers. The use of a combination of organic and inorganic species, such as organics + boric acid or organics + hydrofluoric acid enabled the preparation of several novel materials. On the opposite, much effort was made to decrease the amount of organic species in synthesis mixtures and even to perform organic-free syntheses.

Effort was also made to develop specific synthesis routes such as conversion of an organic-free zeolite to an organozeolite, nonaqueous procedures and fast heating of reaction mixtures.

From the Proceedings of the 9th International Zeolite Conference, Montreal 1992, Eds. R. von Ballmoos et al.,

The paper will conclude with some comments regarding the perspectives in molecular sieve synthesis.

NEW COMPOSITIONS AND STRUCTURES OF TETRAHEDRAL FRAMEWORKS

All-silica structures

Currently, numerous silica-only molecular sieves are known. They were all prepared in the presence of an organic species. In some cases the organics were the same as for the corresponding Si,Al zeolites, in others, specific templates were used. The silica forms of the following structure types were obtained so far by direct synthesis : AFI [1], BEA*[2], FER [3], MEL [4,5], MFI [6-8], MTT [9], MTW [10], SOD [11,12], TON [13] and ZSM-48 [2]. The types BEA*, MTT and MTW crystallized with one single template whereas for the others several organics were efficient. It can be expected that more porous silica forms will be obtained, perhaps by finding new templates with efficient host-guest interactions. One way could be to use an association of organics and other species such as H_3BO_3, which was used respectively with 4,4'-trimethylenedipiperidine and ethylene diamine in the synthesis of silica-Beta [2] and silica-ferrierite [3] . Both, the organic species and H_3BO_3 are occluded in the structure and thus contribute to its stabilization.

Metallosilicates with divalent heteroatoms

A few microporous materials with Be, Co, Cu and Zn as heteroatoms beside Si have been published.

A synthetic analogue of the beryllosilicate mineral lovdarite, $K_4Na_{12}Be_8Si_{28}O_{72}\cdot 18H_2O$, was described recently [14]. Be [15] and Co [16,17] were incorporated in limited amounts in the MFI-type framework. A copper-containing analcime $Cs_2CuSi_5O_{12}$ has been made by high-temperature synthesis [18]. A microporous sodium zincosilicate, VPI-7 with a novel structure comprising 3-membered rings besides larger rings, was synthesized from gels containing tetraethylammonium cations as help species [19]. Another zincosilicate, $K_2ZnSi_4O_{10}$, with 10-membered rings was reported [20].

Metallosilicates with trivalent heteroatoms

Currently numerous B, Fe and Ga-containing materials were obtained by direct synthesis. The synthesis of the analogues of ZSM-5 has been the most investigated. They were generally obtained by a procedure derived from that patented by Argauer and Landolt for Al-ZSM-5 [21]. A large number of other structures with these T(III) elements as framework elements are now known too.

The reported B-containing materials are of the types AFI [22,23], BEA* [23,24], FER [25], MEL [24,26], MFI [24,26-29], MTW [24,30,31], NU-1 [24,32], SOD [33] and ZSM-48 [34]. There is good evidence of B incorporation into the tetrahedral framework. It was mostly based on X-ray difraction data (unit cell dimensions variation as a function of B content), and infrared and ^{11}B NMR spectroscopies.

The published Fe analogues are of the types BEA* [35], MEL [36], MFI [37-42], MOR [43], MTT [44], MTW [31], NU-1 [32], SOD [45] and TON [46]. Techniques such as ESR, EXAFS, UV-visible diffuse reflectance and Mössbauer spectroscopy were generally used to demonstrate the tetrahedral coordination of iron.Extra-framework iron species may be present in the as-synthesized materials, therefore, care must be taken to avoid the formation of insoluble brown-colored ferric hydroxides on mixing the reactants [42].

The reported Ga-containing materials are numerous too, the corresponding structure types are ABW [47-49), ANA [50], ERI [51], FAU [52-57], FER [58], LTL [59-61], MAZ [50], MFI [62-68], MOR [69], NAT [50], NU-1 [32], SOD [70,71], THO [72] and TON [73]. X-ray diffraction data, EXAFS, infrared and Ga NMR spectroscopies were employed to check the introduction of gallium into the framework.

Metallosilicates with tetravalent heteroatoms

The partial replacement of Si by Ge in silicalite I and in ZSM-48 was reported by Gabelica et al. [74] and by Pang and co-workers [75] respectively. A substitution degree Ge/(Ge+Si) as high as 0.33 could be reached for Ge-silicalite I.

The isomorphous substitution of Ti for Si is more limited, it was reported for frameworks of the type BEA* [76], MEL [77] and MFI [78,79]. Ti incorporation in the MFI-type framework is now well established, evidence was obtained from X-ray diffraction data, infrared spectroscopy and EXAFS. For Ti-silicalite II (MEL type), unit cell expansion and specific infrared absorption spectra led to the conclusion that Ti was incorporated in the framework [77].

Metallosilicates with other heteroatoms

The direct synthesis of microporous metallosilicates based on Zr [80,81], V [82,83], Cr [83,84] and Mo [85] corresponding to high silica structures has been reported. The amount of element present in the materials is generally very small, therefore it is difficult to get a sound proof of the incorporation of these elements into the framework.

Beryllo-, zincophosphates and arsenates

Harvey and Meier [86] have synthesized five new beryllophosphates from mixtures containing inorganic cations and tetraethylammonium hydroxide at 100-200°C for 7 to 21 days.

One of them shows a new structure type, BPH, that has no aluminosilicate analogue, the others are iso-structural with zeolite RHO (Li), gismondine (Na), edingtonite (K) and analcime (Cs).

Gier and Stucky [87] reported a great number of beryllo-, zincophosphates and arsenates obtained by a low temperature synthesis, 4 to 100°C, in a wide range of pH values, 2 to 12, and crystallization times, 5 to 72 hrs. The new phases are given in table 1.

Table 1. Beryllo-, zincophosphates and arsenates synthesized by Gier and Stucky[87].

Structure type	Chemical formula
FAU	$Na_{96}Be_{96}P_{96}O_{192} \cdot 192H_2O$ $(Na,TMA)_{96}Zn_{96}P_{96}O_{192} \cdot 192H_2O$
Li ABW	$Li_4Be_4T_4O_{16} \cdot 4H_2O$ (T = P, As) $Li_4Zn_4T_4O_{16} \cdot 4H_2O$ (T = P,As)
RHO	$Li_{24}Be_{24}T_{24}O_{96} \cdot 40H_2O$ (T = P, As) $Rb_{24}Be_{24}As_{24}O_{96} \cdot 40H_2O$
SOD	$Na_6Zn_6T_6O_{24} \cdot 8H_2O$ (T = P,As)

All phases, except the Na,TMA FAU-type zincophosphate were prepared in the presence of inorganic species only. In particular, the RHO analogue $Li_{24}Be_{24}P_{24}O_{96} \cdot 40H_2O$ which was prepared by Harvey and Meier [86] in the presence of TEAOH was obtained from an organic-free mixture by operating at 70°C. The structures are generally unstable to water loss which occurs at 140-350°C, depending on the structure. The RHO-type $Rb_{24}Be_{24}As_{24}O_{96} \cdot 40H_2O$ material is stable up to at least 600°C.

Aluminophosphates and derived materials

Wilson has reviewed very recently the synthesis of $AlPO_4$-based molecular sieves [88]. Most of these materials were discovered by the UOP researchers. The synthesis of VPI-5, the first 18-membered ring material, was reported by Davis et al. in 1988 [89]. More than 27 different structures are known and 14 elements (Si, Co, Fe, Mg, Mn, Zn, Be, Sn, B, Ga, Ge, Ti, As, Li) have been incorporated into the frameworks besides Al and P. Over 100 organic additives have been reported, these are generally quaternary ammonium cations or amines.

Wilson [88] has given a list of selected $AlPO_4$-based structures with typical organic additives. In Table 2 this list is extended to include the structure codes newly adopted by the IZA.

Table 2. $AlPO_4$-based structures and typical organics used for their synthesis

Structure	Organic	IZA Code
5	n-Pr_3N	AFI
8	n-Bu_2NH	AET
11	n-Pr_2NH	AEL
12	Ethylenediamine	
14	t-$BuNH_2$	
16	Quinuclidine	AST
17	Quinuclidine	ERI
18	Et_4N^+	AEI
20	Me_4N^+	SOD
21	Pyrolidine	
22	Quinuclidine	AWW
23	Pyrrolidine	
24		ANA
25	Calc.of 21 $\rightarrow$ 25	ATV
28	Calc.of 23 $\rightarrow$ 28	
31	n-Pr_2NH	
33	Me_4N^+	ATT
34	Et_4N^+	CHA
35	Quinuclidine	LEV
36	n-Pr_3N	
37	Me_4N^++Pr_4N^+	FAU
39	n-Pr_2NH	ATN
40	Pr_4N^+	
41	n-Pr_2NH	
42	Me_4N^+	LTA
43	n-Pr_2NH	GIS
44	Cyclohexylamine	CHA
46	n-Pr_2NH	AFS
47	N,N-diethanolamine	CHA
50	n-Pr_2NH	AFY
52	n-Pr_3N+Et_4N^+	AFT
VPI-5	n-Pr_2NH or Et_4N^+ or organic-free	AFI

The two aluminosphophates $AlPO_4$-C and $AlPO_4$-D with structure types APC and APD, which are obtained by controlled dehydration of the hydrate $AlPO_4 \cdot 1.5H_2O$ H3 should be added to this list.

Alumino- and galloarsenates

Xu Ruren et al. reported recently a novel aluminoarsenate $AlAsO_4$-1 [90] and two novel galloarsenates $GaAsO_4$-1 and -2 [91] synthesized hydrothermally in the presence of organic aminocompounds. $AlAsO_4$-1 was obtained with ethanolamine as a template at 150-200°C for 4 days. Its structure comprises a one-dimensional 8-MR channel system. $GaAsO_4$-1 and -2 were synthesized in the presence of tetramethylammonium hydroxide and dimethylamine respectively. The synthesis procedure of $GaAsO_4$-2 is particular. First, a mixture of dimethylamine, $Ga(OH)_3$, pyroarsenic acid and water was heated at 180°C for 5 days. Hydrofluoric acid was added to the mixture (HF/Ga_2O_3 = 1.2) and heated at 200°C for 10 days. The structure shows a one-dimensional 10-MR and a two-dimensional 8-MR channel system. All the As atoms are tetracoordinated whereas 1/3 of the Ga are hexacoordinated and 2/3 are pentacoordinated by 4O and 1F atoms.

Tetrahedral sulfide frameworks

A new field was opened recently with the hydrothermal synthesis of microporous germanium and tin sulfides in the presence of tetraalkylammonium and amine additives [92]. The elements Mn, Fe, Co, Ni, Cu, Zn, Cd and Ga could be incorporated into the framework of the thiogermanates. Typically, precipitated GeS_2 and SnS_2 were used as the Ge and Sn sources, the templates were added as bicarbonate solutions and the extra elements as acetates. Heating was performed at 100-150°C for 18 to 162 hrs. It is worthy to note that the starting pH was 7-9 and the final pH 0.5-3.

Eleven structure types, with no oxide molecular sieve analogue, were isolated for the GeS_2 and GeS_2-derived materials, and one for SnS_2. The structure of the GeS_2-derived material TMA-CoMnGS-2 was solved, it comprises Ge_4S_6 adamantane-type units. Three such units are linked by a MeS_3SH tetrahedron. The organic species were found occluded in the solids, their removal with the usual heating procedure led to a collapse of the structures.

The use of the bigger S atom as a bridging element instead of O should enable the isomorphous substitution of larger atoms which cannot be incorporated in tetrahedral oxide frameworks.This would expand the tetrahedral framework compositions. Secondly the Me-S bonds are more covalent than the corresponding bonds involving O, therefore new secondary building units are possible. This is illustrated by the adamantane-type Ge_4S_6 units which consist of 3 MRs.

MIXED FRAMEWORKS

In this section will be described oxide materials with mixed tri- + tetracoordination, tetra- + pentacoordination and tetra- + hexacoordination which were reported recently.

Sodium calcium aluminoborates

Xu Ruren and co-workers reported four novel types of microporous aluminoborates named B-Cn (n = 7,8,9,10) with a 3-dimensional framework [93]. The synthesis was performed in the presence of tetraalkylammonium cations or amines in a narrow pH range (~one pH unit) below 4.5 at 160-210°C for 5-20 days. On the basis of infrared absorption and X-ray photoelectron spectroscopies, the authors concluded to the presence of AlO_4,BO_4 and BO_3 units for B-C7, B-C8 and B-C10, and AlO_4 and BO_4 units only for B-C9. No organics were found occluded in the structures. The pore size determined by using gauge molecules such as n-hexane and cyclohexane was found to be about 6.0 Å, 4.3 Å - 6.0 Å, 6 Å and 4.3 Å for B-C7, B-C8, B-C9 and B-C10 respectively. They are stable up to 425-733°C, depending on the compound.

Microporous zinc phosphate hydrate

A microporous zinc phosphate with a channel structure, $Zn_3(PO_4)_2 \cdot H_2O$, was synthesized by Corcoran et al. [94] in the presence of tripropylamine at 200°C for 5 days. No organic was found in the structure. The same hydrate was obtained previously at 150°C from an organic-free mixture by Riou et al. [95]. Both groups solved the crystal structure on a single crystal. The three-dimensional framework comprises a one-dimensional channel system which can be built up by combining 3-, 4- and 5-membered rings. The 3-membered rings consist of two zinc and one phosphorus atoms. Zinc is present in ZnO_4 and ZnO_5 units. The fifth oxygen of the ZnO_5 unit is the O atom of a water molecule, thus, one might consider that the framework is in fact a tetrahedral framework and not a true mixed framework. Such a situation appears too in aluminophosphates and even more in gallophosphates in which Al or Ga may be penta- or hexa-coordinated through bonding with OH groups or water molecules.

The loss of the hydration water of $Zn_3(PO_4)_2$ occurs in the range 200-400°C. The solid obtained at 700°C was α-$Zn_3(PO_4)_2$ [96].

Mixed tetrahedral-octahedral frameworks

Microporous phosphomolybdates

Five novel types of microporous phosphomolybdates based on a tetrahedral-octahedral framework were reported by Haushalter and co-workers [97-99]. Typically, the synthesis was performed by heating a mixture of MoO_3, Mo(particle size < 2μm), H_3PO_4, H_2O and possibly an organic species (Me_4N^+ or amine) at 200°C for 3 to 6 days. The chemical formula of the materials is reported in Table 3 with the mean oxidation state of molybdenum.

Table 3. Novel phosphomolybdates with mixed tetrahedral-octahedral framework

Formula	Mean Mo oxidation state	Ref.
$(Me_4N)_{1.3}(H_3O)_{0.7}[Mo_4O_8(PO_4)_2]\cdot 2H_2O$	+5	97
$Mo_8O_{12}(PO_4)_4(HPO_4)_2\cdot 13H_2O$	+5	98
$CH_3NH_3[Mo_2O_2(PO_4)_2(H_2PO_4)]$	+5	99
$(CH_3)_2NH_2[Mo_2O_2(PO_4)_2(H_2PO_4)]$	+5	99
$NH_4Mo_2P_2O_{10}\cdot H_2O$	+4.5	99
$(NH_4)_3Mo_4P_3O_{16}$	+3.5	99

The framework consists of MoO_6 octahedra and phosphate tetrahedra. In $(Me_4N)_{1.3}(H_3O)_{0.7}[Mo_4O_8(PO_4)_2]\cdot 2H_2O$ for example, edge-sharing between four MoO_6 units results in Mo_4O_8 tetramers which are linked by PO_4 tetrahedra. The thermal stability of the materials is variable, the hydrates are stable to dehydration but the organic template removal by heating results in a deterioration of the structure, or even a collapse, as observed for the Me_4N^+-containing solid.

Microporous titanosilicates with a mixed tetrahedral-octahedral framework

Chapman and Roe [100] prepared a novel microporous titanosilicate named GTS-1 by heating strongly alkaline titanium silicate gels. It is a strutural analogue of the mineral pharmacosiderite. Its ideal formula would be $M_4(Ti_4Si_3)O_{16}$ (M = K, Cs) and its framework consists of TiO_6 and SiO_4 units whose arrangement generates a three-dimensional channel system bounded by 8-membered rings of alternating octahedra and tetrahedra. The same synthesis route yielded synthetic analogues of the zeolitic minerals zorite and vinogradovite.

Other types of microporous titanosilicates ETS-4 and ETS-10, consisting of octahedral Ti and tetrahedral Si, were reported in a recent patent [101]. They were obtained by heating aqueous mixtures containing Na_2SiO_3,$TiCl_3$ or $TiCl_4$, NaOH and KF (pH ca. 10) at 150°C for 7 days. According to adsorption measurements the pore sizes are ca. 4 Å and 8 Å for ETS-4 and ETS-10, respectively. The O_2 pore volumes (at 77 K) for various exchanged forms of ETS were found to range from 8.39 cc/100 g to 11.95 g/100 g for the K- and as-synthesized forms, respectively. They were found stable up to 200-300°C (ETS-4) and 600°C (ETS-10).

NOVEL ULTRALARGE-PORE ALUMINOSILICATES

Mobil researchers reported recently in a patent [102] the synthesis of aluminosilicate materials with ultralarge pores (greater than 15 Å) and extremely promising properties such as very high adsorption capacity of organic molecules. The solids show generally a one-dimensional hexa-gonal arrangement of open channels as observed by transmission electron microscopy.

They were synthesized from mixtures containing quaternary ammonium cations with one of the four groups having 6 to 18 carbons such as cetyltrimethyl ammonium (CTMA), and smaller cations such as Me_4N^+, Et_4N^+ or Pr_4N^+. Thus, in example 4 of the patent [102] a mixture of molar composition $6.1(CTMA)_2O : 5.2(Me_4N)_2O : 0.23Na_2O : Al_2O_3 : 33.2SiO_2 : 789H_2O$ prepared from CTMA hydroxide, Catapal alumina, tetramethylammonium silicate and precipitated silica was heated at 150°C for 48 hrs. The authors report that the solid obtained, which was calcined at 540°C, showed a very high surface area, 1043 $m^2\ g^{-1}$, and a benzene adsorption capacity of 66.7 wt % (at 25°C, benzene pressure = 50 torr).

ORGANIC ADDITIVES

It is well known that the use of organic additives brought about a dramatic increase of the number of synthetic molecular sieves. Starting with simple species such as tetramethylammonium cations, more and more complex organic molecules, essentially nitrogen compounds, have been used : di- and triquaternary ammonium cations, linear polyelectrolytes, long chain quaternary ammonium cations such as cetyltrimethylamonium, etc. Recently a new family of templates, crown ethers, was used in the synthesis of silica-rich faujasite and silica-rich hexagonal EMT-type aluminosilicate [103]. A combination of structure directing agents such as organic species and boric acid [2] or organic species and hydrofluoric acid [104] enabled the synthesis of novel materials. On the other hand, much effort was made to decrease the amount of organic species in starting mixtures by seeding with a structure directing mixture. In some cases organic-free syntheses became possible without seeding, by tuning the synthesis conditions or by adding seeds to the starting mixture. Finally, number of materials could be obtained in the presence of what may be called help species which influence the gel chemistry but are not occluded in the crystalline solids.

These various points will be developed in this section.

Synthesis of high-silica faujasite and of silica-rich hexagonal EMT-type material by using crown ether templates

Recently, Delprato et al. [103] reported the synthesis of silica-rich faujasite using Na^+ and the crown ether 15-crown-5 (1,4,7,10,13 pentaoxacyclopentadecane) as an organic structure directing agent. With Na^+ and the crown ether 18-crown-6, EMT-type material was obtained.

Good reproducibility and high yields were obtained with weakly basic aluminosilicate hydrogels of composition 1.8-2.2Na_2O : Al_2O_3 : 10SiO_2 : 1crown ether : 140H_2O. The gels were aged for 24 hrs at 20°C and heated for 8 days at 110°C. The framework Si/Al ratio was 4.5 for both materials.

The crown ether templates appear to be very selective and efficient in these syntheses. In particular, the material obtained with 18-crown-6 is of a virtually pure hexagonal EMT-type structure with almost no intergrowth with the cubic FAU-type structure.

Usual tetraalkylammonium templates were used for the synthesis of the high-silica zeolites ZSM-20 [105] and ECR-30 [106]. The selectivity of these organics is lower than that of the crown ethers, as a matter of fact both materials are silica-rich polytypes of faujasite.

Combination of organic and inorganic additives

As mentioned previously in the section on all-silica molecular sieves, Gies and co-workers [2] were successful in the synthesis of silica-ferrierite and silica-Beta by using respectively ethylenediamine and 4,4'-trimethylenedipiperidine as organics and boric acid as an inorganic additive. Both organic and inorganic species were found occluded in the structure [3].

The use of organic templates and fluoride for the synthesis of silica-based zeolites and $AlPO_4$-derived molecular sieves, has led to several materials in which the fluoride anion is trapped besides the organic species. This is particularly the case for the all-silica AST-type octadecasil obtained in the presence of quinuclidine and HF [107] and the two novel cubic microporous gallophosphates which were only obtained when HF was present. The first, of the LTA type, crystallized from mixtures containing di-n-propylamine and HF [104], and the second which is the first 20-membered ring molecular sieve of the-CLO type was obtained in the presence of quinuclidine and HF [108]. The crystal structure of the three materials was determined, the fluoride anions were found in the double-four rings whereas the organic cations were located in the larger cages of the structures. It can be assumed that fluoride plays a co-structuring role beside the organic species, by templating the small double-four ring cages and thus contributes efficiently to the stabilization of the structure.

Two other examples of cation and anion templating are found in $AlPO_4$-22 [109] and in the LTA-type silicoaluminophosphate SAPO-42 [110]. The former was obtained with quinuclidine which is occluded in the rpa cage of the structure, whereas hydrogen phosphate was found in the aww cage. The latter was synthesized in the presence of Me_4N^+, the ^{31}P MAS NMR spectrum was consistent with phosphate anions being trapped in the sodalite cage of the LTA-type structure. Phosphate anions were also found occluded in zeolites ZK-21 and ZK-22 [110].

Organic-free and almost organic-free syntheses

Organic-free synthesis of MFI-type zeolites

Several patents have been reported on the organic-free synthesis of ZSM-5 a few years after Argauer and Landolt's discovery [21]. Recently there have also been publications in the open literature [111-114]. There is agreement that pure ZSM-5 can be obtained at 190°C, 24-40 hrs, in the absence of seeds, from organic-free gels having the composition (5-14)Na_2O : Al_2O_3: (50-70)SiO_2 : (1000-3200)H_2O. Dai et al. [112,113] performed the syntheses in PTFE-lined autoclaves which were carefully cleaned by HF before each run, in order to eliminate any seeds left from a preceding experiment. By heating a gel of composition 11.2Na_2O : Al_2O_3 : 70SiO_2 : 3213H_2O, at 190°C for 24 hrs without stirring, pure ZSM-5 crystallized. The Si/Al ratio in the zeolite was 22.

On increasing the heating time, ZSM-5 transformed into mordenite, which converted to quartz (Figure 1).

For a starting SiO_2/Al_2O_3 = 20-25 or Na_20/Al_2O_3 = 14, mordenite was the major phase and for SiO_2/Al_2O_3 ~200, quartz was obtained as a major phase [112,114].

As a conclusion, it appears that the range of framework Si/Al ratios obtained from organic-free gels is rather narrow (Si/Al = 20-25). This corresponds presumably to the composition for which enough framework charges are present to interact with the Na^+ cations and stabilize the structure.

On the other hand, numerous recent patents and publications in the open literature report on the organic-free synthesis of ZSM-5 with deliberate seeding. Usually, the seeds were crushed ZSM-5 crystals, either prepared in the presence of an organic or recycled from an organic-free synthesis.

Almost organic-free synthesis

Recently Fajula et al. [115] carried out an almost organic-free synthesis of zeolite Omega by adding a structure directing mixture to an organic-free crystallization mixture.

The structure directing mixture was prepared by heating at 50°C for 20 days a mixture of composition 3.2Na_2O : 0.8$(Me_4N)_2O$: Al_2O_3 : 8.2SiO_2 : 160H_2O. About 25-30 vol. % of this seeding mixture was then added at 20°C to the organic-free crystallization mixture whose composition was 3.2Na_2O ; Al_2O_3 : 8.2SiO_2 : 160H_2O. Heating was performed at various temperatures with stirring. As can be seen on Figure 2, the induction period can be reduced drastically.

The same type of procedure was used for the synthesis of offretite [116]. There are several advantages for such a procedure :

(i) the induction period is decreased with respect to that of the usual synthesis procedure

(ii) the consumption of Me_4N^+ is drastically reduced

(iii) separation of the nucleation and growth steps is possible, therefore a very homogeneous crystal size distribution can be achieved and well crystallized zeolites can be obtained.

Synthesis with help species

As already mentioned above, numerous materials were prepared in the presence of organic species which are not found occluded in the structure. Nevertheless they seem to make the synthesis easier.

This was the case for the synthesis of the 18-membered ring aluminophosphate VPI-5 which was first synthesized from mixtures containing di-n-propylamine or Bu_4N^+ cations [89]. The crystal structure of the as-synthesized form was published recently [117] The only guest species are water molecules linked by hydrogen bonds into a triple helix filling the channels. The helix interacts with Al atoms of the framework. The role of the organic species is not quite clear, they may influence the gel chemistry or participate in the structuration of the water molecules. The organic-free synthesis of this aluminophosphate turned out to be possible [118].

Other examples of syntheses in the presence of help species are the preparations of beryllophosphates [86], VPI-7 [19] and the zinc phosphate hydrate $Zn_3(PO_4)_2{\cdot}H_2O$ [94].The latter was also successfully obtained from wholly inorganic systems [95], but under narrow conditions.

SPECIFIC SYNTHESIS ROUTES

In this section, three routes of interest will be addressed, i. e., the transformation of an organic-free zeolite into an organozeolite, the synthesis in nonaqueous media and the fast crystallization technique.

Transformation of an organic-free zeolite into an organozeolite

The transformation under various conditions such as prolonged heating during the synthesis or heating in an aqueous solution of given composition is well documented. Recently Zones and co-workers performed a rather systematic study on the conversion of zeolites to other zeolites. Particularly, they were successful in preparing Si,B and later Si,Al AFI-type materials by heating calcined B-Beta in an aqueous solution of the appropriate template, i. e., N,N,N-trimethyl-1-adamantammonium [23]. Such a route using a microporous source material may offer interesting possibilities in the future.

Synthesis in nonaqueous media

The first material synthesized from a nonaqueous medium was silica sodalite [11]. It was made from mixtures of ethylene glycol, fumed silica and solid NaOH or solid Na_2CO_3 at 170°C. Great care was taken to remove completely any water present, however a catalytic role of traces of H_2O could not be ruled out. In fact, the addition of small amounts of water had no observable effect on the formation of silica-sodalite, only large amounts hindered its formation.Later, silica-sodalite could be prepared from normal aqueous mixtures with trioxane as a template [12].

Up to now, the use of nonaqueous media has not produced any other novel material. Only the synthesis of ZSM-5 [119], ZSM-35 [119], ZSM-39 [120] and ZSM-48 [121] was reported. Therefore this route using nonaqueous mixtures does not seem very promising. Moreover, it is not always clear if the starting mixtures were really water-free.

Fast crystallization

The synthesis of silica-based zeolites and $AlPO_4$ materials was performed recently by fast crystallization. Zeolites A and ZSM-5 were obtained after 10 min. microwave heating [121], VPI-5 was made in 45 min. by heating the autoclave in an oil bath [122] and $AlPO_4$-5 crystallized by direct heating at 600°C for 5-10 min.[123]. Such routes may be promising because novel metastable materials which would no more be found after a longer heating may be obtained.

PERSPECTIVES

The past of zeolite synthesis is bright, it can be expected that its future will be so too. More progress will probably be made using a large variety of organic species of different natures and shapes or a combination of species. The choice of a template will presumably be made easier by using computor generated models of guests and hosts and looking for the best geometrical fit, or even better, by computing the interactions between the frameworks and the included species. The framework model may be that of an already known framework or of an hypothetical one. It can also be anticipated that more materials, which are presently synthesized in the presence of organics only, will be obtained from organic-free media. This could be at least the case for zeolites which are synthetic counterparts of natural zeolites.

ACKNOWLEDGMENT

I am grateful to Professor J.L. Guth for fruitful discussions.

REFERENCES

1. Zones, S.I., U.S. Patent 4,665,110 (1987) ; Bialek, R., Meier, W.M., Davis, M.E. and Annen, M.J., Zeolites **11**, 438 (1991).
2. Gies, H., in : *Inclusion Compounds*, Vol. 5 : Inorganic and Physical Aspects of Inclusion, Eds. J.L. Atwood, J.E.D. Davies and D.D. Mac Nicol, Oxford University Press, pp. 1-36 (1991).
3. Gies, H. and Gunawardane, R.P., Zeolites **7**, 442 (1987).`
4. Bibby, D.M., Milestone, N.B. and Aldridge, L.P., Nature **280**, 664 (1979).
5. Mostowicz, R., Nastro, A., Crea, F. and B. Nagy, J., Zeolites **11**, 732 (1991).
6. Flanigen, E.M., Bennett, J.M., Grose, R.W., Cohen, J.P., Patton, R.L., Kirchner, R.M. and Smith, J.V., Nature **271**, 152 (1978).
7. Price, G.D., Pluth, J.J., Smith, J.V., Benett, J.M. and Patton, R.L., J. Am. Chem. Soc. **104**, 5971 (1982).
8. Patarin, J., Soulard, M., Kessler, H. Guth, J.L. and Baron, J., Zeolites **9**, 397 (1989).
9. Franklin, K.R. and Lowe, B.M., in: *Zeolites: Facts, Figures and Future*, Studies in Surface Science and Catalysis, Vol. 49A, Eds. P.A. Jacobs and R.A. van Santen, Elsevier, Amsterdam, pp. 179-188 (1989).
10. Gies, H., Marler, B., Fyfe, C.A., Kokotailo, G. T. and Cox, D.E., J. Phys. Chem. **94**, 3718 (1990).
11. Bibby, D.M. and Dale, M.P., Nature **317**, 157 (1985).
12. Keijsper, J., den Ouden, C.J.J. and Post, M.F.M., in: Ref. 9, pp. 237-247.
13. Marler, B., Zeolites **7**, 393 (1987).
14. Ueda, S., Koizumi, M., Baerlocher, Ch., McCusker, L.B. and Meier, W.M., in: *Proceedings of the 7th International Zeolite Conference*, Tokyo, Poster Paper 3C-3 (1986).
15. Romannikov, V.N., Chumachenko, L.S., Mastikhin, V.M. and Ione, K.G., J. Catalysis **94**, 508 (1985).
16. Rossin, J.A., Saldarriaga, C. and Davis, M.E., Zeolites **7**, 295 (1987).
17. Mostowicz, R. , Dabrowski, A.J. and Jablowski, J.M., in: Ref. 9, pp. 249-259.
18. Heinrich, A.R. and Baerlocher, Ch., Acta Crystallogr. **C47**, 237 (1991).
19. Annen, M.J., Davis, M.E., Higgins, J.B. and Schlenker, J.L., in: *Synthesis/Characterization and Novel Applications of Molecular Sieves*, Materials Research Society Symposium Proceedings, Vol. 233, Eds. R.L. Bedard, T. Bein, M.E. Davis, J. Garces, V.A. Maroni and G.D. Stucky, Materials Research Society, Pittsburgh, Pa., pp. 245-254 (1991).

20. Kohara, S. and Kawahara, A., Acta Crystallogr.**C46**, 1373 (1990).

21. Argauer, R.J. and Landolt, G.R., U.S. Patent 3,702,886 (1972).

22. Zones, S.I., Holtermann, D.L., Innes, R.A., Santilli, D.S., Yuen, L.T. and Ziemer J.N., PCT Int. Appl. WO 91-844 (1991).

23. Van Nostrand, A. and Zones, S.I., in: *Zeolites and Pillared Clays Synthesis*, Eds. M.L. Occelli and H.E. Robson, Van Nostrand Reinhold, New York, in press.

24. Kuehl, G.H., US Patent 4,661,467 (1987).

25. Shavoki, M.S. and Schimandle, J.J., Eur. Patent Appl. 234,766 (1987).

26. Perego, G., Bellusi, G., Carati, A., Millini, R. and Fattore, V., in: *Zeolite Synthesis*, ACS Symposium Series, Vol. 398, Eds. M.L. Occelli and H.E. Robson, ACS, Washington DC, pp. 360-373 (1989).

27. Coudurier, G. and Vedrine, J.C., in: *New Developments in Zeolite Science and Technology*, Proc. 7th International Zeolite Conference, Eds. Y. Murakami, A. Iijima and J.W. Ward, Elsevier, pp. 643-652 (1986) and references therein.

28. Kutz, N.A., in: *Perspectives in Molecular Sieve Science*, ACS Symposium Series, Vol. 368, Eds. W.H. Flank and T.E. Whyte, Jr., ACS, Washington DC, pp. 532-541 (1988) and references therein.

29. Cichocki, A.,Parasiewiz-Kaczmarska, J., Michalik, M. and Buś, M., Zeolites **10**, 577 (1990).

30. Zhao Yanan, Xiang Shouhe and Li Hexuan, Cuihua Xuebao **11**, 323 (1990).

31. Zhao Yanan and Li Xe Hexuan, Shiyou Xuebao, Shiyou Jiagong **6**, 33 (1990).

32. Bellusi, G. , Millini, R., Carati, A., Madinelli, G. and Gervasini, A., Zeolites **10**, 642 (1990).

33. Hoelderich, W. and Lermer, H., DE Patent 3,537,998 (1987).

34. Jiao , Q. and Pang W., Huaxue Xuebao **48**, 761 (1990).

35. Kumar, R., Thangarij, A., Bhat, R.N. and Ratnasamy, P., Zeolites **10**, 85 (1990).

36. Reddy, J.S., Reddy, K.R., Kumar, R. and Ratnasamy, P., Zeolites **11**, 553 (1991).

37. Kotasthane, A.N., Shiralkhar, V.P., Hedge, S.G. and Kulkarni, S.B., Zeolites **6**, 253 (1986).

38. Calis, G., Frenken, P., de Boer, E., Swolfs, A. and Hefni, M.A., Zeolites **7**, 319 (1987).

39. Borade, R.B., Zeolites **7**, 319 (1987).

40. Meagher, A., Nair, V. and Szostak, R., Zeolites **8**, 3 (1988).

41. Meagher, A., Zeolites **9**, 87 (1989).

42. Ulan, J.G., Gronski, R., Szostak, R., Zeolites **11**, 466 (1991) ; Zeolites **11**, 472 (1991).

43. Chandwadkar, A.J., Bhat, R.N. and Ratnasamy, P., Zeolites **11**, 42 (1991).

44. Kumar, R. and Ratnasamy, P., J. Catal. **121**, 89 (1990).

45. Szostak, R. and Thomas, T.L., J. Chem. Soc., Chem. Comm., 113 (1986).

46. Borade, R.B., Adnot, A. and Kaliagine, S., Zeolites **11**, 710 (1991).

47. Newsam, J.M., J. Chem. Soc., Chem. Comm., 1295 (1986).

48. Newsam, J.M., J. Phys. Chem. **92**, 445 (1988).

49. Yang, J., Xie, D., Yellon, W.B. and Newsam, J.M., J. Phys. Chem. **92**, 445 (1988).

50. (a) Jarman, R.H., Jacobson, A.J. and Melchior, M.T., J. Phys. Chem. **88**, 5748 (1984) ; (b) Yellon, W.B., Xie, D., Newsam, J.M. and Dunn, J., Zeolites **10**, 553 (1990).

51. Occelli, M.L., U.S. Patent, 4,931,266 (1990).

52. Selbin, J. and Mason, R.B., J. Inorg. Nucl. Chem. **20**, 222 (1961).

53. Kuehl, G.H., J. Inorg. Nucl. Chem. **33**, 3261 (1971).

54. Suzuki, K., Kiyozumi, Y., Shin, S. and Ueda, S., Zeolites **5**, 11 (1985).

55. Newsam, J.M., Jacobson, A.J. and Vaughan, D.E.W., J. Phys. Chem. **90**, 6858 (1988).

56. Occelli, M.L., U.S. Patent 4,803,060 (1989).

57. Delprato, F., Guth, J.L. and Zivkov, C., Fr. Demande FR 2,650,582 (1991).

58. Sulikowski, B. and Klinowski, J., J. Chem. Soc., Chem. Comm., 1289 (1989).

59. Xu, R. and Liu, X., Chimica Sinica **42**, 227 (1984).

60. Newsam, J.M., J. Chem. Soc., Chem. Comm., 1295 (1986).

61. Occelli, M.L., U.S. Patent 4,919,907 (1990).

62. Chu, C.T. and Chang, C.D., J. Phys. Chem. **89**, 1569 (1985).

63. Seive, A., Guth, J.L., Raatz, F. and Petit, L., Eur. Patent Appl. 342,075 (1989).

64. Axon, S.A., Huddersmann, K. and Klinowski, J., Chem. Phys. Lett., **172**, 398 (1990).

65. Ione, K.G., Vostrikova, L.A., Petrova, A.V. and Mastikhin, V.M., in : *Structure and Reactivity of Modified Zeolites*, Eds. P.A. Jacobs, N.I. Jaeger, P. Jirû, V.B. Kazansky and G. Schulz - Ekloff, Elsevier, Amsterdam, p. 151 (1984).

66. Weitkamp, J., Beyer, H.K, Borbély, G., Cortés-Corberán, V. and Ernst, S., Chem.-Ing. Tech. **58**, 969 (1986).

67. Kentgens, A.P.M., Bayense, C.R., Van Hooff, J.H.C., De Haan, J.W. and Van de Ven, L.J.M., Chem. Phys. Lett. **176**, 399 (1991).

68. Challoner, R., Harris, R.K., Barris, S.A.I. and Taylor, M.J., Zeolites **11**, 827 (1991).

69. Barri, S.A.I. and Young, D., Eur. Patent Appl. 130,013 (1985).

70. McCusker, L.B., Meier, W.M. Suzuki, K. and Shin, S., Zeolites **6**, 388 (1986).

71. Newsam, J.M. and Jorgensen, J.D., Zeolites **7**, 569 (1987).

72. Goldsmith, J.R., Miner. Mag. **29**, 952 (1952).

73. Barri, S.A.I. and Young, D., Eur. Patent Appl. 106,478 (1984).

74. Gabelica, Z. and Guth, J.L., in : Ref. 9, pp. 421-430.

75. Jiao, Q. and Pang, W., Huaxue Xuebao **48**, 761 (1990).

76. Chen, N.Y., McCullen, S.M., Eur. Patent Appl. 325,053 (1989).

77. Reddy, J.S. and Kumar, R., Zeolites **12**, 95 (1992).

78. Perego, G., Bellussi, G., Corno, C. Taramasso, M., Buonomo, F. and Esposito A., in : Ref. 27, pp. 129-136.

79. Qiu, S., Pang, W. and Yao, S., in : Ref. 9, pp. 133-142.

80. Pang, W., Yu, L. and Wu, Y., Geodeng Xuexiao Huaxue Xuebao **19**, 951 (1989).

81. Dongare, M.K., Singh, P., Moghe, P.P. and Ratnasamy, P., Zeolites **11**, 690 (1991).

82. Kornatowski, J., Sychev, M., Goncharuk, V. and Baur, W.H., in : *Catalysis and Adsorption by Zeolites*, Studies in Surface Science and Catalysis, Vol. 65, Eds. G. Öhlmann, H. Pfeiffer and R. Fricke, Elsevier, Amsterdam, pp. 581-590 (1991).

83. Jiao, Q.and Pang, W., Gaodeng Xuexiao Huaxue Xuebao **11**, 797 (1990).

84. Prasado Rao, T.S.R., Borade, R.B. and Halgeri, A.B., Chem. Express **1**, 709 (1990).

85. Yu, L. and Pang, W., Jilin Daxue Ziran Kexue Xuebao, 77 (1989).

86. Harvey, G. and Meier, W.M. in : Ref 9, pp. 411-420.

87. Gier, T.E. and Stucky, G.D., Nature **349**, 508 (1991).

88. Wilson, S.T., in : *Introduction to Zeolite Science and Practice*, Studies in Surface Science and Catalysis, Vol. 58, Eds. H. van Bekkum, E.M. Flanigen and J.C. Jansen, Elsevier, Amsterdam, pp. 137-151 (1991).

89. Davis, M.E., Montes, C. and Garces, J.M., in : Ref. 26, pp. 291-304

90. Yang, G., Li, L. Chen, J. and Xu, R., J. Chem. Soc., Chem. Comm., 810 (1989)

91. Chen, J., Li, L. Yang, G. and Xu, R., J. Chem. Soc., Chem. Comm., 1217 (1989).

92. Bedard, R.L., Wilson, S.T., Vail, L.D. Bennett, J.M. and Flanigen, E.M. in : Ref. 9, pp. 375-387.

93. Wang, J., Feng, S. and Xu, R. in : Ref. 9, pp. 143-150.

94. Corcoran, E.W., Jr., Savadar, J.B. and Bhallar-Chawla, M., in : Ref. 23.

95. Riou, A., Cudennec, Y. and Gérault, Y., Rev. Chim. Minérale **23**, 810 (1986).

96. Cudennec, Y., Lecerf, A., Riou, A. and Gérault, Y., C.R. Acad. Sci. **301** (II), 93 (1985).

97. Haushalter, R.C., Strohmaier, K.G. and Lai, W.L., Science **246**, 1289 (1989).

98. Mundi, L.A., Strohmaier, K.G., Goshorn, D.P. and Haushalter, R.C., J. Am. Chem. Soc. **112**, 8182 (1990).

99. Mundi, L.A., Strohmaier, K.G. and Haushalter, R.C., Inorg. Chem. **30**, 154 (1991).

100. Chapman, D.M. and Roe, A.L., Zeolites **10**, 730 (1990).

101. Kuznicki, S.M., U.S. Patent 4,853,202 (1989).

102. Beck, J.S., Chu, C.T.W., Johnson, I.D., Kresge, C.T., Leonowicz, M.E., Roth, W.J. and Vartuli, J.C., PCT Int. Appl. WO 91-11,390 (1991).

103. Delprato, F., Delmotte, L., Guth, J.L. and Huve, L., Zeolites **10**, 546 (1990).

104. Kessler, H., in : Ref. 19, pp. 47-56.

105. Ciric, J., US Patent, 3,972,983 (1976)

106. Vaughan, D.E.W., Eur. Patent Appl. 315,461 (1989).

107. Caullet, P., Guth, J.L., Hazm, J., Lamblin, J.M. and Gies, H., Eur. J. Solid State Inorg. Chem. **28**, 345 (1991).

108. Estermann, M., McCusker, L.B., Baerlocher, Ch., Merrouche, A. and Kessler, H., Nature **352**, 320 (1991).

109. Richardson, J.W., Jr., Pluth, J.J. and Smith, J.V., Naturwiss, **6**, 467 (1989).

110. Kuehl, G.H. and Schmitt, K.D., Zeolites **10**, 2 (1990).

111. Van Santen, R.A., Keijsper, J. Oms, G. and Kortbeek, A.G.T.G., in : Ref. 27, pp. 169-175.

112. Dai, F.Y., Suzuki, M., Takahashi, H. and Saito, Y., in : Ref. 27, pp. 223-230.

113. Dai, F.Y., Suzuki, M., Takahashi, H. and Saito, Y., in Ref. 26, pp. 244-256.

114. Shiralkar, V.P. and Clearfield, A., Zeolites **9**, 363 (1989).

115. Fajula, F., Nicolas, S., Di Renzo, F., Gueguen, C. and Figueras, in : Ref. 26, pp. 493-505.

116. Moudafi, L., Dutartre, L., Fajula, F. and Figueras, F., Appl. Catal. **20**, 189 (1986).

117. McCusker, L.B., Baerlocher, Ch., Jahn, E. and Bülow, M., Zeolites **11**, 308 (1991).

118. Szostak, R.private communication.

119. Xu, W., Li, J., Li, W., Zhang, H. and Liang, B., Zeolites **9**, 468 (1989).

120. Huo, Q., Feng, S. and Xu, R., Huaxue Xuebao **48**, 639 (1990).

121. Xu, W., Li, J. and Lin, G., Zeolites **10**, 753 (1990).

122. Jansen, J.C., Arafat, A. and van Bekkum, H., in: Ref. 23.

123. Cauffriez, H. and Guth, J.L., Zeolites,**12**, 121 (1992).

124. Xu, R. and Huo, Q., J. Chem. Soc., Chem. Comm., 168 (1992).

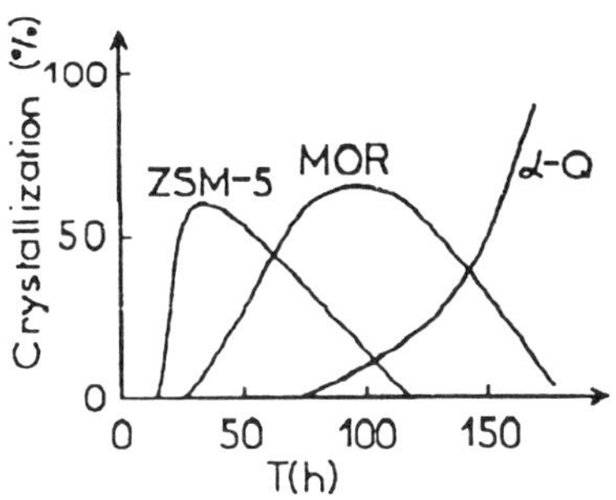

Fig.1.Organic-free synthesis of ZSM-5

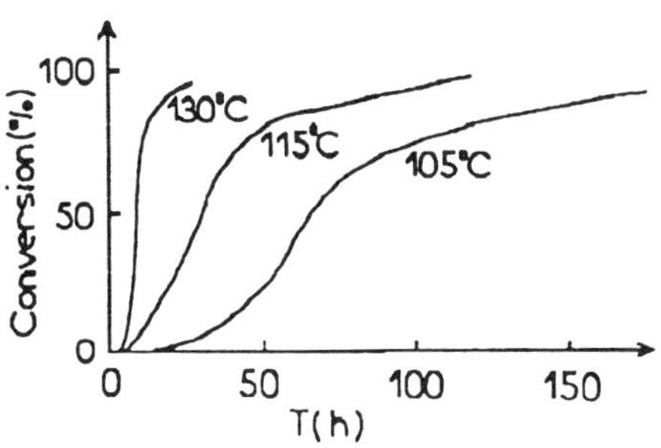

Fig.2. Crystallization curves for zeolite Omega

SODALITE SUPRALATTICES: FROM MOLECULES TO CLUSTERS TO EXPANDED INSULATORS, SEMICONDUCTORS AND METALS

Andreas Stein and Geoffrey A. Ozin

Advanced Zeolite Materials Science Group, Lash Miller Chemical Laboratories
University of Toronto, 80 St. George Street, Toronto, Ontario, Canada M5S 1A1

ABSTRACT

Sodalite is an ancient material with great potential for advanced applications. In the context of Solid-State Chemistry and Condensed Matter Physics, this Federovian framework material is considered to provide a unique opportunity for studying metal-nonmetal transitions, quantum, spatial and dielectric confinement effects, quantum electronic and nonlinear optical phenomena in expanded insulators, semiconductors and metals. One of the major goals of some of our recent research has been to assemble novel nanostructures by encapsulating clusters consisting of the components of insulators, semiconductors and metals, inside the framework aluminosilicate sodalite. This nanoporous host acts as a stabilizing dielectric matrix, capable of organizing single size and shape clusters in perfectly periodic arrays. Such materials, and the related zeolite analogues, have an interesting variety of potential applications such as nanoporous molecular electronic materials, molecular wires, chemical sensors, zeolite electrodes, nonlinear optical materials and high density data storage materials. Advanced materials of this type have been recently reviewed.[1,2]

INTRODUCTION

The distinct physical properties of the materials we now call insulators and metals have been utilized since the early history of mankind. Only in this century two more classes of electrical materials were discovered, namely semiconductors and superconductors. Each new discovery, the insulator ("Stone Age"), the metal ("Iron Age", *ca.* 1200 B.C.), the superconductor (superconducting mercury, 1911 and high-temperature superconductors, 1986) and the semiconductor (transistor action, 1947) has revolutionized technology, with wide effects even on the social structure of the world.[3]

The theory explaining the properties of these materials has, as is usual, lagged in time behind their discovery. Many structural, thermal, electronic and optical properties of bulk insulators, semiconductors and metals are now well understood. A number of theoretical models now exist, which predict the behaviour of these materials reasonably well. More recently, however, new questions have arisen:

1) What happens when a piece of bulk material is divided into smaller and smaller units, approaching the molecular level? Are the present models still valid in the small size regime? This question can also be turned around. Starting from the molecule, what happens when larger and larger clusters are formed? When does a material reach its bulk properties and what happens near this transition?

2) What happens to the properties of a bulk material that has been expanded to increase internuclear separation? Can transitions from metals to semiconductors or insulators be observed?

From the Proceedings of the 9th International Zeolite Conference, Montreal 1992, Eds. R. von Ballmoos et al.,

Both of these points are intensively studied by physicists, chemists and material scientists.

This paper will deal with the questions posed above using the following approach: By confining insulator, semiconductor-type and metallic materials inside a sodalite host matrix, one can create three-dimensional periodic arrays of single-size and shape clusters of the guest components. Their vibrational and electronic interactions with each other and with the sodalite framework may be controlled by selecting appropriate extraframework guest ions, varying their concentrations, and carrying out physical or chemical manipulations on the guest species. Depending on the degree of interaction between clusters, it may be possible to create extended cluster lattices in various size regimes, whose physical properties (bonding, electronic, optical) may be different from both individual clusters and the bulk material composed of the guest ions. These phenomena will be studied by considering the composition/structure/property relationships of synthetic sodalites containing a variety of guest species.

Our approach will be to very briefly review the general properties of insulators, semiconductors and metals, including transitions from one to another. The concept of spatial confinement utilized to produce quantum size materials and the properties of such materials will then be briefly outlined. This is followed by a succinct description of the structure of sodalite, whose framework has played the role of host for the confinement of insulator, semiconductor-component and metal clusters in the work presented here.

A number of questions are addressed throughout this paper. Can insulators, semiconductors and metals be created inside sodalite matrices? How are they stabilized? Do components of insulators, semiconductors and metals behave like their bulk parent materials inside the sodalites? What is the effect of expanding these materials, *i.e.*, increasing their internuclear distances? What is the effect of the dielectric framework? Can a development from molecular structure to band structure be observed? Does a transition occur from one state to another as the occluded guest concentration increases and as the internuclear separation decreases; *i.e.*, can a percolation threshold be observed? Can one observe quantum size effects? What are the properties of such nanocomposites? Do these materials lend themselves to device applications in quantum electronic circuitry, high density optical data storage media and non-linear optical applications?

Metals, Nonmetals and the Metal-Insulator Transition

A distinction between metallic and nonmetallic crystalline solids is given by the states of electron bands. This band description for bulk insulators, semiconductors and metals was first developed by Wilson.[4] A nonmetal is a material in which all bands are full or empty, resulting in localization of electrons, while a metal possesses one or more partially filled bands and electrons become itinerant.[5] If all occupied bands are filled, but the energy difference between the valence band and the conduction band is small, the material is an intrinsic semiconductor. Doping the intrinsic semiconductor with, for example, an n-type donor, creates discrete donor levels containing single electrons just below the conduction band. These electrons are easily promoted across the narrow bandgap into the conduction band. If all bands are filled except for one or two bands which are either nearly filled or nearly empty, the solid is an impurity semiconductor (extrinsic semiconductor). Extrinsic semiconductors have much higher conductivities than similar intrinsic semiconductors at room temperatures. The conductivity of extrinsic semiconductors may be accurately controlled by choosing the dopant concentration, so that materials with desired conductivities may be manufactured. The electrical conductivity of a semiconductor or insulator increases with temperature as the thermal excitation of electrons creates partially filled states. For metals, enhanced electron scattering leads to a decrease in electrical conductivity at higher temperatures.

An approximate criterion for the metal-insulator transition, based on the separation between atoms in an ordered crystalline array, was given by Mott and Hubbard. They proposed that the metallic status is achieved by having a sufficiently high electron concentration so that the ionization energy of any particular "atom" in the system disappears. The Mott-Hubbard theory addresses three regimes of interatomic separation *(a)* or density. For large interatomic separations (low density), Coulombic repulsions between atoms dominate. To avoid significant charge build-up on the atoms, valence electrons are localized in disconnected spatial regions. The material is an antiferromagnetic insulator. For small values of *a*, considerable orbital overlap occurs, relieving any electron excess on any one atom and giving rise to partly-filled bands, itinerant electrons and metallic conductivity. For intermediate values of *a*, the frontier atomic levels broaden into bands with increasing density, until the energy gap between these bands vanishes at a critical density, n. This transition occurs at the condition $n^{1/3} a_H > ca.\ 0.4$, where n is the number of centers per unit volume, and a_H is the hydrogen radius. This criterion is independent of any particular arrangement of constituent atoms. According to Mott, any metal-insulator transition in a crystalline material at zero Kelvin must be a transition from a situation in which bands overlap to a situation when they do not, producing a full valence band and an empty conduction band with an energy gap between them. Such a transition may be induced by a change in parameters, such as composition, concentration, specific volume, pressure and temperature. For example, the expansion of most metals will at first decrease the gap and then enlarge it.[5]

Another criterion for the metal-nonmetal transition in condensed matter was developed by Herzfeld in 1927. The Herzfeld criterion predicts that a material will show metallic behaviour if the molar volume V, is reduced below the molar refractivity, R ($R = 4/3\pi N_A\alpha$ where N_A = Avogadro's number, α = polarizability) at standard conditions. Thus, if $R/V \geq 1$, a material is metallic, if $R/V < 1$ it is an insulator. The power of this criterion has been demonstrated by experimental agreement for a large number of materials.[6]

A different form of the metal-insulator transition must be applied to disordered systems. The appropriate formalism was introduced in 1958 by Anderson, and is based on a random potential energy. In sufficiently random fields the mean free path of electrons is shortened to such an extent that the one-electron wave functions can be localized in discrete regions of space. The material is effectively transformed from a metal to an insulator and will not conduct at absolute zero. When the temperature is slightly raised, conduction occurs by thermally activated hopping. In principle one can always transform a metallic system into an insulating or semiconducting state by expanding it to low densities or by introducing sufficient disorder (*e.g.*, alloying with a nonmetal).[6] In a disordered system containing metallic and insulating regions, metallic conduction may take place through linkages among metallic sites. Percolation occurs when a conducting path spans the complete sample. Tunnelling may provide additional conduction pathways.

Confinement of Materials

The electronic properties of metals, insulators and semiconductors change drastically as one proceeds from individual atoms or molecules, through clusters and aggregates to the bulk materials. On the small scale, electrons are localized in discrete atomic or molecular states, and the terms metal, insulator and semiconductor are not valid. On the bulk scale, continuum band states are formed, and the electrons become itinerant in metals. In insulators and semiconductors electron and hole pairs can form an exciton band. The quantum size regime is a transition region, encompassing microscopically structured solids in the size regime from 5 - 100Å.[7] Particles of this size display electrical, optical and magnetic properties that are essentially independent of their chemical composition, but rather relate to their extremely small dimensions.

The present interest in ultrasmall semiconductor particles can be considered to derive from the original notion of artificially layered structures known as semiconductor superlattices, and from the exciting optical and electrical properties that such quasi two-dimensional (2D) electron systems offer. The formation of modulated chemical composition layers of this kind with a periodicity larger than the natural lattice spacing but smaller than the de Broglie wavelength of electrons of the semiconductor components, allows the realization of bandgap engineering and heterojunctions.[8-11]

Theoretical considerations have indicated that quantum size effects are significantly enhanced when the dimensionality of a quasi-2D semiconductor is further reduced. For example, strongly enhanced electron mobilities have been predicted in the case of quantum wires (1D), whereas quantum boxes (0D) should be insulating. Fundamental alterations in their optical properties are expected by 1D and 0D quantum confinement of excitons and the resultant density of states. A dominant contribution to the non-linear optical property of bulk semiconductors arises from the sharp exciton band found just below the absorption bandedge (VB to CB). The Coulombic attraction between an electron in the CB and hole in the VB leads to formation of a stable hydrogenic quasi-particle, the exciton. The stabilization of the exciton increases with increasing quantum confinement of the bulk semiconductor, so that the exciton resonance is well resolved at room temperature. With the bulk semiconductor oscillator strength concentrated into the few spectral lines of the quantum dot[12], the quantum dot is predicted to show a greater absorption index than the bulk semiconductor, hence enhanced nonlinear optical properties. These effects, however, may be reduced by the interference of phonon coupling, as well as size and shape inhomogeneities. This emphasizes the importance of a narrow particle size distribution, and ideally they should be of a single size and shape.

Because quantum size materials can display remarkable linear and non-linear optical properties (including high second and third order polarizabilites and non-linear refractive indices), they are expected to have many potential applications in areas such as photocatalysis, electro-optics, frequency multiplication and mixing of UV-VIS-IR radiation, optical switching for high speed parallel processing in optical and optoelectronic computers,[2,13] heterostructure lasers,[14,15] quantum well devices such as resonant tunnelling transistors[15,16] and high mobility field-effect transistors.[14] Extremely high electron mobility in quantum-well wires and high performance of quantum well or box lasers and modulators are expected from theoretical predictions.[11,12] Although chemists have had extraordinary success in making quantum dots, none are yet suitable for devices.[15]. Several reviews of the properties, applications and various methods of producing quantum confined materials have been published.[7,17,18]

Sodium Silver Sodalites

In our research, the sodalite framework was used as a host material to confine clusters composed of the components of insulators, semiconductors and metals. Sodalites are aluminosilicates of the type $M_8X_2(SiAlO_4)_6$, M = cation, X = anion. The sodalite structure is shown in Figure 1. The individual cages are stacked in eightfold co-ordination, resulting in a space filling framework (a Federov Solid) with overall cubic symmetry (bcc).

Because of the valence difference between aluminum and silicon, the sodalite lattice possesses a negative charge equal to the number of aluminum atoms. This charge is balanced by exchangeable cations at C_3 sites near the six-rings. Another cation and an anion at the center of the cage are often present as well. In order to maintain a charge balance, divalent anions require the presence of divalent cations or a complementary cage lacking an anion. Certain guest-anions can be used as internal reagents (*e.g.*, for intra-sodalite cage redox reactions). Different kinds of both cations and anions may be mixed throughout the lattice. Thus, a vast range of guests can be incorporated during the synthesis of the sodalites or produced by thermal, photochemical and other reaction processes. These include materials that are insulators, semiconductors, photoconductors and metals in their bulk form. The unit cell dimensions, charge balance requirements and cage-filling can be tuned by incorporating the appropriate ions during the sodalite synthesis and melt cation exchange processes.

Sodalites have been studied mainly as model compounds for larger zeolite structures. Yet because of their small entrance windows, sodalites have not been used in applications involving catalysis or gas separation. However, closer inspection leads to the realization that the sodalite lattice forms a convenient matrix for a variety of encapsulated guests. The sodium cations inside the sodalite cages are exchangeable by small cations, including silver cations. By increasing the silver loading in the parent sodalite, it is possible to form cluster and extended supercluster structures. Such superclusters

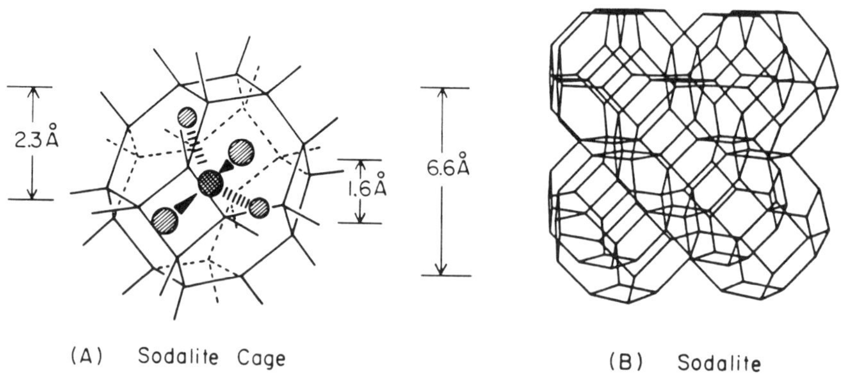

Figure 1. (A) Sodalite cage showing a single cuboctahedron, a central anion, and four cations in the six-ring sites. Each corner represents a TO_4 unit (T = Si or Al). (B) Sodalite framework, emphasizing the close packing of cages.[19]

have been reported in the case of semiconductor materials in faujasites and zeolite A.[20,21,22] One important advantage of sodalites over the other zeolite hosts is that the sodalite lattice is not comprised of co-existing sodalite β-cages and α-supercages, but is instead entirely composed of close-packed sodalite cages. The regular, all-space filling framework of sodalite provides a temperature stable homogeneous nanoporous matrix of sodalite cavities. This is suitable for stabilizing small isolated or interacting molecules, atoms and clusters in the form of neutral species, ions or radicals (all of which may be air-sensitive or reactive otherwise). Only small cations and molecules up to the size of water can pass through the six-rings. Larger species are either excluded at room temperature (*e.g.*, oxygen) or trapped inside the cages (*e.g.*, clusters, anions except OH^-). A number of possible scenarios must be considered for silver and silver halide inclusion in sodalites. Na^+, Ag^+ and Br^- may act as independent ions occupying sites in the unit cell. In this case they are merely a part of the total lattice structure (strong interaction with framework, weak interaction with each other). On the other hand, Na^+, Ag^+ and Br^- may interact, forming $(Na_nAg_{4-n}Br)^{3+}$ clusters with molecular behaviour (only slightly perturbed by the presence of the framework). In this instance the degree of Na^+ involvement could be drastically different from that of Ag^+. Each cluster would behave like an isolated molecule within each cage. The cluster spatial confinement would be limited to a maximum value of 6.6Å. If the $(Na_nAg_{4-n}Br)^{3+}$ clusters interact with each other, possibly mediated through the framework, they can form larger units. In this case, the confinement volume could go up to the sodalite crystallite size. These possibilities are summarized in the scheme shown below (where F = framework, M = cation, X = anion). They will be considered throughout this paper.

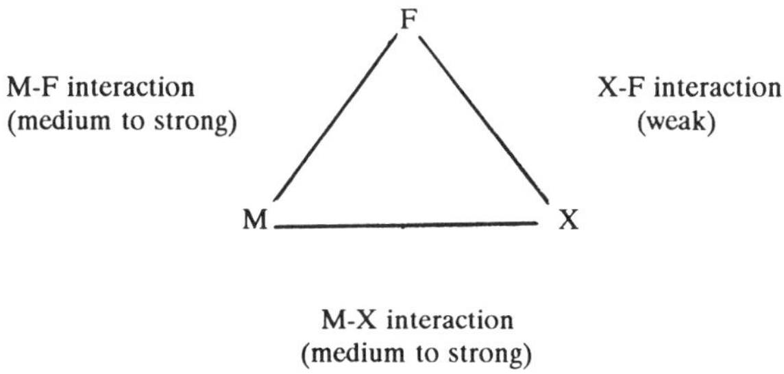

Definition of Terms

In this paper the following definitions will be used to classify sodalites: In Class A materials, all cages are filled with M_4X clusters. Class C sodalites are closely related, except that they can contain two different anions, that is mixed M_4X and M_4Y semiconductor component clusters. Class B sodalites contain isolated M_4X clusters "diluted" in cages with M_3 "spectator triangles" or M_4OH clusters. Diagrams of Class A, B and C sodalites appear in Figure 2. The formal representations, based on a unit cell (u.c.) which consists of two sodalite cages, are given in Table 1.

Table 1: Classification of Sodalites[23,24,25,26]

Class A:	$M_{8-2n}N_{2n}X_2$-SOD
Class C:	$M_{8-2n}N_{2n}X_{2-p}Y_p$-SOD
Class B:	$M_{8-p-2n}N_{2n}X_{2-p}[]_p$-SOD
	$M_{8-2n}N_{2n}X_{2-p}(OH)_p$-SOD

M,N = Ag^+, Na^+
X,Y = Cl^-,Br^-,I^-, (Ox= 1/2 $C_2O_4^{2-}$ for Class B)
[] = Anion-free sodalite cage
2n = 0-8
p = 0-2
SOD ≡ $(SiAlO_4)_6^{6-}$

In all Class B sodalites internal water can be present. These structures will sometimes be abbreviated MNX-SOD, MNXY-SOD, MNX[]-SOD and MNOx-SOD, respectively, and the cation and anion loadings will be stated.

RESULTS AND DISCUSSION

The following represents a synopsis of a large body of knowledge concerning our work on Class A,B and C sodalites. The examples used have been judiciously selected to illustrate the unique quality of sodalite supralattices for probing the transition from molecules to clusters to expanded insulators, semiconductors and metals.

Synthesis[23,27]

The sodalites must be synthesized in their sodium forms according to the recipes summarized in Figure 2. Depending on the choice and relative amounts of X and Y one can generate sodalite compositions described as Class A,B and C, as depicted in Figure 2.

Except for differences in the degree of hydration, only single crystallographic phases were found in as synthesized Class A, B and C sodium sodalites. Analyses of data from the product composition *versus* the feed composition and unit cell size *versus* composition provide some indication of small domain formation, with regions containing predominantly those anions that tend to be occluded preferentially during the hydrothermal synthesis. However, such domains must be very small (< *ca.* 50 Å). Combined results from powder XRD, Rietveld refinement, ^{23}Na DOR/MAS-NMR, far-IR and

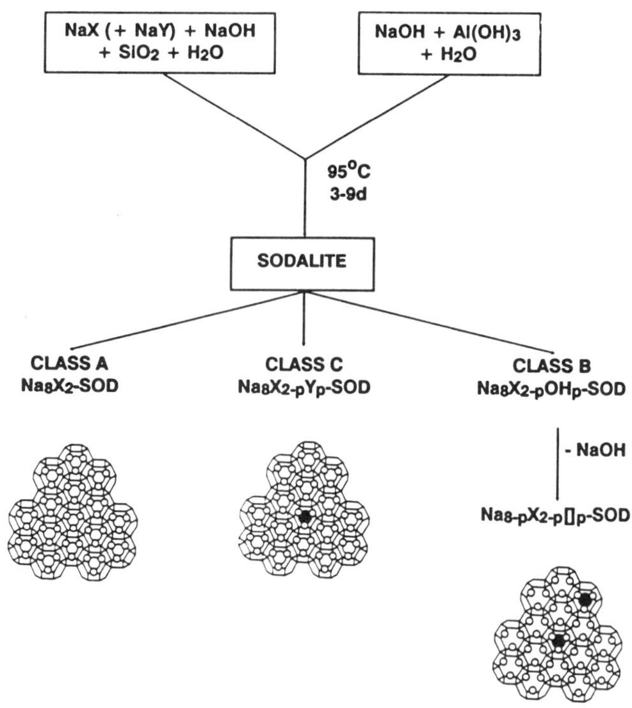

Figure 2. Synthesis of Class A, B and C Sodalite Supralattices[23,27]

mid-IR indicate that the anions (or small domains of anions) are distributed statistically throughout the sodalite lattice, forming a homogeneous solid solution.[24,28,29]

Sodium ions can be exchanged by silver ions, either by silver salt melt ion-exchange or an aqueous ion exchange. The melt exchange is stoichiometric for Class A and C sodalites. In Class A and C sodalites the silver appears to be randomly distributed. However, in Class B sodalites, silver associates preferentially with the halide (and probably oxalate) anions, so that most of the anion-free cages are filled with silver ions only after the remaining cages have been occupied by silver. Yet even in these situations of silver segregation only one crystallographic phase was detected, either because of merely minute differences in the unit cell sizes of silver and sodium-containing regions, or because of the small concentration of such regions.

Class A Sodalites: Isolated Molecules to Expanded Insulators and Semiconductors[24,28]

Rietveld methodology was used to successfully refine many of the sodalite structures presented in this study.[28] Pauling's structure of sodium chlorosodalite[30] and more recent sodalite refinements[31,32] provided suitable starting models. The refinements were carried out using the General Structure Analysis Software (GSAS).[33] As an average space group $P\bar{4}3n$ was used, small disorders in the framework oxygens, due to cages with different content, were not detected by the Rietveld refinement. However, the most important parameters, namely the unit cell sizes, the positions of all guest atoms and thus their interatomic separations, could be refined to a high degree of confidence.

Knowledge of the effect of the sodalite composition on the unit cell size and the interatomic distances is vital to understand the coupling between atoms, and thus the electronic and optical properties of the materials. The unit cell size of Class A sodalites varies with the type of halide and the types and relative concentrations of cations present. The halide effect is shown in Figure 3. The halide effect is mainly a space-filling effect, the large anion causing an expansion of the sodalite cage. Replacement of Na^+ by Ag^+ causes a contraction of the cage. In a series of sodium, silver bromosodalites the unit-cell size decreased slightly as the silver concentration was increased[24,28], forming a tighter, more covalent bond between the guest cation and the central anion (4-coordinate $r(Na^+) = 1.13Å$, $r(Ag^+) = 1.14Å$).[34] At silver concentrations at which most cages contained at least one Ag^+ ion (at loading levels greater then 2.5 - 3 Ag^+/u.c.) additional silver resulted in weakening of the Ag-X bond, preventing further contraction of the unit cell. Abrupt changes or discontinuities can be observed in unit cell size *versus* composition plots at certain compositions if either a change in symmetry of the solid solutions or a change in the solid solution bonding occurs. Taking into consideration that Setter and Depmeier[35] observed an almost linear increase in the unit cell edge with average ionic radius of the cage cations, and that the Na^+ and Ag^+ ions have virtually identical radii, the observed behaviour is best explained by considering the more covalent bonding of Ag^+ with the central halide. This strong interaction results in shortening of the Ag-X distances compared to NaX and an increase in the Ag-O (framework) separations. One is therefore justified in calling a sodalite-encapsulated Ag_4X^{3+} unit a cluster, in spite of its interaction with the framework.

It is interesting to note, that at loading levels near 2.5 - 3 Ag^+/u.c. an abrupt break in magnitude occurred not only for the unit cell edges of the cubic sodalite cage, but also for the peak positions of far-IR absorptions associated with a translational mode of Na^+ near the sodalite 6-ring site.[24,28] Both effects may be related to a percolation threshold for connectivity between AgBr units. In the case of Na^+ far-IR absorptions, the translatory frequency remains essentially constant at lower concentrations, but decreases at higher concentrations when the proportion of cages containing more than one Ag^+ increases. Our present hypothesis is that at low concentrations the AgBr molecules act as isolated defects having no effect on the Na^+ translatory vibrations. Above percolation threshold loadings the Ag^+ ions must be considered part of the whole unit cell for the vibrational problem and, because of their weaker interaction with the support compared to sodium, reduce the vibrational

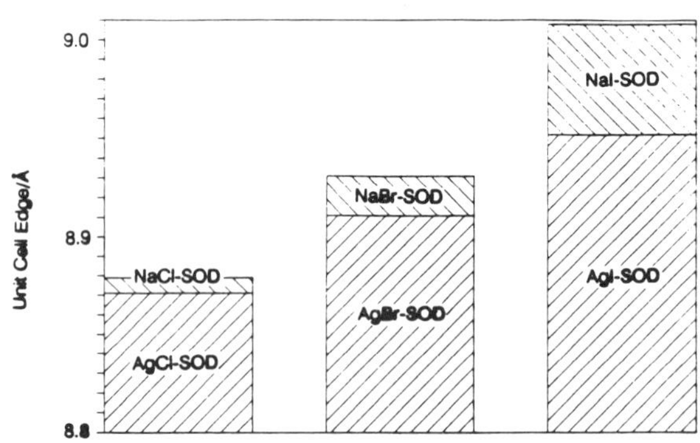

INTERATOMIC DISTANCES, Å UNITS

Sample	M-X	M-O	Ag-Ag	Ag-Ag (next)	M-X (bulk)	Ag-X (mol.)	Ag-X* (next)
NaCl-SOD	2.734	2.372	-	-	2.820	-	-
AgCl-SOD	2.537	2.475	4.142	4.920	2.775	2.28	5.146
NaBr-SOD	2.888	2.356	-	-	2.989	-	-
AgBr-SOD	2.671	2.444	4.361	4.859	2.887	2.393	5.047
NaI-SOD	3.089	2.383	-	-	3.237	-	-
AgI-SOD	2.779	2.576	4.539	4.821	3.040	2.545	4.918

Figure 3: Bar graph showing the unit cell sizes of sodium and silver halosodalites together with a listing of key interatomic distances.[28]

frequency of the Na^+ modes. In contrast to the translational modes in the far-IR, the frequency of the aluminosilicate framework vibrations (mid-IR) varies linearly with the silver concentration.

Figure 3 lists the cation-anion, cation-oxygen, and silver-silver distances in halosodalites, as well as bond lengths in bulk and gas phase silver halides. The silver halide distances inside the sodalite fall between those of the vapour phase molecules and the bulk semiconductor solids. The AgI-distances in AgI-SOD compare remarkably well with other cluster species, such as solvated Ag_4I^{3+} as well as with bulk Ag-I species. The sodalite framework appears to have very little effect on the structure of this cluster. It is notable that, even outside a sodalite cage, Ag_4I^{3+} is claimed to have high stability.[36] For the sodium halosodalites, the bond lengths are only slightly shorter than in the rock salt structures. The distance between silver and a halide anion in the same cage increases going from Cl and Br to I, while the separation of the silver to an anion in an adjacent cage decreases on going through the halide group. The silver-oxygen distances in these sodalites are slightly longer than typical silver-oxygen separations for 4-coordinate silver, and significantly longer than those found in 2-coordinate silver. The Ag-Ag distances within each cage increase from Cl to Br to I. Ag-Ag separations for the closest silver ions in adjacent cages decrease from Cl to Br to I. The intracage Ag-Ag separations are *ca.* 6% longer than in the salts, and Ag-Ag distances between cages are from 25% to 12% longer. The Ag-Ag distances in the intrasodalite Ag_4I^{3+} cluster compare closely with those found in related clusters, such as solvated Ag_4I^{3+} (4.6 Å) and the $Ag_4I_6^{2-}$ cluster, where Ag-atoms are also arranged tetrahedrally (4.55 Å).[37] Extended Hückel molecular orbital (EHMO) calculations have shown that the orbital overlap, and thus the atomic interaction, is significant for silver at these distances (even though the Ag-Ag separations are much larger than in the metal). One can therefore expect electronic interaction between clusters (see below). Taking all these bond lengths into account, one can consider the organized array of Ag_4X units as expanded silver halide semiconductors, although they are influenced by the aluminosilicate host matrix.

The Rietveld refinement of NaAgBr-SOD containing a very low Ag concentration of 0.3 Ag^+ per unit cell (u.c.), yielded an Ag-Br bond length of 2.21 Å (2.0 to 2.4 Å, based on three standard deviations with correction for thermal ion motion).[24,28] This distance compares with the internuclear separation of gaseous AgBr (2.39 Å).[38] The Na-Br distance in this sodalite is 2.94 Å, *i.e.*, slightly longer than in pure NaBr-SOD and only 2% shorter than in the salt. Due to the more extensive covalent bonding of AgBr compared to NaBr, the silver is more closely associated with the halide, while Na^+ remains more strongly coordinated to the framework oxygen. This results in the Na_3AgBr aggregate behaving more like a slightly perturbed AgBr molecule. Based on a binomial distribution of $Na_nAg_{4-n}Br$ moieties, at the silver loading level of this sample one in every eight cages is occupied with an AgBr molecule. The connectivity of AgBr molecules between cages is therefore small, and the AgBr molecules can be considered isolated. At a slightly higher silver concentration (2.5 Ag/u.c.) the AgBr distance (2.519(5) Å) has a value intermediate between the isolated molecule and the fully Ag-loaded sample. This value may in fact be an average of bond lengths for Ag_4Br, Ag_3NaBr,

Ag_2Na_2Br and $AgNa_3Br$ clusters which could not be resolved into individual lengths by the method used.[39]

Figure 4 shows UV-visible reflectance spectra for sodalites containing each of the three different halides Cl, Br and I, at silver loadings of 0, 0.1 and 8 Ag^+ per unit cell. In all cases the positions of the absorption bands show a halide dependence, the most drastic difference occurring for the iodosodalites. The transitions involved in the optical absorptions were assigned by considering the energy band diagrams obtained by extended Hückel molecular orbital (EHMO) calculations, Figure 4. The calculations are based on a single sodalite cage with the composition $M_4XAl_{12}Si_{12}O_{36}{}^{15+}$.[28] The Wolfsberg-Helmholz-Calzaferri[40] weighting formula was used; no charge iterations were carried out. Closely spaced energy levels of the same orbital composition are represented as continuous bands, to simplify the diagram.

The spectra of the sodium chloro- and bromosodalites consist of an absorption at *ca.* 205 nm with a shoulder around 240 - 250 nm. The iodide spectrum is more complex, with bands at 225 - 250 nm, 290 nm and a shoulder near 255 nm. The framework atoms are involved in the optical transitions *via* the O 2p and Si 3s, 3p orbitals. According to the EHMO calculations, the Al 3s, 3p levels lie too high to take part in the transitions. In the absence of silver, the Cl 3p levels mix in with the more energetic O 2p levels, while Na 3s mixes with lower energy Si 3s, 3p levels. No mixed Na/Cl bonding orbitals are present. The bandgap is hardly affected, corresponding mainly to a Si $\leftarrow$ O ligand-to-metal charge transfer (205 nm absorption). The shoulder found at lower energy may correspond to optical transitions within the cluster (Na $\leftarrow$ Cl). The band broadening and greater spectral complexity for the iodide sample may be related to the increased spin-orbit coupling for this heavy halide.

When a small amount of silver ions is present (0.1 Ag^+/u.c.), new features appear, corresponding to transitions from Ag 4d or halide np mixed with Ag 5s, 5p or 4d to mixed {[Na 3s, Ag 5s]Si 3s, 3p} LUMO (lowest unoccupied molecular orbital) levels.[41] They progressively red-shift for the bigger anions in larger unit cells in the series Cl, Br, I. The transitions are similar to the gas phase values of 230 nm and 320 nm for the AgBr monomer.[42] While cation ns orbitals still mix with Si 3s, 3p, the halide np orbitals move above the O 2p in energy, mixing with silver orbitals instead. This mixing is an indication of Ag-X bonding. The fact that the LUMO consists of both cluster and framework orbitals, indicates that the aluminosilicate framework is involved in the electronic interactions between clusters, and affects the optical and electronic properties of silver sodalites.

At higher silver loadings the intensity of the absorptions increases, but in the case of Cl and Br no significant shift in the absorption energy is observed.[24,28] Again, the iodide sample behaves differently, showing a red shift at increased silver loading for the major components. The energy levels due to clusters with 0,1,2,3,4 silver ions are likely to overlap and merge into common bands of the types O 2p, Ag 4d, {X np, Ag 5s, 5p, 4d},{[Na 3s, Ag 5s] Si 3s, 3p}. An average energy level diagram is expected, which dampens any spectral changes related to Ag^+ loading.

After complete silver exchange the spectral features appear simpler again. A sharp absorption

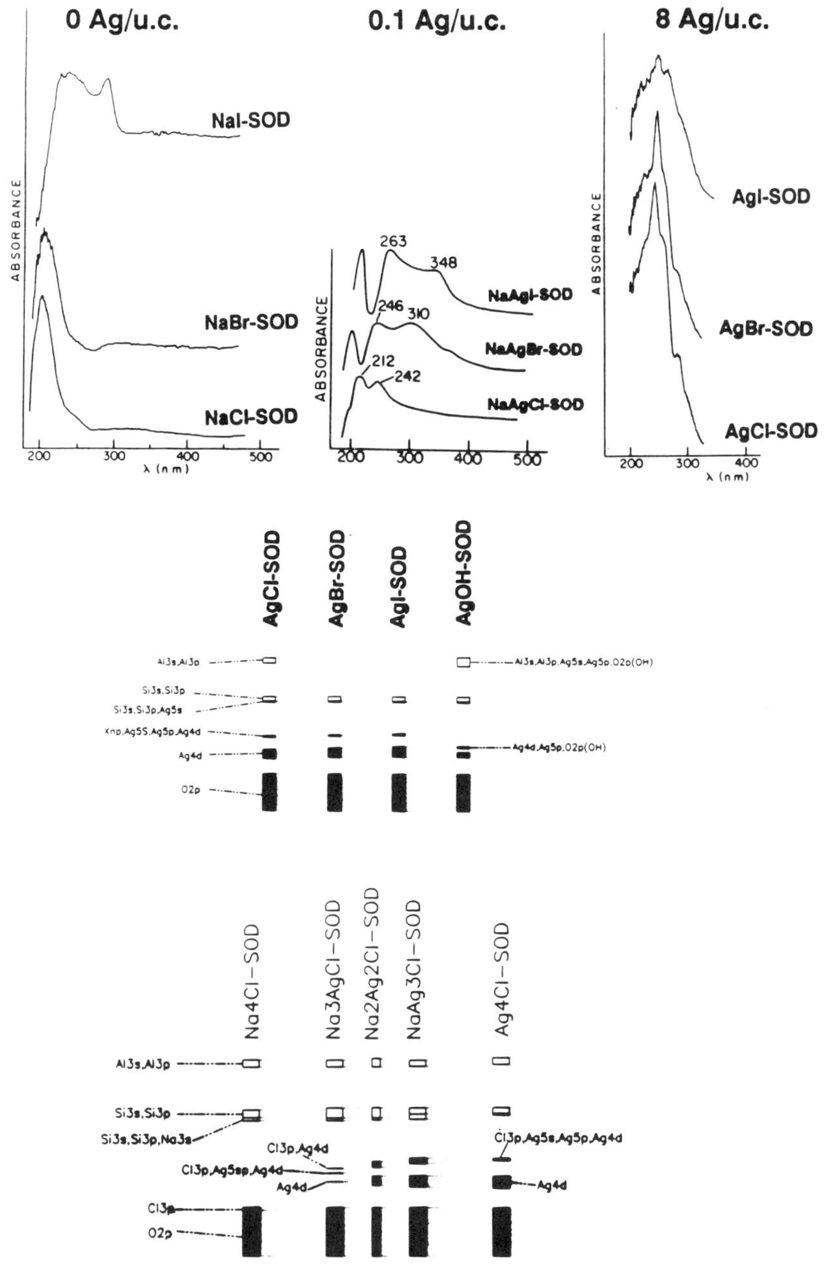

Figure 4. A: UV-visible reflectance spectra of sodium, silver halosodalites. B: Sodalite band diagrams for one cluster of the type indicated, inside a single sodalite cage.[24,28]

between 245-250 nm is superimposed on a broader feature peaking in the same region. Another peak or shoulder is present between 280-300 nm. The absorption bands are broadest for the iodide. For completely silver-exchanged samples, the optical absorptions can be assigned to {Si 3s, 3p} ← {Ag 4d} (broader 245 - 250 nm absorption); {Si 3s, Si 3p, Ag 5s} ← {Ag 4d} (sharp 245 - 250 nm absorption); {Si 3s, 3p, Ag 5s} ← {X np, Ag 5s, 5p, 4d}(280-300 nm). The first two transitions do not involve halide orbitals, and the corresponding optical absorptions show little halide dependence. The energy of the {Si 3s, 3p, Ag 5s} ← {X np, Ag 5s, 5p, 4d} transition exhibits a slight dependence on the type of halide. The almost negligible narrowing of the calculated energy gaps for the series Cl, Br, I is due mainly to a small destabilization of the HOMO (highest occupied molecular orbital) on going down the halogen group. As the LUMO does not involve halide orbitals, it remains virtually unaffected by the halide (except for changes caused by different unit cell sizes). The framework charge transfer transition is still present underneath the more intense silver related absorptions.

In many other host matrices containing quantum size particles, an increase in the loading of the semiconductor material results in a red shift as the particle size increases.[43,44] Inside the sodalites the I-VII cluster nuclearity is limited to five, and no significant absorption band shifts occur at higher loadings for the Cl^- and Br^- series. Significant overlap of atomic orbitals within a cluster, as well as overlap of Ag 5p orbitals between adjacent clusters and through framework atoms (see later) allows electronic communication between cages and the contents of cages. Even after complete silver exchange the absorption edge remains at higher energy than in bulk AgBr, an indication that microcrystalline AgBr is absent.

The absorption edges of silver halides and other semiconductors have been fitted to equations for the energy dependence of the absorption index for both direct and indirect, allowed and forbidden electronic transitions, to determine the allowedness of the interband transitions.[45] Because the nuclearity of the Ag_4X^{3+} clusters is small, the lowest energy absorptions may be simply due to HOMO-LUMO transitions of a molecular nature. However, the observations that: 1) silver ions from adjacent clusters can overlap directly as well as *via* framework atoms; and 2) absorption band broadening occurs for greater connectivity between clusters, suggest that the formation of bands (probably narrow bands) may be possible in silver halosodalites. Narrow bands may occur in insulators and metals when valence electrons have both localized and band-like characteristics. Table 2 lists the fitted bandedges for AgCl-SOD, AgBr-SOD and AgI-SOD above 300 nm. The "Overall"

Table 2: Bandedges of AgX-Sodalites[28]

Sample	Bandedge (eV) (Overall)	(Best)	(EHMO)
AgCl-SOD	3.83 ± 0.05	3.78 ± 0.05 if	2.52
AgBr-SOD	3.85 ± 0.05	3.79 ± 0.08 if	2.52
AgI-SOD	3.75 ± 0.09	3.77 ± 0.14 ia	2.36

value is an average of the values for *df, ia* and *if*; the "Best" value is the one with the least uncertainty in the energy intercept.

These values are significantly smaller than the optical bandgaps in sodium sodalites which show a strong halide dependence: NaCl-SOD, 6.1 eV; NaBr-SOD, 5.9 eV; NaI-SOD, 5.2 eV.

The gaps for the three silver halosodalites are not significantly different from each other. This is in qualitative agreement with the gap calculated by the EHMO method, which also varies only very slightly with the type of halide. The interband transitions appear to be indirect, *i.e.*, the bottom of the conduction band is at a different k-value than the top of the valence band. The lowest energy interband transitions in bulk AgCl and AgBr (allowed) are also allowed indirect.[46,47] β-AgI (wurtzite) is a direct gap ionic semiconductor.[48,49,50] The allowedness of a transition is normally determined experimentally.[45] However, because of the similarity in the fits for *ia* and *if*, our data did not allow us to distinguish between allowed and forbidden processes. Energy band calculations in k-space are required to characterize the interband transitions in silver halosodalites in more detail.

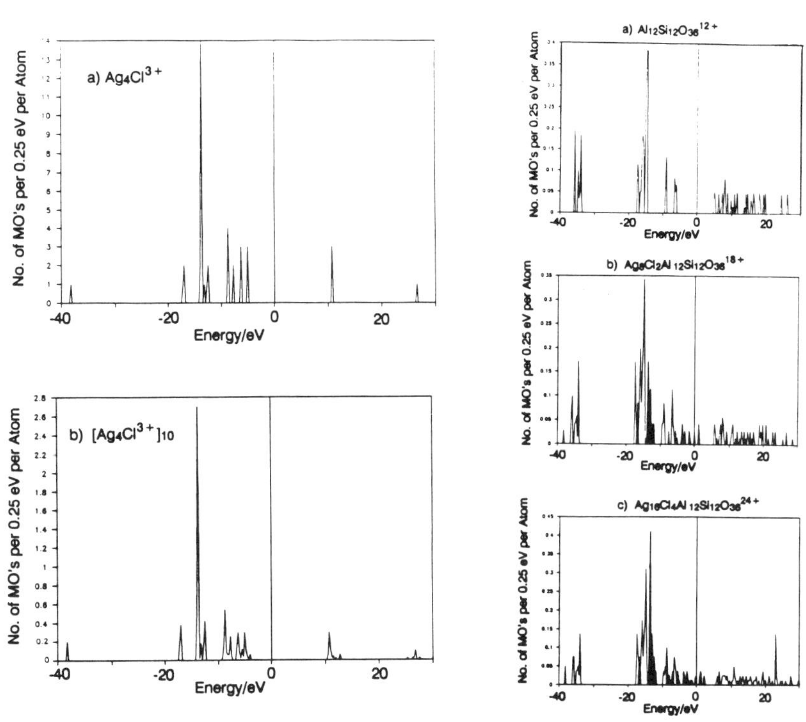

Figure 5. A: Density-of-states diagrams for a) $[Ag_4Cl^{3+}]$ and b) $[Ag_4Cl^{3+}]_{10}$. The graphs show the number of molecular orbitals per atom within 0.25 eV ranges. B: Density-of-states diagrams for an individual cluster-free sodalite cage, $Al_{12}Si_{12}O_{36}^{12+}$, a sodalite cage with two clusters (one inside, one outside), $Ag_8Cl_2Al_{12}Si_{12}O_{36}^{18+}$, and a sodalite cage with four clusters (one inside, three outside), $Ag_{16}Cl_4Al_{12}Si_{12}O_{36}^{24+}$. In the latter, regions pertaining mainly to the cluster orbitals are darkened.[28]

Figure 5 shows density-of-states diagrams for $[Ag_4Cl^{3+}]$ and $[Ag_4Cl^{3+}]_{10}$. One can see that while splitting of levels occurs and the density of levels increases at larger n, the overall pattern remains and levels still appear to be discrete. The question arises: are valence and conduction bands formed upon the addition of more clusters? The overlap matrix, Table 3 columns a and b, shows the orbital overlap (elements of the overlap matrix) between Cl-, Ag- and Ag- orbitals in the same cluster, as well as with the closest Ag-orbitals in an adjacent cluster. One can see that in spite of the relatively large Ag-Ag distances 4.14 Å (a) and 4.92 Å (b)) significant overlap still exists, even between adjacent clusters. As the interaction between atoms is proportional to the overlap integrals,[51] silver atoms are clearly interacting electronically, thus supporting the notion of bands. It is therefore valid to call the arrangement of semiconductor-component clusters an expanded silver halide. Is it an expanded semiconductor? Expansion of the semiconductor components and inclusion in a sodalite matrix has increased its bandgap energy compared to the bulk silver halides, as would be expected. The optically determined bandgaps are of the order of 3.8 eV. This value is at the border between semiconductors and insulators.[52]

Table 3: Percentage Overlap Between Atomic Orbitals Within a Ag_4Cl Cluster, Between Next Neighbours in Adjacent Clusters and with Framework Atomic Orbitals.[28]

Orbital	(a) Same Cluster			(b) Next Cluster			(c) Framework					
	Ag5s	Ag5p	Ag4d	Ag5s	Ag5p	Ag4d	Al3s	Al3p	Si3s	Si3p	O2s	O2p
Cl 3s	19	22	6	-	-	-	-	-	-	-	-	1
Cl 3p	15	11	7	-	-	-	1	2	-	1	-	-
Ag 5s	6	12	3	2	6	1	19	23	14	16	11	10
Ag 5p	12	19	4	6	13	2	30	25	26	22	25	14
Ag 4d	3	4	1	1	2	3	4	6	4	5	7	6

So far, we have treated these clusters as "gas phase species". However, in the sodalite cage they are part of a lattice which is composed of the additional atoms Si, Al and O. Can the framework atoms really be treated as innocent bystanders or are they involved in the interaction between clusters, or even within one cluster? Figure 5B shows the density-of-states diagrams for an individual sodalite cage, a cage with two clusters (one at the centre of a cage, one outside), and a cage with four clusters (one cluster at the centre of the cage, three outside). The structure of the cluster band is clearly visible, even within a sodalite cage (*cf.* Figure 5A). The HOMO pertaining to the cluster is slightly destabilized by the presence of the cage. Mixing of framework and guest levels is also visible, especially in the LUMO region, where it leads to some orbital stabilization. At high cluster loadings, band broadening becomes more pronounced and the orbital mixing is more extensive. While the HOMO still has the Ag 4d, Cl 3p and Ag 5s characteristics, the LUMO now belongs to the framework Si 3s and Si 3p orbitals with some Ag 5s mixed in. An electronic transition between the HOMO and LUMO would therefore involve charge transfer.

Calculations on mixed Ag/Na halide clusters showed that, while the HOMO-LUMO gap is large in the absence of silver, it shifts red as soon as at least one Ag atom is included in the cluster. The red shift continues with an increasing silver content ($[Na_4Cl]_9 \rightarrow [Na_4Cl]_8[AgNa_3Cl] \rightarrow [Na_4Cl]_8[Ag_4Cl] \rightarrow [Ag_4Cl]_9$) but the incremental change is smaller than after addition of the first silver. When the first silver atom is added, the frontier orbitals take on new characteristics (*i.e.,* those pertaining to Ag^+ rather than Na^+). With addition of more Ag^+, communication between the levels leads to stabilization of the LUMO and destabilization of the HOMO, but does not change the identity of these levels drastically. Thus the change in the energy gap is smaller. This behaviour coincides with the observed trend in UV-visible reflectance spectra of silver halosodalites.

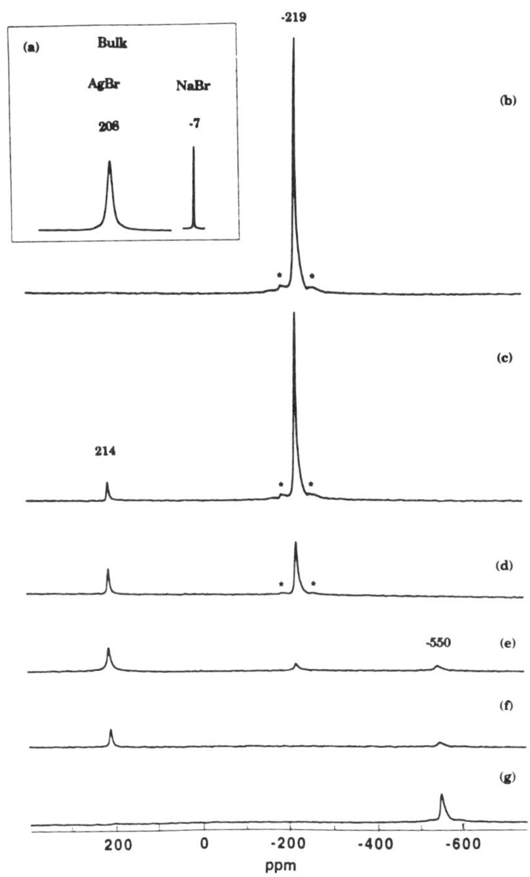

Figure 6: ^{81}Br MAS-NMR at a field of 11.7T of Class A sodalites $Na_{8-n}Ag_nBr_2$-SOD where n = 0, 0.8, 2, 4, 5.2, 8.[53]

Halogen MAS-NMR in a field of 11.7T of $Na_{8-a}Ag_nBr_2$-SOD provides invaluable clues about percolation thresholds and electronic coupling between clusters. Representative ^{81}Br MAS-NMR spectra (referenced against 0.1M NaBr) of Class A $Na_{8-n}Ag_nBr_2$-SOD for n = 0, 0.8, 2, 4, 5.2, 8 are displayed in Figure 6c-6g. The parent material, Na_8Br_2-SOD features a single peak at -219 ppm as shown in Figure 6b. The ^{81}Br peak which is ascribed to the Na_4Br clusters, Figure 6b, appears farther upfield at -219 ppm, than the ^{81}Br MAS signal from bulk NaBr, at -7 ppm, Figure 6a. This shielding indicates a higher charge density around the Br^- anion for the Na_4Br tetrahedra in Na_8Br_2-SOD, compared to the Na_6Br octahedra in bulk NaBr. As Ag^+ cations start replacing Na^+ inside the sodalite cavities, a signal emerges at 214 ppm, Figure 6c-6f. This peak is coincident with the ^{81}Br MAS from bulk AgBr as shown in Figure 6a. The signal at around 214 ppm disappears upon washing the solid sample with diluted $Na_2S_2O_3$, thus it is assigned to nanocrystalline AgBr, probably on the outside surface of the sodalite crystallites, formed through the Ag^+ exchange process (less than 10% of the Br^- content, undetectable by PXRD). The total intensity of the ^{81}Br signal decreases as more Ag^+ cations are loaded within the sodalite framework as shown in Figure 6. The decrease of the overall signal is due to the formation of mixed clusters, $Na_{4-x}Ag_xBr$. The Ag^+ cations in these clusters form a strong covalent bond with the Br^- anion, causing a significant distortion of the charge distribution around the Br^- anion. This might produce substantial quadrupolar interactions which would broaden and spread the ^{81}Br signal over a large frequency range. In the final stages of the Ag^+ exchange, Figure 6e-6g, a distinct signal appears at around -550 ppm, which becomes the dominant peak in the fully Ag^+ -exchanged sample, Ag_8Br_2-SOD, as shown in Figure 6g. The upfield signal at -550 ppm is ascribed to encapsulated Ag_4Br clusters. Note the profound upfield shift observed for the Br^- anions, from -219 ppm in the starting Na_4Br cluster, to -550 ppm in the completely Ag^+ cation exchanged sample. The increased shielding of the Br^- nuclei in the encapsulated Ag_4Br clusters looks extraordinary, in particular since one expects a higher covalent nature for the Ag-Br bonds, compared to the respective Na-Br bonds in the Na_4Br clusters. An increased covalency would reduce the density of electronic charge around the ^{81}Br nuclei, leading accordingly to a deshielding effect. This is indeed demonstrated in the ^{81}Br MAS spectra of the bulk NaBr and AgBr, Figure 6a, where deshielding of the ^{81}Br nuclei is observed in the latter material. The upfield signal at -550 ppm is not detected in Class B sodium, silver, halosodalite samples that intentionally contain a high abundance of Br^--empty cavities, relative to Br^--containing cavities. However, a percolation threshold is observed in a sodalite sample with an intermediate ratio of Br^--containing to Br^--empty cavities. The extraordinary nuclear shielding observed in Figure 6 might be related to intercavity electronic coupling (indirectly through-space and/or directly through the sodalite framework) which introduces orbital overlap between adjacent $Na_{4-x}Ag_xBr$ clusters. Support for this proposal stems from the observation of similar but less pronounced ^{35}Cl MAS-NMR shielding/deshielding chemical shifts for the series $Na_{8-n}Ag_nCl_2$-SOD (-122 ppm, Na_4Cl; -310 ppm, Ag_4Cl; referenced against 0.1M NaCl) most probably caused by less covalent bonding between Ag^+ and Cl^- compared to Ag^+ and Br^-, as well as likely weaker intercavity electronic coupling in the chloro- relative to the bromosodalites.

Is the "isolated AgBr molecule" in the low silver loading NaAgBr-SOD sample fully isolated? Of course not! The sodium cations, framework atoms and even bromide ions in adjacent cages still interact with the AgBr diatomic unit, in spite of its short bondlength. However, at the low silver loading it is isolated from other AgBr units. This directly affects the optical properties of the material resulting in more structure in both the absorption and excitation spectra. In what follows we will deal with Class B sodalites, where it is possible to produce an even more isolated AgBr molecule, by eliminating halide ions in adjacent sodalite cages.

Class B Sodalites: Assembling a Semiconductor Quantum Supralattice One Atom at a Time[25,29]

In this section a series of silver, sodium bromo, hydrosodalites of the type $Na_{8-p-2n}Ag_{2n}Br_{2-p}[]_p$-SOD, where SOD = $Si_6Al_6O_{24}$, [] = an anion free cage, p = 0-2, 2n = 0-8, will be discussed, which are derived from NaBrOH-sodalites. In these materials single size $Na_{4-n}Ag_nBr$ (n = 0-4) clusters are dispersed within a perfectly periodic array of all-space filling β-cages, but in contrast to the previously described Class A sodalites they are isolated from each other by sodalite cages containing no anion, forming a Class B supralattice of nonstoichiometric silver, sodium bromosodalite, Figure 2. The anion-free cages are filled with $Na_{3-n}Ag_n$ (n = 0-3) triangles, and in the hydrated form, with water molecules. Denks[54] has defined a nonstoichiometric sodalite as a solid-phase solution of NaX-SOD and Na[]-SOD, in which overall β-cavities with sodalitic and zeolitic fillings are statistically interspersed. Our definition of Class B sodalites also includes sodalites containing nonstoichiometric amounts of other anions, including the oxalates discussed later.

For Class A sodalites described above, it has been shown that orbital overlap and electronic/vibrational communication is possible between clusters in adjacent cages and therefore throughout the lattice. In Class B sodalites on the other hand, anion deficient cages break the periodicity between Ag_4X^{3+} clusters. This effect can change co-operative effects between clusters, which require the presence of a halide atom (vibrational) or its orbitals (electronic). For a solid solution, variations in the bromide content, p, are expected to alter the mean distance between $Na_{4-n}Ag_nBr$ clusters while changes in the silver content, n, and thus the sodium:silver ratio, shift the constituents of the clusters from those of an insulator to a semiconductor type. At low halide and silver concentrations isolated AgX molecules may be produced in Class B materials. Such controlled tunable volume filling and compositional alteration of the contents of the sodalite β-cages provides an unprecedented opportunity to adjust the extent of electronic and vibrational coupling between fixed nuclearity $Na_{4-n}Ag_nBr$ clusters. It should therefore be possible to chemically manipulate the electronic band structure of the materials, and thus their electronic and optical properties. Some of the key structural, vibrational and optical properties of the Class B sodalites have recently been addressed.[25,29] They will be summarized below.

The $Na_{3-m}Ag_m$ and $Na_{4-n}Ag_nBr$ clusters are statistically distributed throughout the sodalite lattice.

However, silver ions are associated preferentially with halide anions.[29] Rietveld refinements of high resolution powder XRD data were carried out on several samples of the $Na_{8-2n-p}Ag_{2n}Br_{2-p}[]_p$-SOD series for n = 0, p = 0, 1.7, 2 and n = 4, p = 0, 0.8, 1.5, 1.7, 2. Table 4 summarizes the results for the unit cell dimensions and separation of ions from the refinements. The edge lengths of the cubic unit cell of the Class B sodalites studied depended on the silver and bromide concentrations as well as on the degree of hydration.[23,25]

Cluster Interactions: Vibrational Coupling.

A comparison of the trend in atomic separations (Table 4) with the corresponding far-IR spectra reveals a dependence of the vibrational coupling on the bromide content of these samples. The far-IR spectra for samples with varying bromide content (Figure 7) show the effect of dropping the population of Na_4Br clusters from two to zero per unit cell of sodalite. The smoothness of this transformation is demonstrated by the monotonic low frequency shift of the pore opening framework mode (f) with decreasing halide concentration (Figure 7). The splitting of the Br^- anion correlation couplet[55] seen at 161 and 68 cm^{-1} decreases as the average separation between bromides becomes greater (larger p). Its intensity subsequently goes to zero. However, the Na^+ cation correlation couplet observed at 200 (E) and 105 cm^{-1} (A_1) remains, and each band broadens, due to an inhomogeneous distribution of Na_4Br/Na_3 clusters. Even at one Br^- every fourth cavity (p = 1.5), the collective vibrational coupling between Na_4Br clusters has been severely suppressed. The absence of the correlation splitting in Na[]-SOD[55] demonstrates the requirement of Br^- for effective coupling between the Na^+ ions of adjacent β-cages.

The correlation splitting between the sodium translational modes decreases slightly as bromide is added to the sodalite. If the coupling depended solely on the separation between Na^+ ions in adjacent cages one would expect the opposite trend, since the Na-Na separation between adjacent cages was found to decrease at higher Br loadings. However, the Na-Br separation increases at the same time. This provides further evidence that the anion plays a significant mediation role in the coupling between Na^+ ions. Since non-basic hydrosodalite, containing only cationic triangles, exhibits no correlation coupling, it is likely that at dilute bromide concentrations the Na^+ coupling occurs predominantly between cages containing sodium tetrahedra. The overall effect is therefore complicated by the extent of connectivity between M_4Br clusters throughout the lattice. Similar, but less pronounced effects can be discerned in the far-IR spectra of the corresponding fully Ag^+ exchanged $Ag_{8-p}Br_{2-p}[]_p$-SOD.

Cluster Interactions: Electronic Coupling

Evidence for collective electronic interactions stems from both the ^{23}Na NMR spectra and UV-visible absorption spectra. The ^{23}Na MAS NMR spectra of hydrated Class B sodium sodalites show two components, a sharp resonance (full width at half height (FWHH): 310 - 380 Hz) *ca.* 1 ppm

Table 4: Separation of Ions in M,(2-p)Br-SOD, M = Na^+ (n = 0), Ag^+ (n = 4). Values in pm. Standard Deviations are Smaller than Least Significant Digit Shown.[29]

M	p	Unit Cell Dimension (a)	M-X	X-X	M-M (intercage)	M-M (intracage)	M-O (Framework Oxygen)
Na	2	884.8	-	-	508	376	251
Na	1.7	896.2	276	776	484	451	245
Na	0	893.1	289	773	474	472	236
Ag	2	898.0	-	-	485	453	-
Ag	1.7	895.7	280	776	481	457	241
Ag	1.5	895.4	268	775	488	438	249
Ag	0.8	893.1	268	773	487	438	238
Ag	0	891.1	267	772	486	436	244

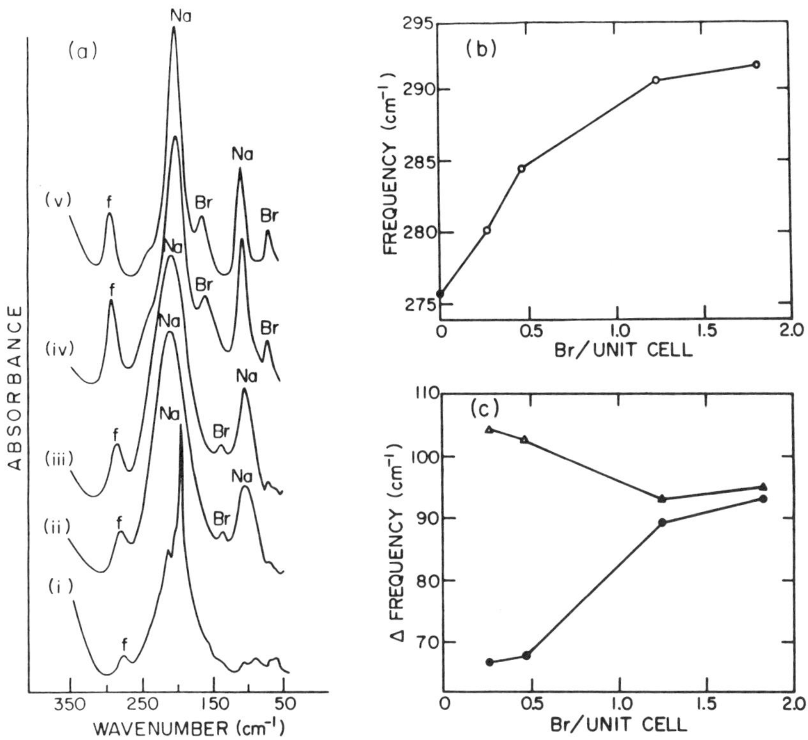

Figure 7. a) Far-IR spectra for the series of dehydrated sodalites of the type $Na_{8-p}Br_{2-p}[]_p$-SOD: i) p = 2, ii) p = 1.7, iii) p = 1.5, iv), p = 0.8, v) p = 0, b) Shift in the far-IR pore opening framework mode (f) with Br^- loading, c) Changes in the far-IR correlation splitting between the high-frequency and low-frequency translatory modes of Na^+ (Δ) and Br^- (•) with Br^- loading.[25]

increases, the unit cell size and upfield from solid NaCl attributed to Na_4Br cages and a broad band (FWHH: 1200 - 1350 Hz) *ca.* 17 ppm upfield from the reference sample assigned to hydrated anion free Na_3 cages. In dehydrated samples only one resonance is observed, which increases in intensity and shifts downfield with higher Br^- loading. This peak is assigned to the ^{23}Na resonance in cages containing Na_4Br. The absence of an observable peak corresponding to Na_3 cationic triangles indicates that intensity loss has occurred through line broadening as a result of greater localization of cations in the cation sites of the []-cages[56] and of locating the bare Na^+ ions in less symmetric environments than hydrated sodium ions.[57] This greatly increases the quadrupolar interactions.[58] Chemical exchange of Na^+ between cages is not expected at room temperature for dehydrated sodalites.[59] In the dehydrated samples the shift of the Na_4Br resonance with increasing Br^- loading is attributed mainly to deshielding of sodium due to electron withdrawal by the relatively electronegative bromide ion, thereby providing evidence for electronic coupling between Na_4Br clusters. Any effects involving charge transfer by framework oxygens must be negligible as they would result in the opposite trend: as the bromide concentration T-O-T angles both decrease,[60,61] allowing more charge density to relocate from the sodalite-cage lattice six-ring oxygens to Na^+ (3s), especially as the Na-O separation is reduced at the same time. As in the case of the unit cell sizes, one has to consider the role of imbibed water to interpret the ^{23}Na chemical shifts of the hydrated bromo, hydrosodalites. In these samples, as the Br^- loading increases the water content correspondingly decreases. The structural water in the cages containing cationic triangles has the effect of deshielding the sodium cations in bromide-containing β-cages. This occurs, presumably, by withdrawing charge density from the framework oxygens *via* hydrogen-bonding.

Optical reflectance data for (8-p-2n)Na,2nAg,(2-p)Br-SOD, where p = 0-2, 2n = 0-8, are shown for the hydrated samples in Figure 8. The series p = 2, 2n = 0-8 serves as a pivotal control group to pinpoint those absorptions associated with m = 0-3 $Na_{3-m}Ag_m$ "ionic triangles", which all fall at 220-240, 300 nm (Figure 8). For one Br^- in every eigth sodalite cage, the line widths of the $Na_{4-n}Ag_nBr$ cluster UV-optical excitations are so narrow compared to the parent case of one Br^- per cavity (Figure 8) that individual components for n = 0, 0.05, 0.5, 1, 4 can be resolved. When these data are compared with those of Class A quantum supralattices (Figure 4), one realizes that all of the resolved components of the Class B supralattice are contained within the spectral envelope of the former.

At low silver and bromide loadings a very sharp UV spectrum resembles that of an isolated gas-phase silver bromide monomer (Br^-(4p), Ag^+(4d) → Ag^+(5s)). The spectral features in the UV region are even sharper than for the "isolated AgBr" (Na_3AgBr^{3+}) in Class A sodalites. The line broadening and red shift of the band edge observed when either the silver or the bromide concentration is increased indicates electronic coupling between $Na_{4-n}Ag_nBr$ clusters (Figure 8). Whereas sodium does not appear to contribute much to this coupling (the absorption bands are sharpest at high sodium concentrations), both bromide and silver must be mediating the communication between clusters. The Ag-Br and Ag-Ag separations (within each cage) change most drastically between p = 1.7 and p =

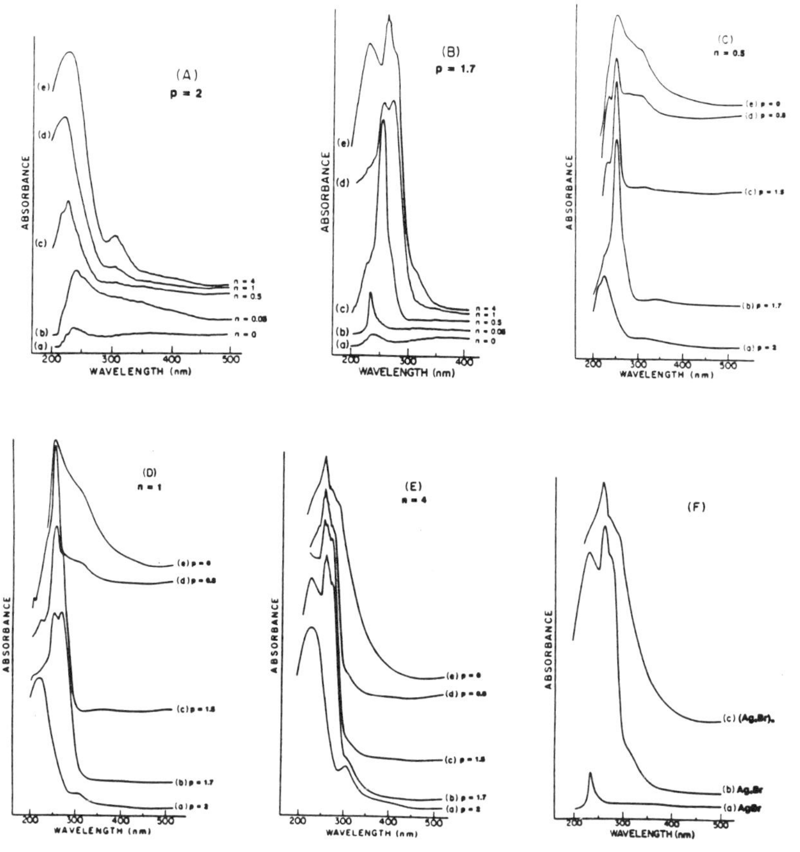

Figure 8. Optical reflectance data for some members of the hydrated sodalite series $Na_{8\text{-}2n\text{-}p}Ag_{2n}Br_{2\text{-}p}[\,]_p$-SOD: A) p = 2; B) p = 1.7, a) n = 0, b) n = 0.05, c) n = 0.5, d) n = 1, e) n = 4; C) n = 0.5; D) n = 1; E) n = 4, a) p = 2, b) = 1.7, c) p = 1.5, d) p = 0.8, e) p = 0; F) Progression of the UV-visible spectra from the sodalite encapsulated isolated AgBr molecule to the isolated Ag_4Br cluster and to the extended $(Ag_4Br)_n$ quantum supralattice.[25]

1.5. However, the most significant broadening of absorption bands in the optical spectra occurs only after the $p=0.8$ loading level. This is an indication that the major mechanism leading to band broadening is not an increase in orbital overlap of the ions within a cage but rather greater communication between clusters in adjacent cages, as the number of bromide centres rises and their separation decreases. Broadening due to phonon coupling is an additional factor; upon cooling Ag,Br-SOD to 27 K three major components could be resolved completely and further components partially. Even though the framework may contribute to long-range through-bond interaction, the gap corresponding to the O(2p) → Al(III)/Si(IV) ligand-to-metal charge-transfer region occurs at higher energy (190-220 mm), and no direct effects were observed.

These data provide compelling evidence for the existence of collective electronic and vibrational coupling interactions between $Na_{4-n}Ag_nBr$ clusters over the full Ag^+ and Br^- loading ranges, as well as, from the isolated silver bromide molecule to the expanded silver bromide quantum supralattice. Furthermore, they reveal the genesis of the $Br^-(4p)$,$Ag^+(4d)$ minivalence band and $Ag^+(5s)$ mini-conduction band on passing from embryonic $Na_{4-n}Ag_nBr$ clusters to a quantum supralattice built of $Na_{4-a}Ag_nBr$ clusters, and ultimately to the parent bulk mixed sodium-silver halides.

Isolated Ag_4^{q+} Clusters to an Expanded $\{Ag_4^{q+}\}_n$ Metal[23,62]

In certain sodalites, containing anions like oxalate or formate, and trapped water, intrasodalite chemical reactions can be carried out in the material's 6.6 Å diameter nanoreactors.[63] For silver sodalites, such redox reactions lead to interesting chromic responses, such as pressure sensitivity, light and X-ray sensitivity, water sensitivity and heat sensitivity, all resulting in changes in the colour or fluorescence. These properties can be exploited, for example, to create materials for high density reversible optical data storage.[64]

Silver oxalatosodalites have been demonstrated to consist of a solid solution of three types of cages containing the following guests: Ag_4Ox^{2+}, Ag_4^{4+}, Ag_3^{3+}. Because of a low oxalate concentration in these materials, the Ag_3^{3+} cages predominate, and the structural properties of the oxalatosodalites strongly resemble those of the hydrosodalites. Oxalate acts as an internal reducing agent for both photoreduction and thermal reduction of silver. During photoreduction Ag_4^{q+} clusters with short Ag-Ag bond lengths are formed, where q = 2 or 3. The Ferroin test,[65] a sensitive and convenient probe of internal versus external silver in nanoporous materials demonstrates that these photogenerated clusters remain encapsulated. (By contrast, heated samples show evidence for external silver after treatment above 250°C). Upon continued UV irradiation, optical spectra suggest the formation of an expanded metal or semiconductor, as the Ag_4^{q+} cluster density increases. In what follows we explore the latter phenomenon in a little more detail.

The Sodium Oxalatosodalite Precursor

To interpret the optical behaviour of silver oxalatosodalites one must first understand their unique structure and the structure of their sodium precursors.[64] Chemical analysis revealed a cation concentration lower than eight cations per unit cell but appropriate for a material containing mainly three cations per cage (NaOx-SOD:$Na_{5.5}Si_{6.3}Al_{5.7}O_{25}C_x$•$nH_2O$; AgOx-SOD: $Ag_{5.9}Si_{6.2}Al_{5.8}O_{25}C_x$•$nH_2O$; x = 0.4 - 0.5). The oxalate concentration was estimated as 0.2 - 0.25 $C_2O_4^{2-}$ (≡ Ox) per unit cell, from the total evolved carbon analysis by thermogravimetric analysis/mass spectroscopy. ^{29}Si MAS-NMR data for NaOx-SOD showed evidence of three types of silicon, supporting the existence of three types of cages, i.e., Si(4Al) tetrahedral groupings corresponding to distinct β-cage contents, namely $Na_4C_2O_4^{2+}$, Na_4^{4+} and Na_3^{3+}. Infrared spectroscopy confirmed the presence of oxalate in both the precursor sodium sodalite and the silver-exchanged product (room temperature, aqueous exchange). X-ray powder patterns showed only a single phase for the fully hydrated sample, eliminating the possibility of significant sodium oxalate deposition on the external surface of the sodalite crystals. Instead, oxalate appears to be randomly distributed throughout the framework.

These results indicate that the hydrothermal synthesis conditions resulted in only *ca.* 10% filling of cages with divalent oxalate anions and four tetrahedrally arranged cations. A corresponding 10% of the cages must contain four cations without an anion, in order to maintain the charge balance. The remaining 80% of the sodalite cages were filled with hydroxide ions, since sodium oxalate has a very low solubility in the basic reaction medium, and sodalite forms preferentially around the hydroxide ions. During hot filtration of the products sodium hydroxide was extracted from the hydroxide cages, leaving 80% of the cages with three cations and no anion. Cages devoid of an anion may also contain water molecules. Thus oxalatosodalites comprise a special case of Class B sodalites.

The results of a full-profile Rietveld analysis of X-ray powder diffraction data for completely silver ion-exchanged AgOx-SOD are illustrated in Figure 9. The data were refined, using the space group $P\bar{4}3n$ and allowing for three types of cages in a solid solution, with appropriate fractional guest occupancies, each statistically distributed throughout the lattice. A good refinement was obtained which is reflected in the small value of χ^2 = 2.16 and final R factors of wR_p = 14.0%, R_p = 10.7%. All framework and silver atoms were located in the refinement. The relative coordinates of two of the silver atoms were further apart than the theoretical resolution of the method; the third type of silver was so close to the other two atoms that it could theoretically not be resolved. However, inclusion of this atom improved the refinement factor. Further support for the validity of including the third atom stems from a refinement of a UV-irradiated silver oxalate sample. In this case the third silver had changed its position so much so that all three atoms could be clearly resolved. In that sample the other two silver atoms were found at nearly the same positions as in the untreated sample.

In accordance with the -2 charge on the oxalate anion, both anion-filled Ag_4Ox and empty Ag_4 cages were observed displaying a tetrahedral disposition of Ag atoms with each Ag coordinated to three oxygen atoms in adjacent six-rings. The intracage Ag-Ag distance is *ca.* 6.7 Å for oxalate ion-containing cages and 5.5 Å for empty Ag_4 cages. For those cages exhibiting loss of OH^- (extracted in a hot washing procedure), residual Ag_3 triangles were observed with an intracage Ag-Ag distance of

4.5 Å. Silver-silver distances between the cages varied from 3.8 to 4.9 Å. In spite of their large magnitudes (normal Ag-Ag bonding distance: 2.88 Å in metallic silver), these interatomic distances still allow direct overlap of the ground and excited state silver orbitals, and they are well within the range of through-bond interaction involving the sodalite framework (see above). Such long range interactions would allow the formation of an expanded silver supercluster. In thoroughly irradiated samples which had undergone a colour change from white to green, those cages containing three silver atoms and one set of four-atom cages remained unaffected, while the silver atoms in the other tetrahedral set moved within bonding distance (2.5 Å) forming a cluster, Figure 9.

The photoactivation of silver oxalatosodalites and the optical changes observed, in general, were related to the ease of silver reduction and the ability of the silver to form small clusters or expanded superclusters within the sodalite framework. In silver oxalatosodalites this was shown to be facilitated by an intrazeolite redox reaction of the following type:

$$2Ag^{+}\ (Z) + C_2O_4^{2-}\ (Z) \dashrightarrow 2Ag^{\circ}(Z) + 2CO_2$$

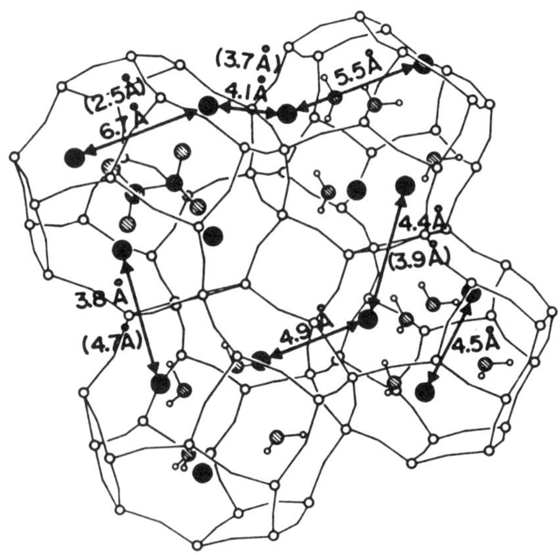

Figure 9: Crystal structure of silver oxalatosodalite, showing cages containing four silver atoms and oxalate, four silver atoms and water, as well as three silver atoms and water. Silver-silver distances are indicated. Values in parentheses correspond to silver-silver distances after 32 hours of UV irradiation.[62]

This reaction can be effected both thermally and photolytically. Note that this equation is intended to illustrate a simple stoichiometric redox reaction and is not meant to represent a balanced unit cell reaction. Depending on the level of silver exchange, aggregates of the type $(Ag_nNa_{4-n})^{q+}$, n = 0 to 4, appear to be formed within the cages, where charge variations of q in the range q = (4-n) to 4 can lead to the observed optical responses. Sodalites containing other anions capable of undergoing intracage redox reactions include OH^-, HCO_2^-, SCN^-, NO_2^-.[63]

A

B

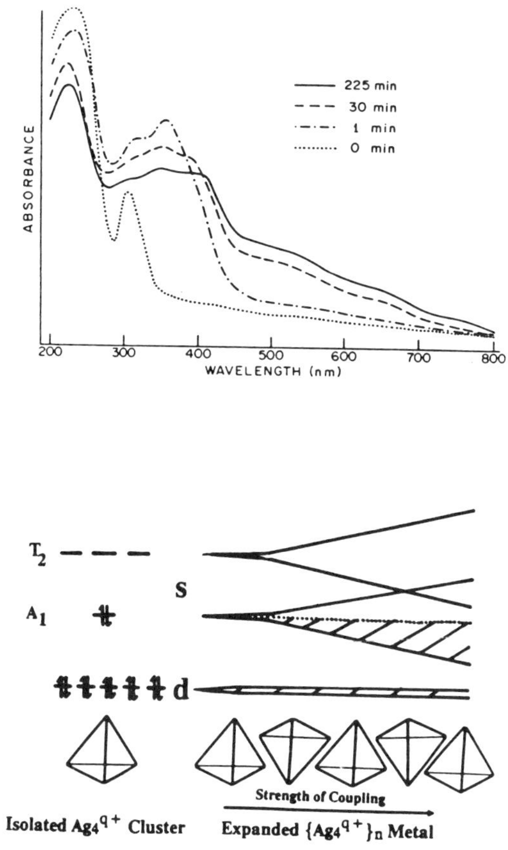

Figure 10. A: Development of the optical absorption bands of silver oxalatosodalites after various UV irradiation times; B: Schematic energy level diagram showing the formation and broadening of bands as the degree of coupling between Ag_4^{q+} silver clusters increases.[23,62]

Figure 10 shows the optical absorption spectra for untreated and UV irradiated AgOx-SOD after various irradiation times. In the initial spectrum two major absorptions were observed with peaks at *ca.* 230 nm and *ca.* 305 nm. After irradiation, both of these absorptions were still present. An additional absorption appeared at *ca.* 360 nm. With longer treatment a plasmon-like background absorption developed, with some structure, covering the whole visible region and tailing into the near IR.

From the combined results of a multiprong *in situ* diffraction, spectroscopy and thermal analysis study the following mechanism is proposed to explain the above observations. Irradiation with a low power UV light results in the dissociation of oxalate to produce CO_2 and two electrons per oxalate anion. Depending on their mobility, the electrons reduce silver ions in the cages containing oxalate or any adjacent cage. Trapped water can react with the CO_2 to produce carbonic acid or carbonate. Since water has a smaller kinetic diameter than CO_2 it is more likely to diffuse through the six-rings into the cages containing CO_2, forming two trapped H_2CO_3 molecules in the former oxalate cages. The silver atoms move within bonding distance forming a small tetrahedral Ag_4^{q+} cluster in the empty Ag_4 cage. A larger Ag_4^{p+} cluster occupies the carbonic acid or carbonate cage. Since the Ag_3 cages seemed to remain unaffected, the triangular units are likely to still consist of Ag_3^{3+} groups of ions rather than reduced clusters.

In order to test this mechanism, extended Hückel molecular orbital (EHMO) calculations were carried out, and the results were compared with the experimental data. Figure 11 shows the band diagrams for Ag_3^{3+}, Ag_4^{4+} clusters and a small Ag_4^{2+} cluster inside a single sodalite cage. {Ag 4d} orbitals form a band within the framework gap, while the {Ag 5s, 5p} orbitals remain as narrow levels, both within the gap and between the Si and Al bands. Addition of a Ag^+ ion to Ag_3^+ to form Ag_4^{4+} stabilizes the {Ag 5s, 5p} levels and broadens the {Ag 4d} band. Closely spaced empty {Ag 5s, 5p, 4d} levels develop upon silver reduction to form a small Ag_4^{2+} cluster with Ag-Ag bond lengths taken from the Rietveld structural analysis. Transitions from the HOMO to these levels would result in a series of low energy absorptions, as observed in photoreduced AgOx-SOD. The optical absorption spectra for untreated and UV-irradiated AgOx-SOD, Figure 10, can therefore be assigned as outlined below (water, oxalate, and carbonate have been ignored in this calculation).

Untreated AgOx-SOD containing 80% cages with Ag_3^{3+}, 10% $Ag_4^{4+}{}_{large}$, 10 % $Ag_4^{4+}{}_{small}$ is very similar in composition to Ag[]-SOD and exhibits a nearly identical absorption spectrum. Based on the energy level diagrams, the two major absorptions are attributed to the {Si 3s, 3p} ← {Ag 4d} (*ca.* 230 nm) and {Ag 5s, 5p} ← {Ag 4d} (*ca.* 305 nm) transitions of the silver triangles and tetrahedra respectively.

After UV irradiation both of these absorptions are still present, which is consistent with the model that the Ag_3^{3+} and half of the Ag_4^{4+} clusters remain unchanged in the photolytic treatment. A new absorption at *ca.* 360 nm is probably due to the {Si 3s, 3p} ← {Ag 4d, 5s, 5p} transition in cages containing small Ag_4^{2+} clusters. The red-shift with respect to the unreduced cluster is due to a destabilization of the {Ag 4d} orbitals upon cluster contraction and concomitant mixing-in of {Ag 5s, 5p} orbitals. A series of additional weaker absorptions reaching down to the near-IR region is likely due to the transitions from the HOMO to the empty {Ag 5s, 5p, 4d} levels mentioned above, Figure 11.

Under reaction conditions where silver is not restricted inside an inclusion compound such as the sodalite framework, continued silver reduction, while proceeding through small clusters such as Ag_2^+ and Ag_4^{2+}, eventually leads to the formation of larger aggregates and metallic silver particles. As long as only mild photolytic treatment is involved, the sodalite framework restricts such agglomeration, keeping the maximum cluster size at four occluded silver atoms. Silver clusters consisting of a small number of atoms do not yet possess metallic properties. However, as the number and density of reduced Ag_4^{q+} clusters increases with irradiation time of silver oxalatosodalites, a broad background feature resembling a silver plasmon absorption develops in the absorption spectra and the Ag_4^{q+} cluster features become less pronounced, Figure 10. In small colloidal silver particles having metallic properties an absorption band is caused by surface plasmon excitation of the electron gas in the particles. In a transition region of growing particles (agglomeration numbers around 10) the plasmon band develops, first peaking at shorter wavelengths (360 nm) and being strongly broadened.[66] A similar process may occur in the sodalites. During the sample irradiation, more and more cages are filled with partially reduced silver clusters. Although they are surrounded by a sea of Ag_3^{3+} triangles, they may communicate electronically, possibly forming bands.

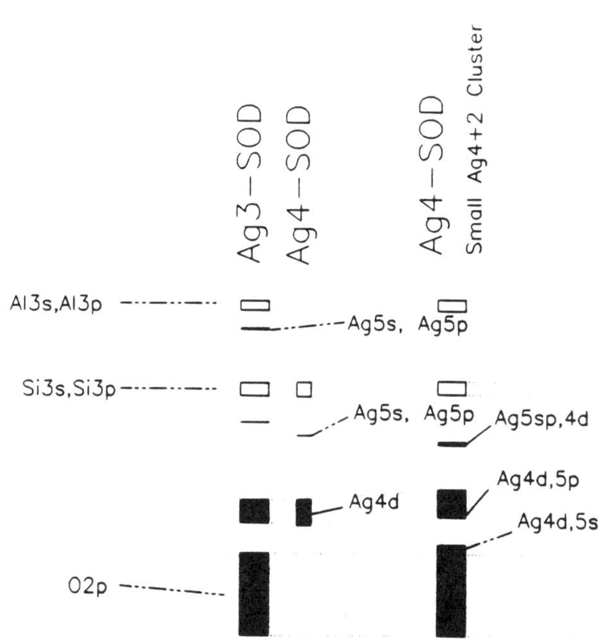

Figure 11: Sodalite band diagrams for one cluster of the type indicated, inside a single sodalite cage. Ag_4-SOD and Ag_4-SOD (Small Ag_4^{2+} Cluster) refer to cages found in silver oxalatosodalites.[23]

The schematic band diagram in Figure 10 illustrates what can occur during this process. Initially the picture consists of isolated Ag_4^{q+} clusters. As more clusters are produced they start to connect, and if the interaction is strong enough they form a band. The bandwidth and band overlap would depend on the degree of coupling. Since after silver reduction some electrons are present in the conduction band, one can consider this arrangement of silver atoms an expanded metal within an insulator framework. In fact, the plasmon absorption does not tail all the way to zero energy, but stops at about 1200 cm^{-1}. This implies that a small bandgap is still present. We may in fact be close to an expanded semiconductor to metal transition. Detailed conductivity studies are required to test this model.

It is notable that with increased irradiation time the quasi-plasmon absorption red shifts and becomes more intense as the number density of reduced silver clusters increases. The Maxwell-Garnett effective medium theory[67] describes plasmon absorptions for small metallic particles in a dielectric medium at low volume concentrations (NV) of the particles. In a typical AgOx-SOD with 10% $C_2O_4^{2-}$ cages the maximum filling is NV ≈ 0.04. The prerequisite of sufficiently low metal volume concentration is therefore fulfilled. In the dielectric regime, where NV << 1 it predicts a red shift, broader line width and a greater intensity of the plasmon absorption as the filling factor NV increases. This agrees with our observations and supports the validity of the expanded metal model. In the sodalites, the particle size is given by the size of a domain of "connected" silver clusters. The domains may have a large size and shape distribution, resulting in a variation of light scattering and absorption for the different domains. With a large particle size distribution one can expect a shoulder in the low energy region of a plasmon.[67] A shoulder with superimposed structure is in fact observed, Figure 10.

Even if photoreduced AgOx-SOD is not metallic, electron tunnelling between clusters could lead to the observed plasmon-like absorption. Zhu *et al.*[68] showed in a quantum mechanical calculation that for a semiconductor superlattice composed of quantum dots with thin barrier layers between them, tunnelling between the quantum dots can cause plasmon excitations. In the case of photoreduced AgOx-SOD, the Ag_4^{q+} clusters are separated from each other by thin aluminosilicate barriers. After brief irradiation times these clusters are likely to be isolated, so that no collective modes can exist. However, after longer irradiation times tunnelling could occur between weakly coupled clusters, and lead to collective electron motions resulting in the observed plasmon modes.

Proposed Future Studies and Applications

Where can one go from here? Traditional or previously suggested applications of sodalites have included pigments (lapis lazuli, a blue sodalite mineral, has been used since antiquity as a rare blue pigment), gas storage materials, and the use of photochromic and cathodochromic sodalites for information storage, display and filter optics. New applications for sodalites would be based on the confinement of small clusters in organized arrays, for example, waveguides, non-linear optical materials, optical computing, quantum electronics[69].

In order to examine the viability of sodalite nanostructures as non-linear optical materials to be used for optical transistors in optical computing, frequency doubling and other all-optical applications, both the linear and non-linear properties of these materials must first be quantified and then optimized. Sodalites fulfill one prerequisite for second-harmonic generation, namely, they are non-centrosymmetric. However, the non-centrosymmetry pertains only to the guest-portion of the material. It may be necessary to increase the dipole moment of the guest by e.g., combining cations with large differences in electronegativity (the sodium/silver system is a good pair), and aligning them by carrying out the ion exchange in a strong electric field, (i.e., a type of "poling" experiment).

For applications of sodalites in the area of quantum electronics and optics (fast switching resonant tunnelling transistors, high gain heterojunction laser arrays) it is necessary to study their electronic properties (including ion and electron mobility) in more detail. It may be important here to increase the electron density throughout the guests, by employing internal redox reactions, similar to those referred to above. Possible n-doping routes of these intrinsic sodalite supralattices involve electron injection from $C_2O_4^{2-}$ or OH^- (photolytic), ion exchanged Cu^+ (thermal) or impregnated Na (thermal, chemical or electrochemical). A feasible p-doping pathway of these intrinsic supralattices involves hole injection from, for example, ion exchanged Cu^{2+} [69].

The optical spectra of the photo-developed AgOx-SOD system showed development of an insulator to semiconductor or quasi-metal transition. This should be further probed by electronic conductivity measurements. A major difficulty to be considered here is making proper electrical contacts with the sodalite, ideally with high geometric selectivity. Several possible approaches to molecular wires have recently been reviewed[1]. The availability of thin sodalite films would probably facilitate the development of sodalite quantum electronic and optical circuitry.

The possibility of using silver sodalites for applications in high density imaging and reversible optical data storage has been discussed elsewhere.[70] In this area, compositional fine-tuning is required to increase "image" stability and contrast, as well as optimize reading, writing and erasing response times, threshold powers and wavelengths. Clearly, all possible device applications for sodalites require much further study. However, the unique structural properties of the sodalite host and the variety of guests that can be included in controlled ways certainly make this ancient material an ideal "model system" for probing the physicochemical properties of assemblies of encapsulated clusters built up of the components of bulk insulators, semiconductors and metals (expanded materials), as well as a promising candidate for advanced materials research and development in the 21st century.

ACKNOWLEDGEMENTS

We wish to thank the Natural Science and Engineering Research Council (NSERC) of Canada's Operating and Strategic Grants Programmes, Alcan Canada and Optical Recording Corporation Toronto for generous financial support of this work. AS would like to thank NSERC for a 1967 Science and Engineering Postgraduate Scholarship.

REFERENCES

1. Ozin, G.A.; Kuperman, A.; Stein, A., *Angew. Chem.* **101**, 373-390(1989).
2. Stucky, G.D.; MacDougall, J.E., *Science* **247**, 669-678(1990).
3. Greenwood, N.N.; Earnshaw, A., *Chemistry of the Elements,* Pergamon Press: N.Y, (1984).
4. Wilson, A.H., *Proc. Roy. Soc. Lond.* **A133**, 458; Wilson, A.H. *Proc. Roy. Soc. Lond.*, **1931**, **A134**, 277(1931).
5. Mott, N.F., *Metal-Insulator Transitions,* Taylor and Francis Ltd., (1974).
6. Edwards, P.P.; Sienko, M.J., *Int. Rev. Phys. Chem.* **3**, 83-137(1983).
7. Halperin, W.P., *Rev. Mod. Phys.* **58**, 533-606(1986).
8. Narayanamurti, V., *Science* **235**, 1023(1987).
9. Ploog, K., *Angew. Chem. Int. Ed. Engl.* 27, 593(1988).
10. Reed, M.A.; Randall, J.N.; Aggarwal, R.J.; Matyi, R.J.; Moore, T.M.; Wetsel, A.E., *Phys. Rev. Lett.* **60**, 535-537(1988).
11. Tsuchiya, M.; Gaines, J.M.; Yan, R.H.; Simes, R.J.; Holtz, P.O.; Coldren, L.A.; Petroff, P.M., *Phys. Rev. Lett.* **62**, 466-469(1989).
12. Reed, M.A.; Kirk, W.P. (eds.) *Nanostructure Physics and Fabrication, Proceedings of the International Symposium*, College Station, Texas, March 13-15, 1989; Academic Press: Boston, (1989).
13. Herron, N. in Marder, S.R.; Sohn, J.E.; Stucky, G.D. (eds.) in *Materials for Non-Linear Optics,* A.C.S. Symposium Series **455**,(1991) A.C.S.: Washington.
14. Kubena, R.L. Joyce, R.J.; Ward. J.W.; Garvin, H.L.; Stratton, F.P.; Brault, R.G., *Appl. Phys. Lett.* **50**, 1589-1591(1987).
15. Coreoran, E., *Scientific American* 122-131 November(1990).
16. Bate, R.T., *Scientific American* **258**, 96-10 March(1988).
17. Morokhov, I.D.; Petinov, V.I.; Trusov,L.I.; Petrunin, V.F., *Sov. Phys. Usp.* **24**, 295(1981).
18. Perenboom, J.A.A.J.; Wyder, P.; Meier, F., *Phys. Rep.* **78**, 173-292(1981).
19. Meier, W.M.; Olson, D.H., *Atlas of Zeolite Structure Types*, Structure Commission of the International Zeolite Association, 81(1978).
20. Wang. Y.; Suna, A.; Mahler, W.; Kasowski, R., *J. Chem. Phys.* **87**, 7315-7322(1987).
21. Wang, Y.; Herron, M., *J. Phys. Chem.* **92**, 4988-4994(1988).
22. Herron, N.; Wang, Y.; Eddy, M.M.; Stucky, G.D.; Cox, D.E.; Moller, K.; Bein, T., *J. Amer. Chem. Soc.* **111**, 530(1089).
23. Stein, A., Ph.D. Thesis, University of Toronto, (1991).
24. Stein, A.; Ozin, G.A.; Stucky, G.D., *J. Amer. Chem. Soc.* **112**, 904-905(1990).
25. Stein, A.; Ozin, G.A.; Stucky, G.D.; Macdonald, P.M., *J. Phys. Chem.* **94**, 6943-48(1990).
26. Stein, A.; Ozin. G.A.; Stucky, G.D.; Macdonald, P.M.; Meszaros, M., *Angew. Chem. Adv. Mater.* **3**, 306-09(1991).
27. Stein, A.; Ozin, G.A., *Advances in the Synthesis and Reactivity of Solids*, (in press).
28. Stein, A.; Ozin, G.A.; Macdonald, P.M.; Stucky, G.P.; Jelinek, R.; Pines, A., *J. Amer. Chem. Soc.*, (in press)
29. Stein, A.; Ozin, G.A.; Stucky, G.D., *J. Amer. Chem. Soc.*, (in press).
30. Pauling, L., *Proc. Nat. Acad. Sci.* **16**, 453(1930); Pauling, L., *Z. Kristallogr.* **74**, 213(1930).
31. Löns, J.; Schulz, H., *Acta Cryst.* **23**, 434(1967).
32. Hassan, I., Grundy, H.D., *Acta Cryst.* **C39**, 3-5(1983). Hassan, I.; Grundy, H.D., *Acta Cryst.* **B40**, 6-13(1984).
33. Larson, A.C.; Von Dreele, R.B., *Generalized Structure Analysis System,* LANSCE, MS-H805, Los Alamos National Laboratory, Los Alamos, NM 87545, (1988).
34. Shannon, R.D., *Acta Crystallogr.* **A32**, 751-767(1976).
35. Setter, N.; Depmeier, W., *Ferroelectrics* **56**, 45-48; 49-52(1984).
36. Holmberg, B., *Acta Chem. Scand.* **A30**, 797-807(1976).
37. Yamaguchi, T.; Johansson, G.; Holmberg, B.; Maeda, M.; Ohtaki, H., *Acta Chem. Scand.* **A38**, 437-451(1984).
38. Krisher, L.; Norris, W.G., *J. Chem. Phys.* **44**, 974-976(1966).
39. A word of caution should be given for those results, where both the occupancy of the silver ion

is quite low, and the separation between the silver and the sodium cations is very small. For a scan to 100°2*θ* (or 80°2*θ*) the maximum theoretical resolution was 1.0 Å (or 1.2 Å). In the above mixed sodium, silver sodalites the separation between cation sites is 0.25 - 0.42 Å, *i.e.*, smaller than the resolution and approaching the range of the thermal vibrations. However, including both sites improved the R-factor over leaving one site out. The fact that the cation of lower occupancy, silver, is a much stronger scatterer than sodium played to our advantage.

40. Calzaferri, G.; Forss, L.; Kamber, I., *J. Phys. Chem.* **93**, 5366-5371(1989); Calzaferri, G.; Forss, L.; Hugentobler, T.; Kamber, I., *ICONC* and *INPUTC*, Fortran software for extended Hückel molecular orbital calculations, Institute for Inorganic and Physical Chemistry, University of Bern, Freiestrasse 3, CH-3009 Bern, (1989).
41. Braces around orbitals imply orbital mixing, brackets imply the presence of either one or more orbitals included in the list.
42. Huber, K.P.; Herzberg, G., *Molecular Spectra and Molecular Structure, IV. Constants of Diatomic Molecules,* Van Nostrand Reinhold: New York, (1979).
43. Brus, L.E., *Nouv.J.Chim.* **11**, 123(1987).
44. Henglein, A., *Top. Curr. Chem.* **143**, 113(1988).
45. Burns, G., *Solid State Physics*, Academic Press, Inc.: Orlando, (1985).
46. Kawate, E.; Masumi, T., *J. Phys. Soc. Japan* **57**, 1814-1819, 1820-1825, 1826-1833(1988).
47. Strurmer, D.M.; Marchetti, A.P. in Sturge, J.; Walworth, V.; Shepp, A. (eds.) *Imaging Processes and Materials*, 8th ed., Van Nostrand Reinhold: New York, 71-109(1989).
48. Rossetti, R.; Hull, R.; Gibson, J.M.; Brus, L.E., *J.Chem. Phys.* **83**, 1406-1410(1985).
49. Micic, O.I.; Meglic, M.; Lawless, D.; Sharma, D.K.; Serpone, N., *Langmuir* **6**, 487-492(1990).
50. Vucemilovic, M.I.; Micic, O.I., *Radiat. Phys. Chem.* **32**, 79-83(1988).
51. Calzaferri, G.; Forss, L., *Helv. Chim. Acta* **70**, 465-479(1987).
52. Strehlow, W.H.; Cook, E.L., *J. Phys. Chem. Ref. Data* **2**, 163-193(1973).
53. Stein, A.; Ozin, G.A.; Jelinek, R.; Pines, A.; *J. Am. Chem. Soc.*, (submitted).
54. Denks, V.P.,*Trudy Instituta Fizika Akad. Nauk Estonskoi SSR* **55**, 14-71(1984).
55. Godber, J.; Ozin, G.A., *J. Phys. Chem.* **92**, 2841-2849, 4980-4987(1988).
56. Welsh, L.B.; Lambert, S.L. in Bradley, S.A.; Gattuso, M.J.; Bertolacini, R.J. (eds.) *Characterization and Catalyst Development*, A.C.S. Symposium Series **411**, American Chemical Society, Washington, D.C., 262-272(1989).
57. Engelhardt, G.; Michel, D., *High Resolution Solid-State NMR of Silicates and Zeolites*, John Wiley & Sons, (1987).
58. Tijink, G.A.H.; Janssen, R.; Veeman, W.S., *J. Am. Chem. Soc.* **109**, 7301-7304(1987)
59. Kundla, E.; Samoson, A.; Lippmaa, E., *Chem. Phys. Lett.* **83**, 229-232(1981).
60. Weller, M.T.; Wong, G., *J. Chem. Soc., Chem. Commun.* 1103-1104(1988).
61. Engelhardt, G. in "Recent Advances in Zeolite Science", *Stud. Surf. Sci. Catal.* **52**, 151(1989).
62. Stein, A.; Ozin, G.A.; Stucky, G.D., *J. Photographic Sci. (Japan)* **53**, 322-28(1990).
63. Weller, M.T.; Wong, G.; Adamson, C.L.; Dodds, S.M.; Row, J.J.B., *J. Chem. Soc. Dalton Trans.*, 593-597(1990).
64. Ozin, G.A.; Stein, A.; Stucky, G.; Godber, J., in *Inclusion Phenomena and Molecular Recognition*, J.L. Atwood (ed.), Pub. Plenum, N.Y.; p.379-393(1990).
65. Stein, A.; Ozin, G.A., *J. Chem. Soc. Chem. Commun.*, 1455(1991).
66. Linnert, T.; Mulvaney,P.; Henglein, A.; Weller, H., *J. Am. Chem. Soc.* **112**, 4657-4664(1990).
67. Papavassiliou, G.C., *Prog. Solid St. Chem.* **12**, 185-271(1979).
68. Zhu, Y.; Huang, D.; Feng, S., *Phys. Rev.B*, **40**, 3169-3175(1989).
69. Ozin, G.A.; Kirkby, S.; Meszaros, M.; Özkar, S.; Stein, A.; Stucky, G.D. in Marder, S.R.; Sohn, J.E.; Stucky, G.D. (eds.), *Materials for Nonlinear Optics,* A.C.S. Symposium Series **455**, A.C.S. Washington, 554-581(1991).
70. Ozin, G.A.; Godber, J.; Stein, A., *United States Patent* 4,942,119(1990).

ZEOLITE STRUCTURAL PROBLEMS FROM A COMPUTATIONAL PERSPECTIVE

J. M. NEWSAM

Biosym Technologies Inc.
9685 Scranton Road, San Diego, CA 92121-2777, USA

ABSTRACT

Computer modeling techniques can be a substantial aid in zeolite structure solutions or refinements, and a means of extracting structural insight from the averaged structural results that diffraction experiments provide. Comparison against databases of hypothetical zeolite structure types, model building or editing subject to space group and translational symmetry constraints, and simulated annealing are three effective computational routes to framework structure solution. Simulated annealing, a Monte Carlo based optimization procedure, is used to adjust an initially random arrangement of the required number of T-atoms in the unit cell (T= tetrahedral species, Si, Al etc.) so as to best match prescribed geometrical and diffraction pattern matching constraints. Simulated annealing has also been exploited in studying aluminum distributions in zeolite frameworks, based on comparison against measured ^{29}Si nmr data. The effects of finite temperature on zeolite structures have been simulated by crystal dynamics, using force field parameters taken from interpretations of zeolite vibrational spectra. Fluctuations in the effective aperture sizes with temperature are found to depend on the framework connectivity, consistent with experimental observation. The frequency spectra of the O–O distances across the apertures reveal generally well-defined periodicities to the pore window motion; a breathing motion is manifested in the LTL-framework as a change in cross-sectional area of the 12-ring window with time during the dynamics simulations.

INTRODUCTION

Substantial successes in new composition fields and advances in analytical methods have both contributed to the steady growth in the number of known zeolite structure types. The first collections of structures by Meier in 1967 [1] and by Meier and Olson in 1970 [2] included 25 and 27 types respectively. The first edition of the Atlas published in 1978 included 38 [3], the number increasing to 64 in the second edition published in 1987 [4]. The list has since grown to a current number of 85 distinct types with approved codes in the third edition to be published in 1992 and included, together with 5 related structures, in ZeoFile, a computerized data base of information on the known zeolite structure types initially derived from the Atlas [5].

Recent papers have described the structures of montesommaite (framework code MON) [6], ZSM-57 (MFS) [7], AlPO4-31 [8], AlPO4-41 (AFO) [9], MAPO-36 (ATS) [10], MAPO-39 (ATN) [11], SAPO-40 (AFR) [12] and AlPO4-18 (AEI) [13]. Several of the new framework structures

From the Proceedings of the 9th International Zeolite Conference, Montreal 1992, Eds. R. von Ballmoos et al.,

have interesting features. Three-rings occur in ZSM-18 (MEI) [14], VPI-7 [15] and zinc tetrasilicate [16] (Figure 1). NU-87 (NES) [17] has a 2-dimensional 10-ring pore system. Large pores are found in the structures of the mineral boggsite (BOG) [18] with 12 and 10-membered rings, AlPO4-8 (AET) [19,20] with 14-rings, and the gallophosphate cloverite (-CLO) [21] with 20-rings and very large supercages.

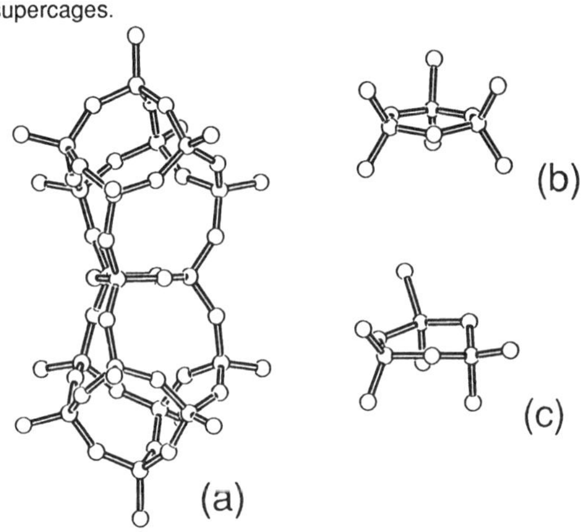

Figure 1. Representations of the 3-ring units in the structures of ZSM-18 (a - the full 5^34^33 cage is shown), lovdarite (b) and zinc tetrasilicate (c).

In all of these studies computational techniques have played a significant role, perhaps in the early phase of model development, in optimizations of initial models based on distance constraints or interatomic potentials, or in structure refinement against measured diffraction data. With the perspective of recent developments, it can be anticipated that computational techniques will play a substantially increased role in future zeolite structural characterization work.

The first step towards characterizing the structure of a new zeolite is the cross-checking of measured powder diffraction data (PXD) against databases of standard measured patterns for known materials, or patterns simulated based on known [5,22] or hypothetical structures. After indexing the PXD pattern, the unit cell size and symmetry deduced are checked against available hypothetical structure data. The number of hypothetical 4-connected 3-dimensional nets that have been described is, however, growing rapidly (see, e.g., [23,24]) and with these various data available only in fragmented form it is possible to miss a prior description of the structure type adopted by a new material. Further, an automated comparison of lists of observed and model PXD peak positions and intensities, even when the model data are available in a computerized database, is not necessarily simple. The flexibility of zeolite frameworks, accomodating changes in unit cell volumes of up to 25% in extreme cases, are such that gross changes can occur in PXD peak positions. Further, dramatic shifts in the diffracted peak intensities can accompany changes in hydration level, alteration of non-framework cation type, content or placement, or sample effects such as preferred orientation. An intelligent database searching strategy might accomodate these types of complications.

FRAMEWORK STRUCTURE SOLUTIONS

When comparisons against the library of known zeolite structure types [4,5,25] or against lists of hypothetical structures do not yield an appropriate model for a new material, other routes to structure solution must be explored. If a single crystal of sufficient size and quality is available, as in the case of boggsite [18] and cloverite [21], conventional single crystal X-ray diffraction methods can be used. Most synthetic zeolites, however, cannot by synthesized in particle sizes much larger than some 10μm and there are often other sample-related complications [26-28]. Improved diffraction instrumentation, the use of synchrotron X-ray sources and better data analysis methods are all helping to combat these problems. Examples of structure solutions by direct powder diffraction methods remain, however, few in number.

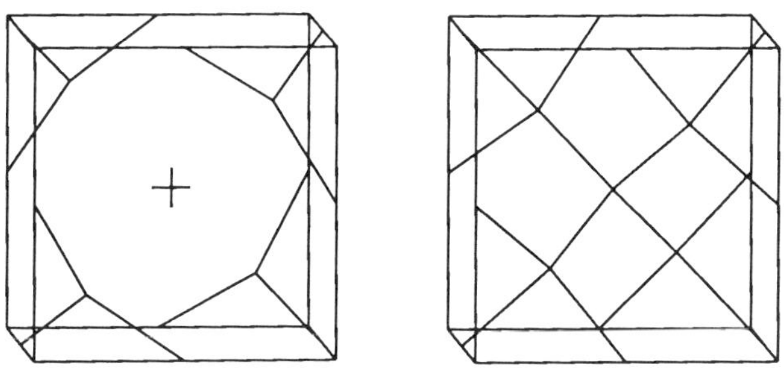

Figure 2. To explore possibilities for 4-connected nets in a 7.8Å cubic unit cell, symmetry-constrained structure editing is used to translate one T-atom along the 3-fold axis on the body diagonal. The symmetry-related atoms move in consort; two more unique T-atoms lie on special positions with no translational degrees of freedom.

The traditional means of developing new zeolite frameworks has been the building of physical models from individual tetrahedra. This approach has both aided in a large proportion of the 85 zeolite framework structure determinations, as well as providing large numbers of interesting hypothetical models. Physical models capitalize directly on the defined geometrial constraints that zeolite structures obey. The selection of those models that might be viable for the material in question is based initially on the character of the model pore architecture and the matching of the model unit cell size and symmerty with the observed data available. Thus, although the extraction of the information on the atomic positions contained within the diffracted intensities is, as above, taxing, a unit cell size can often be derived quite straightforwardly by indexing the PXD pattern or by electron diffraction. The likely space group symmetry or, more typically, a small number of possible space group symmetries can usually also be determined based on those reflections that are systematically absent from the PXD or electron diffraction patterns. The framework density or number of T-atoms contained within the unit cell can, at least approximately, be deduced from sorption experiments or thermogravimetric analyses.

The physical model building approach does not allow the known symmetry or unit cell size to be used directly as a constraint; it can be slow and it is also difficult to avoid an intrinsic bias towards the particular preconceptions of the builder. Mathematical or computational model building methods potentially have substantial advantages. The combination of geometrical constraints and a known unit cell size and space group can, for high symmetry cases, limit greatly the possible T-atom positions (T = tetrahedral species, Si, Al etc.). Such symmetry arguments were used in some early zeolite structure determinations [29]; they have been further exploited by Bennett and Schomaker [30]. Symmetry constraints can be programmed in a quite general way so that as an atom or fragment is moved in the structure, the symmetry-related atoms or fragments are moved in consort; recognizing the character and location of the symmetry elements that are present in the defined space group allows the motion to be constrained so that atoms lying on special positions such as 2-fold rotation axes remain on those symmetry elements (Figure 2). When several symmetry elements are present it can be possible to specify all of the topologically distinct 4-connected nets that might be formed [31]; only a subset of the total number of such topologies represent viable models for new zeolite materials.

The first phase of framework structure determination requires definition only of the approximate T-atom positions. The geometrical constraints on the T-atom configurations obeyed by zeolites and related 4-connected framework structures can be quantified by constructing cost or 'energy' functions based on the T-T distances, and the T-T-T angles and average angles observed in known zeolite structures [32], and a coordination number term that reflects the requirement of framework 4-connectedness The degree to which these combined constraints are satisfied by a given arrangement of T-atoms within the unit cell then provides a measure of the reasonableness of that arrangement. Simulated annealing, a general Monte Carlo based optimization method, can then be used to adjust an initially random set of unique T-atom coordinates so as to best satisfy the specified constraints and hence to generate likely structural models for the material in question (Figure 3). At each Monte Carlo step the unique atom coordinates are displaced randomly and the 'energy' of the new configuration is computed. The new configuration is accepted if its energy is lower than the original or if the energy difference, ΔE, is less than a random 'thermal fluctuation' (that is, if $\exp\{-\Delta E/T\}$ is greater than a random number between 0 and 1) at the current simulation 'temperature', T. The simulation is commenced at high T and the T is gradually lowered to anneal the system into a low energy state. This new approach to structure solution [32] has already been applied to a large number of known zeolite structures with very encouraging results [24] (Figure 4). Some framework structures have also been solved using the method [32,33]. Extensions of the approach to other material types, using different descriptions of the geometrical constraints are also now being explored [34,35].

The simulated annealing method can generate large numbers of hypothetical structures in a short time frame. A non-exhaustive series of simulations based on the unit cell sizes and symmetries of 64 known zeolite structures produced over 5000 hypothetical structures [24]. Many of these structures are not 3-dimensional frameworks, instead forming discrete cages, chains or layers. However, for each different set of input data, each derived structure has a unique set of coordination sequences for the inequivalent T-atoms and therefore has a distinct topology. Use of the degree of match between simulated and target powder diffraction patterns as an additional contribution to the total cost function is a substantial aid for practical structure solution [24]. With this additional term, the success of the method in reproducing known structures is improved. More importantly however, with appropriate weighting the number of structures yielded by the procedure is dramatically reduced. Often only one or two structures result. These test cases indeed indicate that if a result is produced using this diffraction pattern matching term it is most likely the correct structural model for the material in question.

More generally, the success of the simulated annealing approach suggests that diffraction data alone might be used as a basis for adjustment of atomic coordinates from an initially random configuration. Athough zeolites are a rather challenging test case, a significant fraction of the known structures were reproduced using only diffraction pattern matching in the cost function [24]. A still higher success rate is to be expected for more condensed crystal structures.

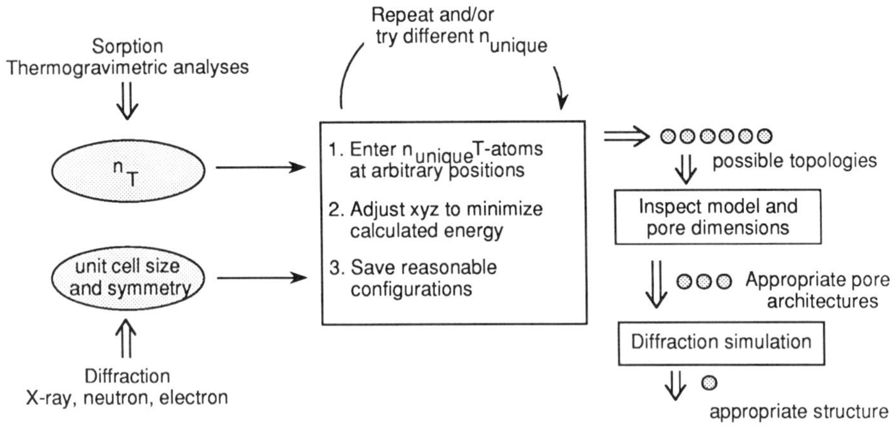

Figure 3. Schematic flow diagram of the use of simulated annealing in framework structure determination when only geometrical terms are used in the 'cost' function.

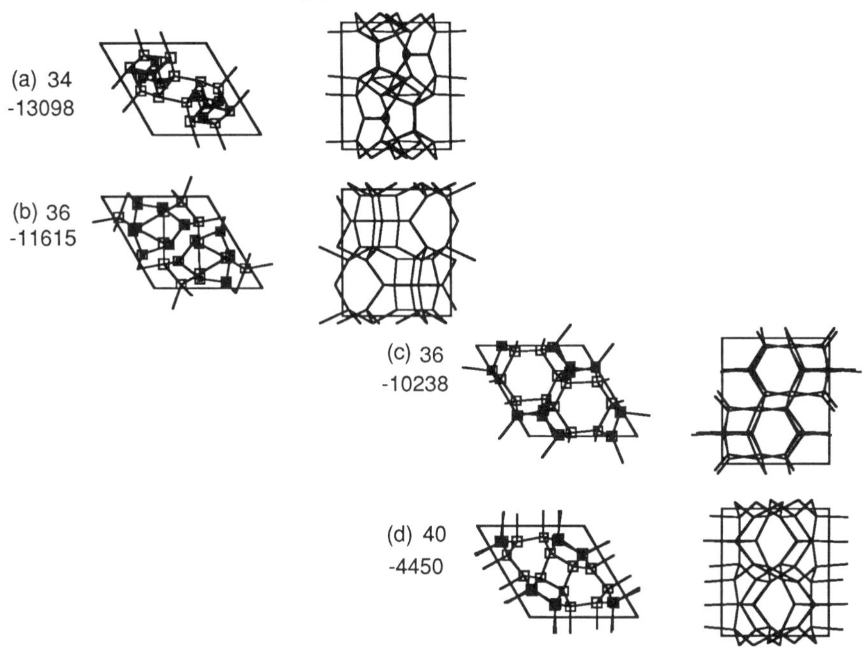

Figure 4. Some of the results of the simulated annealing procedure applied to the known crystallographic data for ZSM-18 ($P6_3/m$, a = 13.2, c = 15.8Å, target number of T-atoms per unit cell, n_T = 34) [24]; the values of n_T for each model, and the calculated 'energies' are shown. The actual MEI-framework is (a).

The simulated annealing method, although initially conceived as an aid to structure solution, can also be used as a true structure design tool. In addition to the geometrical and diffraction terms mentioned above, we have used a term capable of providing bias towards structures that are 3-connected in projection along a defined crystallographic direction. Many known zeolite frameworks have this property (see, e.g., [36]). Another term allows accomodation of one or more cyclindrical or spherical pores. In principle, a contribution to the cost function could be made based on whether or not rings containing a prescribed number of T-atoms are present. However, for large rings this computation can be time-consuming and therefore incovenient to perform at each of the many thousands of steps used in the Monte Carlo simulated annealing procedure. Simplistically, successive shells around each T-atom to the shell corresponding to the defined ring size, N, would need to be sampled to determine whether closure to form a ring had occurred, a calculation scaling as a power of N. Even with substantially improved algorithms, this calculation is expensive. We therefore use a term that adds a positive contribution to the total cost function for each T-atom that protrudes into the defined cylindrical or spherical pore volume. This simple device neatly allows structures to be *designed* with particular pore geometries and, for example, facilitates speculative studies of structures with very large pores.

Although this design of hypothetical structures is purely an academic exercise without the prospect of a rational synthesis approach, computational methods can also assist with this difficult problem. One approach that has recently shown promise is to design a template, stable under the required hydrothermal crystallization conditions, that will optimally fill the cages or channels within the desired structure. A Monte Carlo docking procedure has been developed to automate this template selection process [37].

DISTORTIONS AND FINITE TEMPERATURE EFFECTS

Often, even with a framework topology that appears viable for a new material, the completion and refinement of the structure is difficult. Small structural distortions caused by framework or non-framework cation ordering, or by framework relaxation around the non-framework species can give rise to deviations from the symmetry of the ideal structure. In refinement, subtle deviations from a higher symmetry can cause problems even when high quality single crystal diffraction data are used. With powder diffraction data, the manifestation of slight deviations from an ideal symmetry may be difficult to resolve in the presence of other sample contributions to the measured powder diffraction peak shapes. One approach to this pseudo-symmetry problem, sometimes used for organic crystal structures that are proving troublesome in refinement, is to attempt optimizations in triclinic symmetry, P1, and, after convergence, to inspect the optimized structure for the likely true symmetry. In solution of the structure of NU-87 (NES-framework) this strategy was particulary effective. An energy minimization of the full framework structure was realized in P1 symmetry using shell model potentials. Careful inspection of the results indicated monoclinic symmetry; this description of the structure was then refined successfully using conventional crystallographic least-squares [17]. Determination of the crystallographic symmetry of a physical or computer model is, it should be noted, often not trivial even for the experienced structural chemist; mistakes in symmetry assignment are quite common.

Structures that display slight symmetry distortions frequently undergo a phase transition at elevated temperature to a structure with higher symmetry. Thermal agitation allows the necessary averaging of the atomic positions on the time scale of the diffraction experiment. Thus silicalite shows a transition at ~320K [38], ZSM-11 at ~300K [39] and ZSM-39 at ~380K [40]. Transitions that are reconstructive, that is in which the framework topology is not conserved, occur in other systems. (Na, TMA)-E(AB) transforms to sodalite at above 633K [41], sodium ZK-14 to sodalite at above 610K [42], $AlPO_4$-C to $AlPO_4$-D at ~623K [43] and VPI-5 or AlPO4-54 to AlPO4-8 [20]. The presence of moisture usually facilitates these types of transition.

Direct experimental study of the atomic-scale dynamics of zeolite structures is difficult. Despite significant progress with simple systems [44], mode assignments of vibrational spectra are complicated. The measured vibrational spectra also provide no direct indication of the manner in which the motions of successive tetrahedra are coupled. Diffraction experiments in principle yield mean squared atomic displacements, although measured Debye-Waller factors are usually excessively large on account of a variety of model deficiencies. This is an area to which modeling can contribute physical or mechanistic insight on an atomic scale, while at the same time providing data to compare against experiment for validation or corroboration of the modeling results.

In applying modeling methods to study zeolite framework dynamics, we have used constant volume crystal dynamics to simulate variations in the aperture dimensions with temperature of the six representative zeolite structure types SOD, RHO, TON, MFI, LTL and *BEA [45]. The framework flexibilities were treated by a crystal mechanics force field using parameters taken from quantitative interpretations of Raman and infrared spectroscopic data. The simulations reveal substantial motion of the framework atoms about their equilibrium positions, with fluctuations in the effective aperture sizes with temperature that depend on the framework connectivity (Figure

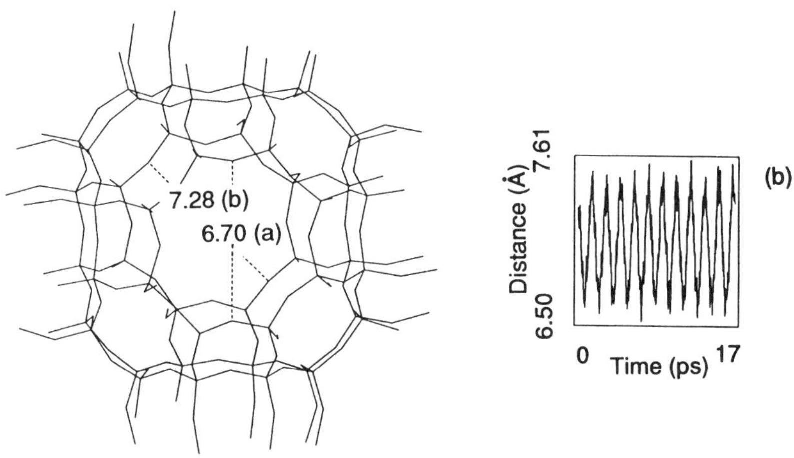

Figure 5. Evolutions of the O-O separations across the 8-ring window in the RHO-framework as a function of time in crystal dynamics simulations [45].

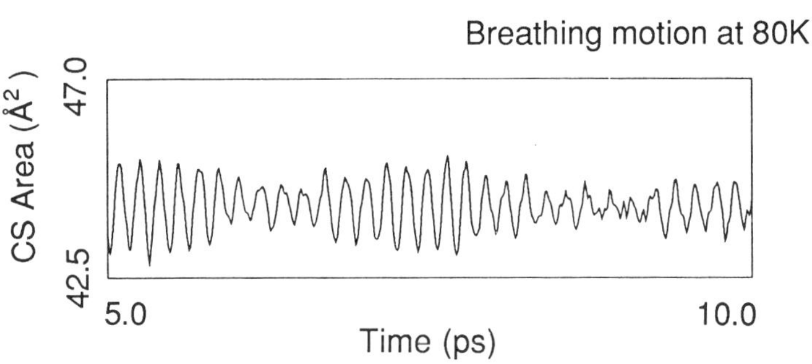

Figure 6. Change in cross-sectional area (CS area) of the 12-ring window in the LTL-framework as a function of time in crystal dynamics simulations at 80K [45].

5). This is consistent with experimental observation. The frequency spectra of the O–O distances across the apertures reveal generally well-defined periodicities to the pore window motion. The definition, extent, and period of the motion depend on the framework connectivity. It is most pronounced in the SOD and RHO-frameworks, previously known from experiment to be most susceptible to static framework distortion. Pore-mouth breathing motion is manifested directly in the LTL-framework in the change in cross-sectional area of the 12-ring window with time during the dynamics simulations [45] (Figure 6).

FRAMEWORK CATION PLACEMENT

The local aluminum distributions are of substantial interest given their role in governing Brønsted site acidities and the selectivities seen in certain catalytic processes. Except in composition regimes where long-range coherence to the Si-Al (or Ga, B, Fe etc) distribution occurs, however, diffraction experiments sample an averaged T-atom site distribution. For structures with crystallographically inequivalent T-sites diffraction experiments can, in favorable cases, indicate the partitioning of aluminum over the inequivalent sites (see, e.g., [46]), but the aluminum site populations are again the averages of all of the local configurations.

Direct information on the relative populations of ^{29}Si nuclei with differing local aluminum distributions is provided by solid state ^{29}Si nmr. Interpretation of these experimental data can then reveal the factors that are at play in governing the aluminum siting. Some early Monte Carlo simulations, that also employed the powerful simulated annealing optimization technique, were based on use of the square planar and diamond lattices as simple models of the more complex zeolite framework topologies [47]. To give reasonable agreement with early measurements of the Si-nAl; n=0-4 distributions as a function of Si:Al ratio, these simple simulations required only a strong first neighbor Al-Al repulsion term (Loewenstein's rule) and a weaker second neighbor Al-Al repulsion term (Dempsey's rule) [47].

Similar simulations have now been performed for the real zeolite framework topology. A model containing several thousand T-sites and with periodic boundary conditions is taken as representative of the bulk crystal structure. The required number of aluminum atoms is introduced at random into the all-silica lattice. The Metropolis Monte Carlo method is then used to adjust this intially random arrangement (by attempting, in each step, substitution of a randomly chosen Al atom onto a randomly chosen Si site) so that the total system energy, that is the sum of the Al-Al repulsion terms, is minimized. Simulated annealing is used as the route towards global energy minimization, and the simulations are repeated for different values of the Al-Al repulsion terms. Without the power of the simulated annealing protocol, convergence is not achieved at the lower Si:Al ratios even for the simpler case of solely Loewenstein constraints [48].

The main result of extensive simulations of Al placement in the FAU-framework topology is that random insertion of Al into the structure, subject to Loewenstein's rule and to a weaker second neighbor Al-Al repulsion term, does not reproduce the measured Si-nAl distribution

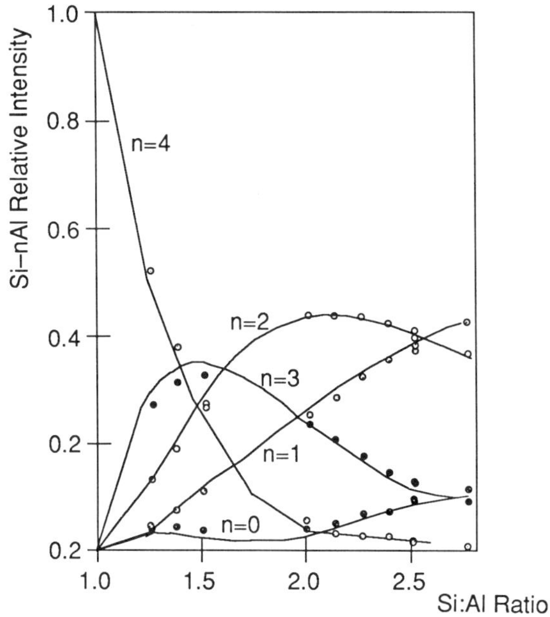

Figure 7. Si-nAl populations in FAU-framework materials measured by ^{29}Si nmr (smoothed continuous lines) and best values calculated using the Monte Carlo procedure (See text). Note the inability of the simulations to reproduce the measured n=3 data and Si:Al < 1.5.

patterns [49] (Figure 7). The aluminum distributions are therefore determined by additional or different factors. This is consistent with Melchior's model of FAU-framework construction from pre-formed 6-ring units [50]. The simulation results also highlight the likely limitations of quantum mechanical studies of aluminum T-site preferences. If the factors controlling the aluminum distributions in zeolites X and Y are also at work in other systems, purely energetic arguments will likely lack direct relevance for application to real materials.

NON-FRAMEWORK SPECIES

Disorder of the non-framework cation configurations is also a common result of diffraction studies [25]. In contrast to the framework cation distributions that would appear, at least in certain frameworks, as the 'petrified' manifestation of formation conditions during synthesis, the non-framework cations are usually highly labile in the hydrated state. They are therefore more likely to adopt configurations governed by equilibrium statistics. Purely energetic arguments [51] or statistical mechanics treatments [52] can therefore be used to rationalize and interpret observed patterns of non-framework cation site occupancies. For example, van Dun and Mortier have had some success in interpreting quantitatively the observed patterns of cation site occupancies measured in sodium and potassium zeolites X and Y [53]. These studies yield the relative cation site energies than could, in turn, be used as a basis on which to improve or validate atomistic simulations of these systems.

At high framework Si:Al ratios when the total number of non-framework cations is small, diffraction methods may have insufficient sensitivity to permit determination of which non-framework cation sites are occupied. If viable interatomic potentials are available, energy minimization, Monte Carlo or molecular dynamics techniques can each be used to probe the preferred non-framework cation configurations. Amongst a small number of results to date, molecular dynamics studies of sodium and calcium cations in zeolite A [54-57] and investigation of possible sites for divalent nickel cations in siliceous mordenite by energy minimization methods [58] are noteworthy.

CONCLUSION

The aforementioned examples are but a few selected from of a growing number of applications of computer modeling techniques as aids in zeolite structure solution or structure refinement, or as means of gaining greater insight into zeolite structural chemistry than the available experimental data afford directly. Zeolite structural problems have traditionally been amongst those driving developments in diffraction methods or amongst those first studied using new diffraction techniques. Several of the computational methods mentioned here have been developed specifically with zeolite structural problems in mind; they will likely now also find broader application in other areas of crystallography. The level of the successes to date and the increased availability of powerful computational capabilities will almost inevitably lead to a substantially increased role for modeling methods in future zeolite structural characterization work.

ACKNOWLEDGEMENTS

The Biosym Catalysis and Sorption Project is supported by a consortium of industrial and academic institutions.

REFERENCES

1. Meier W. M. In: *Molecular Sieves*; Ed. R. M. Barrer, Society of Chemical Industry, London, pp 10-27 (1968).
2. Meier W. M. and Olson, D. H. In: *Molecular Sieve Zeolites – I (ACS Adv. Chem. Ser. No. 101)*; Vol. 1971; Ed. E. M. Flanigen and L. B. Sand, American Chemical Society, Washington, DC, pp 155-170 (1971).
3. Meier W. M. and Olson, D. H. *Atlas of Zeolite Structure Types*, First edition: Structure Commission of the International Zeolite Association (Polycrystal Book Service) (1978).
4. Meier W. M. and Olson, D. H. *Atlas of Zeolite Structure Types*, Butterworths, Guildford, UK (1987).
5. Newsam J. M. and Treacy, M. M. J., Zeolites in press (1992).
6. Rouse R. C., Dunn, P. J., Grice, J. D., Schlenker, J. L. and Higgins, J. B., Amer. Mineral. **75**, 1415 (1990).
7. Schlenker J. L., Higgins, J. B. and Valyocsik, E. W., Zeolites **10**, 293-296 (1990).
8. Bennett J. M. and Kirchner, R. M., Zeolites submitted (1992).
9. Bennett J. M. and Kirchner, R. M., private communication (1990).
10. Smith J. V., Pluth, J. J. and Andries, K. J., Zeolites submitted (1992).
11. McCusker L. B., Brunner, G. O. and Ojo, A. F., Abstracts of 15th. Congress of IUCr, Bordeaux C-59 (1990).
12. McCusker L. B., private communication (1992).
13. Simmen A., McCusker, L. B., Baerlocher, C. and Meier, W. M., Zeolites **11**, 654 (1991).
14. Lawton S. L. and Rohrbaugh, W. J., Science **247**, 1319 -1322 (1990).
15. Annen M. J., Davis, M. E., Higgins, J. B. and Schlenker, J. L., J. Chem. Soc. Chem. Comm. 1175 (1991).
16. Kohara S. and Kawahara, A., Acta Cryst. **C46**, 1373-1376 (1990).
17. Shannon M. D., Casci, J. L., Cox, P. A. and Andrews, S. J., Nature **353**, 417-420 (1991).
18. Pluth J. J. and Smith, J. V., Amer. Mineral **75**, 501-507 (1990).
19. Dessau R. M., Schlenker, J. L. and Higgins, J. B., Zeolites **10**, 522-524 (1990).
20. Vogt E. T. C. and Richardson, J. W., J.Solid State Chem. **87**, 469-471 (1990).
21. Estermann M., McCusker, L. B., Baerlocher, C., Merrouche, A. and Kessler, H., Nature **352**, 320-322 (1991).
22. von Ballmoos R. and Higgins, J. B. *Collection of Simulated XRD Powder Patterns for Zeolites (special edition of Zeolites, Vol. 10, No. 5)*, Butterworths, Guildford, UK (1990).
23. Smith J. V., Chem. Rev. **88**, 149-182 (1988).

24. Deem M. W. and Newsam, J. M., J. Amer. Chem. Soc. in press (1992).
25. Mortier W. J. *Compilation of Extra Framework Sites in Zeolites*, Butterworths, Guildford, UK (1982).
26. Baerlocher C., Zeolites **6**, 325-333 (1986).
27. McCusker L. B., Acta Cryst. in press (1992).
28. Newsam J. M. In: *Zeolite Microporous Solids (NATO ASI Series)*; Ed. E. G. Derouane, F. Lemos, C. Naccache and F. R. Ribeiro, Kluwer, Dordrecht, pp. 167-192 (1992).
29. Taylor W. H., Z. Kristallogr., Kristallgeometrie, Kristallphys., Kristallchem. **74**, 1-19 (1930).
30. Bennett J. M. In: *Perspectives in Molecular Sieve Science (ACS Symp. Ser. 368)*; Ed. W. H. Flank and T. E. Whyte, American Chemical Society, Washington, DC, pp 162-176 (1988).
31. Treacy M. M. J., Rao, S. and Rivin, I. In: *Proc. 8th. Int. Zeolite Conference*; Ed. R. A. von Ballmoos, J. B. Higgins and M. M. J. Treacy, Elsevier, Amsterdam, to be published (1992).
32. Deem M. W. and Newsam, J. M., Nature **342**, 260-262 (1989).
33. Nenoff T. M., Harrison, W. T. A., Nicol, J. M., Stucky, G. D. and Newsam, J. M., submitted (1992).
34. Pannetier J., Bassas-Alina, J., Rodriguez-Carvajal, J. and Caignaert, V., Nature **346**, 242 (1990).
35. Freeman C. M. and Catlow, C. R. A., J. Chem. Soc. Chem. Comm. 89-91 (1991).
36. Newsam J. M. In: *Solid State Chemistry: techniques*; Ed. A. K. Cheetham and P. Day, Oxford University Press, Oxford, in press (1992).
37. Freeman C. M., Catlow, C. R. A., Thomas, J. M. and Brode, S., Chem. Phys. Lett. 137-142 (1991).
38. Hay D. G., Jaeger, H. and West, G. W., J. Phys. Chem. **89**, 1070-1072 (1985).
39. Fyfe C. A., Gies, H., Kokotailo, G. T., Pasztor, C., Strobl, H. and Cox, D. E., J. Amer. Chem. Soc. **111**, 2470-2474 (1989).
40. Strobl H., Fyfe, C. A., Kokotailo, G. T., Pasztor, C. T. and Bibby, D. M., J. Amer. Chem. Soc. **109**, 4733-4734 (1987).
41. Meier W. M. and Groner, M., J. Solid State Chem. **37**, 204-218 (1981).
42. Cartlidge S. and Meier, W. M., Zeolites **4**, 218-225 (1984).
43. Keller E. B., Meier, W. M. and Kirchner, R. M., Solid State Ionics **43**, 93-102 (1990).
44. Creighton J. A., Deckman, H. W. and Newsam, J. M., J. Phys. Chem. **95**, 2099-2101 (1991).
45. Deem M. W., Newsam, J. M. and Creighton, J. A., J. Amer. Chem. Soc. in press (1992).
46. Newsam J. M., J. Phys. Chem. **93**, 7689-7694 (1989).
47. Soukoulis C. M., J. Phys. Chem. **88**, 4898-4901 (1984).

48. Beagley B., Dwyer, J., Fitch, F. R., Mann, R. and Walters, J., J. Phys. Chem. **88**, 1744-1751 (1984).
49. Herrero, C. P. and Ramirez, R., Chem. Phys. Lett. **194**, 79-83 (1992).
50. Melchior M. T. In: *Intrazeolite Chemistry (ACS Symp. Ser. 218)*; Ed. G. D. Stucky and F. G. Dwyer, American Chemical Society, Washington, DC, pp 243-265 (1983).
51. Newsam J. M., Jacobson, A. J. and Vaughan, D. E. W., J. Phys. Chem. **90**, 6858-6864 (1986).
52. van Dun J. J. and Mortier, W. J., J. Phys. Chem. **92**, 6740-6746 (1988).
53. van Dun J. J. and Mortier, W. J., J. Phys. Chem. **92**, 6747-6754 (1988).
54. Shin J. M., No, K. T. and Jhon, M. S., J. Phys. Chem. **92**, 4533-4540 (1988).
55. Cohen de Lara E., Kahn, R. and Goulay, A. M., J. Chem. Phys. **90**, 7482-7491 (1989).
56. Demontis P., Suffritti, G. B., Quartieri, S., Fois, E. S. and Gamba, A., J. Phys. Chem. **92**, 867-871 (1988).
57. Miyamoto A., Matsuba, K., Kubo, M., Kawamura, K. and Inui, T., Chemistry Letters 2055-2058 (1991).
58. den Ouden C. J. J., Jackson, R. A., Catlow, C. R. A. and Post, M. F. M., J. Phys. Chem. **94**, 5286-5290 (1990).

DISCUSSION - Plenary Lectures

"Host/Guest Chemistry and Catalysis In Zeolites" (J. Weitkamp)

P. Dutta: The examples on photochemistry in zeolites is fascinating. However, considering the particle size and the wavelength of uv-visible light, considerable fraction is scattered, thus leading to low quantum yields. Do you see this as a problem; if so, how can one alleviate this problem?

J. Weitkamp: This is, no doubt, an important limitation of host/guest photochemistry in zeolites. I am not aware of systematic work in which quantum yields were measured. The utilization of a given zeolite, e.g., ZSM-5 with different crystal size, could perhaps furnish valuable information.

P. A. Jacobs: 1.) You were addressing the issue of stability of ship-in-a-bottle complexes in zeolites and were stating this was not clear at the moment. I believe that indeed the initial stability was not far beyond stoichiometry. On the recently published systems, however, turnovers over $6x10^3$ have been mentioned for room temperature oxidation of alkanes.

J. Weitkamp: Your comment clearly demonstrates that authors who report on catalysis over ship-in-a-bottle catalysts ought to pay special attention to the number of turnovers achieved in their systems.

P. A. Jacobs: You also mentioned that Co-SALEN were true encapsulates in zeolite Y. Given the molecular dimensions of the complex, I doubt whether irreversible encapsulation will occur. I would appreciate if you could express your views on both issues.

J. Weitkamp: So far, we have not prepared salen complexes in faujasite in our laboratory, so there is at this time no in-house experience. Bleeding of faujasite entrapped salen complexes should however be carefully monitored by those who are working on the system.

P. J. Chong: 1.) What type of photon source and wavelength is used for photolysis of 2-phenylcyclo-dodecanone? 2.) Considering photons hardly interact with adsorbed molecules inside the pores due to opaqueness, surface-scattering, thermal effects, etc., how to overcome the prevailing problem of quantum yield, etc? Have you ever used an excimer laser for the photolysis?

J. Weitkamp: 1.) Conventional mercury lamps, usually of the medium pressure type, ca. 200 to 400 nm, for details, consult ref. [44]. 2.) See question of P. K. Dutta. 3.) It was referring to work by others. So the question is not applicable.

"Ceramic Membranes for Separations and Reactions" (A. J. Burggraaf, et al.)

R. von Ballmoos: 1.) How do you rate the prospects that we will have shape selective (i.e., separating n-from iso-alkanes) microporous membranes soon? 2.) How long do you think their development will take, and which are the critical steps?

A. Burggraaf: 1.) No strong opinion; developments are in very early stage, so certainly not "soon" for shape selective (higher alkanes); however alkanes are certainly possible and partially realized (laboratory). 2.) Development of ceramic membrane reactors of small (laboratory) soon - a few yeas; larger reactors (industrial) - 5 to 10 years.

"Sodalite Supralattices: From Molecules to Expanded Insulators, Semiconductors and Metals" (G. A. Ozin)

G. Schulz-Ekloff: For silver sodalites transitions between semiconductor states to conductor states are mentioned. Both states might exhibit quite different temperature dependences. Are there indications for this, e.g., from impedance spectroscopy i.e., measurements of electron mobilities in isolated clusters?

G. Ozin: Yes, I agree with you and we have been attempting to perform these kinds of measurements on large single crystals of sodium, silver halo sodalites, looking for insultor-semiconductor transitions and resonant tunneling phenomena in the current voltage plots. We are continuing to work on these experiments.

"Zeolite Structural Problems From a Computational Perspective" (J. M. Newsam)

T. Gaffney: In simulated annealing, are T-T distance and T-T-T angle energies minimized independently? If you limit the calculations to aluminosilicates, can a correlation between T-T distance and T-T-T angle be used to reduce the number of structures simulated annealing generates?

J. Newsam: The total cost function includes T-T distance and T-T-T angle contributions as separate terms, appropriately weighted. The two terms are computed independently. As illustrated by the gallosilicate structural results, an increase in T-O distance is usually accompanied by a reduction in the T-O-T angle. This does often not translate into a direct correlation between T-T distance and T-T-T angle. We find that a single set of constraints can be applied successfully to a broad range of framework compositions.

Question: We know that simulated annealing and other techniques may generate a number of structures which meet the constraints and exhibit a locally minimum penalty function. How do you know when to stop generating these structures with your simulation?

J. Newsam: There are two approaches: the sprinter and the jogger. For the sprinter seeking to solve one framework structure of a new material, one goal is one framework structure that matches one experimental data and provides the basis for interpreting them. The SA procedure would be terminated immediately when this solution is achieved. The jogger is more concerned with obtaining many structures and one SA procedure would be continued until one number of new topologies generated asymptotes.

E. Cohen DeLara: Is the structure which corresponds to the minimum of energy dependent of the choice of the potential?

J. Newsam: Depending on how much the potential is changed, yes. We actually find that we have good success with a broad range of framework compositions using a single set of geometrical cost functions. Success for other structure classes, such as mixed tetrahedral-octahedral coordinations found for example, in stannosilicates does require some adjustment of the geometrical cost terms.

Synthesis

THE ROLE OF ALKALI CATIONS IN THE SYNTHESES OF SILICALITE-1 IN FLUORIDE MEDIUM

F. CREA*, R. MOSTOWICZ**, F. TESTA*, R. AIELLO*, A. NASTRO* AND J. B.NAGY***

*Dipartimento di Chimica, Università della Calabria, 87030 Arcavacata di Rende, Rende (CS), Italy

**Industrial Chemistry Research Institute, 01-793 Warsaw, Poland

***Laboratoire de Catalyse, Facultés Universitaires de Namur, 5000 Namur, Belgium

ABSTRACT

MFI Zeolites are synthesized from batches of molar composition 10MF-1.25TPABr-10SiO_2-330H_2O with M = Li, Na, K, Cs and NH_4. Regularly sized 70, 80 or 90 μm crystals are obtained. The aspect ratio (i.e. the ratio of length to width) depends on the crystallization rate. The lower the crystallization rate, the higher the aspect ratio. The formation of crystalline K_2SiF_6 and Cs_2SiF_6 does increase both the induction period and the crystallization rate.

INTRODUCTION

The synthesis of MFI zeolites in fluoride media is particularly interesting [1,2]. Fluoride ions act as solubilizing agent, participate in polycondensation processes and crystallization. Several MFI, MTT, MTN, TON, FER [1,3] and MEL and ZSM-48 [4] structures have been synthesized by the use of fluoride ions instead of or together with hydroxide ions. The so-obtained zeolite crystals exhibit a regular morphology and larger than usual crystal sizes.

On the other hand, there is an increased interest in studying the crystal shape and size in different reaction conditions. The crystal size is strongly dependent on the pH of the reaction medium [5-7], which governs both the rate of nucleation and of crystal growth. The nature of the cations also greatly influences the crystal size of ZSM-5 [8,9]. The crystal shape described by the aspect ratio, i.e. the ratio of length to width, is also dependent on the alkalinity [6,7,10]. This ratio is in reasonable agreement with the growth rate ratios for the crystal length and width. The problem of intergrowth in ZSM-5 crystals has also been recently examined [11,12]. Most frequently adjoining crystals are rotated by 90° around a common c-axis. This form of intergrowth is nucleated from small areas on (010) faces of growing crystals [11]. The formation of additional small crystals [13] or ramps [11] is also linked to the seeding effect in the reaction medium [13].

In this paper, we examine the crystallization rate of silicalite-1, its size and shape as a function of cations, synthesized from an initial batch of composition 10MF-1.25TPABr-10SiO_2-330H_2O with M = Li, Na, K, Cs and NH_4.

From the Proceedings of the 9th International Zeolite Conference, Montreal 1992, Eds. R. von Ballmoos et al.,

EXPERIMENTAL

The MFI zeolites (silicalite-1) were prepared in modified Morey-type PTFE-lined 20 cm^3 autoclaves. The reactants were mixed in the following order: MF (M = Li, Na, K, Cs or NH_4), distiled water, tetrapropylammonium (TPA) bromide and fumed silica. The final gels were characterized by a molar composition of $10MF\text{-}1.25TPABr\text{-}10SiO_2\text{-}330H_2O$. The autoclaves were heated at 170±2°C for various times without stirring under autogeneous pressure. Products were identified and their crystallinity was determined by powder X-ray diffraction, using CuKα radiation. Thermogravimetric analyses of dried gels and crystalline products were scanned from room temperature to 700°C at a rate of 10°C/min under 15 cm^3/min flow of nitrogen.

The M-content (M=Li, Na, K and Cs) was determined with Atomic Absorption. Morphology and crystal size were determined by SEM microscope, and the aspect ratios by an optical microscope during the synthesis.

RESULTS AND DISCUSSION

The variation of crystallinity -as determined from X-ray data- is reported as a function of time in Figure 1. The crystallization rate is higher in presence of K and Cs, than in presence of Na and NH_4. The Al-free MFI zeolite crystallizes at a much lower rate from the Li containing system. Indeed, only 90% crystallinity is obtained after 144 h in the latter, while 100% crystalline phases are formed after 18 h from the K-system, 20 h from Cs-system, 36 h from the Na-system and some 48 h from the NH_4-system. Long crystals of ca. 95 μm are obtained from the NH_4-system and ca. 80 μm from the Na- and Cs-systems and ca. 68 μm from the, K and Li-systems. All crystals have similar size and all the systems show monodispersity. The crystal habit is of particular importance. While the Li, Na, Cs and K-systems all lead to the generally characteristic intergrowth, the NH_4-system essentially shows the main crystals without any significant intergrowth. Figure 2 illustrate the increase of length, width and thickness for the Li, Na, K , Cs and NH_4-systems as a function of time.

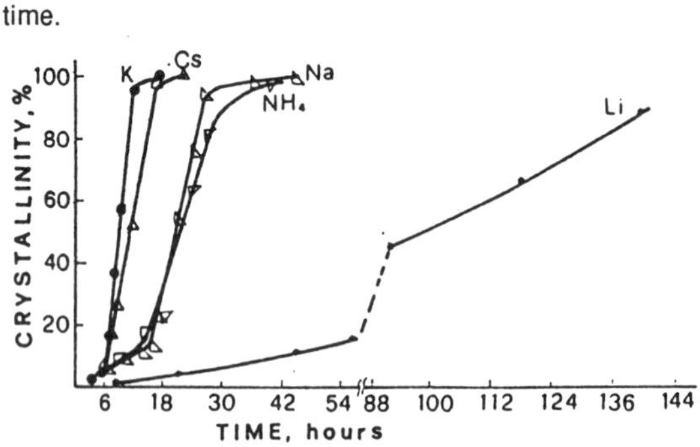

Figure 1. Crystallization curves for the formation of silicalite-1 from gels of molar composition: $10MF\text{-}1.25TPABr\text{-}10SiO_2\text{-}330H_2O$ with M = Li, Na, K, Cs and NH_4

The fastest increase is observed for the K-system, but a *plateau* is reached after ca. 18 h and the size of the crystals remains constant (68 μm). The rate of increase is smaller for the Na and NH_4-systems, but a higher value is obtained at the end of crystallization for the latter (95 μm).

The Li-system rate of increase is even smaller for the and a *plateau* is also reached at ca. 6 days.

In order to get some insight into the nucleation and crystallization mechanism of MFI zeolite in fluoride medium, the chemical analysis of the samples, the rate of crystallization, the induction period, the length, width and thickness of the final crystallites, the aspect ratio (length to width and width to thickness ratios) and the pH of the initial and final gels are systematically examined. The corresponding values are reported in Table 1. The initial values of linear growth rates of crystal length (R_L in μm h^{-1}), width (R_W in μm h^{-1}) and thickness (R_T in μm h^{-1}) are reported in Table 2.

The induction period defined as the first moment of detection of a crystalline phase by XRD increases in the series: K<Cs<NH_4<Na<Li. The crystallization rate R (in % h^{-1}) and the linear crystal growth rate R_L (in μm h^{-1}) both increase monotonously as a function of pH. Note, that a good linear relationship is obtained between R_L and R for the Li, Na and Cs-systems (Fig.3).The NH_4-system shows a higher rate of crystal length growth than that predicted by the relationship while the K-systems increases with a much lower rate R_L. Similar relationships are also obtained for R_W and R_T (Fig.3).

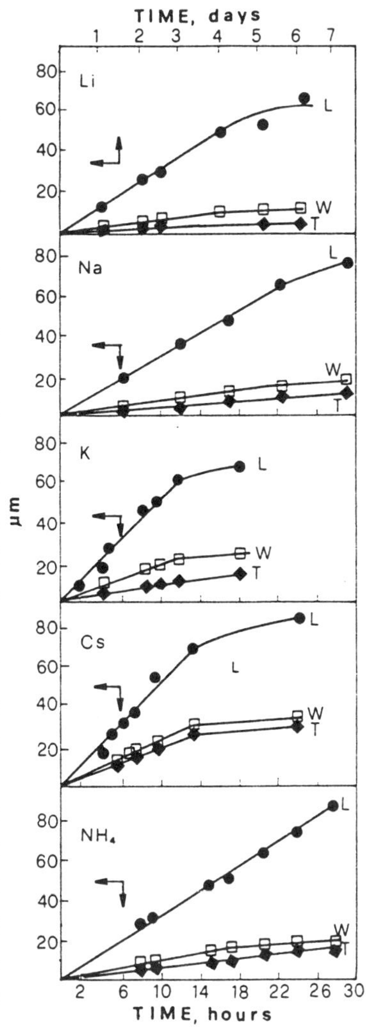

Figure 2. Variation of crystal length, width and thickness as a function of time for Li, Na, K, Cs and NH_4

The initial increase of logR as a function of pH (initial slope of ca. 1) is much higher than that obtained in presence of 1 mole of MF (slope of 0.3) [2]. Indeed, if the F^- amount is varied in a series of gels of molar composition xNaF-yNaCl-1.25TPABr-10SiO_2-330H_2O with x+y=10 and x=1, 2, 3, 5, 7 and 10, at pH of the gels in the range 6.3-7.2, an initial slope at 1.0 is obtained from the logR versus log [F^-] plots [14].This shows clearly that in addition to the effect of OH^- concentration, the F^- ions do play a definite role in the synthesis.

Table 1 Physico-chemical characteristics of silicalite-1 samples during and after the synthesis from gels of molar composition $10MF\text{-}1.25TPABr\text{-}10SiO_2\text{-}330H_2O$.

M	pH_i^a	pH_f^a	$L_{\mu m}^b$	$W_{\mu m}^b$	$T_{\mu m}^b$	L/W^c	W/T^c	$I.P(h)^d$	$R\%h^{-1e}$
Li	5.9	5.8-6.0	70	13	5.6	5.2	3.3	20	0.8
Na	7.2	7.0-7.2	80	17	15	4.7	1.1	9.0	4.9
K	8.0	8.5-8.7	69	24	13	2.9	1.8	2.5	15.0
Cs	8.5	8.5-8.7	80	33	31	2.5	1.1	5.0	9.0
NH_4	7.1	7.6	95	22	19	4.3	1.2	7.0	4.0

a) pH_i=initial pH; pH_f=final pH
b) L = length, W = width, T = thickness of the crystals;
c) Aspect ratios: L/W= ratio of length to width, W/T=ratio of width to thickness
d) Induction period: appearance of ca. 4% crystallinity; e) Rate of crystallization

Table 2.Initial values of linear growth rates of crystals

M	R_L^a	R_W^b	R_T^c	R_L/R_W	R_W/R_T
Li	0.5	0.1	0.1	5.5	1.7
Na	2.9	0.7	0.6	4.0	1.2
K	5.0	2.0	0.9	2.5	2.1
Cs	6.0	2.4	2.1	2.5	1.1
NH_4	3.2	0.8	0.7	4.0	1.1

a) R_L=length growth rate; b) R_W=width growth rate;
c) R_T=thickness growth rate

The accessible amount of F^- ions in the liquid phase does depend evidently on the solubility of the fluorides. The solubility of the fluorides increases in the series: $LiF<NaF<KF<NH_4F<CsF$ [15]. However, if logR is plotted as a function of logS, where S is the solubility of the fluoride, a curve with a maximum is obtained for KF and CsF. Indeed, this solubility is further modified by the presence of the solid part of the gel, where the various cations and anions are adsorbed at the interface [16]. Little is known about the structure of the gels in these fluoride systems. However, the presence of the maximum in the logR vs logS plot could be explained by an interesting behaviour of the K-system and Cs-system. If these gels are dried at 100 °C during 12 h, crystalline K_2SiF_6 and Cs_2SiF_6 are detected, together with some potassium and cesium silicate. It is indeed possible, that these crystalline centers do favour the

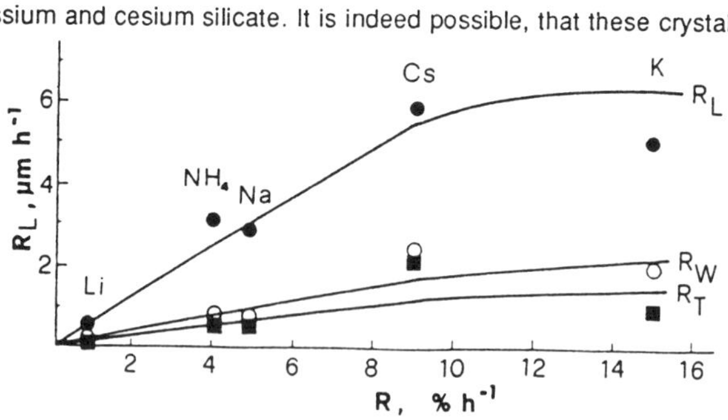

Figure 3. Linear growth rate R_L as a function of crystallization rate R

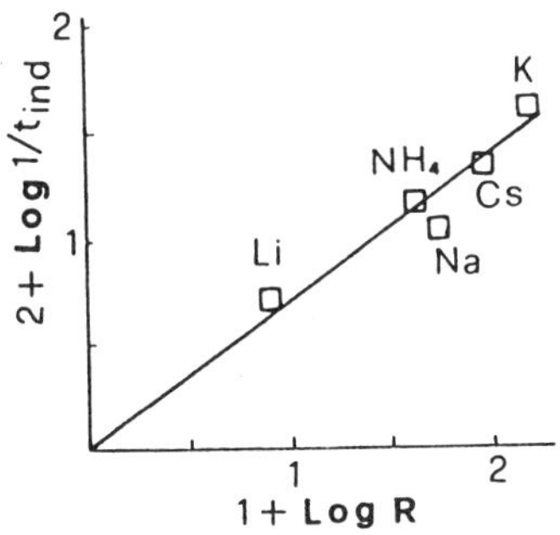

Figure 4. Log R vs. log of the inverse of the induction rate ($1/t_{ind}$)

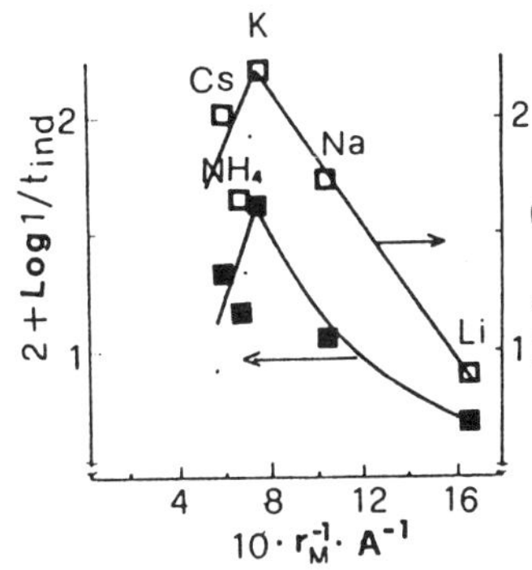

Figure 5. Log inverse of the induction time ($1/t_{ind}$) and log R vs. the inverse of the ionic radii

nucleation, leading faster to the final crystalline phase. Such crystalline centers detected by high resolution electron microscopy where previously supposed to induce nucleation [17,18]. The presence of K_2SiF_6 and Cs_2SiF_6 is also clearly shown by the thermal analysis results, where the different dried gels are analysed in presence and in absence of TPABr. In addition to peaks due to the thermal decomposition of TPABr, a peak at high temperature (ca. 430 °C) is also detected in both TPABr containing and TPABr-free system. These peaks correspond to the decompositions of K_2SiF_6 and Cs_2SiF_6 into 2KF and 2CsF, respectively, and into SiF_4, and to the elimination of the latter. The evolution of SiF_4 during calcination has been confirmed by mass spectroscopy. In fact, the masses of HF, F, SiF^{+3}, SiF_3^+ have been detected. If logR is plotted as a function of the log (induction rate) (i.e., the inverse of the induction period, $1/t_{ind}$), a linear relation is obtained for Li, NH_4, K and Cs ions (Fig. 4).

It means, that similar factors do influence both the induction and the crystallization rate. The Na-system shows a rather lower induction rate (or a higher crystallization rate) than expected from the linear relationship. The influence of the cation can be considered as being due essentially to electrostatic interaction with the gel negative charges, leading to better salting-out effect with increasing electrostatic interaction [19]. For this purpose the radius of the bare cations [20] or better their inverse [19] can be used and the logR and $\log(1/t_{ind})$ as a function of $1/r_M$ are reported in Figure 5.

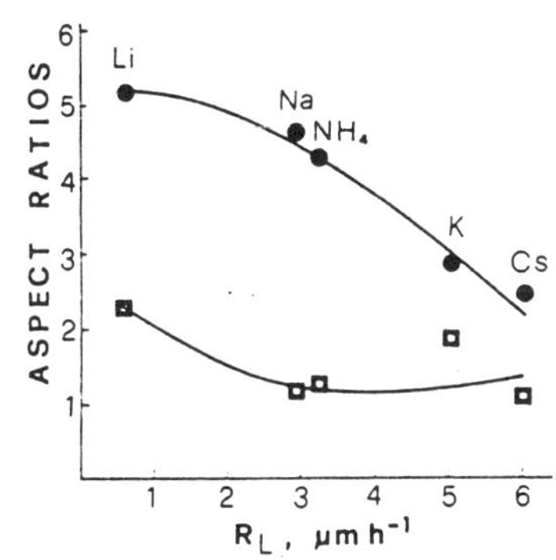

Figure 6. Aspect ratios (L/W and W/T) vs. length growth rate R

It is seen, that oppositely to the effect found in alkaline media, the crystallization or the induction rate both are decreasing with increasing electrostatic potential. A maximum is observed for the K-gel for both induction and crystallization rates. It is very probable that the important factor governing the nucleation and crystallization in the F^--containing medium is not the above-mentioned salting-out effect due to the interaction of the alkali cations with the gel negative charges, but the interaction of the alkali cations with the mineralizing agent F^-. The stronger this interaction, the lower the induction and crystallization rates, at least in a first approximation. More systematic work is necessary in order to better assess the role of the alkali cations. Let us only note that in the Na system the energy of activation for the nucleation (10.8 Kcal mol^{-1}) is higher than the one in the alkaline media (4.8 Kcal mol^{-1}) and the energies of activation for crystallization are equal (10.1 Kcal mol^{-1} in fluoride and 10.0 Kcal mol^{-1} in alkaline medium).

The length to width (L/W) ratio decreases with increasing reaction rate. Figure 6 shows the variation of the aspect ratios (L/W and W/T) as a function of R_L (in μm h^{-1}). The W/T ratio reaches a *plateau*, which can suggest a different phenomenon controlling the growth of length, width and thickness. In that respect, it is interesting to report the variation of the aspect ratios as a function of time. Figure 7 illustrates the variation of both L/W and W/T ratios for the Li, Na, K, Cs and NH_4-systems. It can be seen, that these ratios do change during the crystallization and the increase for length, width and thickness is not linear as a function of time (Fig. 2).

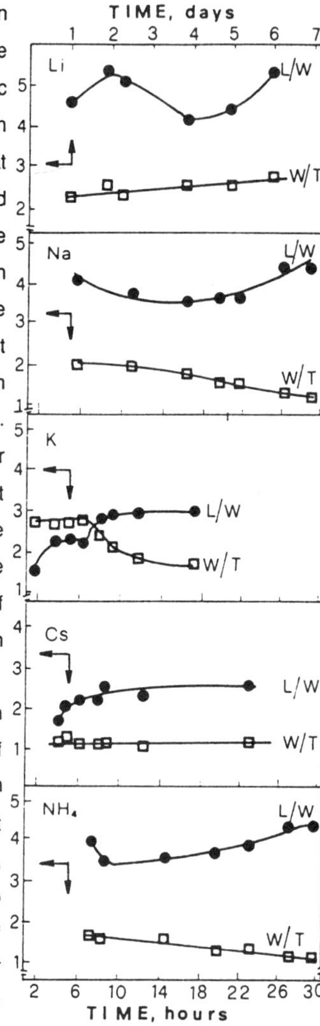

Figure 7. Aspect ratios (L/W and W/T) vs time for Li, Na, K, Cs and NH_4 systems

The initial aspect ratios as computed from the ratios of the corresponding rates (Table 2) are therefore different from the final aspect ratios (Table 1). In the case of Na and NH_4, the L/W and W/T ratios vary quite monotonously. After an initial small decrease, the L/W ratio increases to its final value. The length remains always much greater than the width. The W/T ratio decreases slowly, showing less difference in the width and thickness of the final samples.

The K-system shows a peculiar behaviour (Fig. 7). The initial L/W ratio is rather small, showing that the rate of increase of the crystals in both the c and a directions are similar, while that of the b direction is smaller (high W/T ratio). During the crystallization, the L/W ratio increases, and the W/T ratio decreases to their final values. The peculiar behaviour of the K-system could be linked to the fast initial rate induced by the presence of the solid crystalline K_2SiF_6 phase (*vide supra*). However, a regular variation characterizes the Cs-system, despite the fact, that a solid Cs_2SiF_6 phase is also formed in this case. It is possible, that the aspect ratios are governed by subtle changes in the adsorption at the different faces of the crystallites. The previously proposed method to explain the variation of the aspect ratios with the crystallization rate can also explain the present results [6,7]. It is based on the hypothesis of MFI structure formation from double-five-membered rings (D5R) [6]. The D5R condense into chains which then link together to form the silicalite structure. In this case, however, a perfect structure cannot be obtained leading to as many as 16 missing tetrahedral sites per idealized unit cell ($Si_{96}O_{192}$). Each missing site corresponds to four SiOX defect groups where X=H, M or TPA. As such a high number of defect groups is never observed (64/u.c.), at least half of the missing sites should be filled during the synthesis [6,7]. The missing sites are supposed to be completed by monomers present in the solution. As previously depicted [6], the growth of the D5R chain is along the c-axis, while the increase of the width is along the a-axis and that of the thickness along the b-axis. The increase along the a-axis (i.e. along the zig-zag channels) is made by condensing the preformed D5R chains and by replacing, at least partially, the missing T sites by monomeric species. It can be supposed that the number of defect sites should decrease as a function of increasing rate along the width of the crystals. In samples obtained in alkaline media, however, the same amount of ca. 16 SiOR/u.c. was found for two samples characterized by aspect ratios of 1.8 and 2.7 [6]. In the samples synthesized in fluoride media, even the opposite behaviour is found. The higher the crystallization rate, the higher the number of defect sites in the samples. The variation of the aspect ratios in these F^--containing systems seems to be more complicated than that reported in alkaline medium [6,7]. For a general explanation, the relative stabilization of the crystal faces should be envisaged and the energies of activation for their growth should be determined. These data are lacking at present and more work is needed for the understanding of the aspect ratios.

The importance of the adsorbed layer on the different crystal surfaces is also shown by the absence of intergrowth in the NH_4-system, oppositely to the other three systems, where the usual intergrowths are observed. Note, that the effect of the adsorbed layer was also evoked to explain the seeding effect or the secondary nucleation or existing crystal faces [13].

CONCLUSIONS

MFI silicalite is prepared from fluoride containing media in presence of various alkali cations and ammonia. Long monodisperse crystals of ca. 70, 80 or 90 μm are obtained.The aspect ratios (or length to width and the width to thickness ratio) do depend on the crystallization rate. The latter essentially depends on the solubility of the fluorides. The aspect ratio increases with decreasing reaction rate. A model of MFI formation, based on the condensation of chains formed previously from double-five-membered rings, adequately explains, in a first approximation the observed aspect ratios.

ACKONOWLEDGMENT

Work carried out with financial support of CNR (Progetto Finalizzato Chimica Fine e Secondaria 2)

REFERENCES

1) Guth, J.L., Kessler, H. and Wey, R., Stud Surf Sci. Catal. **28** 121, 1986.

2) Tavolaro, A., Mostowicz, R., Crea , F.,Nastro, A., Aiello, R. and B.Nagy, J., Zeolites, in press.

3) Guth, J.L., Kessler, H., Higel, J.M., Lamblin, J.M., Patarin, J., Sieve, A., Chezeau, J.M. and and Wey, R., ACS Symp. Series **398**, 176, 1989.

4) Mostowicz, R.,Nastro, A., Crea , F.and B. Nagy, J., Zeolites **11**, 732, 1991.

5) P.A. Jacobs and J.A. Martens, "Synthesis of High-Silica Aluminosilicate Zeolites", Eds. Elsevier, Amsterdam, 1987.

6) Hayhurst, D.T., Aiello, R., B.Nagy, J., Crea , F., Giordano, G.,Nastro, A. and Lee, J.C., ACS Symp. Series **368**, 277, 1988, *and references therein.*

7) Hayhurst, D.T, Nastro, A., Aiello, R., Crea , F. and Giordano, G., Zeolites **8**, 416, 1988.

8) Nastro, A., Colella, C. and Aiello, R., Stud. Surf. Sci. Catal.**24**, 39, 1985.

9) Fajula, F., NATO ASI Series B **221**, 53, 1989.

10) Fegan, S.G. and Lowe, B.M., J., Chem. Soc., Faraday Trans1 **82**, 785, 1986.

11) Hay, D.G., Jaeger, H. and Wilshier, K.G., Zeolites **10**, 571, 1990.

12) Handreck, G.P. and Smith, T.D., Zeolites **11**, 356,1991.

13) Warzywoda, J., Edelman, R.D. and Thompson, R.W., Zeolites **11**, 318.,1991.

14) Mostowicz, R.et al., in preparation.

15) Handbook of Chemistry and Physics, Eds. CRC Press, Boca Raton, 1987.

16) Iler, R.K., The Chemistry of Silica, Eds. Wiley, New York, 1979.

17) Bursill, L.A. and Thomas, J.M., in: *Recent Progress Report, Fifth International Conference Zeolites*, Eds. Giannini, Napoli, 1981, p. 25.

18) Zandbergen, H.W. and van Dyck, D., Stud. Surf. Sci. Catal.**49A**, 599, 1989.

19) Jacobs, P.A. and Martens, J.A., Stud. Surf. Sci. Catal.**33**, 65, 1987.

20) Nastro, A., Gabelica, Z., Bodart, P. and B.Nagy, J.,Stud. Surf. Sci. Catal. **19**, 131, 1984.

SYNTHESIS OF NH_4-FORM HETEROATOM-SUBSTITUTED MFI ZEOLITES IN THE PRESENCE OF FLUORIDE ANIONS

JOHN DWYER*, JINGPING ZHAO**, and DAVID RAWLENCE***

*Chemistry Department, UMIST, P.O Box 88, Manchester M60 1QD, UK.

**Research Institute of Petroleum Processing, Beijing 100083, People's Republic of China.

***Crosfield Catalysts, P.O. Box 26, Warrington, Cheshire WA5 1AB, UK.

ABSTRACT

NH_4-form high-silica MFI zeolites with heteroatom frameworks substituted by boron, (aluminum) and gallium have been directly synthesised from a sodium-free system in the presence of fluoride ions. In this synthesis environment, fluoride anions play a critical function as a "catalyst". In general, high-silica products are easily obtained, maximum heteroatom substitution takes place in near-neutral reaction mixtures, and very high limits to heteroatom substitution with very efficient incorporation are observed in the gallium-silica system. The compositions of products can be easily controlled by modifying fluoride concentration and pH of reaction mixtures. Moreover, an effective seeding technique can significantly reduce the synthesis period of these zeolites.

INTRODUCTION

High-silica MFI zeolites with heteroatom substituted frameworks are widely used as catalysts after ion-exchange to the NH_4-form, because of the excellent shape-selectivity and activity in several reactions including conversion of methanol to gasoline (MTG)[1], alkylation of toluene[2], isomerization of xylene[3] and low temperature oxidation and epoxidation[4-6] of organic compounds with H_2O_2 over [Ti]ZSM-5. However, these zeolites are usually synthesised in alkaline media, having a pH value greater than 10. The products are typically in sodium-form (although this can sometimes be avoided in templated or special syntheses) which have to be pre-treated by ion-exchange and calcination procedures, before use, in order to transform them into the NH_4- or H-form and to reduce sodium levels. Residual sodium ions can poison acid-catalytic reactions and can affect the stability of zeolite structures. As a result, pre-treatment of a sodium-form zeolite to generate a finished catalyst can be complex and costly.

On the one hand, direct synthesis of NH_4-form zeolites is very valuable in the study of

*Where Correspondence should be addressed.

From the Proceedings of the 9th International Zeolite Conference, Montreal 1992, Eds. R. von Ballmoos et al.,

zeolite crystallisation because the synthesis environment is fundamentally different from that of a conventional alkaline system. For example, in the direct synthesis system, fluoride ions substitute for hydroxyl anions, during the synthesis[7,8] and as a result, the pH of reaction mixtures can vary from acid to alkaline. On the other hand, direct synthesis of NH_4-form MFI zeolites from a sodium-free system in the presence of fluoride ions can obviously simplify pre-treatment[7] as well as increase the flexibility in synthesis because framework composition and particle size are easily controlled by modifying fluoride concentration in reaction mixtures and/or by changing the pH of these mixtures[8,9]. Heteroatom substitution into frameworks, for some elements, is higher and more efficient in the fluoride system. These advantages may be of benefit in commercial processes. This paper examines, in detail, the incorporation of B, Al and Ga into pentasil frameworks in systems containing fluoride ions, and demonstrates some unusual characteristics of the new synthesis system when compared to a conventional alkaline system.

Moreover, an effective seeding technique is developed which can considerably reduce the synthesis period.

EXPERIMENTAL

The raw materials used in the syntheses of NH_4-[B]ZSM-5, NH_4-[Al]ZSM-5 and NH_4-[Ga]ZSM-5 were boric acid (Aldrich Chemicals), aluminium chloride hexahydrate (Aldrich Chemicals), gallium nitrate solution (prepared at UMIST), ammonium fluoride (BDH Chemicals), Tetrapropylammonium bromide,[TPABr] (Aldrich Chemicals), silica sol [Ludox 40 AS, $SiO_2 = 40$ wt%].

The reaction mixtures were prepared by mixing the solutions containing the above components in the order, silica sol, TPABr, substituted elements and NH_4F. The mixtures were stirred for at least 2 hours to make the gel homogeneous and then hydrothermally crystallised at 448 K under spontaneous nucleation conditions. The crystallisation period depended on the composition of the reaction mixtures.

Zeolites were characterised by XRD, SEM and MAS NMR according to the conditions described previously[8,9].

RESULTS AND DISCUSSION

Very crystalline heteroatom pentasils (NH_4-[M]ZSM-5) where M is B, (Al) and Ga, were synthesised in the presence of fluoride ions. X-ray diffraction patterns show that these as-synthesised products are typical orthorhombic MFI structures[8,9]. MAS NMR spectra of substituted elements (^{11}B, ^{27}Al and ^{71}Ga) give clear signals at -3.78, 50.9 and 147ppm[8,9] respectively, which correspond to these heteroatoms in a tetrahedral coordination to oxygen within the zeolite framework. The evidence shows that very crystalline NH_4-[M]ZSM-5 with heteroatoms substituted into the framework can be synthesised directly in the presence of fluoride ions. However, some synthesis characteristics are similar and others depend greatly upon the type of heteroatom.

1. Catalytic function of fluoride ions in the formation of NH_4-[M]ZSM-5

The most important characteristic of the new synthesis system is that fluoride ions are used to substitute for hydroxyl anions during the synthesis. In order to investigate the effects of fluoride concentration, optimal reaction mixtures containing the different heteroatoms, were used. The compositions of these optimal reaction mixtures depend on the particular substituting element and, in terms of oxide components, can be described as;

xNH_4F - $1.0B(OH)_3$ - $10.0SiO_2$ - 2.0TPABr - $410H_2O$

yNH_4F - $1.4AlCl_3$ - $10.0SiO_2$ - 3.0TPABr - $300H_2O$

zNH_4F - $0.2Ga(NO_3)_3$ - $10.0SiO_2$ - 2.0TPABr - $300H_2O$

The reaction mixtures were crystallised at 448 K for varying periods depending on the substituted elements and chosen composition. The crystallinity of the products for the three systems increases rapidly with increased fluoride content after a "critical" concentration (Figure 1) and thereafter does not change much with further increase in fluoride concentration, or decreases slightly.

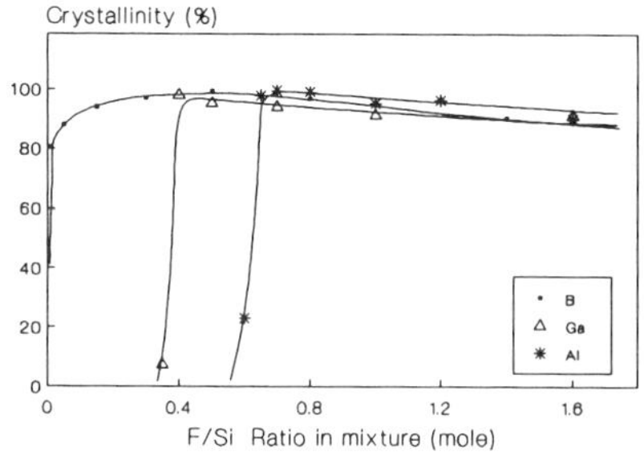

Figure 1. Crystallinity changes with F/Si ratio in reaction mixtures

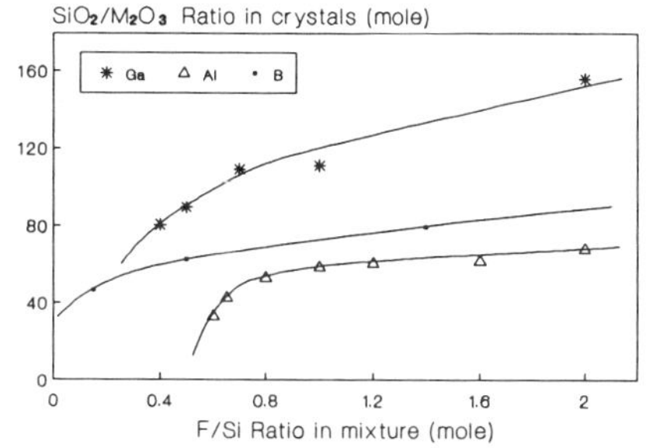

Figure 2. Changes of product Si/M ratio with content of fluoride ions in mixtures

The critical concentration of fluoride ions varies with the individual synthesis parameters and substituting elements. However, the pattern of crystallinity increase with the F/Si ratio in mixtures is very similar for the three different substituting elements. The rapid increase of crystallinity at a certain critical fluoride content suggests that there may be a minimum concentration of fluoride ions which keeps the reactants sufficiently soluble in the crystallisation conditions. Further increase of fluoride ion content, in contrast, slightly restricts the crystallisation of the starting gels. These changes suggest that a critical concentration of fluoride ions appears to be required for gel crystallisation, and that the function of the fluoride ions is similar to that of a "catalyst".

The increase of fluoride content in the reaction mixtures usually leads to an increase in the Si/M ratio in the zeolite framework (Figure 2). This is probably associated with the stability of fluoride complexes in the reaction liquid phase, and in particular to the stability of complexes of the substituting heteroatoms. At high fluoride concentration, fluoride complexes are stabilised in the liquid phase under the reaction conditions, so that high-silica products are usually obtained.

Obviously, fluoride concentration in the reaction mixtures strongly influences crystallisation rate and directly affects the composition of the zeolite products.

2. Effect of starting pH of reaction mixtures

NH_4-[M]ZSM-5 can be directly synthesised, over a wide pH range, in the presence of fluoride ions. The pH range of the reaction mixtures depends on the substituting elements under the given synthesis conditions and composition of the reaction mixtures (Figure 3).

However, the composition of the as-synthesised products has a minium region for Si/M ratio, when the pH of their parent gels is near to neutral. According to general principles of complex chemistry, these substituting elements have lower solubility and the fluoride complexes are relatively unstable around this pH. These results demonstrate that isomorphous substitution into zeolite frameworks depends on the pH of mixtures and that efficient substitution usually occurs under conditions in which the substituting elements are in relatively unstable complexes and have somewhat low solubility in the liquid phase.

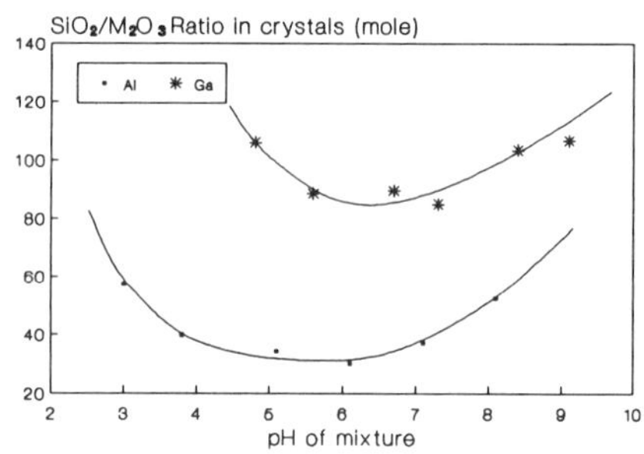

Figure 3 Changes of product Si/M ratio with pH of starting mixtures

3. The influence of substituting elements in the direct syntheses of NH_4-[M]ZSM-5

In order to investigate the effects of heteroatom type on the formation of zeolites, a carefully chosen composition of reaction mixture is used. The composition of the mixtures is as follows,

$$7.0NH_4F - 10.0\ SiO_2 - 1.0M - 3.0TPABr - 350H_2O$$

where the M is $B(OH)_3$, $AlCl_3$ and $Ga(NO_3)_3$ respectively. The synthesis parameters and product characteristics are summarised in Table 1. These results (Table 1) show that the minimum crystallisation periods increase significantly in the order B<Al<Ga implying that larger atoms require more time for reaction under comparable conditions. However, the composition of the products shows that the substitution of heteroatoms is also increased and becomes more efficient in the order given above. The most efficient substitution was observed in the Ga-Si system. These results are in agreement with the decreased solubilities of fluoride complexes from boron to gallium. Gallium fluoride is the least soluble of these fluorides so that the Ga-Si system gives the most efficient substitution. In contrast, a high solubility for the fluoride, as observed for NH_4BF_4 leads to a low substitution for silicon in the [B]ZSM-5 framework. This suggests that the solubility of the fluoride complexes is one of main factors affecting the efficiency of heteroatom substitution into the framework and controls the crystallisation rate in the synthesis system containing fluoride ions.

^{29}Si MAS NMR spectra of the three products show a decrease in a shoulder at -115ppm (Figure 4), which is related to resonance of the $Si(OSi)_4$ peak in high-silica pentasil zeolites[10], and an increase in the $Si(1M)(OSi)_3$ peak at -105ppm when the elements change from B to Ga. This also shows that Ga is incorporated into the framework more efficiently than boron from reaction mixtures having the same starting Si/M ratio.

Table 1. The synthesis parameters and product characteristics

System	Minimum Time For Crystallisation (day)	Si/M Ratio		Crystal Size (μm)
		Mixture	Product	
B-Si	2	10	31.0	120x20
Al-Si	7	10	23.4	55x30
Ga-Si	14	10	10.4	ϕ 31.8

Different framework elements also result in differences in morphology of the crystals (Figure 5). [B]ZSM-5 crystals are long and narrow with an average size 120x20 μm. For [Al]ZSM-5, the average size is approximately 55x30 μm and crystals are shorter and wider than those observed in boron substitution. The Ga-substituted crystals are spherical with a diameter of about 32 μm. Electron micrographic results show that the growth of crystals along the length direction is restricted, in the presence of fluoride ions, as the radius of the substituting heteroatoms is increased. Because in typical ZSM-5 crystals, the length direction corresponds to the **C** axis (Coffin-lid type)[11,13], these differences in crystal morphology suggests that the growth of crystals along the **C** axis is increasingly difficult in the order B, Al, Ga.

A composition diagram for the reaction mixtures containing the different substituting elements is given in Figure 6. The region for very crystalline products decreases from B-Si to Ga-Si. It shows that the synthesis becomes more difficult in the order B, Al, Ga.

4. Special seeding technique for reducing crystallisation period

Although the direct synthesis presents many advantages, as mentioned above, it is also noticed that the crystallisation period is usually longer than a conventional synthesis system. However, a special seeding technique, crystallisation-directing-agent gel (CDA-gel), was developed[9]. The seeding technique can significantly reduce crystallisation period in the synthesis system. An example is the application of the CDA-gel in synthesis of NH_4-[B]ZSM-5. In the example, the CDA gel was prepared from a modified reaction mixture which had a composition similar to that of an ordinary reaction mixture but the H_2O/SiO_2 and F/SiO_2 ratios could be 35-100 and 0.3-3.0 respectively. For this typical sample, the CDA-gel is heated at 448K for 10 hours with $H_2O/SiO_2 = 45$ and $F/SiO_2 = 1.4$. The preparation of a CDA-gel was given in detail in reference[9,12]. The key points in the preparation of the CDA gel slurry concern the choice of a suitable composition and an appropriate heating treatment to achieve zeolite nuclei of the correct size.

CDA gel (4%), based on $SiO_2 + B_2O_3$, was added to an ordinary reaction mixture which was made homogeneous by vigorous stirring. The final reaction mixtures with the CDA gel were then allowed to crystallise under standard conditions[8]. Using the CDA gel, crystalline products could be obtained in only 6 hours at 448 K, a much shorter period than in the case without a CDA gel (2.5 days). X-ray diffraction patterns of the comparable samples are shown in Figure 7. This technology is more efficient in a slower synthesis system such as NH_4-[Al]ZSM-5.

ACKNOWLEDGEMENTS

We appreciate the financial support for this research from the Research Institute of Petroleum Processing, China, the ORS award Scheme, Shell Research Amsterdam, Laporte Inorganics Limited, ICI Limited and SERC vs. The Link Scheme (GRF 99380).

REFERENCES

1 Meisel, S.L. and McCullough, J.P., Chem. Tech. 686, (1976).

2 Paparatto, G., Moretti, E., Leofanti, G., and Gatti, F., J. Catal. **105**, 227, (1987).

3 Young, L.B., Butter, S.A. and Kaeding, W.W., J. Catal. **76** 418, (1982).

4 Kraushaar-Czarnetzki, B.,"Characterisation and Modification of Zeolites and Related Materials", Tech. Univ. Eindhoven, Eindhoven (1989).

5 Thangaraj, A., Kumar, R., Mirajkar, S.P. and Ratnasamy, P., J. Catal., **130** 1, (1991).

6 Clerici, M.G Applied Catalysis, **68**, 249, (1991)

7 Guth, J.L., Kessler, H. and Way, R., Proc. 7th Int. Conf. Zeolites, Tokyo, 121, (1986).

8 Dwyer, J. and Zhao, J., J. Material Science, February No.2, 235 (1992).

9 Zhao, J. Ph D Thesis, Chemistry Department, UMIST (1990).

10 Fyfe, C.A., O'Brien, J.H. and Stroble, H., Nature **Vol. 326**, 281, (1987).

11 Flanigen, E.M., Bennett, J.M et al. Nature, **Vol.271**, 512, (1978).

12 Dwyer, J. Zhao, J.P. Zeolites in press (1992)

13 Chao, K. and Chern, J., Zeolites, **Vol. 8**, January, 83, (1988).

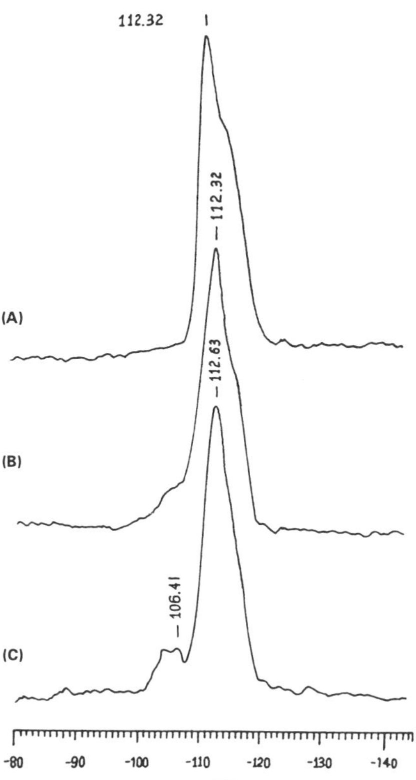

Figure 4. ^{29}Si MAS NMR spectra of samples (A) [B]ZSM-5 (B)[Al]ZSM-5 (C)[Ga]ZSM-5

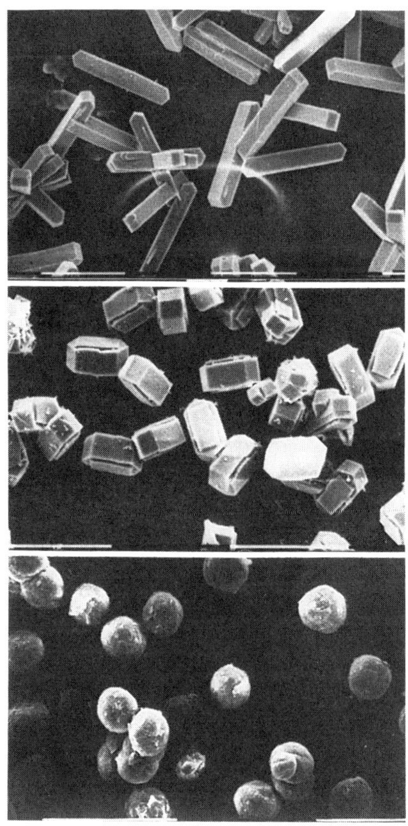

Figure 5. SEM photographs of NH_4-[M]ZSM-5 substituted by B, Al and Ga (Scale = 0.1mm)

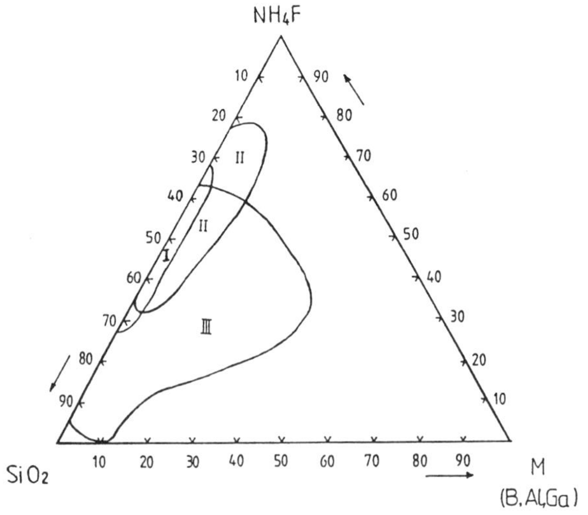

Figure 6. Composition diagram of reaction mixtures for synthesis of NH_4-[M]ZSM-5 in the presence of fluoride anion
Region I = Ga-Si, Region II = Al, Region III = B-Si

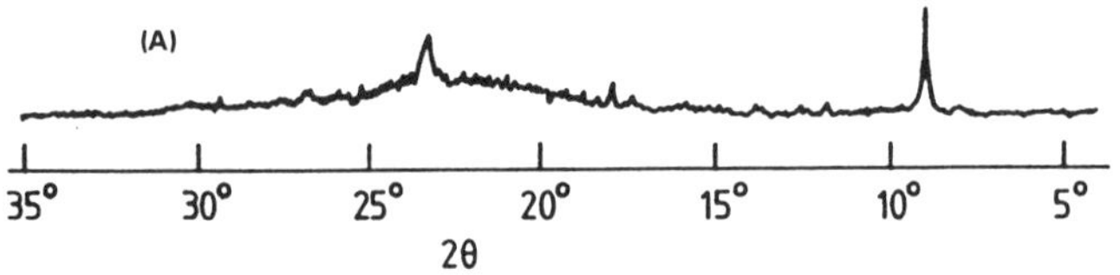

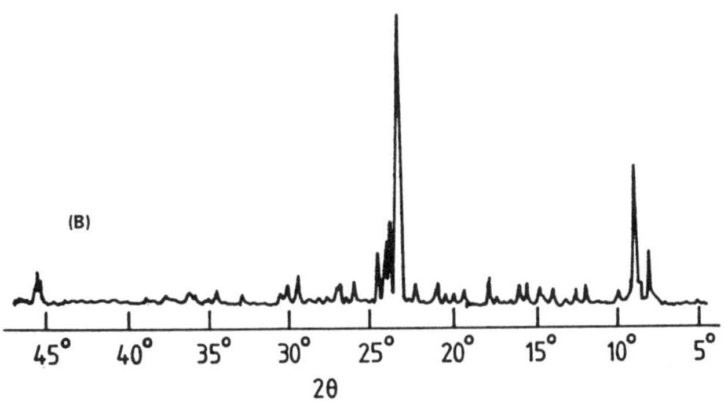

Figure 7. XRD patterns of typical NH_4-[B]ZSM-5
obtained from an optimised reaction mixture in 6 hours at 448K
(A) Without CDA gel (B) With 4% CDA gel in mixture

AN UNEXPECTED AND HIGHLY VERSATILE NEW ZEOLITE SYNTHESIS ROUTE LEADING TO LARGE PORE ALUMINO AND BOROSILICATE SIEVES

S. I. ZONES, L. T. YUEN, Y. NAKAGAWA, R. A. Van NORDSTRAND, and S. D. TOTO

Chevron Research and Technology Company, Richmond, California

ABSTRACT

Use of calcined boron beta zeolite in zeolite syntheses provides a highly reactive and versatile reagent. Large pore borosilicate sieves such as BSSZ-24 and BSSZ-31 can be rapidly crystallized, can use a larger variety of organo-cations to direct their synthesis, and can be made at higher substitutions than seen in conventional syntheses. In general, the use of boron beta zeolite leads to the rapid synthesis of open pore products.

INTRODUCTION

Since observing the intermediacy of organo-free zeolites along the pathway for synthesis of some organo-zeolites, we have been interested in studying how the inorganic components are being used [1]. In reactions involving reasonably high OH^-/Si values ($\geq$0.50), it can be seen that aluminum becomes a mass-limiting component. Thus in studies concerning cubic P zeolite it was found that all of the aluminum in the starting gel was found in either P zeolite or the organo-zeolite product at any point in time [2-4]. Surprisingly, starting the synthesis with P zeolite, as aluminum source, faster rates for organo-zeolite crystallization were observed [4]. The use of Y zeolite (with the same Si/Al ratio as P) resulted in even faster conversion rates. Y is a higher surface area material than P. This prompted a more general search for reactive zeolites. Boron beta zeolite was discovered to be a remarkably reactive and versatile reagent in this capacity.

EXPERIMENTAL

Boron beta zeolite was prepared by a novel method which allows selectivity over ZSM-5 in the synthesis when Al^{+3} is absent and B^{+3} is present [5]. Boron-containing beta was calcined in nitrogen to 1100°F with a slight air bleed introduced. The calcined product was used as the entire metal oxide reactant source or in conjunction with Cabosil M-5 silica. Reactions were carried out using low OH^-/SiO_2 conditions, 150°C and 0 rpm, as have been previously described for SSZ-24 [6]. Products were analyzed by X-ray diffraction. Aluminum replacement of boron was carried out using $Al(NO_3)_3$ solutions [5].

RESULTS AND DISCUSSION

1. Synthesis of Boron-SSZ-24 (AFI Structure)

A. Initial Reactions

When boron beta zeolite was calcined in nitrogen with only a slight oxygen bleed into the furnace, the organic components were decomposed to give an optimal microporosity and retention of framework boron. Direct air calcination leads to greater boron loss. This had been previously demonstrated for the MFI type borosilicate [7].

When the open-pore material was placed into a teflon cup of a Parr 4745 reactor with a 10/1 mass ratio of water and heated to 170°C for 4 days the borosilicate molecular sieve lost 90% of its crystallinity. The aluminosilicate beta zeolite loses some crystallinity under the same experimental treatment, but to a much lesser degree. Thus the borosilicate version of this multi-dimensional, large-pore molecular sieve appears to be hydrothermally unstable.

This observation, in conjunction with our success in converting aluminosilicates into other zeolitic products, led us to test the utility of using calcined boron beta zeolite as a source of boron

From the Proceedings of the 9th International Zeolite Conference, Montreal 1992, Eds. R. von Ballmoos et al.,

and silica in zeolite syntheses. Initially, we tried using the conditions for synthesizing all-silica SSZ-24 (N,N,N-trimethyl-1-adamantammonium hydroxide as template) and replacing some of the reactant silica with boron beta zeolite. This experiment was successful in producing boron SSZ-24. Table I shows the reaction conditions for both the all-silica and borosilicate reactions.

A control experiment using a dissolved borate source ($Na_2B_4O_7$) produced the chabazite material, SSZ-13. This was analogous to the problems we observed previously in efforts to introduce significant quantities of aluminum into the SSZ-24 structure to achieve SiO_2/Al_2O_3 values below 100 [8].

A porous borosilicate glass (Vicor) was also tried as a boron source. Although boron is the next abundant metallic element after silicon in the glass, there is also some aluminum in this material. We found that if the glass was used as a minor component such that $SiO_2/Al_2O_3 + B_2O_3$ remained >100 for the overall reaction, B(Al)SSZ-24 would be produced in about 2 weeks. When the glass was used as the exclusive tetrahedra source, the expected SSZ-13 product was obtained.

The most striking finding was that boron beta zeolite can be used as the only tetrahedrally coordinated element source and still be converted to boron SSZ-24. The verification of borosilicate SSZ-24 (AFI structure) and the corresponding lattice contraction due to boron substitution was first reported by Van Nordstrand [9].

B. Reaction Rates

The synthesis of boron SSZ-24 from boron beta zeolite was remarkably fast, the product forming in 1 day at 150°C. By comparison, it has been reported that the formation of the all silica zeolite requires approximately 1 week [6], or longer [10]. In many reaction systems, the introduction of a lattice substitution component slows down the crystallization rate. This has been shown for aluminum in TON synthesis [11], for example. It was therefore surprising that the substituted SSZ-24 forms so much faster than the all-silica material.

Figure 1 compares the crystallite size achieved by the boron beta zeolite and the all-silica pathways. The faster, new method produces crystallites 5-10 times smaller than the original all-silica case. It appears that use of boron beta zeolite as our reactant allows favorable nucleation to occur much earlier in the reaction and at a high rate. Even when boron beta zeolite comprises

Table 1

Reaction Conditions for Introducting Boron Into Lattice Positions for Zeolite Syntheses With N,N,N Trimethyl 1-Adamantammonium Hydroxide (1-ADA OH^-)[1]

1-Ada OH^-/SiO_2	$NaOH/SiO_2$	H_2O/SiO_2	B/SiO_2	B Source	Product
0.15	0.10	44	0.00	None	SSZ-24
0.15	0.10	44	0.02	B Beta	BSSZ-24
0.15[2]	0.10	44	0.06	B Beta[3]	BSSZ-24
0.15	0.10	44	0.02	$Na_2B_4O_7$	BSSZ-13
0.15	0.10	44	0.02	Vicor[4] Glass	BSSZ-24
0.15	0.10	44	0.04	Vicor Glass[3]	BSSZ-13

[1]Cabosil M-5 as SiO_2 source;150°C, 0 rpm 3-14 days.

[2]N,N,N Trimethyl-2-adamantammonium hydroxide used.

[3]Borate source is also entire silica source.

[4]Vicor is a product of Dow Corning Glass. (Si/B = 25)

only part of the total silica content, with the remainder from cabosil, the crystallization rates exceed the all-silica case. Therefore, interaction of the organo-cation with pure silicates, which are derived from dissolving Cabosil, is not sufficient to retard the more favorable nucleation promoted by the presence of boron beta zeolite in the reaction.

Tetrapropyl ammonium (TPA) bromide is an excellent and versatile template for ZSM-5 synthesis in a wide variety of inorganic conditions. Borosilicate versions of ZSM-5 have been described previously [12]. Conversion of boron beta zeolite to ZSM-5 occurs slowly if the original template is left in the pore system. At several days boron beta zeolite is still present, demonstrating that the hydrothermal transformation is slow without porous calcined material. These experiments indicate that access to the pores is needed for this transformation and that the high surface area leaves boron beta zeolite quite unstable in the absence of a pore-filling organic. Dissolved components may start nucleation near the beta zeolite surface.

C. Boron Concentration

During ammonium exchange of calcined boron beta zeolite, boron can be removed from the lattice. This is largely a rate phenomenon accelerated by increasing temperature and proton concentrations. A similar behavior was recently described for boron ZSM-5 [7]. The proton concentration can be controlled by the choice of ammonium counterion and we found that little boron was lost using ammonium acetate (buffering capacity), but significant loss occurs with the corresponding nitrate.

By using a combined mixture of ammonium acetate and nitrate salts in an ion-exchange step, we have prepared boron beta samples with varying boron content but without loss in crystallinity. Boron retention was related to the exchange solution pH. Table 2 lists the boron concentrations in these materials, which were then used as the only tetrahedra source in

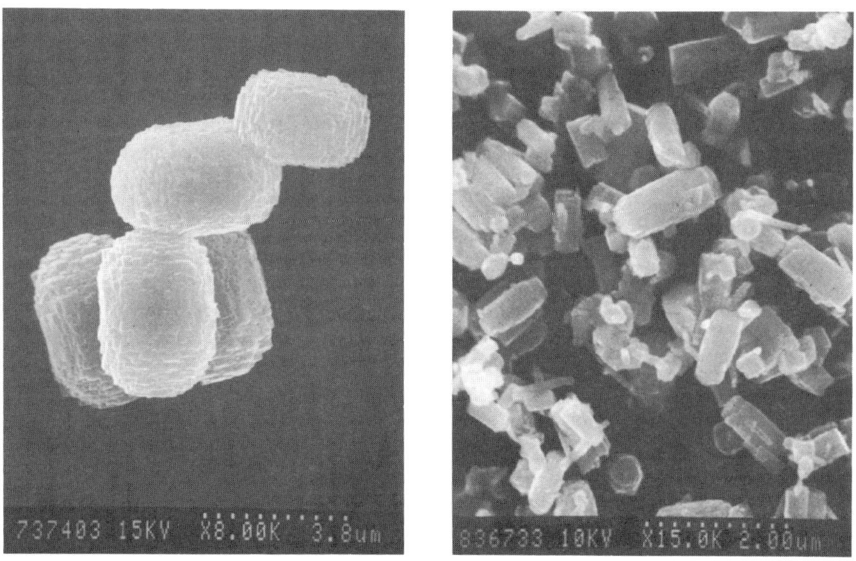

Figure 1. Scanning election micrographs of products with and without boron substitution. The average crystallite size is significantly smaller for BSSZ-24 indicating a greater extent of nucleation in the synthesis.

reactions to produce BSSZ-24. In all cases, BSSZ-24 is the product formed. Some cristobalite appeared as an impurity as the starting SiO_2/B_2O_3 ratio became larger. Table 2 also shows data for transfer of boron to the new zeolite product. Boron does not have low solubility under high pH conditions [13], therefore, unlike cases with aluminum, complete incorporation into the product is not guaranteed. As shown in the table, less than half of the boron in beta is being transferred to the BSSZ-24 product. Subsequent formation of an aluminosilicate version is also given. In the aluminosilicate form more than 90% of the boron has been removed. Catalytic performance of the aluminosilicate has been reported [9].

Table 2

Transfer of Boron From Reactant To Product in BSSZ-24 Synthesis and Subsequent Aluminum Incorporation

Wt % Boron in Beta	Wt % in BSSZ-24	XRD	No. of Al^{+3} Ex.[1]	Wt % Al
1.02	0.34[2]	BSSZ-24	1	0.59[3]
0.60	0.22	BSSZ-24 + Trace Cristobalite	-	-
0.39	0.22	BSSZ-24 + Trace Cristobalite	-	-

[1]This denotes the number of $Al(NO_3)_3$ solution exchanges. (See Reference 5 for method.)
[2]SiO_2/B_2O_3 = 90 in product.
[3]SiO_2/Al_2O_3 = 110 in product.

D. Template Flexibility

One of the intriguing features in the discovery of the all-silica SSZ-24 was that only one template produced it [14]. This was in sharp contrast to the large variety of organic amines which could be used to synthesize the isostructural AlPO-5 under weakly acidic or neutral conditions [15]. We have found that in converting boron beta zeolite to BSSZ-24, there is a greater degree of template flexibility.

Table 3 shows that three other closely related adamantane or noradamantane derivatives also worked to give BSSZ-24. For comparison we also show data for the products obtained by these templates under more conventional silicate or aluminosilicate reaction conditions. No other templates have worked to make SSZ-24 except N,N,N-Trimethyl-1-Adamantammonium hydroxide. (Recently, we have found some additional organo-cations which do work) [16].

2. Synthesis of Other Borosilicate Zeolites

A. Boron SSZ-31

The usefulness of the boron beta zeolite as a reactant can be extended to other reaction systems. A variety of organo-cations produced borosilicate zeolites with relatively rapid reaction rates. One of the most interesting cases was the synthesis of boron SSZ-31.

SSZ-31 was recently described as a novel, very high silica zeolite [17]. Like SSZ-24, it appears to be a large pore (12 ring) structure. Direct synthesis with significant aluminum substitution is difficult. When boron is the substituting component, SSZ-33 is produced instead [18]. The initial SSZ-31 discovery, like SSZ-24, used only one template molecule successfully. Later, Nakagawa described a related [5.2.1.0] tricyclodecane derivative which could produce SSZ-31 [19].

Figure 2 shows that using boron beta zeolite, five different organo-cations can produce boron SSZ-31. The synthesis conditions are the same as in Table 1 for boron beta zeolite conversion.

Table 3

Use of Adamantane Derivatives in the Synthesis of BSSZ-24

Template Structure	XRD			
	SiO_2/B_2O_3 = 30	SiO_2/Al_2O_3 = 50	100	∞
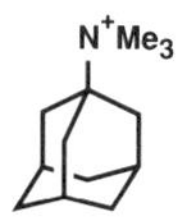	BSSZ-24	SSZ-13 (or SSZ-25)	SSZ-23	SSZ-24
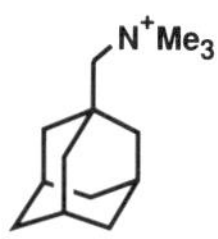	BSSZ-24 ①	ZSM-5	ZSM-5	No Product
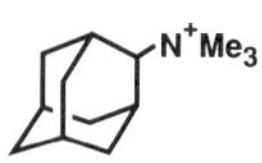	BSSZ-24	SSZ-13 (or SSZ-25)	SSZ-23	No Product
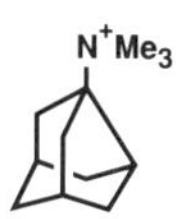	BSSZ-31 ②	SSZ-13	SSZ-13/23	Layered
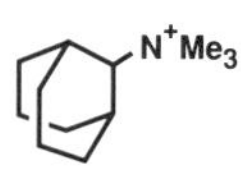	BSSZ-24	SSZ-13	SSZ-23	——

① AT SiO_2/B_2O_3 = 100
② AT SiO_2/B_2O_3 = 70

Aluminosilicate versions can be made by post-synthetic substitution. What is equally striking is that in the absence of boron beta zeolite, none of these templates make boron SSZ-31. If sodium borate is provided to a reaction mixture and the overall B/Si ratio is established to be the same as in Table 1, the various templates produce either layered or clathrate type structures as depicted in Figure 2.

B. Open Pore Systems

As demonstrated above in Figure 2, the use of boron beta zeolite can lead to the open pore (12 ring) boron SSZ-31. A conventional synthetic approach in using soluble reagents like sodium borate and any of the five organo-cations was unsuccessful in synthesizing <u>any</u> open pore structures. As more organocations were tested with boron beta zeolite, two emerging trends were noted: (1) the conversion is generally rapid, and (2) the products were predominantly 10 ring and 12 ring products.

Table 4 lists a variety of organo-cation structures and the resulting borosilicate molecular sieve products. Frequently encountered phases were ZSM-12 and ZSM-48, both open, one-dimensional pore systems. In some instances boron beta was only successfully converted if it had a reduced boron content. This was often true when ZSM-48 was produced and is consistent with previous work where ZSM-48 was found to crystallize at low substitution of silica [20].

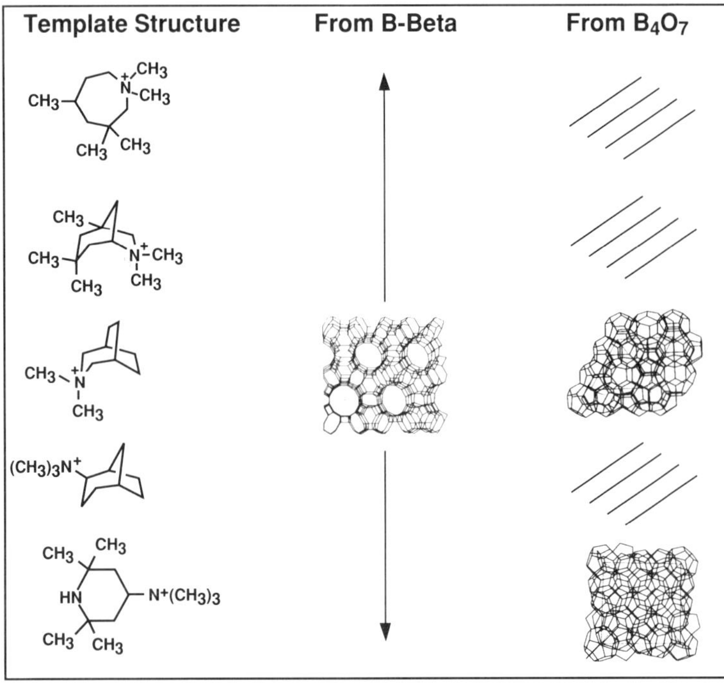

Figure 2. Highly substituted mono and bicyclic organic cations which can convert boron beta zeolite to BSSZ-31. Under conventional syntheses using dissolved borate sources, the product becomes different in all cases. The drawing under boron beta conditions depicts an open pore product only representing BSSZ-31; its structure has not been published.

A few of the organic molecules which we have tested still have selectivity for forming cage structures. When the free adamantanamine was used instead of the N,N,N-Trimethyl adamantammonium derivative, BSSZ-24 was not the product: the small pore product Sigma-1 was made instead [21]. Analogously, a smaller propellane, 1-aza [3.3.3.0] undecane produced a clathrasil structure. The propellane can be considered to be a higher homologue of quinuclidine. Both quinuclidine and N-methyl quinuclidine are known to stabilize a variety of cage or cavity type structures instead of producing open pore systems.

CONCLUSIONS

Boron beta zeolite is relatively unstable in solution at elevated temperatures once the organocations are removed. It can be used as the inorganic reactant in zeolite synthesis and converted very rapidly to BSSZ-24. The product obtained is normally very difficult to make by conventional synthesis routes.

The synthesis using boron beta zeolite also showed a remarkable flexibility in terms of using a variety of organocations to make a range of open pore borosilicates. A number of organocations functioned to make BSSZ-31 as well a BSSZ-33, BSSZ-25, BZSM-12 and BZSM-48. While a few organic amines or quaternary ammonium compounds still produced cage products with 8 ring or smaller pores, this remained very much the exception in this unusual and versatile zeolite synthesis reaction.

Table 4

Some Representative Boron Beta Conversions Demonstrating the Open Pore Products Obtained

Organo-Cation	SiO_2/B_2O_3	XRD
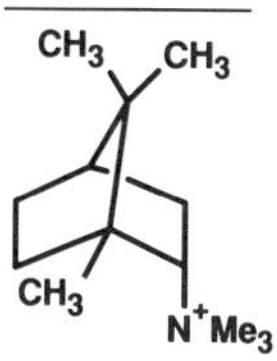	30	SSZ-25
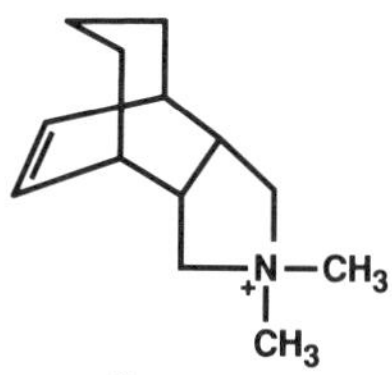	30	SSZ-33
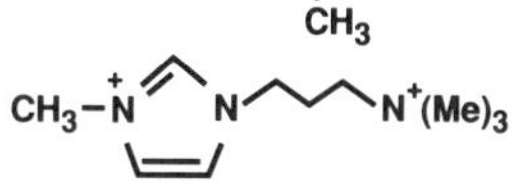	100	ZSM-48
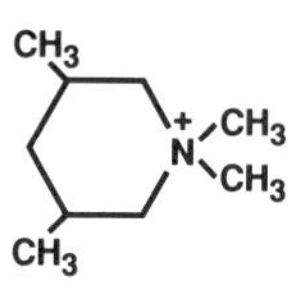	50	ZSM-12
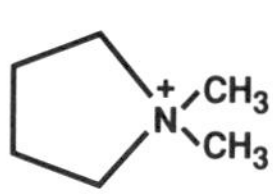	130	ZSM-12 + 48
$(CH_3)_3N^+-(CH_2)_x-N^+(CH_3)_3$		
X = 4	130	ZSM-12
X = 7	130	ZSM-48

ACKNOWLEDGEMENTS

We thank J. A. Gill and G. S. Lee for synthetic expertise in preparing a number of the organocations used in this study. We thank D. J. O'Rear for his enthusiastic support of this project and Chevron Research and Technology Company for continued support.

REFERENCES

1. Zones, S. I.; Van Nordstrand, R. A.; Zeolites, **8** 166, 1988.

2. Zones. S. I.; Van Nordstrand, R. A.; Zeolites, **8**, 4, 1988.

3. Zones, S. I.; Van Nordstrand, R. A.; *Novel Materials in Heterogeneous Catalysis*, (ed. R. T. K. Baker, L. L. Murrell) ACS Symp. 437 Chapter 2, (1990).

4. Zones, S. I. J. Chem. Soc. Farad. Trans. **86** 3467, (1990).

5. Zones, S. I.; Holtermann, D. L.; Jossens, L. W.; Santilli, D. S.; Rainis, A.; Ziemer, J. N.; Pat Appl. WO 91/00777, 1991.

6. Zones, S. I.; U.S. Patent 4,834,958, (1989).

7. Unger, B.; Wendlant, K. P.;, Tougar, H.; Schuieger, W.; Bergk, K. H.; Brunner, E.; Reschetilowski, W.; J. Chem. Soc. Farad. Trans. **87**, 3099 (1991).

8. Zones, S. I.; Van Nordstrand, R. A.; Santilli, D. S.; Wilson, D. M.; Yuen, L. T.; Scampavia, L. D.; *Zeolites: Facts, Figures, Future*, pg 299, 1989.

9. Van Nordstrand, R. A.; Santilli, D. S.; Zones, S. I.; *Synthesis of Microporous Materials Vol. II*, Chapter 25, ed. by M. Occelli and H. Robson (Van Nodstrand Reinhold, 1992)

10. Bialek, F.; Meier, W. M.; Davis, M. E.; Annen, M. J.; Zeolites **11**, 438, (1991).

11. Zones, S. I.; Zeolites **9**, 458, (1989).

12. Klotz, M. R.; U.S. Patent 4,268,420, (1981).

13. Belussi, G.; Perego, G.; Clerici, M. G.; Giusti, A.; Eur. Pat. Appl. 293,031, (1988).

14. Van Nordstrand, R. A.; Santilli, D. S.; Zones, S. I.; *New Perspective in Molecular Sieve Science ACS Symp. 368*, Chapter 15, 1988.

15. Wilson, S. T.; Lok, B. M.; Messina, C. A.; Cannan, T. F.; Flanigen, E. M.; J. Am. Chem. Soc. **104**, 1146, (1982).

16. Nakagawa, Y., to be published.

17. Zones, S. I.; Harris, T. V.; Rainis, A.; Santilli, D. S.; Ep Appl. No. WO 90/04567 (1990).

18. Zones, S. I.; U.S. Patent 4,963,337 (1990).

19. Nakagawa, Y.; Zones, S. I.; *Synthesis of Microporous Materials Vol. II*, Chapter 16, ed. M. Occelli and H. Robson, (Van Nodstrand Reinhold, 1992).

20. Araya, A.; Lowe, B. M.; J. Catal. **85**, 135 (1984).

21. Stewart, A., G.B. 2193202A (1988).

ORGANIC GUEST STABILIZATION DURING SYNTHESIS OF HIGH SILICA ZEOLITES WITH MULTIDIMENSIONAL CHANNELS AND LARGE PORES: SSZ-26 FROM A PROPELLANE-BASED ORGANOCATION

S. I. ZONES and D. S. SANTILLI

Chevron Research and Technology Company, Richmond, CA 94802, U.S.A.

ABSTRACT

A new large pore zeolite, SSZ-26, is produced by systematic design of the organo-cation used in the synthesis. A large propellane diquaternary ammoniun compound was used and circumvents the formation of ZSM-12 or clathrate-type products. These latter two products emerge as favorable products for a variety of silica-based zeolite syntheses. Micropore data suggests the channel system is multidimensional, consistent with the design of the organo-cation used. Multidimensionality may also be related to aluminum substitution for silica-based zeolites.

INTRODUCTION

The history of the use of organic molecules in molecular sieves has taken pathways of increased complexity. Tetramethylammonium, the smallest of the tetraalkyl ammonium compounds, was the first to be used and then higher homologues were used to discover Beta [1], ZSM-5 [2], and ZSM-12 [3], three of the most useful molecular sieve materials found. The use of amines and amino alcohols provided an alternative to the quarternary compounds and the discovery of important high silica one-dimensional materials such as Theta-1 [4], ZSM-23 [5], and ZSM-48 [6] followed. These basic organics have also proven indispensable in the aluminum-phosphate based molecular sieve syntheses [7].

Eventually, the research moved back towards more complex mono- and diquaternary ammonium compounds and new materials have continued to be discovered. Synthetic conditions favor cage- or cavity-based materials, or one-dimensional channel systems over multidimensional channel systems. On the other hand, the relatively few multidimensional channel systems have become the commercially most important zeolites. Also, having discovered the high silica large pore zeolite, SSZ-24 (AFI) using a rigid polycyclic organocation [8], we turned our attention towards efforts to enhance the formation of multidimensional channel systems. The more favorable diffusion characteristics seemed a key to their success as catalysts.

EXPERIMENTAL

The synthesis of the appropriate diquaternary ammonium propellane used to make SSZ-26 is described in the patent [9]. The patent also gives experimental conditions for synthesizing SSZ-26. The X-ray diffraction work was carried out on a Siemans Kristalloflex Diffractometer. Zeolite samples were calcined to 1100°F using a Maxim programmable system. Ion-exchange was performed using a 1:1 mass ratio of NH_4NO_3:zeolite in 10-20 mL H_2O/g zeolite for 2 hours at 100°C. This process was repeated three more times.

Adsorption was studied on a vacuum microbalance system or using the dynamic pore probe technique previously described by Santilli [10]. Hydrocarbon adsorbates were of greater than 99+% purity and used as received from Aldrich Chemical Company. Partial pressure values for the micropore filling measurements using the micro balance are given with the experimental data.

Initial TEM work was carried out on a Jeol 100 CX and then using a JEM 4000 Ex Microscope [11].

From the Proceedings of the 9th International Zeolite Conference, Montreal 1992, Eds. R. von Ballmoos et al.,

RESULTS AND DISCUSSION

A. Comparative template studies based upon quaternized pyrrolidines

In our efforts to produce new zeolite structures and to develop the correct elements for multidimensional channel formation we paid close attention to how template molecules lead to specific products. The role of organic-cations, organic amines, hydrated cations and other polar molecules as templates around which the host lattice will develop is still debated. It is not a requisite system for some of the acid-based aluminum phosphate sieve syntheses [12]. Still, what we observed in the product was a strong guest/host relationship for the organic component used and the lattice developed. There were no surprises in comparing the void regions of the host lattice and the space-filling characteristics of the organic component. In addition, for high silica zeolite materials the correlation was also high for the available void volume of the lattice and the amounts of organic component found encaged within. In an earlier study it was shown that the guest template could become the limiting component in the amount of zeolite product produced [13].

We also observed something of a competition in synthesis reactions between making parallel one-dimensional channel systems and materials with large cavities or cages and little or no channel system. Both types seemed favored over multidimensional channel systems. In terms of one-dimensional materials a number of organo-cations or dications lead to ZSM-12. We were particularly impressed with the behavior of four features: quaternized pyrrolidines, quaternized piperidines, ethyl substituents on charged nitrogen, or a single long axis in the organic guest all favor ZSM-12.

Polycyclic systems favor clathrate until the guest becomes too large. This was recently shown for the Nonasil structure by Nakagawa [14]. Even when one went from amino adamantane to the larger N,N,N-trimethyl ammonium adamantane, while the latter is too large to make clathrates, the favored synthesis was still materials like SSZ-13 or SSZ-23 with cavities larger than the interconnecting pores [8].

Figure 1 shows the contrasting use of quarternized pyrrolidines built into subsequent organic structures and the resulting zeolites produced. The simplest N,N-dimethyl quarternary ammonium compound produced 12-ring ZSM-12. But incorporation into a bicyclic [3.2.1] tropane system now favors cage formation as evidenced by the zeolite structures produced. ZSM-58, SSZ-29, and Σ-2 all have cages with much smaller pores. Even pyrrolidinium cation as part of the larger tricyclic [5.2.1.0] decane produced the caged nonasil [14].

If the two 5 rings are fused end-to-end, and nitrogen moved outside the ring, ZSM-12 or EU-1 are the products. This moves us back to the linear orientation. The template has now become too large to simply make a cage material.

B. SSZ-26 Based Upon a Propellane Candidate

Figure 2 shows a representation of disrupting the linear 12-ring orientation found in ZSM-12 by disrupting the principal axis in the diquaternary organo-cation shown in Figure 1. A third fused ring was added generating a [4.3.3.0] tricyclododecane system. This is an example of materials known as propellanes. SSZ-26 can be made from two isomeric configurations [15]. While these configurations work to produce a zeolite which we believe is neither one-dimensional or a cage with small pore (vide infra) some other propellanes were unsuccessful [15]. It is interesting to note that the propellane based upon the fusion of three pyrrolidine rings and then quaternized is a poor zeolite template in a variety of synthesis reactions. This particular tricationic template is too crowded sterically. Space-filling models show that methyl groups from each nitrogen are very crowded between nitrogens. The much older discovery of ZSM-18 utilized a tris pyrrolidinium template which was sterically less crowded and also much more planar. (See Figure 1.) ZSM-18 would also qualify as an exception from the unidimensional versus cage dilemna [16]. A secondary channel system intersects the large pores.

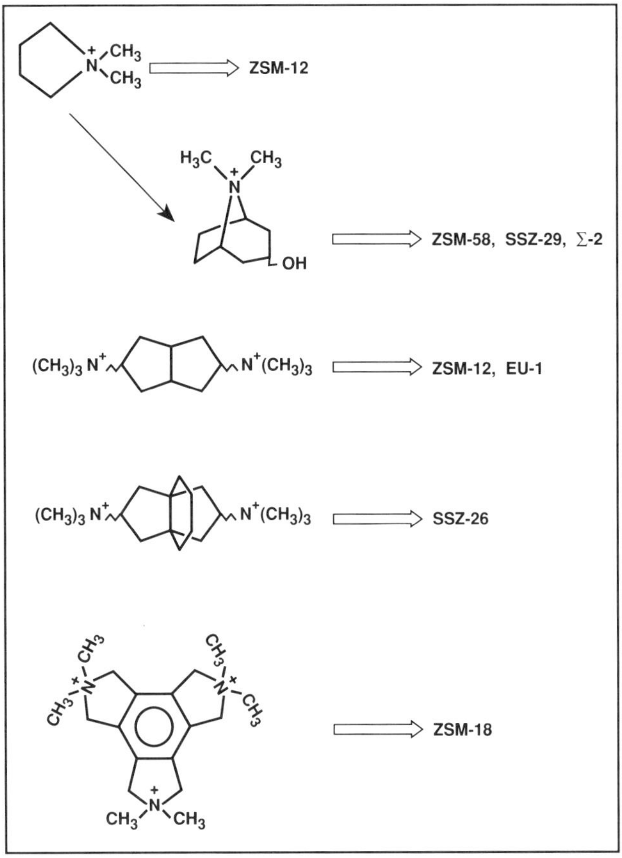

Figure 1. Organo-cations based upon elaborating the size and position of substituents on pyrrolidine. The zeolites made from using such organo-cations are also given.

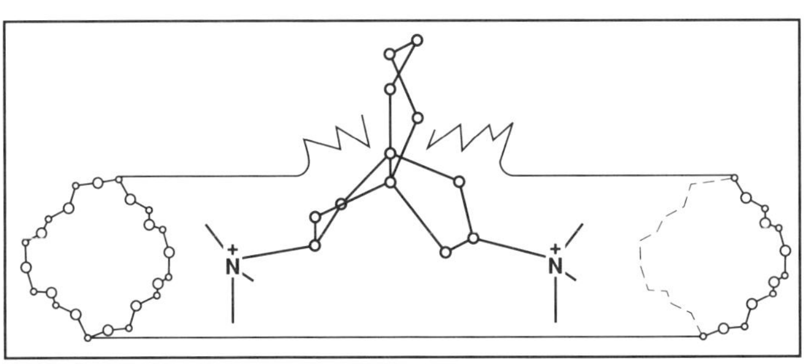

Figure 2. A representation of how the diquaternary ammoniun compound based upon the [4.3.3.0] tricyclododecane disrupts the linear channel formation observed in ZSM-12.

Table 1 gives the key inorganic ratios and reaction parameters for synthesizing SSZ-26. The overall OH^-/SiO_2 and Al/Si ratios are most like those for synthesizing beta zeolite or ferrierite. The latter can in fact be an impurity problem in the SSZ-26 synthesis even though it forms without using the organic. However, its formation is greatly suppressed by using Y zeolite as the aluminum source in the synthesis [17].

Conversely, the relatively large propellane diquaternary ammonium compound has a very high specificity for making SSZ-26. When producing SSZ-26, the pore-filling by the organo-cation is substantial, usually reaching about 15 wt % in the as-made zeolite. This value is higher than those for a variety of one-dimensional 10 and 12 ring structures which usually show organic contents closer to 8-10 wt % of the as-made zeolite.

Table 1

Key Ratios for Synthesizing SSZ-26

Parameter	Ratio Value
Al/Si	0.04-0.06
N^+/Si	0.15
Na^+/Si	0.15
OH-/Si	0.30
H_2O/Si	25-35
Temp.	160-175°C
RPM	30-75

C. Implications About SSZ-26 Being A Zeolite With a Multidimensional Pore System

The open-pore SSZ-26 material with SiO_2/Al_2O_3 ~30 showed considerable microporosity. Table 2 shows the hydrocarbon uptake for SSZ-26 and several other materials. The values are lower for SSZ-26 than for the other two known zeolites with three-dimensional pore systems, beta, and Y. On the other hand the values for SSZ-26 are almost double any known one-dimensional material.

Table 2

Hydrocarbon Uptake for Zeolites at P/PO = 0.14

Zeolite	Cyclohexane Uptake (cc/gm)
Beta	0.25
Y	0.27
SSZ-26	0.20
SSZ-24	0.12
ZSM-12	0.11
LTL	0.12
ZSM-5	0.18 (n-hexane)
EU-1	0.014 (0.144 for n-hexane)[1]

[1]From Casci, J. L.; Lowe, B. M.; Whittam, T. V. in Eur. Pat. Appl. 42,226 (1981).

Two other zeolites are worth comparison in attempting to establish an opinion about SSZ-26. ZSM-5 which has intersecting 10 rings shows a nitrogen pore-filling behavior about 25% below SSZ-26. EU-1, with 10 rings (one-dimensional) but deep side pockets, also shows a value of 25% below SSZ-26. As the porosity increases, the framework density decreases. At some minimal value it would no longer be possible to have pores with side pockets and not have them connect and thus become a multidimensional pore system. The micropore filling of SSZ-26 by nitrogen, leading to a calculated surface area (BET) of 590 m^2/g also suggests a framework density too low to be one-dimensional [18].

Santilli has introduced a method for analyzing the dynamic, steady-state pore-filling by three hexane isomers of varying size [10]. A stream of n-hexane, 3-methyl pentane and 2,2-dimethyl butane is introduced into the dried evacuated zeolite held at elevated temperature. Data for SSZ-26 and several other zeolites appears in Table 3.

Table 3

Hydrocarbon Uptake Using the Dynamic Pore Probe Technique[1]

	mg/g			
Zeolite	n-C_6	3MP	2,2 DMB	ΣC_6
Y (Y82)	38	37	26	101
Beta	57	37	20	114
SSZ-26	31	22	18	71
SSZ-24	12	15	26	53
MORD	10	11	13	34
LTL	13.2	11.6	4.2	29
ZSM-5	25	7	5	37

[1]130°C

Again SSZ-26 does not contain the microporosity of beta or Y zeolites. But the values are significantly higher than for ZSM-5 or for some of the known one-dimensional large pore zeolites. In the case of Y, beta, and SSZ-26 zeolites, although 2,2 dimethylbutane is readily admitted, there is not the unusual preference for it observed for the AFI structures (SSZ-24, AlPO-5, SAPO-5) [19].

VPI-5 has a very large micropore volume but only parallel one-dimensional channels [20]. Is the SSZ-26 an aluminosilicate of this type? Table 4 shows the tabulated X-ray diffraction lines for the calcined SSZ-26. The largest d-spacing observed is 11.36 Å. For VPI-5 it is 16.4 Å. The value for SSZ-26 (and also seen by electron diffraction) is much more consistent with the large pore behavior being a result of a 12-ring pore. Preliminary HREM work also indicates that there is more than one pore system in SSZ-26 [11].

Table 4

X-Ray Powder Diffraction Lines for SSZ-26[1]

2Θ	d/n	$I/I_o \times 100$
7.78	11.36	100
20.33	4.389	63
21.37	4.158	25
21.99	4.042	53
22.85	3.890	46 Sh
23.00	3.867	64
26.49	3.365	33

Sh = Shoulder

[1]For more complete sets of lines see ref. 9

D. Trends in Dimensionally of the Pore System of Aluminosilicate Materials

SSZ-26 crystallizes most efficiently in the SiO_2/Al_2O_3 range of 20-50. This is also true of beta zeolite. What is proposed is that to achieve multidimensional channel systems a requisite amount of substituting Al^{+3} must be present. Al^{+3} substitution provides longer bonds and in instances, more favorable bond angles relative to SiO_2 alone [21].

Figure 3 shows a collection of materials with multidimensional and unidimensional channel systems. We have selected zeolites with at least one 10 or 12 ring for this grouping. The zeolites are placed in the approximate SiO_2/Al_2O_3 range in which they form. With the exception of ZSM-5 two readily discernable groups can be seen: materials which are difficult to produce below SiO_2/Al_2O_3 = 50, and those which become more difficult above that value.

We have grouped a number of zeolites clustered in the $SiO_2Al_2O_3$ range of 5-20 as multidimensional. From experience in adsorption and catalysis some will seem surprising entries. But zeolites like L, Ferrierite, Mordenite, Omega, and offretite all have interconnecting 8-ring channels as well as a main 10- or 12-ring channel. This feature can be contrasted with the candidates in the group best formed beyond SiO_2/Al_2O_3 = 50. These materials all have one-dimensional, parallel channel systems with no secondary intersecting channels. Most of the materials in this latter group can be made out to SiO_2/Al_2O_3 > 250. Some become most favorable with no aluminum substitution [22].

ZSM-5 (MFI Structure) remains the lone exception. In the absence of organo-cation it can be made in the SiO_2/Al_2O_3 = 30-40 range. The branching nature of ZSM-5 is certainly consistent with relatively high aluminum substitution. But this structure can be made aluminum-free (silicalite [23]). This may be a situation where the structure-directing aspect of tetrapropyl ammonium cations (TPA) for the MFI structure is great enough to overcome an increased energy barrier for the all-silica crystallization. It does seem to be the case that formation of MFI at

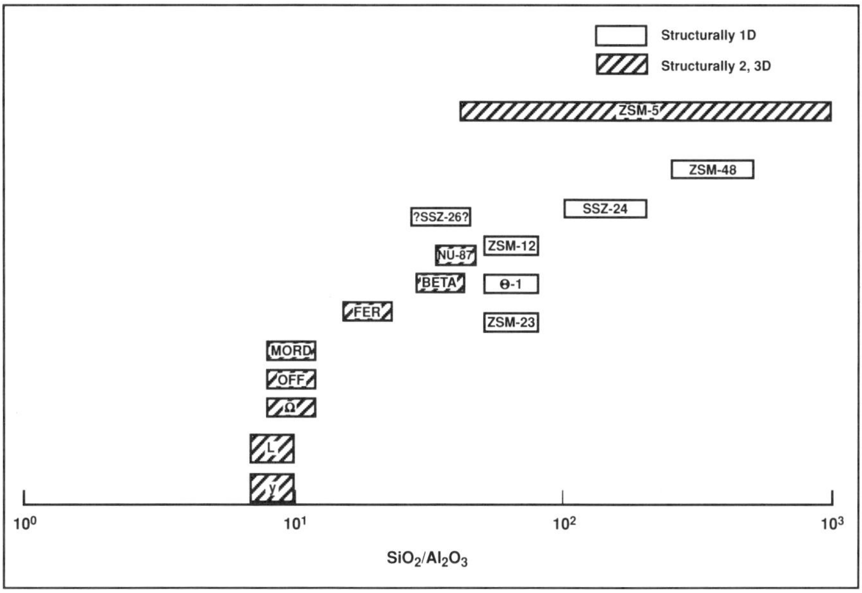

Figure 3. A schematic representation of zeolite type versus SiO_2/Al_2O_3 values which are most favorable. Materials with known multi-dimensional pore systems are cross-hatched.

increasingly higher SiO_2/Al_2O_3 ratios becomes less favorable for the wide variety of organocations and amines which can stabilize the MFI structure. Without Al^{+3}, TPA is the best structure director. Ranges for other zeolite structures may be extended as highly selective templates are found.

CONCLUSIONS

Organic guest molecules stabilize silicate-based microporous lattices at high solution pH. A good structure/fit relationship is observed in almost all cases. Polycyclic hydrocarbon organocations tend to favor cavity of clathrate formation until the guest molecule becomes too large. Molecules with a long principal axis stabilize ZSM-12. The appearance of ethyl groups on charged nitrogen also promotes ZSM-12. The large propellane diquaternary ammonium cation used in this study only produces SSZ-26.

Indications of multidimensionality of the pore system in SSZ-26 is mainly from micropore volumes observed for various adsorbates. The values are greater than one dimensional materials and beyond that for EU-1, a material with extensive side pockets. Multidimensionality may also be greatly influenced by inorganic factors like the extent of lattice substitution. With the exception of MFI, multidimensional pore systems are best made with significant aluminum substitution.

ACKNOWLEDGMENTS

We thank R. A. Van Nordstrand for all X-ray diffraction work and much useful discussion. Lun-Teh Yuen and L. D. Scampavia contributed to propellane syntheses. A. Rainis and T. F. Finger provided micropore volume values for experiments different than pore probe studies. We thank Chevron Research and Technology Company for continued support during these studies and permission to publish.

REFERENCES

1. Wadlinger, R.L.; Kerr, G. T.; Rosinski, E. J.; U.S. Patent 3,308,069 (1967).

2. Argauer, R. J.; Landolt, G. R.; U.S. Patent 3,702,886 (1972).

3. Rosinski, E. J.; Rubin, M. K.; U.S. Patent 3,382,449 (1974).

4. Barri, S. A. I.; Phillip, H.; Telford, C. D.; Eur. Pat. Appl. 57,049 (1982).

5. Plank, C. J.; Rosinski, E. J.; Rubin, M. K.; U.S. Patent 4,076,842 (1978).

6. Rollmann, L. D.; Valyocsik, E. W.; U.S. Patent 4,423,021 (1983).

7. Wilson, S. T.; Lok, B. M.; Messina, C. A.; Cannan, T. R.; Flanigen, E. M.; J. Am. Chem. Soc. **104**, 1146 (1982).

8. Zones, S. I.; Van Nordstrand, R. A.; Santilli, D. S.; Wilson, D. M.; Yuen, L. T.; Scampavia, L. D.; *Zeolites, Facts Figures Future*, pg 299, Elsevier 1989.

9. Zones, S. I.; U.S. Patent 4,910,006 (1990).

10. Santilli, D. S.; J. Catal. **99**, 335 (1986).

11. Pan, M.; Cruzier, P. A.; Chan, I. Y.; Zones, S. I.; *Proc. 49th Meeting of Electron Micro Soc. Amer.*, pg 1132 (1991).

12. Szostak, R.; Duncan, B.; Sorby, K.; Ulan, J. G.; Catt. Lett. **8**, 367 (1990).

13. Zones, S. I.; Zeolites **9**, 458 (1989).

14. Nakagawa, Y., Zones, S. I.; *Synthesis of Microporous Materials Vol. II*, ed. by M. Occelli and H. Robson (Van Nodstrand-Reinhold, 1992), Chapter 16.

15. Zones, S. I.; Olmstead, M. M.; Santilli, D. S.; J. Am. Chem. Soc., **114** 4195 (1992).

16. Lawton, S. L.; Rohrbaugh, W. J.; Science, 1319 (1990).

17. Zones, S. I.; Van Nordstrand, R. A.; Santilli, D. S.; to be submitted.

18. Breck, D. W.; *Zeolite Molecular Sieves* Chapter 5 (J. Wiley and Sons 1974).

19. Van Nordstrand, R. A.; Santilli, D. S.; Zones, S. I.; Ref. 14, Chapter 25 (1992).

20. Davis, M. E.; Saldarriaga, C.; Montes, C.; Garces, J.; Crowder, C.; Zeolites **8** 362 (1988).

21. Alvarado-Swaisgood, A. E.; Barr, M. K.; Hay, P. J.; Kedondo, A.; J. Phys. Chem. **95** 10031 (1991).

22. Araya, A.; Lowe, B. M.; J. Catal. **85** 135 (1984).

23. Flanigen, E. M.; Bennett, J. M.; Grose, R. W.; Cohen, J. P.; Patton, R. L.; Kirchner, F. M.; Smith, J. V.; Nature **271** 512 (1978).

RAMAN SPECTROSCOPIC STUDIES IN ZEOLITE SYNTHESIS

P.K. DUTTA

Department of Chemistry, The Ohio State University, Columbus, Oh 43210, USA

ABSTRACT

The development of specific zeolite frameworks by interaction of organic structure directing molecules and aluminosilicate species is an important area of research. Raman spectroscopy, by providing vibrational and thereby structural information about the framework as well as the organic molecule helps elucidate the mechanism of zeolite evolution. In this study, we contrast the role of tetramethylammonium (TMA^+), tetraethylammonium (TEA^+), tetrapropylammonium (TPA^+) and pyrrolidine (Pr) in the synthesis of ZK-4, mordenite, ZSM-5 and ferrierite, respectively.

INTRODUCTION

Formation of novel zeolitic frameworks, as well as the synthesis of zeolites with varying Si/Al ratios has been made possible by the introduction of organic molecules in the reactant system.[1] The specific role that such molecules can play is an area of active research.[2] An intriguing observation has been the trapping of the organic molecule by the zeolite during synthesis, suggesting the possible role of the organic as a templating agent around which the zeolite forms.[3] Direct experimental evidence for this process is lacking. Spectroscopic methods provide a direct structural probe for examining organic-aluminosilicate interactions. We have been developing the use of Raman spectroscopy to study these processes. As compared to our previous studies on totally inorganic compositions,[4] in this study we bring together our work on role of organic molecules. Four zeolitic systems: ZK-4, mordenite, ZSM-5 and ferrierite are compared. The vibrational spectroscopic features that make it possible to deduce the organic - framework interactions are quite different for the four zeolite systems and form the basis of discussion in this study. The role of Raman spectroscopy in elucidating the nature of the complex zeolite synthesis process is thereby illustrated.

EXPERIMENTAL

The synthesis procedures were adopted from the published and patent literature. The gel compositions used in the synthesis of zeolites were: ZK-4 - 8.6 Na_2O • Al_2O_3 • SiO_2 • TMACl • 556H_2O;[5] mordenite - 7.38 Na_2O • Al_2O_3 • 24.7 SiO_2 • 3.41 TEABr • 597H_2O[6]; ZSM-5 - 45.7 Na_2O • Al_2O_3 • 85 SiO_2 • 8.6 TPABr • 17H_2SO_4 • 2571 H_2O[7]; ferrierite - 1.2 Na_2O • Al_2O_3 • 20SiO_2 • 12.6 Pr • 398H_2O.[8] Synthesis using the same composition were carried out in multiple batches using Teflon lined containers. The syntheses were interrupted at various times, the solution and solid phases separated

From the Proceedings of the 9th International Zeolite Conference, Montreal 1992, Eds. R. von Ballmoos et al.,

and examined by Raman spectroscopy. The spectra were obtained using visible excitation lines from Spectra Physics Ar ion laser by methods described elsewhere.[4] Powder diffraction patterns were obtained with a Rigakiu Geigerflex D/Max 2B diffractometer using a nickel filtered CuK_α source.

RESULTS AND DISCUSSION

This section has three parts focusing on framework vibrations, specific motions of structure directing agents that are sensitive to organic-aluminosilicate interactions, and the possible roles that have been discovered for organic molecules in framework by Raman spectroscopy.[9-11]

a) Framework vibrations: Based on all the published Raman data of zeolitic frameworks, it appears that the most prominent band occurs in the region between 300-500 cm^{-1}, followed by weaker bands in the 700-900 cm^{-1} and 1000-1200 cm^{-1} regions.[12] These latter bands do provide structural information, however it is the prominent low frequency band that has been directly correlated with framework architecture.[13,14] This band is assigned to a ν_s (T-O-T) mode, and involves primarily the motion of the oxygen atom. For a series of frameworks, consisting primarily of even membered 4,6,8,10 and 12 membered rings as in zeolite A, faujasite, chabazite, gismondine and edingtonite, this band occurs in the region between 486-521 cm^{-1}.[13] Moreover the frequency has an inverse dependence on the T-O-T angle. With the introduction of five membered rings, this band shifts to a lower frequency. For ferrierite-like structures with 5,6,8 and 10 membered rings the band is at ~430 cm^{-1}, whereas in the absence of 8-membered rings as in ZSM-23 and Nu-10, the band is at 450 cm^{-1}. With the introduction of 4-membered rings as in ZSM-5, mordenite and ZSM-48, this band appears at 384, 396 and 386 cm^{-1} respectively.[14] Thus, Raman spectra during zeolite evolution can provide information about framework development, as is illustrated below.

b) Vibration of Structure-Directing Agents: Tetramethyl ammonium ion (TMA^+) has been used as a structure directing agent in the synthesis of zeolites such as sodalite, ZK-4 (analogue of zeolite A), gmelinite and cancrinite.[1,2] Four vibrational bands characterize the tetrahedral CN_4 unit and occur at 950 cm^{-1} (T_2, (C-N asymmetric stretch), 752 cm^{-1} (A_1, C-N symmetric stretch), 455 cm^{-1} (T_2, C-N deformation) and 370 cm^{-1} (E, C-N deformation).[9] The Raman spectrum of TMACl is shown in Figure 1a. Figure 1b shows the spectrum of TMA^+ trapped in the zeolite sodalite. The most significant changes are the shifts in C-N stretching frequency from 752, 950 cm^{-1} to 761 and 955 cm^{-1}. The increase of frequencies of bands that lead to an increase of molecular volume is characteristic for vibrations that are influenced by high pressures.[15] In the sodalite crystal, the TMA^+ of diameter ~6.4Å are held in the β-cages of diameter 6.5Å, leading to a tight encapsulation. During the stretching vibration, the outermost H atoms meet resistance from the zeolite framework, thus leading to an increase in frequency. This characteristic shift of frequency therefore provides a diagnostic marker of the formation of the sodalite cages. Zeolite ZK-4, whose framework is identical to that of zeolite A but with Si/Al >1, is made up of sodalite units that are connected through four-membered rings. The synthesis of this material provided an opportunity to examine the buildup of sodalite cages. The appearance of the 767 cm^{-1} band coincided with the X-ray diffraction pattern characteristic of ZK-4, clearly establishing that the framework formation is not proceeding through individual sodalite units formed in the nucleation stage connecting with each other. So, the entrapment of TMA^+ is occurring as subunits combine to form the sodalite cage and the zeolite structure at the same time.

Tetraethylammonium ion (TEA^+) has been used to make novel zeolitic structures such as ZSM-20 and ZSM-12.[2] In addition, the presence of TEA^+ helps in stabilizing higher Si/Al ratio frameworks, as in the case with mordenite.[16] The role of the organic species in such a process is not well understood. Figure 2 shows the Raman spectra of the gel in the ν_s(T-O-T) framework region of 300-600 cm^{-1}.[11] The prominent band in the 0-hr unheated gel spectrum at 418 cm^{-1} and the weak band at 387 cm^{-1} are due to the TEA cation. X-ray diffraction shows the presence of mordenite crystals after 168 hrs of heating and is reflected in the Raman bands at 387 and 467 cm^{-1} due to mordenite. What is of interest is that these mordenite-like bands appear after 24 hrs of heating, though crystals are formed at significantly longer times (>168 hours). It appears that the TEA^+ is helping stabilize the mordenite building blocks until these units can connect to form the crystalline framework. The ability of alkylammonium ions to stabilize specific aluminosilicate structures in solution is well known.[17] In this case, the TEA^+ is stabilizing mordenite-like structures in the gel, that would otherwise be unstable at the high Si/Al ratios.

The tetrapropylammonium (TPA^+) ion is best known for directing the synthesis of ZSM-5.[16] We have examined this system and the Raman data in the characteristic C-C stretching and C-H wagging and rocking modes is shown in Figure 3.[10] In contrast to ZK-4 and mordenite discussed above, the TPA^+ appears to be trapped within the alumino-silicate framework after a day of heating and cannot be removed by ion-exchange or washing. Formation of zeolite crystals defined by the X-ray diffraction pattern occurs after four days of heating. However, as shown in the insert, prior to the formation of any zeolite crystals, there is a decrease in intensity of the 1324 cm^{-1} band and the appearance of a band at 1343 cm^{-1}. This change arises from the TPA^+ assuming a gauche conformation as compared to the normal all-trans form.[18] The trapping of the TPA^+ in its all trans-form by the aluminosilicate followed by a conformational change is consistent with a model that ZSM-5 nucleation may proceed through the more symmetric ZSM-11, as schematically shown in Figure 4. Thus the initial trapping of all-trans TPA^+ (1 day of heating) leads to ZSM-11 nuclei, followed by a conformation change, necessary to accommodate the TPA^+ in ZSM-5 nuclei (3 days of heating)

Finally, we focus on the molecule, pyrrolidine which has been used in the synthesis of zeolites ZSM-21, ZSM-23 and ZSM-39.[2] We focus presently on the zeolite ferrierite, which has a framework devoid of four membered rings. Figure 5 shows the Raman spectra during zeolite growth. The band at 902 cm^{-1} in the initial gel arises from the pyrrolidine ring breathing vibration and resembles the spectrum of pyrrolidine in solution. X-ray diffraction indicates crystal formation after 62 hours of heating. Reaction for 11 hours brings about changes in the ν_s(T-O-T) bending region with the appearence of a band at 485 cm^{-1}, assigned to four membered aluminosilicate rings[4] and the splitting of the pyrrolidine ring breathing mode to 913 and 881 cm^{-1}, the latter assigned to NH_2^+ rocking mode.[19] This indicates that pyrrolidine is getting protonated as it incorporates into the aluminosilicate gel. After 38 hours, the band due to 438 cm^{-1} arising from "ferrierite-like" structures is observed. In analogy with the mordenite system, disordered zeolitic-like units are forming prior to formation of crystals. With further heating the NH_2^+ rocking mode splits into several bands. We interpret this observation as indicative of the fact, that, as crystallization proceeds the pyrrolidine molecule finds itself constrained at different sites. This would indicate that zeolite growth occurs by initial ordering of the aluminosilicate gel, followed by the assembly of smaller domain structure and finally the local structure of the zeolite itself.

c) Raman spectroscopy as a probe of zeolite growth: The above studies clearly indicate that Raman spectroscopy can provide structural information relevant to the formation of zeolite crystals. Information about both the framework and the organic species and their mutual interaction can be derived. Based on these studies, we have discovered that:

- zeolitic like structures in a disordered form are present in the gel phase prior to crystallization. How these units link to form the structure is still unclear.

- nucleation of ZSM-5 may proceed through ZSM-11. The generality of such transformations in zeolite synthesis is unclear. The presence of conformation sensitive Raman bands was critical to this discovery.

- ferrierite crystals grow by a global ordering of the aluminosilicate gel, followed by buildup of smaller domains. This leads to increasing entrapment of the pyrrolidine molecules that act as structure directing agents.

- entrapment of organic cations in zeolite cages can lead to vibrational changes that mimic high pressure effects because of the tight fit of the organic molecule inside the cage. In such cases, the Raman bands provide a clear indication of when these cages form, as was evident in ZK-4 synthesis.

References

1. Lok, B.M.; Cannan, T.R.; Messina, C.A., Zeolites, **3**, 282, (1983).

2. Szostak, R. Molecular Sieves, Principles of Synthesis and Identification, Van Nostrand, N.Y. (1989).

3. Daniels, R.H.; Kerr, G.T.; Rollman, L.D., J. Am. Chem. Soc., **100**, 3097, (1978).

4a. Dutta, P.K.; Shieh, D.C. J. Phys. Chem., **90**, 2331, (1986).

b. Dutta, P.K.; Shieh, D.C.; Puri, M., J. Phys. Chem., **91**, 2332, (1987).

c. Dutta, P.K.; Puri, M.; Shieh, D.C. in Microstructure and Properties of Catalysts (Ed. Teacy, M.M.J.; Thomas, J.M.; White, J.M.). Materials Research Society Publication **111**, 101, (1988).

d. Dutta, P.K.; Bowers, C.; Puri, M. ACS Symposium Series, **398**, 98, 1989.

5. Kacirek, H.; Lechert, C. ACS Symposium Series **40**, 244, (1977).

6. Chu, L.B. US Patent 3758593, (1973).
7. Rollman, L.D.; Valyosik, E.W., Inorg. Synth. **22**, 61, (1983).
8. Plank, C.J.; Rosinski, E.J.; Rubin, M.K. US Patent 4016245, (1977).
9. Dutta, P.K.; Shieh, D.C.; DelBarco, B., Chem. Phys. Lett. **127**, 200, (1986).
10. Dutta, P.K.; Puri, M., J. Phys. Chem., **91**, 4329, (1987).
11. Twu, J.; Dutta P.K.,; Kresge, C.; J. Phys. Chem., **95**, 5267, (1991).
12. Twu, J.; Dutta, P.K., J. Phys. Chem., **95**, 2498, (1991)
13. Dutta, P.K.; Shieh, D.C.; Puri, M., Zeolites, **8**, 306, (1988).
14. Dutta, P.K.; Rao, K.M.; Park, J.Y., J. Phys. Chem., **95**, 6654, (1991).
15. Allkins, J.R.; Obrenski, R.J.; Brown, C.W.; Lippincott, E.R., Inorg. Chem., **8**, 1450, (1969).
16. Jacobs, P.A.; Marsten, J.A., Synthesis of High-Silica Aluminosilicate Zeolites, Elsevier, Amsterdam, (1987).
17. Iler, R.K. The Chemistry of Silica, John-Wiley, NY, (1979).
18. Harada, I.; Takeuchi, H.; Sakakibara, M.; Matsura, H.; Shimanouchi, T., Bull Chem. Soc. Jpn., **50**, 102, (1977).
19. Herlinger, A.W.; Long, T.V. J. Am. Chem. Soc., **92**, 6481, (1970).

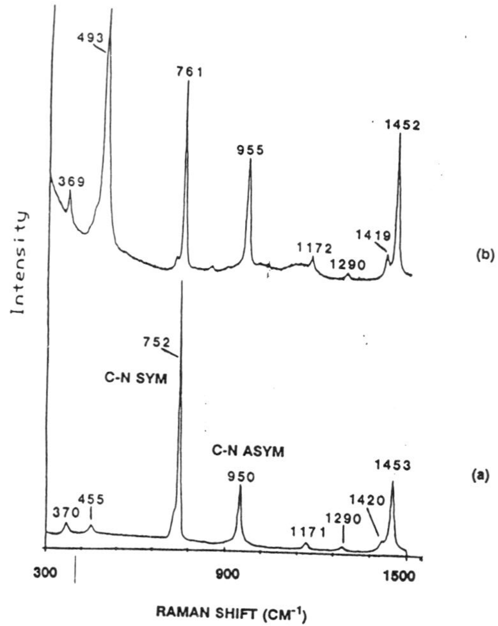

Figure 1. Raman spectrum of TMA^+ (a) in solution (b) trapped in sodalite cages.

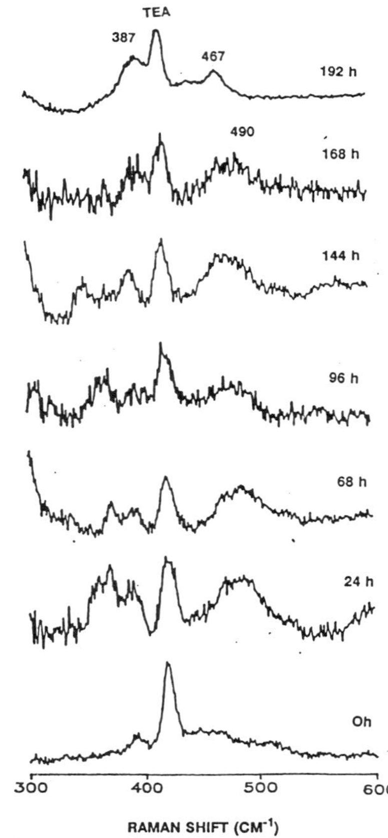

Figure 2. Raman spectra obtained at various times during crystallization of mordenite.

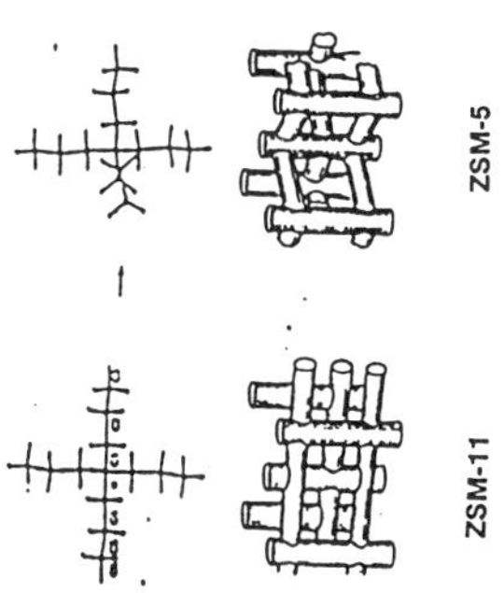

Figure 4. Model of ZSM-5 devlopment proceding through ZSM-11 nucleation.

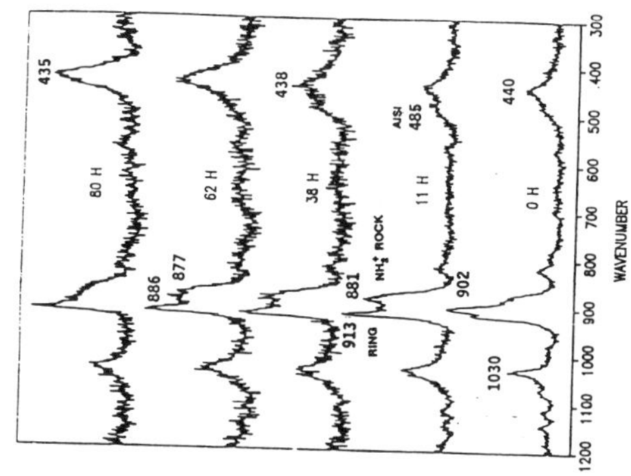

Figure 5. Raman spectra during evolution of ferrierite synthesis.

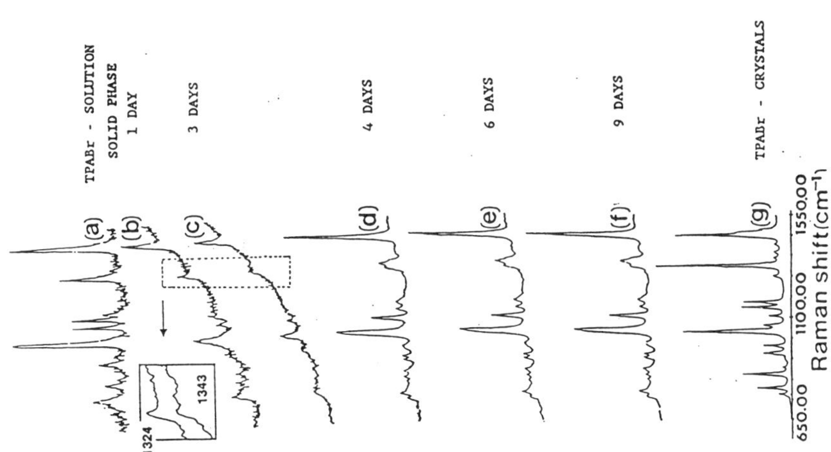

Figure 3. Raman spectra of TPA^+ in solution, at various stages of ZSM-5 synthesis and TPABr crystal.

SYNTHESIS AND CHARACTERISATION OF A NEW HIGH-SILICA, LARGE-PORE ALUMINOSILICATE ZEOLITE, NCL-1

R. KUMAR, K.RAMESH REDDY, ANUJ RAJ and P. RATNASAMY

National Chemical Laboratory, Pune 411 008, India.

ABSTRACT

A new, high-silica (Si/Al = 20 to infinity) large pore zeolite, NCL-1 as well as its Al-free silica polymorph have been synthesised and characterised by XRD, electron microscopy, thermal analysis, framework IR and solid state MAS NMR spectroscopies, adsorption and catalytic methods. Preliminary investigations indicate that NCL-1 possesses orthorhombic symmetry with a = 1.195 $\pm$ 0.003, b = 0.836 $\pm$ 0.003 and c = 2.870 $\pm$ 0.004 nm, respectively. The effective pore diameter of NCL-1 (from adsorption, and diffusivity experiments) seems to be close to that of mordenite (0.7 nm). NCL-1 exhibits activity and shape selectivity in acid catalysed reactions.

INTRODUCTION

High silica, large pore zeolites are of special interest in the field of adsorption and catalysis. Till now, ZSM-12 is the only large-pore, 12 ring, zeolite which can be synthesised in a wide Si/Al range including the Al-free silica polymorph form. Though zeolite beta can also be obtained in a fairly large range of Si/Al = 10 to 120, its silica polymorph is still unknown. Now, we report a new, large pore high silica zeolite, designated as NCL-1, which can be obtained in a very large range of Si/Al ratios (20 to infinity). The silica polymorph of NCL-1, free from Al has also been synthesised. The x-ray diffraction pattern of NCL-1 indicates that NCL-1 has a novel framework structure. The solid state MAS NMR spectra of ^{27}Al and ^{29}Si reveal that both are located at framework positions with tetrahedral symmetry. The adsorption and catalytic properties of NCL-1 are characteristic of aluminosilicate zeolites with pore openings constituted by 12-membered tetrahedra.

EXPERIMENTAL

NCL-1 was synthesised hydrothermally in stainless steel autoclaves (100 ml capacity). Sources of silica (tetraethylorthosilicate or fumed silica) and aluminium (Al^{3+} salts or aluminium isopropoxide) were mixed in the presence of a diquaternary ammonium dibromide salt (R) for the synthesis of NCL-1. Aluminium was excluded in the preparation of the Al-free silica polymorph. Diquaternary salts of the formula (alkyl)$_3$ N-(CH_2)$_6$-N(alkyl)$_3$ were particularly efficient in the synthesis of NCL-1 and its silica polymorph. The molar gel composition yielding phase pure NCL-1 was : 0.2-0.02 Na_2O : 0.05-0.025 RBr_2 : 0-0.025 Al_2O_3

From the Proceedings of the 9th International Zeolite Conference, Montreal 1992, Eds. R. von Ballmoos et al.,

: SiO_2 : 40 H_2O. The OH^-/SiO_2 ratio (0.1-0.15) was adjusted by adding appropriate amount of H_2SO_4. The crystallisation was carried out under agitation at 438 K for 3-14 days depending upon the Al content in the gel. After crystallisation, the solid material was thoroughly washed with distilled water, dried in air (393 K) and finally calcined in flowing dry air at 723 K. The protonic forms of NCL-1 were obtained by ion exchange with ammonium nitrate solution followed by drying and calcination.

The crystallinity of the samples was recorded using Rigaku x-ray diffractometer (Model D-Max III VC) using nickel-filtered CuKα radiation. The XRD powder pattern was recorded at a scanning rate of $2\theta = 0.5^o$/min with silicon as an internal standard. The elemental analysis were performed using plasma emission spectroscopy (Jobin Yvon) and wet chemical methods. Sorption measurements were carried out at 298 K and at $p/p_o = 0.5$ using a Cahn electrobalance. Framework IR spectra (in the region 200-1500 cm^{-1}) were recorded with a Nicolet 60 SXB FT-IR spectrometer. All spectra were recorded with the sample at room temperature. The sample were also characterised by scanning electron microscopy (JEOL, JSM 5200), thermal analysis (Setaram TG/DTA 92), MAS NMR spectroscopy (Bruker, MSL 300) and catalytic methods. The catalytic reactions were carried out in a down-flow tubular silica reactor. Catalyst particles of 35-45 mesh were used in the run. The products were analysed, on-line, using a Shimadzu R1A gas chromatograph with a 5 % bentone 34 + 5% diisodecylpthalate column. The protonic form of the zeolites was used for all catalytic experiments.

RESULTS AND DISCUSSION

X-ray diffraction

Various samples of NCL-1 with Si/Al ratios as low as 20 were synthesised. The Al-free silica polymorph was also prepared. The x-ray diffraction patterns of the calcined forms of NCL-1 with Si/Al = 20, 80 and greater than 1500 (the silica polymorph) are shown in Fig. 1. The XRD pattern of the as-synthesised form of the silica polymorph is also shown, for comparison. The XRD pattern is novel, the principal lines being listed in Table 1. Preliminary studies indicate orthorhombic symmetry with a = 1.195 $\pm$ 0.003 nm, b = 0.836 $\pm$ 0.003 nm and c = 2.870 $\pm$ 0.004 nm. The 'd' values calculated on the basis of the above unit cell dimensions compare reasonably well with those observed experimentally (Table 1). It was also confirmed that the material was not a mixture of two or more crystalline phases (like ZSM-12 and silica X which have some lines in common with NCL-1).

IR spectroscopy

The framework IR spectra (Fig.2) of the calcined samples (723 K) exhibit distinct absorption bands at 1230, 1098, 793, 548, 468 and 610 cm^{-1}, respectively. Based on literature values [1] for crystalline aluminosilicate molecular sieves, these absorption bands are tentatively assigned to external asymmetric stretch (1230 cm^{-1}) internal asymmetric stretch (1098 cm^{-1}), external symmetric stretch (793 cm^{-1}), double rings (548 cm^{-1}) and T-O bending vibrations of TO_4 tetrahedra (468 cm^{-1}) (T = Si or Al). The origin of the band around 610 cm^{-1} is not clear. It may be mentioned that bands around 590-620 cm^{-1} are observed in ZSM-5 zeolites and are usually assigned to external links of complex 5-membered rings [2].

Electron microscopy and thermal analysis

Fig.3 shows the scanning electron micrograph of the silica polymorph of NCL-1. The average crystal size is 3 x 7 μm and the crystal habit is rectangular (Fig. 3). The aluminosilicate analogs were similar in shape. The crystal sizes were, however, smaller. During the thermal decomposition of NCL-1 in air (Fig. 4), two distinct weight losses are observed at 373-390 and around 590 K, respectively. The former is due to the loss of water and the latter weight loss arises probably from the oxidative decomposition of the organic material occluded in the pore system of the zeolite. The exothermic peak observed in this region arises due to the oxidation of the occluded organics in air. The relatively low temperature at which the organics are removed suggests that their major role during the zeolite synthesis is as pore-filling agents. If they had been bound to the zeolite framework as templates, they would have been eliminated at a higher temperature.

NMR spectroscopy.

The solid state MAS NMR spectra of ^{29}Si and ^{27}Al in calcined samples of NCL-1 of varying Si/Al ratios are illustrated in Fig. 5. The ^{29}Si spectrum exhibited only one major signal at -112 ppm (with respect to TMS) attributable to Si(4Si). The samples containing aluminium exhibited only one signal at 52 ppm in the ^{27}Al spectrum, diagnostic for tetrahedral Al^{3+} and no signal due to octahedral Al^{3+}.

Adsorption

One of the distinguishing properties of molecular sieves is their ability to discriminate amongst molecules of different sizes. Those molecules whose critical diameter is smaller than the dimensions of the pore openings are easily adsorbed while those that are larger are excluded/are adsorbed with difficulty. Thus, the rates and extent of adsorption of molecules with different critical diameter provide insight into the size of the pore openings in the zeolite. Fig. 6 illustrates the type of Type I isotherms observed for adsorption in NCL-1. They are typical of adsorption in molecular sieves. The isotherms are reversible. Table 2 shows the amount of adsorption of three molecules of varying critical diameters in the pores of the silica polymorph of NCL-1. For comparison, adsorption values in other zeolites belonging to the small, medium and large pore categories are also shown. These zeolites have pore whose openings are constituted by 8, 10 and 12 TO_4 (T = Si or Al) tetrahedra, respectively. From the adsorption values of mesitylene (critical diameter = 0.78 nm) it is seen that NCL-1 belongs to the large-pore category with pore openings similar to that of mordenite and made up of 12 tetrahedra. The surface area of the sample (by O_2 adsorption) was 355 m^2/g (Omnisorp CX100). The effective pore diameter of NCL-1 (Si/Al = 20, 35, 80) as well as the Al-free sample (silica polymorph) obtained by Omnisorp (Ar adsorption) was close to that of mordenite (0.7 nm). In order to further characterise the pore system of NCL-1, the diffusivities of n-hexane, ortho-xylene and 1,3,5 trimethyl-benzene (mesitylene) were measured from the corresponding rates of adsorption at 298 K by the gravimetric method (Cahn electro-balance)[3]. From the initial, linear portion of the plots of (Q_t-Q_x) versus t (Q_t and Q_x are the amount adsorbed at time t and infinity, respectively) and using values of the average particle size, obtained from scanning electron microscopy, the diffusivity values for n-hexane, ortho xylene and mesitylene were found to be 8.5, 1.5 and

0.6 X 10^{-11} $cm^2/sec.$, respectively. These values seem reasonable in view of those reported in the literature for 10- and 12 membered ring zeolites [4] and suggest that NCL-1 is, indeed, a large pore zeolite with 12-ring pore apertures.

Catalytic activity

The results of acid catalysed test reactions such as constraint index [5] and meta xylene isomerisation over NCL-1 (Si/Al = 35) are compared with those obtained over other zeolites such as ZSM-5, ZSM-12, MOR and Beta (Table 3), categorising NCL-1 as a large pore, 12-ring zeolite. The detailed product distribution in the conversion of meta xylene at various temperatures (Table 4) also confirms the presence of large pores in NCL-1. Further, in ethylbenzene disproportionation (to benzene and diethylbenzene) over H-NCL-1 (Fig. 7), the value of diethylbenzene/benzene molar ratio in the products was 0.87. Both these parameters are characteristic of 12-ring large pore zeolites.

REFERENCES

1. Flannigen E.M., Khatami, H. and Szymanski, H.A., Advances in Chemistry Series, **101**, 201 (1971).
2. Kulkarni, S.B., Shiralkar, V.P., Kotasthane, A.N., Borade, R.B. and Ratnasamy, P., Zeolites **2**, 313 (1982).
3. Ruthven, D.M., Eic, M. and Richard, E., Zeolites **11**, 647 (1991).
4. Wu, P., Debebe, A. and Ma, Y.U., Zeolites **3**, 118 (1983).
5. Frillette V.J., Haag W.O. and Lago R.M., J. Catal. **67** 218 (1981).

Table 1. XRD pattern of NCL-1 (Si/Al>3000).

d(expt.) nm	I/I_o	hkl	d(calc.) nm
1.435	100	002	1.435
1.195	71	100	1.195
1.098	46	101	1.102
0.802	8	110	0.808
0.599	8	014	0.599
0.479	9	006	0.478
0.430	sh	---	---
0.418	85	020	0.418
0.396	56	300	0.398
0.356	18	008	0.358
0.340	vw	---	---
0.287	8	0010	0.287

Table 2. Adsorption (wt%) in zeolites 298 K, $p/p_0 = 0.5$.

	n-hexane	Cyclohexane	Mesitylene
Critical dia., (nm)	0.49	0.61	0.78
NCL-1	7.00	6.10	4.80
Small pore			
Linde 5A	14.90	0.80	0.40
Medium pore			
ZSM-5	13.50	6.80	0.50
ZSM-23	8.50	4.30	1.00
ZSM-48	5.80	3.20	<0.10
Large pore			
Zeolite Y	20.90	21.80	24.00
Mordenite	11.40	11.70	7.20
ZSM-12	8.30	10.30	2.60
Beta	15.90	18.30	17.00

Table 3. Characterization of NCL-1 (Si/Al=20) by catalytic test reactions.

	ZSM-5	ZSM-12	MOR	NCL-1	BETA
1. m-xylene conversion[a]					
p-xylene/o-xylene[b]	1.80	1.10	1.00	1.00	1.00
Isomer./disproport.[c]	58.00	40.00	20.00	15.00	8.50
1,3,5/1,2,4 TMB ratio[d]	---	0.02	0.06	0.11	0.20
2. Constraint index	7.00	3.00	1.20	1.00	0.60

a: The reactions were carried out in a down flow integral reactor at atmspheric pressure b-d: Values are at extrapolated zero m-xylene conversion, obtained by varying temperature and WHSV; c: Molar ratio, (para + ortho) xylenes/trimethyl benzene;

Table 4. m-Xylene isomerisation on NCL-1 (Si/Al = 20)
pressure = atmospheric; LHSV = 4 ml/hr; H_2/Oil = 1.5 moles

Temperature (K)	473	493	518
Conversion (%)	4.88	6.63	12.89
Products (wt%)			
Benzene	0.01	0.01	0.01
Toluene	0.28	0.45	0.89
p-xylene	1.95	2.56	4.75
m-xylene	95.12	93.37	87.11
o-xylene	1.77	2.35	4.44
1,3,5 TMB	0.10	0.21	0.57
1,2,4 TMB	0.43	0.68	1.68
1,2,3 TMB	0.02	0.16	0.27
para/ortho	1.10	1.09	1.07
1,3,5,/1,2,4 TMB	0.23	0.31	0.34
D/I*	0.14	0.19	0.24

* Disproportionation/Isomerisation

Figure 1. XRD pattern of NCL-1 zeolites

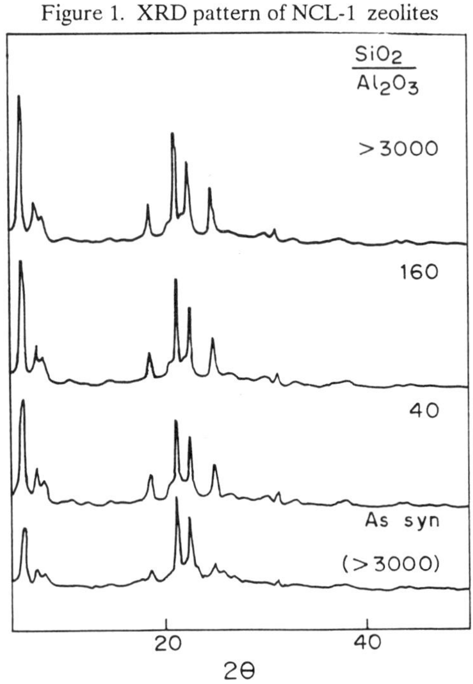

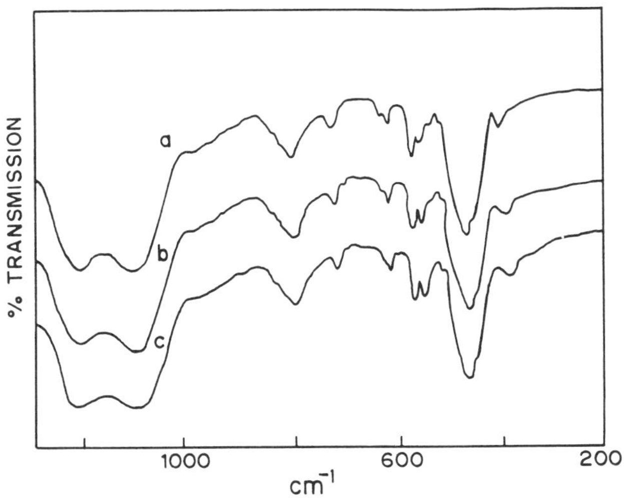

Figure 2. Framework IR spectra. Curves a-c refer to samples with Si/Al>3000, 80 and 20, respectively

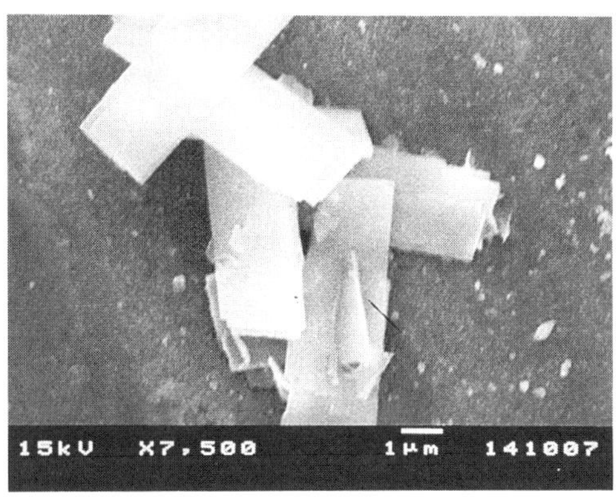

Figure 3. Scannining electron spectroscopy of NCL-1 (Si/Al >3000)

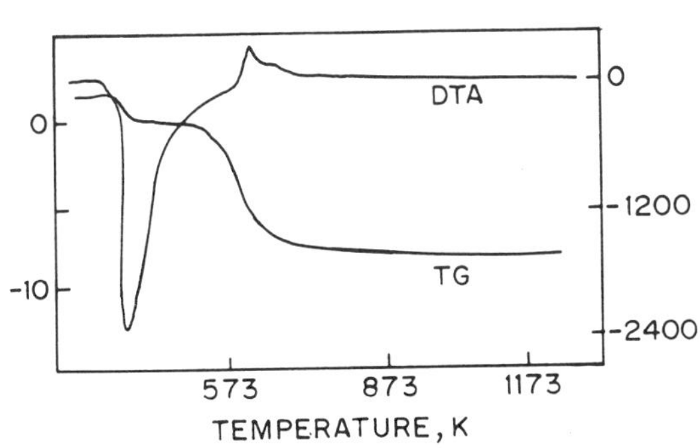

Figure 4. DTA and TG of NCL-1 (Si/Al>3000)

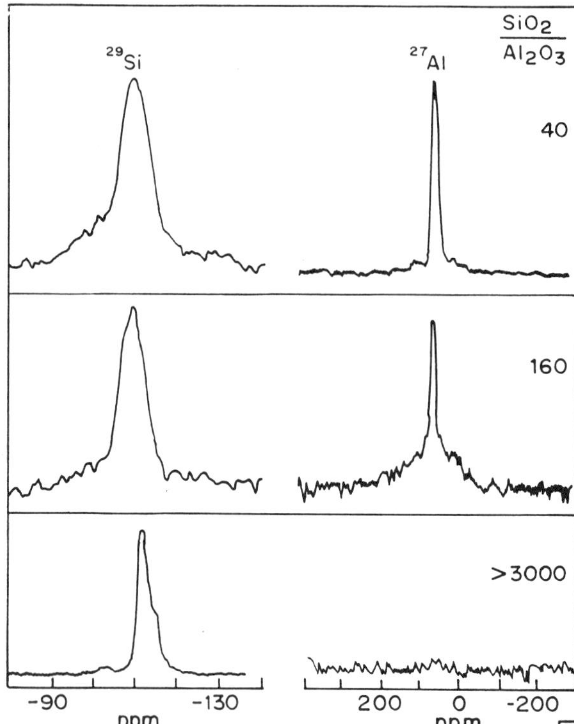

Figure 5. ^{29}Si and ^{27}Al MASNMR of NCL-1 zeolite.

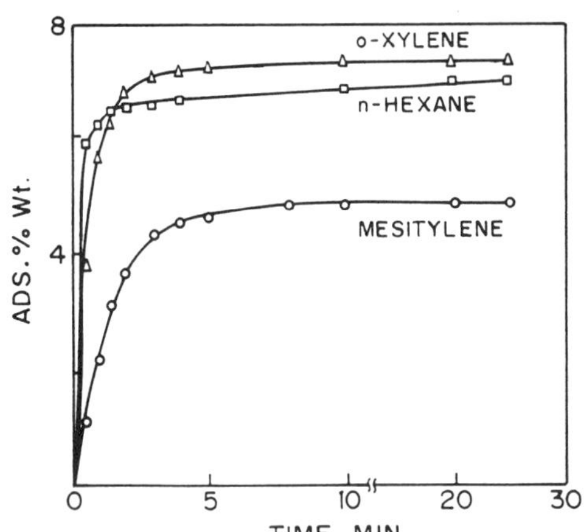

Figure 6. Adsorption isotherms in NCL-1 (Si/Al >3000)

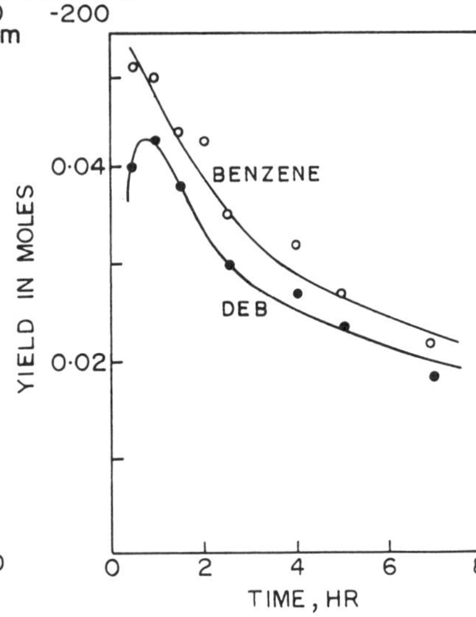

Figure 7. Ethylbenzene disproportionation over NCL-1 (Si/Al =20)

ZEOLITE SYNTHESIS IN AMMONIA AND AQUEOUS AMMONIA SOLVENTS: REACTIVITY OF GROUP IA CATIONS WITH PRE-MADE SILICA-ALUMINA GELS

D.E.W. VAUGHAN and K.G. STROHMAIER

Exxon Research and Engineering Company, Rt. 22 East, Annandale, NJ 08801, USA

ABSTRACT

The synthesis of zeolites using group IA cations in ammonia and aqueous ammonia solvents has been studied. Pressure effects dominate the products of these reactions, yielding dense zeolites or feldspathoids. Differences between aqueous and aqueous ammonia systems are minor except in the sodium system. The sodalite and ECR-5 cancrinite products are unusual in having Si/Al ratios greater than unity and no salt entrainment. In the cancrinites this produces new compositions with sorption properties commensurate with the 12-ring structure and not previously observed in conventionally synthesized materials containing entrained salts.

INTRODUCTION

Water plays a profound part in zeolite syntheses. In combination with alkali it solubilizes the reactant silica and alumina species usually forming a precursor hydrogel from which these species are transported to the growing zeolite crystallites. Water also allows the solubilization of structure directing inorganic cations and organic templates. The hydration sphere of these reactant species is important during the course of the crystallization reaction by effecting the size of the templating cation, the rate of dissolution of the gel and diffusion rates of reactants towards the growing crystallites.

Little work has been reported on zeolite synthesis in non-aqueous solvents. Bibby and Dale synthesized silica-sodalite and aluminosilicate-sodalites over a wide range of Si/Al ratios using ethylene glycol or propanol as a solvent[1]. Ethylene glycol, butyl alcohol, glycerol, amyl alcohol and hexyl alcohol were used in combination with sodium and tetraalkylammonium templates to synthesize the pentasil zeolites silicalite, ZSM-39 and ZSM-48[2]. A study of group IA and IIa cation effects on non-aqueous zeolite synthesis using organic solvents yielded only the dense phases hydroxysodalite, BaT and kaliophilite[3].

Ammonia is similar to water in that it is a dipolar, protonic solvent capable of acid-base chemistry. It has a dipole moment of 3.4×10^{-30} C•m compared to 4.9×10^{-30} C•m for the more polar water molecule. This results in ammonia being a poorer solvent than water for ionic substances, but a better solvent for more covalent compounds. At a pH greater than about 9, the equilibrium $NH_4^+ + OH^- \rightarrow NH_3 + H_2O$ totally favors gaseous ammonia dissolved in the base solution. Pure ammonia has a vapor pressure of over 60 atm. at 100°C compared to 1 atm. for pure

From the Proceedings of the 9th International Zeolite Conference, Montreal 1992, Eds. R. von Ballmoos et al.,

water. In the present paper we evaluate the effect of the less polar ammonia solvent molecules in forming precursor species and cation-solvent complexes as templates in zeolite synthesis. In transition metal systems (a subsequent study) the use of these solvents facilitates the evaluation of ammonia metal complexes at high pH on zeolite synthesis products. The results of the experiments to be described in this paper show that the main influence is that of high pressure on lower temperature zeolite syntheses.

EXPERIMENTAL

A series of silica-alumina gels having SiO_2/Al_2O_3 ratios of 2.7, 5.0, 6.2, 7.2, 7.6, 9.4, 13.6, 17.8, and 27.9 were made by vigorously mixing solutions of sodium silicate(P.Q. Corp. N-brand) diluted to 8% SiO_2, with 12% solutions of aluminium sulfate($Al_2(SO_4)_3 \cdot 17H_2O$) and then adjusting the final stoichiometry to 1.66 Na_2O: Al_2O_3: $xSiO_2$ with either 1<u>N</u> H_2SO_4 or 50% NaOH. This was followed by washing with 10% $(NH_4)_2SO_4$ to remove entrained and exchanged Na, drying at 125°C, and finally calcination at 350°C for 3 hours. The compositions and N_2 surface areas of these reactants are listed in Table 1. For comparison purposes meta-Kaolin(Si/Al = 1, Georgia Kaolin - Hydrite UF-calc. 750°C/3hrs. or Engelhard - Santinone #2), Hi-Alumina FCC(Si/Al = 2.6, W.R. Grace), Lo-Alumina FCC(Si/Al = 5.5, W.R. Grace), and silica gel(Grade 923, W.R. Grace) were also used as reactants, as they are known to be highly reactive in conventional zeolite syntheses.

Table 1
Starting gel compositions and final analyses for silica-alumina gels.

Gel	Starting Gel SiO_2:Al_2O_3	Final Analysis SiO_2	Al_2O_3	Surface Area m^2/g
G2.7*	2.7	40.4%	25.3%	
G5.0	2.0	42.7%	14.6%	134
G6.2	2.5	54.3%	12.4%	169
G7.2	3.0	43.0%	10.1%	
G7.6	3.0	49.6%	11.1%	184
G9.4	4.0	40.4%	7.3%	217
G18.6	6.0	47.2%	5.9%	260
G17.8	8.0	52.6%	5.0%	252
G27.9	16.0	45.4%	2.8%	241

* Used starting gel stoichiometry of 2.13 Na_2O: Al_2O_3: 2.7 SiO_2 and sodium aluminate as alumina source instead of aluminum sulfate.

Measured amounts of silica-alumina gel and IA hydroxides were added to 75 ml. stainless steel sampling cylinders(Hoke 6HD75) and then cooled with dry ice. Aqueous ammonia, or liquid ammonia from an inverted NH_3 cylinder, was added, with simultaneous weighing, and the sampling cylinders quickly capped. The cylinders were then mounted on a rotating shelf(12 RPM) in an air oven and reacted at 90°C. Cylinders were removed from the oven, cooled to room temperature, and then cooled with dry ice before un-capping. The samples were vacuum filtered, washed with distilled water and dried in an air oven at 110°C. Products were analyzed by powder X-ray diffraction (Siemens D-500 diffractometer; Cu target), ICP-AES for multi-element simultaneous chemical analysis (Jarrel Ash), TGA (duPont 1090), ^{29}Si MASNMR (JEOL FX200WB), and hexane sorption using an automated Cahn vacuum microbalance system.

RESULTS AND DISCUSSION

Ammonia Solvent Experiments

Liquid ammonia experiments were performed using Li, Na, and K cations with m-Kaolin, Lo-Alumina FCC, Hi-Alumina FCC, and silica gel having the stoichiometry:

$$2\ M_2O:Al_2O_3:x\ SiO_2(11.5 - 28.5\ NH_3:SiO_2;\ H_2O:SiO_2 < 1).$$

No crystalline product was recovered from these experiments after 8 days at 90°C. This is not surprising since high concentrations of alkali are needed to dissolve alumina and silica species and the solubility of the alkali metal hydroxides in liquid ammonia is very low. Further experiments were then restricted to the aqueous ammonia system.

Aqueous Ammonia - Lithium System

Lithium cation experiments were performed using m-Kaolin, Lo-Alumina FCC, Hi-Alumina FCC, and silica gel in reactions having the stoichiometry:

$$0.7 - 2.0\ Li_2O:Al_2O_3:x\ SiO_2(3.2 - 8.3\ NH_3:SiO_2;\ 8.0 - 20.5\ H_2O:SiO_2)$$

Results from the reaction of lithium hydroxide with silica-alumina gels in aqueous ammonia indicate that the presence of ammonia has little effect in the lithium system. Li-A(ABW) crystallized from m-Kaolin reactions and remained the stable phase even after 7 days at 90°C. Zeolite Linde F (EDI) was recovered from the Hi-Alumina FCC reaction and a phillipsite/merlinoite product was recovered from the reaction with Lo-Alumina FCC. No crystalline phase was seen in the silica gel experiments. Structure codes are reported by Meier and Olson[4].

Aqueous Ammonia - Sodium System

Sodium cation experiments were performed using m-Kaolin, Lo-Alumina FCC, Hi-Alumina FCC, silica gel, and silica-alumina gels in reactions having the stoichiometry:

$$2.0 - 15.3\ Na_2O:Al_2O_3:x\ SiO_2(3.2 - 29\ NH_3:SiO_2;\ 7.4 - 67\ H_2O:SiO_2)$$

Results from the reaction of sodium hydroxide with silica-alumina in aqueous ammonia are listed in Table 2 and indicate that the presence of ammonia has a considerable effect in the sodium system. Whereas sodalite and zeolite P(GIS) are the dominant phases at long crystallization times without ammonia, the combination of both ammonia and a high sodium hydroxide concentration

Table 2
Experimental results from sodium hydroxide-aqueous ammonia syntheses.

Na_2O: Al_2O_3	Na_2O conc.†	gel§	NH_3: SiO_2	H_2O: SiO_2	time (days)	product	Si/Al
0.03*	0.001	Silica Gel	22.3	0.4	8.0	amorph.	
2.04	0.040	m-Kaolin	7.1	17.5	7.0	CAN	1.02
2.11	0.039	m-Kaolin	13.4	14.5	12.0	CAN	0.97
2.11	0.039	m-Kaolin	20.0	7.8	12.0	CAN	0.94
2.04	0.020	m-Kaolin	14.2	34.0	0.9	SOD	
2.00	0.025	m-Kaolin	11.6	26.7	2.0	SOD+CAN	
3.00	0.025	m-Kaolin	17.3	40.1	2.0	SOD+CAN	
5.00	0.025	m-Kaolin	29.0	67.0	2.0	SOD+CAN	
2.00	0.025	G2.7	8.9	20.6	4.0	SOD	
3.00	0.025	G2.7	13.4	30.9	4.0	SOD+CAN	
5.00	0.025	G2.7	22.3	51.5	7.0	CAN	1.09
6.14	0.040	G5.0	9.3	21.9	5.0	CAN	1.15
6.14	0.040	G5.0	0.0	31.1	5.0	GIS	
6.17	0.040	G5.0	9.0	22.1	5.0	CAN	1.23
6.17	0.039	G5.0	15.1	16.4	12.0	CAN	1.94
6.17	0.040	G5.0	9.0	22.1	4.0	CAN	1.24
2.85	0.025	Hi-Al	6.7	15.8	2.0	GIS+CAN	
3.76	0.039	Hi-Al	0.0	18.9	5.0	GIS	
3.76	0.039	Hi-Al	5.6	13.4	5.0	CAN	1.17
4.73	0.025	Hi-Al	11.2	26.2	2.0	CAN	
7.23	0.040	G6.2	8.7	20.6	5.0	CAN	1.77
7.23	0.040	G6.2	0.0	29.3	5.0	SOD	
7.27	0.040	G6.2	8.5	20.9	5.0	CAN	1.25
8.81	0.040	G7.2	0.0	31.0	5.0	GIS	
8.81	0.040	G7.2	9.2	21.8	5.0	CAN	1.25
3.00	0.025	G7.6	4.8	11.0	20.0	amorph.	
5.00	0.030	G7.6	6.6	15.4	2.0	GIS	
8.09	0.040	G7.6	7.7	19.1	5.0	CAN	1.36
8.09	0.039	G7.6	19.5	7.6	12.0	CAN	1.96
8.09	0.039	G7.6	13.0	14.2	12.0	CAN	1.43
8.00	0.027	G9.4	9.6	22.2	14.0	GIS	
12.26	0.040	G9.4	9.8	23.1	5.0	CAN	1.29
12.32	0.040	G9.4	9.5	23.4	4.0	CAN	1.52
2.84	0.025	Lo-Al	3.0	7.2	5.0	GIS	
15.28	0.040	G13.6	8.1	20.0	4.0	ANA+SOD	
5.01	0.025	G17.8	3.4	7.8	5.0	amorph.	
17.94	0.040	G17.8	7.3	18.0	4.0	ANA+CAN	
14.96	0.051	G17.8	4.8	11.6	5.0	GIS+ANA	
14.95	0.043	G27.9	3.7	8.8	14.0	amorph.	

* $Na_2O:SiO_2$
† moles $Na_2O/(NH_3+H_2O)$
§ G + number = silica-alumina gel having $SiO_2:Al_2O_3$ ratio as indicated by number; Hi-Al = Hi-Alumina FCC; Lo-Al = Lo-Alumina FCC.

strongly promotes the crystallization of cancrinite. Silica-alumina gels in the range SiO_2/Al_2O_3 = 2 to 9.4 all produced good cancrinite samples. The silica-alumina gels with SiO_2/Al_2O_3 > 10 no longer form cancrinite; zeolites P, mordenite and analcite become the prominent phases. A minimum Na_2O concentration of 0.04 $Na_2O:(NH_3+H_2O)$ is required for a good cancrinite product. At a Na_2O concentration below 0.04 pure cancrinite is only occasionally seen and zeolites P and sodalite become the dominant products. Good cancrinite products were seen at $NH_3:H_2O$ molar ratios up to 3.

The use of ammonia in preparing cancrinite promotes the synthesis of high silica products. Using conventional synthesis methods, hydroxy-cancrinite has previously been prepared only at Si/Al = 1.0 using high temperatures[5,6] and can be synthesized at temperatures below 100°C only by using nitrate, chromate, and molybdate anions[7] and the mixed bases NaOH plus LiOH[6]. Li-Cs and Li-Tl cancrinites have also recently been synthesized at 200°C[8]. The use of ammonia allows higher silica reactants to be used at lower temperatures, resulting in high silica products as seen in Figure 1. Some of the highest ratio materials, Si/Al = 1.95, were synthesized at $NH_3:H_2O$ levels of 2 to 3. Representative S.E.M. photographs of aqueous ammonia cancrinite samples are compared to meta-Kaolin chromate cancrinite in Figure 2. In general the gel based ammonia cancrintite materials tend to have small crystal sizes(< 0.2μ), confirmed by distinctive X-ray diffraction peak broadening. This new high silica cancrinite is designated ECR-5[9] and it's ^{29}Si-MASNMR, shown in Figure 3, confirms the elemental analyses. The NMR spectra of Si/Al = 1.0 nitrate cancrinite shows the characteristic single peak at -85.7 ppm vs. TMS representing silicon surrounded by 4 lattice Al atoms (Si - 4Al+0Si). (However, this

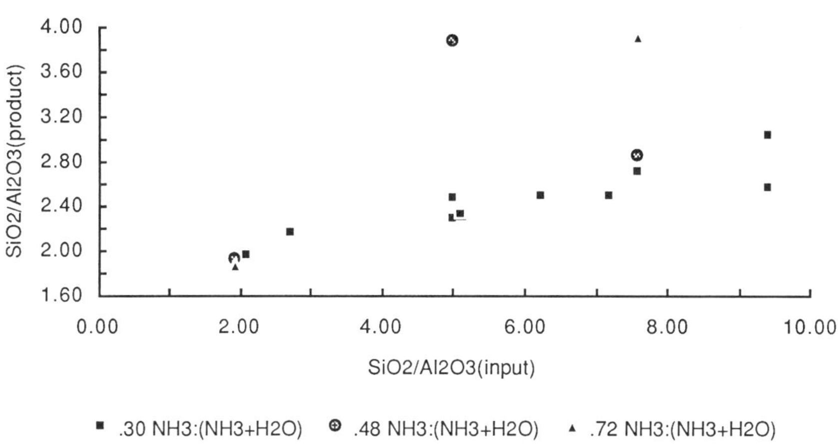

Figure 1. SiO_2/Al_2O_3 ratio of cancrinite products in the NaOH-aqueous ammonia system.

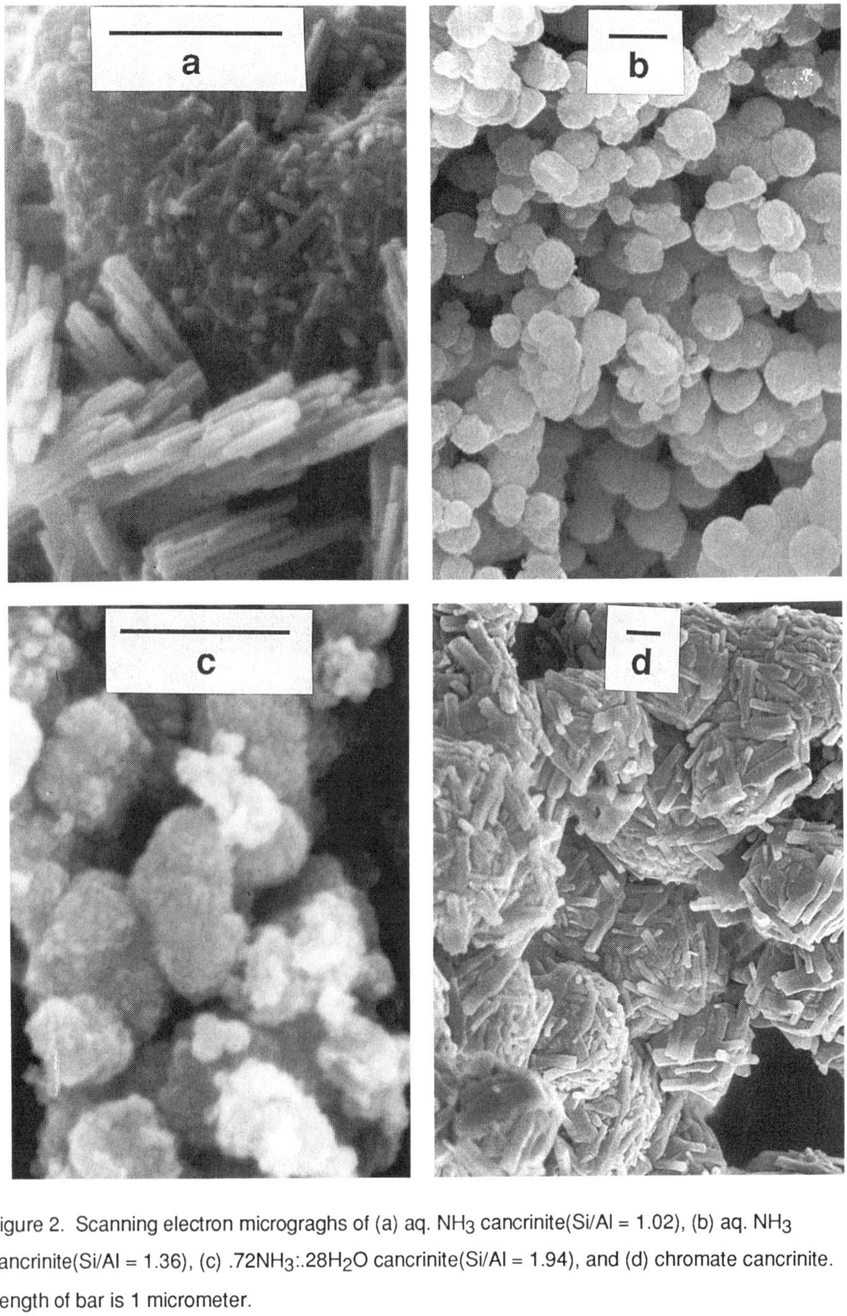

Figure 2. Scanning electron microgragh s of (a) aq. NH_3 cancrinite(Si/Al = 1.02), (b) aq. NH_3 cancrinite(Si/Al = 1.36), (c) $.72NH_3$:$.28H_2O$ cancrinite(Si/Al = 1.94), and (d) chromate cancrinite. Length of bar is 1 micrometer.

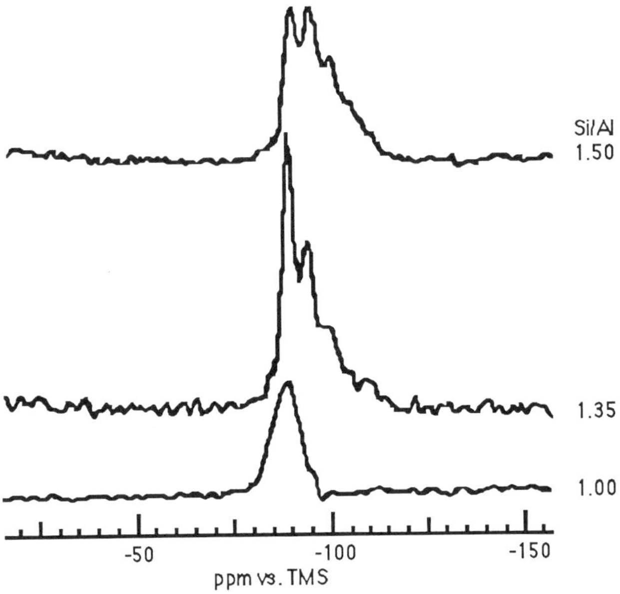

Figure 3. ^{29}Si NMR of aqueous ammonia cancrinite

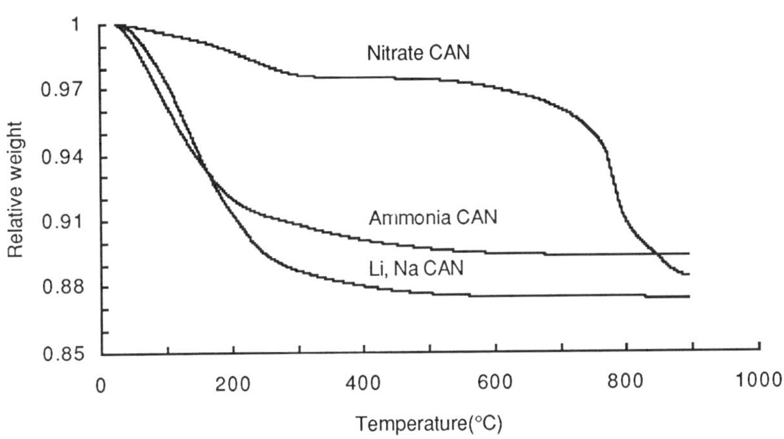

Figure 4. Thermogravimetric analysis of aqueous ammonia cancrinite compared to nitrate and lithium, sodium cancrinite.

chemical shift is sensitive to the trapped anion and is -86.8 ppm for chromate cancrinite and -87.5 ppm for sulfate cancrinite.) ECR-5 samples at Si/Al = 1.36 and 1.50 show NMR peaks at about -87.4, -92.3, -97.7 and -102.8 ppm vs. TMS, representing silicon surrounded by 4 Al atoms (Si-4Al+0Si),3 Al atoms (Si-3Al+1Si), two Al atoms (Si-2Al+2Si) and one Al atom (Si-1Al+3Si) respectively, confirming a higher silica framework composition. Thermogravimetric analysis (Figure 4) shows a much larger water absorption than nitrate cancrinite, but a lower absorption than lithium, sodium cancrinite. Both ammonia cancrinite and lithium-sodium cancrinite absorb about 3.5 wt. % n-hexane at 50 torr, compared to less than 1% for conventional hydroxy-cancrinite.

Aqueous Ammonia - Potassium System

Potassium cation experiments were performed using silica-alumina gels in reactions having the stoichiometry:

$$2.0 - 10\ K_2O{:}Al_2O_3{:}xSiO_2(3.2 - 22\ NH_3{:}SiO_2;\ 7.9 - 51.9\ H_2O{:}SiO_2)$$

Results from the reaction of potassium hydroxide with silica-alumina in aqueous ammonia are listed in Table 3 and indicate that the presence of ammonia has a little effect in the potassium system. In low silica gels the product was zeolite F. In higher silica reactions a merlinoite/phillipsite phase was the major product.

Table 3
Experimental results from potassium hydroxide-aqueous ammonia syntheses.

K_2O: Al_2O_3	gel§	NH_3: SiO_2	H_2O: SiO_2	time (days)	product
.03*	silica gel	3.2	7.9	8	amoprh.
2.0	m-Kaolin	14.2	34.9	8	EDI
2.0	m-Kaolin	7.1	18.4	2	EDI
2.0	G2.7	8.8	20.7	15	amorph.
3.0	G2.7	13.1	31.1	7	GIS
5.0	G2.7	22.0	51.9	4	EDI
2.9	Hi-Al	6.7	15.1	8	PHI/MER
4.7	Hi-Al	16.8	39.2	6	PHI/MER
5.0	G7.6	11.9	27.6	15	PHI/MER
3.0	G9.4	3.8	9.0	20	amorph.
2.7	Lo-Al	3.0	7.4	5+	MER
4.7	Lo-Al	5.0	12.1	8	amorph.
10.0	G17.8	5.0	12.0	14	amorph.
10.0	G27.9	3.2	7.6	14	amorph.

* $K_2O{:}SiO_2$

§ G + number = silica-alumina gel having $SiO_2{:}Al_2O_3$ ratio as indicated by number; Hi-Al = Hi-Alumina FCC; Lo-Al = Lo-Alumina FCC.

\+ Temp. = 100°C

Aqueous Ammonia - Cesium System

Cesium cation experiments were performed using silica-alumina gels in reactions having the stoichiometry:

$$2 - 5\ Cs_2O{:}Al_2O_3{:}x\ SiO_2(2.5 - 28.4\ NH_3{:}SiO_2;\ 6 - 67.6\ H_2O{:}SiO_2)$$

Results from the reaction of cesium hydroxide with silica-alumina in aqueous ammonia are listed in Table 4. Only two products were made, cesium aluminum silicate from m-Kaolin, and pollucite, the cesium analogue of ANA, from a gel having $SiO_2/Al_2O_3 = 7.6$.

Table 4
Experimental results from cesium hydroxide-aqueous ammonia syntheses.

Cs_2O: Al_2O_3	gel§	NH_3: SiO_2	H_2O: SiO_2	time (days)	product
2.0	m-Kaolin	11.4	27.1	2	$CsAlSiO_4$
5.0	m-Kaolin	28.4	67.6	2	$CsAlSiO_4$
2.8	Hi-Al	6.6	16.0	15	amorph.
4.7	Hi-Al	11.0	26.5	15	amorph.
2.0	G7.6	3.1	7.4	15	ANA
5.0	G7.6	7.8	18.6	5	ANA
3.0	G9.4	3.8	9.0	15	amorph.
5.0	G9.4	6.3	15.0	15	amorph.

§ G + number = silica-alumina gel having $SiO_2{:}Al_2O_3$ ratio as indicated by number; Hi-Al = Hi-Alumina FCC.

CONCLUSIONS

The synthesis of zeolites in aqueous ammonia and liquid ammonia using dried silica-alumina gels in the presences of group IA cations produced zeolites characteristic of high pressure/ temperature reactions. More open framework zeolites such as LTA, FAU, LTL and OFF were not observed. Only the highest silica containing gels (silica gel itself and Si/Al=27.8) were unreactive in any of the compositions studied. Reactions in liquid ammonia yielded no crystalline products. Apart from the differences induced by higher pressures, observed differences between aqueous and aqueous-ammonia syntheses were minor. The exception was that in the sodium system new higher silica cancrinites were synthesized and designated ECR-5. Characterization of ECR-5 by X-ray diffraction and ^{29}Si-MASNMR confirmed the high silica /alumina ratio, and n-hexane absorption and thermogravimetric analysis demonstrated the increased accessibility and capacity of this 12-ring framework compared to conventional cancrinites.

REFERENCES

1. D.M. Bibby and M.P. Dale, Nature **317**, 157(1985).
2. H. Qisheng, F. Shouhua and X. Ruren, in: *Zeolites: Facts, Figures, Future*, Eds. P.A. Jacobs, and R.A. van Santen, Elsevier, Amsterdam, pp. 291-298 (1989).
3. W.A. van Erp, H.W. Kouwenhoven and J.M. Nanne, Zeolites **7**, 286(1987).
4. W.M. Meier and D.H. Olson, *Atlas of Zeolite Structure Types*, Butterworths/Intlernational Zeolite Assoc.(1987).
5. R.M. Barrer and E.A.D. White, J. Chem. Soc., 1561(1952).
6. R.M. Barrer and D.E. Mainwaring, J. Chem. Soc. Dalton, 2534(1972).
7. R.M. Barrer and J.F. Cole, J. Chem. Soc. (A), 1516(1970).
8. C. Colella and M. De' Gennaro, in: *Zeolite Synthesis*, A.C.S Symp. Ser. 398, Eds. M.L. Ocelli and H.E. Robson, A.C.S. Washington, DC, 196(1989).
9. D.E.W. Vaughan, U.S. Patent 4,717,560(1988).

STUDY ON THE INSERTION OF VANADIUM IN THE SILICALITE FRAMEWORK

G. BELLUSSI, G. MADDINELLI, A. CARATI, A. GERVASINI and R. MILLINI

Eniricerche S.p.A., 20097 S. Donato M.se, ITALY

ABSTRACT

The presence of Na+ cations plays and important role in the synthesis of V-containing silicalite. It affects the state of the vanadium contained in the crystalline solid and its stability to the thermal treatments. Characterization of V-containing silicalites, synthesized both in the presence and in the absence of Na+ cations, by XRD, ^{51}V-MAS-NMR, ^{29}Si-MAS-NMR, ESR and FT-IR techniques, indicates that the tetracoordinated vanadium is present in both cases in the precursors but only a part of it remains in the same coordination after calcination only in samples prepared in the absence of sodium.

INTRODUCTION

Because of the importance of supported vanadium oxides as selective oxidation catalysts, several attempts have been made with the aim of inserting vanadium in the zeolite framework [1-6]. The vanadium modified zeolite catalysts described in the literature are mainly characterized on the basis of their catalytic activity, while few informations are available about the real situation of vanadium and its transformation upon thermal treatments. Moreover, since vanadium can readily go through different oxidation states, its interaction with silicalite framework must be strongly affected by the preparation method and by the subsequent treatments.

The aim of our study is to investigate the situation of vanadium in vanadium containing silicalites as a function of the synthesis procedures.

EXPERIMENTAL

Synthesis

V-containing silicalites were prepared following two different synthesis procedures.

(A) type synthesis: colloidal SiO_2 (Ludox HS-40, DuPont), aqueous tetrapropylammonium hydroxide (TPA-OH, 30% wt., EniChemSynthesis) and a solution of sodium hydroxide (Carlo Erba, RLE) in water were successively added to an aqueous solution of VCl_3 (Fluka, purum) under vigorous stirring. The crystallization of the resultant gel was performed in a stainless steel shaked autoclave at 170 °C for 6

From the Proceedings of the 9th International Zeolite Conference, Montreal 1992, Eds. R. von Ballmoos et al.,

days under autogeneous pressure.

(B) type synthesis: an aqueous solution of VCl_3 and tetraethylsilicate (Dynasil A, Dynamit Nobel) were added successively to aqueous TPA-OH under vigorous stirring. The resultant slurry was homogenized for two hours with an high speed stirrer. Crystallization occurred in a stainless steel shaked autoclave during 15 hours at 170 °C under autogeneous pressure. After cooling, in both cases the solids were separated from the mother liquors, washed and dried at 120 °C. The dried materials (here inbelow named precursors) were then calcined at 550 °C in air for 5 hours, in order to destroy the trapped organic cations. Mild treatments with aqueous ammonium acetate at 70 °C were performed in order to wash away unstable vanadium species. Samples were successively dried and calcined at 550 °C. The reactant mixture and products composition for significant tests are reported in Table 1.

Physico-chemical characterization

X-Ray powder diffraction patterns (XRD) were recorded on a Philips diffractometer equipped with a pulse-height analyzer; the CuK_α radiation was used ($\lambda = 1.54178$ A). The infrared spectra were recorded on a Perkin-Elmer 1730 FT-IR using the KBr wafer technique (2% wt. concentration). ESR measurements were performed on ESR/Varian spectrometer working at X-band (9000 MHz) at 100 KHz modulation, both at room temperature and at liquid nitrogen temperature. The resonance of DPPH (g=2.0036) was used as standard. Solid state NMR studies were carried out on a CXP-300 Bruker spectrometer using an Oxford Instrument 7.046 T, 89 mm bore superconducting magnet. Samples were rotated within Delrin spinners at a speed between 4 and 5 KHz. ^{51}V-MAS-NMR spectra were obtained at 78.864 MHz using a 0.5 s repetition time and a 125 KHz spectral width. Chemical shifts were referred to external liquid $VOCl_3$. The centerbands were attributed by changing the spinning rate. ^{29}Si-MAS-NMR spectra were recorded at 59.595 MHz and chemical shifts were referred to that of TMS.

RESULTS AND DISCUSSION

In the (A) type synthesis the amount of sodium hydroxide added to the reaction mixture plays an important role on the properties of the final product. For lower NaOH content (see preparations A-1 and A-2, Table 1) a MFI type silicalite is formed which maintains the same structure after calcination in air. Only traces of vanadium are present in the product. Increasing the amount of NaOH in the synthesis gel (see preparation A-3 , A-4 and A-5, Table 1) mixtures of silicalite and amorphous material or an unidentified crystalline phase (hereinbelow named phase X) are obtained. The calcination in air at 550°C of the samples gives rise to the formation of cristobalite. If the same treatment is performed in hydrogen atmosphere, the silicalite structure i s mantained. Figure 1 reports the typical XRD powder diffraction spectra of such a mixture (sample A-4, Table 1). Figure 2 shows the infrared spectra of the same sample dried at 120 °C (a) or calcined at 550 °C in air (b) or in H_2 atmosphere (c). In the IR spectrum of the precursor, beside the typical bands of MFI structure type [7], a new band is present located at 869 cm^{-1} (Figure 2a); the latter is probably due to the unidentified crystalline phase detected by the XRD analysis. In fact a linear correlation is found between the intensity ratio of the diffraction peak at $2\vartheta = 23.15 \pm 0.05$ (reflection 5 0 1 of Silicalite-1) and $2\vartheta = 24.76 \pm 0.05$ (strongest reflection of phase-X) and the area ratios of the 869 cm^{-1} (phase-X) and 796 cm^{-1} (Silicalite-1) IR bands. The factor affecting the formation of the phase-X seems to be the Na+/ TPA-OH molar ratio, nevertheless a pure phase-X was never obtained. The air calcined sample (b) shows the vibration pattern characteristic of cristobalite, while no differences exist between the precursor (a) and the sample calcined in the H_2 atmosphere (c).

Table 1. Reaction mixture compositions and product analyses for several synthesis

Test n.	Composition of starting gel (1)				Product Analyses: XRD		Product Analyses: Composition	
	V_2O_3	TPA+	Na+	H_2O	Precursor (2)	Calcined in air	V_2O_3	Na+
A-1	0.033	0.25	0.15	35	S-1	S-1	(3)	-
A-2	0.025	0.40	0.15	35	S-1 + A	S-1	(3)	-
A-3	0.025	0.40	0.20	35	S-1 + X	C	0.015	0.35
A-4	0.025	0.60	0.30	35	S-1 + X	C	0.019	0.85
A-5	0.016	0.40	0.30	35	S-1 + A	C	0.018	1.10
B-1	0.025	0.50	-	35	S-1	S-1	0.011	-
B-2	0.025	0.70	-	35	S-1	S-1	0.009	-
B-3	0.025	1.20	-	35	S-1	S-1	(3)	-
B-4	0.016	0.50	-	35	S-1	S-1	0.010	-
B-5	0.010	0.50	-	35	S-1	S-1	(3)	-

(1) Moles referred to 1 mole of silica.
(2) S-1= Silicalite-1; A= amorphous; X= phase-X; C= Cristobalite
(3) Vanadium present in traces.

The transformation to cristobalite upon calcination in air seems not to be strictly related to the presence of the phase-X since it occurs also in samples in which this phase is not present (preparation A-3, Table 1). The ESR spectrum of the precursor obtained from the synthesis type (A) shows a very intense signal with hyperfine structure both at room temperature and at liquid nitrogen temperature, due to the nuclear magnetic moment of vanadium. The spin Hamiltonian parameters of the spectra, calculated following the method of Hecht and Johnston [8] are reported in Table 2. The values of the g tensors lead to the conclusion that the V^{4+} species is in fourfold symmetry, either D_{4h} or C_{4v}, in tetragonal compression [9]. The well resolved hyperfine splitting indicates that the V^{4+} species are atomically dispersed in the silicalite framework [9a]. The ESR spectra of the air calcined sample shows a very broad unstructured signal. The air thermal treatment probably causes the oxidation of most of vanadium to diamagnetic V^{5+}. Calcination in hydrogen atmosphere leads to strong modification of the ESR spectra of the precursor. An unstructured singlet line centered at g_o=1.96 with lorentzian shape appears. The line width, 80 G, is much smaller than the one of the precursor (600 G). According to Miyamoto et al. [2] a singlet line derives from agglomerated paramagnetic V centers in dipole- dipole interaction. The behaviour suggests that calcination in H_2 preserves part of vanadium in the V^{4+} form but causes its agglomeration.

Table 2. Magnetic parameters for samples analyzed by ESR.

Test n.	Thermal treat.	$A_{//}$ (G)	$A_{\perp}$ (G)	$g_{//}$	$g_{\perp}$	$H_{//}$(G) $m_I=0$	$H_{\perp}$(G) $m_I=0$	(g_o)	(A_o)
A-4	Dried at 120 °C	191	89	1.93	1.97	3344	3274	1.96	123
B-4	Dried at 120° C	191	55	1.95	2.00	3330	3243	1.98	100
B-4	Calcined in air	205	55	1.95	2.00	3339	3248	1.98	105

^{51}V-MAS-NMR analysis permits to investigate about the presence of vanadium species in the 3+ or 5+ oxidation states. According to previous studies [10] it is possible to obtain information about the symmetry of the vanadium species under investigation by comparing static ^{51}V-NMR spectra with those of model compound such as V_2O_5 or Na_3VO_4. We extended this approach to ^{51}V-MAS-NMR technique. The vanadium nucleus in V_2O_5 is known to be in a distorted octahedral symmetry while Na_3VO_4 is a model compound for vanadium in a distorted tetrahedral symmetry [11].
The ^{51}V-MAS-NMR spectrum of the above mentioned compounds are reported in Figures 3a and 3d respectively. The spectrum of the A-4 sample precursor (Figure 3b) is different from both the model spectra since it is constitued only by a singlet resonance at -570 ppm, the same chemical shift region of the tetrahedral model compound [11], with a linewidth of 2700 Hz. In Figure 3e the static spectrum of the same sample is reported. The line broadening with respect to the previous one is moderate, due to a small anisotropic effect; this indicate the presence of a quite symmetric system. The samples calcined at 550 ° C in air or in H_2 atmospheres show a ^{51}V-MAS- NMR spectrum completely different with respect to the previous. Shape and isotropical chemical shift are similar to that of the model

compound Na_3VO_4 (Figure 3f). This behaviour suggests that the vanadium species present in A-4 precursor is assigned to V^{5+} species.

By the (B) type synthesis it is possible to obtain silicalites containing significant amount of vanadium depending on the molar composition of the reactant mixture (See Table 1). In our experiments, the maximum amount of vanadium detected in the crystalline solids corresponds to 0.01 moles of V_2O_3 per silica mole. All the products obtained from the (B) type synthesis reported in Table 1 are characterized by an XRD powder pattern typical of pure MFI structure type. No structural collapses or damages are detected upon calcination in air at 550 °C. In fact, XRD patterns of air calcined samples B-1, B-2 and B-4 (see Table 1) show the orthorhombic MFI type structure. FT-IR spectra of the above mentioned samples were similar to that of pure silicalite. In some preparation a low intensity band at 970 cm $^{-1}$ was observed which disappeared following washing with aqueous ammonium acetate. This treatment was performed with the aim a of verifying the stability of vanadium contained in the silicalite. The washing with ammonium acetate solution removes most of vanadium; for instance in sample B-2 the SiO_2/V_2O_3 molar ratio was 110 in the calcined form, 439 after 2 washings with fresh ammonium acetate solution, 536 after 4 and 760 after 6 treatments. As the vanadium content decreased a transition toward the monoclinic symmetry was observed. The ESR spectra of B-4 precursor (Figure 4a) and of the air calcined sample (Figure 4b) show paramagnetic signals of V^{4+} species well resolved in their eightfold components (Table 2). The values of the g tensors are similar to the ones of the A-4 precursor sample, a strong decrease in $A_\perp$ parameter is observed indicating more covalency in the axial component. It is worthnotice that the V^{4+} species remain atomically dispersed in the silicalite framework also after calcination in air.

The ^{51}V-MAS-NMR analysis of the B-4 precursor do not show any lines due to the presence of V^{3+} or V^{5+}, while the calcined sample has a pattern and an isotropical chemical shift similar to that of the model compound V_2O_5 (Figure 3c). In order to collect informations about the environment of the silicon atoms we performed the ^{29}Si-MAS-NMR analysis on A-4 and B-4 samples before and after calcination in air. Beside the typical peak at -112 ppm another signal at -102 ppm was present in all the samples with the exception of calcined A-4. A signal at -102 ppm could be due to the presence of SiOH groups in defect centers within the silicalite structure as well as to silicon atoms bound to unlike nuclei by analogy with spectra reported for silicalites containing other nuclei (i.e. aluminium [12]). On the other hand, ^{29}Si Cross Polarization MAS-NMR experiments performed on samples with different SiO_2/V_2O_3 ratios do not provide sufficient informations to discriminate in favour of one of the two hypothesis.

From the results reported above we drew the major points summarized in the following.

For samples obtained through (A) type synthesis and from reaction mixtures having a high NaOH content (Na+/V203 molar ratio >2) vanadium is present in the precursor as V^{4+} and V^{5+} both in fourfold coordination. The calcination in air, performed to remove the organic cations, produces the oxidation of all the V^{4+} species to V^{5+} and the collapse of the structure to the final form of cristobalite. There are not enough informations to discriminate whether there is a consequencial dependence between the two facts. Zhdanov et al. [13] observed the collapse of a silicalite synthesized in the presence of high amount of alkali cations. In our case both the oxidation of all the V^{4+} to V^{5+} and the presence of alkali cations could be responsible for the transformation to cristobalite. The presence of structural vanadium can be excluded also in the product calcined in the H_2 atmosphere although the silicalite structure is maintained. In fact the vanadium present as V^{5+} is in a situation similar to that of the model compound Na_3VO_4 and the remaining V^{4+} is in an aggregated form as indicated from the absence of fine structure of the ESR spectrum. In samples synthesized in the presence of a very low content or in the absence of sodium, vanadium is present in the precursor as V^{4+} atomically dispersed. During the calcination in air a part of vanadium is oxidized to V^{5+} and it aggregates to form a V_2O_5 type species. The part that remains as V^{4+} is still atomically dispersed. The stability of vanadium in our samples is very low since it can be removed almost completely by mild treatments such as washing with aqueous solution of ammonium acetate.

CONCLUSIONS

It is possible to prepare vanadium containing silicalites by direct synthesis, but the amount and the state of vanadium in the final product strongly depends on the synthesis procedure. Especially the presence of sodium in the reaction mixture plays a not negligible role. The presence of vanadium in framework position is uncertain although indication about this possibility exists for the as synthesized samples coming either from (A) type or from (B) type synthesis. After calcination at 550 °C, only for samples prepared through the (B) type procedure there are still indications about the possibility that vanadium atoms could be in framework sites or at least bound to silicon atoms in defective centers. In these compounds a significant amount of vanadium is weakly bound to the structure since it can be removed by mild treatments.

REFERENCES

[1] Inui T., Medhanavyn D., Praserdham P., Fukuda K., Ukava T., Sakamoto A. and Miyamoto A., Appl. Catal. 18, 311 (1985).
[2] Miyamoto A., Medhanavyn D. and Inui T., Appl. Catal. 28, 89 (1986).
[3] Centi G., Tvaruzkova Z., Trifirò F., Jiru P., Kubelkova L., Appl. Catal. 13, 69 (1984).
[4] Cavani F., Trifirò F., Jiru P., Habersberger K., Tvaruzkova Z., Zeolites 8, 12 (1988).
[5] Zatorsky L.W., Centi G., Lopez-Nieto J., Trifirò F., Bellussi G., Fattore V., Stud. Surf. Sci. Catal., Eds. P.A. Jacobs and VanSanten R.A., 49B, 1243 (1989).
[6] Rigutto M.S., VanBekkum H., Appl. Catal. 68, L1 (1991).
[7] Jacobs P.A., Beyer H.K., Valyon J., Zeolites 1, 161 (1981).
[8] Hecht H.G., Johnston T.S., J. Chem. Phys. 46, 23 (1967).
[9] a) Takahashi H., Shiotani M., Kobayashi H., Sohma J., J. Catal. 14, 134 (1969); b) Kucherov A.V., Slinkin A.A., Zeolites **7**, 583 (1987); c) Sharma V.K., Wokaun A., Baiker A., J. Phys. Chem.

90, 2715 (1986).

[10] a) Zamaraev K., Mastikhin V.M., Colloids Surf. 12, 401 (1984) ; b) Taouk B., Guelton M., Grimblot J., Bonnelle J.P., J. Phys. Chem. 92, 6700 (1988); c) Chary K.V.R., Venkt Rao V., Mastikhin M., J. Chem. Soc. Chem. Comm., 92 (1989); ibid. 202; d) Oldfield E., Kinsey R.A., Montez B., Ray T., Smith K.A., J. Chem. Soc. Chem. Comm. 254 (1982).

[11] LeCoustomer L.R., Tauk B., LeMeur M., Payen E., Guelton M., Grimblot J., J. Phys. Chem. 92, 1230 (1988).

[12] Engelhardt G., Michel D." High-resolution Solis State NMR of Silicates and Zeolites", John Wiley & Sons, 147-154 (1987).

[13] Zhdanov S.P., Feoktistova N.N, Titova T.I., Kosheleva L.S., Peuker Ch., Loffler E., Pilz W., Ad.Coll. Int. Sc. 25, 203 (1986).

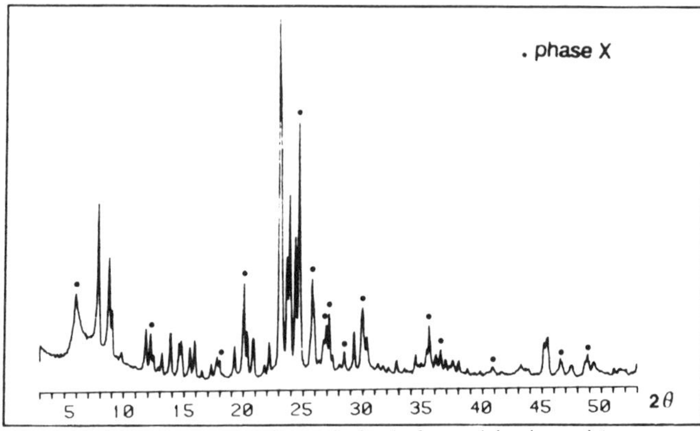

Figure 1. X-Ray spectrum of sample A-4 before calcination: mixture of Silicalite-1 and phase X.

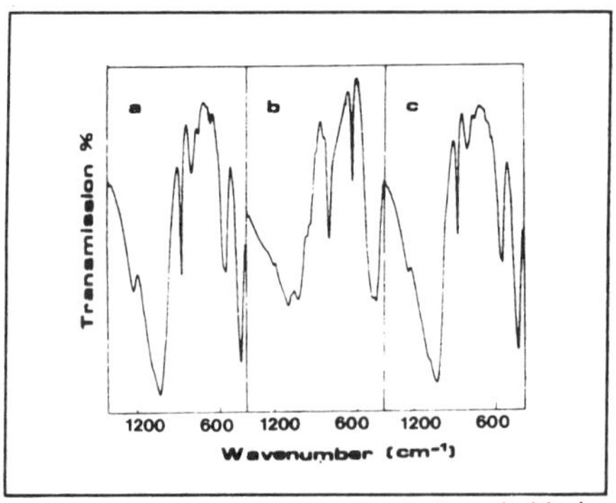

Figure 2. FT-IR spectra of sample A-4: a) dried, b) calcined in air, c) calcined in H_2.

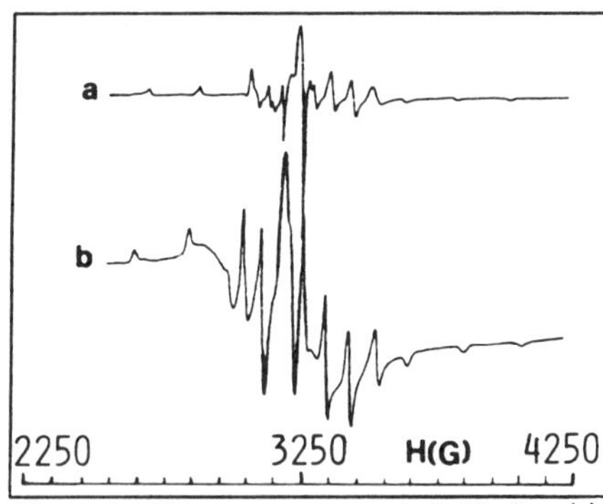

Figure 4. ESR spectra of sample B-4 before (a) and after (b) calcination in air.

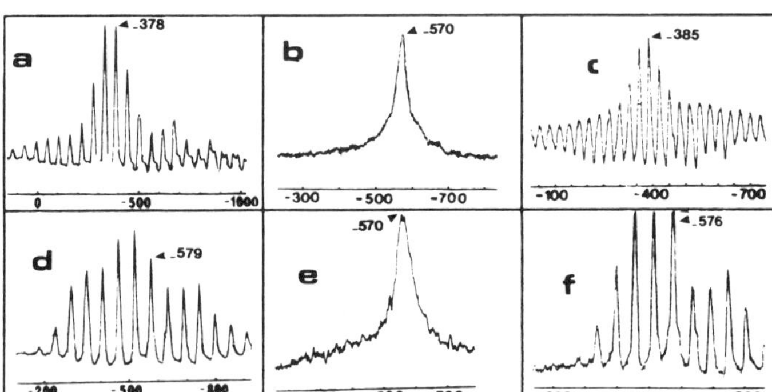

Figure 3. ^{51}V-MAS-NMR spectra of: a) V_2O_5, d) Na_3VO_4, b) sample A-4 dried, e)same as b)taken in static mode, c)sample B-4 and f)sample A-4 both after calcination in air.

F^- : A MULTIFUNCTIONAL TOOL FOR MICROPOROUS SOLIDS a) MINERALIZING, STRUCTURE DIRECTING AND TEMPLATING EFFECTS IN THE SYNTHESIS

J.L. GUTH, H. KESSLER, P. CAULLET, J. HAZM, A. MERROUCHE and J. PATARIN

Laboratoire de Matériaux Minéraux (URA-CNRS 428) ENSCMu-UHA,
3, rue Alfred Werner, 68093 MULHOUSE Cedex (FRANCE)

ABSTRACT

Three effects may be observed if fluoride anions are present in synthesis mixtures of microporous solids. The first one is a mineralizing (or mobilizing) effect : F^- allows crystallizing from a gel through a dissolution process. This effect applies especially for silica-based materials which thus may be prepared in media with pH lower than 10-11. The second one is a structure directing effect which means that its presence is necessary to obtain a material with a given structure and/or composition from a given reaction mixture. By this effect three entirely novel materials could be prepared (the silica form of the AST structure type, LTA-$GaPO_4$ and CLO-$GaPO_4$). The third one is a templating effect : F^- is incorporated into the framework interstices of the material and contributes strongly to the stabilization of the structure as do the organic template molecules. This effect is chiefly observed when the framework has no charges or if its charges are autocompensated.

INTRODUCTION

The mineralizing effect of fluoride anions in hydrothermal synthesis was known by the ancient mineralogists and chemists [1] but until the end of the seventies it was little studied in molecular sieves synthesis. The first clear example was the crystallization of silicalite I in the presence of F^- by Flanigen and Patton [2]. Later the fluoride route was extensively investigated in our Laboratory for silica-based zeolites and $AlPO_4$- or $GaPO_4$-based materials [3-6]. It was observed, that beside its mineralizing role, F^- may show a structure directing and a templating effect. The aim of this paper is to give a general overview of the synthesis results obtained in the presence of fluoride anions, emphasizing especially the two last effects which allowed the formation of new materials.

We define for F^- the three above-mentioned effects as following :

- F^- is a mineralizer(or mobilizer) if it makes the formation of a more stable solid phase from a less stable solid phase possible through a dissolution-crystallization process. Thus it allows the formation of a solution, generally from a gel, supersaturated in polycondensable species. It also often increases the dissolution and polycondensation rates. Like a catalyst, the mineralizer is normally not consumed during the transformation.

From the Proceedings of the 9th International Zeolite Conference, Montreal 1992, Eds. R. von Ballmoos et al.,

- F^- is a structure directing agent if its presence in the reaction mixture was found to be necessary to form a given material, but this does not necessarily imply its incorporation into the formed material.
- F^- is a template when it is incorporated into the formed material.

EXPERIMENTAL PART

Synthesis

All syntheses were performed from aqueous gels (pH = 4-11 for silica-based materials, pH = 3-9 for $AlPO_4$, $GaPO_4$ and derived materials) under hydrothermal conditions in PTFE-lined autoclaves using F^- as a mineralizer.

For silica-based mixtures the fluoride source was chosen according to the nature of the organic template (e.g., HF with an amine or a tetraalkylammonium hydroxide, NH_4F or NaF with alkylammonium salts). NaF was the preferred source in the absence of organic template. Pure silica sources (e.g., Aerosil from Degussa) or silica combined with the other framework elements (e.g., aluminosilicate gels) were used. When not combined with silica, the other framework elements were added as soluble salts (e.g., $FeCl_3$, $GaCl_3$, $GeCl_4$), acids or hydroxides (e.g.,$B(OH)_3$, precipitated TiO_2,xH_2O).

For the $AlPO_4$- and $GaPO_4$- based synthesis mixtures, P was introduced as phosphoric acid (85 % in water). Pseudo-boehmite (Catapal B from Vista) or isopropoxide were used as an Al source, and gallium sulfate was used as a Ga source.As the presence of alkaline or $NH4^+$ cations is to be avoided, F^- was added as HF.

The temperature ranged from 60 to 200°C and the heating times were a few hours to a few weeks. Tables 1 and 2 show typical examples with reaction mixtures, composition, and synthesis conditions. Other examples are described in the papers, theses and patents referred to in the two tables [3,7-27].

Characterization

Identification was performed by X-ray diffraction and purity was checked by optical or electron microscopy. All elements except oxygen were quantitatively determined by classical chemical analysis. Multinuclear MAS NMR spectroscopy (Bruker MSL 300) was used to study structural features such as state of the organic template, number of sites and coordination of the framework elements. Other techniques such as thermal analysis (T.G., D.T.G., D.S.C.) were also employed for the characterization of the materials. Some discussed results are inferred from the structure determination by the Rietveld technique [29-30] and single crystal X-ray diffraction [20,23].

Table 1. Silica-based materials (The main T elements in the framework are underlined)

Molar composition of the starting mixture	T(°C)	Durat. (h)	Struct. type	T	Struct. direct.	Templa-ting	Ref.
0.1-1(Pr_4NBr or Pr_3N):SiO_2:0.0-0.1 $TO_2$0.1-1NH_4F:10-50H_20	200	48-240	MFI	$\underline{Si}$,($\underline{Si}$,T^{IV})	No	Yes	3,7,8
$H_2N(CH_2)_{7\text{-}8}NH_2$:$SiO_2$:2HF:10$H_2O$	170	240	MFI	$\underline{Si}$	Yes	Yes	9
0.1-1(Pr_4NBr or Pr_3N or Bu_3N or Et_4NCl):SiO_2:0.1-0.01T_2O_3:0.1-1NH_4F:10-50H_2O	170	48-240	MFI	($\underline{Si}$,T^{III})	No	Yes	4,10-12
$H_2N(CH_2)_{6\text{-}8}NH_2$ or $(CH_3)_2N(CH_2)_3N(CH_3)_2$:$SiO_2$:0.1-0.01$Al_2O_3$:2HF:10-50$H_2O$	170	300	MFI	($\underline{Si}$,Al)	Yes	Yes	9
0.8-1.8NaF:SiO_2:0.1-0.02Al_2O_3:25H_2O(pH 7-8)	170	48[a]	MFI	($\underline{Si}$,Al)	No	Yes	13
2(Piperidine or $H_2N(CH_2)_{3\text{-}4}NH_2$):$SiO_2$:0.1-0.03$Al_2O_3$:2HF:10-30$H_2O$	170	300	FER	($\underline{Si}$,Al)	No	No	9,14
2R [b]: SiO_2:0.1-0.03Al_2O_3:2HF:10-30H_2O	170	300	FER	($\underline{Si}$,Al)	Yes	No	9,14
3$CH_3(CH_2)_3NH_2$ or1$H_2N(CH_2)_5NH_2$:SiO_2:0-0.01Al_2O_3:3 or 2HF:10H_2O	170	72[a]	TON	$\underline{Si}$,($\underline{Si}$,Al)	No	Yes	9,15
3$CH_3(CH_2)_4NH_2$ or 1$H_2N(CH_2)_{4or6}NH_2$ or 3$Pent_2NH$:SiO_2:0-0.01Al_2O_3:2-3HF:10H_2O	170	72[a]	TON	$\underline{Si}$,($\underline{Si}$,Al)	Yes	Yes	9,15
2-3Pyrrolidine:SiO_2:0-0.01Al_2O_3:2-3HF:10-30H_2O	170	72[a]	MTT	$\underline{Si}$,($\underline{Si}$,Al)	No	Yes	16
2-3(Pr_2NH or i-$PrNH_2$):SiO_2:0-0.01Al_2O_3:2-3HF:10-30H_2O	170	72[a]	MTT	$\underline{Si}$,($\underline{Si}$,Al)	Yes	Yes	16
(1-4diazabicyclo[2,2,2]octane+0.5$MeNH_2$):SiO_2:0.1-0.015Al_2O_3: 2HF:10-40H_2O	200	168[a]	BEA*	($\underline{Si}$,Al)	Yes	No	17
1,4-diazabicyclo[2,2,2]octane: SiO_2:0-0.015Al_2O_3: 1-2HF:10-40H_2O	200	48[a]	MTW	$\underline{Si}$,($\underline{Si}$,Al)	Yes	No	18
Quinuclidine: SiO_2:0.1-0.05Al_2O_3:2HF:10-40H_2O	170	96	LEV	($\underline{Si}$,Al)	Yes	No	19
Quinuclidine : SiO_2:0-0.05Al_2O_3:HF:15-25H_2O	170	360	AST	$\underline{Si}$,($\underline{Si}$,Al)	Yes	Yes	9,20
(Me_4NOH+0.4Pr_2NH):SiO_2:0.002-0.025Al_2O_3:0.4HF:40-60H_2O	170	24[a]	"NU1"	($\underline{Si}$,Al)	Yes	Yes	-
0.5Me_4NCl:SiO_2:0-0.08Al_2O_3:NH4F:1-4H_2O	170	100-15	MTN	$\underline{Si}$,($\underline{Si}$,Al)	No	Yes	9
0.4-4.8NaF:SiO_2:0.1-0.02Al_20_3:10-30H_2O(pH=8-10)	200	120[a]	MOR	($\underline{Si}$,Al)	No	No	21

(a) with seeds and agitation (b) R = Me_2 NH or CH_3 $(CH_2)_{1\text{-}3}NH_2$ or $H_2N(CH_2)_5NH_2$ or piperazine

Table 2. $AlPO_4$-and $GaPO_4$-based materials (The main T elements in the framework are underlined)

Molar composition of the starting mixture	T(°C)	Duration (h)	Structure type	T	Struct. direct.	Templa-ting	Ref.
2 i-$PrNH_2$:Al_2O_3:P_2O_5:0.5HF:100H_2O	200	24	"$AlPO_4$-14A"	Al,P	Yes	No	22
Pr_3N or Et_3NOH:Al_2O_3:P_2O_5:0-0.6SiO_2:0.5-1HF:70H_2O	150-170	4-24	AFI	Al,P,Si	No	Yes	23
n-Pr_2NH:Al_2O_3:P_2O_5:0-0.6SiO_2:HF:70H_2O	170	17-24	AEL	Al,P,Si	No	Yes	24
Quinuclidine: Al_2O_3:P_2O_5:HF:60H_2O	150-170	24	AST-like	Al,P	Yes(a)	Yes	-
1.5Morpholine: Al_2O_3:P_2O_5:HF:50-100H_2O	170-200	258-303	CHA-like	Al,P	Yes(a)	Yes	5,24
1-1.5 R(b):Al_2O_3:0.47-0.75P_2O_5:0.4-2.8SiO_2:HF:50-100H_2O	170-200	144-303	CHA	Al,P,Si	No	Yes	5,24
0.6Me_4NOH:2Pr_4NOH:Al_2O_3:P_2O_5:0.5SiO_2:0.25-1HF:50-100H_2O	200	24	FAU	Al,P,Si	No	Yes	24
x n-Pr_2NH:Ga_2O_3:Pr_2O_5:1-2.66HF:70H_2O(b)	80-200	6-36	LTA	Ga,P	Yes	Yes	6,25
x Quinuclidine: Ga_2O_3:P_2O_5:0.37-1HF:70H_2O(b)	140-170	15-24	CLO	Ga,P	Yes	Yes	6,26;27

(a) Not exactly the same symmetry as the phase obtained in the absence of F

(b) R = Et_4 NOH or morpholine

(c) x is such that the starting pH is equal to 3.5-5

RESULTS AND DISCUSSION

Silica-based materials

According to the low pH values in the reaction mixtures (with the exception of the "NU1" synthesis), the presence of F^- was necessary as a mineralizer for all materials of table 1 in order to allow their crystallization. Four of them (FER, BEA*, LEV, MOR) were obtained only with low Si/Al ratios (5<Si/Al<30), the other being formed generally with Si/Al ratios extending to high values (MFI, MTW, AST, "NU1", MTN with Si/Al>10-15) or only with very high Si/Al ratios (TON, MTT with Si/Al>50). Moreover, for the MFI-structure type, silicon could be substituted by tetravalent elements (extensively with Ge and in a more limited range with Ti, Sn and Zr) or trivalent elements other than Al (B, Fe, Ga). The framework compositions are generally similar to those published for materials prepared in alkaline medium and in the absence of fluoride.

The first nine structure types could be synthesized with new organic templates and in the case of the AST structure type, with an entirely novel composition of the framework, owing to the presence of F^- which is then considered as a structure directing agent.

The templating effect of F^- is observed in the following structure types : MFI, TON, MTT, AST, "NU1" and MTN [31]. In these cases, F^- is incorporated, as ion pairs, with the organic cation. F^- was localized by the structure determination from X-ray diffraction data in the small double four-membered ring cages of the AST-type framework [20] and between two five-membered rings of the MFI-type framework [30]. According to the chemical shift values observed for ^{19}F in the other structures, we can likewise expect a localization in small framework interstices [31]. The decreasing of the templating effect of F^- with increasing amounts of trivalent T elements such as Al in the framework may be explained by the fact that

- the organic cations are used to compensate the negative charges of the framework
- negatively charged frameworks accept more difficultly the insertion of F^- (repulsion interactions).

The last effect could explain the absence of F^- as a template in the other alumina-rich materials. The alumina-rich MFI structure type (5<Si/Al<20) prepared with NaF as a template is a particular case. Chemical analysis shows always the presence of NaF. But its localization is not clearly established yet. The ^{19}F MAS NMR chemical shift ($\delta \cong -220$ ppm/$CFCl_3$)) and the absence of XRD lines of crystalline NaF seem to indicate its presence in the channels of the structure.

$AlPO_4$- and $GaPO_4$-based materials

The mineralizing role of F^- is not so evident here as in the case of the silica-based materials. Indeed the employed pH conditions are the same as those which would be used in absence of F^-. Nevertheless beneficial effects are observed (generally shorter crystallization times, better crystallized products). The substitution degree of Si for P and Al and the corresponding mechanisms are similar to those observed for $AlPO_4$- derived materials synthesized in the absence of fluoride.

On the other hand the structure directing role is here more evident since 5 new (in structure and/or in composition) phases out of 9, could be crystallized owing to the presence of F^- : pure $AlPO_4$-14A [22,32], a tetragonal AST-like [33] and a triclinic CHA-like [5,24] $AlPO_4$ and two new gallophosphates with the LTA [6,25] and CLO [6,26,27] structures.

With the exception of $AlPO_4$-14A, all the other phases incorporate F^-, which is thus a template accompanying the organic cation. As in the case of the silica-based materials, the amount of F is lower for SAPO phases than for $AlPO_4$ phases because in the former, one part of the organic cations compensates the negative charges of the framework. In $AlPO_4$-5 (AFI), F^- was found between two 4 rings in the 4-ring columns of the framework [23]. As the ^{19}F chemical shift value in $AlPO_4$-11 (AEL) is very close to the value observed for $AlPO_4$-5, one may conclude that the environment of F is essentially the same in both structures [31]. By analogy with the silica AST-type phase [20], it can be assumed that in the tetragonal AST-like $AlPO_4$ phase, F is also located in the D4R cages. In the triclinic CHA-like $AlPO_4$ phase two F atoms were found to bridge two Al of a 4-ring which connects two double 6-rings of the chabazite structure [29]. The hexacoordination of two Al out of six was confirmed by ^{27}Al MAS NMR spectroscopy [5,24]. The presence of these additional framework F atoms is compensated by two protonated morpholine molecules located in each chabazite cage.

In the LTA $GaPO_4$ [28] and in the CLO-type $GaPO_4$ [27], fluorine was found in the double four-rings of the structure. For both materials the pentacoordination of Ga (4O, 1F) in a trigonal bipyramidal environment was confirmed by ^{71}Ga MAS NMR.

Thus, it may be assumed that in all structures in which F is a template, it contributes to the stabilization of the structures allowing in some cases their formation.

CONCLUSION

Whereas F^- plays a mineralizing role as does OH, in many cases the synthesis of a given crystalline microporous material with a given organic template will only be possible in the presence of fluoride anions. Two situations then arise, in the first, F is not present in the structure and hence seems to be involved in the gel chemistry only, in the second, F interacts with the framework and thus contributes to the stabilization of the structure. This situation is mainly observed with non charged or charge-autocompensated frameworks.

REFERENCES

1. Barrer, R.M. *Hydrothermal Chemistry of Zeolites*, Academic Press, London (1982).
2. Flanigen, E.M. and Patton, R.L., US Patent 4073865, (1978).
3. Guth, J.L., Kessler, H. and Wey, R. in: *New Development in Zeolite Science and Technology*, Eds Y. Murakami, A. Iijima and J.W. Ward, Kodensha-Elsevier, Tokyo, pp. 121-128 (1986).
4. Guth, J.L. Kessler, H., Higel, J.M. Lamblin, J.M., Patarin, J., Seive, A., Chezeau, J.M. and Wey, R. in: *Zeolite Synthesis*, Eds M.L. Occelli and H.E. Robson, ACS Symp. Series 398, ACS, Washington D.C., pp. 176-195 (1989).
5. Kessler, H. in :*Recent Advances in Zeolite Science*, Eds J. Klinowski and P.J. Barrie, Studies in Surface Science and Catalysis, Vol. 52, Elsevier, Amsterdam, pp. 17-38 (1989).
6. Kessler, H. in : *Synthesis/Characterization and Novel Applications of Molecular Sieves*, Eds R.L. Bedard, T. Bein, M.E. Davis, J. Garces, V.A. Maroni and G.D. Stucky, Material Research Society, Vol. 233, Pittsburg, Pennsylvania, pp. 47-55 (1991).
7. Gabelica, Z. and Guth, J.L., Angew. Chem. **28**(1), 81 (1989).
8. Lopez, A., Thesis, Université de Haute Alsace, Mulhouse (1990).
9. Lamblin, J.M., Thesis, Université de Haute Alsace, Mulhouse (1989).
10. Patarin, J., Kessler, H. and Guth, J.L., Zeolites **10**(7), 674 (1990).
11. Seive, A., Thesis, Université de Haute Alsace, Mulhouse (1989).
12. Kessler, H., Chezeau, J.M., Guth, J.L., Coudurier, G. and Strub, H., Zeolites **7**(4), 360 (1987).
13. Joly, J.F., Caullet, P., Guth, J.L., Faust, A.C., Brunard, N. and Kolenda, F., French Patent Appl. 90/16529 (1990).
14. Guth, J.L., Faust, A.C., Raatz, F. and Lamblin, J.M., French Patent Appl. 86/16362 (1986).
15. Patarin, J., Lamblin, J.M., Faust, A.C., Guth, J.L. and Raatz, F., French Patent Appl. 88/06841 (1988).
16. Patarin, J., Lamblin, J. M., Faust, A.C., Guth, J.L. and Raatz, F., French Patent Appl. 88/08105 (1988).
17. Caullet, P., Guth, J.L., Faust, A.C., Raatz, F., Joly, J.F. and Deves, J.M., French Patent Appl. 89/12556 (1989)
18. Joly, J.F., Caullet, P., Faust, A.C., Baron J. and Guth, J.L., French Patent Appl. 90/16256 (1990).
19. Caullet, P., Guth, J.L., Faust, A.C., Joly, J.F. and Raatz, F., French Patent Appl. 89/17163 (1989).

20. Caullet, P., Guth, J.L., Hazm, J., Lamblin, J.M. and Gies, H., Eur. J. Solid State Inorg. Chem. **28**, 345 (1991).

21. Caullet, P., Guth, J.L. Faust, A.C., Joly, J.F., Travers, C. and Raatz, F., French Patent Appl. 89/13317 (1989).

22. Goepper, M. and Guth, J., Zeolites **11**(5), 477 (1991).

23. Qiu, S., Pang, W., Kessler, H. and Guth, J.L., Zeolites **9**(5), 440 (1989).

24. Guth, F., Thesis, Université de Haute Alsace, Mulhouse, (1989).

25. Merrouche, A., Patarin, J., Kessler, H. and Anglerot, D., French Patent Appl. 91/01106 (1991).

26. Joly, J.F., Merrouche, A., Kessler, H. and Guth, J.L., French Patent Appl. 91/04225 (1991).

27. Estermann, M., McCusker, L.B., Baerlocher, C., Merrouche, A. and Kessler, H., Nature, **352**, 320 (1991).

28. Simmen, A., Patarin, J. and Baerlocher, C., in : *Proceedings of the 9th International Zeolite Conference*, Eds R. von Ballmoos, J. B. Higgins and M.M.J. Treacy, Butterworth-Heinemann, Stoncham M.A. (1992).

29. Simmen, A., Thesis, Eidgenössische Technische Hochschule, Zurich (1992).

30. Mentzen, B.F., Sacerdote-Peronnet, M. Guth, J.L. and Kessler, H., C.R. Acad. Sci. Paris **313(II)**, 177 (1991).

31. Klock, E., Delmotte, L., Soulard, M. and Guth, J.L. in : *Proceedings of the 9th International Zeolite conference*, Eds R. von Ballmoos, J.B. Higgins and M.M.J. Treacy, Butterworth-Heinemann, Stoneham M.A. (1992)

32. Pluth, J.J. and Smith, J.V., Acta Cryst., **C43**, 866 (1987)

33. Bennett, J.M. and Kirchner, R.M., Zeolites **11** (5), 502 (1991)

THE SYNTHESIS AND CHARACTERIZATION OF Ga-MFI AND Ga-MTW ZEOLITES

S. L. LAMBERT

UOP, Des Plaines, IL 60017-5017, U.S.A.

ABSTRACT

The primary synthesis of MFI and MTW zeolites, in which gallium has been substituted for aluminum in the framework, is described. A wide range of gallium substitution was achieved. Characterization, including silicon and gallium NMR, X-ray diffraction, chemical analyses, and XPS, was obtained to verify the framework substitution of gallium. The thermal stability of gallium in the MTW framework is evaluated.

INTRODUCTION

The substitution of elements other than silicon and aluminum in the frameworks of known zeolites has received much attention in recent years [1-2]. The substitution of an element in the framework of a zeolite affects the chemical properties of the zeolite but maintains the molecular sieving capabilities of the structure. The substitution of gallium for aluminum in the framework of the MFI (ZSM-5) and MTW (ZSM-12) zeolites is intended to moderate the strong acidity of the silica-alumina forms of these molecular sieves [3]. The framework of the MFI zeolite contains a three-dimensional system of pores delimited by 10 rings. The two types of pores are approximately 5.3 x 5.6 Å and 5.1 x 5.5 Å in cross-section and are orthogonal. The MTW framework is characterized by one-dimensional, 12-membered ring pores with an effective cross-section of 5.5 x 5.9 Å. These gallium-substituted materials can then find application in processes in which their molecular sieving action is desirable but in which strong acidity is undesirable [4-5]. Other examples of gallium-substituted frameworks have recently begun appearing in the literature [6-9].

EXPERIMENTAL

The synthesis procedures used to produce the target gallium-substituted zeolites were modified from syntheses that the author had previously used to successfully prepare the alumino-silicate versions of ZSM-5 and ZSM-12 zeolites. Sodium gallate solution, obtained by dissolving gallium oxide in sodium hydroxide, was substituted for the aluminum source. This substitution required an adjustment of the gel pH. The adjustment required the use of the bromide salt of the tetraalkylammonium ion acting as the template to maintain a pH in the range of 12 to 13.

A wide range of substitution was achieved for the MFI structure, from 0.5% Ga (Si/Ga = 230) up to 7.52% Ga (Si/Ga = 14). Gallium substitution, in the range of 0.84% to 4.23%, was achieved in the case of Ga-MTW; however, far fewer experiments were done in the case of the MTW framework, and this range may not accurately represent the limits of gallium substitution in this framework.

From the Proceedings of the 9th International Zeolite Conference, Montreal 1992, Eds. R. von Ballmoos et al.,

A sample of gallium-substituted MFI, at a high level of gallium substitution, was prepared by dissolving tetrapropyl-ammonium bromide (TPABr) (62.95 g) in deionized water (180.36 g). This solution was filtered and then added to a sample of Ludox HS-40 (58.23 g) to produce a translucent solution. Separately, sodium hydroxide (23.52 g) was dissolved in deionized water (41.33 g). Gallium oxide (12.31 g) was added to this solution, and the resulting slurry was sealed in a Parr bomb and heated for 20 hours at 110°C to produce a clear sodium gallate solution. A portion of the sodium gallate solution (15.44 g) was added drop by drop to the stirred Ludox-TPABr solution. A minor amount of additional deionized water (5 g) was added to the resulting gel, which had a pH of 12-13. The gel was crystallized hydrothermally in teflon-lined Parr bombs for 114 hours at 150°C. The resulting solids were washed with deionized water and dried at 110°C overnight. This procedure yielded 22.63 g of white product.

The Ga-MTW sample was prepared by first dissolving sodium hydroxide (21.72 g) in deionized water (37.04 g) and adding gallium oxide (11.37 g). The dissolution of the gallium oxide was achieved by heating in a Parr bomb at 110°C for 72 hours. Triethylmethylammonium bromide (TEMABr) (10.78 g) was dissolved in deionized water (29.22 g). A portion of the sodium gallate solution (4.65 g) was added drop by drop to the stirred template solution. A sample of Ludox AS-40 (31.47 g) was diluted with deionized water (11.63 g) and then added drop by drop to the TEMABr-Ga solution. The resulting gel was hydrothermally crystallized at 150°C for 20 days. The solids were recovered by filtration and washed with deionized water. After drying at 110°C, 13.64 g of white powder was obtained.

RESULTS AND DISCUSSION

Data from the X-ray powder diffraction were used to confirm the framework structure of these samples. Both samples are of high purity. The Ga-MFI, whose XRD pattern is shown in Figure 1, contains no additional crystalline phases, but the Ga-MTW (Figure 2) has a minor cristobalite impurity. The crystallinity for these samples was not calculated because an accurate assessment for isomorphously substituted samples is difficult to obtain by X-ray methods. This assessment is difficult because the scattering intensities are element dependent.

The chemical analyses for the two samples are presented in Table 1. Mass balance calculations for both syntheses indicate that all of the gallium in the synthesis gel has been incorporated into the solid product. Other syntheses conducted in this lab have shown that the gallium level can be varied over a wide range of concentrations by altering the gallium content of the synthesis gel. Although gallium appears to be readily incorporated in these frameworks, the substitution of gallium for aluminum in these synthesis resulted in the need for extended crystallization time. The higher the gallium content in the gel, the slower the crystallization occurred. The samples discussed in this paper were deliberately prepared at a high gallium level to facilitate characterization of the gallium in these materials. The Ga-MFI has a bulk Si/Ga ratio of 14, and the Ga-MTW has a bulk Si/Ga ratio of 25.

Table 1
Chemical Analysis Data

Sample	Ga-MFI	Ga-MTW
Ga(VF), %	7.52	4.23
Si(VF), %	42.0	43.7
Al(VF), ppm	435	738
Si/Ga mole ratio	14	25
Si/Al mole ratio	897	576

Note: VF is volatile free.

The framework substitution of gallium was evaluated next. The framework Si/M(III) ratio was determined from the silicon MASNMR data and compared to the bulk Si/Ga ratio determined by chemical analysis. The silicon NMR data for the Ga-MFI sample are shown in Figure 3. The Si/M(III) ratio calculated from these data is 15. This ratio is in good agreement with the calculated Si/M(III) ratio of 14 determined from the chemical analysis data. Insufficient aluminum is present in the sample to account for the distribution of Si(nM(III)) peaks observed in the silicon NMR data.

The gallium-static NMR spectra were obtained for this sample. The spectrum consists of one broad line located at 162 ppm relative to gallium nitrate (Figure 4). The presence of only one line is consistent with gallium in one environment. Data were obtained at two frequencies: 110.8 MHz and 152.5 MHz. The line broadening is not due to quadrupolar interactions. The line width is consistent with gallium in a symmetric environment, such as the zeolite framework.

X-ray photoelectron spectroscopy (XPS) was also used to characterize this Ga-MFI sample. The XPS survey scans for Ga_2O_3 and Ga-MFI were obtained on pressed powders. They indicate that the only significant impurity is carbon. High-resolution scans were obtained for various Ga, O, C, and Si (for the zeolite) peaks to determine if the gallium was present as gallium oxide or had been incorporated into the zeolite framework. The binding energies (BE) corrected for charge shift are given in Table 2. Examination of the BE's clearly indicates that the gallium in Ga-MFI is different from that in gallium oxide. In addition, the differences in energy between a photoelectron peak and a corresponding Auger peak were compared. The ΔBE's are 263.3 and 264.9 for gallium oxide and Ga-MFI, respectively. This difference supports the thesis that gallium does not exist in the Ga-MFI as extra-framework gallium oxide. The line widths were also observed to be narrower for the Ga-MFI sample than for the reference oxide. This result suggests that only one type of environment exists for gallium in the zeolite sample.

Table 2
XPS Data: Binding Energies and Line Widths

		Ga_2O_3		Ga-MFI	
		B.E., eV[a]	Line Width eV[b]	B.E., eV[a]	Line Width eV[b]
Ga	$3d_{5/2}$	19.9	2.0	18.5	1.4
	$3d_{3/2}$	20.4	2.0	19.2	1.4
	$3p_{3/2}$	105.4	3.2	103.8	2.2
	$3p_{1/2}$	108.9	3.1	107.3	2.7
	3s	160.6	3.0	159.4	1.9
	$2p_{3/2}$	1117.8	2.4	1116.5	2.1
	Auger	423.9	3.0	424.3	2.9
	Auger	420.6	3.5	420.8	2.8
O	1s	530.6	2.3	530.9	1.8
	Auger	976.1	5.3	975.8	4.3
Si	2p	—	—	101.7	1.9

[a] Binding energies, which are corrected for sample charging, assume a binding energy of 284.6 eV for adventitious carbon.
[b] Line widths in eV are full widths at half maximum intensity.

The silicon MASNMR data for the Ga-MTW sample are shown in Figure 5. The framework Si/M(III) ratio calculated from these data is 14. This ratio is significantly lower than the value calculated from the chemical analysis data (Si/Ga=25). The gallium NMR data were also obtained for this sample (Figure 6). These data are consistent with all of the gallium being located in one uniform, crystalline environment. The chemical shift is close to that observed for gallium in the Ga-MFI samples. The difference in the bulk versus framework ratios may be due to the presence of nonframework silica in the Ga-MTW sample. In the simulation of the silicon NMR data, a line representing amorphous SiO_2 was not required to achieve an acceptable fit. This line would be difficult to detect at low concentrations because it overlaps with the lines of the zeolite framework. The amount of amorphous silica required to account for this difference in Si/Ga is small, only 4 wt-%. The presence of a significant concentration of silanol groups could also lead to a lower value of Si/Ga for the framework. The silicon NMR intensity for the silanol groups falls in the same location as the Si(1Ga) line. The data were not corrected for the presence of silanol groups.

The Ga-MTW sample was calcined for 10 hours at 500°C in a muffle oven. The resulting sample was white, and X-ray diffraction verified the phase as MTW. Qualitatively, about a 10% loss of crystallinity appears to occur for these calcination conditions. The crystallite size of the sample appears to have decreased because the line widths in the XRD data have increased. The silicon NMR data also support the reduced crystallinity of the sample. The Si/Ga mole ratio for the framework has increased slightly: to 18. The gallium NMR data have a noticeably broader line width after calcination, indicating that gallium now occupies a greater variety of environments.

CONCLUSIONS

The substitution of gallium into the lattices of the MFI and MTW frameworks appears to be quite facile. All of the gallium in the synthesis gel is incorporated into the solid product. The level of gallium in the zeolite can be controlled by varying the amount of gallium in the gel.

Characterization of these materials was done using X-ray diffraction to verify the formation of the framework structure. Chemical analyses were obtained to determine the bulk Si/Ga ratio. The NMR data were then used to characterize both the silicon and the gallium in the samples. The framework Si/M(III) ratio was determined from the silicon NMR data and compared to the bulk Si/M(III) ratio. The gallium NMR data provided support for the presence of only one type of gallium in these samples. The XPS data were obtained for the Ga-MFI sample, and these data also support the substitution of gallium into the framework.

The thermal stability of gallium in the MTW framework is poorer than that observed for Ga-MFI zeolites. After 10 hours, at 500°C calcination, a 10% loss in crystallinity was observed. This loss was accompanied by some loss of framework gallium. The Si/Ga ratio, as determined by silicon NMR, increased from 14 to 18. No change in the Si/Ga ratio was observed for the Ga-MFI zeolite when it was subjected to a similar calcination.

REFERENCES

1. Barrer, R. M., *Hydrothermal Chemistry of Zeolites*, Academic Press, 1982, Chapter 6.
2. Tieler, M., Geelen, M., and Jacobs, P. A., Acta. Phys. Chem. **31**, 1 (1985).
3. Dompas, D. H., Mortier, W. J., Kenter, O. C. H., Janssen, M. J. G., and Verduijn, J. P., J. Catal., **129**. 19-24 (1991).
4. Luo, J., Tan, C., and Dong, Q., Shiyou Haugong **19**, 587-92, (1990).
5. Bayense, C. R., Van der Pol, A. J. H. P., and Van Hooff, J. H. C., Appl. Catal. **72**, 81-98 (1991). 6. Delprato, F., Guth, J. L., and Zivkov, C., Fr. Demande FR 2,650,582 (1991).
7. Occelli, M. L., US Patent 5,303,432, (1991).
8. Occelli, M. L., PCT Int. Appl. WO 91 98,172.
9. Luo, J., Tan, C., and Dong, Q., Ranliao Huaxue Xuebao, **18**, 207-14 (1990).

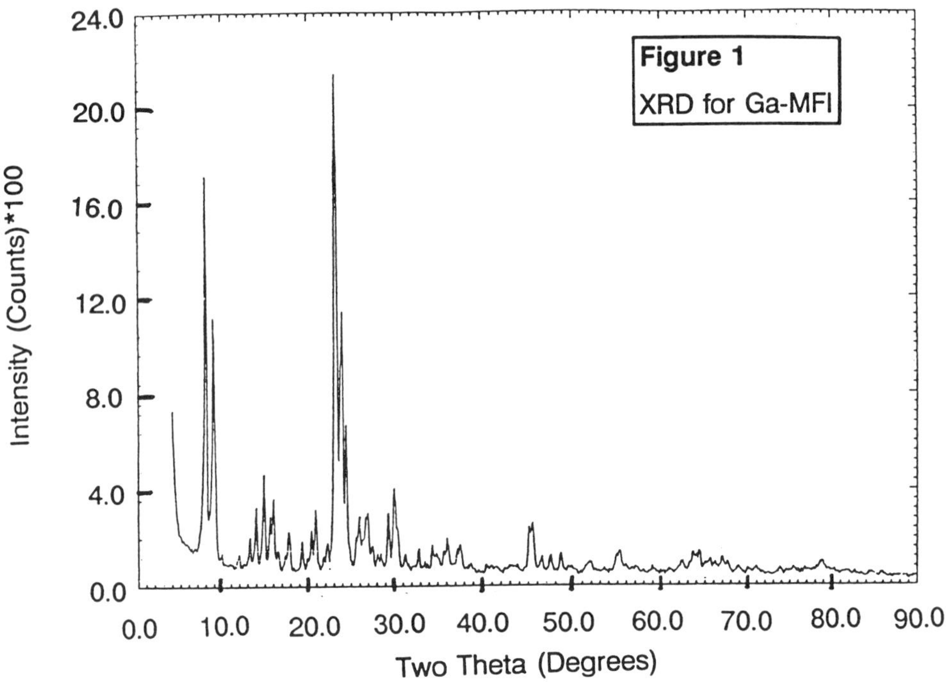
Figure 1
XRD for Ga-MFI
Intensity (Counts)*100
24.0
20.0
16.0
12.0
8.0
4.0
0.0
0.0
10.0
20.0
30.0
40.0
50.0
60.0
70.0
80.0
90.0
Two Theta (Degrees)

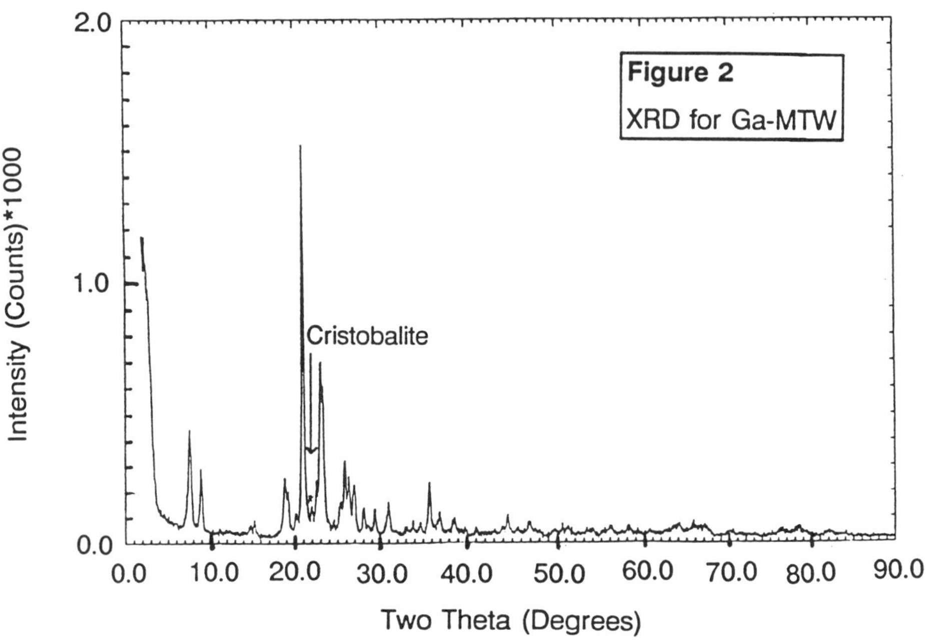
Figure 2
XRD for Ga-MTW
Intensity (Counts)*1000
2.0
1.0
0.0
Cristobalite
0.0
10.0
20.0
30.0
40.0
50.0
60.0
70.0
80.0
90.0
Two Theta (Degrees)

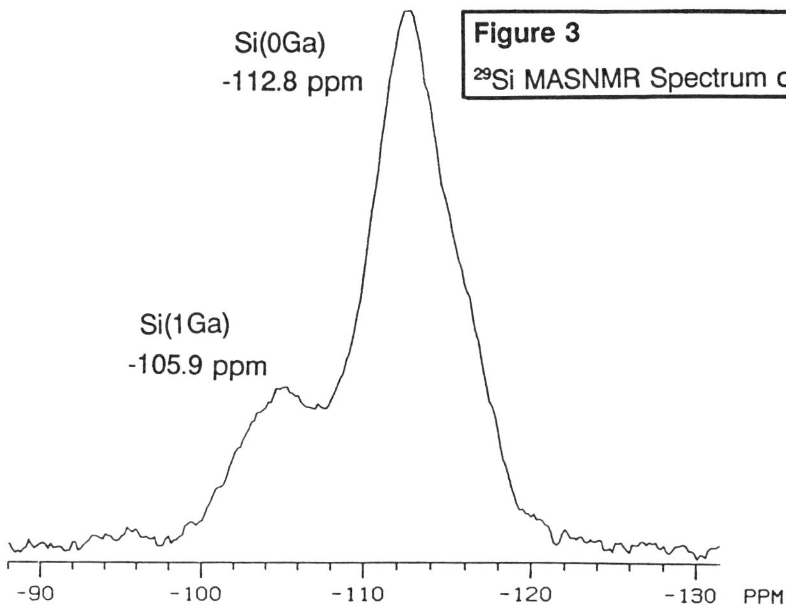

Figure 3
^{29}Si MASNMR Spectrum of Ga-MFI

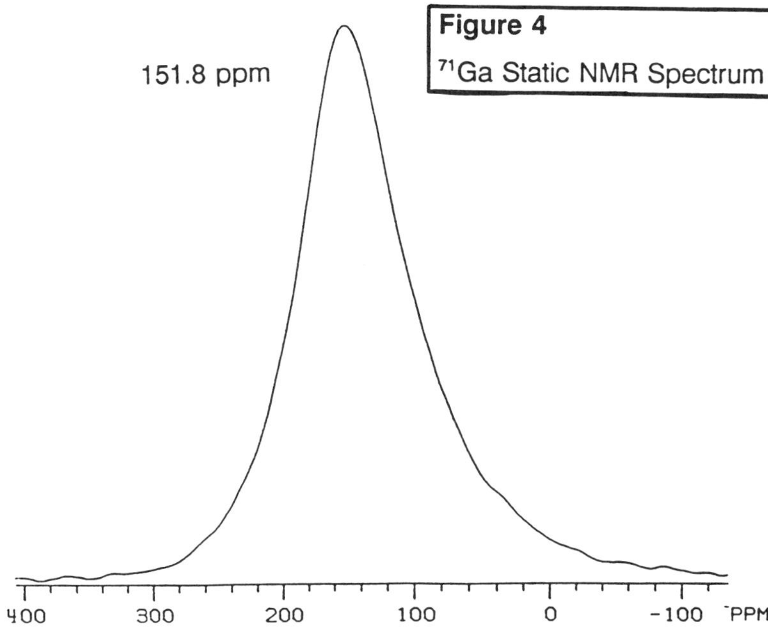

Figure 4
^{71}Ga Static NMR Spectrum of Ga-MFI

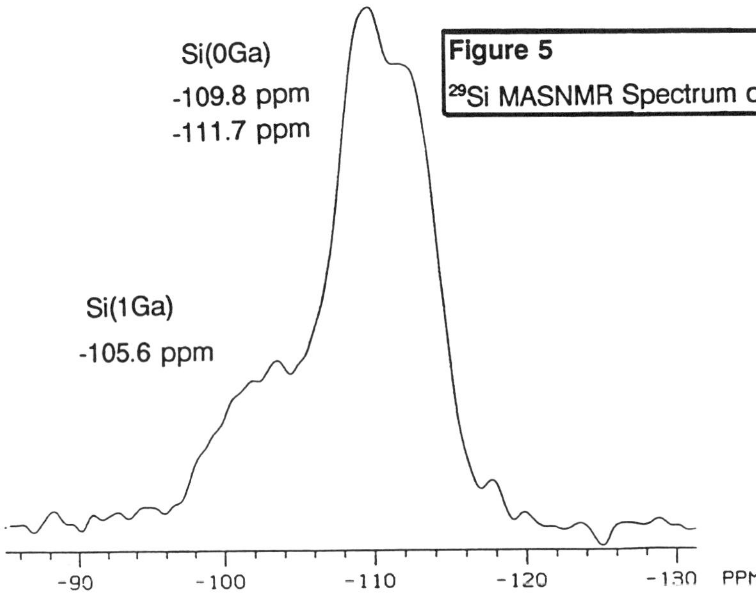

Figure 5
^{29}Si MASNMR Spectrum of Ga-MTW

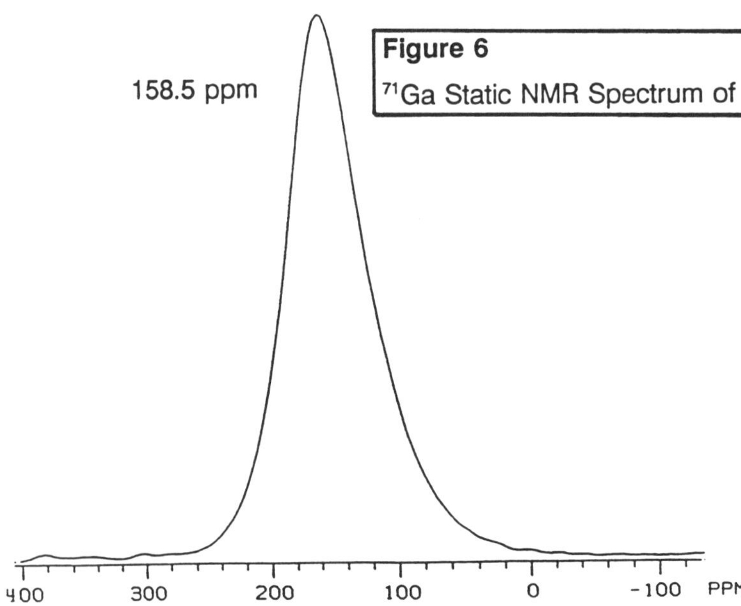

Figure 6
^{71}Ga Static NMR Spectrum of Ga-MTW

SYNTHESIS OF GALLO- AND GALLO-ALUMINO ZSM-5 ZEOLITES WITH VARIABLE AND QUASI-HOMOGENEOUS AL-GA DISTRIBUTION, FROM ALKALI-FREE MEDIA

Z. GABELICA*, G. GIANNETTO**, F. DOS SANTOS**, R. MONQUE***, and R. GALIASSO***

*University of Namur, Dept. Chemistry, 61, Rue de Bruxelles, B-5000, Namur, Belgium
**Universidad Central, Fac. Engen., Los Chaguaramos, Caracas, Venezuela
***INTEVEP A. S., Apdo. 76341, Los Teques, Caracas, Venezuela

ABSTRACT

A new synthesis route leading to a rapid crystallization of alkali-free gallo-, alumino-, or gallo-alumino MFI-type zeolites from methylamine-Pr_4N^+ media, was proposed. Methylamine plays a similar mobilizing-complexing role towards both Al^{3+} and Ga^{3+} ions so that they are quasi-homogeneously distributed in the final zeolite lattice. Mild (hydro)thermal treatments of the as-synthesized materials directly yield the protonated forms involving quite strong Brönsted acid sites. While the Al species remain stable in the framework, gallium is partly extracted from the lattice upon these treatments. The framework Ga stability appears to be markedly dependant on the actual lattice Al concentrations

INTRODUCTION

Pentasil-type gallo- and gallo-aluminosilicates have proved to be very active and selective catalysts in the transformation of lower alkanes to aromatics (1-12). Besides the zeolitic acid sites commonly generated by ion exchange of the calcined materials that catalyze various reaction steps of the conversion such as cracking, oligomerization and cyclization, highly dispersed gallium oxydic species were recognized to catalyze more specifically the initial dehydrogenation of alkanes to olefins (1,4,5,7,9-14), to show a good selectivity to stable aromatics (1,4,7,9-14), and to increase the overall yield (4). Such Ga_xO_y-type species can be prepared and stabilized by calcining under appropriate conditions Ga-exchanged, Ga-impregnated or Ga_2O_3-loaded H-ZSM-5 zeolites (1,2,4-18), but also by heating under various conditions and atmospheres, the as-synthesized MFI gallo- or gallo-aluminosilicates (4,5,15-20). Parameters such as (hydro)thermal temperature and time or the water vapor partial pressure can markedly influence the (in)stability of the framework gallium and, consequently, the amount of non-framework Ga species eventually released upon these treatments (5,11,18-20). No doubt that the formation of these latter species should also depend on the actual zeolite lattice regularity (presence of defects...), composition (nature and vicinity of the close gallium neighbors such as templates, pore filling molecules or counterions...) or homogeneity (Ga or Ga-Al framework distribution), properties that should in principle be controlled by an appropriate optimization of various synthesis parameters.

In this paper, we propose and describe a new synthesis route using Pr_4N^+ ions and methylamine (MA), leading to alkali-free MFI-gallo and gallo-aluminosilicates that readily yield acid catalysts after calcination of the as-synthesized materials. A combination of physicochemical characterization techniques was used to investigate some important characteristics of these syntheses, as well as a series of properties of the resulting crystalline phases: competitive Ga-Al incorporation during crystallization of gels involving variable Ga and Al concentrations, final framework distribution of these ions in the as-synthesized zeolites, their consequent (hydro)thermal stability, as well as the resulting acidity of the calcined phases. Some of these properties are compared with those characterizing analog materials prepared via the so-called fluoride route.

From the Proceedings of the 9th International Zeolite Conference, Montreal 1992, Eds. R. von Ballmoos et al.,

EXPERIMENTAL

Synthesis

A series of MFI zeolites containing Ga, Al or both elements were synthesized by using the following original recipe. Appropriate amounts of gallium and/or aluminum nitrates ($Ga(NO_3)_3.9H_2O$, 99.999% from Aldrich, $Al(NO_3)_3.9H_2O$, P.A., from Merck) pre-dissolved in a small amount of water, were added dropwise, under stirring, to an aqueous solution containing methylamine (CH_3NH_2 40% aq., pure, from Fluka) and predissolved Pr_4NBr (Merck, P.A.). Both trivalent ions can be added simultaneously or successively to the methylamine -Pr_4N^+ solution, in variable order and after variable time intervals. The resulting clear solution (pH of about 13) was stirred for 1h at ambient temperature before addition of solid SiO_2 (Aerosil 200 from Degussa, either with high-surface area of 270 $m^2.g^{-1}$(all the "ZN"-coded samples) or with "low" surface area of about 150 $m^2.g^{-1}$ (samples mentioned but not fully characterized in this work). The final gel (pH of about 12) was stirred for another 1 to 2 h, before transfer to Teflon-coated stainless steel 80 ml autoclaves. These were heated in static or agitated conditions at 185°C, for variable periods of time, typically for 3 to 5 days. The final as-synthesized solids were filtered, washed thoroughly with cold water and dried at 90°C for 12h.

A typical gel molar composition involves 0.96 -x mole of SiO_2, y mole of Ga nitrate, z mole of Al nitrate (with y+z=x, and x varying from 0 to 0.04), 1 to 10 mole of CH_3NH_2, 0.25 mole of Pr_4NBr, optionally 0.25 to 1 mole of HF or NH_4F (added to the mixture at any stage of its preparation), and 40 mole of H_2O. The preferred amount of CH_3NH_2 mostly depends on the initial pH of the solution and 2 to 8 mole of CH_3NH_2 per mole of SiO_2 were shown to be the optimized ratios in other systems (21). It should be noted that methylamine shows irritating properties to eyes, skin and respiratory track.

Characterization

Nature and purity of the samples were checked by the classical X-ray powder diffraction method (Philips P.W. 1349/30 diffractometer) and by SEM (Philips XL 20 microscope).

Water and organic contents were determined by combined thermal analysis, either TG-DTA-DTG (Stanton Redcroft ST-780) or TG-DTG-DSC (Setaram 111). This latter instrument also allows the automatic titration of the gaseous basic species (NH_3, amines...) released upon thermal degradation of the as-synthesized or calcined and NH_4^+-exchanged zeolitic samples, by coupling a Metrohm-type titrimeter to the thermobalance (22).

Crystallinities of the various samples were evaluated from n-hexane sorption capacities measured within the thermobalance (22,23).

The total gallium contents were determined selectively in the presence of Al^{3+} ions, by adapting the complexometric "EDTA-copper PAN" titration method (24) in the case of zeolitic materials, after their pre-dissolution in diluted HF (22).

Spot EDX quantitative analyses of Si, Ga or Al on selected areas (core or edge) accross individual crystallites were obtained using an EDAX P.V. 9800 Philips analyser coupled with the electron microscope. Performances and limitations of such a technique have been discussed elsewhere (25,26).

High resolution solid state MAS ^{27}Al- and ^{71}Ga-NMR techniques have been recently successfully used as accurate tools for a quantitative determination of zeolitic framework Al^{3+} and Ga^{3+} contents. Detailed experimental setup, conditions for quantitative measurements and their limitations, are discussed in a subsequent paper (22).

RESULTS AND DISCUSSION

Crystallization kinetics of (Ga,Al)-MFI zeolites

The use of methylamine in zeolite preparations presents numerous advantages with respect to classical preparations achieved in alkali basic media. Indeed this compound shows, along with F^- ions, potential mobilizing abilities towards many ionic species, among which Si^{4+}, Al^{3+} and Ge^{4+} (21) but also Ga^{3+}, B^{3+}, Cr^{3+}, Fe^{3+}, Ti^{4+}...(22), by forming with them soluble complexes. It also readily brings the pH of the gel to basic values (12-12.5) needed for a rapid and efficient condensation of the various (here Si, Al or Ga) oligomers and the nucleation and growth of the first zeolite crystallites, thus avoiding the disadvantageous use of alkali bases or the costly use of Pr_4NOH, to reach these pH values. Finally, when used along with Pr_4N^+ ions, well known for their ready and specific templating action in the case of MFI-type structures, methylamine would not act as a template itself, essentially because of its reduced size. In specific preparations described elsewhere, it may at the best behave as a stabilizing pore filling molecule (27).

A series of (Ga_yAl_z)-MFI zeolites were prepared from the methylamine medium in the absence of any fluoride or alkali cationic species, in the compositional range varying from y=2.5 for z=0 to y=0 for z=4.

When low surface area silica is used, and under agitation (twisting autoclaves at 185°C), pure, 100% crystalline silicalite (y=z=0) and the $Ga_{2.5}$ -end member (y=2.45, z=0) are formed after 5 days heating , while less cristalline samples are obtained when Al is added to the system, the richer the gel in Al, the lower the crystallinity. For example, for z=1 or 2 (for y=1.88 and 1.22 respectively), the crystallinities measured were about 75% and 8 days of heating are necessary to recover the corresponding 100% crystalline phases. The Al-rich phases (z=3, y=0.61) and (z=4, y=0) show poor crystallinity (e.g. less than 50%), even after 8 days heating. These results show that, like in the case of most of the explored media from which ZSM-5 can crystallize (28,29), the presence of Al^{3+} also impedes the growth of the MFI framework in the presence of methylamine. It is possible that, like $Al(OH)_4^-$ and other low oligomeric hydroxy-Al species formed at lower pH values, soluble methylamine-complexed Al species do not readily react with silicate species as to yield the adequate (Si,Al) oligomers ready to condense in the early nucleation-growth steps. Conversely, gallium yields stable monomeric gallate ions already in weak alkaline media, that would not easily transform into other type of Ga oxy-hydroxy phases, so that their incorporation into a crystallizing silicate structure is expected to be more straightforward than that of Al (29). Such considerations could at least in part explain the faster kinetics observed in the case Ga-rich gallo-aluminosilicates.

Reactivity of silica is certainly another factor influencing this kinetics. Indeed, when high surface area silica is used in the synthesis, the same phases, even the Al end-members, are formed more rapidly, i.e. after about 3 days of heating, both in static or in agitated conditions. A continuous series of (Ga,Al)-MFI materials synthesized under these conditions were therefore selected for a more in depth study of their trivalent composition and their further (hydro)thermal stability.

Composition and stability of as-synthesized (Ga,Al)-MFI zeolites prepared from methylamine media

Table 1 describes a typical series of 9 samples synthesized from methylamine media, in the presence of high surface area silica and variable Al and Ga initial concentrations.

Table 1
Al and Ga contents (mol. per 96 T atoms, T= Si + Al + Ga) in various as-synthesized (Ga,Al)-MFI zeolites crystallized from (TPA^+, MA) media (185°C, 3 days, in static (S) or agitated (A) conditions)

Sample (code)	Al/96 T(1)			Ga/96 T (1)			
	In gel =(a)	In zeolite (2) (NMR) =(b)	Synth. eff. (3) =(b/a) X100	In gel =(d)	In as synth. zeol. Chem.	NMR=(e)	Synth. eff.(3) =(e/d) X100
ZN-61 (A)	0	0	_	2.45	2.38	2.39	96
ZN-58 (A)	1	0.97	97	1.84	1.75	1.74	97
ZN-64 (S)	1	1.05	(105)	1.84	1.83	1.83	100
ZN-59 (A)	2	1.98	99	1.22	1.18	1.18	97
ZN-65 (S)	2	1.96	98	1.22	1.20	1.18	98
ZN-67 (S)(4)	2	2.07	(103)	1.22	1.00	1.00	82
ZN-60 (A)	3	2.90	97	0.61	0.66	0.60	98
ZN-66 (S)	3	2.80	93	0.61	0.59	0.56	92
ZN-62 (A)	4	3.98	100	0	0	0	-

(1) The molecular weight of each unit cell supposed to contain 96 T atoms , was evaluated from known framework composition (recurrent calculation of M^{3+} contents) and from the actual amounts of H_2O and organics, as determined by thermal analysis. The Al and Ga contents were determined from ingredient ratios (in the gel phase), from titrimetric analysis (bulk Ga concentrations in as-synthesized zeolites), from ^{27}Al-NMR or ^{71}Ga-NMR (framework Al and Ga concentrations in as-synthesized zeolites), or from ^{27}Al-NMR and TPD of NH_3 (framework Al and Ga in calcined, NH_4-exchanged zeolites). Detailed analytical conditions are reported elsewhere (22).
(2) The framework Al concentrations in as-synthesized and in calcined zeolites are the same (22)
(3) "Synthesis efficiency", i.e. the percentage of trivalent element incorporated from the gel phase to the zeolite framework.
(4) Ga added to the (MA+Al) solution, in contrast to other samples, for which Al was added to the (MA+Ga) solution during the gel preparation.

One can see that both the total Ga (chemical analysis) and Al (^{27}Al-NMR) contents roughly correspond in each case to the Al or Ga contents per 96 T atoms (T=Si+Al+Ga) introduced in the gel phase, thereby showing that the synthesis efficiency for both trivalent ions is close to 100% (columns 4 and 8). Furthermore, while ^{27}Al-NMR can be considered as a quantitative measurement of framework Al^{3+} concentrations, the chemical analysis data never indicate whether all the Ga is actually incorporated in the zeolitic framework. In the absence of accurate methods to determine framework Ga concentrations in as-synthesized zeolites, we have been able to derive successfully empirical correlations between the normalized ^{71}Ga-NMR intensities of the central resonance line located in the 150-160 ppm range currently attributed to tetrahedral framework Ga^{3+} species in zeolites, and the total Ga contents determined by titrimetric analysis (22). For samples synthesized by following similar recipes (here the MA + Pr_4N^+ route) that are supposed to involve the same potential counterions to framework Ga^{3+} (here Pr_4N^+), linear correlations were obtained, from which the framework Ga^{3+} concentrations could be derived (Table 1, column 7).

Sample ZN-67, the only one prepared from gels in which the Ga^{3+} salt was added to the solution already containing Al^{3+} ions pre-dissolved in methylamine, shows a synthesis efficiency for Ga slightly (but significantly) lower than 100%. The longer the contact time between Al and MA, the lower the synthesis efficiency for Ga (27). This may suggest that when Al is sufficiently "conditioned" in the MA-complexing medium for a rapid further incorporation in the growing MFI crystallite, the subsequent incorporation of Ga is somewhat impeded. However, in the case of the reverse order of addition of the trivalent elements (all other samples), both ions are totally incorporated, even if Al is added at the very end of the gel preparation (27).

No further mechanistic considerations could be inferred from these observations but this behavior is probably closely related to the final Ga-Al gradient concentration in the as-synthesized gallo-aluminosilicate crystallites, that always show an outer rim slightly richer in Al than in Ga (Table 2). Such a gradient is even more pronounced when the zeolites are synthesized from fluoride media, that involve more stable (soluble) Al-fluoro complexes than the corresponding Ga-ones(19). One can therefore imagine that the Al incorporation, favored through a prolongated contact of Al^{3+} ions with the mobilizing agent (MA or F^-) in the gel, impedes to some extent the further incorporation of Ga, at least at the end of the crystallization process, i.e. near the outer surface of the crystallites. By contrast, the reverse addition, namely Al in (MA + Ga), could result in a total and rapid incorporation of Ga, that will be found preferentially located in the crystallite cores. The Al species, of which the framework incorporation is often retarded, especially in diluted, weakly saturated systems (28), will finally integrate the crystal lattice during the last stages of the growth process, and therefore be found located near the outer crystallite surface.

The Al-Ga gradient concentration could indeed be detected by using spot EDX probing of various individual crystallites, either near their edge or in their center (Table 2). The results presented for various systems synthesized either from methylamine or from fluoride media, do confirm the above proposed scheme. However, the gradient is very small for zeolites synthesized in the presence of methylamine, with respect to those crystallized from fluoride media (19). This suggests that a quasi-homogeneous Ga-Al distribution is achieved when MA is used as mobilizer, probably due to a very similar stability of Ga-MA and Al-MA complexes. This trend is also observed for pure Ga and Al end-members of the series (ZN-82 and ZN-92), that also involve a quasi-homogeneous repartition of the trivalent ions throughout the crystallites (Table 2, core/rim molar ratios in bold characters; see also footnote (3)).

Stability of Al^{3+} and Ga^{3+} ions in calcined (Ga,Al)-MFI zeolites

Table 3 gives the actual residual Al and Ga concentrations in the framework of calcined zeolites of the same series. As it has been demonstrated elsewhere (22), ^{71}Ga-NMR is no longer accurate for a quantitative determination of framework Ga in calcined samples. Indeed, for the same samples calcined under comparable conditions, the intensities of the resonance line characterizing the residual framework tetrahedral Ga^{3+} species are random and sometimes even "invisible", thus broadened beyond detection. This is probably due to strong quadrupolar interactions occurring in dry, not water-re-equilibrated samples (30). A convenient method to determine the total amount of trivalent ions (Ga^{3+} and Al^{3+}) is to compute the total amount of acid sites generated by both ions in the framework, by e.g. using the classical technique of TPD of ammonia evolved upon heating the corresponding NH_4^+-exchanged samples. The knowledge of the total trivalent concentration by TPD and of the Al^{3+} concentration by ^{27}Al-NMR, directly yields the residual Ga^{3+} concentration that remains in the lattice after the calcination experiments (Table 3, column 4).

One can see that under our mild (hydro)thermal calcination conditions, namely a slow heating in dry N_2 flow to 550°C followed by an isothermal heating at the same temperature under a weak, possibly slightly moist (ambient) air flow, the zeolites are not dealuminated. Conversely, the framework Ga is found extracted from the lattice of all samples (column 5). Undoubtedly, the amount of Ga^{3+} ions that remain in the lattice after calcination, or, in other words, the general stability of the framework Ga species, is noteworthily influenced by the actual Al^{3+} framework content. The richer the (Ga,Al)-MFI framework in Al^{3+}, the easier the Ga^{3+} extraction from it, upon calcination.

Table 2
Spot EDX analysis of Al and/or Ga concentrations (mol. per unit cell of 96 T atoms, T=Al+Ga+Si) in the core and at the outer rim of individual (Ga)-,(Al)- and (Ga,Al)-MFI crystallites synthesized from fluoride (F^-) or methylamine (MA) media.

Sample (code) (1)	Ga/unit cell (2) Bulk	Core	Rim	Core/Rim(3)	Al/unit cell (2) Bulk	Core	Rim	Rim/Core(3)	Reference
AS-206 (Ga,Al) (F^-)	1.96 (AA)	3.70 (M)	1.35 (M)	2.7	1.02 (AA)	0 (M)	0.66 (M)	(∞)	(19)
	1.85	2.42	0.80	3.0	1.05	0.3	0.91	3.0	This work
ZN-80 (Ga,Al) (MA)	1.30(NMR)	1.10	1.00	**1.10**	2.05 (NMR)	1.9	2.26	**1.19**	"
	1.24	1.80	1.60	**1.13**	1.0	1.17	1.35	**1.15**	"
ZN-83 (Al) (F^-)	–	–	–	–	4.0 (NMR)				
					4.0	2.40	5.80	2.42	"
ZN-82 (Al) (MA)	–	–	–	–	3.90 (NMR)				
					4.10	3.77	3.96	**1.05**	"
ZN-92 (Ga) (MA)	2.95(NMR)				–	–	–	–	
	2.80	2.28	2.50	**0.91**					"
AS-207 (Ga) (F^-)	1.03 (AA)	3.10 (M)	0.70 (M)	4.40 (M)	–	–	–	–	(19)
	1.20	3.31	1.28	2.60					This work

(1) All ZN-type samples are synthesized in the presence of methylamine or NH_4F, as described in the experimental part (185°C, 5 days, static conditions).The AS-type samples, kindly provided by Dr. J. L. Guth (Mulhouse, France) were sythesized in the presence of NH_4F (200°C, 16-23 days, static conditions, as described in ref. (19). All the samples are 100% crystalline MFI zeolites (n-hexane sorption data). Note that pure (Al_4)-MFI crystallizes in the presence of NH_4F after only 5 days (sample ZN-83), while the corresponding gallo(alumino)silicates need a far longer time to be 100% crystalline (samples of AS-type). These features are discussed in a subsequent paper (27).

(2) All analyses by EDX, except "AA " (Atomic Absorption) or "M" (Electron Microprobe) performed in Mulhouse (19), and "NMR" (quantative ^{27}Al- and ^{71}Ga-NMR) as described elsewhere (22).

(3) For the sake of clarity, the core/rim (or rim/core, as to keep positive values) molar ratios for samples synthesized in the presence of methylamine, are printed in bold characters.

Table 3
Framework Al and Ga contents (mol. per unit cell involving 96 T atoms) in various calcined (under N_2 flow, then in ambient air, at 550°C) and further NH_4^+-exchanged (Ga,Al)-MFI zeolites.

Sample (code)	Al (NMR) =(a)	Al+Ga (1) (TPD) =(b)	Ga (c)=(b)-(a)	% Ga extracted upon calcination (2)	% Ga in acid sites =(c/b)X100	Acid site strength (T°C max. /TPD)
ZN-60	0	1.91	1.91	20	100	386
ZN-58	0.97	2.68	1.71	4	64	398
ZN-64	1.05	2.80	1.75	4	63	404
ZN-59	1.98	2.87	0.89	25	31	416
ZN-65	1.96	2.80	0.84	24	30	420
ZN-67	2.07	2.74	0.67	33	24	418
ZN-60	2.90	2.86	0	100	0	438
ZN-66	2.80	2.68	0	100	0	436
ZN-62	3.98	3.84	0	-	-	450

(1) Calculated from the total number of acid sites, as determinated by NH_3-TPD on NH_4^+-exchanged samples.

(2) $\frac{\text{Ga (as synth.)-Ga (calc.)}}{\text{Ga (as synth.)}} \times 100$

Finally, all the Ga^{3+} ions are released from samples that initially involve 3 Al^{3+} and 0.61 Ga^{3+} in the framework, while only about 4% of the 1.84 framework Ga^{3+} ions are released when only 1 Al^{3+} completes the framework composition. By contrast, more gallium is lost from the pure gallosilicate ZN-61 (about 20%), possibly suggesting that small amounts of Al^{3+} in the lattice (in ZN-58 and ZN-64) play a kind of stabilizing role towards framework Ga^{3+}. Such a preliminary observation undoubtely needs confirmation and generalization and is now subject of further investigations (27).

Other interesting data can be derived from these determinations. We can, for example, evaluate the amount of framework Ga^{3+} ions that still (partly) generate the acidity of the calcined samples (column 6), directly related to the percentage of Ga extracted from the lattice upon calcination (column 5).

Another remarkable feature is the regular variation of the temperature of the TPD peaks corresponding to the NH_3 release from the acid sites, generated by both Ga^{3+} and Al^{3+} framework ions (Table 3, column 7). This temperature smoothly increases from 360°C (pure Ga^{3+}-sample), to 450°C for the pure Al^{3+} counterpart. This is a good indication for a relative acid site homogeneity, thus suggesting that Al^{3+} and Ga^{3+} ions are quasi-simultaneously incorporated in the growing zeolite framework during the crystallization process, leading to a rather homogeneous final Ga-Al distribution within the lattice. This observation, further confirmed by monitoring the Pr_4N^+ decomposition in these samples by DSC (27), is also in line with the EDX measurements.

The relatively high TPD temperatures, with respect to those reported for various MFI gallo- or gallo-aluminosilicates synthesized from Na^+-bearing gels (20,27,31), also suggest that the corresponding acid sites are substantially stronger in our series of samples than in zeolites currently prepared by using more conventional methods.

Our investigations on the stability of the lattice Ga species in these materials open new possibilities to improve their catalytic performances, by monitoring the relative strength and concentration of acid sites, as well as the amount and dispersion of the extra-lattice Ga_xO_y oxidic species, through a selective adjustment and a careful control of either some specific synthesis parameters (e.g. Al^{3+} concentration) or post-synthesis (hydro)treatments.

ACKNOWLEDGMENTS

The authors would like to thank Miss C. Mayenez and Miss M. Wiame (Namur, Belgium) for their efficient participation in the synthesis work and their skillful technical help. They are indebted to Dr. J. L. Guth (Mulhouse, France) for providing some zeolite samples synthesized in his laboratory.

REFERENCES

1. Inui, T., Makino, Y., Okazumi, F., Nagano, S. and Miyamoto, A., Ind. & Eng. Chem. Res., **26**, 647 (1987).
2. Petit, L., Bournonville, J. P. and Raatz, F., in: *Zeolites: Facts, Figures, Future,* Eds. P. A. Jacobs, and R. A. Van Santen, (Proc. 8th Int. Zeolite Conf.), Elsevier, Amsterdam, The Netherlands, pp. 1163-1171 (1989).
3. Inui, T., Ishikawa, Y., Kamachi, K. and Matsuda, H., in ref. 2, pp. 1183-1192.
4. Kanai, J. and Kawata, N., Appl. Catal. **55**, 115 (1989).
5. Khodakov, A. Yu., Kustov, L. M., Bondarenko, T. N., Dergachev, A. A., Kazansky, V. B., Minachev, Kh. M., Borbely, G. and Beyer, H. K., Zeolites, **10**, 603 (1990).
6. Price, G. L. and Kanazirev, V., J. Mol. Catal., **66**, 115 (1991).
7. Meriaudeau, P., Sapaly, G. and Naccache, C.,in ref.2, pp. 1423-1431.
8. Fujimoto, K., Nakamura, I. and Yokota, K., Zeolites, **9**, 120 (1989).
9. Meriaudeau, P. and Naccache, C., J. Mol. Catal. **59**, 1312 (1990).
10. Gnep, N. S., Doyemet, J. Y. and Guisnet, M., in *Zeolites as Catalysts, Sorbents and Detergent Builders,* Eds. H. G. Karge and J. Weitkamp, Elsevier, Amterdam, The Netherlands, pp.153-262 (1989).
11. Fujimoto, K., Nakamura, I., Yokota, K. and Aimoto, K., Bull. Chem. Soc. Japn. **64**, 2275 (1991).
12. Corma, A., Goberna, C., Lopez Nieto, J. M., Peredes, N. and Perez, M., Stud. Surf. Sci. Catal. **69**, 409 (1991).
13. Kanazirev, V., Price, G. L. and Dooley, K. M., J. Chem. Soc. Chem. Commun. **1990**, 712.
14. Kitagawa, H., Sendoda, Y. and Ono, Y., J. Catal., **101**, 12 (1986).
15. Simmons, D. K., Szostak, R., Agrawal, P. K. and Thomas, T. L., J. Catal., **106**, 287 (1987).
16. Belussi, G., Millini, R., Carati, A., Maddinelli, G. and Gervasini, A., Zeolites, **10**, 642 (1990).
17. Bayense, C. R., Van Hooff, J. H. C., Kentgens, A. P. M., De Haan, J. W. and Van de Ven, L. M. J., J. Chem. Soc. Chem. Comm., **1989,** 1292.
18. Kucherov, A. V., Slinkin, A. A., Beyer, H. K. and Borbely, G., J. Chem. Soc. Faraday Trans, 1, **85**, 2737 (1989).
19. Seive, A., Ph. D. Thesis, University of Mulhouse, France, 1990.
20. Challoner, R., Harris, R. K., Barri, S. A. I. and Taylor, M. J., Zeolites, **11**, 827 (1991).
21. Gabelica, Z. and Guth, J. L., in ref. 2, pp. 421-430; Gabelica, Z. and Guth, J. L., Angew. Chem. Int. Ed. Engl., **28**, 81 (1989).
22. Gabelica, Z., Mayenez, C., Monque, R., Galiasso, R. and Giannetto, G., in *Advances in Zeolites and Pillared Clays Synthesis,* Eds. M. L. Occelli and H. E. Robson, Van Nostrand Reinhold, New-York, Chap. 15, 1992, in press; ibid, U. S. Patent, disclosed in 1992.
23. Gabelica, Z., B.Nagy, J., Derouane, E. G. and Gilson, J. P., Clay Minerals, **19**, 803 (1984).
24. Wagner, W. and Hull, C. J., *Inorganic Titrimetric Analysis,* Marcel Dekker, Inc., New-York, 1971
25. Gabelica, Z., Blom, N. and Derouane, E. G., Appl. Catal. **5**, 227 (1983).
26. B.Nagy, J., Bodart, P., Collette, H., El Hage-Al Asswad, J., Gabelica, Z., Aiello, R., Nastro, A. and Pellegrino, C., Zeolites, **8**, 209 (1988).
27. Gabelica, Z., Wiame, M., Galiasso, R., Monque, R. and Giannetto, G., in preparation.
28. Rollmann, L. D. and Valyocsik, E. W., Eur. Patent, 21,674 and 21,675 (1981).
29. Sheka, I. A., Chaus, I. S. and Mityureva, T. T., *The Chemistry of Gallium,* Elsevier, Amsterdam, pp. 49 (1966).
30. Kentgens, A. P. M., Bayense, C. R., Van Hooff, J. H. C., De Haan, J. W. and Van de Ven, L. J. M., Chem. Phys. Lett., **176**, 399 (1991).
31. Chu, C. T. W. and Chang, C. D., J. Phys. Chem. **89**, 1569 (1985).

PREPARATION OF ZEOLITE FILM USING CELLULOSE MOULDING

T. SANO, Y. KIYOZUMI, K. MAEDA, M. TOBA, S. NIWA, and F. MIZUKAMI

National Chemical Laboratory for Industry, Tsukuba, Ibaraki 305, Japan

ABSTRACT

The crystallization of silicalite, ZSM-5, ZSM-11 and SAPO-5 was carried out in the presence of the cellulose moulding such as a filter paper. It was found that the zeolite-cellulose composites with the high structural strength are obtained, and that the zeolite films are not disintegrated by the calcination process at 500~600°C. The shape of zeolite film was the same as that of the filter paper used, indicating that the shape of the zeolite film is easily controlled. Nitrogen adsorption experiments suggested that mesopores ($2<r_p<100$ nm) hardly existed in the film. Mercury porosimetry, however, showed the existence of large pores (r_p: ca.10000 nm) in the film.

INTRODUCTION

Zeolites and molecular sieves are widely used in many chemical and physical processes such as heterogeneous catalysis and gas separation. Recently, great interest has been focused on zeolite membranes or films due to their uniform pore sizes and resistance to high temperatures. There are many articles and patents concerning zeolite membranes. Most of the zeolite membranes have been made by embedding the zeolite crystals in polymer or ceramic matrixes or by the in situ crystallization of zeolites on porous substrates [1]. We have studied the synthesis of pure zeolite film from the standpoint of understanding of growth process of zeolite crystals. As the Al distribution and the change in the morphology along the width of film reflect the time course of the crystal growth, we can get much information regarding the crystal growth process by analyzing the film obtained. More recently, we succeeded in preparing pure ZSM-5 zeolite films of 30~100 μm thickness, which are made up of zeolite crystals themselves, by using a Teflon slab or a Teflon sleeve [2-4]. Large differences in the surface morphology and the surface Al concentration between both sides of the film were observed. An amount of Al was hardly detected on the Teflon side. It was concluded that during the initial crystallization of ZSM-5 zeolite, the zeolite crystals with high SiO_2/Al_2O_3 ratios are generated. The similar pure zeolite films were also synthesized by Tsikoyiannis et al. [5] and Anderson et al. [6]. However, the pure zeolite films obtained on the Teflon surface are fragile and the control of the shape of zeolite film is not easy. Much improvement must be made before the zeolite film can be applied in many fields. In this paper, we will describe a new preparation method of silicalite, ZSM-5, ZSM-11 and silicoaluminophosphate(SAPO-5) films using the cellulose moulding such as a filter paper.

EXPERIMENTAL

The hydrothermal synthesis of silicalite, ZSM-5 and ZSM-11 zeolite films was performed as follows. Aluminum nitrate and colloidal silica were added to a stirred mixture of

From the Proceedings of the 9th International Zeolite Conference, Montreal 1992, Eds. R. von Ballmoos et al.,

tetrapropylammonium bromide (TPABr) or tetrabutylammoninum bromide (TBABr) and sodium hydroxide in solution, to give a hydrogel with a composition of 0.1 TPABr(TBABr)-0.05 Na_2O-(0~0.01) Al_2O_3-1.0 SiO_2-(70~100) H_2O. Then the hydrogel was transferred to a stainless steel autoclave with a Teflon sleeve. A piece of filter paper was immersed in the solution and placed vertically along the axis of the autoclave. The autoclave was placed in an air-heated oven at 170 °C for appropriate periods. After the appropriate periods, the autoclave was cool down, and the filter paper was recovered. The zeolite-cellulose composite was washed with deionized water and dried at 120°C for 24 h. The hydrothermal synthesis of molecular sieve SAPO-5 film was conducted in a similar manner described above. Aluminium isopropoxide or boehmite was added to a mixture of 85% phosphoric acid and water, subsequently, colloidal silica and 2-diethylethanolamine (DEA) were added under vigorous stirring to give a hydrogel with a composition of 1.0 P_2O_5-1.0 Al_2O_3-0.30 SiO_2-1.0 DEA-47 H_2O. The hydrogel was aged overnight at room temperature and transferred to the autoclave. The crystallization was conducted at 200°C for appropriate periods.

The identification of the products was achieved by X-ray diffraction (MAC Sci. MXP18). To analyze a film, the incident beam fell at a glancing angle of 1.00°. Concentration of Si, Al and P on the surface of the film was measured by energy-dispersive X-ray analysis (EDX) on a Kevex Delta system attached to a scanning electron microscope (Hitachi S-800) with an accelerating voltage of 15 kV. The determination of the bulk chemical composition of the films were also carried out by wet chemical analysis by means of inductively coupled plasma emission spectrometry (Seiko ICP-SPS-1200VR). The amount of cellulose in the composite was determined by thermal analysis (MAC Sci. TG-DTA 2000). The pore size distribution of the film was measured by a BELSORP 36 nitrogen porosimeter and a Micromeritics AutoPore 9200 mercury porosimeter.

RESULTS AND DISCUSSION

Synthesis of Silicalite, ZSM-5 and ZSM-11 zeolite films

The crystallization of silicalite, ZSM-5 and ZSM-11 zeolites was carried out in the presence of the cellulose moulding such as a filter paper. The synthesis conditions of the zeolites were summarized in Table 1. In the case of silicalite, the zeolite-cellulose composite was easily obtained. Due to gravity, powdery silicalite crystals were also obtained on the bottom of the autoclave. Figure 1 shows the X-ray diffraction diagrams of the composites prepared at varying crystallization times. The intensities of the reflection peaks corresponding to silicalite increased with an increase in the crystallization time. After crystallization for 24 h, the peaks (2θ : ca.15, 16, 23) corresponding to cellulose were not observed in the X-ray diffraction diagram of the composite. Figure 2 shows the scanning electron micrographs of the composites and the filter paper used. Cellulose fibers as well as small silicalite crystals were observed in the micrograph of the composite after crystallization of 3 h. It was also observed that several silicalite crystals grew through cellulose fibers, as shown in Figure 3. Taking into account the existence of OH groups in the cellulose, this seems to suggest that cellulose fibers provide the unique sites for crystallization of zeolite. The surface of the filter paper became gradually covered with densely packed zeolite crystals 5~30 μm in size, and then cellulose fibers could not be observed in the micrograph after crystallization of 24 h. The thickness of the composite at crystallization time of 48 h was ca. 500 μm and was ca. 2 times that of the filter paper used. Scanning electron micrographs of the silicalite film (Run no.5) prepared with a Teflon slab was shown for a reference in Figure 4. It may be noted that the sizes of the silicalite crystals on the solution side of the film were similar to those on the composite. The thermogravimetric (TG) measurement of the composite is illustrated in Figure 5. Two distinct steps were observed in the composite. These steps were exothermic processes. Considering that only one step was observed in the TG curve of the pure silicalite film prepared with the Teflon slab, it seems to be obvious to ascribe the first step to the decomposition of the filter paper and the second one to the destructive desorption of TPA cations occluded in the zeolite framework. The composite was calcined at 500°C for 20 h in order to decompose and burn the cellulose. However, the zeolite film was not disintegrated by the calcination process. This suggests that the silicalite crystals are strongly bonded together in the form of a continuous film. Figure 6 shows the pore size distributions of the

Table 1. Preparation of zeolite films using filter paper under various synthesis conditions[a)]

Run no.	Zeolite	Synthesis conditions			Film
		SiO_2/Al_2O_3	OH^-/SiO_2	Time (h)	
1	Silicalite	∞	0.1	3	×
2		∞	0.1	12	×
3		∞	0.1	24	o
4		∞	0.1	48	o
5[b)]		∞	0.1	48	o
6		∞	0.2	48	o
7		∞	0.3	48	×
8	ZSM-5	200	0.1	48	o
9		100	0.1	48	×
10		100	0.1	96	o
11[b)]		100	0.1	48	o
12	ZSM-11	200	0.1	72	×
13		200	0.1	240	×
14		200	0.1	504	o
15[b)]		100	0.1	216	o

a) $TPABr/SiO_2$=0.1, Temp.=170°C
b) A Teflon slab was used instead of a filter paper.

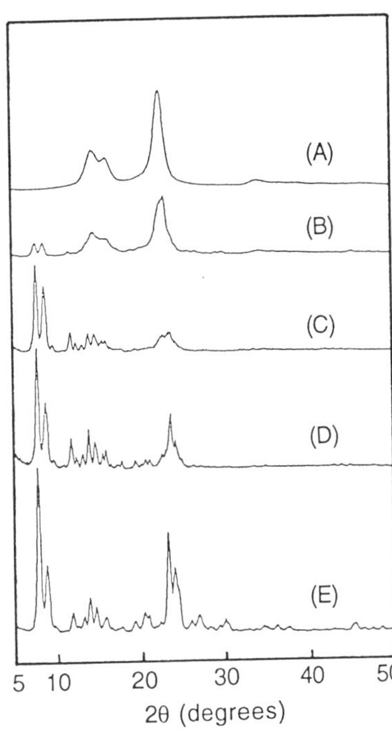

Figure 1. X-ray diffraction diagrams of silicalite-cellulose composites prepared at varying crystallization times : (A) 0h, (B) 3h, (C) 12h, (D) 24h, (E) 48h

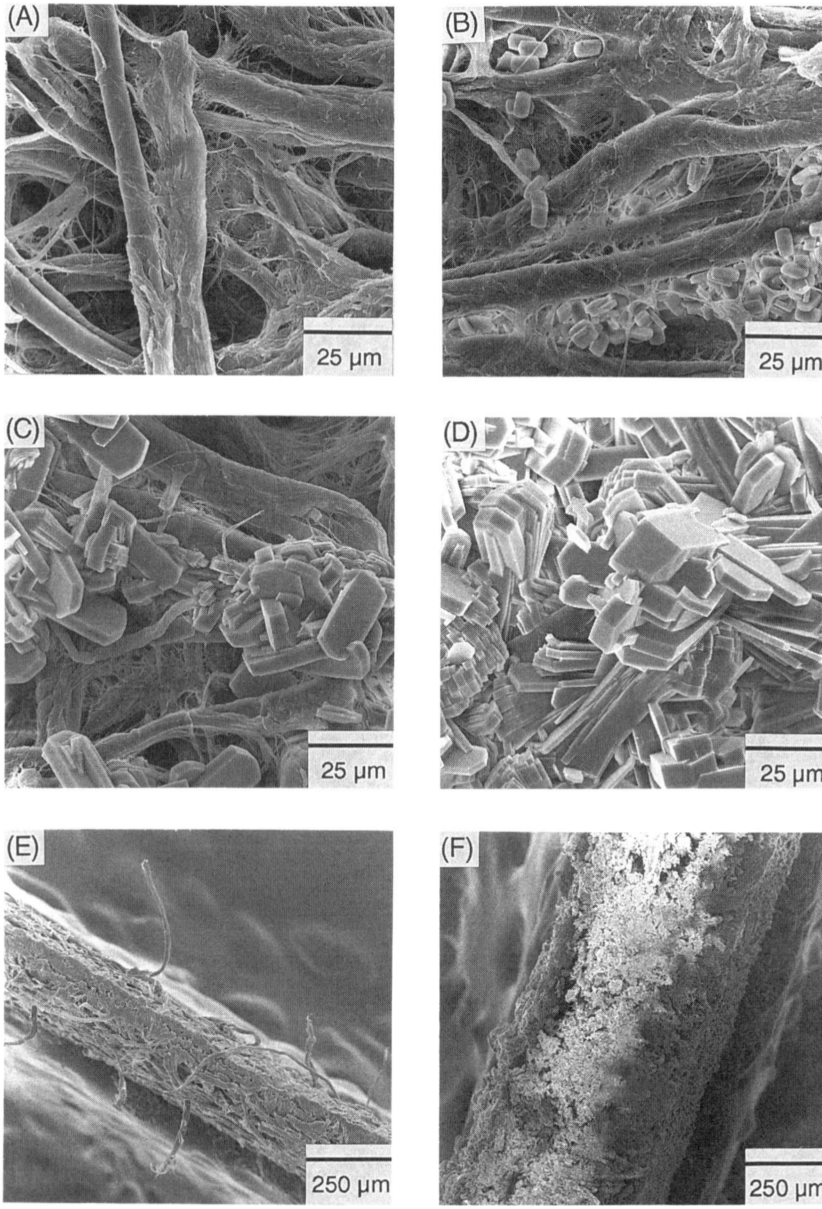

Figure 2. Scanning electron micrographs of silicalite-cellulose composites prepared at varying crystallization times: (A) 0h, (B) 3h, (C) 12h, (D) 48h, (E) Cross section of filter paper, (F) Cross section of the composite after crystallization of 48 h.

films determined by nitrogen adsorption and mercury porosimetry. No peak was observed in the radius range of 2~100 nm. However, mercury porosimetry showed the peak at ca.10000 nm. Considering that the intensity of the peak in the silicalite-cellulose composite before calcining at 500°C was considerably smaller than that in the silicalite film after calcining, the peak seems to be attributable to the void space within the film, which is generated by burning of the cellulose. Our preliminary experiments suggested that the void space is filled up to some extent with amorphous materials by treatment of the film with metal alkoxide solutions. Figure 7 shows an optical photograph of the silicalite-cellulose composites (Run no. 4). The shape of the composite was the same as that of the filter paper used, indicating that the shape of zeolite film is easily controlled. These results suggest that such silicalite films can be applied to many fields.

In the case of ZSM-5 and ZSM-11, the synthesis of the zeolite-cellulose composite needed a longer crystallization time, as compared to the synthesis of zeolite film using the Teflon slab. Scanning electron micrographs of the zeolite-cellulose composites obtained were shown in Figure 8 (Run nos. 10 and 14). The surface of the filter paper consisted of the densely packed zeolite crystals. The surface SiO_2/Al_2O_3 ratios of the two composites were lower than those expected from the synthesis mixtures. This enrichment of Al on the surface is consistent with the results for the zeolite film prepared with the Teflon slab. It is well recognized that the crystallization rates of ZSM-5 zeolites decrease as the Al content increases (below a SiO_2/Al_2O_3 ratio of ca.100) [7,8]. To clarify

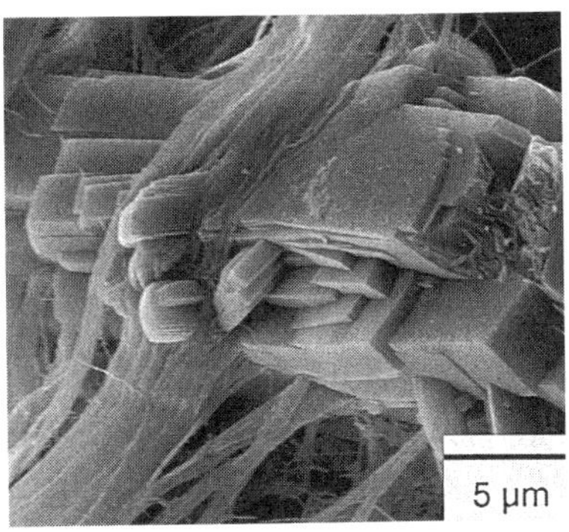

Figure 3. Scanning electron micrograph of crystal growth through cellulose fiber.

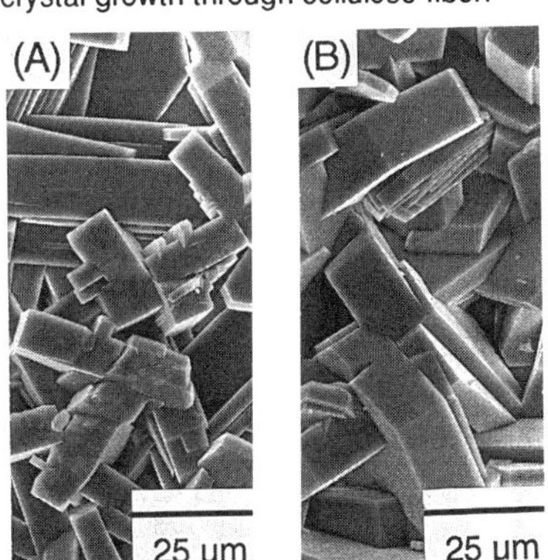

Figure 4. Scanning electron micrographs of the silicalite film prepared with a Teflon slab. (A) Teflon side, (B) Solution side

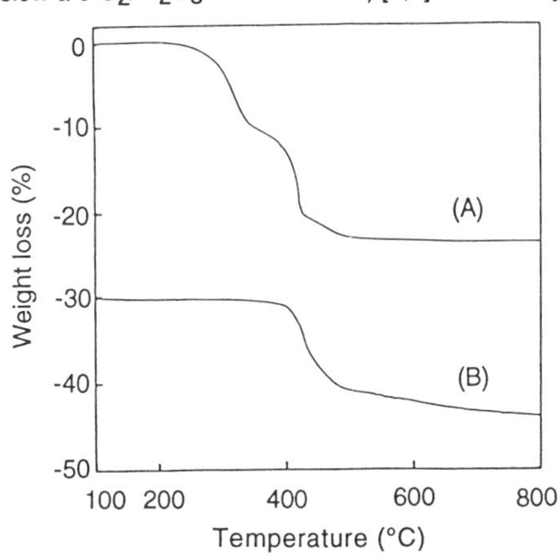

Figure 5. TG curves of the silicalite films prepared with (A) filter paper and (B) Teflon slab at crystallization time of 48h.

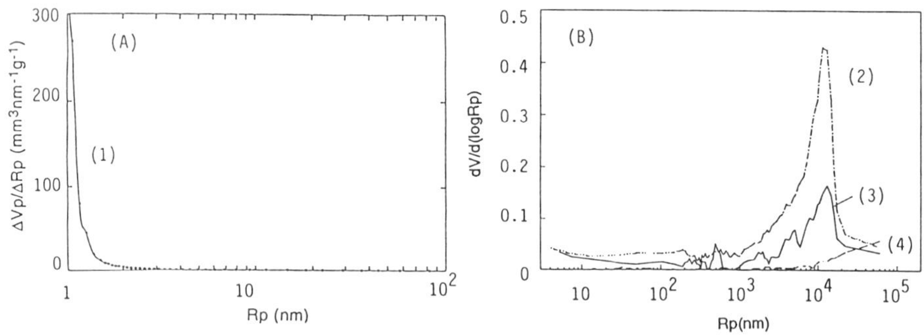

Figure 6. Pore size distributions of the silicalite films determined by (A) nitrogen adsorption and (B) mercury porosimetry.
(1), (2) : after calcination at 500°C, (3) : before calcination, (4) : the silicalite film (after calcination at 500°C) prepared with a Teflon slab.

the reason for the difficulty of preparing the zeolite-cellulose composites with high Al contents, therefore, the filter paper was immersed into the synthesis mixture at 50°C for 3~24 h and the surface concentration of Al was analyzed by EDX. The surface SiO_2/Al_2O_3 ratios of the filter paper were 30~55 and were lower than those expected from the synthesis mixtures (100~200), indicating an enrichment of Al on the surface. This enhancement of Al seems to cause the requirement of the longer crystallization time of ZSM-5 and ZSM-11 on the filter paper.

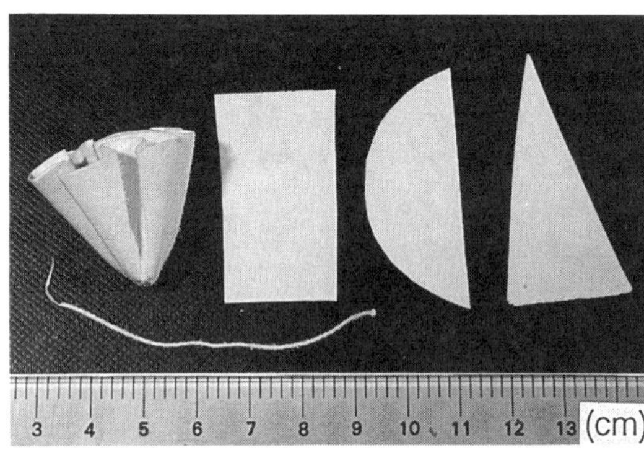

Figure 7. Optical photograph of silicalite cellulose composites.

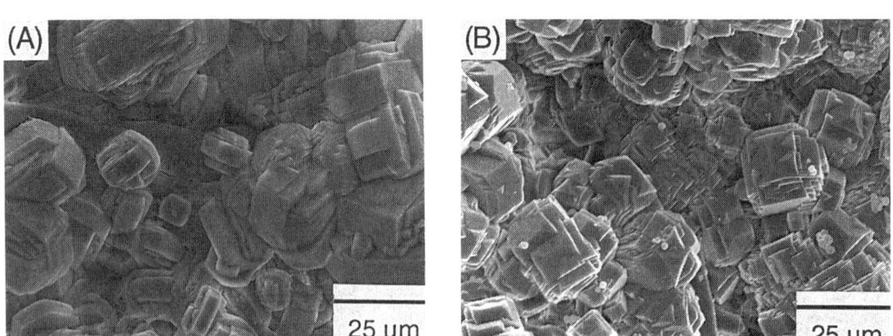

Figure 8. Scanning electron micrographs of (A) ZSM-5- and (B) ZSM-11-cellulose composites.

Synthesis of SAPO-5 film

The synthesis SAPO-5 film was also carried out in a similar method described above. As part of the filter paper was dissolved due to the hydrolysis of cellulose in a phosphoric acid solution, the shape of the SAPO-5-cellulose composite could not be controlled by the filter paper used. The distinct peaks corresponding to SAPO-5 was observed in the X-ray diffraction diagram of SAPO-5-cellulose composite (Figure 9). No peak corresponding to the cellulose was observed. Figure 10 shows the scanning electron micrographs of the composites prepared by using aluminum isopropoxide and boehmite as a Al source. The surface morphology of the composite strongly depended upon the Al source used. The thickness of the composite was ca.500 μm. The film was not disintegrated by the calcination process at 600°C for 24 h. The amount of the cellulose calculated by TG analysis was ca.5 wt%. The surface atomic ratios of P/Al and Si/Al determined by EDX are summarized in Table 2. The pore size distribution of the film was similar to that of the silicalite film, indicating the existence of the void space within the film.

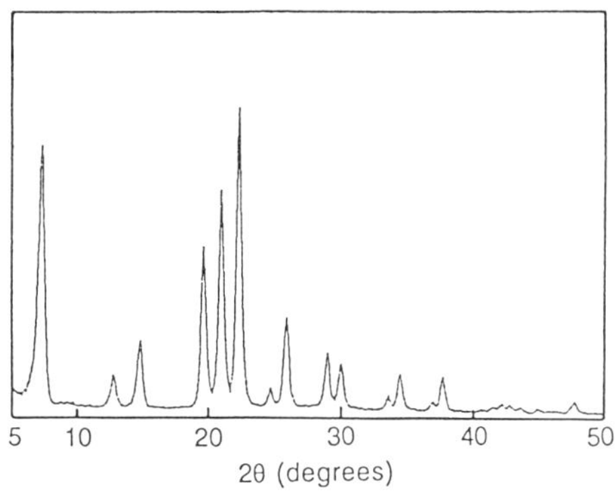

Figure 9. X-ray diffraction diagram of SAPO-5-cellulose composite.

Table 2. Characteristics of SAPO-5-cellulose films

Run no.	Al source	Thickness (μm)	P/Al	Si/Al
16	aluminium isopropoxide	480	1.09	0.201
17	boehmite	520	1.18	0.154

Figure 10. Scanning electron micrographs of SAPO-5-cellulose composites prepared with (A) aluminium isopropoxide and (B) boehmite as a Al source.

CONCLUSION

The crystallization of silicalite, ZSM-5, ZSM-11 and SAPO-5 was carried out in the presence of the filter paper. It was found that the zeolite-cellulose composites with the high structural strength are obtained, and that the zeolite films are not disintegrated by the calcination process. From nitrogen adsorption and mercury porosimetry, it was found that mesopores originated from grain boundary were not observed in the radius range of 2~100 nm, but large pores with the radius of ca.10000 nm were observed, indicating the existence of the void space within the film generated by burning of the cellulose. If the void space is completely filled up with amorphous materials, this suggests strongly that this kind of the film is applied to many fields such as gas separation, water treatment and hydrocarbon processing. Although the exact interpretation of the role of the filter paper in the preparation of the zeolite film could not be given at the present time, we consider that OH groups in the cellulose provide the unique sites for the crystallization of zeolite.

REFERENCES

1. For example ; (1) Wernick,D.L. and Osterhuber,E.J., J.Membrane Sci., **22**, 137 (1985), (2) Demertzis,M. and Evmiridis, N.P., J.Chem.Soc.,Farady Trans., **82**, 3647 (1986), (3) Ishikawa,A., Chiang,T.H. and Toda,F., J.Chem.Soc.Chem.Commun., 764 (1989), (4) Bein,T., Brown,K. and Brinker,C.J., Stud.Surf.Sci.Catal., **49**, 887 (1989), (5) Goldman,M., Fraenkel,D. and Levin,G., J.Appl.Polym.Soc., **37**, 1791 (1989), (6) Michaels, A.S., Chemtech, 162, (1989), (7) Geus,E.R., Jansen,A.E., Jansen,J.C., Schoonman,J. and van Bekkum,H., Stud.Surf.Sci.Catal., **65**, 457 (1991), (8) Sakurada,S., Tagaya,N., Maejima,T. and Yoshida,T., Jpn. Kokai Tokkyo Koho, 59-213,615 (1984), (9) Miyazaki,K. and Mitsumata,T., Jpn. Kokai Tokkyo Koho, 60-129,119 (1985), (10) Suzuki,H., Eur.Pat.Appl., 180,200 (1985), 135,069 (1986), US Pat., 4,699,892 (1987), (11) Ooyama,M., Jpn.Kokai Tokkyo Koho, 63-291,809 (1988), (12) Lachman,I.M. and Patil,M.D., US Pat., 4,800,187 (1989)
2. Sano,T., Kiyozumi,Y., Kawamura,M., Mizukami,F., Takaya,H., Mouri,T., Inaoka,W., Toida,Y., Watanabe,M. and Toyoda,K., ZEOLITES, **11**, 842 (1991)
3. Sano,T., Kiyozumi,Y., Mizukami,F., Takaya,H., Mouri,T. and Watanabe,M., ZEOLITES, **12**, 131 (1992)
4. Sano,T., Mizukami,F., Takaya,H., Mouri,T. and Watanabe,M., Bull.Chem.Soc.Jpn., **65**, 146 (1992)
5. Tsikoyiannis,J.G. and Haag, W.O., ZEOLITES, **12**, 126 (1992)
6. Anderson,M.W., Pachis,K.S., Shi,J. and Carr,S.W., J.Mater.Chem., **2**, 255 (1992)
7. Derouane,E.G., Gilson,J.P. and Nagy,B.J., J.Mol.Catal., **10**, 331 (1981)
8. Rollmann,L.D. and Valyocsik,E.W., Eur.Pat.Appl., 21,674 (1981)

CONTROLLED GROWTH OF THIN FILMS OF MOLECULAR SIEVES ON VARIOUS SUPPORTS

J.C. JANSEN, W. NUGROHO, and H. van BEKKUM

Delft University of Technology, Laboratory of Organic Chemistry and Catalysis,
Julianalaan 136, 2628 BL Delft, The Netherlands.

ABSTRACT

MFI-type zeolite films were grown on different substrates. The zeolite crystals could be oriented either in a lateral or in an axial orientation with respect to the substrate surface depending upon the presence or absence of a continuous gel layer, respectively.

INTRODUCTION

The preparation of layers, films and coatings of zeolites has been reported recently.[1-5] The polycrystalline material with a thickness between 3 and 30 μm was precipitated on metallic-, monolithic ceramic-, teflon-, and silica substrates. The main objectives to prepare these composite products are the potential application as catalyst, membrane and sensor.

For certain catalyst uses the axial orientation of zeolite crystals on a support is of interest in order to expose the zeolite pores as good as possible. In the case of membrane application a lateral orientation of the crystals on a macroporous support is required.

In this work the parameters to achieve either type of orientation of the zeolite crystals have been studied for MFI-type zeolite.

EXPERIMENTAL

Crystal supports and set-ups

Metal supports were horizontally positioned on the bottom of a 3 ml teflon lined autoclave with a diameter of 20 mm which was filled with 2 ml synthesis mixture. The metals chosen were: aluminum foil, aluminum-, nickel-, copper- and titanium platelets of 10×10×2 mm. All metal supports were cleaned with ethyl alcohol before use.

The following supports were vertically positioned against the wall of a 35 ml teflon lined

From the Proceedings of the 9th International Zeolite Conference, Montreal 1992, Eds. R. von Ballmoos et al.,

autoclave:

- Stainless steel gauze (50×50 mm) with a wire diam. of 0.08 mm and a spacing of 0.1 mm.
- Modules of cordierite and mullite, 10×10×40 mm.
- Mica.
- Quartz platelets, diameter 15 mm.

Silicon [100] wafers were used in particular model studies. They were obtained from a silicon single crystal, grown by a crystal pulling method. The wafers were 0.7 mm thick and 98 mm in diameter. Prior to cutting to provide platelets of 10×10 mm, the wafers were chemically and mechanically treated and polished on one side down to atomic scale.

Contaminants will be present on the surface after these operations, so in order to provide a clean hydrophilic silicon surface, the platelet has to be cleaned. Firstly, gross organic contaminants were removed by treating of the platelets in boiling xylene. Solution RCA-1, see Table 1, is thereafter used to remove residual organic contaminants. In this step, the platelets are already placed vertically in teflon holders, see Figure 1. Inorganic contaminants were removed with RCA-2 solution.[6,7] As final step, the platelets were exposed to sonification in water. These cleaning operations are given in Table 1.

Table 1: Cleaning operations data of the silicon platelets.

Cleaning sequence	Composition	Conditions
Xylene	Xylene	Boiling point 30 minutes
RCA-1	H_2O: H_2O_2: NH_4OH 5: 1: 1*	80°C ± 5°C 15 minutes water rinse
RCA-2	H_2O: H_2O_2: HCl 6: 1: 1*	80°C ± 5°C 15 minutes water rinse
Ultra sonic	H_2O	Room temperature 10 minutes water rinse
H_2O : distilled deionized water. NH_4OH : 25%, JT Baker.		H_2O_2 : 30%, stabilized, Merck. HCl : 36 - 38%, JT Baker.

*parts by volume

Without drying (to prevent any contamination by the atmosphere), the holder with the platelet was placed in a teflon-lined autoclave containing 32 ml synthesis mixture in such a way that the platelet was positioned in the upper part of the liquid.

Synthesis

The syntheses are summarized in Table 2 and 3.

Table 2: Synthesis conditions using various supports.

Support	Synthesis mixture molar ratio			Temp.	Time
	SiO_2	TPA_2O	H_2O	(K)	(h)
Al-foil	2	1	5600	453	6-8
Al/Ti/Ni/Cu platelets	2	1	13800	448	10
Stainless steel gauze	2	1	1400	443	8
Cordierite	2	1	1400	443	8
Quartz	2	1	2000	453	5
Si : Aerosil 200, Degussa.			TPA : TPAOH, 40%, CFZ.		

Table 3: Synthesis conditions using Si-platelet support, at 436 K

Code	Synthesis mixture molar ratio					Time
	SiO_2	Al_2O_3	TPA_2O	Na_2O	H_2O	(h)
¼A	10.0	-	1.4	-	16800	20.0*
½A	21.2	-	3.0	-	16800	14.0*
A	39.4	-	5.8	-	16800	12.0*
4A	139.4	-	19.7	-	16800	3.5*
6A	209.1	-	27.3	-	16800	3.5*♦
TNA-1	39.4	-	5.5	5.8^{+}	16800	20.0**
TNA-2	39.4	-	5.5	11.8^{+}	16800	22.5*
NA	39.4	-	-	5.5^{++}	16800	20.0**
NAAl	39.4	0.4	-	5.5^{++}	16800	20.0**
Si : TEOS, 98%, Janssen Chimica.				TPA : TPAOH, 40%, CFZ.		
Na : $NaCl^{+}$, Merck or $NaOH^{++}$, Lameris Pleuger.				Al : $Al(NO_3)_3.9H_2O$, Merck.		

* for complete crystallization ♦ etching of silicon wafer

** no crystallization observed

Characterization

XRD, powder X-ray diffraction data were collected on a Philips PW 1840 diffractometer and a Siemens D500 diffractometer, using the Cu-K α line in both cases.

FTIR, infrared spectra were recorded on a Bruker IFS 66 spectrometer in transmission mode in case of the silicon platelets.

SEM, photographs were taken of samples covered with a thin layer of gold by sputtering. The microscope used was a JEOL JSM-35 scanning electron microscope.

RESULTS AND DISCUSSION

Metal supported films

Figure 2 shows an extensive coverage of about 1 μm thick crystals of the MFI-type with mainly the crystal c-direction perpendicular to the plane of the Al-support. The SEM photograph clearly shows a minimum intergrowth and thus as many crystal faces or zeolite pores as possible are exposed this way. Other metal supports such as Ni-,Ti-, and Cu-platelets gave similar results. The crystallization of MFI-type zeolite on a stainless steel gauze, on a cordierite module and a quartz support resulted in an abundant but randomly oriented crystal layer with elongated prismatic forms axially grown on the support surface. At higher concentration the crystals seem to adopt a more lateral orientation as observed on an Al-foil support. The supports used are supposed to be oxidized and form hydroxyl groups protruding into the synthesis mixture under the prevailing conditions. The zeolite growth on the support surface is then expected to start via a condensation mechanism thus resulting in firm chemical bonding of the crystals on the surface even after calcination and also after performing a reaction in a liquid phase using the composite as the catalyst.

Figure 1. Teflon holder with silicon wafer. The holder was used in 35 ml stainless steel teflon lined autoclaves to adjust the silicon platelet vertically and close under the synthesis mixture level.

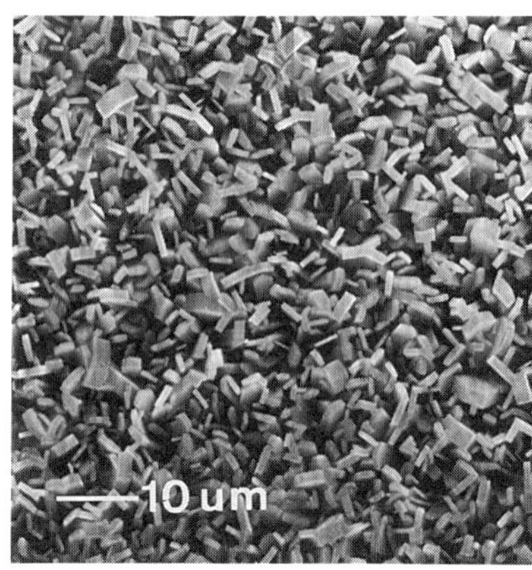

Figure 2. Axially oriented zeolite crystals of the MFI-type on a metal support, mean thickness of the crystals 1 μm.

Silicon supported films

The effect of the concentration of the components in the synthesis mixture on the final growth of a silicon supported zeolite film was studied. It was found that the crystallization time, using the largest observed crystals as a yardstick varies inverse with the concentration, see Figure 3, whereas the final crystal size is smaller at higher concentration, see Figure 4. In the final products the support is for more than 98% covered with a continuous monocrystal film with crystals having a lateral orientation as is shown in Figures 5c and 6b,c for experiments A and 4A, respectively.

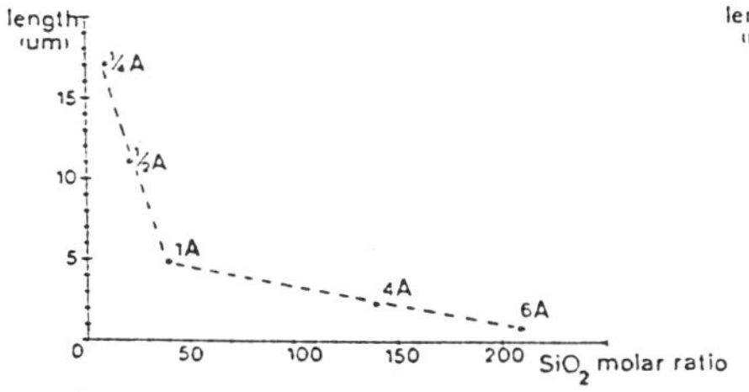

Figure 3. The maximum crystal dimension in c-direction versus the SiO_2 molar ratio of synthesis mixtures ¼A, ½A, A, 4A, 6A.

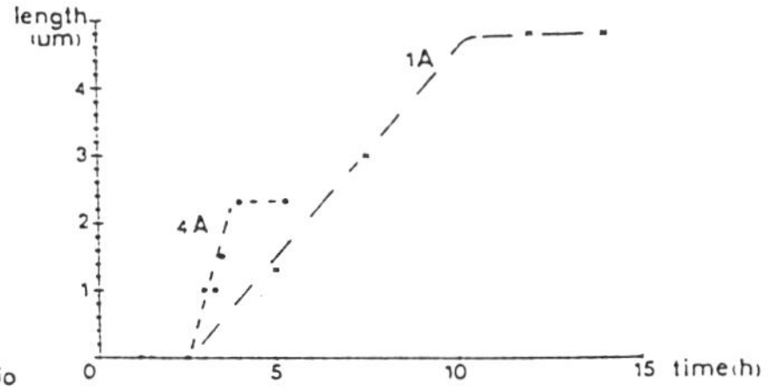

Figure 4. The crystal dimension in c-direction as a function of synthesis time, using synthesis mixtures A and 4A. In both cases a continuous monocrystal film was achieved.

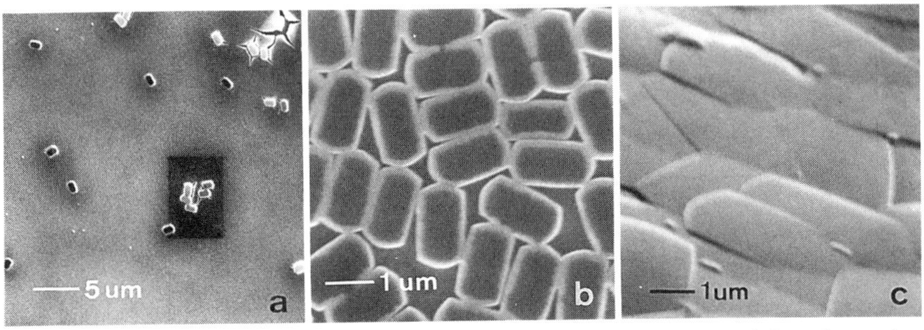

Figure 5. Evolution of the crystal growth of MFI-type with synthesis mixture A, a) first observed crystals after two hours in gel phase, b) crystals apparently associated via surface tension forces after 3¾ hours, oriented with the ac plane parallel to the support surface, c) A 20° view angle on final crystallization of a monocrystal film of 500 nm in thickness, indicating a slight deviation of the crystals with respect to each other and the support surface.

The narrow crystal size distribution of both products indicates an equal growth time for each crystal although Figures 5a and 6a show that the homogeneous nucleation is not an explosive but rather a gradual progressive process. The unexpected homogeneous crystal size distribution which is observed might be due to a drastic change in nutrient concentration during crystallization. Initially small globular shaped fast growing crystals are formed, see Figure 6a, which are embedded in a gel layer of about 0.5 μm thickness which covers the silicon platelets. Accordingly initial growth occurs under conditions of extremely high supersaturations. The smooth rounded and/or cubic form of the initial crystallization, see Figures 5a and 6a, indicates a kinetic roughening of the crystals growing in the gel layer. An analogous observation has been reported earlier for crystallization of zeolites in gel spheres and it was concluded that nucleation occurs at the gel/liquid interface.[8] It is assumed that also in the present experiments nucleation occurs at the gel/liquid interface as shown by the initially observed 'loose' crystallites in the top layer of the gel.

The crystallization then proceeds into the gel (layer thickness about 0.5 μm) which is consistent with the typical crystal form obtained in very high supersaturations.[8] As the gel density (1.6) is lower than the crystal density (2.0), a liquid phase is gradually formed between the growing crystal and the consumed gel. Nuclei started with different orientation on the gel/liquid interface will result in crystals with different orientations. The crystals developed in the direction of the Si-platelet and surrounded by liquid, are then probably affected in their orientation by the Si-platelet via surface tension- and v.d. Waals forces. The crystals are, based on the above, all more or less oriented parallel to the Si-platelet. Hereafter chemical bondings are formed apparently via condensation of hydroxyl groups of the crystals and surface hydroxyls of the Si-platelet which is assumed to possess an oxidized surface. Even after calcination the crystals are still firmly bonded to the platelets. The coverage and orientation of the crystals on the silicon platelets is shown in Figures 5c and 6c. The orientation of the crystals is almost completely parallel. A powder X-ray diffraction scan shows diffraction lines representing 0k0 planes only, see Figure 7, indicating a strong preference of the crystal orientation. Figure 7b gives the intensities of the 0$\underline{10}$0 plane at different ω-offset angles. It can be concluded that after 2° ω-offset, the intensity of the 0$\underline{10}$0 plane has been disappeared. The crystal monolayer is thus within 4° deviation flat and the absence of other diffraction lines shows that of the crystallites are mainly in a parallel orientation.

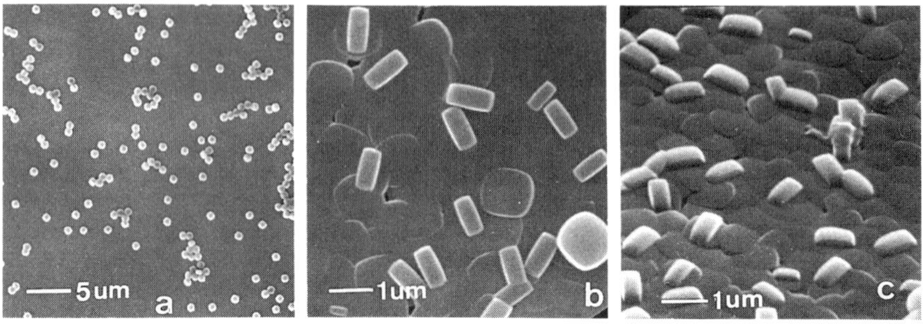

Figure 6. Evolution of the crystal growth of MFI-type with synthesis mixture 4A, a) smallest observed crystals after two hours, b) A 90° view angle on final crystallization showing a continuous monocrystal film and more than 10% additional film, c) A 20° view angle on final crystallization indicating the rather parallel orientation of the monocrystal film of 400 nm in thickness.

In the case of more diluted synthesis mixtures such as ¼A and ½A almost no gel layer is formed and the crystals show the well known elongated prismatic shape. The crystals are larger but less abundantly present. Here the nucleation occurred, according to the SEM-photograph in Figure 8,

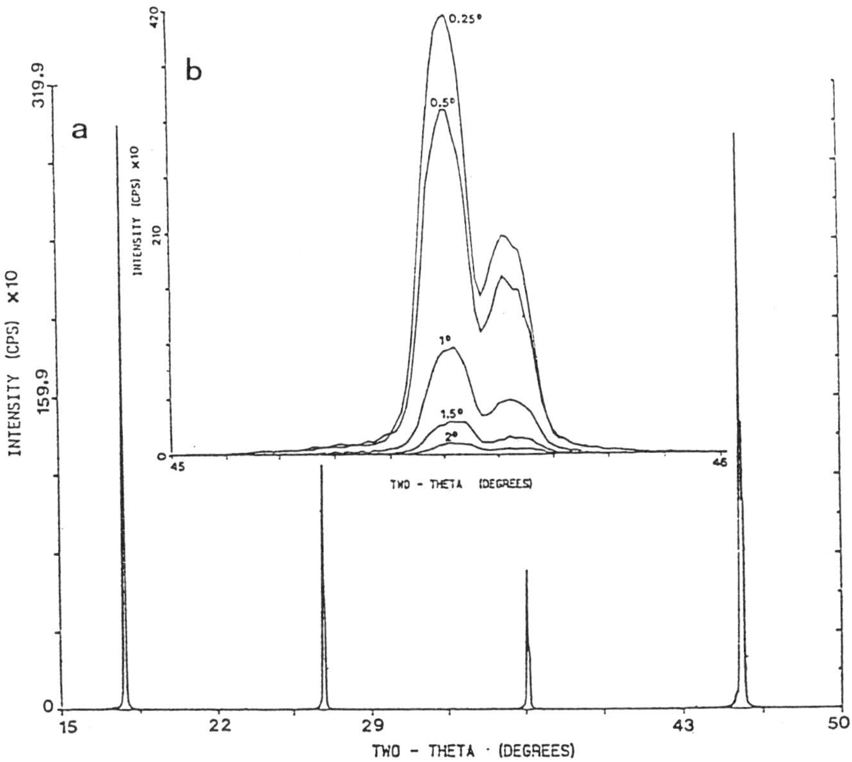

Figure 7. XRD powder pattern of zeolite MFI crystals in lateral orientation, see Figure 6c, a) Only 0k0 diffraction lines with relative intensities of MFI-lattice are observed here, b) 0100 intensities at different ω-offset angles clearly show the unique orientation of the crystals within 4°.

on the surface of the Si-platelet and axial growth with a random orientation of the crystals is clearly obtained. The addition of sodium ions to the synthesis mixture, code NA, resulted in a heterogeneous coverage of the Si-platelet with gel. Crystals were grown partly from the gel/liquid interface and partly from the Si-platelet surface. For this reason a mixture of crystal orientations was observed.

Figure 8. Surface nucleation on the Si-support using ¼A synthesis mixture, resulted in axially grown crystals.

Experiments without template TPA^+ but with NaOH and without and with an aluminum source, mixtures NA and NAAL respectively, gave so far no results under the prevailing synthesis conditions.

CONCLUSION

It can be concluded that crystals of zeolite MFI can be grown on a proper support which appear to be firmly bonded even after calcination. The orientation is mainly axial when low concentrations and temperatures are used with the support on the bottom of the autoclave. The crystal orientation is parallel to the support surface when, *i*) a relatively low temperature, *ii*) support vertical in the upper part of the synthesis mixture and, *iii*) refined synthesis conditions, thus without Na^+, are used.

ACKNOWLEDGMENT

The authors like to thank Ing. N.M. van der Pers, Laboratory of Metallurgy, Delft University of Technology, for XRD measurements of some samples; Drs. H.W. Kouwenhoven, E.T.H. Zürich, and R. de Ruiter, Delft University of Technology, for valuable discussions.

REFERENCES

1. Sano,T., Kawamura,M., Mizukami,F., Takaya,H., Mouri,T., Inaoka,W., Toida,Y., Watanabe,M., Toyoda,K., Zeolites **11** (1991) 842.
2. Davis,S.P., Borgstedt,E.V.R. and Suib,S.L., Chem. Mater. **2** (1990) 712.
3. Lachman,I.M., Patil,M.D., US Patent 4,800,187 (1989).
4. Mercer,W.C., Coughlin,P.K., McLeod Jr.,D., Flanigen,E.M., U.S. Patent 4,860,584 (1989).
5. Mercer,W.C., Coughlin,P.K., Wilson,S.T., Flanigen,E.M., U.S. Patent 4,904,518 (1990).
6. Runyan,W.R. and Bean,K.E., Semiconductor Integrated Circuit Processing Technology, Addison-Wesley Publishing Company, Inc., (1980) 26-31.
7. Kern,W., Puotinen,D.A., RCA Review **31** (1970) 187-206.
8. Jansen,J.C., Engelen,C.W.R. and van Bekkum,H., ACS Symp. Ser. **398** (1989) 257.

MOLECULAR DESIGN OF LAYERED ZIRCONIUM PHOSPHONATE CATALYSTS

K. SEGAWA*, N. KIHARA*, H. YAMAMOTO*, and S. NAKATA**

*Department of Chemistry, Sophia University, 7-1 Kioi-cho, Chiyoda-ku, Tokyo 102, JAPAN

**R & D Center, Chiyoda Corp., 13 Moriya-cho, Kanagawa-ku, Yokohama 221, JAPAN

ABSTRACT

Preparation and characterization of two-dimensional zirconium phosphonate derivatives have been investigated. Two composite zirconium phosphonates in single crystal phase have also been investigated and characterized by XRD, ^{13}C-, and ^{31}P-MASNMR. These compounds are lamellar structures comprising zirconium phosphates or organophosphonates. Each layer consists of planes of zirconium bridged through phosphonate groups which alternate above and below the Zr atom planes, oriented away from the basal surfaces in a bilayered fashion in the interlayer region. The catalytic performance over zirconium phosphonates is evaluated by hydrolysis of ethyl acetate or by esterification of acetic acid. When the composite zirconium phosphonate is composed with an acidic function and with a hydrophobic function in single crystal phase, the catalytic activity showed a higher activity without swelling of the catalyst in the reaction media. The composite materials become accessible to any reactant molecule and improve hydrophobicity.

INTRODUCTION

Zirconium phosphonates are solid materials typically synthesized in amorphous forms under aqueous conditions by the reaction of a soluble salt of a tetravalent metal and a phosphonic acid or an organophosphoric acid:

$$Zr^{4+} \quad + \quad 2(OH)_2\overset{\overset{\displaystyle O}{\|}}{P}R \quad \longrightarrow \quad Zr(O_3PR)_2 \quad + \quad 4H^+ \qquad (1).$$

However, it is possible to synthesize many of them as crystalline compounds. The potential applications of crystalline zirconium phosphonates have significant recent interest. Depending on the R group, different applications for these materials in the field of chromatographic separation, photochemistry, fuel cell, and catalysis are expected. In the case of zirconium phosphates, with the general formula $Zr(O_3POH)_2{\cdot}nH_2O$, the most extensively investigated crystalline zirconium phosphate is an α layered acid salt [1]. α-$Zr(O_3POH)_2{\cdot}H_2O$ crystallizes in the monoclinic system, as assigned by Troup and Clearfield from their single crystal work [2]. For the zirconium

From the Proceedings of the 9th International Zeolite Conference, Montreal 1992, Eds. R. von Ballmoos et al.,

phosphonates, however, the two-dimensional tetravalent metal plane has a structure essentially similar to the zirconium phosphate structure; substituted for hydroxyl groups are the desired organic functional groups, oriented away from the basal surfaces in a bilayered fashion in the interlayer region [3]. Therefore, zirconium phosphonate materials can act as a series of modified surfaces, and become accessible for reactant molecules of catalysis.

EXPERIMENTAL

Preparation of zirconium phosphonates

Addition of a water-soluble Zr(IV) salt to phosphoric acid results in the precipitation of a gelatinous amorphous precipitate. Several types of zirconium phosphonates, $Zr(O_3PR)_2$, either in single component, such as one kind of R, or in composite zirconium phosphonates, such as two different R's, were prepared: here R could be -OH, $-CH_3$, $-C_6H_5$, $-C_{12}H_{25}$, $-C_{22}H_{45}$, $-CH_2COOH$, $-C_2H_4COOH$ and $-CH_2SO_3H$. Most of the zirconium phosphonate derivatives were obtained by the addition of aqueous solution of $ZrOCl_2{\cdot}8H_2O$ (1.17 M) to the (organo) phosphonate reagent, $(OH)_2P(O)R$, (0.75 M, [R]/[Zr]=2-10) to yield small particle, amorphous zirconium phosphonates. For the crystallization process, the amorphous zirconium phosphonate was treated with HF solution. The use of HF as a complexing reagent to inhibit release of free Zr^{4+} had a considerable effect on increasing crystallinity and particle size. The zirconium fluorocomplexes were decomposed slowly by removing the hydrofluoric acid in the presence of phosphonic acid. The crystallinity of composite zirconium phosphonates increased with increasing concentrations of HF. However, some segregation of each function was observed when HF concentrations, [F]/[Zr], exceeded 9. Usually for the crystallization, amorphous zirconium phosphonates were treated with HF solution ([F]/[Zr] = 6 or 8) at 333K [4].

Characterization of zirconium phosphonates

The experimental procedures of TGA, XRD, and measurement of ion-exchange capacity have been described in detail elsewhere [5]. The interlayer d-spacing from XRD pattern was determined by the angle of (001) reflection.

High-resolution solid-state MASNMR spectra were obtained on a Fourier Transform pulsed NMR spectrometer (JEOL, JNM-GX270) equipped with a CP/MAS unit (JEOL, NM-GSH27MU). All ^{31}P-NMR spectra combined with cross polarization (CP) and or with magic angle spinning (MAS) at 109.38 MHz were measured with high-power proton decoupling during data acquisition [6]. The ^{13}C-MASNMR spectra were obtained at 67.94 MHz with CP and proton decoupling.

The catalytic activities of hydrolysis of ethyl acetate at 341 K, and of esterification of acetic acid with ethanol at 343 K were measured. 250 mg of catalyst was suspended in ethanol solution of acetic acid or aqueous solution of ethyl acetate, and the reaction rates were measured by GC (PORAPAK Q, 2-m).

RESULTS AND DISCUSSION

The single component zirconium phosphonates

The structure of zirconium phosphonates is similar to the zirconium phosphate core layer with pendant organic groups attached to this core and extending perpendicular to the plane of the layer [5]. Table 1 gives some physico-chemical properties for several single-crystal zirconium phosphonates. Elemental analysis results were in fairly good agreement with values calculated from the theoretical chemical formulae. The interlayer d-spacing of n-alkyl zirconium phosphonates are increasing linearly with increasing carbon numbers. The results suggest that n-alkyl groups are distributed uniformly between each pair of Zr atom planes and that the interlayer d-spacing obeyed the bulkiness of n-alkyl groups. Solid products consisting of layered sheets with ordered arrays of the pendant organic groups on both sides of the layer were obtained with most phosphonic acids.

Table 1. Physico-chemical properties of crystalline zirconium phosphonates

Compound	d-spacing /nm	Decomposition temp.[a] /K	Ion-exchange capacity /meq g^{-1}
$Zr(O_3PCH_3)_2$	0.78	873	0.00
$Zr(O_3PC_6H_5)_2$	1.55	733	0.00
$Zr(O_3PC_{12}H_{25})_2$	3.24	754	0.00
$Zr(O_3PC_{22}H_{45})_2$	5.62	505	0.00
$Zr(O_3POH)_2 \cdot H_2O$	0.76	823	6.50 (6.67)[b]
$Zr(O_3PCH_2COOH)_2$	1.13	632	6.30 (5.45)
$Zr(O_3PC_2H_4COOH)_2$	1.34	633	5.40 (5.06)
$Zr(O_3PCH_2SO_3H)_2$	1.54	533	5.79 (4.55)
Nafion-H	–	473	0.83 (0.80)

a Decomposition temperature is determined by the first break-point of TGA.
b Numbers in parentheses are calculated from chemical formulae.

The TGA curves of zirconium phosphonates showed multistep decomposition of pendant groups, due to the elimination of alkyl or hydroxy groups. The thermal stability values (decomposition temperatures) of zirconium phosphonates in Table 1, determined by the first break point of the TGA curve, shifted to lower temperatures as the number of alkyl group carbons increases. Although no ion-exchange capacity was associated with the pure alkyl zirconium phosphonates; replacing R with -OH, $-CH_2COOH$, $-C_2H_4COOH$, or $-CH_2SO_3H$ groups increased the ion-exchange capacity in direct proportion to the number of acid hydrogens in the replacing groups.

For the catalytic test reaction, esterification of acetic acid with ethanol at 343 K and hydrolysis of ethyl acetate at 341 K in aqueous solution (0.67 M) showed relatively lower activity except over $Zr(O_3PCH_2SO_3H)_2$ (Figure 1). For this $Zr(O_3PCH_2SO_3H)_2$ catalyst, the activity was higher than that for the hydrogen form of Nafion. However, the reaction proceeded under somewhat homogeneous reaction, because $Zr(O_3PCH_2SO_3H)_2$ is water-soluble or soluble with polar compounds. The test reaction results suggest that, even when the acidic functions are present (R; -OH, $-CH_2COOH$, $-C_2H_4COOH$) in zirconium phosphonates, the reactant molecules are not accessible, because the d-spacings are smaller than those of alkyl zirconium phosphonates.

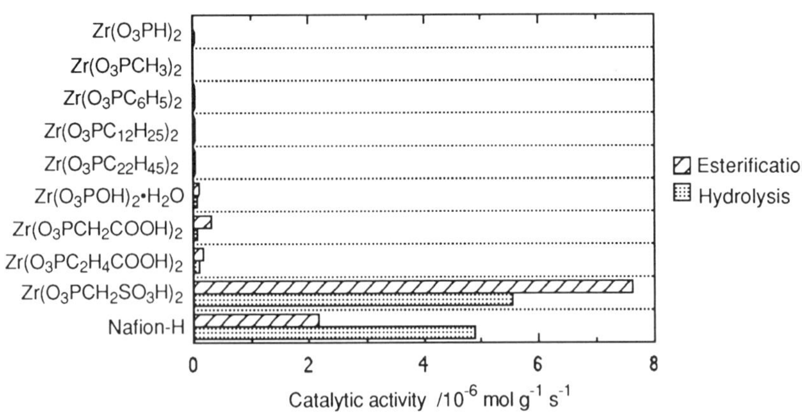

Figure 1. Catalytic activities for esterification of acetic acid and hydrolysis of ethyl acetate on crystalline zirconium phosphonates

As was mentioned previously, the Zr atoms lie very nearly in a plane and are bridged by phosphorous tetrahedra. These tetrahedra are situated alternately above and below the Zr atom plane. Three oxygen atoms of each phosphorous atom are bonded to three different Zr atoms, which form a distorted equilateral triangle. Thus each phosphonate R group in $Zr(O_3PR)_2$ is directed into the interlayer space. ^{31}P-MASNMR and ^{13}C-CP/MASNMR have been used to study the phosphorous and carbon microenvironment of layered zirconium phosphonates. For ^{31}P-MASNMR spectra, each chemical shift of crystalline zirconium phosphonate from phosphoric acid represents the single identical resonance line [5], which shows one kind of phosphorous environment between each pair of layers.

The composite zirconium phosphonates in single crystal phase

The acidic function of single zirconium phosphonate showed rather poor catalytic activities for esterification of acetic acid and hydrolysis of ethyl acetate except over $Zr(O_3PCH_2SO_3H)_2$ (see Figure 1). In addition, over $Zr(O_3PCH_2SO_3H)_2$ catalyst, the reaction proceeds as a homogeneous reaction, even though the catalytic activity is higher than other acidic zirconium phosphonates. To overcome the problem of water solubility or swelling, a second function was introduced to achieve a catalytic advantage in certain applications. The objective of this study is to explore the role of a second phosphonate function in single crystal phase on the catalytic performance of acidic function and hydrophobic function of zirconium phosphonates.

For the preparation of composite zirconium phosphonates, we preferred two different functions: $-CH_2SO_3H$ group for acidic function, and $-C_{12}H_{25}$ group for hydrophobic function, to give for example $Zr[(O_3PCH_2SO_3H)_x(O_3PC_{12}H_{25})_{1-x}]_2$. Here x is the initial mole fraction of the acidic group for preparation. This x is not changed even after the crystallization process; the values are confirmed by elemental analysis and by potentiometric titration.

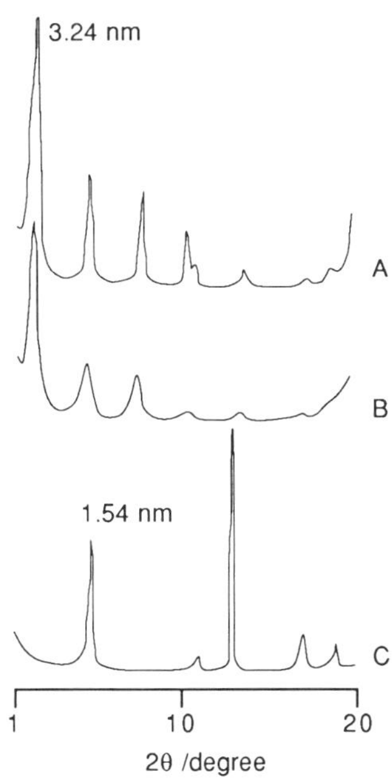

Figure 2. XRD patterns of single and composite zirconium phosphonates:
A ; $Zr(O_3PC_{12}H_{25})_2$, B; $Zr[(O_3PCH_2SO_3H)_{0.5}(O_3PC_{12}H_{25})_{0.5}]_2$, C; $Zr(O_3PCH_2SO_3H)_2$

Interlayer d-spacing of composite material depended on the bulkier functional groups. Figure 2 shows the XRD pattern of $Zr(O_3PC_{12}H_{25})_2$, $Zr[(O_3PCH_2SO_3H)_{0.5}(O_3PC_{12}H_{25})_{0.5}]_2$, and $Zr(O_3PCH_2SO_3H)_2$. The XRD patterns did not show any phase segregation regardless of concentration of each component. In addition, the interlayer d-spacing showed constant (3.24 nm) values similar to that of single zirconium dodecyl phosphonate, $Zr(O_3PC_{12}H_{25})_2$. Ion-exchange capacities increased with increasing the concentration of $-CH_2SO_3H$ groups in direct proportion to the number of acid hydrogens in the replacing groups. The results suggest that the functional groups are distributed uniformly between each pair of Zr atom planes and the interlayer d-spacing obeyed that of the bulkier functional groups, such as $-C_{12}H_{25}$.

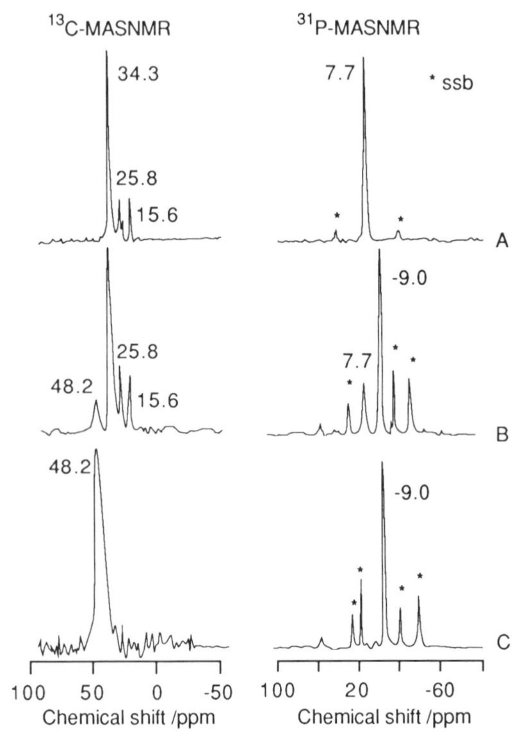

Figure 3. ^{13}C- and ^{31}P-MASNMR spectra of single and composite zirconium phosphonates: A ; $Zr(O_3PC_{12}H_{25})_2$, B; $Zr[(O_3PCH_2SO_3H)_{0.5}(O_3PC_{12}H_{25})_{0.5}]_2$, and C; $Zr(O_3PCH_2SO_3H)_2$

The Zr atoms in zirconium phosphonates lie very nearly in a plane and are bridged by phosphorus tetrahedra. These tetrahedra are situated alternately above and below the Zr atom plane. Thus, each phosphonate R group in $Zr(O_3PR)_2$ is directed into the interlayer space. ^{13}C- and ^{31}P-MASNMR have been used to study the phosphorus and carbon microenvironment of layered zirconium phosphonates. The MASNMR spectra of single and composite zirconium phosphonates are shown in Figure 3. In the case of ^{31}P-MASNMR spectra of single component zirconium phosphonate, the chemical shifts of crystalline zirconium phosphonates represent a single resonance. This indicates the presence of only one kind of phosphorus environment between each pair of layers. For ^{31}P-MASNMR spectra of composite zirconium phosphonate, however, two identical resonance lines are observed: -9.0 ppm for phosphorus of $P\text{-}CH_2SO_3H$, and 7.7 ppm for phosphorus of $P\text{-}C_{12}H_{25}$. The relative intensity of the resonance lines varied as a function of *x*. The interlayer d-spacing of composite $Zr[(O_3PCH_2SO_3H)_x(O_3PC_{12}H_{25})_{1-x}]_2$ was 3.24 nm regardless of *x* values. ^{13}C-CP/MASNMR spectra of composite compounds show the same resonance lines as those of single $Zr(O_3PC_{12}H_{25})_2$ and $Zr(O_3PCH_2SO_3H)_2$. The major resonance line, which is attributed to the methylene carbons is observed at 34.3 ppm from TMS (tetramethylsilane). The other two lines in ^{13}C-CP/MASNMR spectra are attributed to the carbon of $P\text{-}CH_2\text{-}$ (25.8 ppm) and to the terminal methyl carbon (15.6 ppm).

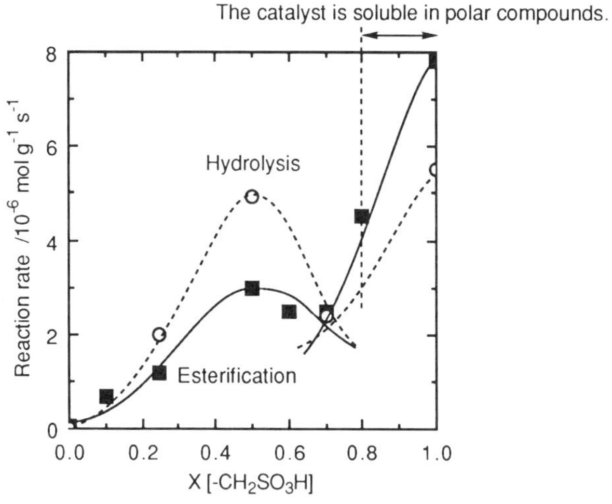

Figure 4. Catalytic activities for esterification of acetic acid and hydrolysis of ethyl acetate on $Zr[(O_3PCH_2SO_3H)_x(O_3PC_{12}H_{25})_{1-x}]_2$ as a function of X

For catalytic esterification of acetic acid in ethanol medium and hydrolysis of methyl acetate in aqueous medium over zirconium phosphonates, the composite materials showed higher catalytic activities than single zirconium phosphonate as illustrated in Figure 4. Above 0.8 of *x* values, the reaction proceeded under somewhat homogeneous reaction, because the catalyst is water-soluble or soluble with polar compounds. It is quite interesting that the catalyst became water insoluble below 0.7 of *x* values. No matter how small the protonic concentrations became, the reaction rate for esterification and hydrolysis was enhanced with increasing x values, and passed through the optimum concentration of P-CH_2SO_3H (x = 0.5). After introduction of hydrophobic functions (-$C_{12}H_{25}$) to $Zr(O_3P\text{-}CH_2SO_3H)_2$, the composite materials became accessible to any reactant molecule and improved in hydrophobicity. The functional groups for catalysis are now incorporated in one crystal phase. These concepts suggest that the composite compounds may be versatile catalysts in variety of reactions in aqueous or polar organic solvents.

In such compounds, with the same α-layered structure as zirconium phosphate [7], two different tetrahedral phosphonate groups are bonded to the plane of zirconium atoms. Thus, it is of great interest to examine the possibility of obtaining various organic derivatives of zirconium phosphonate in which two or more different functional groups are present in the same compounds. Diffraction data confirm the expected correlation of interlayer spacing with the size of the pendant organic group. Solid products consisting of layered sheets with ordered arrays of the pendant organic groups on both sides of the layer are obtained with most phosphonic acids.

REFERENCES

1. Whittingham, M. S. and Jacobson, A. J., Intercalation Chemistry (Materials Science Series) Academic Press, New York, 1982.
2. Troup, J.M. and Clearfield, A., *Inorg. Chem.*, **16**, 3311 (1977) .
3. Alberti, G., Costantino, U., Allulli, S., and Tomassini, N., *J. Inorg. Nucl. Chem.*, **40** 1113(1978).
4. Dines, M. B. and DiGiacomo, P. M., *Inorg. Chem.*, **20**, 92 (1981).
5. Segawa, K., Sugiyama, A., and Kurusu, Y., *Studies in Surface Sci. and Catal.*, **60**, 73 (1991).
6. Segawa, K., Nakajima, Y., Nakata, S., and Asaoka, S., *J. Catal..* **101**, 81 (1986).
7. Segawa, K., Kurusu, Y., Nakajima, Y., and Kinoshita, M., *J. Catal.*, **94**, 491(1985).

SYNTHESIS AND CHARACTERIZATION OF THE NOVEL GALLOPHOSPHATE CLOVERITE WITH 20-MEMBERED RING CHANNELS

J. PATARIN, C. SCHOTT, A. MERROUCHE, H. KESSLER, M. SOULARD, L. DELMOTTE, J.L. GUTH and J.F. JOLY*

Laboratoire de Matériaux Minéraux (URA-CNRS 428) ENSCMu-UHA,
3, rue Alfred Werner, 68093 MULHOUSE Cedex (FRANCE)

* Institut Français du Pétrole, B.P. 311, 92506 RUEIL MALMAISON Cedex (FRANCE)

ABSTRACT

The influence of the synthesis conditions such as HF to quinuclidine ratio, gallium and silicon sources, pH and temperature were determined for the novel gallophosphate cloverite and the corresponding silicogallophosphate. For the latter, little Si was incorporated in the crystals (electron microprobe analysis). The thermal stability of the $GaPO_4$ form was determined by studying the variation of the adsorption capacity of n-hexane as a function of the calcination conditions. It is stable to at least 500°C. It was found that after calcination at 450°C, part of the fluoride was still present in the structure, in a slightly different environment (^{19}F MAS NMR). Presumably due to an amorphization reaction, NH_3 turned out to be an unsuited probe molecule for the determination of the acid properties of the materials.

INTRODUCTION

Beside the dense phase types quartz, tridymite and cristobalite of $GaPO_4$ [1], a few microporous gallophosphates were recently described[2-4]. Subsequently, two novel structures comprising 10- and/or 8-membered ring channels were reported [5-6], the others are counterparts of the aluminophosphates $AlPO_4$-12, 14, 21 and 25. In the same system (Ga_2O_3-P_2O_5), the use of a fluoride medium in the presence of quinuclidine enabled us to prepare selectively the novel gallophosphate named cloverite. The gallium source used was the sulfate. The main channels of the cubic structure [7] are circumscribed by 20-MR (Ga, P atoms). Their four-leafed clover shape is due to the fact that one type of D4R's with dangling OH groups protrudes into the window. Multinuclear NMR spectroscopy (^{13}C, ^{19}F, ^{31}P, ^{71}Ga) confirmed the structure analysis [8].

In the present paper are described the effect of synthesis variables such as starting chemical composition and nature of the reactant sources on the formation of gallophosphate or silicogallophosphate cloverite. The thermal stability, particularly the behavior of the F^- anions, the adsorption capacity after various calcination conditions and the ammonia-STD are also discussed.

From the Proceedings of the 9th International Zeolite Conference, Montreal 1992, Eds. R. von Ballmoos et al.,

EXPERIMENTAL

Synthesis

The reactants were quinuclidine (Aldrich), 85 % phosphoric acid (Prolabo, Normapur), 40 % hydrofluoric acid (Prolabo, Normapur). The silicon sources were colloidal SiO_2, Ludox AS40 (40 wt % SiO_2,60 wt % H_2O), Aerosil 130 (Degussa) and tetraethoxysilane (Fluka, 98 wt %). Four gallium sources were used, a hydrated gallium sulfate prepared as described previously [8], a gallium nitrate solution (Rhône-Poulenc), a gallium oxyhydroxide from gallium metal (Grade 1, Johnson Matthey) and an amorphous source obtained after heating the nitrate source at 250°C for 4 hours.

The order of addition of the reactants under stirring was the following : H_3PO_4, water, gallium source, silica source to prepare a silicogallophosphate (GaPSO) and finally quinuclidine and hydrofluoric acid. The obtained gel (pH= 3-4.5) was mixed and transferred into a PTFE-lined stainless steel autoclave. After heating at 150-170°C for 24 hours, the solid was filtrated, washed with water until the pH of the filtrate was 5.5. and dried at 70°C.

Adsorption measurements

The adsorption capacity measurements were performed on computerized equipment,a TG92 from Setaram. About 100 mg of the as-synthesized solid was calcined by heating to 450°C under vacuum or flowing dry N_2 (heating rate 5°C.min^{-1}). After 1 h at 450°C the solid was cooled to 100°C and reheated in a flowing dry O_2 to either 450°, 500°C or 600°C (heating rate 5°C.min^{-1}).After 1.5 - 2 h at the selected temperature, the sample was cooled to room temperature and subjected to a flow of dry N_2 + sorbate. The relative pressure p/p_o was 0.5 and the sorbate molecules were n-hexane, p-xylene, m-xylene, mesitylene and 1,3,5-triisopropylbenzene. The adsorption capacity for n-hexane was also determined after calcination at 380°C under vacuum for 1.5 h. The calcination conditions are summarized as follows:

Type of calcination	Conditions
Vacuum 380	380°C, vacuum (1.5 h)
Vacuum $\rightarrow O_2$ 450	450°C, vacuum (1 h) + 450°C, O_2 (1.5h)
Vacuum $\rightarrow O_2$ 600	450°C, vacuum (1 h) + 600°C, O_2 (1.5h)
$N_2 \rightarrow O_2$ 450	450°C, N_2(1 h) + 450°C, O_2 (2h)
$N_2 \rightarrow O_2$ 500	450°C, N_2(1 h) +500°C, O_2 (2h)

Solid state NMR spectroscopy

^1H- ^{29}Si CPMAS NMR spectroscopy was used to determine the silicon incorporation into the gallophosphate framework and ^{19}F MAS NMR to study the environment of fluoride after heating. The spectra were recorded on a Bruker MSL 300 spectrometer, the conditions were

^{1}H-^{29}Si CP MAS NMR : frequency 59.6 MHz: 9.5 μs pulse length, 8 s pulse interval, 3 kHz spinning rate and 3 ms contact time

^{19}F MAS NMR : the parameters were 282 MHz, 3 μs, 4 s and 8 kHz.

Stepwise thermal desorption of ammonia (NH_3 STD)

The STD of ammonia was carried out on 100-120 mg of powder. An activation similar to the $N_2 \rightarrow O_2$ 450 procedure used for the adsorption measurements was employed. After calcination the sample was cooled to 100°C and NH_3 was adsorbed at that temperature for 15 min. After a purge at 140°C (30 min. under helium) NH_3 was desorbed by heating the sample by steps of 50°C from 150 to 500°C. The desorbed ammonia was detected by a thermal conductivity detector and neutralized by a HCl solution. The total amount of base was then determined by back-titration.

RESULTS AND DISCUSSION

Synthesis, chemical composition

In table 1 are given some typical synthesis examples having produced cloverite. Among the four gallium sources, sulfate and nitrate only led to a pure sample. With the oxyhydroxide source, traces of residual GaOOH were still present beside cloverite (sample E), whereas a dense $GaPO_4$ phase (type quartz) was obtained as an impurity (traces) with the amorphous source (sample F). In the latter case, a lower temperature or a shorter heating time might lead to cloverite without such an impurity.

Samples A, D, E and F show a similar crystal size close to 1 μm. However, large isolated and/or agglomerated truncated cubes (Figure 1) were obtained when the fluoride amount in the starting mixture was low (comparison between samples A, B and C). Nevertheless, when the fluoride content was too low and especially lower than the stoichiometric value, i.e., 0.5F per 1 Ga_2O_3, traces of the gallophosphate $GaPO_4$-a [3] were detected by X-ray diffraction. As reported previously, with a similar starting composition, but without fluoride anions, only the latter phase was obtained [9]. When the pH value was lower than 3, $GaPO_4$ type quartz crystallized. The chemical composition found for the gallophosphate samples is in good agreement with the unit cell formula $8([Ga_{96}P_{96}O_{372}(OH)_{24}](QF)_{24}(H_2O)_n)$ determined by the single crystal structure analysis [7]. The values of chemical composition for the samples in the dehydrated form are as follows :

- calculated (wt%)	Ga: 34.90,	P:15.51,	C:10.52,	N:1.75,	F:2.38,
- observed (wt%) sample A,	Ga: 34.89,	P:16.17,	C: 10.07,	N: 1.71,	F:2.42
sample D,	Ga: 35.10,	P:15.44,	C: 10.87,	N: 2.25,	F:2.23

Three cloverite-type samples were synthesized from a mixture based on a silicogallophosphate composition with three different silica sources(samples G, H and I, table 1). For each sample, no significant change was observed on the XRD pattern with respect to the one of the gallophosphate phase, neither for the peak intensities nor the peak angles. The Si/(Si+Ga+P) molar ratio inside the crystals found by electron microprobe analysis is equal to 0.02 .

The ^{1}H-^{29}Si CP MAS NMR spectrum of sample G given in Figure 2 consists of two broad resonances at -102.0 and -110.3 ppm. These signals can be assigned to amorphous silica with silanol groups and, maybe, to a few Si(4Si) islands resulting from a $2Si^{4+} = P^{5+} + Ga^{3+}$ substitution.

Table 1. Typical synthesis conditions leading to cloverite and crystal size of the samples

Sample	Reaction mixture	T (°C)	Heating time(h)	Results	Crystal size(μm)
A	$1Ga_2O_3$ (S)*: $1P_2O_5$: $65H_2O$: 6Q : 1HF	150	24	CLO	~1
B	$1Ga_2O_3$ (S)*: $1P_2O_5$: $65H_2O$: 6Q : 0.4HF	150	15	CLO+a[a]	10-20
C	$1Ga_2O_3$ (S)*: $1P_2O_5$: $65H_2O$: 6Q : 0.28HF	150	24	CLO+a	60-100[d]
D	$1Ga_2O_3$ (N)*: $1P_2O_5$: $80H_2O$: 5.8Q : 1HF	170	24	CLO	~1
E	$1Ga_2O_3$ (O)*: $1P_2O_5$: $70H_2O$: 1.5Q : 1HF	170	24	CLO+H[b]	~1
F	$1Ga_2O_3$ (Am)*: $1P_2O_5$: $70H_2O$: 1.1Q : 1HF	170	24	CLO+Q[c]	~1
G	$1Ga_2O_3$ (S)*: $1P_2O_5$: $70H_2O$: 6Q : 1HF : $0.6SiO_2$(L)*	170	24	CLO	10-60
H	$1Ga_2O_3$ (N)*: $1P_2O_5$: $80H_2O$: 5.4Q : 1HF : $0.6SiO_2$(Ae)*	170	24	CLO	1-60
I	$1Ga_2O_3$ (N)*: $1P_2O_5$: $80H_2O$: 5.4Q : 1HF : $0.6SiO_2$ (T)*	170	24	CLO	~1

* Gallium sources : S : sulfate; N : nitrate; O : oxyhydroxide; Am : amorphous

* Silica sources : L : Ludox As-40; Ae : Aerosil ; T : tetraethoxysilane CLO : Cloverite

a traces of gallophosphate type $GaPO_4$-a *b* traces of GaOOH

c traces of quartz- type $GaPO_4$ *d* isolated and agglomerated truncated cubes

Thermal stability, adsorption measurements

The thermal stability of the structure and the behavior of fluoride were detemined by heating the solid at various temperatures. Indeed in the calcined sample, fluoride anions might by compensated by protons resulting in a new type of acid sites. On heating the as-synthesized samples at 100°C dehydration occurred (wt. loss ≈10%). The same amount of water was readsorbed at room temperature in a 80 % relative humidity atmosphere. After removal of the organic species by calcination up to 500°C no significant change of the powder XRD spectrum was observed for the calcined non-rehydrated sample. However when the calcined solid was exposed to moist air at room temperature, the crystallinity decreased significantly, and after a few hours the product was amorphous [8]. After heating the as-made sample at 600°C for 1.5 h, some quartz-type $GaPO_4$ appeared and above 700°C there was a total conversion to tridymite or cristobalite. The adsorption capacity values for n-hexane, m- and p-xylene, mesitylene and 1,3,5-triiso-

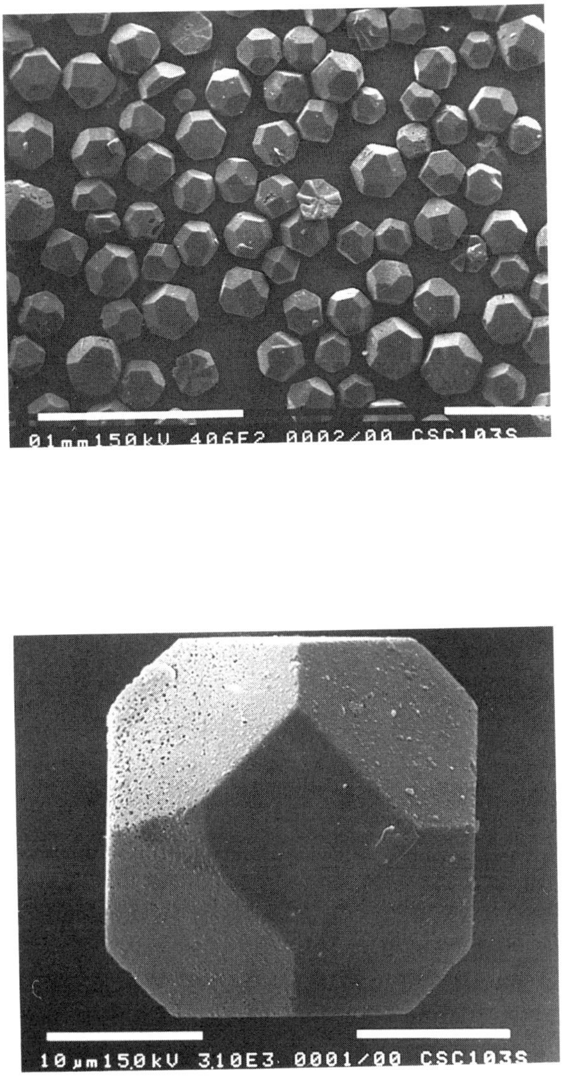

Figure 1 : Scanning electron micrographs of sample B.

propylbenzene after the vacuum $\rightarrow O_2$ 450 calcination procedure were respectively 15.8, 14.3, 13.1, 15.3 and 5.7 wt % [10]. The change in adsorption capacity as a function of the activation conditions was followed by using n-hexane as an adsorbate. The values found (wt %)were 14.9 (vacuum 380), 15.8 (vacuum $\rightarrow O_2$ 450), 15.7 ($N_2 \rightarrow O_2$ 450), 14.8 ($N_2 \rightarrow O_2$ 500) and 9.0 (vacuum $\rightarrow O_2$ 600). The adsorption capacity is essentially the same for activation temperatures between 380°C and 500°C, whereas a sharp drop is observed after heating at 600°C. As mentioned above, the heated solid contained quartz-type $GaPO_4$, presumably as a result of a dehydroxylation process.

Information on the removal of fluoride on heating was obtained by calcining a portion of sample A at 450°C following the $N_2 \rightarrow O_2$ 450 procedure. According to TG and the color (yellow-brown) of the residue most of the organic template was removed. A residual amount of 2-3F per 192 T was found by chemical analysis. The ^{19}F MAS NMR spectrum of the calcined non rehydrated sample is given in Figure 3 with the spectrum of the as-synthesized sample A. In both cases, a single line is observed respectively at -71.1 ppm and -68.1 ppm. Such a chemical shift range was previously found for the LTA gallophosphate synthesized in fluoride medium in which the F^- anions have been also located inside the D4R's of the framework [10]. The small chemical shift difference observed between the as-synthesized and the calcined forms corresponds probably to a slight distortion of the D4R's units occurring on calcination.

Stepwise thermal desorption of ammonia (NH3 STD)

For the NH_3 STD measurements, the cloverite samples A, B, D and G were calcined using the $N_2 \rightarrow O_2$ 450 procedure. The desorption temperature maximum and the total amount of desorbed ammonia (in mmol $NH_3.g^{-1}$) for each sample were: sample A 200°C, 3.0; sample B 400°C, 3.5; sample D 200°C, 2.7 and sample G 400°C, 3.6. Concerning the amount of NH_3 desorbed, which turned out to be large, the values found are higher than the maximum value that could be expected from the chemical formula taking into account all the OH groups and all the F^- of the structure. The overestimated value which would correspond to 192 dangling OH groups and 192 HF pairs is close to 2.9 mmol NH_3 g^{-1}. Moreover, the difference observed between samples B, G and A,D seems to be related to the crystal size rather than to the possible presence of silicon in the framework. Indeed for large crystals (samples B and G), the total amount of NH_3 observed was close to 3.5 mmol $NH_3.g^{-1}$, whereas for small sized samples (samples A and D) the values were 2.7-3.0 mmol $NH_3.g^{-}$) After the STD NH_3 experiment, the solid was amorphous. Nervertheless, after ammonia adsorption at 100 °C, the structure was found intact. The amorphization probably occurred during the desorption process. Hence ammonia is not the right probe molecule to study the acid properties of cloverite. A preliminary n-heptane cracking test (T = 350°C, WHSV = $3h^{-1}$)

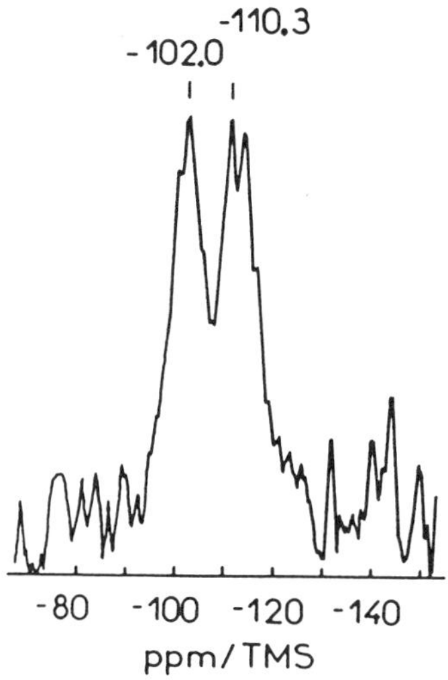

Figure 2 : ^{1}H-^{29}Si CP MAS NMR spectrum of as-synthesized sample G (no. of scans = 6,355)

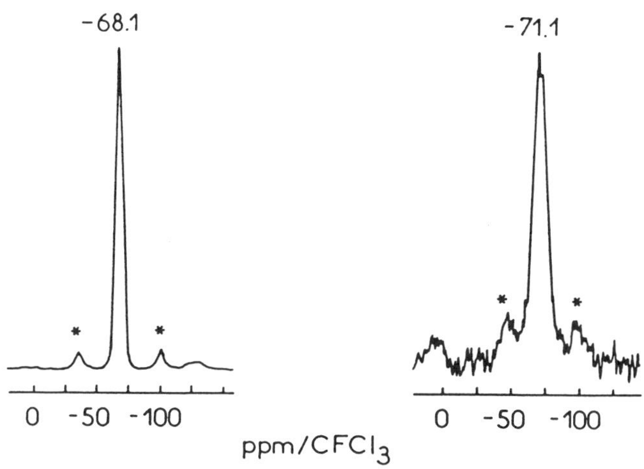

Figure 3 : ^{19}F MAS NMR spectra. Left : as-synthesized sample A (no. of scans = 180).
Right : calcined sample A (no of scans = 180)
* spinning sidebands

was performed. After 10 minutes on stream the product was heavily coked. This could mean that a catalytic cracking occurred, but also a fast deactivation since no cracking products were detected.

CONCLUSION

Cloverite can be synthesized with various gallium sources in the pH range 3.0-4.5 at 150-170°C. By changing the fluoride amount in the starting mixture the crystal size can be controlled. Very little Si was found in the crystals when a silicogallophosphate mixture was used.

After heating at 450°C under a dry atmosphere, only a small amount of fluoride was still present inside the D4R's units of the structure but in a slightly different environment. Such a result shows that the fluoride anions are loosely bound in the as-synthesized solid. Those present in the calcined sample could be compensated by residual organic cations rather than protons. Ammonia does not seem to be a suitable probe molecule to study the acid properties of cloverite.

REFERENCES

1. Goiffon, A., Bayle, G., Astier, R., Jumas, J.C., Maurin, M. and Philippot, E., Rev. Chimie Minér. **20**, 338 (1983).
2. Parise, J.B., J. Chem. Soc., Chem. Comm. 606 (1985).
3. Wilson, S.T., Woodward, N.A., Flanigen, E.M. and Eggert, H.G., Eur. Patent Appl. 226,219 (1987).
4. Feng, S. and Xu, R., Chemical Journal Chinese Universities **3**, 867 (1987).
5. Yang, G., Feng, S. and Xu, R., J. Chem. Soc. Chem. Comm. 1254 (1987).
6. Wang, T., Yang, G., Feng, S., Shang, C. and Xu, R., J. Chem. Soc., Chem. Com. 998 (1989).
7. Estermann, M., McCusker, L.B., Baerlocher, C., Merrouche, A. and Kessler, H., Nature **352**, 320 (1991).
8. Merrouche, A., Patarin, J., Kessler, H., Soulard, M., Delmotte, L., Guth, J.L. and Joly, J.F., Zeolites,in press.
9. Kessler, H., in : *Synthesis, characterization and Novel Applications of Molecular Sieve Materials*, Eds. R.L. Bedard, T. Bein, M.E. Davis, J. Garces, V.A. Maroni, and G.D., Stucky, Materials Research Society, Pittsburgh, pp. 47-55 (1991).
10. Merrouche, A., Patarin, J., Soulard, M., Kessler, H. and Anglerot, D., in : *Zeolites and Pillared Clays Synthesis* ,Vol.1, Eds. M. Occelli, and H. Robson, van Nostrand Reinhold, New York, to be published in 1992.

THE CRYSTALLIZATION OF ALUMINOPHOSPHATE MICROPOROUS COMPOUNDS IN ALCOHOLIC SYSTEM

Ruren Xu, Qisheng Huo, and Wenqin Pang

Department of Chemistry, Jilin University, Changchun, P.R.China,

ABSTRACT

Syntheses of aluminophosphate microporous compounds, both zeolite-type and layered aluminophosphate materials containing organic amines, in the presence of various amines from alcohol media are described. The results of investigation into the crystallization of aluminophosphate from nonaqueous systems are preliminarily summarized. The influences of some main factors (organic amine template, alcohol solvent, crystallization temperature, time, etc.) on the resulting structures are studied.

INTRODUCTION

Since the first extensive systematic study of aluminophosphate-based molecular sieves was reported by Wilson et al. in 1982 [1], to date, a large number of microporous $AlPO_4$ have been synthesized under hydrothermal conditions. Some of the aluminophosphate molecular sieve framework topologies are analogous to those of aluminosilicate zeolites, but some of them are specific [2]. Organic amines are used in the synthesis of all of these materials. The role of organic bases is understood to a certain degree. Recently, we reported the syntheses of aluminophosphate molecular sieves, AlPO-5, 11, and 21 [3]; a chain aluminium phosphate [4]; and a sheet aluminium phosphate anion [5] from alcoholic systems. The synthesis of JDF-20 [6] was more recently described. We are attempting here to give a more detailed discussion of syntheses of aluminophosphates from alcoholic system.

EXPERIMENTAL

The aluminophosphate gel was prepared with aluminium isopropoxide, phosphoric acid (H_3PO_4 85 wt%), organic amine and alcohol. The general procedure of gel preparation was as follows: (1) aluminium isopropoxide was first added in alcohol solvent, (2) then amine was added with stirring, (3) the compound containing the heteroatom was added, if

From the Proceedings of the 9th International Zeolite Conference, Montreal 1992, Eds. R. von Ballmoos et al.,

required, (4) the mixture was stirred until it was completely homogenized, (5) phosphoric acid was slowly dropped into the mixture,(6) the resultting gel was stirred for at least 0.5 h, (7) the crystallization of the reaction gel was carried out in Teflon-lined stainless steel autoclaves, heated in an oven under static conditions at autogenous pressure. The products were recovered by filtration and/or sedimentation. They were then washed with copious amounts of water or organic solvent and dried in ambient air. All samples obtained were initially identified by means of powder X-ray diffraction (Rigaku D/max-IIIA X-ray diffractometer) using CuK$_\alpha$ radiation. The main phases were characterized by single crystal XRD, SEM, IR, NMR and DTA-TG.

RESULTS AND DISCUSSION

As relatively low polarity of alcohol solvent by contrast with aqueous solution, it lead to the low solubility of starting reaction gel, the low polymerization rate, and the low crystal growth rate. The crystallization of aluminophosphate in alcoholic system is slower than that in aqueous solution in general. On the other hand, synthesis in alcoholic system is favorable to grow some single crystals, which might be advantaged for X-ray single crystal structure analysis. We have been obtained more than 20 single crystals. A good route is developed for growing single crystals of microporous compounds.

Our general findings and some conclusions come from a study of a large number of different preparations with 32 kinds of amines and 33 kinds of alcohols as solvent. We found some similarities and differences exist in between aqueous and alcoholic synthetic systems. The solubility and dispersion of starting material in alcohol solvent were taken into account, on the premise of introduction of no large amount of foreign cation and anion. Aluminum triisopropoxide and phosphoric acid were generally chosen as the proper sources of aluminum and phosphorus, respectively.

As compared with the crystallization of aluminophosphate in aqueous systems, many phase in alcoholic systems were crystallized. These phases include both the well known molecular sieves in aqueous system and the novel structures which have not discovered yet from aqueous system. Some of known $AlPO_4$ molecular sieves can not be obtained from nonaqueous systems to date in our laboratory, but various types of APO-n have been synthesized, which include small pore AlPO-20, middle pore AlPO-11 and large pore AlPO-5; and element substituted MgAPO-43. The new structures obtained include the chain, sheet, and framework aluminophosphates (from the 8-ring of APO-JDF to the 20-ring of JDF-20, the largest ring in aluminophosphate structures to date.

The amines and products in ethylene glycol(EG) (as an example of alcohol systems), are shown in Table 1. In EG system, of 32 used amines, only i-Pr_2NH failed to give a crystalline phase; Oct_3N and aniline gave dense phases.

Table 1. Products obtained from the reaction mixture 0.5-20 amine : 1 Al_2O_3 : 1.2-3.0 P_2O_5 : 40 EG (180-200°C)

amine	abbreviation	product
<u>primary amine</u>		
methylamine	$MeNH_2$	AlPO-21,K(3D, 8-ring)
i-propylamine	i-$PrNH_2$	AlPO-41,ZnAPO-43,MgAPO-43
propylamine	$PrNH_2$	AlPO-CC(2D, [5])
butylamine	$BuNH_2$	AlPO-CH(2D),CF(unknown)
<u>secondary amine</u>		
dimethylamine	Me_2NH	AlPO-21
dipropylamine	Pr_2NH	AlPO-11, 41
<u>tertiary amine</u>		
triethylamine	Et_3N	AlPO-5
tripropylamine	Pr_3N	AlPO-5, CoAPO-36
<u>quaternary ammonium</u>		
tetramethylammonium	TMA	AlPO-20
<u>alkanolamine</u>		
ethanolamine	MEA	AlPO-JDF(3D,[4]), MgAPO-43
triethanolamine	TEA	AlPO-5
<u>diamine, cyclic amine</u>		
ethylenediamine	en	AlPO-CA(2D,[5]),CAn(n=1-4)
1,3-propyldiamine	1,3-PDA	AlPO-CL(unknown)
1,6-hexanediamine	1,6-HDA	AlPO-CN,CN1(unknown)
1,4-diazabicyclo(2,2,2)octane	DO	AlPO-5

Many layered (or sheet) aluminophosphates were relatively easily obtained in alcoholic system, such as AlPO-CA: $Al_3P_4O_{16}^{3-}$[5]; AlPO-CA4: $AlP_2O_8^{3-}$; AlPO-CC: $Al_3P_4O_{16}^{3-}$ [5]; AlPO-CH: $HAl_2P_3O_{12}^{2-}$, which supplies a possible basis to development of new pillard interlayered $AlPO_4$. Some amines, particularly diamine and medium size primary amine, as templates were favored for the formation of layered aluminophosphates. These layers consist of macroanionic sheets separated by organic moieties.

Figure 1 gives the structure of AlPO-CA. The organic cations have strong influence on the structure. AlPO-CA and AlPO-CC based on the same $Al_3P_4O_{16}^{3-}$ layers, but the layers stack in AAAA and ABAB sequence in AlPO-CC and AlPO-CA, respectively. Such low-dimensional structures may have the potential application as novel acid catalysts. It may be possible by appropriate use of macrocations to render these materials acidic in a manner not dissimilar to the ways in which certain group IV metal phosphaonates have also been made acidic and catalytically active [7].

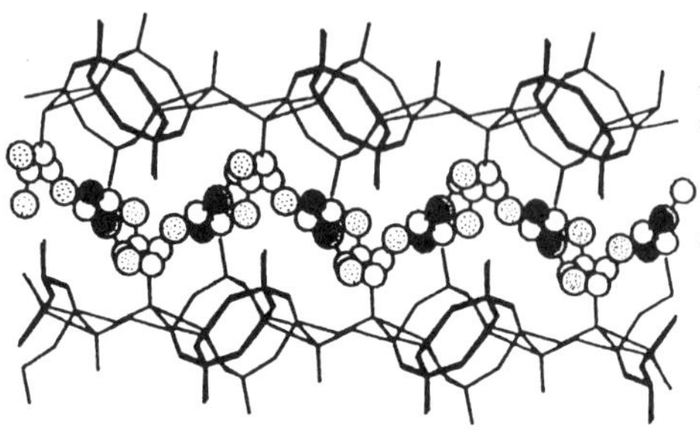

Figure 1 The view of AlPO-CA along the sheet plane (interlamellar layer retained), the interlamellar region consisted of a diprotonated ethylenediammonium cation and two ethylene glycol moieties.

Our results indicate that organic amines have the similar effect in aqueous [8-13] and alcoholic synthesis process. (A) The organic base plays an important role in directing the formation of specific structure in the synthesis (see Table 1). The structure-directing [8-11] effects by the steric and electronic effects of these amine and quaternary ammonium cations. The phenomena of various amines directing same structure was also found. See Table 2. (B) The presence of the organic base has a large influence on the (reaction) gel chemistry [12,15] of aluminophosphate mixture(pH, composition, solubility. and presence of various precursor species, etc.). The templating becomes operative only in the gel with right gel chemistry. Some amines, such as ethylenediamine, can lead to various products with different structures by varying the preparation conditions. The one-template-multiple-structure phenomena[10] are observed (see Table 3). (C) The stabilizing role of the organic base was proposed that the organics act like pore fillers[13]. This is verified by the facts that as-synthesized aluminophosphate microporous crystal includes organic amine, and the amines with large size can not be used as templates (e.g., Oct_3N).

Table 2. selected examples of multiple-template-one-structure phenomena

Product structure	Organic amine[a]
AlPO-21	$MeNH_2$, $EtNH_2$, Me_2NH.
AlPO-41	$iPrNH_2$, Et_2NH, Pr_2NH, Bu_2NH, Am_2NH.
AlPO-C	$EtNH_2$, $iPrNH_2$. $PrNH_2$.
AlPO-5	Et_2NH, CHA, Et_3N, Pr_3N, Bu_3N, DEA, TEA, HMTA, 4-picoline, piperidine, DO.

a CHA=cycle-hexylamine; DEA=diethanolamine; HMTA=hexamethylenetetramine.

Table 3. selected examples of one-template-multiple-structure phenomena

Template	Product structure
$iPrNH_2$	MAPO-43, AlPO-41, AlPO-CC
$BuNH_2$	AlPO-CF, AlPO-CH
en	AlPO-CA, AlPO-CA1, AlPO-CA2, AlPO-CA3, AlPO-CA4
Pr_2NH	AlPO-11. AlPO-41
Pr_3N	CoAPO-36, AlPO-5

Table 4. synthesis in wide compositions

No.	gel composition (molar ratio)				temperature	time	product
	Al_2O_3	P_2O_5	amine	solvent	(/°C)	(/days)	
1	1.0	1.2[a]	$1.2Pr_2NH$	44 EG	180	5	AlPO-11
2	1.0	1.7[b]	$3.3Et_3N$	20 EG	180	15	AlPO-5
3	1.0	2.5	$9.8Et_3N$	60 EG	180	15	AlPO-5
4	1.0	1.8	$2.4Et_3N$	44 EG	180	30	AlPO-5
5	1.0	1.8	$4.7Et_3N$	44 EG	180	30	AlPO-5
6	1.0	1.8	$11.8Et_3N$	44 EG	180	30	AlPO-5

a pseudoboehmite(AlOOH), b $Al(OH)_3$ as aluminium source, respectively.

The crystallization of microporous aluminophosphates in the weak acid medium exhibited a new kind of gel chemistry that is different from that encountered in synthesizing aluminosilicate zeolite in strong basic medium [14]. However, synthesis of aluminophosphate molecular sieves in wide apparent basicity rang, i.e., amine/P_2O_5=1-8 for AlPO-5, from alcoholic system (by contrast, narrow synthetic rang, amine/P_2O_5=0.3-0.7 for AlPO-5, in aqueous system [15]) shows different gel chemistry from that in aqueous system. See Table 4. The further investigation is required making.

It is worth to mention that some regularity, which are lack in aqueous system exist in some alcohol systems. For example, in s-BuOH solvent system, the amines with small size can lead to the formation of three dimensional framework structure, e.g., Me_2NH leads to AlPO-21. The medium size amines, particularly primary amine, can result in production of two dimensional layered structure, e.g., $(BuNH_3)_2HAl_2P_3O_{12}$(APO-CH). The large size amine Et_3N, which is a good template in hydrothermal system, can lead to framework structure in EG; in BuOH it can only lead to one dimensional structure, such as $(C_2H_5)_3NH^+H_2AlP_2O_8^-$(APO-CI) [4]. These probably result from low concentration and less distribution of reactive aluminophosphate species in lower polarity solvent.

In order to understand the role of alcoholic solvent, we selected certain amine Et_3N as template to study the syntheses in more than 20 kinds of alcohols (see Table 5). we found that the structure of product mainly depends on the polarity of solvent. Among many chemical and physical parameters of solvent, the empirical solvent polarity parameter, E_T^N [16], is selected to represent the solvent polarity here. It implies that different precursor species are present in different polarity solvents as the nutrients for crystallization. This was identified by ^{31}P NMR of solution of AlPO-5 and AlPO-CI during crystallization.

In general, the rate of crystallization for a martial increases with temperature. In some cases different structures were formed at different temperatures and some product phases can transform to other phase through a long crystallization time at same temperature. Table 6 gives some examples of these phenomena.

Table 5. Relationship between alcohol solvent and product structure ($Et_3N:Al_2O_3:P_2O_5$=5:1:1.8, 180°)

alcohol[a]	E_T^N	product	alcohol	E_T^N	product
(water	1.000	AlPO-5)	1,4-butanediol	0.704	JDF-20
glycerol	0.812	AlPO-5	1,3-butanediol	0.682	JDF-20
glycol	0.790	AlPO-5	tEG	0.664	JDF-20
methanol	0.762	AlPO-5	butanol	0.602	AlPO-CI
1,3-propanediol	0.747	AlPO-5	s-butanol	0.506	AlPO-CI
DEG	0.713	JDF-20	c-hexanol	0.500	AlPO-CI
TEG	0.704	JDF-20	t-amyl alcohol	0.321	amorphous

a DEG=diethylene glycol; TEG=triethylene glycol; tEG=tetraethylene glycol.

Table 6. effect of temperature and time

No	gel composition (molar ratio) Al_2O_3	P_2O_5	amine	solvent	temperature (/°)	time (/days)	product
1	1.0	1.8	4.8 Pr_2NH	18 TEG	180	1	AlPO-11
2	1.0	1.8	4.8 Pr_2NH	18 TEG	180	7	AlPO-41
3	1.0	2.4	10 $BuNH_2$	26 HexOH	180	2	AlPO-CF
4	1.0	2.4	10 $BuNH_2$	26 HexOH	180	10	AlPO-CH
5	1.0	1.8	4.7 Et_3N	18 TEG	180	5	JDF-20
6	1.0	1.8	4.7 Et_3N	18 TEG	200	5	AlPO-5

CONCLUSIONS

In alcoholic system, new microporous, chain, and layered aluminophosphates can be synthesized. Organic amine takes the similar effect in alcoholic similar to that in aqueous synthesis process. Some similarities and differences exist in between aqueous and alcoholic synthetic systems. The formation mechanisms of aluminophosphate are not the quite same as in aqueous solution. The novel chemistry of microporous aluminophosphate solids in nonaqueous systems is developed.

ACKNOWLEDGE

We thank the National Natural Science Foundation of China for financial support.

REFERENCE

1.Wilson,S.T., Lok,B.M., and Flanigen, E.M., US Pat. 4310440, (1982).

2.Meier,W.M. and Olson,D.H. *Atlas of Zeolite Structure Types,* Butterworths, London, (1987).

3.Huo Qisheng and Xu Ruren, *J.Chem.Soc.,Chem.Commun.,* 783, (1987).

4.Jones,R.H., Thomas,J.M., Xu,R., Huo,Q., Xu,Y., Cheetham,A.K., and Bieber,D., *J.Chem.Soc.,Chem.Commun.,* 1170, (1990).

5.Jones,R.H., Thomas,J.M., Xu,R., Huo,Q., Cheetham,A.K., and Powell,A.V., *J.Chem.Soc.,Chem.Commun.,* 1266, (1991).

6.Huo,Q., Xu,R., Li,S., Ma,Z., J.M.Thomas and R.H.Jones, *J. Chem. Soc., Chem. Commun.,* in press, (1992)

7.Dines,M.B., and Griffith,P.C., *Polyhedron,* **2**, 607, (1983)

8.Flanigen,E.M., Patton,R.L., and Wilson,S.T., in *Innovation in Zeolite Materials Science* (eds. P.J.Grobet et al.), Elsevier, Amsterdam, p.13, (1988).

9.Wilson,S.T., Lok,B.M., Messina,C.A., Cannan,T.R., and Flanigen,E.M., *ACS Symp. Ser. 218,* Am. Chem. Soc., Washington, DC, 79, (1983).

10.Wilson,S.T., Lok,B.M., Messina,C.A., and Flanigen,E.M., *Proceedings of the 6th International Zeolite Conference,* Butterworths, Guildford, UK, 97, (1985).

11.Tapp,N.J., Milestone,N.B., and Bibby,D.M., *Zeolites,* 8, 183, (1988).

12.Lok,B.M., Cannan,T.R., and Messina,C.A., *Zeolites,* 3, 282, (1983).

13.Barrer,R.M., *Stud. Surf. Sci. Catal.,* 24, 1, (1985)

14.Stringham,B., *Econ.Geol.,* 47, 661, (1952).

15.Ren,X., Komarneni,S., and Roy,D.M., *Zeolites.,* **11**, 142,(1991).

16.Reichardt,C., *Solvents and Solvent Effects in Organic Chemistry,* Verlag Chemie, Weinheim, Basel, Cambridge, New York, 2nd edn, (1988).

SYNTHESIS, CHARACTERIZATION AND PHASE TRANSITION OF THE 20-MEMBERED RING $AlPO_4$: JDF-20

Qisheng Huo*, Ruren Xu*, Shougui Li*, Yihua Xu*, Zhanguo Ma*, Yong Yue** and Liyun Li**

*Department of Chemistry, Jilin University, Changchun, P.R.China

**Wuhan Institute of Physics, The Chinese Academy of Sciences, Wuhan, P.R.China

ABSTRACT

The synthetic procedure used to crystallize JDF-20 is described. The predominantly nonaqueous reaction mixture is capable of crystallizing JDF-20. JDF-20 was characterized by X-ray powder diffraction, MAS NMR, IR, Scanning electron microscopy, Chemical analysis and thermoanalysis. JDF-20 can transform or re-crystallize to AlPO-5 under thermal and hydrothermal conditions.

INTRODUCTION

Microporous compounds, of which zeolites are the best known, have been studied for a long time. Since 1982, scientists of Union Carbide Corporation have synthesized a series of aluminophosphates (AlPO-n) [1] and element-substituted aluminophosphate molecular sieves such as SAPO and MeAPO [2]. Zeolites with pores consisting of 12-membered rings were known many years ago, but molecular sieves with larger pores more than 12-membered rings have been synthesized only in recent years. In 1988, the first molecular sieve with larger pores, VPI-5 was reported [3], which is an aluminophosphate containing one-dimensional, circular pores made of 18-membered rings with free diameters of about 13 Å. More recently, Estermann et al [4] reported a new gallophosphate, Cloverite, containing larger pores, which has a 20-membered ring pore that includes four terminal hydroxyl groups which intrude into the opening and has a cloverleaf shape. The supercage formed at the intersection of the channels has a body diagonal of 29-30 Å, significantly larger than any other known cages. More recently, we have synthesized [5], for the first time, a microporous crystalline aluminophosphate possessing 20-membered ring, designated JDF-20 (JDF represents Jilin David Faraday) [5]. The

From the Proceedings of the 9th International Zeolite Conference, Montreal 1992, Eds. R. von Ballmoos et al.,

crystal structure of JDF-20 has been solved with X-ray single crystal analysis by R.H.Jones, J.M.Thomas and their coworkers at Davy Faraday lab. [6]. JDF-20 structure consists of elliptically shaped main 20-membered ring channels intersected by smaller 10- and 8-membered rings. The entrance of the main channel has oxygen atoms projecting into it (see Figure 1). The structural parameters of JDF-20, $[Al_5P_6O_{24}H]^{2-}.2[Et_3NH]^+ .2H_2O$, are space group C2/c, a=32.035, b=14.308, c=8.852A, and beta=104.65^o.

The JDF-20 was synthesized in a predominantly nonaqueous system. In general, hydrothermal synthesis is a typical method for the syntheses of zeolite and aluminophosphate-based molecular sieve. For example, $AlPO_4$-n were generally prepared from aqueous synthetic systems through crystallization of active aluminophosphate gel containing organic amine or quaternary ammonium templating agents. A few zeolites can be synthesized from organic solvent or mixed solvent systems [7,8] in which water is a minor component. We have obtained a series of aluminophosphate molecular sieves and aluminophosphate crystals with specific structures [9-11] in the presence of organic amine templates in alcoholic synthetic systems. In this paper, we report the detailed synthetic procedures, which were used to crystallize the the aluminophosphate JDF-20, characterization of JDF-20, and, particularly, transformation of JDF-20 to AlPO-5 at thermal and hydrothermal conditions, which enrich the chemistry of microporous aluminophosphates.

EXPERIMENTAL

The JDF-20 was synthesized from predominantly nonaqueous systems in which some glycols were used as the solvent. A typical procedure is in following: (i) an aluminium triisopropoxide-solvent slurry was formed; (ii) the amine was added to the slurry; (iii) after 10 min. stirring, phosphoric acid (H_3PO_4, 85 wt.%) was dropped slowly to the mixture; (iv) the precursor mixture was aged under ambient conditions; (v) the final gel was charged into the Teflon-lined autoclave and heated in an oven; (vi) at specified time, the autoclaves were removed from the oven and allowed to cool and products were recovered by filtration with copious amounts of water and dried in ambient air.

The crystalline products were characterized by XRD, IR, ^{27}Al-, ^{31}P-, ^{13}C-MAS NMR, DTA-TG and MS techniques. The X-ray diffraction pattern was recorded by a Rigaku D/max-IIIA X-ray diffractometer using CuK radiation. The X-ray powder diffraction data and their indexing results obtained with the TREOR program show that JDF-20 has a unique structure. Figure 2 illustrates the X-ray powder diffraction pattern for JDF-20. ^{31}P MAS NMR spectrum was recorded on a Varian XL-200 spectrometer, ^{13}C and ^{27}Al MAS NMR spectra were measured on a Bruker MSL-400 spectrometer. Thermogravimetric analysis (TGA) was performed in air on a Perkin-Elmer TG7 thermogravimetric analyzer.

RESULTS AND DISCUSSIONS

Synthesis

We have synthesized JDF-20 in the presence of triethylamine template with a variety of organic alcohols such as triethylene glycol, ethanol and 1,4-butanediol as solvents. Table 1 gives the effect of gel composition on the final product crystallized. Notice that relative wide compositional variations are allowable. VPI-5, by contrast, was synthesized in small compositional variations. The aging time of reaction mixture has no obvious influence on the resulting structure. The aim of ageing is to homogenize reaction mixture. JDF-20 was synthesized from a reactive aluminophosphate gel that contains the triethylamine template. To date, we have not been able to crystallize JDF-20 in the absence of the triethylamine template. Other organic amines led to the formation of different molecular sieves, e.g., triethanolamine and dipropylamine result in APO-5 and APO-11, respectively.

The crystallization of JDF-20 occurs in the temperature range of 140-190°C. At relatively low temperatures, a longer crystallization time is required, and at high temperatures, JDF-20 forms rapidly with small amounts of AlPO-5 impurity. The formation of AlPO-5 is favored at 200°C from the reaction mixture with the same composition by autogeneous pressure crystallization, or by direct high temperature crystallization: heating at 300-700° in air [12]. Using H_2O or EG instead of TEG as solvent, the same gel mixture results in the formation of AlPO-5. AlPO-5 can also be obtained in TEG system under the different crystallization conditions by varying the composition of the reaction mixture. JDF-20 is stable at 180^oC in the mother liquor for many days.

Table 1. Gel composition and crystallization conditions

No.	gel composition Al_2O_3	(molar ratio) P_2O_5	amine[a]	solvent[b]	temperature (/°)	time (/days)	product
1	1.0	1.8	4.7Et_3N	18 TEG	180	5	JDF-20
2	1.0	3.0	7.1Et_3N	18 TEG	180	30	amorphous
3	1.0	1.8	2.5Et_3N	18 TEG	180	5	AlPO-5
4	1.0	1.8	4.7Et_3N	26 DEG	180	10	JDF-20
5	1.0	1.8	5.9Et_3N	14 tEG	180	10	JDF-20
6	1.0	1.2	5.9Et_3N	14 tEG	180	50	JDF-20
7	1.0	1.8	4.7Et_3N	28 BG	180	15	JDF-20
8	1.0	1.8	6.1TEA	18 TEG	180	15	AlPO-5
9	1.0	1.8	4.8Pr_2NH	18 TEG	180	3	AlPO-11
10	1.0	1.8	4.7Et_3N	18 TEG	200	5	AlPO-5

a TEA=triethanolamine, b TEG=triethylene glycol; DEG=diethylene glycol; tEG=tetraethylene glycol; BG=1,4-butanediol.

Heteroatoms, such as Li, Mg, Mn, Ga, B, Si, Ti, and Sn can be introduced into the JDF-20 lattice to form so called M-JDF-20, e.g., Sn-JDF-20 for tin subsitution. The introduction of heteroatoms has an influence on the shape and size of the product crystals.

Characterization

The chemical composition of the as-synthesized sample of JDF-20 is Al_2O_3 31.4 wt%,P_2O_5 43.7 wt%, C: 13.9wt%, H:3.4 wt%, and N: 2.3 wt%. The atomic Al/P ratio is 1. The organic molecule occluding in the structure was triethylamine (Et_3N), which was also identified by MS and ^{13}C MAS NMR. Notice that JDF-20 contains a high organic content. For comparison, AlPO-5 was synthesized with tetrapropylammonium and the analysis of the organic template gives C: 7.5 wt% and N: 0.67% [1].

It was observed by using optical and scanning electron microscopy that JDF-20 crystallized as a pure phase. The JDF-20 crystals have 'sheet' or 'needle-like' morphology, and these crystals are aggregated into bundles. Figure 3 shows scanning electron micrographs of JDF-20.

The MAS NMR spectra of JDF-20 samples are given in Figure 4. The ^{27}Al NMR spectrum of JDF-20 gives a broad resonance at 33 ppm with respect to an external $Al(NO_3)_3$ aqueous solution. This result is typical of tetrahedral Al in microporous $AlPO_4$-type materials [13]. Presumably because of quadrupole effects on the ^{27}Al quadrupole nucleus,the line is broad. In the ^{31}P spectrum of JDF-20, two lines are observed at -22 ppm and -29 ppm with respect to 85 wt.% H_3PO_4. This indicates that two types of tetrahedral P environments [14] exist in JDF-20. The JDF-20 gives two ^{13}C NMR signals at 8.2 and 46 ppm, the chemical shifts were relative to $SiMe_4$. The two resonance lines belong to the two kinds of carbon atoms of the triethylamine molecule, respectively. The line for $-CH_3$ at 8.2 ppm is nearly 2 ppm lower than that (10.8 ppm) of free Et_3N and very close to the value(8.4 ppm) of Et_3N in water. this indicates that the Et_3N in the structure is protonated.

Thermogravimetric analysis of a bulk sample indicates that the weight loss (up to ca.25% at 100-400^oC) is caused by the removal of triethylamine and water molecules in the structure. The result has been verified by elemental analysis and ^{13}C MAS NMR. The removal of

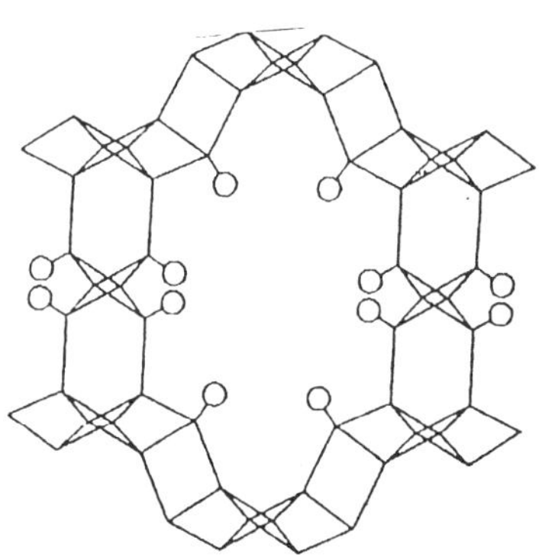

Figure 1 Main channel structure of JDF-20

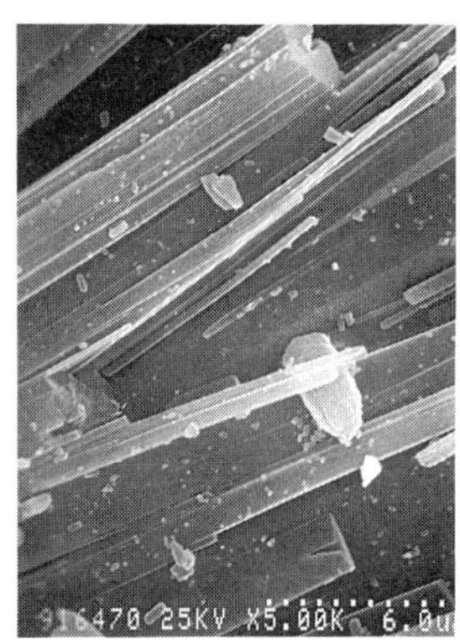

Figure 3 Scanning electron micrograph of JDF-20 crystals

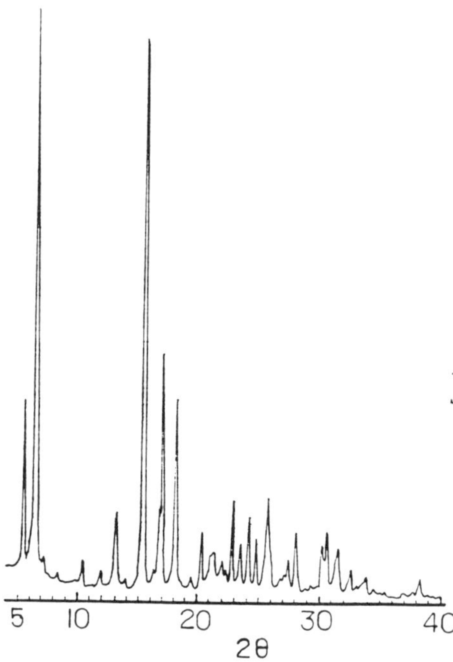

Figure 2 X-ray powder pattern of as-synthesized JDF-20

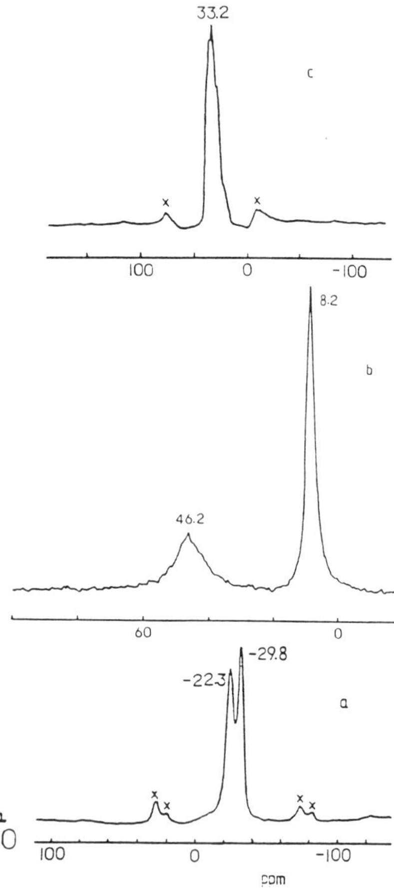

Figure 4 ^{31}P(a), ^{13}C(b) and ^{27}Al(c) MAS NMR spectra of JDF20, x=spinning side bands.

triethylamine at relative high temperature, ca 200°C (triethylamine boiling point 98°C), implies the interaction between triethylamine and the framework. This is also in agreement with the result of ^{13}C MAS NMR. No further weight losses were observed on heating to above 800°C.

The IR spectrum of JDF-20 shows absorption bands at 1237, 1188,

1177, 1089, 759, 649, 618, 582, 561, 477, and 449 cm^{-1}, which are similar to these of 4- and 6- membered ring aluminophosphate. The spectrum of JDF-20 is more complex than that of AlPO-5 or AlPO-11, etc.. The bands around 1240-1000cm^{-1} are somewhat similar to those of VPI-5 [14].

Phase Transition

It is interesting that JDF-20 can transform or recrystallize to AlPO-5 under thermal and hydrothermal conditions. Phase transitions occurring in aluminophosphate molecular sieves have been described before (e.g., the transitions of AlPO-21 to AlPO-25 [1, 15], AlPO-23 to AlPO-28 [1], and VPI-5 to AlPO-8 [16,17]). The phase transition of JDF-20 were observed in two ways. First, the JDF-20 transforms to APO-5 at above 300°C. The transformation of JDF-20 to AlPO-5 depends on removal of the triethylamine template, heating rate (e.g.,during calcination), post-synthesis treatments, and probably lattice defects. Some of the specific conditions under which JDF-20 transforms into AlPO-5 and the various factors that affect such a transformation will be delineated elsewhere. Secondly, JDF-20 digested in water transforms to pure AlPO-5 phase. While the mixture (2-10 ml water per gram. as-synthesized JDF-20) was heated at 180° under autogeneous pressure, after 3 hr, the AlPO-5 was obtained. The introduction of amine into this mixture influences the resulting product structure, (e. g. propylamine results in the AlPO-CC: $Al_3P_4O_{16}{}^{3-}$, a layered aluminophosphate [10]). The results imply that the process of phase transformation involves the recrystallization of JDF-20. Study on the possible mechanism of this phase transformation is in process.

CONCLUSIONS

JDF-20 is a new member of family of aluminophosphate-based microporous materials, and the largest ring aluminophosphate (20-membered rings), JDF-20, can be synthesized from alcoholic system using Et_3N as template and can transform to AlPO-5 in two routes under different conditions. JDF-20 contains protonated triethylamine in its channels.

Acknowledgment. The work was supported by the National Natural Science Foundation of China.

REFERENCES

1.Wilson,S.T., Lok,B.M., and Flanigen,E.M. U.S.Patent 4,310,440. (1982).

2.Lok,B.M., Messina,C.A., Patton,R.L., Gajek,R.T., Cannan,T.R., and Flanigen,E.M., *J.Am.Chem.Soc.,* 106,6092, (1984).

3.Davis,M.E., Saldarriage,C., Montes,C., Garces,J., and Cronder,C., Nature(London), 331, 698, (1988).

4.Estermann,E., McCusker,L.B., Baerlocher,C., Merrouche,A., and Kessler,H., *Nature,* 352, 320, (1991).

5.Huo,Q., Xu,R., Li,S., Ma,Z., J.M.Thomas, and R.H.Jones, *J. Chem. Soc., Chem. Commun.,* in press,(1992).

6.Jones,R.H., Xu,R., Huo,Q., Li,S., Ma,Z., Chen,J., Thomas,J.M., and Chippendale,A.M., submitted to *J.C.S., Chem. Commun.,*

7.Bibby,D.M., and Dale,M.P., *Nature,* 317, 157, (1985).

8.Huo,Q., Feng,S., and Xu,R., *J.Chem.Soc.,Chem.Commun.,* 1486, (1988).

9.Huo,Q., and Xu,R., *J.Chem.Soc.,Chem.Commun.,* 783, (1990).

10.Jones,R.H., Thomas,J.M., Xu,R., Huo,Q., Cheetham,A.K., and Powell,A.V., *J.Chem.Soc.,Chem.Commun.,* 1266, (1991).

11.Jones,R.H., Thomas,J.M., Xu,R., Huo,Q., Xu,Y., Cheetham,A.K., and Bieber,D., *J.Chem.Soc.,Chem.Commun.,* 1170, (1990).

12.Huo.Q., and Xu.R., *J.Chem.Soc.,Chem.Commun.,*, 168, (1992).

13.Blackwell,C.S., and Patton,R.L., *J.Phys.Chem.,* **92**, 3965, (1988).

14.Davis,M.E., Montes,C., Hathaway,P.E., Arhancet,J.P., Hasha,D.L., and Garces,J.M., *J.Am.Chem.Soc.,* 111, 3919, (1989).

15.Richardson,J.W.,Jr., Pluth,J.J., and Smith,J.V., *J.Phys.Chem.,* **94**,3365,(1990).

16.Vogt,E.T.C., and Richardson,J.W.,Jr. *J.Solid State Chem.,* 87, 469, (1990).

17.Annen,M.J., Young,D., Davis,M.E., Cavin,O.B., and Hubbard,C.R., *J.Phys.Chem.,* **95**, 1380, (1991).

HYDROTHERMAL TRANSFORMATION OF ZEOLITE ZSM-39 INTO ZSM-23

S. ERNST, R. KUMAR*, and J. WEITKAMP

Institute of Chemical Technology I, University of Stuttgart, Pfaffenwaldring 55, D-W-7000 Stuttgart 80, Germany

*On leave from National Chemical Laboratory, Catalysis Division, Pune - 411 008, India

ABSTRACT

The hydrothermal transformation of the clathrasil-like zeolite ZSM-39 (MTN) into the 10-membered ring zeolite ZSM-23 (MTT) is observed in aluminosilicate systems containing pyrrolidine as organic base under specific synthesis conditions. Initially, ZSM-39 grows from a gel having the starting chemical composition: 27.5 Na_2O - 50 pyrrolidine - Al_2O_3 - 110 SiO_2 - 4950 H_2O - 38.5 H_2SO_4 at 180 °C under continuous agitation. During the course of crystallization, the ZSM-23 phase with a needle-like morphology grows at the expense of the large octahedral ZSM-39 crystallites and ultimately, almost pure ZSM-23 is obtained. TGA/DTA experiments reveal clear differences in the thermal behavior of the pure and the mixed phases and can be used for a rough estimation of the quantitative composition of the mixtures.

INTRODUCTION

The transformation of a given zeolite structure into another one during conditions of hydrothermal synthesis in the presence of organic cations ("templates") is a well known phenomenon [e. g., 1-10]. Pertinent examples of such transformations are (i) the appearance of faujasite as an intermediate crystalline species during the synthesis of zeolite ZSM-4 from aluminosilicate gels containing tetramethylammonium cations [2]; (ii) the conversion of cubic zeolite P [3-8] or faujasite [9] into a high silica chabazite-like structure (SSZ-13) in the presence of N,N,N-trimethyl-1-adamantammonium cations; and (iii) the tetraethylammonium-mediated transformation of gmelinite into analcime [10]. As pointed out by Zones [7,9], such transformations can be exploited to achieve a faster overall synthesis of a desired material, provided that a less stable zeolite structure is used as silica and alumina supplying reactant. Moreover, the investigation of such structural transformations is of fundamental interest because the knowledge emerging from these studies may contribute to a more detailed understanding of the extremely complex mechanisms involved in hydrothermal synthesis of zeolites. In most of the known examples for such transformations, a more open structure is converted into a less open one, i. e. into a structure with a higher density. Furthermore, hydrothermal transformation of one structure into another one is usually occurring in aluminosilicate systems in which one particular organic base can favor the formation of various structures depending on the gel composition. An example for such an organic

From the Proceedings of the 9th International Zeolite Conference, Montreal 1992, Eds. R. von Ballmoos et al.,

base is pyrrolidine [11-14]: Depending on the starting materials, the exact gel composition and the crystallization conditions, an impressive number of different high silica zeolites can be synthesized, viz. ZSM-5, ZSM-23, ZSM-35, ZSM-39, ZSM-48 and ZSM-51. So far, however, no transformations of these zeolites into each other have been reported. In this paper, we report on the transformation of the relatively dense 6-membered ring clathrasil ZSM-39 into the more open 10-membered ring zeolite ZSM-23 occurring in aluminosilicate gels containing pyrrolidine.

EXPERIMENTAL SECTION

All synthesis experiments were carried out in rotating stainless steel autoclaves with an approximate volume of 250 cm^3. The synthesis mixtures were prepared as follows: Fumed silica with a specific surface area of ca. 200 m^2/g (CAB-O-SIL M-5) was mixed with 90 % of the total amount of distilled water. To this mixture were added stepwise under stirring: NaOH pellets (Merck, p. a.), pyrrolidine (Fluka, p. a.) as structure directing agent, $Al(NO_3)_3 \cdot 9\,H_2O$ (Merck, p. a.) dissolved in the remaining 10 % water, and concentrated sulfuric acid (98 wt.-%, Merck, p. a.). The composition of the reaction mixtures was calculated as described by Rollmann and Valyocsik [15]. The following composition of the synthesis gel turned out to be particularly suitable to study the structural transformation: SiO_2/Al_2O_3 = 110, Na^+/SiO_2 = 0.5, pyrrolidine/SiO_2 = 0.45, H_2O/SiO_2 = 45, OH^-/SiO_2 = - 0.2 and SO_4^{2-}/SiO_2 = 0.35. The mixtures were thoroughly homogenized, filled into the autoclaves and rotated inside a thoroughly thermostated furnace. The transformation was typically observed at 180 °C. Periodically, autoclaves were withdrawn from the oven and quenched with cold water. The crystalline products were filtered, washed with distilled water, dried at 120 °C in air and afterwards characterized by X-ray powder diffraction (Siemens D 5000 diffractometer; CuK α-radiation; 40 kV, 30 mA) and scanning electron microscopy (Cambridge Stereoscan S 4-10). The chemical compositions of the pure zeolites ZSM-23 and ZSM-39 as well as of the solid recovered from the autoclave during the transformation (mixtures of ZSM-23 and ZSM-39) were determined by dissolving the zeolites in HF followed by atomic emission spectroscopy using inductively coupled plasma (AES/ICP). Selected samples were further characterized by simultaneous thermogravimetry and differential thermal analysis (Setaram TGA/DTA 92) using a heating rate of 5 K/min from room temperature to 1 000 °C and a purge of dry air ($\dot{V}_{air}$ = 30 cm^3/min).

RESULTS AND DISCUSSION

Table 1 summarizes the results of a set of crystallization experiments which were originally intended to synthesize zeolite ZSM-23. The alkalinity of the gel (OH^-/SiO_2) was varied, keeping the SiO_2/Al_2O_3-ratio, the template concentration (pyrrolidine/SiO_2) and the water content (H_2O/SiO_2)

Table 1. Influence of OH^-/SiO_2 on the crystalline product obtained from a gel with the composition SiO_2/Al_2O_3 = 110, pyrrolidine/SiO_2 = 0.45 and H_2O/SiO_2 = 45. Crystallization temperature: 180 °C.

Run No.	Na^+/SiO_2	SO_4^{2-}/SiO_2	OH^-/SiO_2	time, d	product
1	0.5	0.15	0.2	1 - 4	ZSM-23 + cristobalite
2	0.5	0.25	0.0	3 - 5	ZSM-39/ZSM-48
3	0.5	0.35	- 0.2	2 - 7	ZSM-39 ⟶ ZSM-23
4	0.4	0.10	0.2	1 - 5	ZSM-23(70%)
5	0.4	0.20	0.0	4	ZSM-23
6	0.4	0.30	- 0.2	6	ZSM-23(30%)
7	0.4	0.35	- 0.3	7 - 10	ZSM-51

constant. Since OH^-/SiO_2 can be principally varied by either adjusting the amount of NaOH (Na^+/SiO_2) or the amount of H_2SO_4 (SO_4^{2-}/SiO_2) added to the gel, both possibilities were explored: For Na^+/SiO_2 = 0.5 the OH^-/SiO_2-ratio was varied between 0.2 und - 0.2 by the addition of increasing amounts of sulfuric acid. In run No. 1, ZSM-23 was obtained along with a considerable portion of cristobalite. After reducing the alkalinity (to OH^-/SiO_2 = 0.0), a mixture of ZSM-39 and ZSM-48 crystallizes. If the amount of sulfuric acid added to the gel is further increased (run No. 3) pure ZSM-39 crystallizes during the first two days and then gradually converts into zeolite ZSM-23 in the following five days. After this time, almost pure ZSM-23 is obtained. From the results of runs No. 4 to 6 it can be concluded that the concentration of Na^+-cations plays an important role in the formation of ZSM-39 from these systems: Even if the overall alkalinity (OH^-/SiO_2) is unchanged, ZSM-39 is completely absent, if Na^+/SiO_2 is reduced by only 20 %. Now, ZSM-23 is the only crystalline product observed. Under very specific conditions (run No. 7), an additional pure material is formed (pyrrolidine-ZSM-51, [14]). In conclusion, it results from Table 1 that the transformation of ZSM-39 into ZSM-23 critically depends on the OH^-/SiO_2 and Na^+/SiO_2 ratios and can only be observed for a very specific composition of the synthesis gel.

The course of the transformation observed during run No. 3 can be seen in Figure 1. After a crystallization time of approximately two days, the typical X-ray powder pattern of zeolite ZSM-39 is observed. The evolution of the strongest diffraction lines of zeolite ZSM-23 begins after ca. 3 days at the expense of zeolite ZSM-39 without the intermediate formation of amorphous material. After a crystallization time of ca. 7 days, almost pure ZSM-23 of good crystallinity is obtained. The absence of significant amounts of amorphous residues suggests that zeolite ZSM-39 was completely consumed during the transformation.

The observed transformation of zeolite ZSM-39 into zeolite ZSM-23 was also followed by scanning electron microscopy. The change in crystallite size and morphology with increasing crystallization time is documented in Figures 2 to 5. Initially, well shaped octahedral crystallites with a size of 5 - 10 μm are formed (not shown). This morphology is typical for ZSM-39. With increasing crystallization time, the octahedra are more and more covered with tiny crystallites of a needle-like shape (Figure 2). This latter morphology is typical for zeolite ZSM-23 crystallized from aluminosilicate gels containing pyrrolidine [11-13]. As the transformation proceeds, the ZSM-39 crystallites are more and more covered by the small needles of ZSM-23 and finally completely wrapped (Figure 3). When ZSM-23 has formed after ca. 7 days, sphere-like agglomerates of the tiny crystallites are observed. Some of these spheres are broken and contain either, like a nest, smaller spheres of the same material (Figure 4) or they are hollow (Figure 5). This probably indicates that the transformation of large octahedral ZSM-39 crystallites into ball-shaped agglomerates of needle-like crystallites of zeolite ZSM-23 begins at the outer surface of the ZSM-39 octahedra and progressively consumes the lattice constituents from the rim to the core. Chemical analysis of the samples withdrawn after different times from the autoclave during run No. 3 reveals that the composition of the solids is almost equal during all steps of the transformation (Si/Al $\approx$ 50). Hence, no further conclusion with respect to the reasons and the exact mechanism of the transformation can be drawn from this information. However, a tentative mechanism could be as follows: During crystallization of ZSM-39, pyrrolidine is occluded in the crystallites which leads to a depletion of template from the mother liquor. The reduced pyrrolidine concentration possibly favors the formation of zeolite ZSM-23 instead of ZSM-39. Hence, ZSM-23 starts to crystallize. Concomitantly, ZSM-39 dissolves, thereby supplying further silicate and aluminate species for the growth of zeolite ZSM-23.

In the TGA/DTA experiments under the conditions applied in this study, the weight loss of ZSM-39 was negligible. By contrast, the total weight loss of pure ZSM-23 produced in run No. 3 amounted to ca. 10 wt.-%. It occurred in two distinct exothermic steps, one at temperatures between 250 and 300 °C (ca. 7 wt.-%), the other one between 440 and 520 °C (ca. 3 wt.-%), respectively. From the absence of an endothermic weight loss at low temperatures (viz. at 100 to 200 °C) it can be concluded that almost no water is adsorbed on the as-synthesized zeolite. Rather, pyrrolidine occupies two energetically different positions in zeolite ZSM-23 and is therefore burnt-off in two steps.

Qualitatively, the TGA/DTA curves of the ZSM-39/ZSM-23 mixtures closely resemble those of pure ZSM-23, with the exception that the higher the amount of (inert) ZSM-39 present in the mixture, the lower is the observed weight loss. Hence, TGA/DTA measurements can be used as a rough estimate for the composition of the ZSM-39/ZSM-23 mixture and are therefore a valuable supplement for the characterization by X-ray powder diffraction.

CONCLUSIONS

The hydrothermal transformation of zeolite ZSM-39 into zeolite ZSM-23 can occur under very specific conditions from aluminosilicate gels containing pyrrolidine as organic base. According to the results obtained in this study, the structural transformation starts at the outer rim of the ZSM-39 crystallites and proceeds progressively to the core. From the experimental results obtained so far, no conclusions with respect to the mechanism of this transformation can be drawn. Characterization of the pure zeolites ZSM-39 and ZSM-23 and their mixtures by thermal analysis is a valuable supplement to the routine characterization by X-ray powder diffraction. TGA/DTA measurements can be used to roughly estimate the composition of the mixtures of the two zeolites. This principle can possibly be applied to determine the composition of other zeolite mixtures consisting of two materials which differ significantly in their thermal behavior.

ACKNOWLEDGEMENTS

Financial support by Deutsche Forschungsgemeinschaft, Fonds der Chemischen Industrie and Max Buchner-Forschungsstiftung is gratefully acknowledged. R. K. is grateful to Alexander von Humboldt-Stiftung for a Humboldt Research Fellowship.

REFERENCES

1. Barrer, R.M., *Hydrothermal Chemistry of Zeolites*, 360 p., Academic Press, London (1982).
2. Dwyer, F.G. and Chu, P., J. Catal. **59**, 263 (1979).
3. Zones, S.I. and Van Nordstrand, R.A., in: *New Developments in Zeolite Science and Technology,* Prepr. of Poster Papers, 7th International Zeolite Conf., Japan Association of Zeolite, Tokyo, Japan, August 17 - 22, 1986, pp. 39-40 (1986).
4. Zones, S.I. and Van Nordstrand, R.A., Zeolites **8**, 166 (1988).
5. Zones, S.I. and Van Nordstrand, R.A., Zeolites **8**, 409 (1988).
6. Chan, I.Y. and Zones, S.I., Zeolites **9**, 3 (1989).
7. Zones, S.I., J. Chem. Soc., Faraday Trans. **86**, 3467 (1990).
8. Zones, S.I. and Van Nordstrand, R.A., in: *Novel Materials in Heterogeneous Catalysis*, Eds. R.T.K. Baker and L.L. Murrell, Am. Chem. Soc. Symp. Series, Vol. 437, American Chemical Society, Washington, D.C., pp. 14-24 (1990).
9. Zones, S.I., J. Chem. Soc., Faraday Trans. **87**, 3709 (1991).
10. Joshi, P.N., Thangaraj, A. and Shiralkar, V.P., Zeolites **11**, 164 (1991).

11. Suzuki, K., Kiyozumi, Y., Shin, S., Fujisawa, K., Watanabe, H., Saito, K. and Noguchi, K., Zeolites **6**, 290 (1986).

12. Ernst, S., Kumar, R. and Weitkamp, J., Catalysis Today **3**, 1 (1988).

13. Ernst, S., Kumar, R. and Weitkamp, J., in: *Zeolite Synthesis,* Eds. M.L. Occelli and H.E. Robson, Am. Chem. Soc. Symp. Series, Vol. 398, American Chemical Society, Washington, D.C., pp. 560-573 (1989).

14. Ernst, S., Kumar, R. and Weitkamp, J., in: *Zeolites for the Nineties,* Eds. J.C. Jansen, L. Moscou, and M.F.M. Post, Recent Research Reports, 8th International Zeolite Conf., Amsterdam, The Netherlands, July 10-14, 1989, pp. 45-46 (1989).

15. Rollmann, L.D. and Valyocsik, E.W., Inorganic Syntheses **22**, 61 (1982).

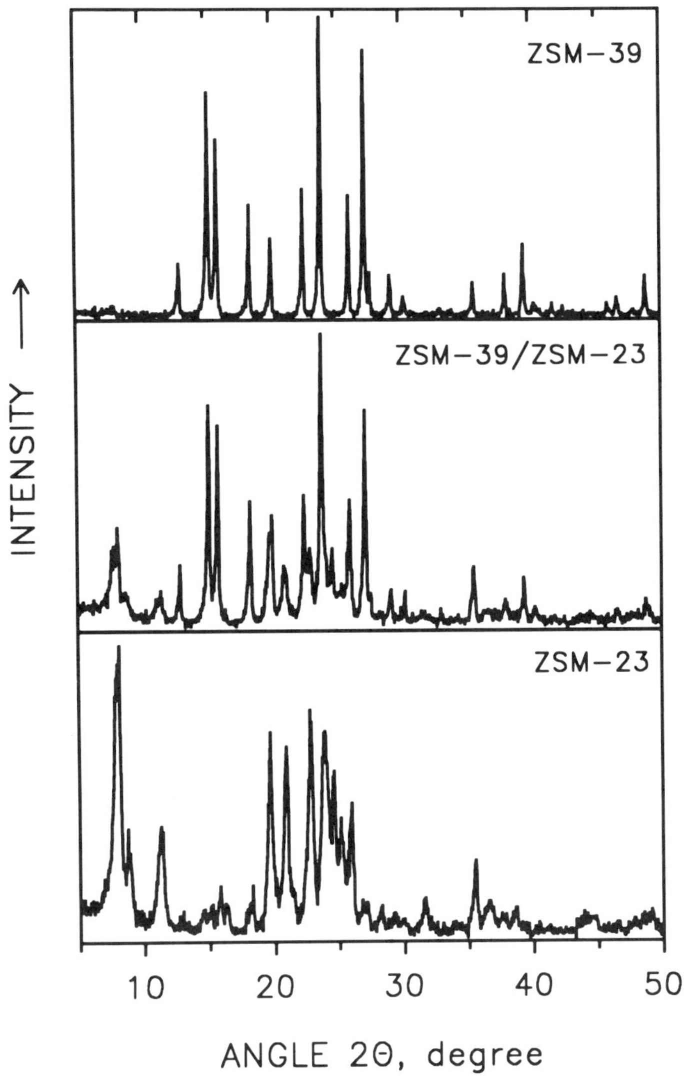

Figure 1. X-ray powder patterns for the transformation of zeolite ZSM-39 into zeolite ZSM-23 observed during run No. 3 after 2 (top), 3 (middle) and 7 (bottom) days.

Figure 2. Scanning electron micrograph of the product from run No. 3 after ca. 3 days.

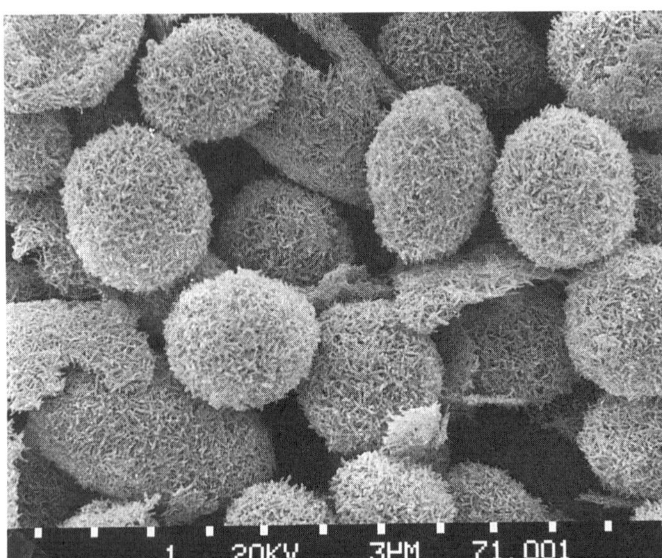

Figure 3. Scanning electron micrograph of zeolite ZSM-23 produced from run No. 3 after ca. 7 days.

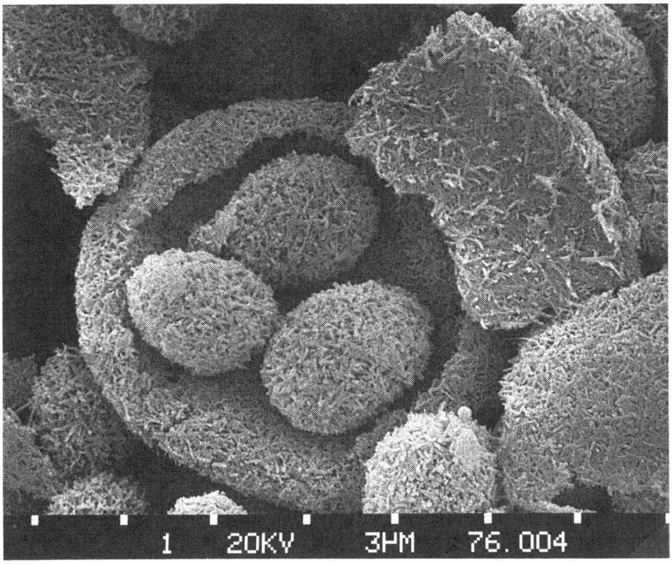

Figure 4. Spheres of zeolite ZSM-23 in larger nests of the same material.

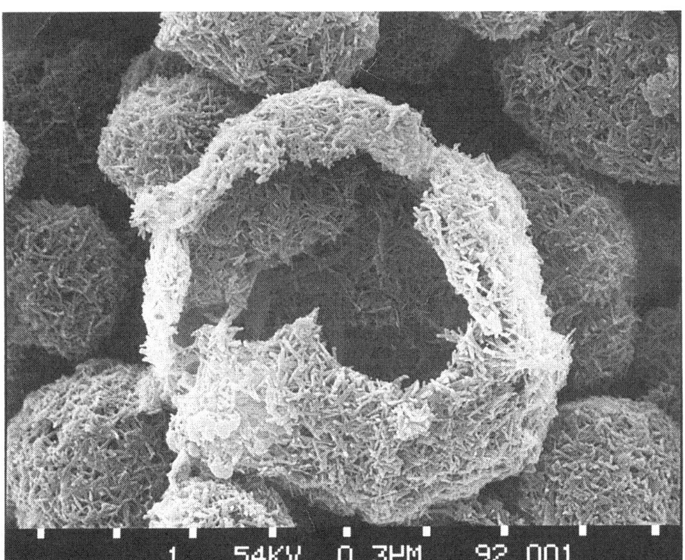

Figure 5. Scanning electron micrograph showing the broken hollow spheres after transformation of ZSM-39 into ZSM-23.

SYNTHESIS OF ULTRAFINE ZEOLITE L

XIANPING MENG, YAN ZHANG, CHANGGONG MENG and WENQIN PANG

Institute of Synthetic and Catalytic Chemistry, Jilin University
Changchun 130023, P.R. China

ABSTRACT

Ultrafine zeolite L with particle size ~ 30nm in diameter was synthesized from the initial mixture of K_2O-Al_2O_3-SiO_2-H_2O system under hydrothermal conditions. The crystallite sizes of products were measured by X-ray powder diffractometry and transmission electron microscopy. The effects of crystallization temperature and time, SiO_2/Al_2O_3, H_2O/Al_2O_3 and K_2O/Al_2O_3 ratios on the crystallite size of products were investigated. The results show that the crystallite size and shape can be controlled by varying synthesis parameters. The ultrafine zeolite L is well-crystallized particles. The optimal ranges of SiO_2/Al_2O_3, H_2O/Al_2O_3 and K_2O/Al_2O_3 ratios (mol. ratio) in initial mixture are 16-40, 250-450 and 10-12.5, respectively.

INTRODUCTION

Ultrafine powders are series of particles with a diameter generally less than 100nm. One of the characteristics of ultrafine powder is that the ratio of external atomic numbers to internal atomic numbers increases rapidly as the particle size decreases. Therefore, ultrafine powders have large external surface area and high surface activity. On the other hand, zeolite molecular sieves have very large internal surface areas. The proportion of external surface area to internal surface area of ultrafine powder differs from that of zeolite molecular sieve because the internal surface area of zeolite is majority. Recently, investigators pay much attention to searching for large pore molecular sieves so as to increase the surface area, but, large pore molecular sieves are generally less thermostable[1-4]. An attempt to increase both external and internal surface of materials was made by preparing ultrafine zeolite molecular sieves. The distribution of secondary pore in such ultrafine zeolites can be controlled in sintering or shaping. Both of the external and internal surfaces of ultrafine zeolite molecular sieves can be sufficiently utilized and the high catalytic activity of such materials can be expected.

In 1986, K. Rajagopalan et al.[5] investigated the selectivity of Y type zeolites in fluid catalytic cracking. The selected zeolites have three particle

From the Proceedings of the 9th International Zeolite Conference, Montreal 1992, Eds. R. von Ballmoos et al.,

sizes with average diameter 60～650nm. Their results showed that smaller particle zeolite exhibits improved activity and selectivity to intermediate cracked products like gasoline and light cycle oil, and the selectivity differences can be explained by the diffusion resistance of species in larger particles. Qin Guanlin et al.[6] early reported that the catalytic activity of 50nm ZSM-5 molecular sieve is much better than that of those with bigger particle size for the isomerization of xylene. V.P. Shiralkar et al.[7] studied the influence of ZSM-5 with different crystallite size on the physicochemical properties and their results proved that longer diffusional path and inhomogeneous aluminum distribution in big particle reduces product yield and selectivity. Zeolite L is a microporous potassium aluminosilicate and effective catalyst in use. The synthesis of normal zeolite L with average particle size 300-1000nm and its applications were early reported[8-17].

EXPERIMENTAL

The pretreatments of raw materials: Calculated amounts of aluminium hydroxide(AR Grade, Beijing Chemical Factory) and amorphous silica(99.95%, Shenyang Chemical Factory) were separately added in KOH solution to form two suspensions. One of the resulting suspension(aluminium hydroxide + KOH solution) was heat-treated in a sealed vessel at 453K for 12 hr. and another one(amorphous silica + KOH solution) was heat-treated in a sealed vessel at 413K for 12 hr. Two solutions were obtained after the treatments.

The two solutions formed above were mixed together and stirred until homogeneous. The homogeneous mixture was placed in a sealed stainless steel pressure vessel lined with polytetrafluoroethylene and heated in an oven at desired crystallization temperature for 1-60 hr. The vessel was rotated at a speed of 20 r.p.m. during crystallization. The solid products were centrifuged by using a high speed centrifuge unit, washed with deionized water, and dried in vaccum at low temperature(77K). The phase analysis of the products was performed by using a X-ray powder diffractometer at 30KV and 20MA (D/max-IIIA, Rigaku Denki Co. Ltd.). Silicon powder was used as an internal standard. Crystallite sizes were measured from the half-width of (221) diffracion lines of zeolite L and calculated by Scherrer equation, $D=0.9\lambda(\beta\cos\theta)^{-1}$, where D is the crystallite size, λ the wavelength (Cu $K\alpha$), β the corrected half-width obtained from the (101) line of 5-15μm low quartz as a standard, and θ the Bragg angle of the diffraction line. The crystallite appearance was observed by a transmission electron microscope(TEM) at 80KV(JEM-100CXII, JEOL Co. Ltd.). The ratios of silica to alumina of products were determined by chemical analysis.

RESULTS AND DISCUSSIONS

Effect of crystallization temperature on the crystallite size

The relationship between crystallization temperature and crystallite size of products was investigated under the conditions of dynamic crystallizing for 4 hours with compositions of initial mixture $10.0K_2O:1.0Al_2O_3:20SiO_2:400H_2O$, and the results are presented in Figure 1. At 393K only amorphous product was obtained, and at 413-463K, crystal zeolite L was produced, but the crystallite size was not changed obviously. Above 473K the particle size was varied evidently.

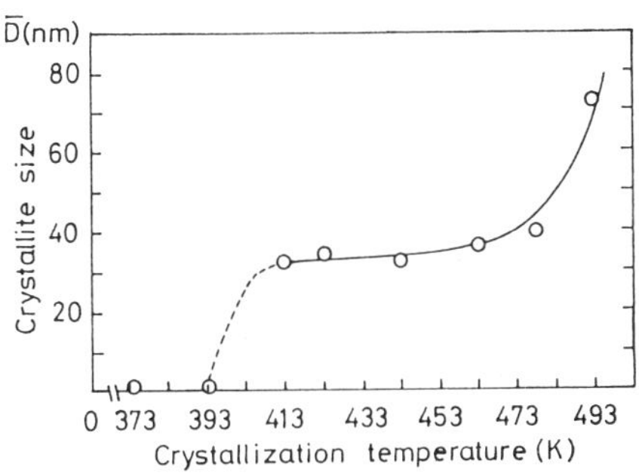

Figure 1. The effect of crystallization temperature on the crystallite size.

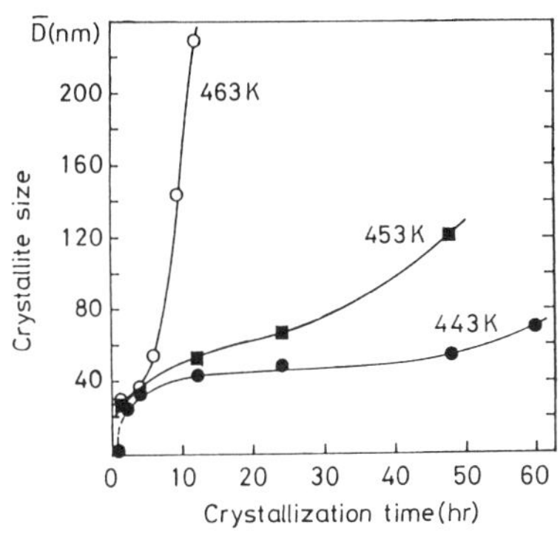

Figure 2. The effect of crystallization time on the crystallite size.

Effect of crystallization time on the crystallite size

Figure 2 shows the effect of crystallization time on the particle size of zeolite L under the conditions of the same compositions as mentioned above but at variable crystallization temperatures. At 463K, the particle size of products were obviously variable with increasing crystallization time, which is different from that at 443K and 453K. It was shown that nucleation and crystal growth of zeolite L is fast at higher temperature, and the perfect crystal of zeolite L can be synthesized at 443K. The results at 463K as shown in Figure 2 are similar to those reported by P.N. Joshi, but, the results at other temperatures are different from his report[18]. Under the same conditions of crystallization, the rate of nucleation and growth at high temperature is much higher than that at low temperature.

Effect of the SiO_2/Al_2O_3 ratio in initial mixture on the crystallite size

The effects of the SiO_2/Al_2O_3 ratio in initial mixture on the crystallite size of zeolite L were investigated by using dynamic crystallizing at 443K for 4 hours, and the results of variable SiO_2/Al_2O_3 ratio (H_2O/Al_2O_3=400, K_2O/Al_2O_3=10) are presented in Figure 3. The crystallite size of products increases slowly with increasing SiO_2/Al_2O_3 ratio of initial mixture. The appearance of products was dependent on SiO_2/Al_2O_3 ratio of initial mixture, as shown in TEM photographs of products (Figure 4).

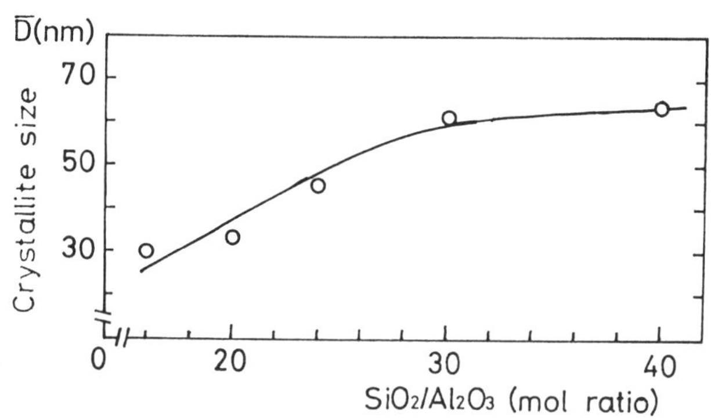

Figure 3. The relationship of crystallite size of products and SiO_2/Al_2O_3 ratio of initial mixure.

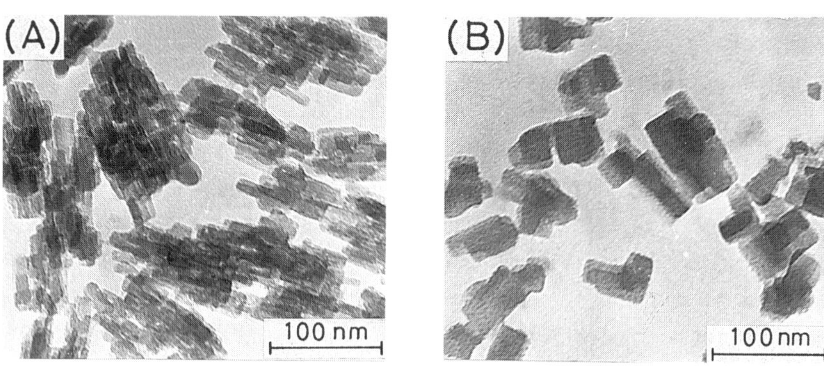

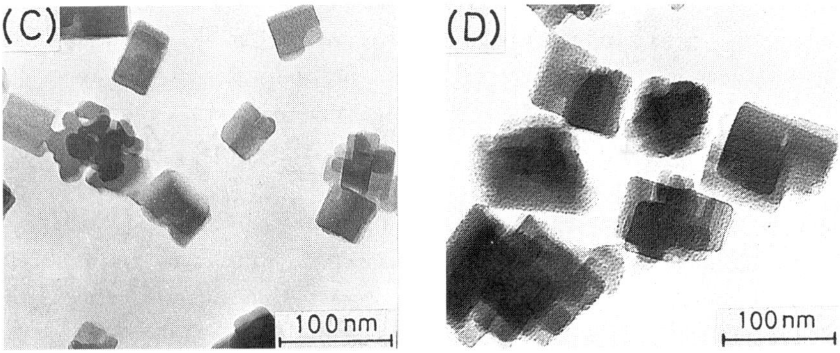

Figure 4. TEM photographs of products synthesized with variable SiO_2/Al_2O_3 ratio of initial mixture (H_2O/SiO_2=400 and K_2O/SiO_2=10) at 443K for 4 hours. (A)SiO_2/Al_2O_3=16; (B)SiO_2/Al_2O_3=20; (C)SiO_2/Al_2O_3=26; (D)SiO_2/Al_2O_3=30.

With increasing the SiO_2/Al_2O_3 ratio in initial mixture, the crystallite size of zeolite L became larger and wider gradually. The SiO_2/Al_2O_3 ratio of products is directly related to the SiO_2/Al_2O_3 ratio of initial mixture as shown in Figure 5.

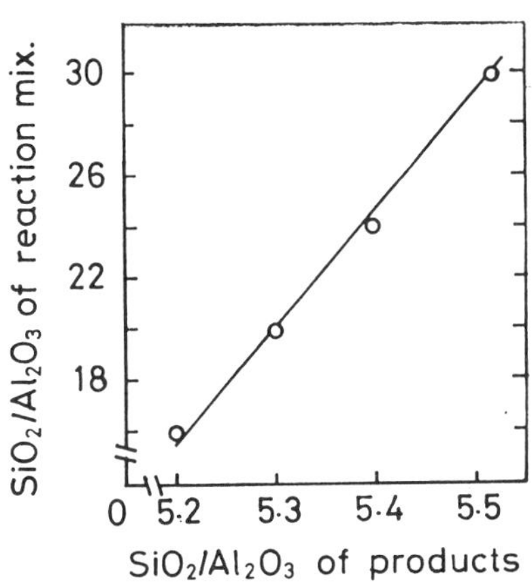

Figure 5. The relationship of SiO_2/Al_2O_3 ratio of initial mixture and products.

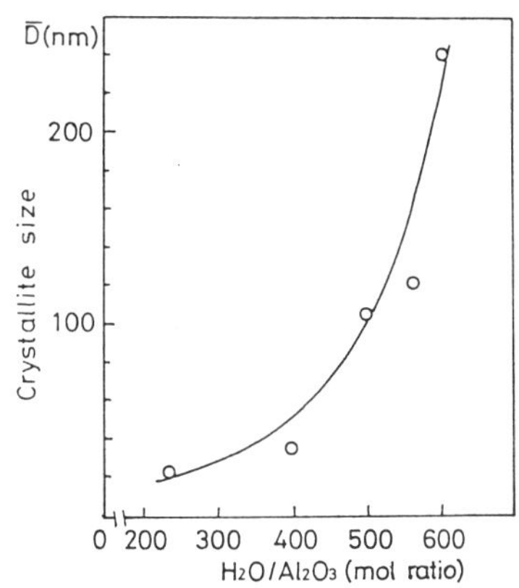

Figure 6. The effect of H_2O/Al_2O_3 ratios of initial mixture on the particle size.

Effect of the H_2O/Al_2O_3 ratio in initial mixture on the particle size

The H_2O/Al_2O_3 ratios of initial mixtures (SiO_2/Al_2O_3=20, K_2O/Al_2O_3=10) influence the crystallite sizes of products synthesized at 443K for 4 hours as shown in Figure 6. The variation of particle size is large in the range of H_2O/Al_2O_3 ratios from 400 to 600, and impurity appears while the H_2O/Al_2O_3 ratios are more than 700.

Effect of the K_2O/Al_2O_3 ratio in initial mixture on the particle size

The effect of the K_2O/Al_2O_3 ratios in initial mixture on synthesis of ultrafine zeolite L is important. Figure 7 shows the XRD patterns of products synthesized in variable K_2O/Al_2O_3 ratio(SiO_2/Al_2O_3=20, H_2O/Al_2O_3=400) at 443K for 4 hours. At the K_2O/Al_2O_3 ratio$>$15, zeolite L can not be formed and at the K_2O/Al_2O_3 ratio$<$8, impurity appears. The particle sizes decrease with increasing K_2O/Al_2O_3 ratio. The optimal range of K_2O/Al_2O_3 ratio is 10-12.5.

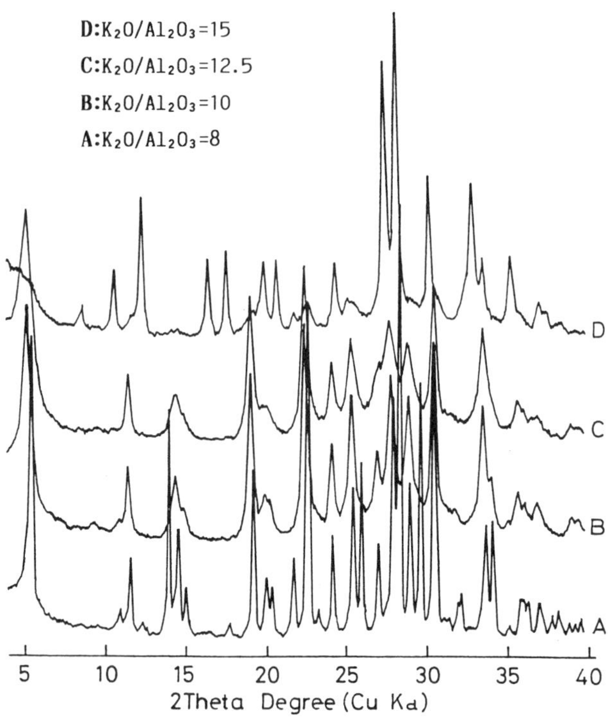

Figure 7. XRD patternts of products synthesized with variable K_2O/Al_2O_3 ratio in initial mixture.

Effect of the dynamic crystallization and the pre-treatment of raw materials on synthesizing ultrafine zeolite L

The dynamic crystallization plays an important role in synthesizing ultrafine zeolite L. Under the same conditions, crystallization time under dynamic condition is much more shorter than that under static condition as shown in Figure 8. Another important factor on the synthesis of ultrafine zeolite L is pre-treatments of raw materials. After pre-treatments of raw materials, pure ultrafine zeolite L is obtained, otherwise particle sizes of products are larger, or/and impurities are formed. This can be clearly seen from the results of TEM and XRD of products in the same systems.

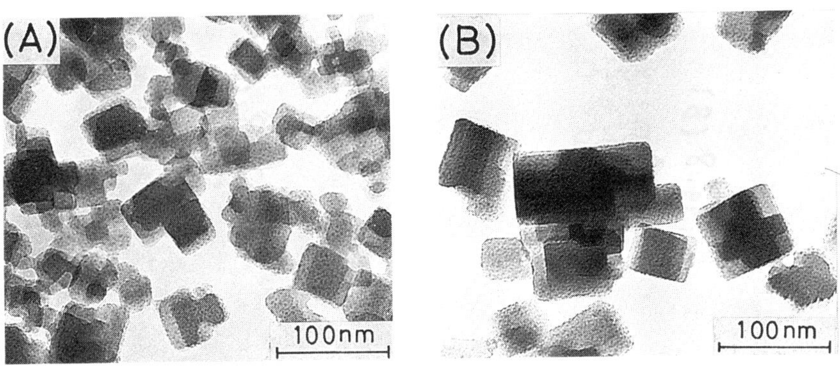

Figure 8. TEM photographs of products synthesized with dynamic and static crystallizing, Composition in initial mixture: $10.0K_2O:1.0Al_2O_3:30SiO_2:400H_2O$, at 443K, (A)static 168 hours; (B)dynamic 6 hours.

CONCLUSIONS

The ultrafine zeolite L with various particle sizes can be obtained by changing the compositions of initial reaction mixture and the conditions of crystallization. The optimal ratios of SiO_2/Al_2O_3, H_2O/Al_2O_3 and K_2O/Al_2O_3 are 16-30, 250-450 and 10-12.5, respectively, and crystallization temperature and time are 403-493K and 1-60 hours, respectively. The dynamic crystallization and the pre-treatment of raw materials shorten crystallizing time, reduce size of well-distribution particle and increase the purity of product.

ACKNOWLEDGEMENTS

This work was supported by the National Natural Science Fundation of China and the fundation of State Key Laboratory for Physical Chemistry of the Solid Surface, Xiamen University.

REFERENCES

1.Davis, M.E., Saldarriaga, C., Montes, C., Garces, J. and Crowder, C., Nature **331** , 698(1988).

2.Dessau, R.M., Schlenker, J.L. and Higgins, J.B., Zeolites **10**, 522(1990).

3.Estermann, M., McCusker, L.B., Baerlocher, C., Merrouche, A. and Kessler, H., Nature**352** 320(1991).

4.Qisheng Huo, Ruren Xu et al., in: Proceedings of the 9th International Zeolite Conference, Eds. R.von Ballmoos, J.B. Higgins, and M.M.J. Treacy, Butterworth-Heinemann, Stoneham, MA,(1992).

5.Rajagopalan, K., Peters A.W., et al., Appl. Catalysis **23**, 69 (1986).

6.Qin Guanlin et al., Petroleum Refining, (in Chinese) **11-12** , 85(1973).

7.Shiralkar, V.P., Joshi, P.N., Eapen, M.J. and Rao, B.S., Zeolites **11**,511(1991).

8.Breck, D.W., "Zeolite Molecular Sieves: structure, chemistry and use", London: John Wiley and Sons. chap. 4, (1974).

9.Barrer, R.M., Beaumont, R.K. and Collela, C., J. Chem. Soc., Dalton, 934(1974).

10.UCC of American, British Patent Specification 1 202 511(1970).

11.Thomas, V.W., British Patent Specification 1 393 365(1975).

12.Young, D. A. and Linda, Y., U.S. Patent 3,867,512(1975).

13.Bernard, J.R. et al., U.S. Patent 4,104,320(1978).

14.Olavesen, C., British Patent Specification 1 601 655(1981).

15.Durez, D., et al., U.S.Patent 4,238,368(1980); Waldeen, C.B., et al., U.S. Patent 4,721,695(1988).

16.Wortel,T.M. and Verduijn, J.P., Eur. Pat. Appl.220 881(1987).

17.Verduijn, J.P., Eur. Pat. Appl.288 294(1988); 323 892(1989); 323 893(1989); 357 252(1990).

18.Joshi, P.N., Kotasthane, A.N. and Shiralkar, V.P., Zeolites**10**, 598(1990).

NEW SILICA-ALUMINA WITH NANO-SCALE PORES PREPARED FROM KANEMITE

S.INAGAKI*,Y.FUKUSHIMA*,A.OKADA*,T.KURAUCHI*,K.KURODA**,C.KATO**

*Toyota Central R&D Labs., Inc., 41-1, Yokomichi Nagakute-cho Aichi-gun 480-11 JAPAN

**Department of Applied Chemistry, Waseda University, Ohkubo-3, Shinjuku-ku, Tokyo 169 JAPAN

ABSTRACT

New silica-alumina with uniform nano-scale pores was prepared from a layered silicate, kanemite ($NaHSi_2O_5 \cdot 3H_20$). After the formation of hexadecyltrimethylammonium-kanemite complex, it was treated with an aqueous aluminum chloride solution and the resulting product was calcined. The XRD and MAS NMR results indicated that the original SiO_2 layers were condensed with each other to form three dimensional SiO_2 networks with Al^{3+} substitution in the frameworks. The calcined products have uniform pore-size distribution. The acidic property was similar to that of an amorphous silica-alumina. The surface area was about 600 m^2/g, and the porosity did not decrease by heat treatment up to 800 °C.

INTRODUCTION

A microporous material whose pore size is larger than that of zeolite group has been of great interest[1)], because of their availability as catalysts for many useful reactions such as cracking and syntheses of molecules having relatively large kinetic diameters. Pillared clays are known as typical wide opening microporous materials. However, their maximum pore size is about 2.0 nm and their thermal stabilities are not enough for cracking reactions[2)]. Amorphous silica-alumina as well as zeolites has widely been used as catalysts for cracking. The pore size distribution of the amorphous silica-alumina is not so uniform as zeolite, which would cause low selectivity in the products. No material has been reported to have uniform and relatively large pores[1)].

Kanemite ($NaHSi_2O_5 \cdot 3H_2O$) is one of the layered polysilicates. The layers of kanemite are composed of single-layered sheets of SiO_4 tetrahedra, although the precise crystal structure has not yet been determined. Recently, Yanagisawa et al.[3)] have reported that highly porous silica can be prepared from kanemite by the intercalation of alkyltrimethylammonium ions in the interlayer space and the calcination of the resulting products. During the organic intercalation, the SiO_2

From the Proceedings of the 9th International Zeolite Conference, Montreal 1992, Eds. R. von Ballmoos et al.,

layers are condensed with each other to form three dimensional SiO_2 networks containing the organic cations. After calcination, the product shows uniform pore size distribution which can be controlled by the kind of the organic cations used.

However, the porous silica had only a little acidity, although surface acidity is essential to the reactions. Therefore, the attempt to prepare similar porous materials with surface acidity is quite interesting. The surface acidity of zeolites is attributable to an isomorphous substitution of Al^{3+} for Si^{4+}. We have tried to prepare porous materials with uniform nano-scale pores and surface acidity from kanemite by means of the substitution with Al^{3+}.

EXPERIMENTAL

Sample preparation

Kanemite was synthesized from an aqueous solution of sodium silicate (water-glass); $SiO_2/Na_2O=2.07$. The water glass was vacuum-dried at room temperature for 2d and calcined at 700 °C for 6h to crystallize δ-$Na_2Si_2O_5$. One hundred grams of δ-$Na_2Si_2O_5$ were dispersed in distilled water (1 L) by stirring for 3h. Kanemite powder was obtained by filtration of the suspension and dried at 25°C. The formation of kanemite was checked by means of X-ray powder diffraction.

A hexadecyltrimethylammonium-kanemite complex was prepared by cation exchange reaction using an autoclave. The mixture of 3 g of kanemite and a 0.1 N aqueous solution of hexadecyltrimethylammonium chloride was heated in a teflon autoclave at 65 °C under shaking for a week. The resulting product was filtered, washed with distilled water repeatedly and air-dried to yield the cation-exchanged kanemite.

The cation-exchanged kanemite was mixed with 50 ml of an aqueous solution containing 0.3–1.5 g of $AlCl_3{\cdot}6H_2O$ and dried, followed by heating in air at 700 °C for 6h. During this heat-treatment the organic cations were burned out and we got a porous silica-alumina. The Al_2O_3 contents in the materials were 5.3, 8.6, 15.8, 22.0 wt% respectively, which were estimated from the concentrations of the $AlCl_3{\cdot}6H_2O$ aqueous solutions used.

A porous silica without Al^{3+} which was reported by Yanagisawa et al.[4)] was prepared by calcination of the cation-exchanged kanemite obtained by the same conditions.

Characterization

The structure of kanemite, cation-exchanged kanemite and the porous materials were examined by X-ray powder diffraction (XRD) and solid state MAS ^{29}Si and ^{27}Al-NMR. XRD spectra were recorded on a Rigaku RAD-B instrument with

Co–Kα radiation. Solid state NMR spectra were obtained on a Bruker MSL–300WB spectrometer.

Pore–size distributions (Cranston–Inkley method) and B.E.T. surface areas were calculated using the results of N_2 adsorption isotherms at temperature of liquid N_2. The isotherms were measured by means of volume method using conventional vacuum–line with Baratron pressure gauge 127A[4]. Samples were preheated at 250°C for 2h.

Thermal stability of porous materials was estimated from the change in B.E.T. surface areas of samples by calcination at 700°C to 1000 °C for 6h in air.

Ammonia–TPD spectra were also obtained with a conventional TPD apparatus, in which the desorbed materials were detected by a thermal conductivity detector. About 0.1 g of samples was vacuum–dried at 500 °C for 1h. The dried samples were exposed to 13.3 kPa of ammonia gas at 100 °C for 45 min, followed by evacuating at 100 °C for 1h. The TPD spectra were measured from 100 °C to 1000 °C with heating rate of 10°C/min. in a helium flow as carrier gas controlled at 4.0×10^{-3} of W/F.

Powder sample was compressed into a disc of 20 mm in diameter and about 10 μm in thickness. The disc vacuum–dried at 200 °C was exposed to pyridine vapor at room temperature, followed by evacuation. IR spectra of the disc sample before and after the adsorption of pyridine were recorded on a Jasco FT/IR–5M spectrometer in the range 4000cm^{-1} to 1200 cm^{-1}. The vacuum cell for IR measurement was used for the pretreatment of the sample and the IR measurements.

RESULTS AND DISCUSSION

Pore structure

The pore–size distributions of the porous silica–alumina and the porous silica are shown in Figure 1. The porous silica had a uniform pore–size distribution as reported by Yanagisawa et al[3]. Although the pore–size distribution was not changed virtually by the addition of Al^{3+}, the B.E.T surface area and the pore volume of the porous silica–alumina were decreased.

Figure 2 shows the relationship between B.E.T. surface area of the porous materials and calcination temperature. These results indicated that these porous materials could stand at high temperature up to 800 °C in air without decreasing the surface area.

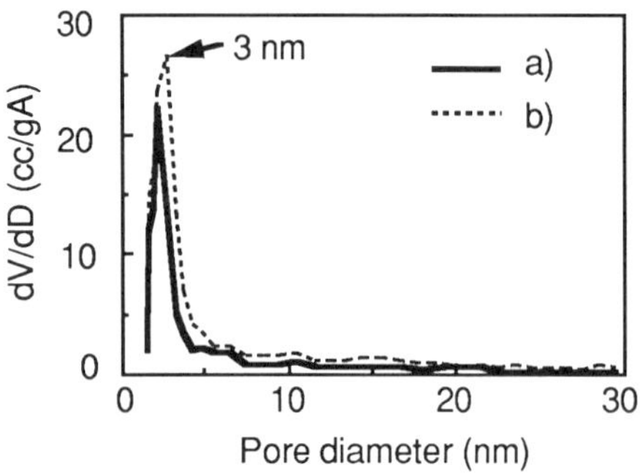

Fig. 1. Pore-size distributions

a) porous silica-alumina(Al_2O_3;15.8wt%)
b) porous silica

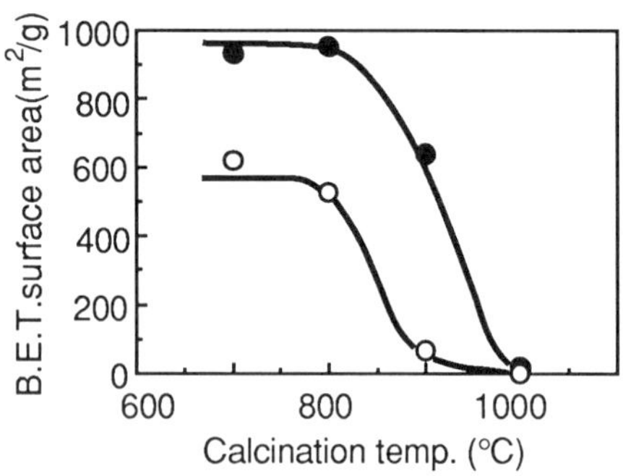

Fig.2 Relationship between B.E.T. surface area and calcination temp. (○) porous silica-alumina (Al_2O_3;15.8 wt%), (●) porous silica.

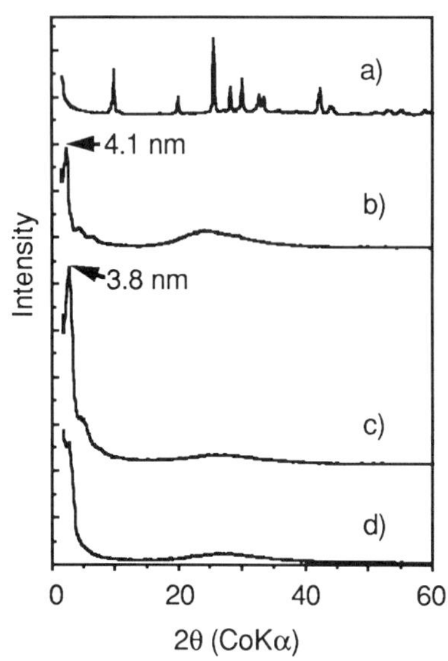

Fig.3. XRD patterns of a)kanemite, b) cation-exchanged kanemite, c) porous silica, d) porous silica-alumina(Al_2O_3;15.8wt%).

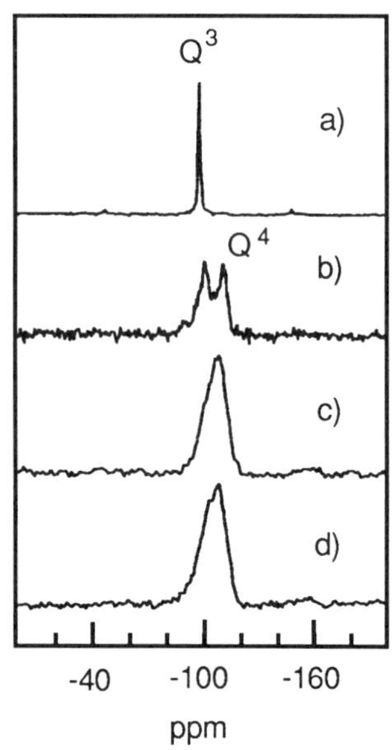

Fig.4. ^{29}Si-MAS NMR spectra a) kanemite, b) cation-exchanged kanemite, c) porous silica, d) porous silica-alumina (Al_2O_3;15.8wt%).

Crystal structure

The XRD patterns of kanemite, cation-exchanged kanemite and the porous materials are shown in Figure 3. The characteristic peaks for kanemite were not observed for the cation-exchanged kanemite and the porous silica and only a few peaks in low angle region were observed for them. These results suggested that the basal spacing of kanemite was increased by the ion-exchange with hexadecyltrimethylammonium ions for Na^+ ions of kanemite and that the SiO_2 layers were deformed and condensed with each other to form SiO_2 networks, as reported by Yanagisawa et al.[3] The peaks at d=4.1 nm and 3.8 nm were observed in the XRD patterns of the cation-exchanged kanemite and the porous silica, respectively. Beneke et al.[5] reported that the long-chain trimethylalkylammonium ions (nc > 12, nc is the number of carbon atoms in the alkyl-chain) formed gauche block-type bilayers. Although the peak at d=3.8 nm was also observed for the porous silica-alumina, the peak intensity was weaker than that for the porous silica. These results suggested that the structure of the porous silica-alumina was somewhat degraded by the addition of Al^{3+} though the structure was similar to the porous silica.

Figure 4 shows the ^{29}Si-MAS NMR spectra of kanemite, cation-exchanged kanemite and the porous silica. Only Q^3 peak was observed for kanemite, which indicated the single layered silicate structure. Q^4 peaks with overlapped Q^3 peaks were also observed in Figures 4 b) and c). These results supported that three-dimensional SiO_2 networks were formed in kanemite during the cation exchange as has already been reported by Yanagisawa. et al.[3] The spectrum of the porous silica-alumina as shown in Figure 4 d) was similar to that of the porous silica except the peak broadening. The broadening can be attributed to the isomorphous substitution of Al^{3+} for Si^{4+} in the SiO_2 frameworks. The isomorphous substitution was supported by the ^{27}Al-MAS NMR spectra (Figure 5). The porous silica-alumina contained both 4-coordinated and 6-coordinated Al^{3+} similar to the usual amorphous silica-alumina. The existence of 4-coordinated Al^{3+} suggested the isomorphous substitution of Al^{3+} for Si^{4+} in the SiO_2 framework.

Acidic properties

NH_3-TPD profiles of the porous silica and the porous silica-alumina are shown in Figure 6. The relationship between the amounts of acid and Al_2O_3 contents for the porous silica-alumina are shown in Figure 7, in which the results for ZSM-5 (SiO_2/Al_2O_3=40) and the amorphous silica-alumina (JRC-SAL-2[6]) are also shown. The amounts of acid increased with the Al_2O_3 contents. This

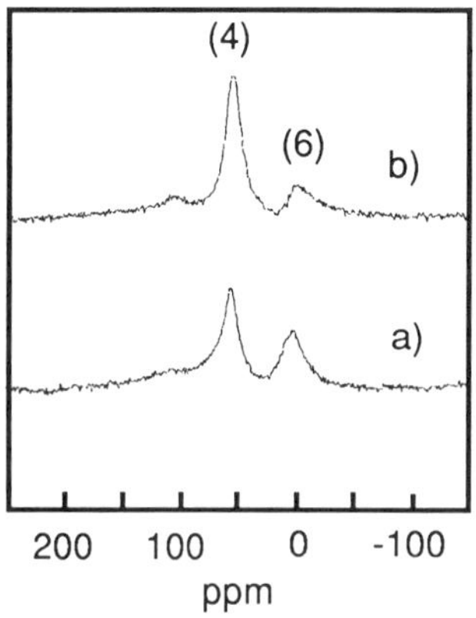

Fig.5. ^{27}Al-MAS NMR spectra
a) porous silica-alumina (Al_2O_3 ;15.8wt%)
b)amorphous silica-alumina (JRC-SAL-2[6]).

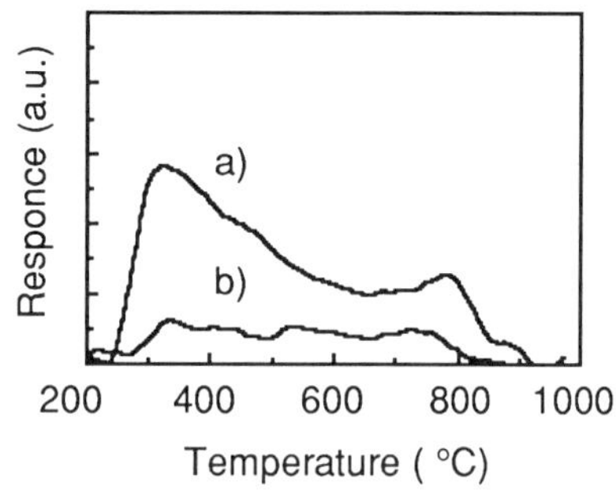

Fig.6 TPD profiles of NH_3 on
a) porous silica-alumina (Al_2O_3; 22.0 wt%, b) porous silica.

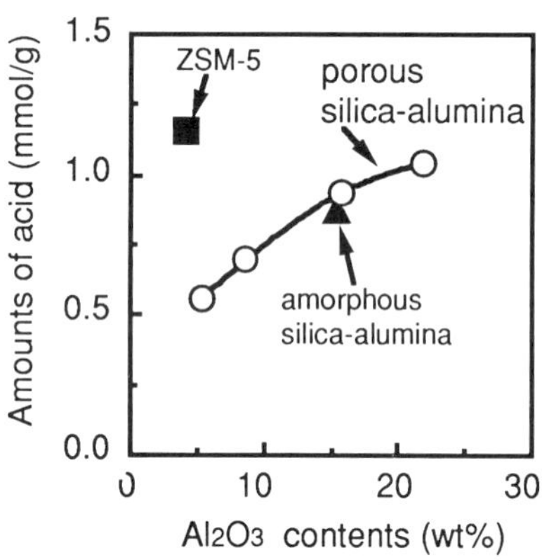

Fig.7 Relationship between amounts of acid and Al_2O_3 contents.
(O) porous silica-alumina, (■) ZSM-5 (SiO_2/Al_2O_3=40), (▲) amorphous silica-alumina (JRC-SAL-2[6]).

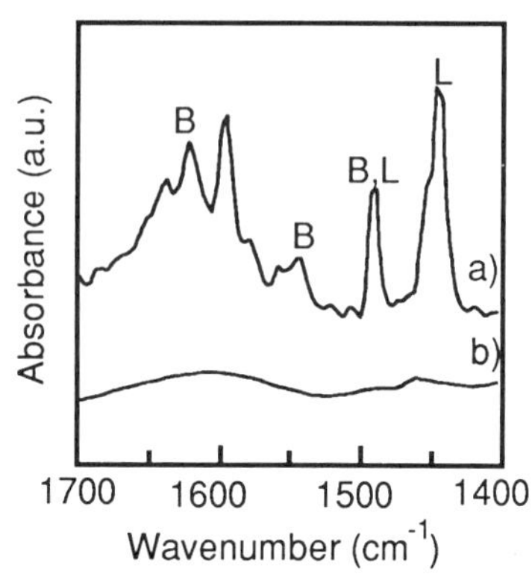

Fig. 8 Infrared spectra of porous silica-alumina (Al_2O_3;15.8 wt%)
a) after and b) before adsorption of pyridine.

B; Brφnsted acid, L; Lewis acid

result suggested again the isomorphous substitutions. A small amount of acid was observed in the porous silica, because of contamination of Al^{3+} in water-glass. The broad TPD profiles were observed for the porous silica–alumina, which was similar to the amorphous silica–alumina (JRC–SAL–2[6]). Although the amounts of acid on the materials prepared from kanemite agreed well with that on the amorphous silica–alumina, those amounts were 1/2 than on ZSM–5 with the same Al_2O_3 contents.

The infrared spectra shown in Figure 8 suggested that the porous silica-alumina had both Lewis and Brϕnsted acid sites and the result was similar to the usual amorphous silica–alumina[7]. It was confirmed by TPD spectra and infrared spectra that the acidic properties of the porous silica–alumina was similar to that of an amorphous silica–alumina.

CONCLUSION

New porous materials which possessed uniform nano–scale pores, high surface area, high thermal stability, and acidic surface were obtained. They were expected to be useful for cracking reactions and so on.

ACKNOWLEDGMENTS

One of the present authors (S.I.) is grateful to A. Koiwai, Toyota Central R&D Labs. Inc., for his cooperation in the NMR study and to N. Takahashi, ibid, for his NH_3–TPD measurements.

REFERENCES

1) D.E.W. Vaughan, Proc. 8th. Intl. Zeolite Conf., Ed. P.A. Jacobs and R.A. van Santen, ELSEVIER, p95 (1989).
2) F. Figueras, Catal. Rev.–Sci. Eng., 30 (3), 457–499 (1988).
3) T. Yanagisawa, T.Shimizu, K. Kuroda and C. Kato, Bull. Chem. Soc. Jpn., 63, 988–992 (1990).
4) S. Inagaki, Y. Fukushima, H. Doi and O. Kamigaito, Clay Minerals, 25, 99–105 (1990).
5) K. Beneke and G. Lagaly, Am. Mineral., 62, 763 (1977).
6) Reference catalyst from Catalysis Society of Japan.
7) K. Tanabe, "Solid Acids and Bases", p60, KODANSHA.

A MICROPOROUS TETRAETHYLAMMONIUM PERMUTITE AS SYNTHESIS INTERMEDIATE OF THE ZEOLITE BETA

M.A. NICOLLE*, F. DI RENZO*, F. FAJULA*, P. ESPIAU* and T. DES COURIERES**

*CNRS URA 418, Ecole Nationale Supérieure de Chimie, 8 Rue de l'Ecole Normale, 34053 Montpellier, France (fax 33/67144353)

**Centre de Recherche ELF-France, P.O. Box 22, 69360 Solaize, France

ABSTRACT

Tetraethylammonium permutites are XRD-amorphous silicoaluminates metastable towards the formation of several zeolites. They present a short-range organization around the organic agent, whose decomposition leaves a micropore system open. Their properties allow to consider the solids of this class as non-crystalline equivalents of the high-silica zeolites.

INTRODUCTION

The amorphous solid present in the synthesis system influences the crystallization of zeolites in several ways: acting as reagent reservoir, buffer for concentrations in solution, hydrodynamic baffle at a micro-scopic level, potential site for heterogeneous nucleation. In which way the presence of organics affects the properties of the synthesis gel? The influence of tetraalkylammonium cations on the colloidal properties of silicates has been early assessed [1-3]. Their ability of stabilizing amorphous solid phases before the formation of any crystal would modify the partition coefficients between solid and solution, which often control the selectivity of zeolite formation [4].

Some recent papers have discussed the incorporation of tetraalkyl-ammonium cations in amorphous zeolite precursors. Most of them were devoted to the characterization of the tetrapropylammonium (TPA) silicate gels, intermediates in the ZSM-5 synthesis. Henderson and White [5], using small-angle neutron scattering, found some evidences for the presence of short-range structures around tetrapropylammonium (TPA). Ito et al. [6] studied the microporosity of calcined gels by ^{129}Xe-NMR spectroscopy. They observed cavities from 15 to 25 Å, which they supposed to have been opened by the decomposition of occluded TPA. Chang and Bell [7] observed by ^{29}Si NMR a rapid cross-linking of the gel at the beginning of the hydrothermal treatment. They attributed the gel reorganization to the template effect of the organic cation.

From the Proceedings of the 9th International Zeolite Conference, Montreal 1992, Eds. R. von Ballmoos et al.,

Literature results on the incorporation of TEA in the amorphous pre-cursors of zeolite beta or ZSM-20 are more split. The source of silica seems to be an important parameter. It has been claimed that, when tetra-ethylorthosilicate is used, almost no TEA is associated with the amorphous solid. This has been the conclusion of Perez-Pariente et al. [8], on the basis of thermal analysis and charge balance data, and of Ito et al. [6], who did not detect any microporous system in the calcined amorphous gels. On the contrary, in the case of gels prepared from amorphous silica, Camblor and Perez-Pariente [9] reported the presence of TEA^+ associated with aluminate sites in the amorphous solid.

In this communication we describe the properties of some amorphous precursors of zeolites formed in the presence of TEA. The final crystal-lization products were zeolite beta, mordenite, ZSM-12 or ZSM-5, depending on the experimental conditions.

EXPERIMENTAL.

Hydrothermal syntheses have been carried out at 150°C in a 500 ml stirred autoclave, equipped for the withdrawal of representative samples throughout the preparation without altering the experimental conditions. Temperature was risen from 25 to 150°C in 2 hrs.

The compositions of the synthesis batches are given in Table 1. The source of silica was a silica sol (Cecasol 30C, SiO_2 25%, Na 0.14%, Al 100 ppm, pH 8.8, grain size 12-18 nm) for the experiments A, B, and C, and tetraethylorthosilicate (Aldrich, Al 1 ppm, Na 1.3 ppm, K 1.3 ppm) in the case of experiment D. Tetraethylammonium (TEA) hydroxide (Aldrich, 40% solution) and bromide (Merck) were used in the ratio needed to obtain an OH^-/SiO_2 ratio of 0.39. The other reagents were sodium aluminate (Carlo Erba RLE), sodium hydroxide (Prolabo RP Normapur), and deionized water. For experiments A, C, and D the order of mixing of the reagents was: alka-line solution, organic agent, aluminate (when appropriate), source of si-lica. In the case of experiment B, the silica sol was stirred for 12 hrs in the alkaline solution before the addition of aluminate and TEA. Before the hydrothermal treatment, the synthesis mixture was stirred for 14 hrs in the case of experiments A, B, and C, and for 15 minutes in the case of experiment D.

Samples of 10 ml were periodically withdrawn. The solid fraction was recovered by filtration, washed with deionized water up to pH 9 and dried at 70°C in air. Solids were characterized by X-ray powder diffraction (CGR Theta 60 diffractometer, Cu K radiation), electron microscopy (Cambridge Stereoscan 100), IR spectroscopy (Nicolet 320 FT-IR, KBr pellets), atomic absorption spectroscopy, adsorption of N_2, cyclohexane and trimethylben-zene. ^{27}Al MAS-NMR data were collected on a Brüker AM 300 spectrometer at 78 MHz, spin frequency 4 MHz, pulse time $\pi/6$, 1 sec recycle, $Al(OH)_6^{3+}$ reference. Thermal gravimetry was carried out under air flow on a Setaram TG 85 balance, using 20 mg samples and a heating rate of 5 K/min. Cation exchange capacity was measured on the solids calcined at 550°C under air flow and ammonium-exchanged. The amount of ammonium cations displaced by potassium was measured by colorimetry.

RESULTS AND DISCUSSION.

Compositions of the solid phases are reported in Table 2. For each experiment sample 1 refers to the solid fraction of the synthesis mixture just before hydrothermal treatment, sample 2 to the solid withdrawn at 150°C, and sample 3 to the final product. For experiments A and B, the evolution of the gel took place during the temperature rise, allowing samples A2 and B2 to be withdrawn immediately after reaching the final temperature. For experiments C and D, the synthesis mixture at room tempe-rature was a clear solution, hence no solid sample 1 is available.

The data of Table 2 indicate that the hydrothermal treatment deeply modifies the composition of the solid phase. The preferred cation is sodium at room temperature (samples 1) and TEA after heating at 150°C (samples 2). The TEA content in samples 2 varies in a quite narrow range. It corresponds to one organic cation per 16-20 tetrahedron-forming atoms, whatever the method of preparation or the Si/Al ratio.

Infrared spectroscopy of samples 2 does not show any zeolitic bands, nor X-ray diffraction allows to detect any crystal line. Characteristic diffraction profiles are depicted in Figure 1. The silica sol used as a reagent and both samples 1 feature one wide band at 3.9 Å (Figure 1a), corresponding to local cristobalite-like structures. This peak is still present in samples 2 (Figure 1b), but it features a wide shoulder at 3.05 Å, at a position expected for quartz-like structures. Another shallow band is present at 12 Å in all samples 2, indicating another type of local ordering.

Cation exchange capacities are 86 and 85 mmol NH_4^+/100 g for samples A2 and B2, respectively, in fair agreement with their aluminium contents, respectively 84 and 82 mmol/100 g. As amorphous ion exchangers, samples 2 correspond then to the definition of permutites [10], and will be denoted as TEA-permutites henceforth.

All specimens of TEA-permutite present a similar habit: spheres 20-25 nm in diameter. The grains are loosely held together, and may be easily separated by ultrasonic treatment. The room temperature amorphous solids (samples A1 and B1) feature a different morphology: irregular particles some tens of μm in size, in which the grains of the original silica sol are no more detectable.

The ^{27}Al MAS-NMR data for experiments A and B, summarized in table 3, indicate that tetrahedral aluminium is present in all samples. No band characteristic of a different coordination can be observed.

The area of the 52-55 ppm peak allows to evaluate the amount of aluminium in tetrahedral coordination. In the case of sample A1, only a small fraction of the aluminium present can be detected by NMR. In this case most aluminium is adsorbed at the surface of the silica sol, in a configuration inconsistent with the spectrometer setting optimized for quantitative analysis of tetrahedral aluminium. On the contrary, most aluminium can be detected in the case of sample B1. This solid has been precipitated at room temperature by adding aluminate to a silicate solution obtained by dissolution of the silica sol. TEA-permutites (samples 2) feature all aluminium in tetrahedral coordination, in the same way as final zeolites (samples 3).

The width of the NMR peaks is related to the dispersion of the alu-minium configurations, a narrower band indicating more uniform geometry. Data of Table 3 indicate that the band width decreases from the room temperature amorphous solid (samples 1) to the TEA-permutite (samples 2), and further to the zeolite (samples 3).

The decomposition of the occluded organic cation liberates a pore system. Surface area and micropore volume of samples calcined at 550°C under air flow, as measured by N_2 adsorption, are reported in Table 4. Cyclohexane adsorbed on the sample B2 occupies the same micropore volume as N_2. No trimethylbenzene is able to enter the microporosity of the same sample.

The DTG curve obtained by heating the TEA-permutite A2 under air flow is reported in Figure 2, compared to the DTG curve of the zeolite beta A3. The decomposition patterns of TEA for the two samples are very much alike, suggesting a similar environment of the organic cation in the two solids. DTG curves of the TEA-permutites withdrawn from the other experiments are not exactly alike, but they all largely differ from the DTG curves of the corresponding final zeolites.

CONCLUSIONS

The amorphous solids formed in hydrothermal conditions in the pre-sence of TEA differ from the synthesis gels prepared at room temperature. The differences are related to the bulk properties of the solid, and they could hardly be accounted for by local rearrangements of the initial gel. The simultaneous incorporation of tetrahedral aluminium and organic cation is in good agreement with a mechanism implying the dissolution of the initial solid in hydrothermal conditions and the formation of a permutite around the organic cation acting as a precipitating agent.

TEA-permutites are organic silicates featuring several properties of the high-silica zeolites: in both cases clusters of silica or aluminate tetrahedra are organized around a bulky molecule, whose decomposition forms a micropore system. The essential difference between the two classes of solids is represented by the dispersion of the cluster geometry: zeolites feature well-defined and repeatable site geometries, whereas permutites present irregular, hence aperiodical, local organization.

Templating effect of an organic cation does not imply the formation of a periodic pattern. Exactly as when no organic agent is used, the con-densation of silicoaluminate species at high supersaturation may bring to the precipitation of an amorphous solid. This precipitate is metastable towards zeolite formation, and a promising field for future research is the way in which permutites affect the crystallization process. They can either enhance zeolite formation, by providing heterogeneous nucleation sites, or hinder the crystallization, by competing with zeolites for common growth units.

Another point deserving further research is the possible parallelism between the easy formation of aperiodic networks in the presence of TEA and the formation of defects during the crystallization of zeolite beta. It is tempting to consider TEA-permutite as the extreme example of the structural disorder typical of zeolite beta.

REFERENCES

1. Weldes, H.H., and Lange, K.R., Ind. Eng. Chem. **61** (4), 29 (1969).
2. Iler, R.K., *The Chemistry of Silica*, Wiley, New York, pp. 150, 237, 376 (1979).
3. Marsman, H.C., Z. Naturforsch. **29B**, 495 (1974).
4. Barrer, R.M., in: *Molecular Sieves*, Society of Chemical Industry, London, p. 69 (1968).
5. Henderson, S.J., and White, J.W., J. Appl. Cryst. **21**, 744 (1988).
6. Ito, T., Fraissard, J., B.Nagy, J., Dewaele, N., Gabelica, Z., Nastro, A., and Derouane, E.G., Stud. Surface Sci. Catal. **49**, Elsevier, Amsterdam, p. 579 (1989).
7. Chang C.D., and Bell, A.T., Catal. Lett. **8**, 305 (1991).
8. Perez-Pariente, J., Martens, J.A., and Jacobs, P.A., Appl. Catal. **31**, 35 (1987).
9. Camblor, M.A., and Perez-Pariente, J., Zeolites **11**, 202 (1991).
10. Iler, R.K., *The Chemistry of Silica*, Wiley, New York, p. 11 (1979).

Table 1. Composition of the crystallization batches.

exp.	Na/(Si+Al)	TEA/(Si+Al)	Al/(Si+Al)	H_2O/SiO_2
A	0.04	0.34	0.040	21
B	0.42	0.34	0.040	17
C	0.06	0.32	0.003	19
D	0.45	0.37	0.042	25

Table 2. Composition of the solid phases.

sample	time at 150°C	Si (w%)	Na/(Si+Al)	TEA/(Si+Al)	Al/(Si+Al)	phase XRD
A1	-	38.7	0.033	0.006	0.035	amorphous
A2	0	33.8	0.015	0.060	0.058	amorphous
A3	3 d	34.2	0.001	0.097	0.058	BEA
B1	-	38.1	0.049	0.005	0.050	amorphous
B2	0	33.6	0.016	0.056	0.055	amorphous
B3	24 h	34.9	0.068	0.049	0.062	MOR
C2	24 h	29.5	0.005	0.050	0.008	amorphous
C3	9 d	40.0	0.010	0.049	0.006	MTW
D2	40 h	34.8	0.028	0.056	0.051	amorphous
D3	10 d	36.6	0.015	0.069	0.056	MFI

Table 3. ^{27}Al MAS-NMR data.

sample	band position ppm	band width at half height, ppm	tetrahedral Al / total Al
A1	55	16	0.08
A2	52	14	0.99
A3	53	11	1.00
B1	54	16	0.84
B2	52	13	0.95
B3	54	7	0.94

Table 4. Surface area and microporosity by N_2 adsorption.

sample	surface area m^2/g	micropore volume ml/g
A1	100	-
A2	430	0.16
A3	700	0.26
B1	20	0.01
B2	440	0.13
C2	410	0.17

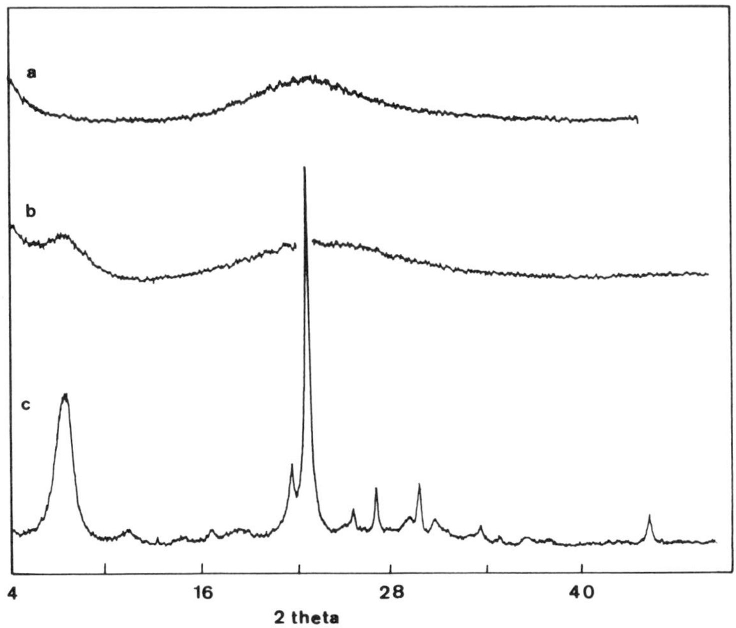

Figure 1: XRD patterns of a) sample A1, gel formed at room temperature, b) sample A2, gel formed at 150°C and c) sample A3, zeolite beta.

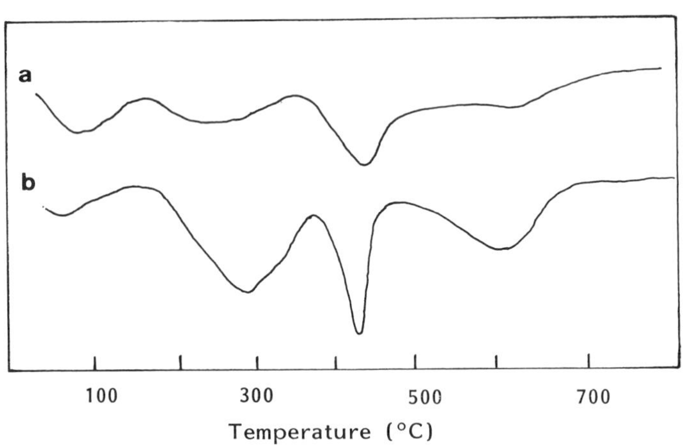

Figure 2: DTG profiles for the decomposition of tetraethylammonium from a) sample A2 and b) zeolite beta, sample A3.

MODELING AND SIMULATION OF ZEOLITE CRYSTALLIZATION

B. SUBOTIĆ and J. BRONIĆ

Ruđer Bošković Institute, P.O.Box 1016, Zagreb, Republic of Croatia

ABSTRACT

The changes of characteristic parameters during the crystallization of zeolite A from differently aged gels were simulated by population balance method, using appropriate mechanisms of gel dissolution, formation of primary zeolite particles (nucleation) and their crystal growth. The correlation between simulated values and experimental data showed that a combination of heterogeneous and autocatalytic nucleation is responsible for the formation of primary zeolite particles. The mechanisms of gel dissolution and crystal growth of zeolite microcrystals were also considered.

INTRODUCTION

Among different models of zeolite crystallization, only population balance models are based on fundamental theories of particulate processes which occur during the crystallization [1,2]. Hence, the population balance methodology enables the simulation of crystallization processes using different mechanisms of gel dissolution and nucleation and the crystal growth of zeolite particles. The objective of this work was the analysis of critical particulate processes which occur during the crystallization of zeolite A from differently aged gels. The analysis was performed by the simulation of crystallization processes using the population balance method and by the correlation of simulated and experimental results taking into consideration earlier defined and/or assumed mechanisms of zeolite nucleation [3,4] and crystal growth [5,6], as well as the mechanism of gel dissolution [7].

EXPERIMENTAL

Zeolite A was crystallized at 80° C from differently aged (at 25° C) gels of 2.04 $Na_2O \cdot Al_2O_3 \cdot 1.9\ SiO_2 \cdot 212\ H_2O$ batch composition. X-ray diffraction, particle size distribution and chemical analysis of the solid and the liquid phases were applied to measure the changes in fraction of zeolite crystallized size of largest crystals, average crystal size, number of crystals and chemical composition of the solid and the liquid phase during the crystallization. Details about gel preparation, its ageing, crystallization procedure sampling and sample characterizations were described earlier [8].

From the Proceedings of the 9th International Zeolite Conference, Montreal 1992, Eds. R. von Ballmoos et al.,

THEORETICAL APPROACH

The population balance for a well mixed, isothermal, constant volume batch crystallizer can be defined by a set of ordinary differential equations [1,2], i.e.,

$$dm_o/dt = B \tag{1}$$

$$dm_1/dt = Q\ m_o \tag{2}$$

$$dm_2/dt = 2\ Q\ m_1 \tag{3}$$

$$dm_3/dt = 3\ Q\ m_2 \tag{4}$$

where m_i ($i = 0, 1, 2$ and 3) are the moments of particle size distribution at crystallization time t, defined as:

$$m_i = \int_0^{\infty} L^i \left[dN(L,t)/dL\right] dL \tag{5}$$

$$B = dN/dt \tag{6}$$

is the nucleation rate and,

$$Q = dL/dt \tag{7}$$

is the crystal growth rate. In accordance with Eq. (5), the mass m_z, of zeolite formed up to crystallization time $t = t_c$ is proportional to the 3rd moment of particle size distribution established at time t_c, and can be expressed as:

$$m_z = G\ \rho \int_o^{\infty} L^3 \left[dN(L,t)/dL\right] dL = G\ \rho\ m_3 \tag{8}$$

where G and ρ are geometrical shape factor and density of growing zeolite particles. Now, taking into consideration that the starting growth of heteronuclei present in the liquid phase of the crystallizing system causes the dissolution of gel and releasing the particles of quasi-crystalline phase distributed through the gel matrix [3,4,9], the rate B, of the formation of new primary particles can be expressed as [4,6,9]:

$$B = \tilde{N}_a\ dm_z/dt_c = G\ \rho\ \tilde{N}_a\ dm_3/dt_c = 3\ G\ \rho\ \tilde{N}_a\ Q\ m_2 \tag{9}$$

where $\tilde{N}_a$ is the number of particles of quasi-crystalline phase released from the mass of gel needed for the crystallization of unit mass of zeolite. Based on findings that the rates of crystal growth and dissolution of zeolites depend on concentrations of both silicon and aluminum in the liquid phase [5,6,10,11], parallelism between the dissolution and crystal growth [12] and Davies and Jones model of crystal growth and dissolution [13], the kinetics of crystal growth of zeolite A microcrystals can be defined as:

$$dL/dt_c = k_g \left[C_{Al} - C^*_{Al}(z)\right]\left[C_{Si} - C^*_{Si}(z)\right] \qquad (10)$$

where k_g is the growth rate constant, C_{Al} and C_{Si} are concentrations of aluminum and silicon in the liquid phase during the crystallization, $C^*_{Al}(z)$ and $C^*_{Si}(z)$ are concentrations of aluminum and silicon in the liquid phase which correspond to the solubility of zeolite at given experimental conditions. Finally, the rate of dissolution of aluminosilicate gels can be expressed as [7]:

$$dm^*_G/dt_c = K_s\,(m^o_G - m^*_G)^{2/3}\left[C^*_{Al}(G) - C_{Al}\right] \qquad (11)$$

where dm^*_G is a differential mass of gel dissolved in a differential time t_c, K_s is a lumped constant, m^o_G is the initial mass of gel, m^*_G is the mass of gel dissolved up to time t_c and $C^*_{Al}(G)$ is the concentration of aluminumin the liquid phase which correspond to the solubility of gel. Since the kinetics of both the gel dissolution and crystal growth of zeolite particles depend on the concentrations of silicon and aluminum in the liquid phase, the differential equations:

$$dC_{Al}/dt_c = a\, dm^*_G/M_G\, dt_c - 2\, dm_z/M_z\, dt_c \qquad (12)$$

$$dC_{Si}/dt_c = b\, dm^*_G/M_G\, dt_c - 2\, dm_z/M_z\, dt_c \qquad (13)$$

also to be included to the calculation procedure. Here a = 2 and b = 2.106 are moles of aluminum and silicon in a mol of gel and M_G and M_z are molecular weights of gel and zeolite A. Equations (1)-(7), (10) and (12)-(14) completely specify the system. The system behavior was simulated by simultaneous solution of those equations using a fourth-order Runge-Kutta method and corresponding numerical values of appropriate constants and initial conditions.

RESULTS AND DISCUSSION

Symbols in Figure 1A show the kinetics of crystallization of zeolite A from the gels aged for 0, 3, 9 and 17 days. The rate of crystallization increases with incensing time of gel ageing as discussed and explained earlier [8]. Chemical analysis of the experimental systems at the beginning, during and at the end of the crystallization process showed that 78.8 g/dm^3 (= m^o_G) of amorphous aluminosilicate having the composition: 1.17 $Na_2O \cdot Al_2O_3 \cdot$ 2.106

$SiO_2 \cdot 3.9\ H_2O$ (M_G = 317.54 g/mol) has been transformed to 85.45 g/dm^3 (= $m_A(eq)$) of zeolite A of the composition: $Na_2O \cdot Al_2O_3 \cdot 2\ SiO_2 \cdot 4.5\ H_2O$ (M_z = 365.17 g/mol). Starting concentrations of aluminum ($C^*_{Al}(G)$ = 0.1 mol dm^{-3}) and silicon ($C^*_{Si}(G)$ = 0.0509 mol dm^{-3}) in the liquid phase remained almost constant during the main part of the crystallization process, thereafter gradually decreased and attained equilibrium values $C^*_{Al}(z)$ = 0.056 mol dm^{-3} and $C^*_{Si}(z)$ = 0.0299 mol dm^{-3} (see Figure 1B). In all examined cases, the size of largest particles increased linearly with the rate $dL/dt_c = K_g$ = 2.74 μm/h, at constant supersaturation (see Figure 1C). The number $\bar{N}_o$, of microcrystals formed by the crystal growth of nuclei present in the liquid phase at t_c = 0 (nuclei-I = heteronuclei + part of particles of quasi-crystalline phase released from the gel during its recrystallization at 25° C [8])and the number $\bar{N}_a$ of microcrystals formed by the growth of particles of quasi crystalline phase released from the gel during the crystallization at 80° C, both contained in 1 g of the final product of crystallization, were determined by the analysis of the kinetic data using the equation [6,14]:

$$f'_A = (\bar{N}_o/3 \cdot \bar{N}_a)\left[\exp(a1 \cdot t_c) + 2\exp(-a1\ t_c/2)\cos(\sqrt{3}a1 \cdot t_c/2) - 3\right] \quad (14)$$

where f'_A is fraction of zeolite A at time t_c and $a1 = (G \cdot \rho \cdot \bar{N}_a K_g^3)^{1/3}$.The numerical values of $\bar{N}_o$, $\bar{N}_a$ and $N_o = m_z(eq)\ \bar{N}_o$ (number of nuclei-I distributed in the liquid phase of a dm^3 of the system) are presented in Table 1 as functions of the time t_a, of gel ageing. The numerical value of the growth rate constant, k_g = 0.2965 cm h^{-1} mol^2 dm^{-6} was calculated by Eq. (10) using the numerical values of the constants K_g, $C^*_{Al}(z)$, $C^*_{Si}(z)$, $C^*_{Al}(G)$ and $C^*_{Si}(G)$. The values of fraction of zeolite A, $f_A = m_z/(m_z + m_g)$ (m_z and m_g are the mass of zeolite A and gel in the solid phase) and $f'_A = m_z/m_z(eq)$, concentrations C_{Al} and C_{Si} of aluminum and silicon in the liquid phase, size L_M of the largest crystals, average crystal size $\bar{L} = m_1/m_o$ and the nucleation rate B were simulated by simultaneous solution of the system of the moment equations (1)-(4) in which B and Q are defined by equations (9) and (10) and the auxiliary equations (11)-(13), using the numerical values of the constants G = 1 (cubes), ρ = 2 g/cm^3 (density of zeolite A), k_g, K_s (see Table 1), $C^*_{Al}(G)$, $C^*_{Si}(G)$, $C^*_{Al}(z)$, $C^*_{Si}(z)$, m^o_g, $m_z(eq)$, M_G and M_z. The initial conditions required for the numerical solution of the mentioned system of differential equations by the Runge-Kutta method, were defined as: $m_i(0) = N_o L^i_o$, where i = 0, 1, 2, 3, N_o is the number of nuclei-I at t_c = 0, L(0) = 10^{-8} cm is the size of nuclei [15], $m^*_G(0)$ = 0 , $C_{Al}(0)$ = 0.1 mol dm^{-3} and $C_{Si}(0)$ = 0.0509 mol

dm^{-3}. The results of simulation, presented in Figure 1. by curves are in good (Figure B) or even excellent (Figures A and C) agreement with corresponding measured values (symbols). Due to low supersaturation of silicon and aluminum in the liquid phase, the differences between the fractions f_A and f'_A are neglected (see Figure 1A) showing that Equation (14) can be used for the analysis of the kinetic data and providing the numerical values of $\bar{N}_o$ and $\bar{N}_a$. The increase in the rate of crystallization with increasing time of the gel ageing is caused by the increase in the number of both nuclei-I ($\bar{N}_o$) and nuclei-II ($\bar{N}_a$) (see Table 1) at constant rate of the crystal growth (see Figure 1C). The increase of $\bar{N}_a$ during gel ageing was explained by simultaneous formation of both distorted (Si,Al,O)- network and simplest structural blocks (for example single and double 4 and 6 membered rings of (Si-O)- and (Al-O)-tetrahedra) by polymerization-depolymerization processes during the gel formation and ageing [3]. The increase of number of nuclei-I can be explained by releasing of particles of quasi-crystalline phase from the gel particles during its recrystallization (dissolution of small particles and the growth of large ones) via Ostvald's ripening at ambient temperature [8]. Starting linear increase of average particle size (see Figure 2A) corresponds to to the growth of nuclei-I with the liner growth rate $Q = dL_M/dt_c = 2.74\ \mu m/h$. The change in L_{av} at extended crystallization time is governed by simultaneous growth with constant rate of increasing number of nuclei. Although the change in L_{av} cannot be measured during the crystallization process, the very well agreement between simulated (curves) and measured (symbols) average crystal size in crystalline end products (see Figure 2A) indicates that simulated results give an realistic feature of the change in L_{av} during the crystallization of zeolites from gels. Finally, the changes in nucleation rate (see Figure 2B) have the shape similar to those earlier predicted by Zhdanov [3] and experimentally confirmed by several authors [5,16,17], but with the maximum shifted toward the crystallization time at which the crystallization rate is greatest, as predicted by the proposed model [6].

CONCLUSION

By using assumed kinetics of gel dissolution and zeolite nucleation and the crystal growth, the population balance method enables simulation of specific events which occur during the crystallization of zeolite, and hence the evaluation of assumed model(s) of crystallization by comparing simulated and measured data. The very well agreement between measured values of f_A , C_{Al}

, C_{Si} , L_M and L_{av} and the same values simulated by population balance method clearly indicates that crystallization of zeolites from aluminosilicate gels is governed by particulate processes (gel dissolution, nucleation and the crystal growth of zeolite) defined by Equations (9) - (11) and that the influence of gel ageing on the kinetics of crystallization of zeolite A is in close relation with low-temperature recrystallization of gel, as explained earlier [5,8].

REFERENCES

1. Randolph, A.D. and Larson, M.A., Theory of Particulate Processes, Academic Press, London (1971).
2. Thompson, R.W. and Dyer, A., Zeolites **5**, 292 (1985).
3. Zhdanov, S. P., Adv. Chem. Ser. **101**, 20 (1971).
4. Subotić, B., ACS Symp. Ser. **398**, 110 (1989).
5. Zhdanov, S.P. and Samulevich, N.N., in: ***Proceedings of the 5th International Zeolite Conference***, Ed. L.V.C. Rees, Heyden, London-Philadelphia-Rheine, pp. 75-84 (1980).
6. Katović, A., Subotić, B., Šmit, I. and Despotović, LJ.A., Zeolites **10**, 634 (1990).
7. Čižmek, A., Antonić, T. and Subotić, B., to be published.
8. Bronić, J., Subotić, B., Šmit, I. and Despotović, LJ.A., Stud. Surf Catal. **37**, 107 (1988).
9. Subotić, B. and Graovac, A., Stud. Surf. Sci. Catal. **24**, 199 (1985).
10 Čižmek, A., Komunjer, Lj., Subotić, B., Široki, M. and Rončević, S., Zeolites **11**, 258 (1991).
11. Čižmek, A., Komunjer, Lj., Subotić, B., Široki, M. and Rončević, S., Zeolites **11**, 810 (1991).
12. Buckley, H.E., Crystal Growth, Wiley, New York, p.304 (1950).
13. Davies, C. W. and Jones, A.L., Trans. Faraday Soc. **51**, 812 (1955).
14. Warzywoda, J. and Thompson, R.W., Zeolites **9**, 341 (1989).
15. Cook, J.D. and Thompson, R.W., Zeolites **8**, 322 (1988)
16. Katović, A., Subotić, B., Šmit, I. and Despotović, LJ.A., Zeolites, **9** 45 (1989)
17. Plee, D.and Biver, C., in: ***Zeolites for Nineties, Recent Research Reports of 8th International Zeolite Conference***, Eds. J.C.Jensen, L. Moscou and M.F.M. Post, Amsterdam, pp. 17-18 (1989).

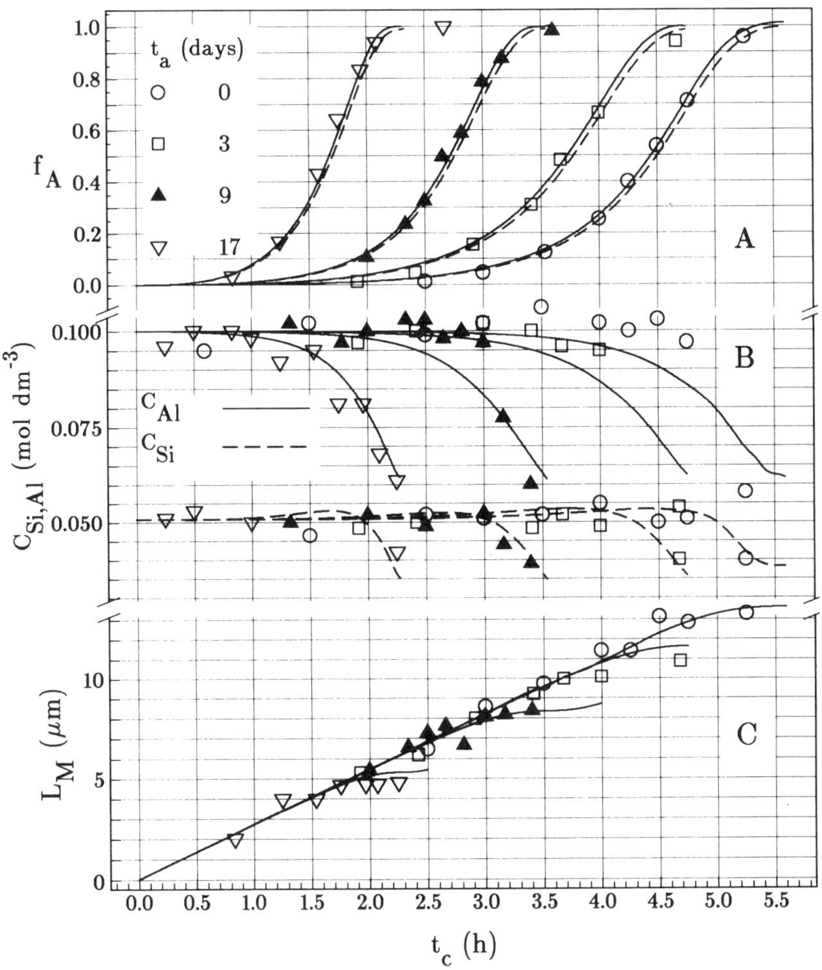

Figure 1. (A) Correlation between simulated (curves) and measured (symbols) kinetics of crystallization of zeolite A from differently aged gels. Solid curves represent simulated fractions f_A and dashed curves represent simulated fractions f'_A. (B) Correlation between simulated (curves) and measured (symbols) changes of concentrations C_{Al} of aluminum (solid curves) and C_{Si} of silicon (dashed curves) in the liquid phase during the crystallization of zeolite A from differently aged gels. (C) Correlation between simulated (solid curves) and measured (symbols) in the size of largest crystals during the crystallization of zeolite from the gels aged for time t_a. t_c is time of crystallization.

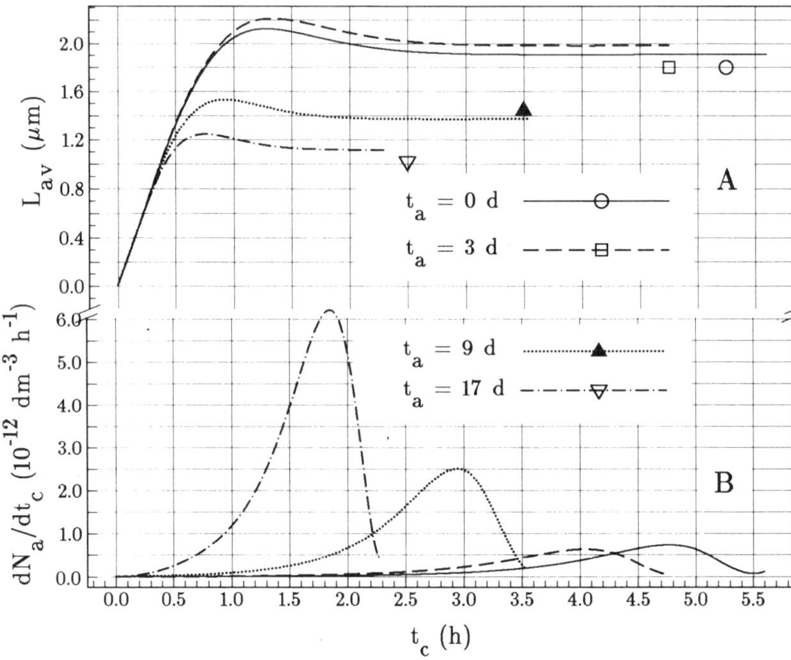

Figure 2. (A) Simulated changes (curves) in average crystal size of zeolite A during its crystallization from differently aged gels. Symbols represent the values of L_{av} measured in crystalline end products. (B) Simulated changes of the nucleation rates (releasing of particles of quasi-crystalline phase) during the crystallization of zeolite A from differently aged gels.

Table 1. Numerical values of $\bar{N}_o$, $\bar{N}_a$ and K_s which correspond to the crystallization of zeolite A from the gels aged for various times t_a .

t_a (d)	$\bar{N}_o$ (particles/g)	$\bar{N}_a$ (particles/g)	K_s ($g^{1/3}$ mol^{-1} dm^3 h^{-1})
0	2.857×10^7	1.200×10^{10}	600
3	9.017×10^7	1.164×10^{10}	500
9	2.210×10^8	3.208×10^{10}	600
17	1.559×10^9	5.294×10^{10}	850

PROBING THE SYNTHESIS OF CoAPO-n MOLECULAR SIEVES BY ANALYSIS OF WET PRECURSOR GELS.

M.G. UYTTERHOEVEN and R.A. SCHOONHEYDT

Katholieke Universiteit Leuven, Centrum voor Oppervlaktechemie en Katalyse (COK), Material Research Center, Kardinaal Mercierlaan 92, B-3001 Leuven, Belgium

ABSTRACT

CoAPO-n with chabazite (CHA) topology crystallizes competitively with CoAPO-5. Insight in the co-crystallization is obtained by Diffuse Reflectance Spectroscopy (DRS). Wet gels with variable template content are analyzed as such and after centrifugation. Increasing template concentrations lead to typical changes of the gel properties: (1) increasing interaction between the cobalt ion and the gel; (2) transformation from an octahedral to a tetrahedral cobalt complex; (3) distortion of the environment of the cobalt atom. Drying of the gel induces important structural modifications which are not necessarily representative for a real crystallization process.

INTRODUCTION

Gel preparation and ageing is an important step in molecular sieve crystallization. Minor changes in conditions can lead to structures with different topologies. Co^{2+} has been used successfully as a probe ion to follow the crystallization of CoAPO's [1]. The question arises whether it is also a probe for the study of the wet precursor gels. This is investigated in this paper by the study of the competitive crystallization of CoAPO-5 and chabazite-like materials.

EXPERIMENTAL

Synthesis gels, having the general composition r(triethylamine)/$(Co_{0.08}Al_{0.92}P_{1.00})O_4$/w H_2O, were prepared and autoclaved for 24 h. Pseudoboehmite and phosphoric acid were used as Al and P source respectively. Details on the procedures of mixing and crystallization can be found in a previous publication [1]. Values for r and w are given in table 1, together with synthesis temperatures and gel pH values.

Name	r	w	Temp.	pH, initial	pH, 4 washings	Topology
G1	0.60	20.7	423 K	2.45	3.60	AFI
G2	0.86	15.4	463 K	4.13	4.54	AFI+CHA
G3	1.00	15.4	423 K	5.40	5.79	CHA
G4	1.00	2.1	423 K	7.07	-	-

Table 1: Gel composition, synthesis conditions and topology of end products.

From the Proceedings of the 9th International Zeolite Conference, Montreal 1992, Eds. R. von Ballmoos et al.,

Crystal phases were identified with X-Ray Diffraction (XRD). Co^{2+} was studied with Diffuse Reflectance Spectroscopy (DRS) on the wet gels, the dried gels and on the crystalline end products.

The gels were analyzed by DRS as such and after washing with water followed by centrifugation. The washing was repeated four times. The pH was measured on the gels. Aliquots of the washed gels were dried at room temperature and at 333 K for 20 h. DRS spectra of these dried samples were also recorded.

RESULTS

XRD spectra of well-crystallized samples derived from G1 and G3 are given in figure 1. Table 1 reports the topology of the crystalline end products. It is seen that a lower template concentration (G1) produces CoAPO-5 (AFI-topology) [2a], while the higher template concentration (G3) produces samples with chabazite topology (CHA) [2b]. For intermediate template concentrations (G2) a mixture of AFI and CHA was obtained. The gel G4, having an extremely low water content did not produce a crystalline phase. A DRS spectrum of well-crystallized G1 is given in figure 4, indicated as CoAPO-5. The two triplets, one in the visible (VIS) region with maxima at 542, 580 and 624 nm, the other in the near infra-red (NIR) region with maxima at 1329, 1492 and 1719 nm were previously assigned to tetrahedral cobalt isomorphically substituted into the framework [1].

The other spectra of figures 2 to 4 relate to DRS spectra of gels. The symbols used include the gel number (G1-G4) as in table 1, the number of centrifugations (C0-C4) and the indication wet or dry. Figure 2 relates to the wet gels as-prepared. The spectrum of G3 was recorded after one centrifugation because the viscosity of G3 as-such was too low. The spectrum was therefore recorded after one centrifugation. At low template content (G1) an intense band is present with maximum at 519 nm and a shoulder around 470 nm. These bands are typical for octahedral Co^{2+} [3]. In the UV-region, sharp absorption bands are located at 304 nm and 214 nm, the latter being very intense. The same features are present in the spectrum of the sample with intermediate template content (G2). In addition, a new band is observed at 630 nm. At high template content (G3), the spectrum shows a broad asymmetric triplet with positions at 527 nm, 574 nm and 634 nm. The VIS-region of G4 has a symmetrical triplet with maxima located at 540, 583 and 628 nm. These are very similar to those of the crystalline end product and are typical for tetrahedral Co^{2+}. In the UV-region, three bands appear at 215, 256 and 304 nm.

The effect of washing and centrifugation on G1 and G2 is presented in figure 3. For gel G1 the characteristic Co^{2+} bands can be progressively washed out, until complete disappearance. In the course of this process, the 470 nm shoulder gains some intensity with respect to the 519 nm band. As the template content of the gels increases (G2, G3) Co^{2+} is more retained, the octahedral as well as the tetrahedral form. Only the 304 nm band disappears upon washing.

Drying at room-temperature for 20 h does not remove all the water and does not cause changes in the DRS spectra. Drying at 333 K induces remarkable effects which are dependent on the number of washings. This is shown in figure 4 for G1. The DRS spectra reflect nicely the colour change from blue over pink to almost white. After one washing the spectrum of the dried gel is dominated by the VIS triplet of tetrahedral Co^{2+}, very similar to that of Co^{2+} in CoAPO-5. When the number of washings increases, the amount of Co^{2+} decreases and, after drying, the 470 nm band (of octahedral Co^{2+}) becomes the most intense one. This is accompanied by a progressive increase of the 1452 nm and 1938 nm bands of adsorbed water.

DISCUSSION

The synthesis of molecular sieves generally involves a gel phase which is digested under the hydrothermal synthesis conditions. Several attempts have been made to elucidate the chemistry of the gel and its hydrothermal digestion. A variety of techniques have been used such as liquid phase NMR [4], IR and Raman spectroscopy [5] and probing of transition metal ions [1, 6, 7, 8].

In our work we have chosen Co^{2+} as a probe ion and DRS as characterization technique. The advantages are: (1) Co^{2+} is easily incorporated into the framework during synthesis; (2) the DRS spectra have an excellent quality; (3) DRS is applicable to wet gels and to dried solids, amorphous and crystalline as well. By investigating the wet gels we try to understand the behavior of an ion which incorporates into the structure and its interaction with the template molecules.

The particles of pseudoboehmite have a pH-dependent surface charge [9,10]:

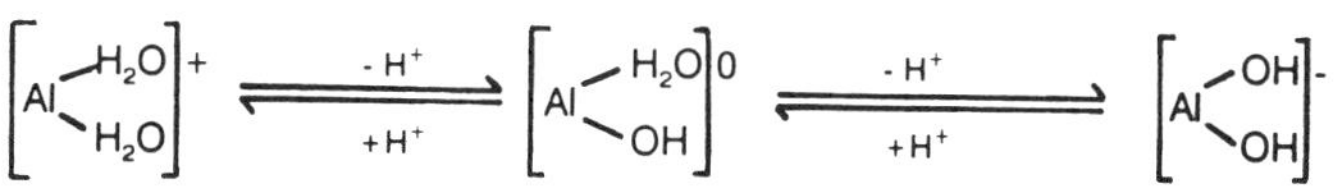

(S1) (S2) (S3)

Contacting this solid with the phosphoric acid solution under the given pH conditions does not result in complete dissolution, but a part of the phosphate species are adsorbed at the surface of the pseudoboehmite particles.

At low pH (gel G1) the positive aquo-site (S1) reacts with the monomeric $H_2PO_4^-$ according to the ligand exchange $(Al\text{-}OH_2)^+ + H_2PO_4^- \rightleftharpoons (Al\text{-}H_2PO_4)^0 + OH_2$ [10,11]. At the pH of 2.4 the triethylamine (Et_3N) is fully protonated. Since cobalt has no tendency to bind to this protonated Et_3N^+ nor to the neutral surface complex, it is detected mainly as the $Co(H_2O)_6^{2+}$ complex in the solution phase, which produces the band at 519 nm.

However, progressive centrifugation of G1 shows that a second cobalt species is present as indicated by the shoulder at 470 nm, which becomes more apparent with washing. It must be assigned to a species fixed to the gel surface. We have previously [1] assigned this band to an octahedral complex $[CoO_4{\cdot}2H_2O]$ in a structural entity with four framework oxygens and two additional water ligands. The bands at 1452 and 1938 nm, assigned to overtones or combination bands of water or hydroxy groups are related to this entity. The data in figure 4 are in full agreement with this interpretation. Going from G1/C1/DRY to G1/C3/DRY, an increase of the relative intensity of the 470 nm band is observed together with an increase of the relative intensity of the bands at 1452 and 1938 nm. We can not give the number of Co^{2+} present as $[CoO_4(H_2O)_2]$. However, since the absorption coefficient of octahedral species is generally 100 to 1000 times lower than that of tetrahedral species, we estimate that after washing and drying $[CoO_4.2H_2O]$ is the predominant species.

At high pH (gel G3, G4) adsorption takes place on the neutral hydroxy site S2 with release of the hydroxyl: $(Al\text{-}OH)^0 + H_2PO_4^- \rightleftharpoons (Al\text{-}H_2PO_4)^0 + OH^-$. At high H_3PO_4 concentrations, reaction also occurs with the internal bonds of the pseudoboehmite [10,11]. This results in the formation of smaller polymeric entities and the creation of additional surface for reaction with the $H_2PO_4^-$. These reactions can be represented as follows:

$$\begin{matrix} Al \\ Al \end{matrix}\!\!>\!OH + H_2PO_4^- \rightleftharpoons [Al\text{-}H_2PO_4]^{1/2-} + [Al\text{-}OH]^{1/2-}$$

$$[Al\text{-}OH]^{1/2-} \underset{+H^+}{\rightleftharpoons} [Al\text{-}H_2O]^{1/2+} \overset{H_2PO_4^-}{\rightleftharpoons} [Al\text{-}H_2PO_4]^{1/2-} + H_2O$$

The negative surface charges generated in this way can react with positively charged cobalt species. A tetrahedral surface complex is formed which can not be removed by washing. Oxygen ligands of the aluminum phosphate surface complex are thought to be part of this complex. The presence of N-ligands provided by the template molecules is unlikely. Indeed, tetrahedral cobalt species ligating simultaneously to O and to N, have adsorption bands at different wavelengths [12]. It is striking that the band positions observed for G3 and G4 are almost identical to those of Co^{2+} in the lattice of a well-crystallized sample.

Sample G2 appears as a situation intermediate between G1 and G3. This is because at the intermediate pH of this gel, the two mechanisms are important. Free $Co(H_2O)_6{}^{2+}$, $CoO_4(H_2O)_2$ as well as tetrahedral Co^{2+} are present.

Let us define a nucleus as the smallest chemical entity in which the ions have the same coordination as in the final crystal. In the case of Co^{2+} this means that on nucleation it is transformed from octahedral to tetrahedral and its spectral features provide information on the nucleation step. The following conclusions can be drawn. (1) Both in G1 and in G3 there is a cobalt species strongly adsorbed at the surface of the gel particles. In G1 this is the octahedral $CoO_4(H_2O)_2$. In G3 this cobalt has the same tetrahedral coordination as in the final crystals. At least the tetrahedral complex is thought to be a precursor in the crystallization of CoAPO-n. This indicates that nucleation can take place at the surface of the gel particles. (2) In the gel stage the organic molecules regulate the pH but are not part of the coordination sphere of cobalt and hence are not an essential part of the nuclei. The pH, along with other factors, determines the nature of the structure building entities and their complexation behavior. In the present system the pH influences the surface properties of the pseudoboehmite particles and their rate of dissolution or disaggregation.

The transformation of G1 into AFI is likely to occur via a solution phase, since most of the cobalt which ultimately ends up in the crystals is present as a soluble aquo-complex. On the contrary, in G3 - leading to CHA - all the cobalt is associated with the solid phase even in the early stages of mixing. A solid-solid transformation cannot be ruled out a priori.

In the literature of competitive co-crystallization of AFI and CHA structures, the following observations were reported: (1) high concentrations of Mg and Si in the gel favor the formation of MAPO-34 [13] and SAPO-44 [14] respectively; (2) low temperature leads to MAPO-34 instead of MAPO-5 [13]; (3) an Fe(III)-salt results in FAPO-5, Fe(II) gives preferably FAPO-34 [15]. (4) longer synthesis times favor the formation of FAPO-34 [15]. In our experiments the template concentration is another structure directing parameter: high Et_3N concentrations favor the formation of a CHA-like material instead of CoAPO-5.

Drying of the gels produces drastic changes in the spectra, due to removal of water. The octahedral $Co(H_2O)_6^{2+}$ complex (519 nm) is transformed into a tetrahedral Co^{2+} complex (VIS triplet), similar to that observed in a well-crystallized sample. However, there is no evidence that the structural changes caused by drying are representative for changes taking place during real crystallization. This clearly demonstrates the importance of a method which permits the analysis of wet gels.

CONCLUSIONS

DRS is a sensitive tool to monitor the coordination of Co^{2+}. It can be applied to wet gel systems and to dry solids. Therefore DRS is well suited to follow the fate of cobalt during synthesis and to investigate aspects of crystallization mechanism.

In this paper we have shown that at low template concentration (and therefore low pH) phosphoric acid forms surface complexes with pseudoboehmite particles, while most of the Co^{2+} remains in solution with a small amount present as a $CoO_4(H_2O)_2$ surface complex. As the template concentration (and thus the pH) increases, pseudoboehmite particles are digested by H_3PO_4 and tetrahedral surface Co^{2+} complexes are formed. The first case leads to CoAPO-5, the latter to chabazite-like structures.

ACKNOWLEDGEMENTS

A research grant from the IWONL is gratefully acknowledged by M.G.U. Many thanks to P.M. Uytterhoeven and to E. Heykants for their technical assistance and to J.B. Uytterhoeven for many discussions.

REFERENCES

1. Uytterhoeven M.G., Van Keyenberg W.A., Schoonheydt R.A., MRS Symp. Proc., Synthesis, Characterization and Novel Applications of Molecular Sieve Materials, **233**, 57 (1991)
2. a) R. von Ballmoos, J.B. Higgins, Zeolites, **10**, 330S-331S (1990). b) Ibid., 368S-371S (1990).
3. Lever, A.B.P., Inorganic Electronic Spectroscopy, Stud. Phys. Theor. Chem., 2nd ed., Elsevier, Amsterdam, **33**, (1984)
4. (a) Mortlock R.F., Bell A.T., Radke C.J., J. Phys. Chem., **95**, 372 (1991). (b) ibid., 7847 (1991)
5. Twu J., Dutta P.K., Kresge C.T., J. Phys. Chem., **95**, 5267 (1991)
6. Okabe K., Matsubayashi N., Shimada H., Hagiwara H., Nishijima A., Chemistry Letters, 955 (1990)
7. Padovan M., Genoni F., Leofanti G., Petrini G., Trezza G., Zecchina A., Stud. Surf. Sci. Cat., Preparation of Catalysts V, (Poncelet G. et al., eds), **63**, 431 (1991)
8. Li H., PhD thesis, K.U.Leuven, 62-82 (1989)
9. Parks G.A., Chem. Rev., **65**, 177 (1965)
10. Rajan S.S., Perrott K.W., Saunders W.M., J. Soil Science, **25**, 439 (1974).
11. Rajan S.S., Nature, **262**, 45 (1976)
12. Banci L., Bencini A., Benelli C., Gatteschi D., Zanchini C., Structure and bonding, **52**, 37 (1982)
13. Martens J.A., Mertens M., Grobet P.J., Jacobs P.A., Stud. Surf. Sci. Catal., Grobet P.J. et al. (ed), **37**, Elsevier, Amsterdam, 97 (1988)
14. Wilson S.T., Flanigen E.M., ACS Symp. Ser., Zeolite Synthesis, Occelli M.L. and Robson H.E., **398**, 329 (1989)
15. Wohlrab S., Rademacher O., Scheler H., Extended Abstracts of the third German Workshop on Zeolite Chemistry, H.G. Karge, J. Volter (eds), Berlin FRG, P2 (1991)

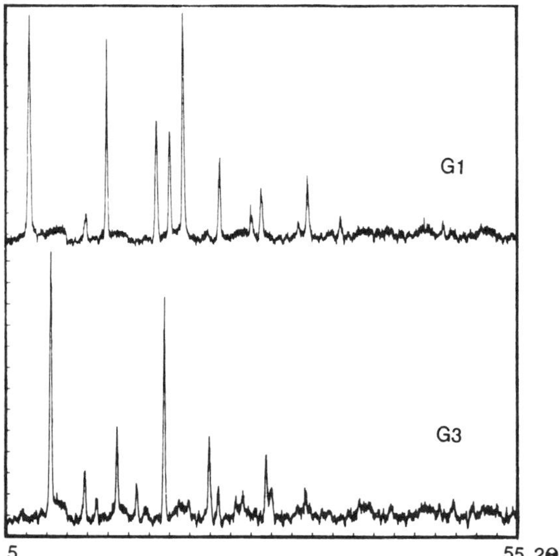

Figure 1: XRD spectra of crystallized G1 and G3.

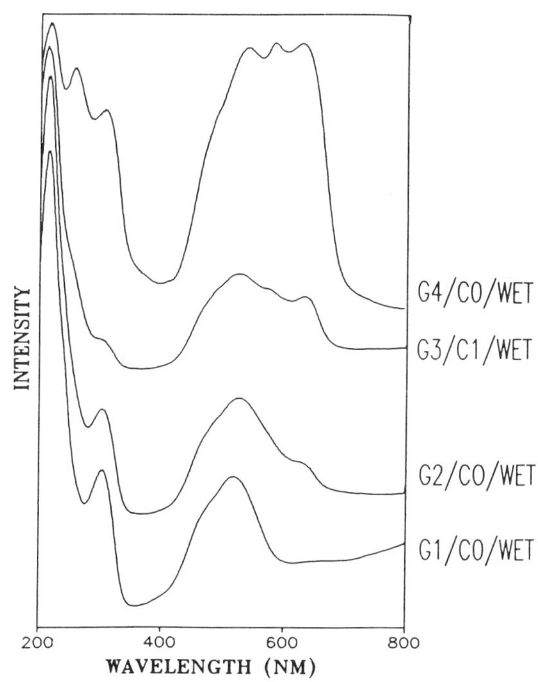

Figure 2: DRS spectra of wet gels as-such: influence of the template content.

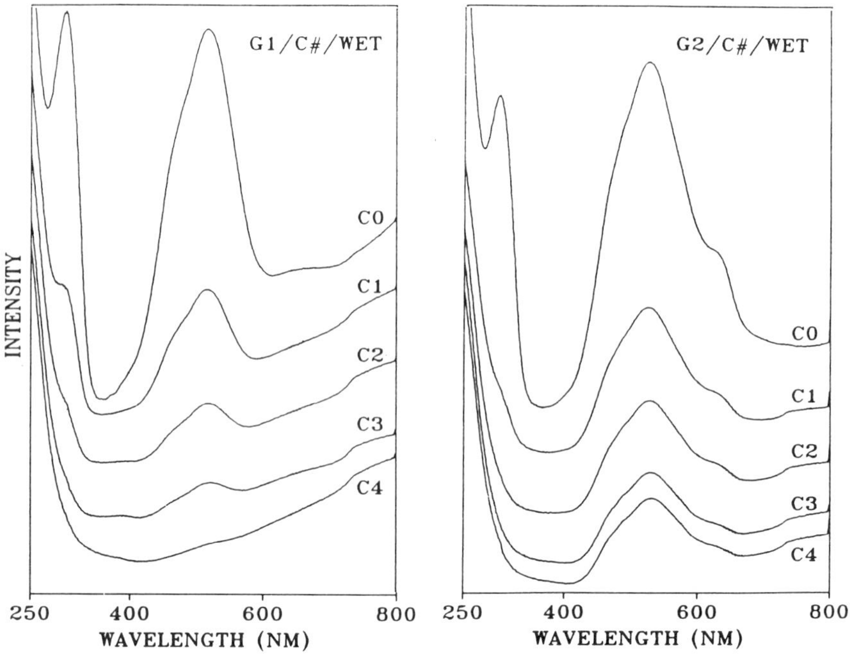

Figure 3: DRS spectra of wet G1 and G2 after different centrifugation steps.

Figure 4: DRS spectra of crystallized CoAPO-5 and centrifuged G1, dried at 333 K

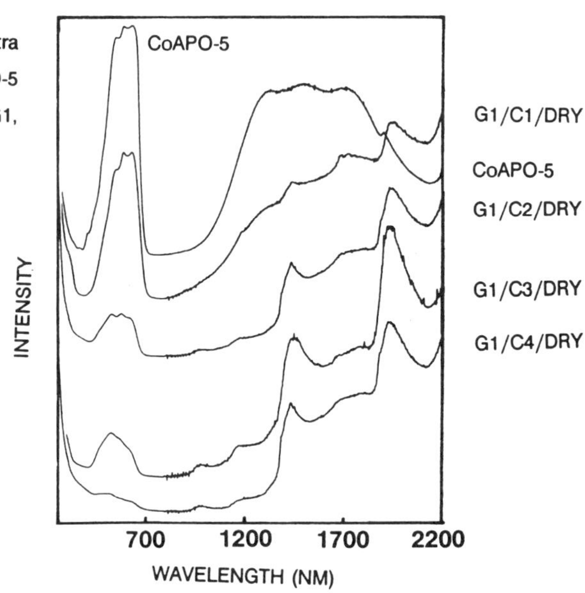

ZEOLITES SYNTHESIZED FROM NONALKALINE MEDIA

Zhao Daqing, Qiu Shilun and Pang Wenqin

Department of Chemistry, Jilin University, Changchun, P. R. of China

ABSTRACT

Hydrothermal crystallization of zeolites and clathrass, at 100～200℃ from nonalkaline media containing F^- anions and organic template has been investigated. Under such conditions Zeolite P, ZSM-39, Theta-1, ZSM-5, a novel zeolite(CJS-1) and two novel clathrasils(CJS-2 and CJS-3) could be prepared. The conditions and various effects on the synthesis were systematically investigated. The large single crystals of Zeolite P, ZSM-39, CJS-2 and CJS-3 were obtained. The ZSM-5 single crystals were synthesized using seven kinds of organic amines.

INTRODUCTION

As a general rule, silica-based zeolites are obtained by hydrothermal crystallization from an alkaline reaction medium [1,2]. In recent years a new route to prepare pentasil zeolites has been developed by Guth and his coworker [3] by using a nonalkaline medium in the presence of fluoride ion. In this way, the MFI-, FER-, MTT-, MTN- and TON-structure types of materials have been prepared [4,5]. This route is especially suitable for high-silica zeolites synthesis. So far, there have been no reports on the synthesis of zeolites with novel structure and clathrasils from the nonalkaline medium.

The morphology of zeolite crystals is an important factor in many studies and industrial applications [6,7]. Great effort has been devoted to the synthesis of large crystals of ZSM-5[3,8,9]. However, previous authors reported the synthesis of large crystals of ZSM-5 using TPA^+(tetrapropylammonium cation) as a template. There has not been any report in the literature on the synthesis of large crystals of ZSM-5 with other organic templates.

Here we report the hydrothermal crystalization of zeolite(CJS-1) with a novel structure, two novel clathrasils(CJS-2 and CJS-3) and four zeolites with known structure(Zeolite P, ZSM-39, Theta-1 and ZSM-5) from nonalkaline fluoride media using various organic templates.

EXPERIMENTAL

A typical synthesis began with combination of water and templates, and colloidal

From the Proceedings of the 9th International Zeolite Conference, Montreal 1992, Eds. R. von Ballmoos et al.,

silica was added to this solution. Then the required amount of NH_4F or aqueous HF (40% in water) and $Al_2(SO_4)_3 \cdot 18H_2O$ were added to the above solution and the mixture stirred until completely homogeneous. The mixture was transferred to the reaction vessel. Static reactions were carried out in 50mL PTFE-lined steel autoclaves. The autoclaves were placed in thermostated ovens at the crystallization temperature. At regular intervals, the atuoclaves were removed, cooled rapidly with running water, and then a small amount of solid product was taken for inspection by optical microscopy. They were then reheated to the reaction temperature. Although this procedure may have reduced the nucleation time, there is no evidence that this occurred[10]. In most cases, the heating was stopped only when optical microscopy showed the product was free from amorphous material and was fully crystalline. The final solid products were recovered by filtration, washed with water, and dried at 110℃.

X-ray powder diffraction patterns were recorded on a Rigaku D/MAX Ⅲ A diffractometer using CuK_α radiation. Electron micrographs were obtained by using a Hitachi SEM X-650B. Thermogravimetry and differential thermal analysis were performed on WCT-2 thermoanalyzer. The measurements were made in flowing N_2 (50mL · min^{-1} with heating rate of 10℃ · min^{-1}). The spectra of ^{13}C CP MAS NMR were recorded on a Bruker MSL-400 spectrometer, and the frequency was 100.6 MHz. The compositions were determined by chemical analysis and electron probe microanalysis(on a Shimadzu EMS-SM7 microanalyzer).

RESULTS AND DISCUSSION

Crystallization of Zeolite P, ZSM-39 and Theta-1

Zeolite P and ZSM-39 were obtained with the same amine (tetramethylamimonium cation). A low Si/Al ratio leads to the Zeolite P, whereas a high Si/Al ratio favours the ZSM-39 formation. The composition limits of the starting mixtures for Zeolite P and ZSM-39 were $1.0SiO_2 \cdot (0.05\sim0.5)Al_2O_3 \cdot (0.5\sim3.0)TMA^+ \cdot (0.5\sim2)NH_4F \cdot (10\sim200)H_2O$ and $1.0SiO_2 \cdot (0\sim0.05)Al_2O_3 \cdot (0.4\sim2.5)TMA^+ \cdot (0.5\sim3)NH_4F \cdot (10\sim200)H_2O$, respectively. For Zeolite P, the mixture was digested at 100~200℃ for 2~6 weeks and ZSM-39 was obtained with heating 5~14 days at 150~200℃.

A remarkable characteristic of this synthesis method is the formation of large crystals. Large single crystals of Zeolite P and ZSM-39 were obtained. The typical synthesis examples of large single crystals are listed in Table 1.

Table 1 Typical synthesis examples of large single crystals of Zeolite P and ZSM-39

starting mixture compositions	time (days)	Temperature (℃)	product
1.0SiO_2 · 0.1Al_2O_3 · 1.0TMA^+ · 1.0NH_4F · 42H_2O	21	190	P
1.0SiO_2 · 1.1TMA^+ · 1.8NH_4F · 40H_2O	7	190	ZSM-39

The scanning electron microscopy analysis (Figure 1) shows that both the single crystals of Zeolite P and ZSM-39 have the typical polyhedral shape and the size, ca. 150 and 400μm in length, respectively. DTA-TG analysis show that Zeolite P loses organic species at 548℃ and the organic template could not be completely removed from the silica-ZSM-39 framework by heating in air at 1200℃ for 9h.

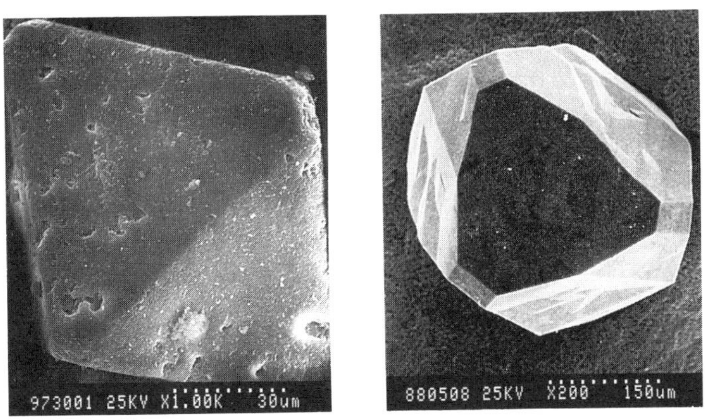

(a) (b)

Figure 1 SEM photographs of large single crystal of (a) Zeolite P and (b) ZSM-39

The composition limits of the starting mixtures for Theta-1 were: 1.0SiO_2 · (0 ~ 0.1) Al_2O_3 · (0.2 ~ 2.0) HF · (10 ~ 200) H_2O · (0.5 ~ 4.0) [1, 6-diaminohexane]. The mixture was digested at 140 ~ 190℃ for 4 ~ 9 weeks. The pure Theta-1 was obtained in all cases when the temperature was below 170℃. At a high temperature or long crystallization time, however, ZSM-5 and more condensed phases such as cristobalite were formed. At a low temperature (below 140℃), no zeolite phase was formed. It is concluded that the optimum temperature for synthesizing Theta-1 is this system is 150℃.

Crystallization of ZSM-5 using various organic templates

A large number of ZSM-5 zeolites could be obtained in F^- media[11～14] according to the described route[3] with $Pr_xNH^+_{(4-x)}$ (Pr=n-propyl; x=1 to 4) templates. The TPA^+ is the best template. The use of seven amines instead of TPA^+ yields ZSM-5 zeolite from F^- media. In Table 2 are given the composition limits of starting mixtures and crystallization temperature for ZSM-5. The silica-ZSM-5 was not obtained by using TEA^+, Et_2NH, or EDA as template. For CH^+-, PIP- and DO-ZSM-5, substitution of silicon by Al was impossible. The pure phase of ZSM-5 was synthesized when the temperature was below 150℃ using Et_2NH, EDA and DO, respectively. At a higer temperature, however more condensed phases were formed.

Table 2 The composition limits of starting mixture and crystallization for ZSM-5

template	SiO_2/Al_2O_3	template/Si	H_2O/Si	F^-/Si	temperature (℃)
tetraethylammonium(TEA^+)	200～400	0.5～6	5～300	0.3～6	140～200
tripropylamine(Pr_3N)	20～∞	0.3～4.5	5～300	0.3～5	150～200
diethylamine(Et_2NH)	20～400	0.5～5	10～200	0.3～4	150
ethylenediamine(EDA)	20～400	0.5～4	10～200	0.5～4	150
choline(CH^+)	20～∞	0.5～4	10～300	0.5～4	150～200
1,4-diazabicyclo-(2,2,2)octane(DO)	∞	0.5～4	10～200	0.5～3	150
piperazine(PIP)	∞	0.5～4	10～200	0.5～3	150～200

We were successful in preparing the large single crystals of ZSM-5 with above seven organic templates, respectively. The typical synthesis examples of large single crystals of ZSM-5 are listed in Table 3.

SEM photographs of the single crystals of ZSM-5(Figure 2) show that they are uniform and with no gel or other crystalline impurities present in the products. Except for PIP-ZSM-5, all the crystals have an average size of about 100μm.

Table 3 The typical synthesis examples of large single crystals of ZSM-5

starting mixture compositions	time(days)	temperature(℃)
$1.0SiO_2 \cdot 0.04Al_2O_3 \cdot 1.2Pr_3N \cdot 1.0HF \cdot 60H_2O$	31	190
$1.0SiO_2 \cdot 0.02Al_2O_3 \cdot 1.1TEA \cdot 1.0NH_4F \cdot 50H_2O$	35	200
$1.0SiO_2 \cdot 0.02Al_2O_3 \cdot 1.2Et_2NH \cdot 1.1HF \cdot 60H_2O$	61	150
$1.0SiO_2 \cdot 0.02Al_2O_3 \cdot 1.1EDA \cdot 1.2HF \cdot 60H_2O$	65	150
$1.0SiO_2 \cdot 0.5CH^+ \cdot 1.0NH_4F \cdot 50H_2O$	21	190
$1.0SiO_2 \cdot 0.5DO \cdot 1.5HF \cdot 50H_2O$	52	150
$1.0SiO_2 \cdot 1.0PIP \cdot 1.2HF \cdot 50H_2O$	47	190

(a)

(b)

(c)

(d)

Figure 2 SEM photographs of single crystals of (a) TEA-ZSM-5, (b)Pr_3N-ZSM-5, (c) CH-ZSM-5 and (d) DO-ZSM-5.

Crystallization in zeolite CJS-1

A novel aluminosilicate zeolite, denoted as CJS-1[15], has been synthesized from nonalkaline media by using three of heterocyclic compounds (piperazine, pyrrolidine and piperidine) as template. The composition limits of starting mixture were $1.0SiO_2 \cdot (0.05\sim0.5)Al_2O_3 \cdot (0.5\sim4.0)R \cdot (0.5\sim3.0)HF \cdot (20\sim200)H_2O$(R

= template). The digested temperature and time were 150～200℃ and 6～10 weeks, respectively.

The X-ray powder diffraction data is given in Table 5. By comparison with the characteristic X-ray powder diffraction data of known zeolite, it was found that CJS-1 has a novel structure. Determination of the N_2 adsorption isotherm of CJS-1 showed that it has characteristic micropore adsorption properties. The apparent surface area was 327 m^2g^{-1}. The langmuir type adsorption isotherms are similar to those of other zeolites. DTA-TG analysis showed that CJS-1 loses organic template at about 500℃. Scanning electron microscopy showed that the crystals of CJS-1 have the typical sheet shape. ^{13}C CP MAS NMR study showed that the decomposition of templates into fragments is responsible for crystallization of zeolite CJS-1.

Table 5 The X-ray powder diffraction data for CJS-1

2θ	d	I/I_0	2θ	d	I/I_0
9.21	9.594	100	26.87	3.315	1
12.83	6.894	2	28.22	3.160	31
13.49	6.558	22	29.19	3.056	4
15.32	5.779	4	37.08	2.422	1
18.71	4.738	21	38.02	2.365	2
22.28	3.987	11	44.57	2.031	3
22.51	3.946	9	45.38	1.997	1
23.11	3.845	5	46.52	1.951	1
23.62	3.763	7	47.23	1.923	2
24.19	3.675	25	47.92	1.897	3
25.24	3.525	10	48.81	1.864	1
25.73	3.459	8	49.59	1.836	1
26.48	3.363	7	51.47	1.774	1
			8.24	1.583	14

Crystallization of two clathrasils

The two clathrasils, named as CJS-2 and CJS-3, were obtained by using the fluoride route at pH = 7 in the presence of pyrrolidine and N, N, N′, N′-tetramethylethylenediamine, respectively. The Si/Al ratio of starting mixtures was infinity and crystallization temperature was at 150～200℃. The SEM analysis showed that both CJS-2 and CJS-3 have not been containinated with gel or other species. The large single crystals of CJS-2 and CJS-3 were obtained from starting mixtures with molar composition $0.5R \cdot 1.0SiO_2 \cdot 0.5HF \cdot 40H_2O$(R=pyrrolidine and N,N,N′, N′-tetramethylethylenediamine, respectively) at 190℃ for 23 and 54 days, respectively. The SEM photographs of single crystals are shown in Figure 3.

(a) (b)

Figure 3 SEM photographs of single crystals of (a) CJS-2 and (b) CJS-3.

The single crystal structure determination of CJS-2 by X-ray diffraction (Rw= 0.107) shows that the topological structure of CJS-2 is consistent with the nonasil [16], but the space group (Pnnb) is not the same. CJS-3 is probaly a novel clathrasil. The X-ray powder diffraction data of CJS-3 is given in Table 6, and the unit cell parameter is a=19.4795(3) Å.

Table 6 The X-ray powder diffraction data of CJS-3

2θ	d	I/I_0	2θ	d	I/I_0
13.01	6.794	16	26.31	3.384	100
15.22	5.816	37	28.12	3.171	22
15.89	5.550	32	29.18	3.058	8
18.41	4.813	24	31.05	2.877	5
20.03	4.429	26	31.53	2.834	11
21.62	4.107	36	36.12	2.485	13
22.88	3.883	63	37.50	2.396	16
25.04	3.553	22	39.95	2.255	1

The ^{13}C CP MAS NMR study of the templates shows the protonated form of the two kinds of organic amines are occluded in the framework of clathrasil. Both pyrrolidine and N,N,N′,N′-tetramethylethylenediamine could not be completely removed from the clathrasil framework by heating in oxygen at 900℃ for 4h.

CONCLUSIONS

It is possible to obtain zeolites and clathrasils with novel structures with fluoride anions in zeolite synthesis. We were successful in preparing a novel zeolite (CJS-1) and two novel clathrasils (CJS-2 and CJS-3). The large zeolite crystals with fewer defects could be prepared from this system because of better controlled nucleation and slower growth rate. The large single crystals of Zeolite P, ZSM-39, CJS-2 and CJS-3 were obtained. The use of seven amines instead of TPA^+ yields the large single crystals of ZSM-5. In nonalkaline fluoride media, the protonated forms of the organic amines replace the usual alkai cations, the materials obtained are therefore alkali-free and a simple calcination yields the acidic form of zeolite.

ACKNOWLEDGEMENT

The authors thank for the financial support by the National Natural Science Foundation of China.

REFERENCES

1. Zhdanov, S. P. "Molecular Sieve Zeolites- I ", ACS, Washington. p. 20(1971).
2. Breck, D. W. "Zeolite Molecular Sieve, Structure, Chemistry and Uses", Wiley, New York, p. 245(1974).
3. Guth, J. L. , Kessler, H. and Wey, R. in "Proc. 7th Int. Conf. Zeolites", eds. Y. Murakami. A. Iijima and J. W. Ward, Kodansa, Tokyo, p. 121(1986).
4. Guth, J. L. , Kessler, H. , Higel, J. M. et al, in"Zeolite synthesis", ACS, Washington, p. 176(1989).
5. Qiu, S. , Pang, W. and Yao, S. in"Zeolite, Facts, Figure, Future", eds. P. A. Jacobs and R. A. Van Santen, Elsevier, Amsterdam, p. 133(1989).
6. Sand, L. B. Preceedings of the 5th International Conference on Zeolites, p. 1(1980).
7. Chen, S. Z. Huddersman, K. Keir, D. and Lees, L. V. C. Zeolites, 8,106(1988).
8. Lermer, H. Zeolites, 5,131(1985).
9. von Ballmoos, R. and Meier, M. W. Nature(London), 289,782(1981).
10. K. R. Franklin and B. M. Lowe, Zeolites, 8,495(1988).
11. Guth, J. L. Kessler, H. Bourgogne, M. Wey, R. and Szabo, G. Fr. Pat. Appl. , 84-07773, 1984.
12. Guth,J. L. Kessler, H. Wey, R. and Faust, A. C. Fr. Pat. Appl. , 85. 07978(1985).
13. Patarin, J. Soulard, M. Kessler, H. Guth, J. L. and Baron, J. Zeolite, 9,397(1989).
14. Gabelica, Z. and Guth, J. L. Fr. Pat. Appl. , 88-04367(1988).
15. Zhao, D. and Pang, W. J. Chem. Soc. , Chem. Commun. , 884(1990).
16. Marler,B. Dehnbostel, N. Ewert, H. H. Gies, H. and Lieban, F. J. Inclusion Phenon. 4, 339(1986).

A NEW FAMILY OF MICROPOROUS GERMANIUM DIOXIDE CRYSTALS

Shougui Li, Ruren Xu, Yuqin Lu and Yihua Xu

Department of Chemistry, Jilin University,Changchun 130023,P.R.China

ABSTRACT

A new family of microporous germanium dioxide crystals have been obtained hydrothermally from organic template-containing reaction systems. Some characterization and structure determination have been made. In the framework structure, the coordination numbers of germanium atoms can be four, five and six and the GeO_n-polyhedra can share corners or edges. There are singly bridging, triply bridging oxygen and/or hydroxyl group in the framworks. The GeO_2 crystals are materials with intermediate and/or small pores.

INTRODUCTION

The high interest in the application potential of zeolite molecular sieves as catalysts, absorbents and ion-exchange agents has resulted in rapid development of the synthesis of crystalline microporous materials with novel structure and compositions. Over the last decade a series of microporous crystals, such as $AlPO_4$-n,SAPO, MeAPO, MeAPSO, ElAPSO[1-3],silica molecular sieves[4] and microporous metal sulfides[5] have been synthesized, followed by the synthesis of many microporous crystals of $GaPO_4$[6,7] $AlAsO_4$[8,9] and $GaAsO_4$-n[8,9]. Up to now ,there are a lot of investigations on aluminosilicate zeolite and $M(III)X(V)O_4$ type microporous crystals, but few investigations on oxide type microporous crystals other than microporous SiO_2 crystals were carried out. More than ten kinds of microporous SiO_2 crystals have been synthesized hydrothermally from organic template-containing reaction systems [4]. Ge is below Si in the periodic table. So it is an extremely attractive question whether GeO_2 forms microporous crystals from organic template-containing reaction systems. The aim of this paper is to describe the synthesis and characterization of a new family of microporous GeO_2 crystals obtained from organic template-containing reaction systems.

EXPERIMENTAL

GeO_2 powder(99.999%, 60 mesh) was used as the Ge source. All the syntheses are carried out in the presence of organic template. The

From the Proceedings of the 9th International Zeolite Conference, Montreal 1992, Eds. R. von Ballmoos et al.,

templates used were following: tetramethylammoniun hydroxide(TMAOH), dimethylamine (DMA), ethylenediamine (EDA),1,3-propanediamine(1,3-PDA), 1,2-propanediamine(1,2-PDA) and 1,6-hexanediamine (1,6-HDA).

The procedure of a typical synthesis was following: appropriate amounts of GeO_2 and organic template were added to certain amount of distilled water, then mixture was stirred until homogeneous. The reaction was allowed to proceed under hydrothermal (static) condition in a Teflon-lined autoclave at 180-200°C for 1-7 days. The crystalline products were filtered, washed with distilled water and dried at ambient temperature.

The crystalline products were characterized by a Rigaku D/max-3A X-ray diffractometer, using Cu Kα radiation. The single crystal X-ray structural determination was made on a Nicolet XRD R3 diffractometer[10,11]. The IR spectrum was recorded on a Nicolet 5DX spectrometer by using a KBr pellet. The differential thermal analysis was carried out on a Perkin-Elmer PC/DTA 1700 instrument in a flow of N_2 at a rate of 10°C min^{-1}.

RESULTS AND DISCUSSION

Synthesis

The reaction compositions and products are listed in Table 1 .

Table 1. Reaction compositions and products

Reaction compositions	Products
(0.4-0.7)TMAOH:GeO_2:(30-50)H_2O	Ge-C_1(novel phase)
(5-6)DMA:GeO_2:(50-60)H_2O	Ge-C_2(similar to $M_3HGe_7O_{16}\cdot 6H_2O$)
(2.5-13)EDA:GeO_2:(30-60)H_2O	Ge-C_3(novel phase)
(2.8-5.6)1,2-PDA:GeO_2:(38-44)H_2O	Ge-C_3
(2.8-5.6)1,2-PDA:GeO_2:(44-52)H_2O	Ge-C_2
(2.0-10)1,3-PDA:GeO_2:(30-60)H_2O	Ge-C_3
(0.3-0.9)1,6HDA:GeO_2:(40-50)H_2O	Ge-C_4(novel phase)

Four kinds of microporous GeO_2 crystals designated by Ge-C_n (n=1-4) were obtained. The X-ray powder diffraction patterns of Ge-C_n are given in Fig.1. By comparing the X-ray powder diffraction patterns of Ge-C_n with those of the germanates known so far, it is found that the Ge-C_1,

Ge-C_3 and Ge-C_4 have novel structures and the Ge-C_2 has similar structure to $M_3HGe_7O_{16}6H_2O$(M=Li,Na,K,Rb,Cs and NH_4^+) [12]. The large single crystals of Ge-C_1 were obtained by using TMAOH as template in the presence of 10% ethylene-glycol [10] and the large single crystals of Ge-C_3 were also obtained in the reaction system of EDA-GeO_2-H_2O [11]. Similar to the crystallization of microporous SiO_2 crystals the pH of the crystallization systems of Ge-C_n were larger than 12. But difference from the syntheses of microporous SiO_2 crystals is that the reaction

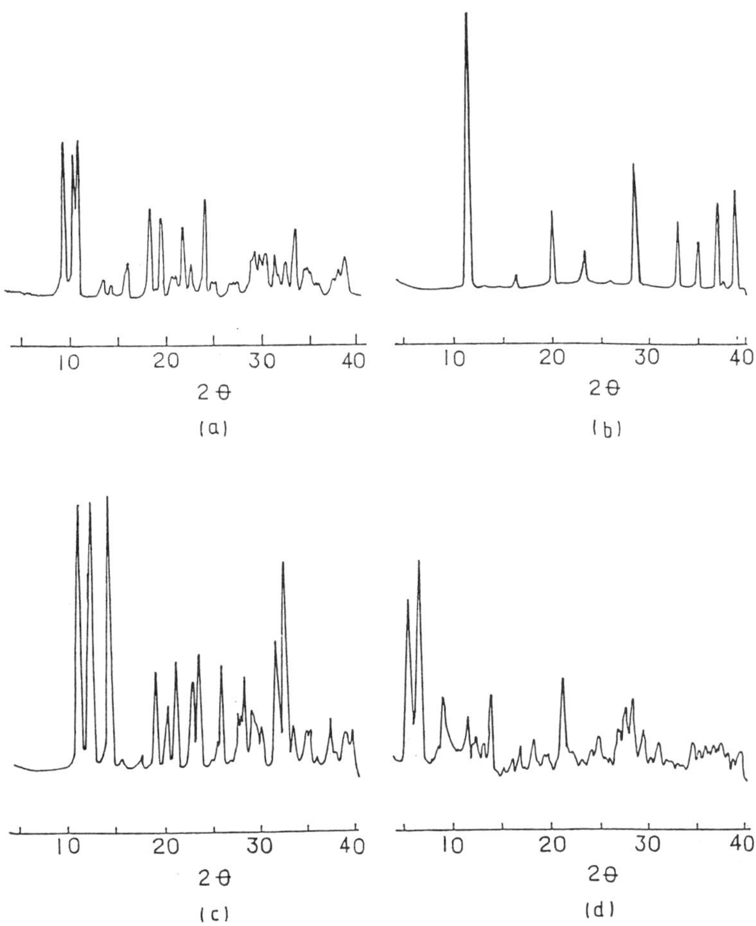

Figure 1. X-ray powder diffraction patterns of Ge-C_1(a), Ge-C_2(b), Ge-C_3(c) and Ge-C_4(d)

mixtures of Ge-C_n became transparent solutions after stirring for enough time. The crystals of Ge-C_n were formed from the solution at 180-200^oC. As the molar fraction of organic templates increased, the crystallization processes of Ge-C_n speeded up, but higher content of organic template often resulted in irregular and smaller crystal morphology. As crystallization temperature was increased ,the crystallization rates of Ge-C_n were increased. The phenomena of metastable phase transformation of Ge-C_n crystals happened. They transformed to dense phase after longer time crystallization.

Structure

The crystal structure of Ge-C_1 has been determined by single crystal XRD [10]. It indicates that the structure of Ge-C_1 consists of Ge atoms with different coordination numbers. Fig. 2a shows the independent building block of Ge-C_1 in which one Ge atom is coordinated by five O atoms and the others are coordinated by four O atoms. There are two types of framework O atoms in the structure of Ge-C_1. In the building block, one O atom is O atom of hydroxyl group attached framework Ge atom and others are singly bridging O atims. The three dimensional framework viewed along the c axis is shown in Fig. 2b. The largest one-dimensional open channel is constructed by packing of eight membered rings and also there are four and six membered ring channels along the c axis. There is no channel in any other directions. The cations of TMA^+ are located in the six-membered ring channels, playing the structure-directing role.

The X-ray powder diffraction pattern of Ge-C_2 is similar to the X-ray powder diffraction pattern of $M_3HGe_7O_{16}6H_2O$ which structure have been determined by means of powder X-ray diffraction [12]. In the structure, Ge atoms exhibit fourfold and sixfold coordination. Four GeO_6-octahedra share edges with each other and six GeO_4-tetrahedra share corners with the four GeO_6-octahedra to give rise to a building block (see Fig. 3a). Such four building block construct an eight-membered ring (see Fig.3b). The eight-membered ring channels run in three perpendicular directions.

The crystal structure of Ge-C_3 was determined by means of single crystal XRD [11]. Fig. 4a shows the independent building block of Ge-C_3 in which one of these Ge atoms is bonded to six O atoms, the four Ge atoms have fivefold coordination and the remaining four Ge atoms are tetrahedrally coordinated. In the structure there are three types of

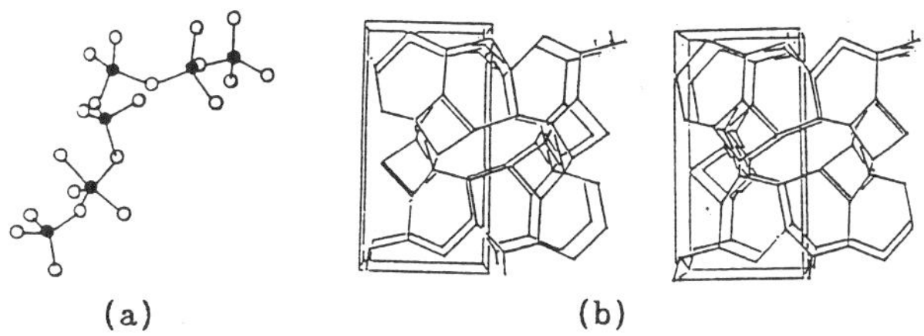

Figure 2. Structure of Ge-C_1, (a) independent building block(● Ge atom, o O atom), (b) stereoview plot along the c axis

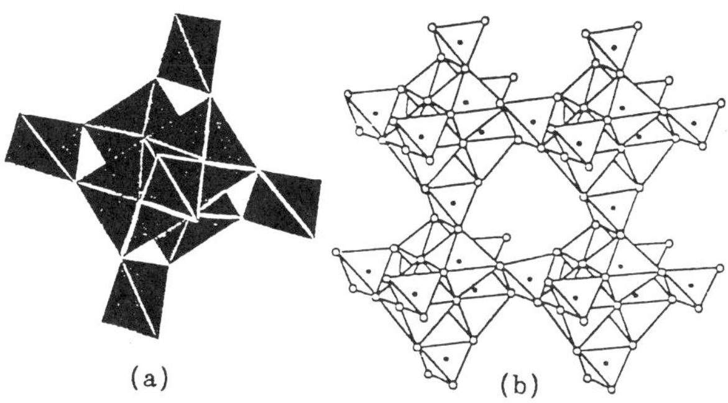

Figure 3. Structures of $M_3HGe_7O_{16}6H_2O$ [12], (a) independent building block (● Ge atom, o O atom), (b) eight-membered window ring

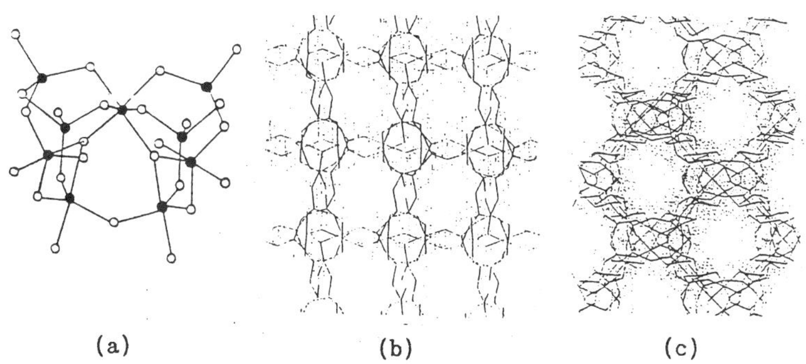

Figure 4. Structures of Ge-C_3,(a)independent building block (● Ge atom, o O atom), (b) stereoview plot along the a axis, (c) streoview plot along the b and c axis

channels and two of them have same ring aperture. The smaller one has an eight membered ring and runs along a axis (see Fig.4b). Intersecting this channel are the two with oval-shaped ten-membered window (see Fig. 4c). Chemically there are three distinct types of O atoms in the Ge-C_3 framework. The most numerous type are singly bridging O atoms. The second type are hydroxyl group attached to the framework Ge atoms. The final type are triply bridging O atoms.

The crystal structure of Ge-C_4 remains to be solved.

In comparison with microporous SiO_2 crystals the framework structures of microporous GeO_2 crystals are complex. For the former the coordination number of Si atom is always four and the SiO_4-tetrahedra only share corners with each other. For the latter the coordination numbers of Ge atoms can be four, five and six and GeO_n (n=4-6) polyhedra not only can share corners, but also can share edges. In the structures of microporous SiO_2 crystals there are only single bridging O atoms, whereas there are singly bridging O atoms, triply bridging O atoms, and/or hydroxyl groups in the structures of Ge-C_n. The GeO_2 crystals are materials with intermediate and/or small pores.

Characterization

The differential thermal analysis results of Ge-C_n in a flow N_2 are shown in Fig.5. The temperature of endothermic peaks for Ge-C_n is much higher than the boiling point of template, implying that the molecule of template exists in the channels of Ge-C_n and there are some interactions between the molecule of template and the frameworks of Ge-C_n. The escaping temperatures of templates are not correspondent with the boiling point. For example, the 1,6-HDA has highest boiling point among these six templates, but its escaping temperature is lowest in the Ge-C_n. This may be associated with the aperture size of channel. In the structure of Ge-C_1 the template molecules are located in smaller six-membered ring channel [10], so the escaping temperature of the template is higher. On the basis of the lower escaping temperature of template in the Ge-C_4, we believe the Ge-C_4 has larger channel aperture. As the template molecule escapes, the framework of Ge-C_n collapse.

The IR spectra of Ge-C_n are shown in Fig.6 which are more complex.

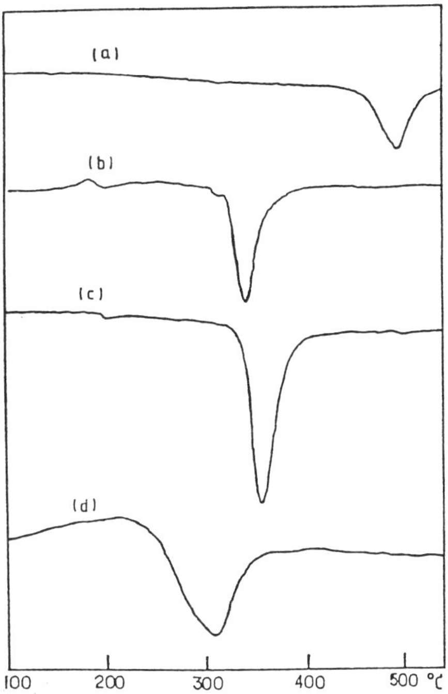

Figure 5.DTA curves of Ge-C_1(a), Ge-C_2(b), Ge-C_3(c) and Ge-C_4(d)

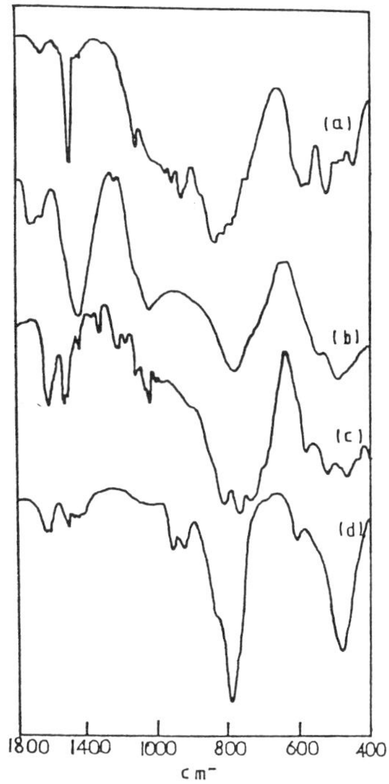

Figure 6.IR spectra of Ge-C_1(a), Ge-C_2(b),Ge-C_3(c) and Ge-C_4(d)

Detail assignment for the IR spectra is difficult, but the absorptions at 700-900 cm^{-1} should be assigned to the asymmetric stretching vibration of Ge-O and at 400-600 cm^{-1} to the symmetric stretching and bending vibration of Ge-O [10]. According to ref. [13], in the tetragonal form of GeO_2 the Ge atom has a sixfold coordination with oxygens, whereas in the quartz-type hexagonal GeO_2 the Ge atom has fourfold coordination. In former the frequency of the Ge-O-Ge asymmetric stretching vibration is at 720 cm^{-1}, whereas in the latter it is found at 885 cm^{-1}. In the structure of Ge-C_1 the Ge atoms mainly exist in the form of fourfold coordination, so the main absorption (837 cm^{-1}) is close to 885 cm^{-1}. The main absorption for Ge-C_2 and Ge-C_3 is at around 790 cm^{-1} owing to the increasing proportions of Ge atoms with five and six coordinations. The main absorption of Ge-C_4 is at around 790 cm^{-1}, so we can deduce that six coordination Ge atoms may exist in the structure of Ge-C_4. The absorption at 1400-1600 cm^{-1} belongs to the vibration of template molecule.

CONCLUSIONS

Four kinds of microporous GeO_2 crystals designated by Ge-C_n (n=1-4) have been obtained hydrothermally from organic template-containing reaction systems. The template molecule exist in the channels of the Ge-C_n crystals. The Ge atoms can have four, five and six coordination with O atoms, and the O atoms can exist in the forms of single bridging, triply bridging oxygen and hydroxyl groups. GeO_n-polyhedra can share corner or edge. The crystals of Ge-C_n have low thermal stability.

ACKNOLEDGE We thank the National Natural Science Foundation of China For financial support.

REFERENCES

1. Wilson,S.T., Lok,B.M. and Flanigen,E.M., U.S. patent, 4310440.(1982)
2. Wilson,S.T., Lok,B.M., Messina,C.A.,Cannan,T.R. and Flanigen,E.M., J. Am. Chem. Soc., 104, 1146 (1982).
3. Flanigen,E.M.,Lok,B.M., Patton,T.R. and Wilson,S.T., Proc. 7th Int. Zeolite Conf., Tokyo, p103 (1986).
4. Flanklin,K.R. and Lowe, B.M., Studies in Surface Science and Catalysis, Elsevier, Amsterdam, Part A, p179(1989).
5. Bedard,R.L. Wilson,S.T., Vail,L., Bennett,J.M., and Flanigen,E.M., Studies in Surface Science and Catalysis, Elsevier, Amsterdam, Part A, p379,(1989).
6. Parise, J.B., J.Chem.Soc.,Chem.Commun.,606(1985).
7. Xu,R. and Feng,S. Chem.J.Chinese Univ., Ser.A, 11, 1335 (1990).
8. Chen,J. and Xu,R., J. solid State Chem., 80, 149(1989).
9. Xu,R., Chen,J. and Feng, S., Chemistry of microporous Crystals, Eds. Invi,T. et al. Kodansha (Tokyo) p63(1990).
10. Cheng,J.,Xu,R. and Yang G., J, Chem. Soc. Dolton trans, 1537(1991).
11. Jones,R.H., Chen, J. et al, J.Solid State Chem., in press.
12. Wittmann, A., Fortschr, Miner., 43(2),230(1966).
13. Lippincott, E.R., Valkenburg,A.V.,Weir,C.E. and Bunting,E.N., J. Res. Nat. Bur. Standards, 61, 61(1958).

SYNTHESIS, DEALUMINATION AND PHYSICO-CHEMICAL CHARACTERIZATION OF ZEOLITE EMT

J. WEITKAMP and R. SCHUMACHER

Institute of Chemical Technology I, University of Stuttgart, Pfaffenwaldring 55, D-W-7000 Stuttgart 80, Germany

ABSTRACT

The synthesis of zeolite EMT was studied by systematically varying the Si/Al- and Na/Al-ratios. EMT, faujasite and a mixture of EMT and faujasite were obtained. By treatment with $SiCl_4$ vapor after the method of Beyer and Belenykaja, EMT zeolites with a wide range of Si/Al-ratios were produced. The effect of steaming dealuminated EMT was investigated. It was, moreover, found that extra-lattice aluminum can be efficiently removed from the pores by washing with aqueous EDTA solution.

INTRODUCTION

In the past, there has been considerable interest in the synthesis and characterization of zeolite ZSM-20 [1,2] which is an intergrown hybrid of the zeolites faujasite and EMT. EMT is the new code assigned to the hexagonal variant of faujasite by the Structure Commission of the International Zeolite Association. Very recently, Delprato et al. [3] succeeded in synthesizing pure EMT with a crown ether, viz. 18-crown-6 as template. The synthesis was modified by mixing the templates 18-crown-6 and 15-crown-5, and this resulted in a mixture of EMT and faujasite [4]. Using high-resolution electron microscopy it could be proven, that mixing of the two templates leads to intergrowths of EMT and faujasite [5]. The behavior of pure EMT in chemical modifications, e.g. by dealumination [6], is of particular interest. The present paper reports on a systematic investigation of the influence of the Si/Al- and Na/Al-ratio in the synthesis gel. Furthermore, dealumination of EMT with $SiCl_4$ vapor and the effect of steaming after the $SiCl_4$ treatment were studied.

EXPERIMENTAL SECTION

The synthesis experiments were carried out in autoclaves from stainless steel (150 cm^3). The standard gel was prepared by dissolving sodium aluminate (Riedel-de Haën, 41 wt.-% Na_2O, 54 wt.-% Al_2O_3), 18-crown-6 (Fluka) and an aqueous solution of NaOH (50 wt.-%) in distilled water. To this solution, a colloidal silica sol (30 wt.-% SiO_2, 70 wt.-% H_2O) was added under vigorous stirring. The resulting gel was aged for one day at room temperature.

From the Proceedings of the 9th International Zeolite Conference, Montreal 1992, Eds. R. von Ballmoos et al.,

For crystallization, the system was allowed to stand at 110 °C without agitation. After 12 days the solid product was filtered, carefully washed, dried at 120 °C and finally calcined at 540 °C. The standard gel composition was the one employed by Delprato et al [3]. Starting from this composition, the Si/Al- and the Na/Al-ratios were varied. Furthermore, other silica sources were tried. Scaling-up experiments were performed in an autoclave with a volume of 750 cm^3.

For dealumination of pure zeolite EMT, the method of Beyer and Belenykaja [7] was applied: The materials were first dried for 2 hours in a moisture-free N_2 stream. After cooling to the reaction temperature, a N_2 stream which was saturated at 25 °C with $SiCl_4$ was passed over the activated zeolite powder. To avoid excessive heat evolution, the samples were first contacted with $SiCl_4$ at low temperatures, viz. 250 °C or lower. In some experiments, this first dealumination step was followed by a second one. In these cases, the temperature was increased at a rate of 5 °C/min up to its final value under continuous treatment with $SiCl_4$. This temperature was held constant for a certain time. The zeolite was then purged in a pure N_2 stream to remove excess $SiCl_4$. After cooling to room temperature, the samples were carefully washed chloride-free with hot water. The detailed dealumination conditions are given in Table 1.

The samples were steamed in pure water vapor for 6 hours at 800 °C. Lattice Si/Al-ratios were determined by standard ^{29}Si-MAS-NMR spectroscopy (Bruker MSL-400) [8]. ^{27}Al-MAS-NMR spectroscopy was employed to collect information about the lattice and extra-lattice aluminum. Bulk Si/Al-ratios of the zeolites were determined after dissolution in HCl/HF/HNO_3 mixtures by atomic emission spectroscopy with inductively coupled plasma (ICP/AES). The X-ray powder diffraction patterns were taken on a Siemens D 5000 instrument using Cu Kα radiation (40 kV, 30 mA).

RESULTS AND DISCUSSION

The nature of the silica source was found to be very critical for a crystallization of pure EMT. Various silica sources like Cab-osil M5 (Fluka), tetraethylorthosilicate, or SiO_2 (Roth) were tested but failed to give EMT. The only silica source which led to highly crystalline EMT was colloidal silica sol, purchased from Bayer AG (type No. VP 4039, 30 wt.-% SiO_2).

With this silica source, the scale-up experiments were successful and EMT could be easily synthesized in a mass of 60 g per batch using the standard recipe. The resulting material was essentially of the same quality as EMT synthesized in the small autoclaves.

Varying the gel composition revealed that pure EMT can only be synthesized in a very narrow range of the Si/Al- and the Na/Al-ratios, in agreement with the results of Delprato et al [3]. The optimum gel composition reported by these authors led to EMT with a good

Table 1. Si/Al-ratios, water uptake over saturated $Ca(NO_3)_2$ solution at room temperature and conditions of dealumination with $SiCl_4$ vapors (T: temperature; t: time; act.: activation; deal.: dealumination; f.: final).

Sample designation	Lattice Si/Al-ratio		Bulk Si/Al-ratio	Adsorption capacity H_2O, wt.-%	$T_{act.}$, $T_{deal.}$, t_{deal}, $T_{deal.,f.}$, $t_{deal,f.}$
	not steamed	steamed		steamed	
EMT-1	> 100	> 100	10.2	2.4	360 °C 250 °C, 20 min 480 °C, 60 min
EMT-2	> 100	> 100	9.2	2.3	480 °C 250 °C, 20 min 480 °C, 60 min
EMT-3	12	> 100	6.1	5.7	360 °C 250 °C, 20 min –
EMT-4	> 100	> 100	7.5	3.7	360 °C 250 °C, 20 min 480 °C
EMT-5	12	> 100	5.2	4.8	360 °C 250 °C, 60 min –
EMT-6	15	> 100	5.1	4.6	360 °C 250 °C, 20 min 300 °C
EMT-7	6.5	50	6.1	6.6	360 °C 200 °C, 20 min –
EMT-8	5.5	29	6.3	12.6	360 °C 150 °C, 20 min –
EMT-9	4.5	n.d.	24.5	n. d.	360 °C 100 °C, 20 min –
EMT (starting material)	3.78	–	3.9	28.2 (not steamed)	– – –

crystallinity and a crystal size of 4 to 5 μm. These syntheses were fully reproducible. The X-ray powder pattern of pure zeolite EMT is shown in Figure 1A. EMT crystals with a size above 4 to 5 μm were never observed. It can be seen in Figure 2 that at very low Na/Al-ratios the resulting material is inevitably amorphous. If the Na/Al-ratio is increased either EMT with contributions of amorphous material or EMT/faujasite mixtures result. A typical X-ray pattern of such mixtures is presented in Figure 1B. At this time we are trying to tailor the amount of EMT and faujasite in such mixtures by appropriate synthesis conditions. It is still unknown whether these synthesis products are physical mixtures of pure EMT and faujasite crystals or intergrown crystals. At high Na/Al-ratios, pure faujasite crystallizes. Its X-ray pattern is shown in Figure 1C. The lattice Si/Al-ratio of this material was found to be between 2.4 and 2.5. At an Si/Al-ratio of 5.0 and Na/Al-ratios above 4, faujasite crystallizes in admixture with another crystalline material the nature of which remains to be determined. Upon further increasing the Na/Al-ratio, this unidentified material could be obtained in a pure form.

The ^{29}Si-MAS-NMR spectrum of calcined EMT (Figure 3A) consists of three intense peaks at -93.5, -99.8 and -105.6 ppm along with a small peak at -87.5 ppm. From right to left, these peaks can be assigned as Si(0Al), Si(1Al), Si(2Al) and Si(3Al). The standard evaluation [8] of this spectrum leads to a lattice Si/Al-ratio of 3.78 which is in good agreement with the bulk Si/Al-ratio of 3.9 (Table 1) compared to the Si/Al-ratio in the gel of 5.0. ^{27}Al-MAS-NMR reveals that in the pure EMT sample (Figure 4A) aluminum occurs on tetrahedral lattice positions only as indicated by the signal at 61.3 ppm.

By treatment with $SiCl_4$ vapors, zeolite EMT can be readily dealuminated. Depending on the dealumination temperature and time, a wide range of lattice Si/Al-ratios can be achieved, as shown in Table 1. Also given in Table 1 are the adsorption capacities of the steamed EMT samples for water vapor. With increasing severity of the $SiCl_4$ treatment, the water uptake decreases drastically and essentially parallels the lattice Si/Al-ratio. Before steaming, only those EMT samples exhibit a lattice Si/Al-ratio above 100 which have been dealuminated at 480 °C. Figure 3D shows a typical ^{29}Si-MAS-NMR spectrum for highly dealuminated EMT. There is only one peak at the Si(0Al) position. Presumably, the broadening at the bottom of this peak is due to some amorphous parts in the lattice generated during removal of aluminum. It has been repeatedly suggested that during the $SiCl_4$ treatment, removal of aluminum from the lattice and insertion of silicon into the lattice occur at different rates, hence defect sites may be formed [7,9]. Upon steaming the same sample, the broadening at the bottom of the peak vanishes and the signal becomes very sharp (Figure 3E). This indicates that healing of the lattice defect sites has taken place. It has been argued for faujasites [10] that the silicon contained in amorphous debris inside the pores can indeed heal lattice defects at high temperatures in the presence of steam. In the steamed EMT samples the lattice and the bulk Si/Al-ratios deviate considerably from each other. This indicates that the extra-lattice aluminum could not be removed efficiently from the pores even by excessive washing with hot water. Figure 4C clearly shows the extra-lattice aluminum present after the dealumination, washing and steaming procedures. Three broad peaks occur in the ^{27}Al-MAS-NMR spectrum. The signal at 55 ppm is assigned to tetrahedrally coordinated aluminum, whereas the peak at 0.4 ppm stems from octahedrally coordinated aluminum outside the lattice. The peak at 29 ppm cannot be assigned unambiguously at this time. It has been speculated for faujasites that this signal is indicative of strongly distorted aluminum in a tetrahedral coordination [11]. Alternatively, aluminum in a five-fold coordination could also be envisaged.

In a further experiment the steamed EMT-1 sample was washed with a concentrated solution of EDTA in water. This way, most of the extra-lattice aluminum could be removed (Figure 4D), although the differences between spectra D and C cannot be evaluated quantitatively.

The ^{29}Si-MAS-NMR spectrum for sample EMT-7, a mildly dealuminated zeolite, exhibits three peaks (Figure 3B), which are assigned to Si(0Al), Si(1Al) and Si(2Al). A comparison with Figure 3C shows that the steaming procedure has led to a strong increase in the lattice Si/Al-ratio (from 6.5 to 50, cf. Table 1). From the ^{27}Al-MAS-NMR (Figure 4B) it is seen that beside the tetrahedrally coordinated aluminum, there is only one additional signal representing octahedral aluminum.

CONCLUSIONS

The synthesis procedure disclosed by Delprato et al. [3] gives zeolite EMT in a good crystallinity and turned out to be very reproducible in our studies. Varying the Na/Al- and the Si/Al-ratio in the synthesis gel for the given amount of template leads either to EMT with an impurity of faujasite or to pure faujasite.

EMT can be readily dealuminated with $SiCl_4$ vapors. While the lattice Si/Al-ratio in the starting zeolite is 3.78, this value after dealumination can be tailored within a wide range. Removal of the extra-lattice aluminum generated in the dealumination step is very difficult as long as the materials are simply washed with hot water. Using aqueous EDTA solution for washing makes the removal of extra-lattice aluminum much more efficient. Steaming after $SiCl_4$ treatment brings about a further increase in the lattice Si/Al-ratio. Concomitantly, the lattice defects in the EMT lattice produced during the reaction with $SiCl_4$ are healed.

ACKNOWLEDGEMENTS

Financial support by Deutsche Forschungsgemeinschaft, Max Buchner-Forschungsstiftung and Fonds der Chemischen Industrie is gratefully acknowledged.

REFERENCES

1. Ernst, S., Kokotailo, G.T. and Weitkamp, J., in: *Innovation in Zeolite Materials Science*, Studies in Surface Science and Catalysis, Vol. 37, Eds. P.J. Grobet et al., Elsevier, Amsterdam, Oxford, New York, Tokio, p. 29-36 (1988).

2. Vaughan, D.E.W., Treacy, M.M.J., Newsam, J.M., Strohmaier, K.G. and Mortier, W.J., in: *Zeolite Synthesis*, Am. Chem Soc. Symp. Series, Vol. 398, Eds. M.L. Occelli and H.E. Robson, pp. 544-559 (1989).

3. Delprato, F., Delmotte, L., Guth, J.L., Huve, L., Zeolites **10**, 546 (1990).

4. Arhancet, J.P. and Davis, M.E., Chem. Mater. **3**, 567 (1991).

5. Anderson, M.W., Pachis, K.S., Rébin, F., Carr, S.W., Terasaki, O., Ohsuna, T. and Alfreddson, V., J. Chem. Soc., Chem. Commun. 1660 (1991).

6. Li, H.-X., Annen, M.J., Chen, C.-Y., Arhancet, J.P. and Davis, M.E., J. Mater. Chem. **1**, 79 (1991).

7. Beyer, H.K., Belenykaja, I., in: *Catalysis by Zeolites*, Studies in Surface Science and Catalysis, Vol. 5, Eds. B. Imelik et al., Elsevier, Amsterdam, Oxford, New York, pp. 203-210 (1980).

8. Engelhardt, G. and Michel, D., *High Resolution Solid-State NMR of Silicates and Zeolites*, Wiley, New York, 504 p. (1987).

9. Martens, J.A., Geerts, H., Grobet, P.J. and Jacobs, P.A., J. Chem. Soc., Chem. Commun. 1418 (1990).

10. Fejes, P., Kiricsi, I., Hannus, I. and Schöbel, G., in: *Zeolites - Synthesis, Structure, Technology and Application*, Studies in Surface Science and Catalysis, Vol. 24, Eds. B. Drzaj, S. Hocevar and S. Pejovnik, Elsevier, Amsterdam, Oxford, New York, Tokio, pp. 263-270 (1986).

11. Engelhardt, G., in: *Introduction to Zeolite Science and Practice*, Studies in Surface Science and Catalysis, Vol. 58, Eds. H. van Bekkum, E.M. Flanigen and J.C. Jansen, Elsevier, Amsterdam, Oxford, New York, Tokio, pp. 285-315 (1991).

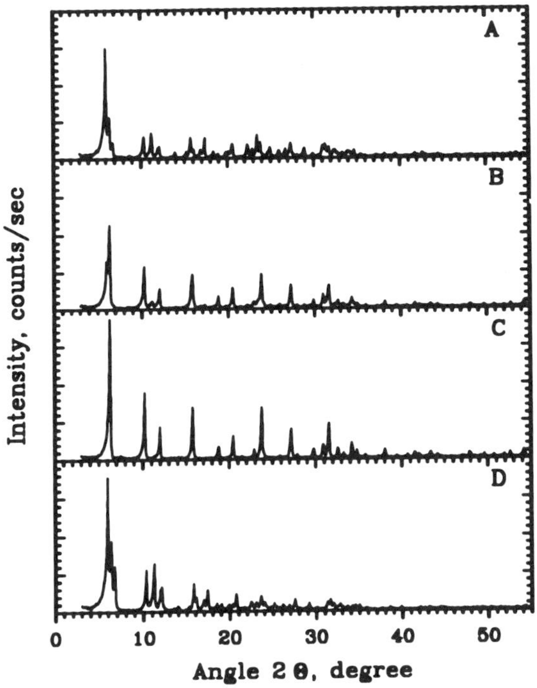

Figure 1. X-ray diffraction pattern of pure EMT (A), a mixture of EMT and faujasite (B), pure faujasite (C) and sample EMT-1 dealuminated and steamed (D).

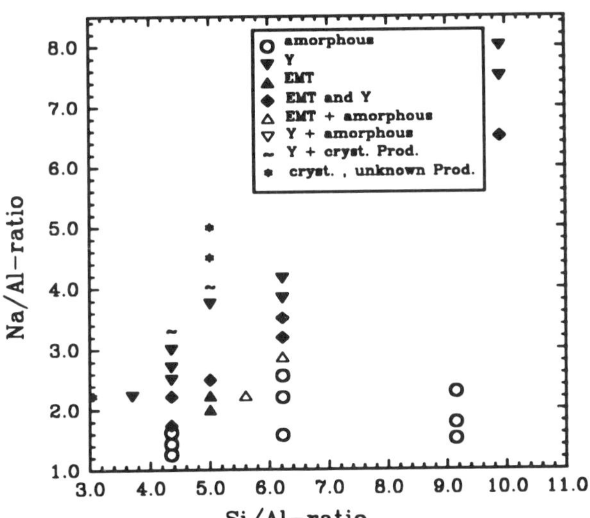

Figure 2. Influence of the Si/Al- and Na/Al- ratio of the gel on the material resulting in the hydrothermal synthesis.

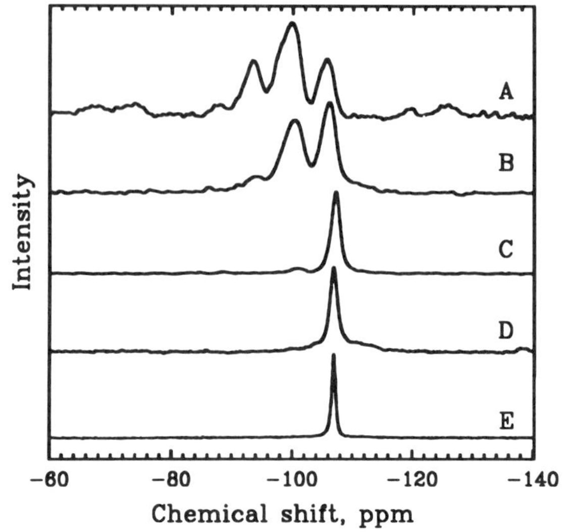

Figure 3. ^{29}Si-MAS-NMR spectra of pure EMT (A), EMT-7 (B), EMT-7 after steaming (C), EMT-1 (D) and EMT-1 after steaming (E).

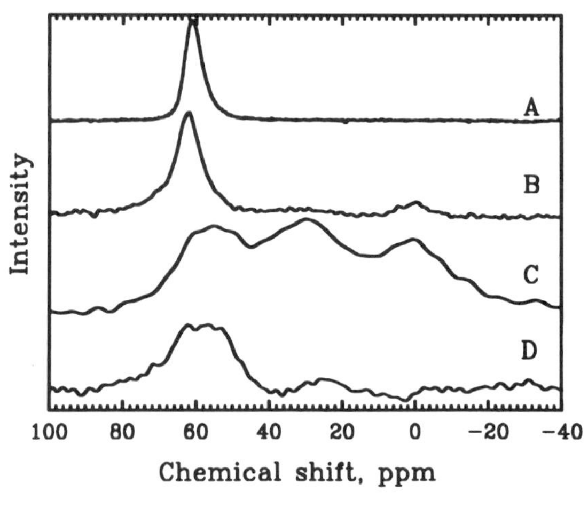

Figure 4. ^{27}Al-MAS-NMR spectra of pure EMT (A), EMT-7 after steaming (B), EMT-1 after steaming (C), EMT-1 after steaming and treatment with EDTA (D).

DISCUSSION - Synthesis

"The Role of Alkali Cations in the Synthesis of Silicalite-1 In Flouride Medium" (F. Crea, et al.)

Harry Robson: You did not include Rb^+ in your group of alkali metal cations. Also Tl^+ is chemically very similar to the alkali metal cations. Would you anticipate any deviations for Rb^+ or Tl^+?

Answer: With RbF, the formation of Rb_2, SiF_6 is also expected in that case due to the presence of small Rb_2 SiF_6 crystallites in the amorphous gel, the rate of both the nucleation and the crystallization is expected to be enhanced. For TlF, we have to carry out the experiments to be able to give a reasonable answer.

F. DiRenzo: Both pH and silica species concentration can affect crystallization kinetics. Do you have any actual measurement of the solution composition in order to tell the effect of each of these factors?

Answer: For the same F^- content, we have found a linear increase of the rate of crystallization as a function of increasing pH. We did not analyze the liquid phases by multi NMR for those samples, but for similar synthesis batches, we have observed by ^{19}F NMR, the presence of F^- ions and of SiF_6^{2-} complexes.

K. Unger: When one discusses parameters which affect the particle morphology dissolution and nucleation play a dominant role in synthesis, both processes take place at the initial period of synthesis. Dissolution and nucleation are much influenced by the heating role and the control of temperature. How did you control these operation parameters in your work?

Answer: The autoclaves were put in the oven at 170°C. The necessary time for the inner part of the autoclave to reach that temperature is ca. 15 minutes. All the samples were prepared following the same procedure.

A. Nastro: Were all dry gels prepared the same way?

Answer: Yes, all the gels were dried at 100°C overnight.

J. L. Guth: Are the alkaline cations present in the crystals after synthesis?

Answer: In our reaction conditions, the alkali cations are not incorporated in the crystals. For example, zeolite obtained from the NaF batch only contained traces of Na, as determined by proton induced x-ray emissions.

"Synthesis of NH_4-Form Heteroatom-Substituted MFI Zeolites in the Presence of Fluoride Anions" (J. Dwyer, et al.)

K. Unger: Did you check the distribution of aluminum and gallium across the MFI crystals? In other words, did you observe a homogeneous or heterogeneous distribution?

J. Zhao: We have checked the Al distribution of Al-MFI by EDAX and SIMM. These methods indicate that two types of Al distribution have been observed; (A) homogeneous (B) surface Al-rich. Other evidence shows that the Al-distribution type depends on preparation method of the products.

Question: What is the yield in your reaction based upon SiO_2? Is there a stoichiometric ratio for Si/F in solution at the end of the reaction?

J. Zhao: We have not calculated the mass balance, but the liquid analysis shows that Si concentration is not high. Therefore, the yield is expected not to be low. The Si/F ratio in the mother liquid is quite near to the ratio in the original reaction mixtures.

H. Robson: Seeding accelerated crystallization, did it also influence substitution of B for Si, Al for Si, or Ga for Si?

J. Zhao: Due to the seeding technology applications [limitations], the sizes of crystals are very small which considerably increases the difficulty. We are going to do some investigation in this field.

H. Kessler: You have performed ^{71}Ga MAS NMR on the as-synthesized material. Have you also looked at the calcined Ga MFI material in order to have an idea of possible Ga removed from the framework?

J. Zhao: The framework Ga is easy to change into non-framework Ga during calcination, particularly in the high Ga products. Therefore, calcination conditions are very important for reduction of transfer from framework Ga to non-framework Ga.

"An Unexpected and Highly Versatile New Zeolite Synthesis Route Leading to Large Pore Alumino- and Borosilicate Sieves" (S. I. Zones, et al.)

J. H. C. van Hooff: Do you have a chemical explanation for the fact that the synthesis of B-SSZ-24 is much more successful with B-Beta as starting material? Is it also possible to use silica-boria gels instead of B-Beta?

Answer: Although the mechanism of this transformation is uncertain, we believe that by using the B-Beta zeolite as a starting material in these reactions, nucleation of B-SSZ-24 becomes more favorable than in reactions using Carbosil and $Na_2B_4O_7$. The high surface area of the calcined beta may provide many local environments where the concentrations of Si, B and other reactants are right for crystallization to occur.

M. L. Occelli: Using B-Beta as a starting reaction, several adamantine derivatives can be used as templates to induce crystallization of the desired phase. Do you observe crystallization when you remove the template? If yes, which phase did you obtain?

Answer: If you remove the template, crystallization does not occur. The beta zeolite goes to amorphous material.

K. Unger: What is the amount of template (adamantyl-derivative) relative to silica in the starting reaction mixture to prepare SSZ-13 and how much remains in the framework of the crystals?

Answer: The ratio of template to silica in these reactions typically is 0.15. Typically, there is ~ 15-20% by weight of the organic template in the SSZ-13 crystals.

Question: Have you tried other borosilicate sources?

Answer: We have tried other borosilicate zeolites like ZSM-5 and amorphous borosilicate gels, but they do not work either very well or at all in this transformation.

"Organic Guest Stabilization During Synthesis of High Silica Zeolites with Multidimensional Channels and Large Pores: SSZ-26 From A Propellane-Based Organocation" (S. I. Zones, et al.)

Question: Could you comment on the general role of organic molecules in zeolite synthesis? The second part of this question deals with preformed species in the solutions, favoring the formation of zeolites. Could you precise your ideas? Also see previous communication by Dr. Nakagawa.

S. I. Zones: I believe the general role of organic molecules in zeolite syntheses follows the pattern of either space-fillers or templates as previously described by Araya and Lowe. In the template case, the interaction is much more specific and I believe the organo-zeolite then achieves a much lower energy stage, making it more stable and less susceptible to subsequent Ostwald ripening phenomena. With the high silica materials, it appears that favorable wall contacts (SiO_2) and hydrophobic solution of the template by the zeolite are part of what makes the product more favorable, particularly in relation to the relatively high ionic strength surroundings.

In answer to your second part, I believe the relevant pre-formed species are still difficult to detect. We published some work a few years ago using an ion-selective electrode for hydrophobic cations. At high ionic strength the organo-cations were not so tethered or clathrated by silicate that we couldn't get the expected, correct electrode sensitivity for organo-cations. Only loss of organo-cation through crystallization of the zeolite produced signal change.

M. Shannon: Do you have information on the unit cell size or space group of this new phase?

S. I. Zones: I do not have the data on hand, but I believe that Crozier and Pan of ASU may have published it in last year's EMSA proceedings.

A. McCormick: What has been the level of your success at using molecular mechanics calculations to predict whether a template will be successful?

S. I. Zones: Dr. Nakagawa, from our lab, presented a paper last year at the New York ACS meeting describing the use of Diels-Alder chemistry to control template dimensions. Since that time, she and Tom Harris have been involved in calculating energies for those molecules in different host structures. The correlations are good, but also lack any framework charge. But I think we are getting closer to using this method to predict what a given template molecule might do.

"Raman Spectroscopic Studies in Zeolite Synthesis" (P. Dutta)

M. Bulow: I find your result on the gauche-trans conformation state of the alkyl chains in the templates with respect to the MFI synthesis extremely interesting. It is well known that the constant of these gauche-trans equilibria is temperature-dependent. Might that be a way to be followed in molecular sieve synthesis or to explain appropriate results which may be available but not explored?

P. Dutta: This is clearly an intriguing question. I think that at the temperatures of synthesis, gauche conformers of TPA might exist, which could directly nucleate ZSM-5. In order to prove this, we will have to do in-situ spectroscopic experiments at the temperature of synthesis. This is, at present, a difficult experiment because of the corrosive condition of synthesis.

R. Schoonheydt: 1. Did you check on decomposition of the templates during synthesis? 2. How do you explain protonation of pyrolidine in alkaline synthesis mixture?

P. Dutta: 1. We only checked on the integrity of the template that was retained in the crystal. Based on amines that are found in the solution at the end of synthesis, there is clearly some decomposition of template in solution. However, spectroscopic and analysis of the crystal indicates that the template integrity in the crystal is maintained.

2. The pH of the synthesizing solution is close to the pKa of the pyrrolidine in the gel as arising from Al incorporation and lowering of pH in the gel as compared to solution.

F. DiRenzo: You did show incorporation of organic agent in the gel in early stages of the synthesis. Which is the amount of organics incorporated? Does it correspond to complete transformation of the silica source into an intermediate organic-rich amorphous solid or only to the formation of local structures in a small portion of the synthesis gel?

P. Dutta: This depends if the gel is washed prior to analysis. In the case of ZSM-5, which was washed, the organic incorporation is small, indicating that the local structures are formed and are stable towards hydrolysis. In the case of ferrierite, the gel was not washed and based on the signal, it appeared that incorporation was extensive. However, if these samples are washed, the gel structure changes and the organic is not retained.

"Synthesis and Characterization of a New High Silica, Large Pore Aluminosilicate Zeolite, NCL-1" (R. Kumar, et al.)

H. van Bekkum: As the template of your interesting new zeolite, you mentioned $(alkyl)_3N^+(CH_2)_6N^+(alkyl)_3$. Could you indicate which alkyl substitutions can be applied and which one is to be preferred?

Answer: Alkyl substitutions with more than one carbon atom like ethyl, propyl, isopropyl, etc.

G. Bellussi: In the oxidation of xylenes with H_2O_2 how was the selectivity with respect to hydrogen peroxide?

Answer: Selectivity (with respect to H_2O_2) for conversion to (xylenes + methyl benzyl alcohol + methyl benzaldehyde) was about 60%.

Question: Did you check the surface area for the new material?

Answer: The surface area was around 400 m^2/g.

"Zeolite Synthesis in Ammonia and Aqueous Ammonia Solvents: Reactivity of Group IA Cations with Pre-Made Silica-Alumina Gels" (D. E. W. Vaughan & K. G. Strohmaier)

P. J. Chong: The cancrinite crystals you have grown appear to be single-crystalline, irregular in size but consistently hexagonal in shape. What would be the max. size of the crystallites experimentally attainable free of defects? Could you indicate some views on impurity inclusion, crystal transparency and/or electronic property whatever assessed additionally?

Answer: The cancrinite crystals in this study were up to about 5 microns in length. Crystals several millimeters in length and a millimeter diameter were grown by E.A.D. White and co-workers in the 1960's at the General Electric Co. in the UK.

J. Garces: Have you performed synthesis with NH_3 above the critical point (of NH_3)? If yes, I wonder if you observe dense or open phases.

Answer: No, work has been limited to 90°C or less. All phases are dominated by pressure, not solvation of cations; at even higher pressures I would expect no new phases.

L. Guth: Have you tried to replace the alkaline cations of the synthesis mixture by cations of transition details?

Answer: As discussed in the introduction, these were the blank experiments for a study of transition metal ammonia complex effects on zeolite synthesis. No phases of interest were observed in the transition metal experiments. Most transition metal experiments resulted in dense oxide phases.

W. Harrison: Have the crystallographic consequences of the non 1:1 Si/Al ration in the novel SOD and CAN phases been investigated? Is ammonia (ammonium cation??) incorporated in the products?

Answer: The Si/Al ordering problem has been done but not published (Melchior and Vaughan). The ammonium cation does not exist at the pH of these experiments (pH>9). Ammonia and water can coexist in the channels, though NH_3 concentrates in the vapor phase in the autoclave at 90°C.

H. Robson: Your synthetic cancrinite shows adsorption capacity where natural cancrinite does not. Is there an analogy with natural/synthetic mordenite or natural/synthetic offretite?

Answer: It is possible that sodalite intergrows in the cancrinite crystals, so blocking the channel. Both are ABC-6 ring structures, as are offretite and erionite, and the suggested analogy is a good one. Millwood, Ramdas and Thomas evaluated this problem and showed that intergrowths of CAN and SOD could be possible in OFF/ERI. The low ratio of CAN/SOD makes them difficult to lattice image (low beam stability) but it is a worthy study.

"Study on the Insertion of Vanadium in the Silicalite Framework" (G. Bellussi, et al.)

J. B. Nagy: Did you observe any difference between the Si-NMR spectra of silicalite-1 and the V-silicalite? If yes, how do you explain the difference?

Answer: ^{29}Si-MAS-NMR analysis was performed on A-4 and B-4 samples before and after calcination in air. Beside the typical peak at -112 ppm, another signal at -102 ppm was observed for all the samples with the exception of calcined A-4. The latter peak can be attributed to the presence of SiOH in defective sites or to silicon atoms bound to vanadium atoms. ^{29}Si cross polarization NMR analysis performed on samples with different vanadium content do not provide sufficient information to exclude one of the two hypotheses.

R. Schoonheydt: 1. Explain the oxidation of V^{3+} to V^{4+} + V^{5+} during synthesis. 2. Why does V^{4+} + V^{5+} not reduce in H_2?

Answer: In the precursors from B type synthesis, V^{4+} was the only V species detected. All V^{3+} added during the synthesis was oxidized to V^{4+} probably during the preparation of the reaction mixture by the atmosphere O_2. In the case of precursors from a type syntheses, ^{51}V-NMR analysis indicate the presence of diamagnetic V species (V^{3+} or V^{5+}). Since samples calcined in air or in H_2 showed the same NMR spectrum which was similar to that of model compound Na_3VO_4, it appears reasonable to assume the V^{5+} as the diamagnetic species. It is very difficult to answer the second question; just to speculate, beside the fact that V^{3+} is less stable than V^{4+} or V^{5+}, it is possible that the interaction with the surrounding plays a role in the stabilization of different oxidation states.

"F^-: A Multifunctional Tool for Microporous Solids a) Mineralizing, Structure Directing and Templating effects in the Synthesis" (J. L. Guth, et al.)

R. Xu: Do you consider the complex effect of F^- in the crystallization of microporous crystals; if so, did you try to investigate the other types of complex ions instead of F^- for crystallizing the SiO_2 - $AlPO_4$ and $GaPO_4$ based microporous crystals?

J. L. Guth: Depending on the nature of the "T" element, other liquids than F- may be used to solubilize those elements (e.g., oxalate anions for Al^{3+} or pyrocatechol complex for Si). We have tried to use such species in the synthesis of zeolites but until now without success.

J. B. Nagy: Following your suggestion, the fluorosilicate complexes do not initiate the zeolite synthesis? Or do they?

J. L. Guth: We believe that the octahedral fluorosilicate complexes (e.g., SiF_6^{2-}) are not directly at the origin of the zeolite crystallization but rather tetrahedral coordinated species which are in equilibrium with them.

D. E. W. Vaughan: Extensive synthetic geochemical studies by Wyllie and Tuttle in the 1960's and on high F pegmatites by Johns and Burnham (also in the 1960's) indicated that F^- is incorporated into micas as terminal groups replacing OH^-. Is it possible that in the zeolites, particularly the phosphates, F^- is a terminal group and not a trapped anion in a cavity?

J. L. Guth: According to the structure and composition of the microporous solids obtained in the presence of F^- we observed three different situations:

- terminal F (=T - F)
- bridging F
- trapped F^-

The last situation seems to be the most frequent.

Z. Gabelica: Did you try to investigate the "structure" (coordination of F species) of the soluble [Si(Al)F] complexes, for example, by using ^{19}F-NMR?

J. L. Guth: This work is in process but for the moment the results are rather disappointing. ^{19}F and ^{29}Si NMR show the presence of $Si\text{-}F_6^{2-}$ only. We believe that other species are in equilibrium with SiF_6^{2-} but difficult to evidence (too low concentration, rapid exchange). Some weak signals appear when the temperature is lowered.

"The Synthesis and Characterization of Ga-MFI and Ga-MTW Zeolites" (S. L. Lambert)

W. Schwieger: What will happen if you reduce the amount of the organic compound in the synthesis mixture, or can you produce template-free Ga-silicates?

S. Lambert: Crystalline Ga-MFI was obtained with template levels as low as 60% of that specified in my data. At lower levels of organics, only amorphous material was obtained at constant crystallization times. We did not try to prepare a template-free Ga-MFI.

J. Kornatowski: 1.) What is the upper limit of the amount of Ga incorporated into the framework of MFI? 2.) What is the influence of content of Ga on morphology of crystals?

S. Lambert: 1.) 14:1 Si/Ga is the highest level achieved for both MFI and MTW frameworks (~7.5% Ga content on a volatile-free basis). 2.) No change in crystal morphology was observed in the range of 0.8% Ga --> 7.5% Ga; however, crystal size increased as the gallium content was decreased with no other changes in the Gel composition.

G. Gianetto: Why did you say that Gallium content reduced the crystallization rate of gallosilicates MFI and MTW zeolites? Where are the evidences for this conclusion?

S. Lambert: I only showed the synthesis and characterization data for 2 samples of the many we prepared. I based my conclusion as the basis of having to increase the crystallization time from that for the analogous Si/Al gel; for instance, the Ga-MTW discussed required at 20 day synthesis (we examined 2 day intervals) to achieve good crystallinity while the analogous gel in the Si/Al system crystallized in 7 days. Other synthesis in the Ga-MFI system showed shorter times were required for crystallization when less gallium was present in the gel. Synthesis work we are doing with gallium in other zeolite frameworks show a similar effect.

D. Meriaudeau: Are XPS and SS NMR adequate techniques to probe the presence or the absence of extra framework Ga? In our opinion, SS NMR is not sensitive enough to probe the presence of traces of extra framework Ga. The best experience to probe the absence of extra framework Ga is to run catalytic test with nC_4 or C_3 as indicated in the literature.

S. Lambert: The XPS and solid state NMR experiments would not be sensitive enough to detect trace levels of extra framework gallium - our experiments were designed to show that a majority of the gallium was in the framework. Catalytic testing (not, however nC_4 or C_3) showed these materials to be less acidic than the aluminum containing analogs.

R. Howe: Did you measure the Ga:Si ratio by XPS to check on the uniformity of Ga distribution in your crystals?

S. Lambert: The Ga:Si ratios were not measured in the XPS experiment. The gallium uniformity across the crystals was not evaluated.

"Synthesis of Gallo- and Gallo-Alumino ZSM-5 Zeolites with Variable and Quasi Homogeneous Al-Ga Distribution, From Alkali-Free Media" (Z. Gabelica et al.)

R. Le Van Mao: Upon calcination, some Ga species are expulsed form the framework and this Ga explusion effect is stronger at higher concentration of Al. Do you have any explanation for that?

Z. Gabelica: We believe that when only a few Ga^{3+} ions are incorporated in the framework (e.g., less than one per unit cell) with respect to Al^{3+}, the structure is locally "expanded" (defected?) because of the larger size of Ga^{3+}. The latter is therefore more readily extracted as to recover a more stable MFI lattice. The same is observed for silicalites with low Ga content (<1/u.c.) obviously for the same reasons. This also explains whey dealumination does not occur at normal calcination temperatures (air 550°C), but starts only above 650-700°C.

F. DiRenzo: Do you consider your synthesis method as easily applicable to other zeolite systems?

Z. Gabelica: The research in this direction appears very promising. A series of other structures has been prepared using the "methylamine/ethylamine route". They are now being fully characterized (papers in preparation). The most promising result was obtained in the case of (Al,Ga)-MTW type zeolites that involve either Al or Ga in variable proportions. Pure Ga and Al-MTW end-members, with extended Ga or Al framework incorporation were obtained.

"Preparation of Zeolite Film Using Cellulose Moulding" (T. Sano, et al.)

K. Unger: Did you investigate the relationship between the crystal size of zeolite and the pore size and mechanical stability of the final product?

T. Sano: The pore size distributions of silicalite, ZSM-5 with SiO_2/Al_2O_3 ratios of 100 and 200, and SAPO-5 films were evaluated by nitrogen adsorption and mercury porosimetry. Except for the void space generated by burning of cellulose fibers, mesopores did not exist within the film regardless of zeolite crystal size.

Question: What is your evidence that the OH groups of the cellulose film are essential for crystallization since we know that we can crystallize films on metals?

T. Sano: At the present time we do not have any evidence. However, from the facts that the thickness of the zeolite film prepared using a filter paper was considerably bigger than that using a Teflon slab, and
that an enrichment of Al on the surface of filter paper occurred, we have considered that the cellulose having many OH groups provides the unique sites for crystallization of zeolite.

"Molecular Design of Layered Zirconium Phosphonate Catalysts" (K. Segawa and N. Kihara)

R. von Ballmoos: Two questions regarding the in-plane distance between the pillared or pendant phosphonate groups: 1.) what is this distance; 2.) is it temperature independent?

K. Segawa: Interlayer d-spacing has been determined by the first line (001) of XRD pattern. In the case of pillared type, if you divided by 2 for the carbon numbers of alkyl group, it becomes the same as pendant ZrP. (If you have C_{10} ZrP, the interlayer d-spacing is the same as Cr pendant ZrP.) Zirconious phosphonate may decompose at higher temperature (300 - 500°C), and end up with zirconium diphosphate structure.

"Synthesis and Characterization of the Novel Gallophosphate Cloverite with 20-Membered Ring Channels" (J. Patarin, et al.)

J. Kornatowski: Have you a hypothesis how to explain the difference in TPD of ammonia for small and large crystals?

J. Patarin: Up to now, we have no explanation to explain the difference observed in STD of ammonia.

P. Malla: Do you have any clue as to why cloverite is not stable against water (vapor)? As you know, SAPO-37 has the same problem. But many other $AlPO_4$'s and SAPO's are fairly stable. What will determine the hydrolysis of Ga or Al in the framework? Is the hydrolysis uniform or does it occur at some specific sites?

J. Patarin: Up to now, we have no real explanation about the amorphization of cloverites after cooling the calcined product under moist air. Nevertheless, one might think that the Ga-O-P linkages are very sensible to the water, since a similar result is observed for LTA-type gallophosphate.

G. Ozin: Concerning the NH_3 desorption and collapse of cloverite we found the $NH_3(g)$ adsorption-desorption process to be reversible under the conditions used in our "in situ" experiments of cloverite, post-treated under vacuum thermal situations using self-supporting wafers. What were the conditions of your STD experiments that led you to your conclusions? Specifically, the post-treatment state of cloverite when the $NH_3(g)$ is introduced, as this appears to be a key point with respect to the stability of the material.

Answer: The calcination procedure used consists of a calcination of an as-made sample under dry N_2 at 450°C (heating rate 5°C.min^{-1}). After that, the solid is cooled to 100°C and reheated under dry O_2 at 450°C (heating rate 5°C.min^{-1}. For the NH_3 STD measurement, the solid is then cooled to 100°C and ammonia is adsorbed for 15 min. (1.5 lh^{-1}).

Question: Was pulsed NH_3 microcalorimetry used in studying the acidity?

Answer: Not yet.

Question: Over what time is cloverite stable in moist atmosphere?

Answer: After removal of organic species by calcination amorphization is observed in moist air after one hour.

"The Crystallization of Aluminophosphate Microporous Compounds In Alcoholic System" (R. Xu et al.)

J. L. Guth: Have you observed templating effects for some alcohols used as solvent (this means with incorporation of alcohol molecules in the cavities or channels of the crystals)?

W. Pang: For 3-D framework microporous $AlPO_4$, not yet. But we observed templating effect in layered $AlPO_4$ for example in AlPO-CA along the sheet plane, the interlamellar region consisted of $^{+}H_3N\text{-}CH\text{-}_2\text{-}CH_2NH_3^{+}$ and two $HO\text{-}CH_2\text{-}CH_2\text{-}OH$ (solvent) moieties.

J. L. Guth: Have you tried to work with complete absence of water in the reaction mixture?

W. Pang: We try to use the P_2O_5 instead of H_3PO_4 as P source, but the crystallization time increased too much, so maybe the H_2O can increase the crystallization rate.

J. Kornatowski: What is the influence of alcoholic solvents on the size of crystals and on the yield of reaction? Can you give some examples in figures? Can you observe some by-products in your batches?

W. Pang: We did not study the influence of alcoholic solvents on the size of crystals. Only the crystallization of $AlPO_4$ in alcoholic solution for a long time, so generally speaking, there are very small impurities contained in the products.

"Synthesis, Characterization, and Phase Transition of the 20-Membered Ring $AlPO_4$: JDF-20" (Q. Huo, et al.)

H. Kessler: Can the organic additive be removed from JDF-20 without transformation into $AlPO_4$-5? If yes, what are the adsorption properties of the emptied material?

R. Xu: No. We have just tried to remove the template with other route or technique.

R. Szostak: 1. You have prepared metal substituted $AlPO_4$'s from alcoholic systems; however, you do not report the preparation of SAPO's. Can SAPO's be made by your synthesis route? 2. When JDF-20 converts to $AlPO_4$-5, do you observe any pore blockage in the $AlPO_4$-5? This transformation appears very similar to that of VPI-5 converting to $AlPO_4$-8. The $AlPO_4$-8 obtained contains many stacking faults which limit the adsorption of molecules larger than 0_2.

R. Xu: 1. Yes, we have synthesized some kinds of SAPO (3D and layered structures) from this route. 2. We did not investigate this transition in more detail for the appearance of stacking faults; we will investigate further.

Structure

CATION DISTRIBUTION AS A STRUCTURAL-CHEMICAL PROBE

J.L. LIEVENS*, and W.J. MORTIER*,**

* Centrum voor Oppervlaktescheikunde en Katalyse, Materials Research Centre, K.U.Leuven, Kardinaal Mercierlaan 92, B-3001 Heverlee, BELGIUM

** Exxon Chemical Holland/BCT, P.O. Box 7335, NL-3000 HH Rotterdam, THE NETHERLANDS

ABSTRACT

The cations generate their own site energy by polarising the framework. This polarisation is easiest for the sites in the open parts of the framework (channels, cages) and for Si-rich frameworks (as opposed to materials with a high degree of Al and Ga isomorphous framework substitution). This has been experimentally demonstrated by determining the site-energy differences for FAU, EMT and KFI-type structures with varying cation loading and Si/Al (Si/Ga) ratio. This is consistent with the direct computation of the global "hardness" and local "softness" of zeolite-type frameworks. With the support of an "exact" statistical thermodynamical model for the cation distribution in zeolite structures with hexagonal prisms, the experimentally determined cation distribution can be used as a unique structural chemical probe for the direct measurement of the site energies, and the parameters influencing these, such as the Si/Al ratio, Al-zoning at the level of the unit cell, site geometry, substitution of Ga, ... etc.

INTRODUCTION

It has been demonstrated recently that a systematic study of the cation distribution can be used as a unique physicochemical probe for experimentally testing the polarisability ('reactivity') of zeolite frameworks. To enable this, two types of investigations have to be successfully completed : (i) the formulation of a theoretical model relating the site energies to the structure type (which can be done via statistical thermodynamics if the partition function for the cations can be evaluated) and (ii) a parametric study of the cation-site occupancies (preferably by direct structure determination). An analysis of a series of such combined studies has enabled the formulation of some 'rules of thumb' regarding the cation distribution (the models of which are predictive), and the framework polarisability, consistent with theoretical predictions by the semi-empirical density functional based Electronegativity Equalisation Method (EEM). The consistency of these experimental and theoretical findings is the subject of this paper.

SITE ENERGY DIFFERENCES RELATED TO THE STRUCTURE TYPE

A statistical thermodynamical model for the cation distribution in dehydrated faujasite-type zeolites was developed by Van Dun et al. [1]. The model involves the site energies ($-e_i$) and the repulsive interaction energy (-w) between the sites I (centre of the hexagonal prism) and the site I' (in the sodalite cage opposite the six-ring window of the hexagonal prism). For all structures with isolated hexagonal prisms, the evaluation of the partition function of interacting cations is reduced to a one-

From the Proceedings of the 9th International Zeolite Conference, Montreal 1992, Eds. R. von Ballmoos et al.,

dimensional problem, which can be solved exactly. The relations between the site population (P_i) and the site energies are then directly derived from the partition function by classical thermodynamical methods [1]. The relations for site I, I' and II are :

$$\left(\frac{1-P_I}{P_I}\right)\left(\frac{n^*}{2P_{I'}-n^*}\right)^2 = j_I \exp-\alpha \exp\frac{(e_I+2w)}{kT}$$

$$\left(\frac{n^*}{2P_I-n^*}\right) = j_{I'} \exp-\alpha \exp\frac{(e_{I'}+w)}{kT}$$

$$\left(\frac{P_{II}}{1-P_{II}}\right) = j_{II} \exp-\alpha \exp\frac{e_{II}}{kT}$$

with

$$n^* = P_I + P_{I'} + \frac{x}{1-x} \pm \left(\left(P_I+P_{I'}+\frac{x}{1-x}\right)^2 - 2\frac{N_I N_{I'}}{1-x}\right)^{\frac{1}{2}}$$

and $x = \exp(-\frac{w}{kT})$

where j_i is the partition function for the three vibrational degrees of freedom of a cation at site i (these are taken as identical for all sites; when calculating population ratios j_i and $\exp(-\alpha)$ disappear). A fourth equation relates the total number of cations per unit cell to the site populations :

$\overline{N} = \sum m_I P_I$ with m_i the multiplicity of site i.

The cation populations as these can be derived from the crystal structures, can be used directly for the estimation of the energy level differences and the interaction energy between the cation exchange sites. Conversely from these it is possible to calculate the site population as a function of the temperature and cation loading. This "exact" model has been extended to other framework structures with similar cation sites such as EMT and KFI.

EXPERIMENTAL AND RESULTS

The cation distribution in a series dehydrated Na_xH-FAU (x = 13, 23, 42 and 54) was studied. To monitor the influence of the framework composition several high silica FAU-type zeolites (both as synthesized and dealuminated) and a series of Na_xH-GaFAU (x = 21,36 and 54) were studied. The influence of a changing framework topology was investigated by studying the cation distribution in NaEMT and NaKFI. These framework topologies and the cation sites are given in figure 1 (FAU), 2 (EMT), 3 (KFI). All structures were dehydrated and studied under a dynamic high vacuum at high temperature (>623 K). Both conventional peak decomposition [11] and Rietveld [12] refinement methods were used to process the data. An overview of sample details is given in Table 1.

The observed distribution of Na cations in dehydrated zeolite-type frameworks of type FAU, EMT and KFI are given in Table 2. The energy level differences are estimated and used to calculate the cation distribution. The observed and calculated cation distribution are in excellent agreement (see

Table 2).

The site energy differences (relative to site I) and the repulsion energy are summarized in table 3. For dehydrated Na-FAU a linear relation between the site energy differences and the aluminum content was found.

Table 1 : Studied structures (sample name, chemical composition, dehydration temperature, structure type, refinement method and reference).

		Si/Al	Na/Al	Temperature	Structure	Method	Ref.
Na_xH-Y	x = 13	2.46	0.23	623 K	FAU	POWLS [11]	[2]
	x = 23		0.41	623 K			
	x = 42		0.75	823 K			
	x = 54		0.98	723 K			[10]
Na-LZ210		4.8	0.96	623 K	FAU	POWLS	[4]
Na-ECR30		4.9	0.76	623 K			
NaH-Y6		5.8	0.50	623 K			
NaH-Y10		9.9	0.40	623 K			
NaH-Y21		20	0.49	623 K			
Na_xH-GaY	x = 21	2.29	0.36	623 K	FAU	DBW3.2 [12]	[5]
	x = 36		0.62	623 K			
	x = 54		0.93	623 K			
Na-EMT		3.31	0.95	723 K	EMT	DBW3.2	[8]
NaH-EMT			0.43	723 K			
Na-KFI		3.29	0.94	723 K	KFI	DBW3.2	[9]
			0.16	723 K			
NaH-KFI							

Table 2 : Experimental and calculated site populations.

			site I	site I'	site II
Na_xH-Y	x = 13	exp.*	0.36(5)	-	0.23(3)
		QC**	0.35	0.04	0.19
	x = 23	exp.	0.57(6)	0.03(4)	0.42(4)
		QC	0.55	0.06	0.38
	x = 42	exp.	0.82(5)	0.18(5)	0.67(5)
		QC	0.61	0.24	0.77
	x = 54	exp.	0.48(6)	0.42(5)	0.92(5)
		QC	0.53	0.48	0.93
Na-LZ210		exp.	0.09(9)	0.38(9)	0.55(6)
		QC	0.061	0.320	0.624
Na-ECR30		exp.	0.08(6)	0.22(6)	0.47(6)
		QC	0.012	0.22	0.489
NaH-Y6		exp.	0.02(8)	0.12(8)	0.23(7)
		QC	0.012	0.096	0.257
NaH-Y10		exp.	0.01(5)	0.09(5)	0.17(6)
		QC	0.002	0.070	0.195
NaH-Y21		exp.	-	0.04(7)	0.06(6)
		QC	0.000	0.025	0.075

			site I	site I'	site II
Na_xH-GaY	x = 21	exp.	0.76(1)	0.13(1)	0.29(1)
		QC	0.78	0.06	0.29
	x = 36	exp.	0.80(1)	0.11(1)	0.65(1)
		QC	0.81	0.12	0.62
	x = 54	exp.	0.54(1)	0.67(1)	0.85(1)
		QC	0.48	0.60	0.96
Na-EMT	type a	exp.	0.24(1)	0.37(1)	0.50(1)
		QC	0.29	0.38	0.45
	type b	exp.	-	1.15(5)	0.72(3)
		QC	0.0	0.99	0.73
NaH-EMT	type a	exp.	0.427(6)	0.296(6)	0.198(8)
		QC	0.25	0.21	0.21
	type b	exp.	-	0.84(1)	0.41(2)
		QC.	0.0	0.97	0.47
Na-KFI		exp.	0.049(8)	0.574(5)	0.972(6)
		QC	0.09	0.54	0.98
NaH-KFI		exp.	0.027(7)	0.043(7)	0.216(6)
		QC	0.005	0.013	0.27

* standard deviation of the least significant digit between brackets
** calculated population factors from the Quasi-Chemical Method (QC)

Table 3 : Site energy level differences (e_i-e_I)* and repulsion energy (-w).

	e_I-e_I (kJ/mol)	$e_{I'}-e_I$ (kJ/mol)	$e_{II}-e_I$ (kJ/mol)	-w (kJ/mol)
Na-FAU	0	-38.23+0.93 Al/UC	-45.23+0.93 Al/UC	50
Si/Al=2.46	0	12.0	5.0	
Si/Al=3.29	0	3.3	-3.7	
Na-GaY	0	15.0	15.0	50
Na-EMT** type a	4	7	9	50
type b	0	-20	2	
Na-KFI	0	-5	-25	50

*a negative number means increased stability,** energy level differences relative to site I_b

DISCUSSION

The experimental cation population for a series of structures, with constant framework composition (i.e. Si/Al ratio) but varying cation content (i.e. Na/Al ratio) can be rationalized (at all temperatures [2]) using only three parameters : the site energy level differences ($e_{I'}-e_I$ and $e_{II}-e_I$) and the cation-cation interaction energy. At high cation loadings the repulsion term can dominate; an evaluation of the site energy level differences is best done at lower cation loadings. Only short range interactions are important. The cation generates its own site energy by inducing charge shifts; these shifts are limited to a sphere with a diameter of about 7 Å [14]. This effect is demonstrated in Figure 4 where the variation of the electrostatic potential (for a SiO_2 composition) along the symmetry axis through the different cation sites is presented. The v_1 curve represents the electrostatic potential without any cation in the structure, while v_2 takes the charge shifts induced by the cation at this site into account. In the perturbed system the difference in electrostatic potential between the sites becomes clearly more pronounced. More details about the calculation of charge shifts and potentials are found in ref. [14].

A variation with the degree of isomorphous substitution certifies that there is a linear relation with the framework aluminum content. This linear variation of the site energy level is valid over the whole composition range, irrespective of the sample preparation (as synthesized or dealuminated) or extra-framework material. This again indicates that the cation-framework interactions are short-range in nature. It is clear that the number of aluminum atoms in the neighbourhood of the cation site influences the site energy. The site preference I>II>I' at high aluminum content reverses to II>I'>I at low aluminum contents (below 48 Al/UC, which is contradictory to conventional wisdom that the octahedrally coordinated site I is *always* the most preferred site). The difference between sites I' and II (both six-ring sites) is due to the different topology of these sites : site I' is located in the sodalite cage, i.e. in a more dense part of the framework compared to the supercage site II. The oxygen atoms of the supercage are indeed softer [14]; i.e. easier to polarise.

From these results we can probably generally state that in dehydrated zeolites, univalent cations will increasingly favour the large cage/channel sites with increasing Si/Al ratio.

The isomorphous substitution of aluminum by gallium reduces the difference between the site energy levels of site I' and II. This indicates a reduced polarisability of the gallosilicate compared to the aluminosilicate. The higher site energy levels for the GaY structure compared to AlY caused by the lower electronegativity of gallium [5,13].

Zeolite Na-KFI, with a Si/Al ratio of 3.3 has similar site preference as Na-FAU (with the same Si/Al ratio) : i.e. II>I'>I. The most stable sites are again located in the least dense parts of the structure [8] : i.e. soft and therefore easy to polarise. The electrostatic potential of the perturbed system (Figure 5, v_2) is the most pronounced at these sites.

The cation distribution in Na-EMT is more complex. The differences between sites of type a and type b are clearly not due to the different framework topology : the electrostatic potential calculations show the same behaviour for both type of sites (Figure 6). Compared to the site energy level differences in FAU-type materials, type a sites have characteristics of a high aluminum FAU, while type b sites have energies for a low aluminum content. Therefore the complex cation distribution pattern in Na-EMT is most probably dominated by a zoning of aluminum distribution, i.e. a higher concentration of aluminum in the twin planes [9].

CONCLUSION

An over-all assessment of the data enables to conclude that (i) the cation-framework interactions are indeed short-range in nature, (ii) the cations generate their own site-energy; the framework charges as well as the charge polarisability play a dominant role; this polarisability is most pronounced in the more open parts of the framework [14], and (iii) the cation location can be used as a structural chemical probe for testing the polarisability of the framework, and for the detection of Al-zoning.

REFERENCES

1. Van Dun, J.J., Mortier, W.J., J. Phys. Chem., **92**, 6740 (1988).

2. Van Dun, J.J., Dhaeze, K., Mortier, W.J., J. Phys. Chem., **92**, 6747 (1988).

3. Van Dun, J.J., Dhaeze, K., Mortier, W.J., Vaughan, D.E.W, J. Phys. Chem. Solids, **50**, 469 (1989).

4. Lievens, J.L., Mortier, W.J., submitted for publication.

5. Lievens, J.L., Mortier, W.J., Verduijn, J.P., accepted for publication in J. Phys. Chem.

6. Meier, W.M., Kokotailo, G.T., Z. Kristallogr., **135**, 374 (1972).

7. Delprato, F., Delmotte, L., Guth, J.L., Huve, L., Zeolites, **10**, 546 (1990).

8. Lievens, J.L., Verduijn, J.P., Bons, A.-J., Mortier, W.J., accepted for publication in Zeolites.

9. Lievens, J.L., Mortier, W.J., Verduijn, J.P., accepted for publication in Zeolites.

10. Mortier, W.J., Van den Bossche, E., Uytterhoeven, J.B., Zeolites, **4**, 41 (1984).

11. Mortier, W.J., N.B.S. Special Publication (U.S.A.), No. 567, 315 (1980).

12. Young, R.A., Wiles D.B., J. Appl. Cryst., **15**, 430 (1982).

13. Dompas, D.H., Mortier, W.J., Kenter, O.H.C., Janssen, M.J.G., Verduijn, J.P., J. Cat., **129** ,19 (1991).

14. Baekelandt, B.G., Mortier, W.J., in press, *'Modelling of Structures and Reactivity in Zeolites'*, R. Catlow and R. Vetrivel, Eds.

Figure 1 : Idealized stereoplot of the FAU structure type, together with the cation sites (I, I' and II).

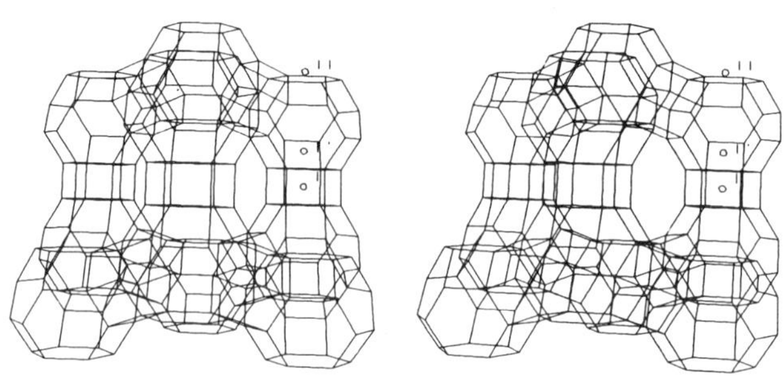

Figure 2 : Idealized stereoplot of the EMT structure type, together with the cation sites (I_a, I_a', II_a, I_b, I_b' and II_b).

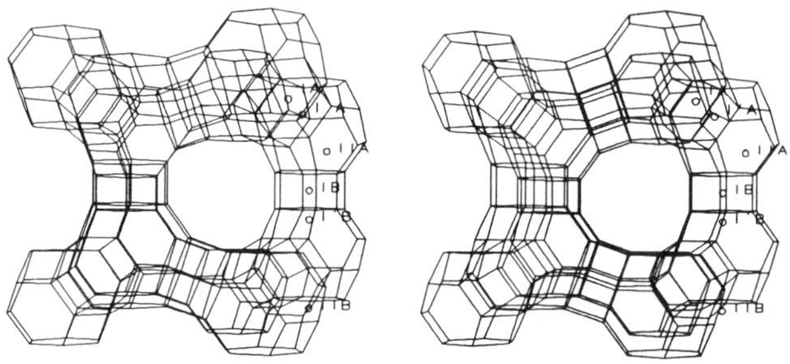

Figure 3 : Idealized stereoplot of the KFI structure type, together with the cation sites (I, I' and II).

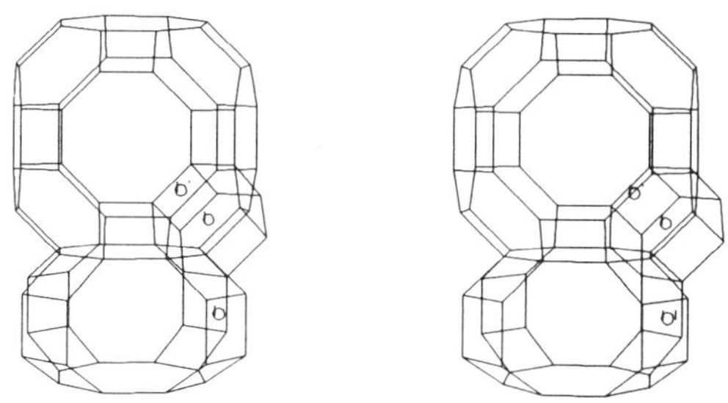

Figure 4 : Electrostatic potential as a function of the distance to the centre of the hexagonal prism, along a path through the sites I, I' and II of FAU (hypothetical SiO_2 composition). (o) v_1 = potential without any cation, (▵) v_2 = potential with cation present.

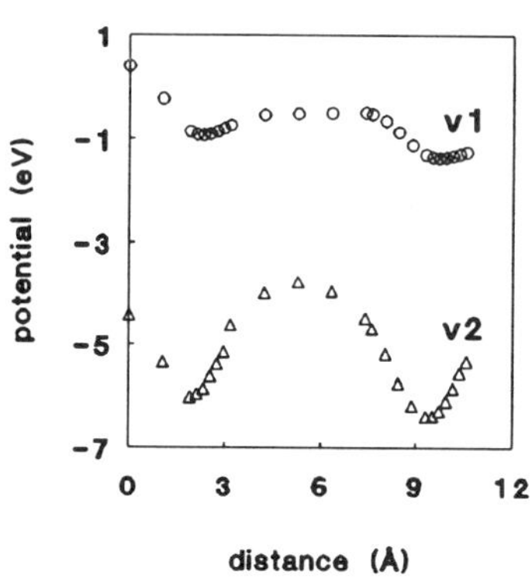

Figure 5a : Electrostatic potential as a function of the distance to the centre of the hexagonal prism, along a path through the sites I and I' of KFI. 5b : Electrostatic potential as a function of the distance to site II of KFI (hypothetical SiO_2 composition). (o) v_1 = potential without any cation, (Δ) v_2 = potential with cation present.

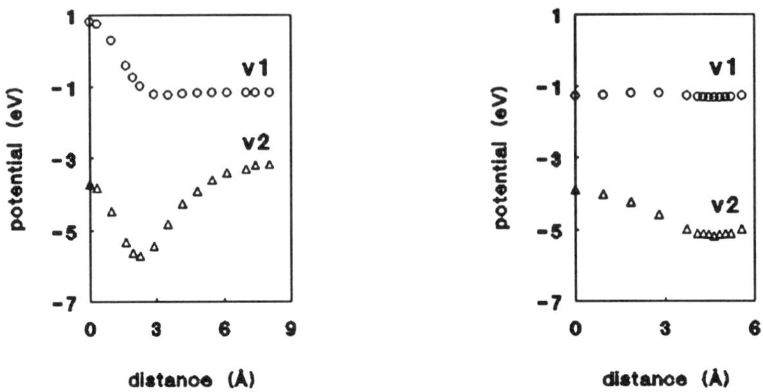

Figure 6 : Electrostatic potential as a function of the distance to the centre of the hexagonal prism, along a path through the sites I_a, I_a' and II_a (full symbols) and along a path through the sites I_b, I_b' and II_b (open symbols) of EMT (hypothetical SiO_2 composition). (o) v_1 = potential without any cation, (Δ) v_2 = potential with cation present.

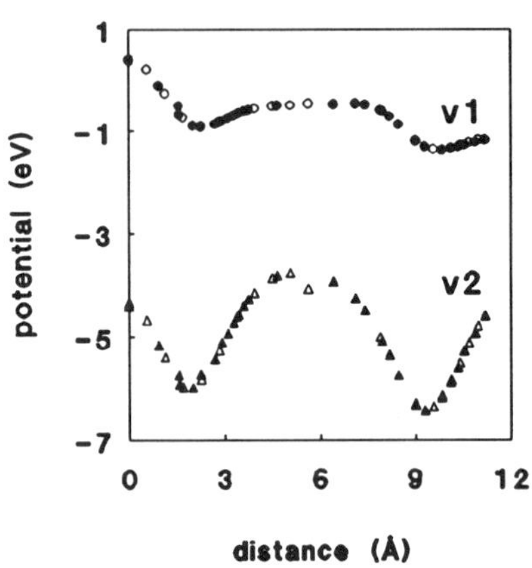

A COMBINATORIAL METHOD FOR GENERATING NEW ZEOLITE FRAMEWORKS

M. M. J. TREACY, S. RAO and I. RIVIN

NEC Research Institute, Inc., 4 Independence Way, Princeton, NJ 08540, *U. S. A.*

ABSTRACT

We show that by applying crystallographic and topological constraints to 2 unique tetrahedrally coordinated atoms (T-atoms, such as Si or Al) within the space group *P6/mmm*, 46 fully-connected frameworks can be generated. Of these, 29 can be refined to give acceptable interatomic distances and bond angles. This combinatorial approach to enumerating frameworks can, in principle, be generalized to other space groups with any number of basis atoms, and provides a systematic method for identifying and classifying new zeolite frameworks.

INTRODUCTION

Direct experimental determination of zeolite structures is a difficult process. This is partly because of the structural complexity of typical zeolite frameworks, but is also because most zeolites form sub-micron crystallites that are too small for structure determination by single-crystal diffraction methods.

Another approach is to systematically derive all of the theoretical frameworks whose computed physical properties, such as unit cell dimensions, framework density and diffraction patterns match those of the unknown zeolite. This approach allows structural identification based on powder diffraction data and porosimetry measurements. However, for this to work reliably, it is important that all appropriate theoretical frameworks be identified.

There are several approaches to framework enumeration, ranging from the intuitive to the fully computerized. For example, if a known zeolite framework can be partitioned into two-dimensional sheets, new frameworks can be generated by iterating over all the sheet conformations [1]. Similarly, known polyhedral building units can be connected in new ways by systematically applying symmetry operators [1][2]. Alternatively, simulated annealing programs, given the number of unique T-atoms and the space group, can search for low-energy configurations [3][4]. In simple systems involving one uniqueT-atom, steric restrictions (annealing) can be computed by hand [7].

In this paper we report on a combinatorial method that, given the unique T-atoms, the space group, generates all the sterically feasible 4-connected graphs. The list can then be pruned further by using computer programs such as DLS76 to identify energetically unfavorable configurations.

We demonstrate the method for the case of 2 unique T-atoms in space group *P6/mmm*, of which zeolite L (the LTL framework) is an instance.

From the Proceedings of the 9th International Zeolite Conference, Montreal 1992, Eds. R. von Ballmoos et al.,

METHODOLOGY

Figure 1 shows an oblique view of the hexagonal unit cell of space group *P6/mmm.* All points in the unit cell can be related to points within the wedge-shaped fundamental region by applying combinations of the symmetry operations belonging to space group *P6/mmm.* Thus the contents of the fundamental region describe the contents of the unit cell uniquely. One way to envisage the action of the symmetry operators in this space group is to note that the five faces of the fundamental region are bounded by mirror planes. Acting much in the same way as a kaleidoscope, these mirrors will fill space with images of the contents of the fundamental region. These mirror images, however, must interconnect in a self-consistent manner. All atomic bonds, including those to mirrored atoms, must obey the rules of chemistry. Thus, there are constraints on the disposition of atoms in the fundamental region. The 6 unique oxygen atoms and 2 unique T-atoms that generate the LTL framework are shown within the fundamental region in Figure 1. The constraints for generating the LTL framework are such that one of the T-atoms, and five of the oxygen atoms lie on mirror planes. To establish how many other distributions of 2 unique T-atoms are possible within space group *P6/mmm*, we first need to establish the rules that constrain atom positions.

Topological constraints

TO_4 tetrahedra are strongly constrained by the symmetry operations that act on the fundamental region. For space group *P6/mmm* this list of constraints is;

(1) No T-atom can lie on a 6-fold axis.

(2) No T-atom, or T-atom vertex, can lie on a vertex of the fundamental region.

(3) If a T-atom lies on a face of the fundamental region, then two (and only two) of the T-atom vertices lie on the same face.

(4) Connections to atoms outside the fundamental region must involve either a T-atom, or one of its vertices, that lies on a mirror or on an edge defined by perpendicular mirrors.

(5) All T-atoms are connected to four other T-atoms.

(6) Tetrahedra are denied edge- and face-sharing connectivities.

(7) Each of the five faces of the fundamental region must have at least one bond connecting through it.

Constraints (1) and (2) arise because these positions are incompatible with tetrahedral symmetry. Constraint (3) occurs because we exclude from consideration those tetrahedra that have coplanar apices. Constraint (4) ensures that an atom cannot lie arbitrarily close to a mirror plane. Otherwise, it will lie too close to its mirror image. Constraints (5) and (6) are specific to the generation of corner-sharing tetrahedral (zeolitic) frameworks. Constraint (7) is required if 3-dimensionally connected frameworks only are to be considered, and could be relaxed if ball, sheet, tubular and interlocked structures are desired.

Nomenclature

Figure 2 shows the fundamental region of space group *P6/mmm*, and presents the nomenclature used in this paper for the faces. The notation of Figure 2 differs from the Wyckoff notation in the International Tables for Crystallography [5] because of ambiguities associated with sites on faces *c* and *d*, both faces being assigned Wyckoff letter *o*. Thus, in this work, the symbol T_a represents a T-atom located on face *a*; O_g an oxygen atom in the general position (within the volume of the fundamental region); and O_{ac} an oxygen atom on the edge defined by the intersection of faces *a* and *c*. Capital letters represent mirror images of atoms. Thus O_{eC} represents the image of the oxygen atom O_e, as reflected in face *c*. Using this notation, the tetrahedral connectivity of the LTL framework, which has one T-atom on face *a* and a second T-atom in the general position *g*, could be described by the sequence $T_aO_gO_{gA}O_{ac}O_{ae}\,T_gO_gO_bO_dO_c$.

Combinatorial analysis

Given the topological constraints listed above, we can now enumerate the topologies by listing the permitted connections. As an example, we analyze the set of configurations that has one T-atom on the bottom face, T_a, and the other in the general position T_g. The LTL framework is a member of this set. According to restriction (3), two of the oxygen atoms connected to T_a must also lie on face *a*. Restriction (4) requires that these two oxygens can occur only on the edges *ac*, *ad* or *ae*. To connect to atom T_g, one of the oxygens connected to T_a must be of type O_g, which is also connected to T_g. The fourth oxygen connected to T_a will be, by symmetry, the image of O_g as reflected in face *a*, i.e. O_{gA}. Thus, only the two oxygens lying on face *a* have any freedom of choice in how they connect. There are only three possible sites for each of these oxygen atoms. Furthermore, restriction (6) means that these oxygens cannot both lie on the same site. Thus the total number of configurations for the T_a atom is therefore given by the quantity $\binom{3}{2}= 3$. The T_a atom can be connected to a T_g atom, with oxygens located at *g*, *ac* and *ae*, or *g*, *ac* and *ad*, or *g*, *ad* and *ae*. These configurations are indicated in Figure 3.

The T_g atom must be connected to O_g. This leaves three other oxygens to be accounted for. If T_a is connected to *ac* and *ae*, constraint (7) dictates that two of these oxygens must lie on faces *b* and *d*, since these faces are not contacted by T_a. The fourth oxygen cannot connect to faces *b* and *d* as well (restriction (6)), but it is free to connect to any of the faces *a*, *c* and *e*. The number of configurations of the T_g atom is therefore $\binom{3}{1} = 3$. This is true for the other two configurations for T_a. This gives the combined number of configurations for the T_aT_g pair as $\binom{3}{2}\binom{3}{1} = 3\times3 = 9$. These nine configurations are found to refine well using DLS76, and are described in Table I. Figures 5.1 - 5.9 show the *c*-axis views of these nine structures. Three of these frameworks contain 3-rings. Figure 5.1 is the LTL framework.

There are a number of other starting possibilities for the T atoms such as T_bT_g, T_cT_g etc. It should be noted that the triangular faces *a* and *b*, being parallel, are equivalent for most configurations. The nine configurations found for T_bT_g are mirror images of those found for T_aT_g, and are not included in this analysis.

In total, 46 3-dimensionally connected frameworks are found. These are listed in Table I. If sheet, tube, ball and interconnected frameworks are allowed, the possibilities increase to 416. The distance-least-squares refinement program DLS76 [6] was applied to all of these 46 topologies. 29 were found to be refinable to *R*-values less than 0.01 and with acceptable interatomic distances. The connectivities of all 46 frameworks are listed in Table I. The *R*-values, hexagonal unit cell constants *a* and *c* in Å, and framework density in $g.cm^{-3}$ are presented for the 29 refinable structures. For the DLS76 calculation the composition was taken to be pure SiO_2 with preferred Si-O bond length of 1.605Å and preferred T-O-T angle of 144°. The weights for T-O, O-O and T-T bonds were 2, 1, 0.1 respectively. Sketches of the rejected frameworks are presented in Figure 4.

DISCUSSION

It is clear from Table I, that a significant number (17) of the 46 topologically realizable frameworks are rejected on geometric grounds. The TO_4 tetrahedra may be unacceptably distorted from the ideal conformation, or atoms are forced too close together. Inspection of the 17 structures rejected by DLS76 analysis shows that they all have a feature in common - namely that one of the T-atoms has a connection to both the *a* and *b* faces. This has the effect of constraining the *a* and *b* faces to be close together (no more than about 5.3Å), which limits the volume available to the second T-atom, inhibiting its ability to make satisfactory bonds. This volume consideration alone, if it had been implemented as a geometric constraint in conjunction with the 7 topological constraints, would have enabled correct identification of all 29 viable frameworks without having to rely on time-consuming algorithms such as DLS76.

As the number of basis T-atoms increases, the number of topologically realizable configurations increases rapidly. However, the geometric constraints may also become more severe and serve to eliminate many of the possibilities. Since the DLS76 program was the rate-limiting step in our analysis, it is desirable to improve the present algorithm so as to eliminate geometrically infeasible structures in the early stages of the analysis. We are currently addressing this issue.

REFERENCES

[1] Smith, J. V., in: Zeolites: Facts, Figures, Future. Eds. P. A. Jacobs, and R. A. van Santen, Elsevier Science Publishers, Amsterdam, pp. 29-47 (1989).
[2] Akporiaye, D. E. and Price, G. D., Zeolites, 9, pp23-32 (1989)
[3] Deem, M. W. and Newsam, J. M., Nature, 342, pp. 260-262, (1989).
[4] Deem, M. W. and Newsam, J. M., submitted to J. Appl. Chem Sci.
[5] International Tables for Crystallography, Vol. A, Ed. T. Hahn, Kluwer Acad. Press p585 (1989)
[6] Baerlocher, Ch., Hepp, A. and Meier, W. M. DLS-76 – A program for simulation of crystal structures by geometric refinement. ETH, Zurich Report, (1977).
[7] Bennett, J. M., Perspectives in Molecular Sieve Science, Eds. W. H. Flank & T. E. Whyte, ACS Symp. Ser 368, pp162–176 (1988).
[8] Chung, S.J. and Hahn, Th., Acta. Crystallogr. Suppl., A31, S1 (1975).

TABLE I

N	ok?	Coordination		Rings	*a* (Å)	*c* (Å)	R_{DLS}	ρ(g/cm^3)
1	√	$T_aO_gO_{gA}O_{ac}O_{ae}$	$T_gO_gO_bO_dO_c$	4,6,8,12	18.056	7.544	0.0008	1.696
2	√	$T_aO_gO_{gA}O_{ac}O_{ae}$	$T_gO_gO_bO_dO_e$	d4r,d6r,8,12	18.078	7.720	0.0053	1.653
3	?	$T_aO_gO_{gA}O_{ac}O_{ae}$	$T_gO_gO_bO_dO_a$	3,4,6,8,12	18.650	5.259	0.0100	2.280
4	√	$T_aO_gO_{gA}O_{ac}O_{ad}$	$T_gO_gO_bO_eO_c$	4,6,8,12	18.046	7.624	0.0014	1.680
5	√	$T_aO_gO_{gA}O_{ac}O_{ad}$	$T_gO_gO_bO_eO_d$	4,d6r,8,24	20.092	7.755	0.0022	1.332
6	√	$T_aO_gO_{gA}O_{ac}O_{ad}$	$T_gO_gO_bO_eO_a$	3,6,8,24	21.504	5.307	0.0061	1.700
7	√	$T_aO_gO_{gA}O_{ad}O_{ae}$	$T_gO_gO_bO_cO_d$	4,6,d8r,24	21.401	7.555	0.0018	1.206
8	√	$T_aO_gO_{gA}O_{ad}O_{ae}$	$T_gO_gO_bO_cO_e$	4,6,d8r,12	17.832	7.584	0.0021	1.730
9	√	$T_aO_gO_{gA}O_{ad}O_{ae}$	$T_gO_gO_bO_cO_a$	3,4,6,8,12,24	22.017	5.258	0.0056	1.637
10	×	$T_cO_gO_{gC}O_{ac}O_{bc}$	$T_gO_gO_eO_dO_a$	d4r,8,18	—	—	—	—
11	×	$T_cO_gO_{gC}O_{ac}O_{bc}$	$T_gO_gO_eO_dO_c$	3,4,6,8,18	—	—	—	—
12	√	$T_dO_gO_{gD}O_{bd}O_{de}$	$T_gO_gO_cO_aO_d$	3,4,6,8,24	20.799	9.267	0.0031	1.041
13	√	$T_dO_gO_{gD}O_{bd}O_{de}$	$T_gO_gO_cO_aO_e$	4,d6r,8,9,12	15.932	9.463	0.0020	1.737
14	×	$T_dO_gO_{gD}O_{bd}O_{de}$	$T_gO_gO_cO_aO_b$	4,9,24	—	—	—	—
15	×	$T_dO_gO_{gD}O_{bd}O_{ad}$	$T_gO_gO_cO_eO_d$	3,6,8,12	—	—	—	—
16	×	$T_dO_gO_{gD}O_{bd}O_{ad}$	$T_gO_gO_cO_eO_a$	6,9,12	—	—	—	—
17	×	$T_eO_gO_{gE}O_{be}O_{de}$	$T_gO_gO_cO_aO_b$	4,12,18	—	—	—	—
18	√	$T_eO_gO_{gE}O_{be}O_{de}$	$T_gO_gO_cO_aO_d$	4,d6r,8,18	18.147	9.391	0.0017	1.349
19	√	$T_eO_gO_{gE}O_{be}O_{de}$	$T_gO_gO_cO_aO_e$	3,4,8,12	17.924	9.227	0.0028	1.407
20	×	$T_eO_gO_{gE}O_{be}O_{ae}$	$T_gO_gO_cO_dO_b$	6,4,18	—	—	—	—
21	×	$T_eO_gO_{gE}O_{be}O_{ae}$	$T_gO_gO_cO_dO_e$	3,4,6,8,18	—	—	—	—
22	√	$T_{g1}O_gO_bO_{c1}O_d$	$T_{g2}O_gO_eO_{c2}O_a$	4,6,8,12	18.546	9.948	0.0009	1.626
23	√	$T_{g1}O_gO_bO_cO_{d1}$	$T_{g2}O_gO_eO_{d2}O_a$	4,6,d8r,24	22.014	9.992	0.0020	1.149
24	×	$T_{g1}O_gO_bO_{c1}O_a$	$T_{g2}O_gO_eO_{c2}O_d$	4,6,8,24	—	—	—	—
25	×	$T_{g1}O_gO_bO_cO_{a1}$	$T_{g2}O_gO_eO_dO_{a2}$	d4r,12,24	—	—	—	—
26	?	$T_{g1}O_gO_bO_{e1}O_c$	$T_{g2}O_gO_{e2}O_dO_a$	4,8,12	14.805	10.422	0.0084	2.435
27	×	$T_{g1}O_gO_{b1}O_eO_c$	$T_{g2}O_gO_{b2}O_dO_a$	d4r,8,12	—	—	—	—
28	×	$T_{g1}O_gO_bO_{e1}O_a$	$T_{g2}O_gO_{e2}O_cO_d$	4,6,8,24	—	—	—	—
29	×	$T_{g1}O_gO_{b1}O_eO_a$	$T_{g2}O_gO_{b2}O_cO_d$	4,6,8,24	—	—	—	—
30	×	$T_{g1}O_gO_bO_{d1}O_a$	$T_{g2}O_gO_eO_cO_{d2}$	4,8,12	—	—	—	—
31	√	$T_{c1}O_cO_{ac}O_eO_{ec}$	$T_{c2}O_cO_{bc}O_dO_{dc}$	3,4,8	12.699	10.003	0.0020	1.724
32	√	$T_aO_gO_{ad}O_{ac}O_{gA}$	$T_bO_gO_{be}O_{bd}O_{gB}$	4,d6r,24	19.409	5.242	0.0023	1.408
33	√	$T_aO_gO_{ad}O_{ac}O_{gA}$	$T_bO_gO_{be}O_{bc}O_{gB}$	8,4,6,12	17.503	5.242	0.0016	1.732
34	√	$T_aO_gO_{ad}O_{ae}O_{gA}$	$T_bO_gO_{be}O_{bc}O_{gB}$	4,d6r,12	17.230	5.242	0.0039	1.787
35	√	$T_aO_gO_{ad}O_{ae}O_{gA}$	$T_cO_gO_{ac}O_{bc}O_{gC}$	3,4,8,9,18	17.248	5.259	0.0060	1.778
36	×	$T_aO_gO_{ac}O_{ae}O_{gA}$	$T_dO_gO_{bd}O_{ad}O_{gD}$	3,6,8,9,12	—	—	—	—
37	√	$T_aO_gO_{ac}O_{ad}O_{gA}$	$T_dO_gO_{bd}O_{de}O_{gD}$	3,4,6,8,24	19.252	7.151	0.0046	1.049
38	√	$T_aO_gO_{ac}O_{ae}O_{gA}$	$T_dO_gO_{bd}O_{de}O_{gD}$	d4r,8,9,12	15.402	7.080	0.0036	1.656
39	√	$T_aO_gO_{ac}O_{ad}O_{gA}$	$T_eO_gO_{be}O_{de}O_{gE}$	d4r,6,8,18	16.719	7.161	0.0025	1.389
40	√	$T_aO_gO_{ac}O_{ae}O_{gA}$	$T_eO_gO_{be}O_{de}O_{gE}$	3,6,8,12	17.154	7.154	0.0047	1.321
41	√	$T_aO_gO_{ac}O_{ad}O_{gA}$	$T_eO_gO_{be}O_{ae}O_{gE}$	3,6,18	17.008	5.259	0.0060	1.828
42	×	$T_cO_gO_{ac}O_{bc}O_{gC}$	$T_dO_gO_{de}O_{bd}O_{gD}$	4,6,18	—	—	—	—
43	×	$T_cO_gO_{ac}O_{bc}O_{gC}$	$T_eO_gO_{de}O_{be}O_{gE}$	4,6,9,12	—	—	—	—
44	√	$T_{ac}O_cO_{ae}O_{ca}O_{aeC}$	$T_cO_cO_{bc}O_dO_{dC}$	d3r,4,6,8	12.136	7.690	0.0027	1.842
45	√	$T_{ac}O_cO_{ad}O_{ca}O_{adC}$	$T_cO_cO_{bc}O_eO_{eC}$	3,4,6,8	12.289	7.458	0.0020	1.852
46	√	$T_{ac}O_cO_{ad}O_{adc}O_{cA}$	$T_{bc}O_cO_{be}O_{bec}O_{cB}$	3,6,8	11.661	5.242	0.0027	1.951

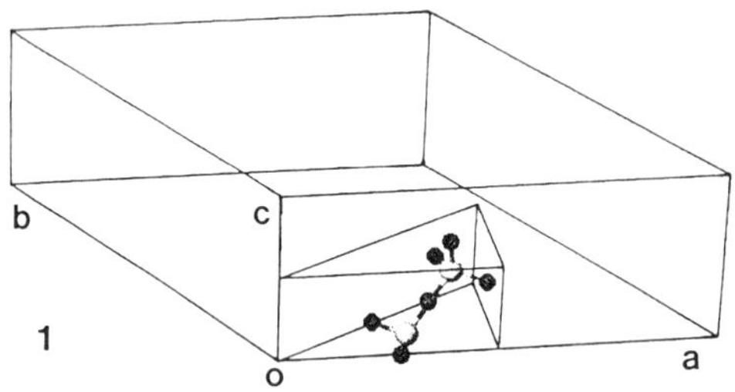

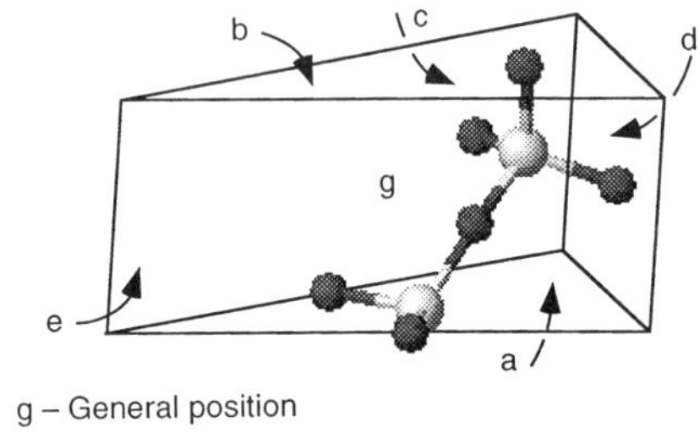

Figure 1. Hexagonal unit cell showing the location of the fundamental region. The basis atoms for zeolite L (LTL) are shown.

Figure 2. The fundamental region for space group *P*6 / *mmm*. The five faces are labelled *a* to *e*.

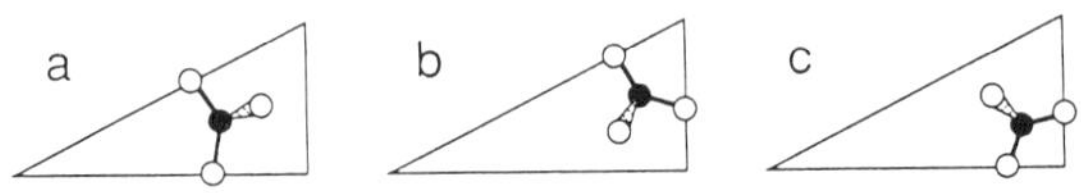

Figure 3. The 3 configurations available to a T-atom, T_a, located on face *a* of the fundamental region and connected to an oxygen atom in the general position, O_g.

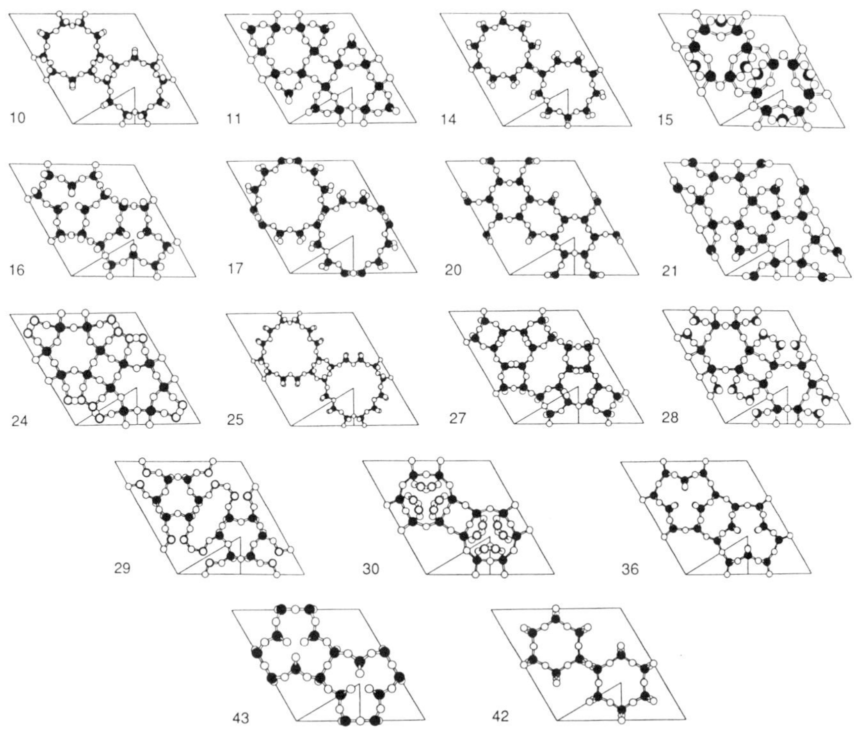

Figure 4. *c*-axis views of the 17 rejected frameworks.

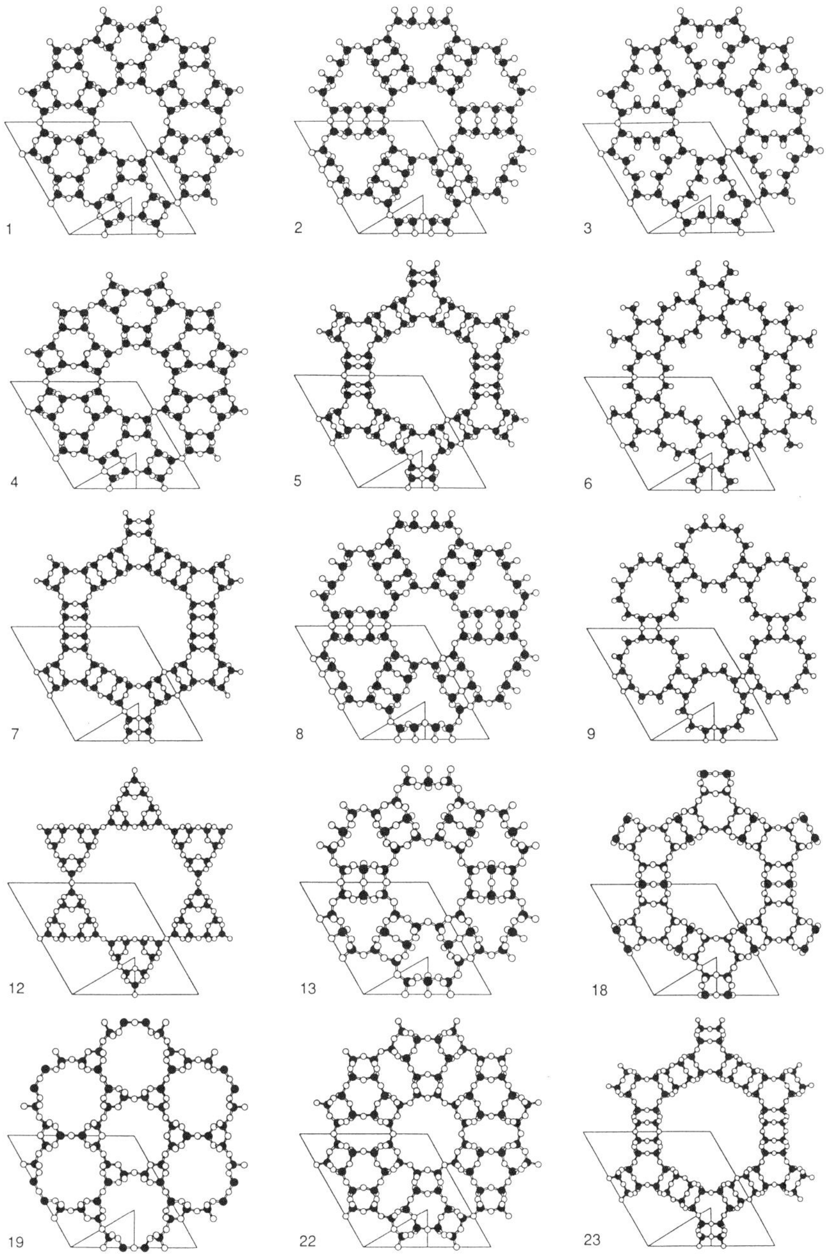

Figure 5. *c*-axis views of the 29 geometrically plausible frameworks.

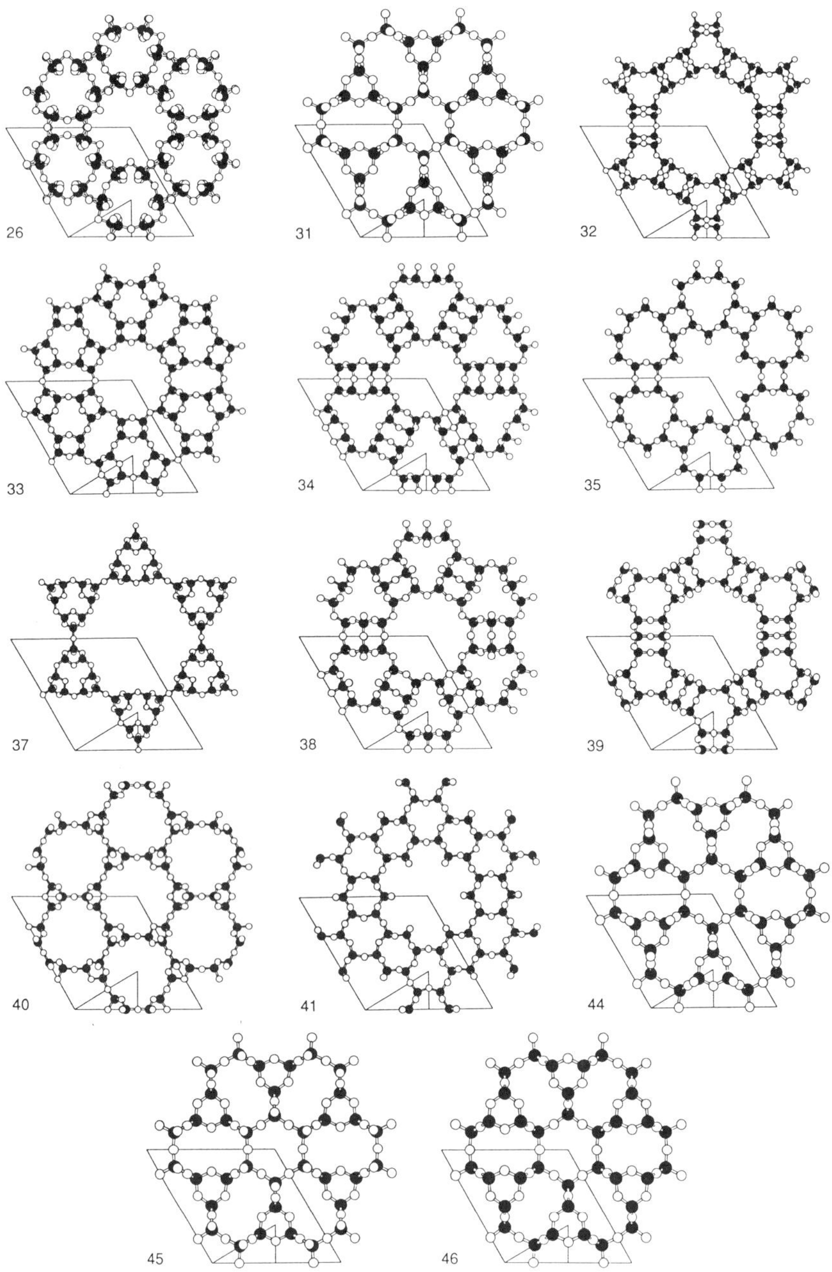

Figure 5 (*continued*). *c*-axis views of the 29 geometrically plausible frameworks.

METHOD OF SOLUTION AND STRUCTURE OF 3 NOVEL HIGH-SILICA MEDIUM-PORE ZEOLITES WITH MULTI-DIMENSIONAL CHANNEL SYSTEMS

M D SHANNON

ICI Chemicals & Polymers Limited, PO Box 8, Runcorn WA7 4QD, England

ABSTRACT

The method of Akporiaye and Price for the enumeration of (some) zeolite structures has been extended and new operators employed in the structure solution of three high-silica zeolite frameworks with multi-dimensional channel systems: NU-87, beta and NU-86.

INTRODUCTION

Of the many classification schemes for zeolite frameworks, one of the most powerful is that introduced by Akporiaye and Price[1]. It provides a simple description of a large fraction of the known frameworks in terms of a "corrugated" sheet and a number of stacking operators that act iteratively on each sheet to produce the next until the unit cell is complete. Structures that can be described using the formalism are generally easily recognised by having one of a characteristic set of lattice parameters, eg 5.0, 10.0 or 20.0±0.1 Å, that represent the repeat normal to the sheets. The specific, characteristic lattice dimension will often dictate just two or three possible sets of stacking operators that can produce that repeat.

In terms of solving an unknown structure, if the unit cell determined from electron diffraction or otherwise contains one of these marker lattice parameters, the problem can be reduced to finding a sheet which, when appropriately corrugated, has the correct additional lattice parameters, is consistent with the symmetry data available and can sustain at least one of the stacking sequences associated with that layer repeat. That is, the problem has effectively been reduced to one in two-dimensions: determining the sheet. This can be straightforward if it involves recognition of a known sheet with a different corrugation or stacking sequence or both, or it can be difficult if a totally new sheet has to be identified.

In solving a number of zeolite structures an extension to the Akporiaye and Price formalism was recognised that enabled it to describe the framework of NU-87[2] which, at that time, had only been found in zeolite NU-85[3] as an intergrowth with EU-1. Then, with an additional concept, zeolite beta was solved quickly early in 1987. More recently this was applied to solve the structure of a novel, highly-faulted zeolite NU-86[4,5] with a 3-D channel system and 30 T atoms in its asymmetric unit.

From the Proceedings of the 9th International Zeolite Conference, Montreal 1992, Eds. R. von Ballmoos et al.,

All unit cells and space group data were initially determined by electron diffraction and models confirmed by simulation of powder X-ray diffraction patterns. For zeolite beta and NU-86 models were conceived within this extended Akporiaye-Price-Shannon formalism and large super-cells were used to approximate the effect of the extensive faulting on these patterns. Latterly the DIFFaX program[6] has been used for NU-86.

RESULTS

The 3 frameworks referred to here have in common a lattice parameter that is a multiple of 12.5Å. This is not one of the characteristic lattice parameters noted by Akporiaye and Price and required the introduction of a new stacking operator.

Zeolites NU-85 and NU-87

NU-87 is related to zeolites nonasil (15Å repeat) and EU-1 (20Å). It shares not only the same 3-connected sheet of T-atoms with them but also (essentially) the same corrugation (Figure 1). But, of course, the stacking operator sequences differ. In all figures of corrugated sheets, T atoms at the nodes at either end of an "arrowed" link are said to have "zig-zag" bonding to the sheet above (arrow end) or below. T atoms at other nodes must either bond directly "up" or "down" to an adjacent sheet. The conventions adopted here are that the starting configuration for a stacking sequence has these bond downwards and that stacking operators, in general, build beneath the previous sheet. The T atom at the top of a zig-zag link is 2.5Å above that at the bottom and the "down" nodes are 1.0Å below the nearest node in a zig-zag link.

A chain of T-atoms from the NON sheet is also shown in Figure 1 in elevation. The effect of stacking on this chain is followed for nonasil, EU-1 and NU-87 in subsequent figures. This does not give a full representation of the corresponding projection. For nonasil and EU-1 it suggests that there are 8-ring windows. In fact these are from minor cages ($5^4 6^4$) and 6-rings would be seen in the full projection. However these simpler diagrams are used to highlight the essential features of the M_o, M_z and M_i operators.

The structure of nonasil is produced (from Figure 1) by the operator sequence $M_oG_aM_oG_a$, where M_o indicates a mirror through the outer T atoms of the sheet (the lowest in our convention) and G_a indicates a glide with translation in the a direction (Figure 2a). EU-1 on the otherhand is produced by the sequence $M_zG_aM_zG_a$ where M_z is the "mould" operator (Figure 2b). A simple description of this operator is: translate the whole sheet downwards by 5.0Å and then any "down" nodes are converted to "up" nodes and raised 2.0Å (or vice versa for "up" nodes). Now

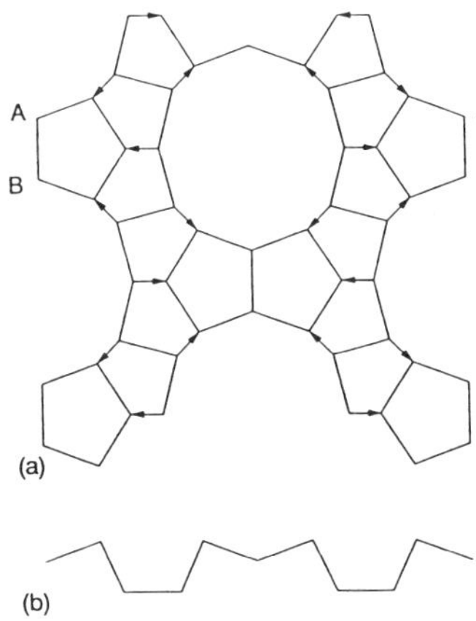

Figure 1. (a) NON sheet and corrugation for nonasil and EU-1 and (b) elevation along AA' (shortest path) or BB'.

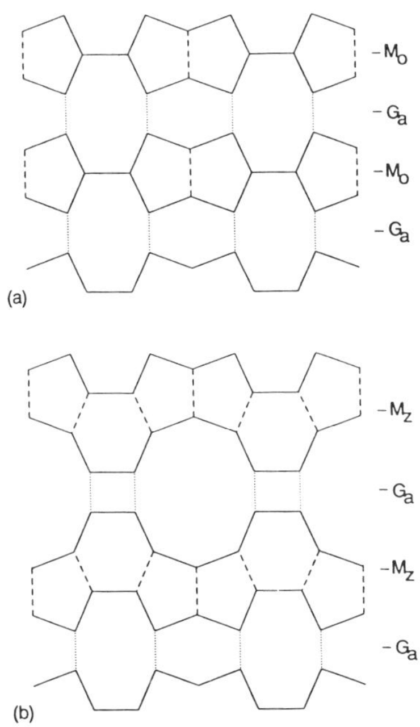

Figure 2. AA' elevation for (a) nonasil and (b) EU-1. Solid lines from sheet, dashed lines link AA' chains in neighbouring sheets, dotted lines link AA' to BB' in neighbouring sheets. a out of paper.

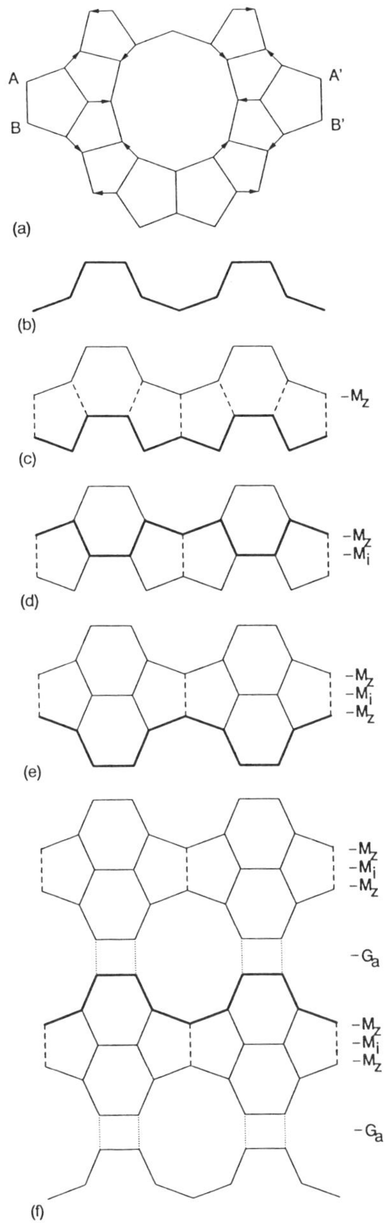

Figure 3. (a) NU-87 sheet, (b) AA' elevation cf Figure 1, (c)-(e) development of stacking operator sequence (operator results shown as broad line) to produce (f) NU-87.

where there were 6-rings crossing the glide planes in nonasil there are 10-rings at alternate glide planes. One could view this as adding additional material to one side of the symmetrical "double sheet" produced by the M_o operator in nonasil. When a 25Å repeat was found for a phase intergrown with EU-1 in zeolite NU-85[3] it seemed likely that this material could be added to both sides, giving 10-ring windows straddling every glide plane.

To describe this a new operator, M_i, is required. It is defined as a mirror through the innermost T atoms of a sheet (Figure 3 where it is more convenient to start with a sheet in which the arrows are reversed). First M_z is applied, then M_i (which produces no new T-atoms) and finally M_z to produce a symmetrical elevation (Figure 3(e)). Thus the stacking sequence $M_zM_iM_zG_aM_zM_iM_zG_a$ was proposed for the phase intergrown with EU-1 in NU-85. This is the structure of NU-87[2] which was made as a single phase at a later date - after we had solved beta. NU-87 has a 2-dimensional channel system controlled by 10-ring windows.

The repeat in the stacking direction for NU-87 would be 12.5Å if the glides were replaced with mirrors to give $M_zM_iM_zM$ as the stacking operator sequence (a hypothetical structure). So by an extension to the AP formalism a means has been introduced to model structures with repeats that are multiples of 12.5Å.

Zeolite beta

In trying to solve the structure of zeolite beta, N A Briscoe (private communication) recognised that there was a strong probability that the two principle projections of the structure (normal to one another) had the same topology as the principle projection in ZSM-12 (Figure 5). The closeness of the a parameters and monoclinic angles, beta, demanded consideration. After allowing for the doubling of the c dimension of the sheet for ZSM-12 due to its corrugation, it was clear that the sheet for beta needed to be less tightly corrugated. This was achieved by the corrugation in Figure 6. Now the repeat along b (as a) was known to be 12.5Å. So, ignoring the faulting, the $M_zM_iM_zM$ stacking operator sequence was applied with the result in Figure 7. This structure contains cubes which, it was immediately recognised, were removed by shear of +1/3 or -1/3 of b as indicated in Figures 7 and 8 to recover the same projection as Figure 6. By starting with the sheet of Figure 7, applying the corrugation from Figure 6 and the $M_zM_iM_zM$ operator sequence Newsam and Treacy's polymorph C[7] was obtained. Appropriate choices of shear directions for the sheet and, following application of the operator sequence, for the elevation led to polymorphs A and B.

The two concepts of the M_i operator and of shear planes normal to the sheet in combination with $M_zM_iM_zM$ are powerful in reducing the description of these complex structures to a 2-D sheet plus simple operators.

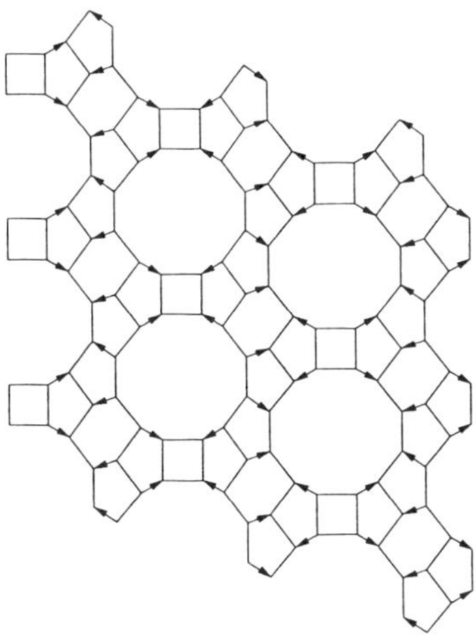

Figure 5. ZSM-12 sheet and corrugation.

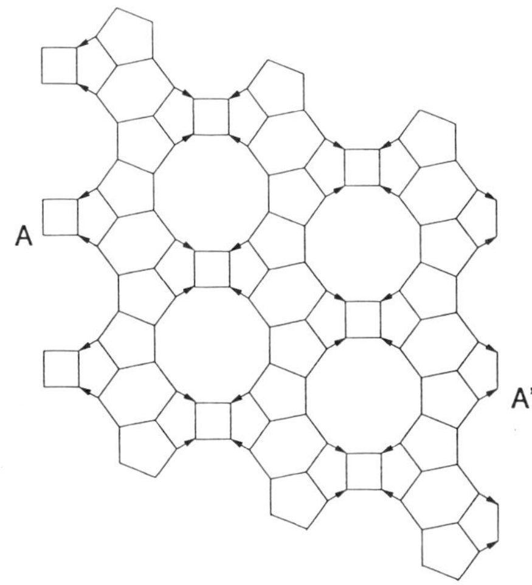

Figure 6. ZSM-12 sheet and beta corrugation.

Figure 7 (right). AA' elevation and projection normal to Figure 6 following $M_ZM_iM_ZM$ operations. Lines at top and bottom indicate positions of "shear" planes.

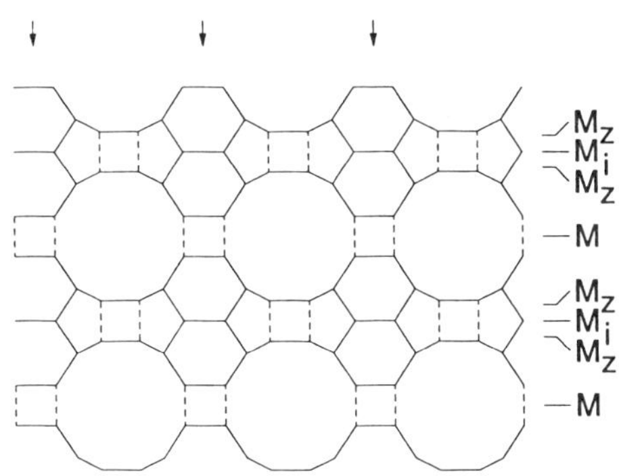

Figure 8 (below right). Result of "shear".

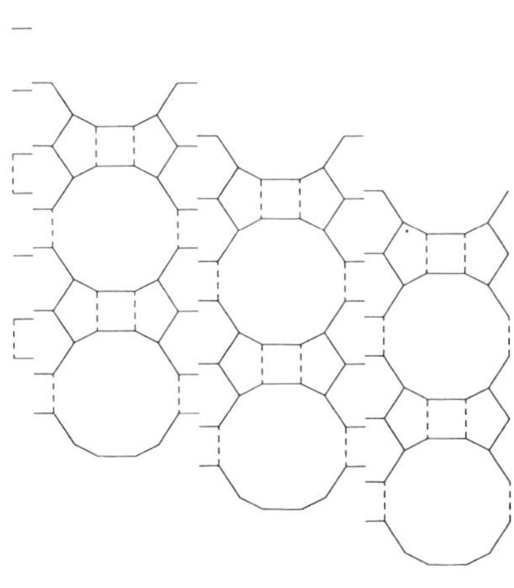

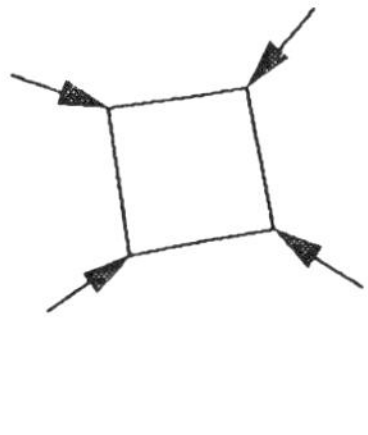

Figure 9. 8 T-atom component of asymmetric unit for NU-86 sheet.

Zeolite NU-86

NU-86 has a principle projection that gives an almost identical electron diffraction pattern and lattice image (at 4Å resolution) to that for beta, with the exception that the fault plane separation is 12.5Å compared with 13.2Å for beta. However this view is just the elevation in NU-86. Consideration of all the diffraction evidence, showed that a rather large sheet 23.3 by 25.1Å with pgg symmetry was needed. The structure would then be completed by application of the $M_ZM_iM_ZM$ operator sequence plus positive or negative shear through the resulting chains of 6,6,4-rings, as in beta, to account for the faulting. The problem was to find the sheet. Density and symmetry data determined that the asymmetric unit contained 28-32 T atoms which meant that the 2-D asymmetric unit for the sheet itself contained at least 12 T atoms. The measured density was intermediate between that for NU-87 and beta, suggesting that the structure had a 3-D channel system. For this reason the corrugation of beta (Figure 6) was assumed rather than that for NU-87 (Figure 3) which would lead to a 2-D channel system. Thus the moiety in Figure 9, containing 8 T-atoms, was taken to be part of the 2-D asymmetric unit. The path from here to the final solution was still difficult. No trial structures that satisfied all the symmetry data could be found until 11-rings were introduced into the sheet. Once that was appreciated the third trial structure (Figure 10) proved to be correct[4]. As with beta 3 ordered polymorphs could be formed; C with no shears following $M_ZM_iM_ZM$, B with all shears in the same direction and A in which they are alternately +1/3 and -1/3 of $\underline{a}$.

The zeolite has a 3-D channel system. One set of channels is straight with 11-ring windows. The striking aspect of the elevation (Figure 11) is that there are columns of 10-ring windows alternating with those of 12-rings. Indeed the alternation of 10- and 12-rings also occurs along the length of these straight channels. The third set of channels crosses the fault planes and so zig-zags, again with alternating 10- and 12-ring apertures. So, although there are 12-ring entrances into the pore system, diffusion is controlled by the 11- and 10-ring windows of the straight channels. These have dimensions intermediate between ZSM-5 and beta.

Given the complexity of NU-86, confirmation that the proposed framework is correct could not proceed via a Rietveld-type refinement. A DLS refinement of the framework positions gave an R-factor of less than .002 and METAPOCS lattice energy calculations showed no significant changes [4]. Good agreement between lattice image and simulation was also obtained. Using DIFFaX [6] the calculated (kinematical) zone-axis electron diffraction patterns showed good qualitative agreement with the experimental (dynamical) data. Simulation of the powder X-ray diffraction pattern by DIFFaX is complicated by the particle morphology of thin platelets with the shear planes normal to the platelet face - a situation that DIFFaX cannot handle. The particles are of a size where high resolution powder data can be recorded at synchrotron sources. Such data from the

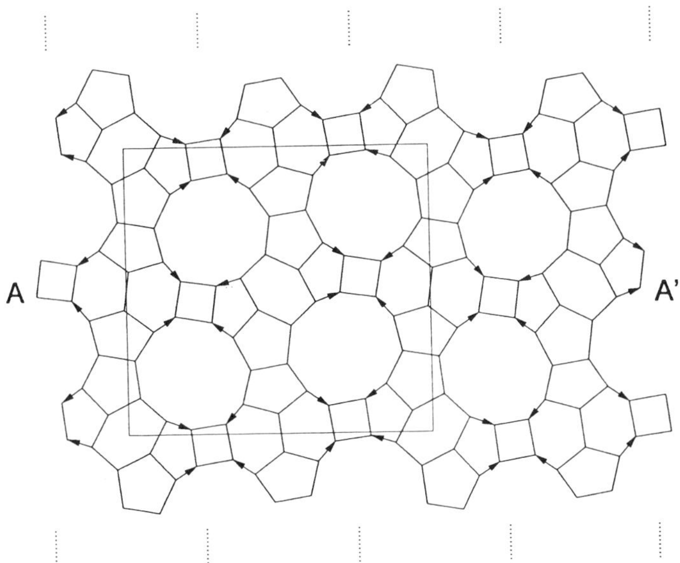

Figure 10. NU-86 sheet and corrugation with position of shear "planes" marked. Shear direction out of or into paper.

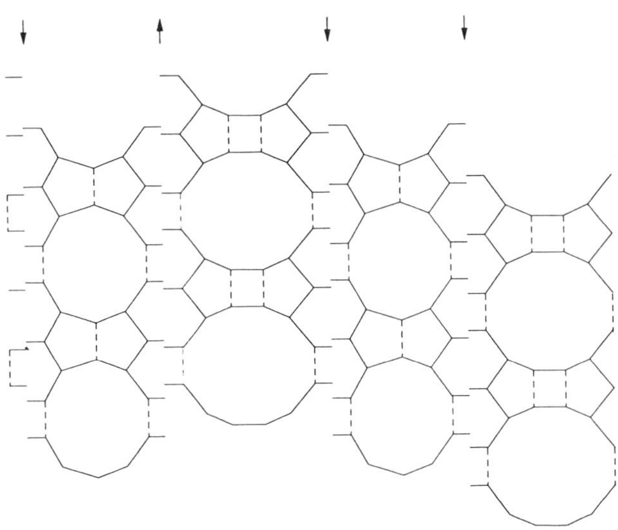

Figure 11. Projection of AA' elevation after $M_ZM_iM_ZM$ plus random shear as marked by arrows.

Daresbury SRS were used to refine the lattice parameters. Figure 12 shows the range 0-35° 2-theta for a wavelength 1.5416Å from the synchrotron in comparison with a DIFFaX simulation. This range is substantially affected by the particle size and shape. Reflections not marked with a dot in the simulation have indices that lead to sustantial (c. 2-fold) broadening and reduction in peak height, as observed. Calculations including size and shape for the ordered polymorphs A and B do account for this effect, but of course introduce other lines. Laboratory diffractometer data in the range 30-60° 2-theta show much closer agreement with simulation. Further high resolution data are being recorded and super-cell calculations are in hand to incorporate the shape effects [4].

CONCLUSIONS

Given the extensive disorder and the large unit cell it is the author's view that NU-86 could not have been solved without these additions to the Akporiaye-Price formalism. They are quite distinct from the extension introduced by Akporiaye[8] to enable him to describe beta. That could not describe NU-87 nor the C polymorphs of beta and NU-86.

The extensions described involve simple concepts which those actively solving structures will find easy to grasp and apply. The APS formalism has facilitated the solution of 2 novel zeolite phases, NU-87 and NU-86. NU-85, NU-86 and NU-87 have multi-dimensional medium-pore channel systems, are high in silica and contain additional novel and potentially important structural features in relation to acid catalysed reactions.

The author is particularly grateful to N A Briscoe, D W Johnson, D E Akporiaye and G D Price for discussions, to P A Cox for lattice energy calculations, S J Andrews for synchrotron data and to ICI PLC for permission to publish.

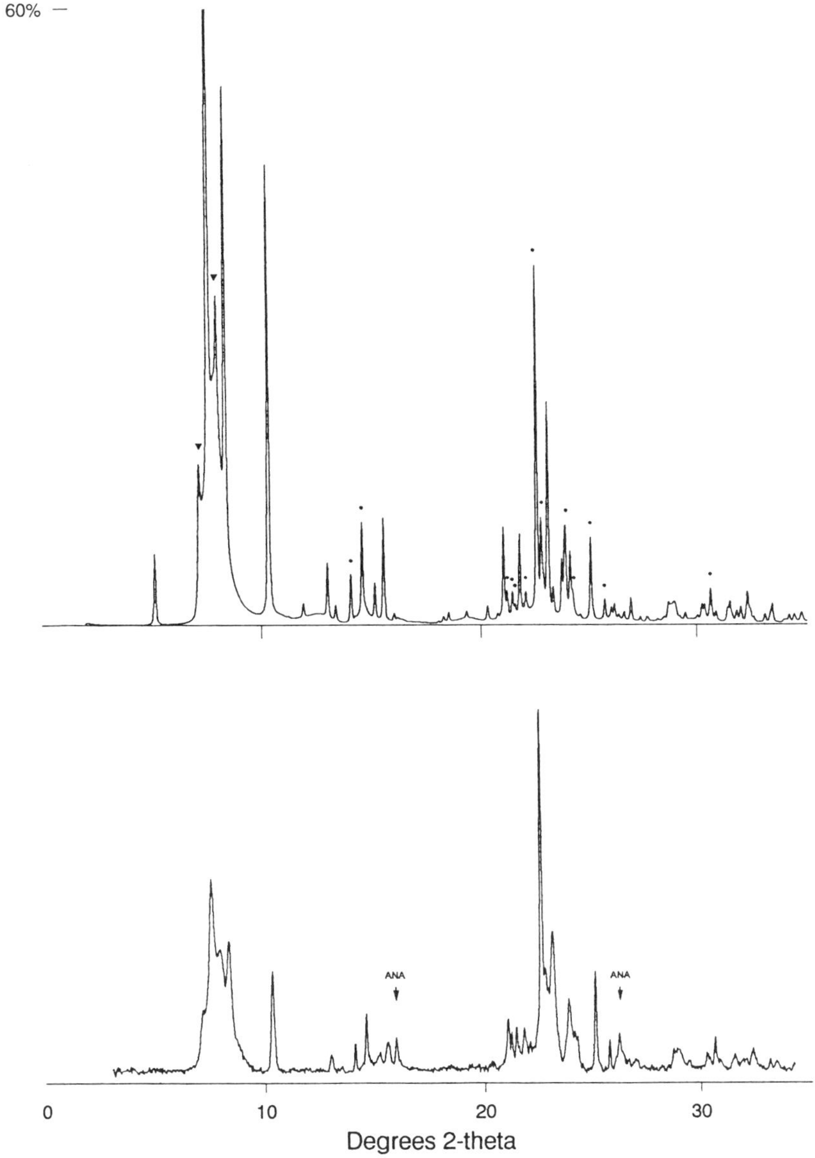

Figure 12. DIFFaX simulation (upper trace) and experimental powder X-ray diffraction patterns for NU-86 (data recorded at wavelength of 1.003Å and transformed to 1.5416Å). See text for explanation of dots. Triangles signify principle "streaked" reflections and should not be affected by shape broadening.

REFERENCES

1. Akporiaye, D.E. and Price, G.D., Private communication (1986) and Zeolites 9, 23-32 (1989).
2. Shannon, M.D., Casci, J.L., Cox, P.A. & Andrews, S.J., Nature 353, (1991).
3. Casci, J.L., Shannon, M.D. & Lake I.J. European Patent Appl. No. 462 745 (ICI plc, 1991).
4. Shannon, M.D., Andrews, S.J., Briscoe, N.A., Casci, J.L. & Cox, P.A. In preparation.
5. Casci, J.L., Shannon, M.D. & Cox, P.A. Proceedings of the 9th International Zeolite Conference, Eds. R. von Ballmoos, J.B. Higgins & M.M.J. Treacy, Butterworth-Heinemann, Stoneham, MA, (1992).
6. Treacy, M.M.J., Deem, M.W. and Newsam, J.M., Proc. Roy. Soc. (London) A433, 499-520 (1990).
7. Newsam, J.M., Treacy, M.M.J., Koetsier, W.T. and de Gruyter, C.B., Proc. Roy. Soc. Lond. A 420, 375-405 (1988).
8. Akporiaye, D.E., Z. Kristallogr. 188, 103-120 (1989).

GROUP 2(Be)/12(Zn)–15(P,As)O_4 MOLECULAR SIEVES

WILLIAM T. A. HARRISON,* THURMAN E. GIER, TINA M. NENOFF and GALEN D. STUCKY

Department of Chemistry, University of California, Santa Barbara, CA 93106-9510, USA

* now at: Department of Chemistry, University of Houston, Houston, TX 77204-5641, USA

ABSTRACT

The crystal structures of three new materials, $Zn_5(PO_4)_2(HPO_4)_4 \cdot H_2O \cdot 2H_2N_2C_6H_{12}$, $Li_8(HPO_4)(BePO_4)_6 \cdot H_2O$ and $LiZnPO_4 \cdot H_2O$ are briefly described. The open, anionic frameworks of each are made up of Be/ZnO_4 and PO_4 tetrahedra, with extra-framework cations in the pores for charge balance. Some systematics of the beryllo(zinco)-phosphate(arsenate) structure field are outlined.

INTRODUCTION

Non-aluminosilicate molecular sieve structures are currently of great interest. Aluminophosphates [1] and gallophosphates [2] are now well characterized, and more recently beryllophosphate (BePO) phases have been discovered in nature [3,4] and synthesized [5].

The group 2/12/15 (Be/Zn)(P/As)O_4 beryllo(zinco)-phosphate(arsenate) framework ion has the same formal charge (but not the same formal *individual* atomic charges) as the $AlSiO_4$ aluminosilicate framework ion, and appears to be a natural candidate for forming tetrahedrally-connected framework structures, since all its component elements (Be, Zn, P, As) form stable tetrahedral species.

Synthesis of microporous beryllo(zinco)-phosphate(arsenate)s has proven to be a productive field [6–11]—here we briefly describe three new structures [$Zn_5(PO_4)_2$-$(HPO_4)_4 \cdot H_2O \cdot 2H_2N_2C_6H_{12}$ (ZnPO-dab), $Li_8(HPO_4)(BePO_4)_6 \cdot H_2O$ (BePO-LOS) and $LiZnPO_4 \cdot H_2O$ (ZnPO-ABW)] which exemplify some of the general features of these zeolite-like materials.

EXPERIMENTAL

Reaction conditions varying from "mild" (*e.g.*, ambient to 70 °C, a few minutes to a few hours) to "high pressure" (650 °C, $\sim$ 30,000psi) have successfully prepared new group 2/12/15 open frameworks [6,10]. Here, $Zn_5(PO_4)_2(HPO_4)_4 \cdot H_2O$-$\cdot 2H_2N_2C_6H_{12}$ was prepared by a moderate-condition hydrothermal reaction (200 °C,

From the Proceedings of the 9th International Zeolite Conference, Montreal 1992, Eds. R. von Ballmoos et al.,

4 days) from soluble precursors, including *DABCO* molecules [12]. $LiZnPO_4{\cdot}H_2O$ was prepared at ambient pressure [6] from lithium-, zinc- and phosphorus-containing precursors, and the beryllophosphate $Li_8(HPO_4)(BePO_4)_6{\cdot}H_2O$ was made in a high pressure bomb using similar conditions to those described in ref. [10].

The crystal structures of these materials were established by standard methods. The atomic arrangements in $Zn_5(PO_4)_2(HPO_4)_4{\cdot}H_2O{\cdot}2H_2N_2C_6H_{12}$ and $Li_8(HPO_4)$-$(BePO_4)_6{\cdot}H_2O$ were determined using single crystal X-ray methods as described earlier [10,11]. The structure of $LiZnPO_4{\cdot}H_2O$ was refined using the Rietveld method with laboratory X-ray powder data [7,9]. Some salient crystallographic parameters for these phases are listed in Table 1. Full experimental details will be published later.

RESULTS

These three structures are built up from anionic, 3-dimensional networks of tetrahedra, enclosing pores containing extra-framework species—metallic cations, organic cations, water molecules, and extra-framework anions. Full details will follow later; here the main structural features for each material are denoted.

• $Zn_5(PO_4)_2(HPO_4)_4{\cdot}H_2O{\cdot}2H_2N_2C_6H_{12}$ is a complex, anionic, framework (Figure 1) consisting of a variety of tetrahedral groups (ZnO_4, ZnO_3OH_2, PO_4, PO_3OH and PO_3O'), where the PO_3O' group contains an "unsaturated" P–O link. Charge-balance is accomplished by extra-framework, protonated, *DABCO* (1,4-diazabicyclo[2.2.2]octane) cations. The tetrahedral network is "interrupted", *i.e.*, some potential Zn–O–P links are replaced by $Zn{-}OH_2$ and P–OH groups, which partake in a structurally-important *intra*-framework H-bond network. The *DABCO* guest also bonds to the framework, *via* $N{-}H{\cdots}O_f$ bonds (O_f = framework oxygen), and appears to play a significant structural role. Complete framework-cation alternation is maintained—there are no Zn–O–Zn or P–O–P bonds in the crystal, despite the non 1:1 Zn:P framework atomic ratio. The average framework atom bond distances (d_{av}(Zn–O) = 1.939 (3) Å, d_{av}(P–O) = 1.516 (3) Å, d_{av}(P–OH) = 1.577 (5) Å) are typical of the species concerned [12].

The tetrahedral connectivity in $Zn_5(PO_4)_2(HPO_4)_4{\cdot}H_2O{\cdot}2H_2N_2C_6H_{12}$ leads to a 3-dimensional network of 4- and 8-rings, as shown in polyhedral representation in Figure 1. In particular, there are three different 8-ring channels, propagating in the **a**-direction, two of which contain the *DABCO* guest species. One channel is regular in shape, with an approximate window dimension of 4.5 × 4.5 Å, one is squashed (dimension ≈ 3.5 × 5.5 Å), and the other distorted channel contains most of the "hanging" tetrahedral bonds which form an H-bonding network. This H-bond array appears to be important in stabilizing the 3-dimensional structure, as well as helping to offset the high framework-atom (Zn/P/O) charge.

Table 1: Crystallographic Parameters

	ZnPO-dab	BePO-LOS	ZnPO-ABW
formula wt.	596.25	245.43	185.31
habit	needle	"chunk"	powder
color	clear	clear	white
crystal system	triclinic	trigonal	orthorhombic
a (Å)	9.366(1)	11.504(2)	10.512(1)
b (Å)	9.882(1)	11.504(2)	8.1320(8)
c (Å)	19.183(2)	9.273(2)	5.0273(5)
α (°)	85.406(5)	90	90
β (°)	85.029(5)	90	90
γ (°)	114.485(5)	120	90
V (Å^3)	1596.5	1062.8	429.7(2)
Z	2	2	4
space group	$P\bar{1}$	$P31c$	$Pna2_1$
space group #	2	159	33
T (°C)	25(2)	25(2)	25(2)
λ(Å)	0.71073	0.71073	1.54178
ρ_{calc} (g/cm^3)	2.386	2.30	2.86
μ (cm^{-1})	41.98	7.0	110.8
observed data	4095§	1321§	1498¶
refinement on:	F	F	powder data
software	CRYSTALS [20]	CRYSTALS	GSAS [21]
R (%)	7.11^a	4.63^a	9.26^c
R_w (%)	7.55^b	4.88^b	12.59^d

§ $I > 3\sigma(I)$, ¶ number of powder points.

$R^a = \Sigma||F_o| - |F_c||/|F_o|$, $R_w{}^b = [\Sigma w(|F_o| - |F_c|)^2/\Sigma w|F_o|^2]^{1/2}$.

$R_p{}^c = \Sigma|y_o - \mathbf{C}y_c|/\Sigma|y_o|$, $R_{wp}{}^d = [\Sigma w(y_o - \mathbf{C}\ y_c)^2/\Sigma w y_o{}^2]^{1/2}$, where **C** is a scale factor.

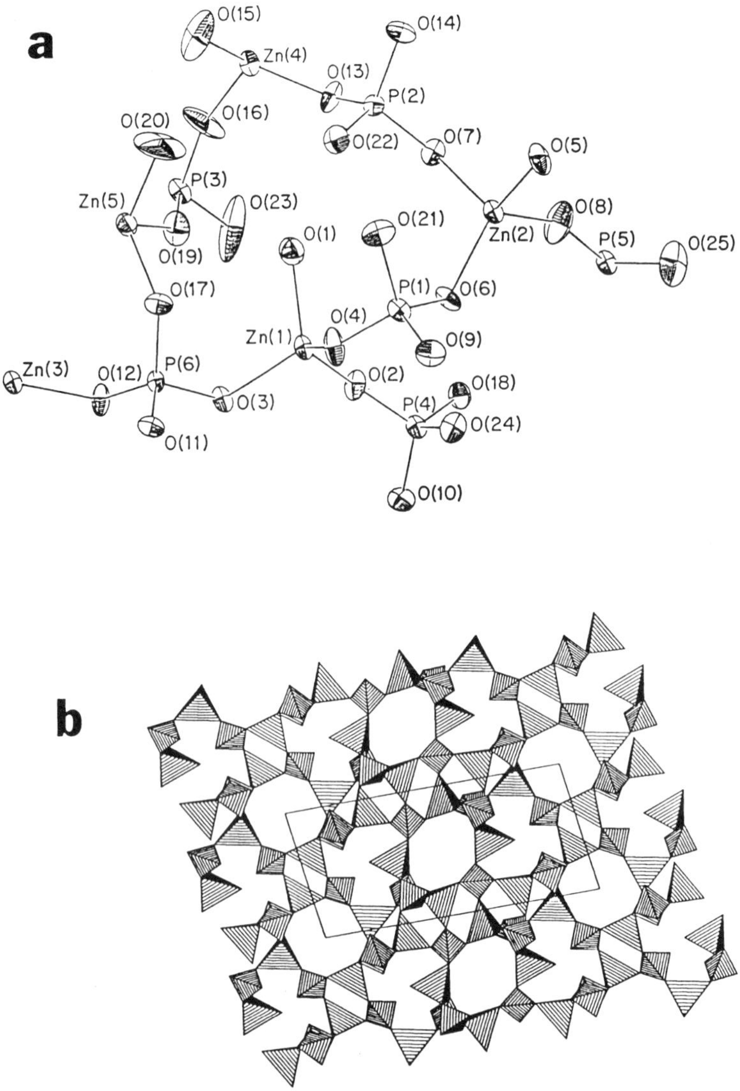

Figure 1: The crystal structure of $Zn_5(PO_4)_2(HPO_4)_4 \cdot H_2O \cdot 2H_2N_2C_6H_{12}$, showing (a) the bonding scheme in the framework asymmetric unit and (b) the complete crystal structure in polyhedral representation [22], viewed down **a**.

• $Li_8(HPO_4)(BePO_4)_6{\cdot}H_2O$ has the same framework configuration as the sodium aluminosilicate *losod* zeolite [13,14]. Complete tetrahedral-atom alternation is maintained *via* Be–O–P links, and all the tetrahedral vertices of each T-atom are connected to 4 distinct neighbors of the opposite type. Two cage types, both assembled from 4- and 6-rings, are formed by this topology: the barrel-shaped 30-membered (30 tetrahedral T-atoms) *loso* cage and the 18-membered *canc* cage, named for its occurence in cancrinite, as illustrated in Figure 2. In Smith's notation, [15] they are the $4^6 6^2 6^3 6^6$- and $4^6 6^2 6^3$-type cages respectively, this terminology indicating the numbers of different types of 4- and 6-rings forming each cage. The *loso* and *canc* cages are assembled into one-dimensional stacks, which propagate along the polar, crystallographic **c**-direction. Both types of cages are interconnected *via* 1-dimensional 6-ring channels in the **c**-direction, forming independent stacks of *loso* and *canc* cages—there are no stacks containing moieties.

The guest species consist of 4 distinct Li^+ cations, a water molecule and a hydrogen phosphate anion. Each of the lithium cations is tetrahedrally-coordinated to oxygen atoms—in each case, 3 Li–O_f and 1 Li–O_g (O_g = "guest" oxygen atom) bonds are formed. *All* of the 6-ring windows are occupied by Li^+ species. The *loso* cage contains the incorporated hydrogen phosphate species, akin to anion incorporation in aluminophosphate cancrinites [16]. This anion-incorporation in lithium beryllophosphates helps to balance the high charge:size ratio of the Li^+ cation, and is consistent with its strong preference for tetrahedral coordination.

• $LiZnPO_4{\cdot}H_2O$ (Figure 3) is an exact analogue of the hydrated lithium aluminosilicate zeolite-ABW, $LiAiSiO_4{\cdot}H_2O$ [17]. It consists of a 3-dimensional framework of alternating, corner-linked, ZnO_4 and PO_4 tetrahedra, formed into 4-, 6- and 8-ring windows (in Smith's notation this topology may be decomposed to 2-d 6^3 or 4.8^2 nets [15]). The predominant channel system is formed of puckered, 1-dimensional 8-rings in the **c**-direction. The extra-framework lithium cations appear to occupy the 6-ring windows parallel to the main 8-ring channels, by making 3 Li–O bonds, in pyramidal geometry, with the fourth Li tetrahedral vertex supplied by a water molecule. This coordination behavior is also found for the extra-framwork species in $LiAlSiO_4{\cdot}H_2O$ [17]. Powder neutron diffraction experiments are in progress to elucidate the complete structural details of $LiZnPO_4{\cdot}H_2O$.

CONCLUSIONS

The existence of an extensive group-2/12/15 molecular sieve chemistry has been demonstrated here and elsewhere [1–11]. Some of the structures of these new phases are isomorphous with known aluminosilicate zeolite frameworks, while others are new. Here, we have described a novel material ($Zn_5(PO_4)_2(HPO_4)_4{\cdot}H_2O{\cdot}2H_2N_2C_6H_{12}$), a lithium

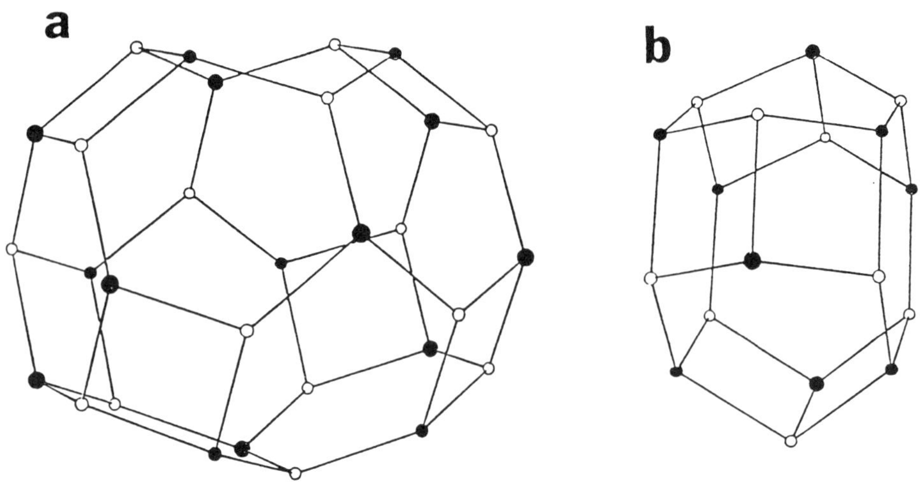

Figure 2: Schematic ORTEP [23] view of the two cage topologies [(a) *loso*, (b) *canc*]in the *losod*-type $Li_8(HPO_4)(BePO_4)_6 \cdot H_2O$, showing only Be–P connectivity. Be atoms are ilustrated as filled circles.

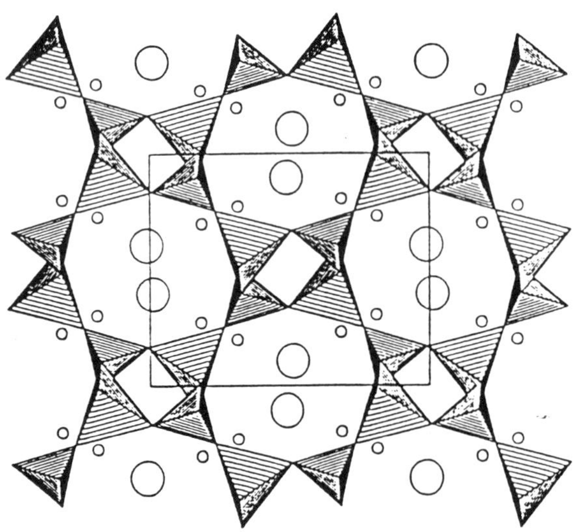

Figure 3: Polyhedral drawing of the $LiZnPO_4 \cdot H_2O$ crystal structure viewed down the c-direction. Probable Li-cation positions are illustrated as small circles, water-molecule oxygen atom sites as large circles.

beryllophosphate analogue of a sodium aluminosilicate ($Li_8(HPO_4)(BePO_4)_6 \cdot H_2O$) and a lithium zincophosphate analogue of a lithium aluminosilicate ($LiZnPO_4 \cdot H_2O$). Some general structural principles are outlined here:

• large ($>$ 0.5 mm) single crystals of these materials may often (but not always) be prepared, for single-crystal X-ray structure determination, and other physical characterization. Aluminosilicate and aluminophosphate materials often pose a much more serious challenge in terms of crystal growth.

• Be, Zn, P and As form stable, tetrahedral, framework ions with oxygen atoms, commensurate with their known crystallochemical behavior, in terms of bond distances and angles.

• these (Be/Zn)(P/As)O phases show a strict avoidance of M–O–M (M = Be, Zn) and X–O–X (X = P, As) links, akin to the "Lowenstein's rule" Al–O–Al avoidance in aluminosilicates. This militates against odd-membered-ring-containing systems—all the systems described above are based on 4-/6-/8-ring topologies.

• "interrupted" frameworks are more commonplace in microporous (Be/Zn)(P/As)O frameworks than in aluminosilicate structures (*e.g.*, roggianite [18]). This is expected due to the high charge density of these group 2/12/15 frameworks.

• dehydration/rehydration and ion-exchange reactions are feasible for some of these (Be/Zn)(P/As)O-based structures [19]. In general, however, they appear to be less stable than aluminosilicates to these "zeolitic" reactions.

ACKNOWLEDGEMENTS

We thank our colleagues at UCSB for their assistance in this work. This work is partially funded by the National Science Foundation (Division of Materials Research).

REFERENCES

1. Flanigen, E. M., Patton, R. L. and Wilson, S. T., In *Innovation in Zeolite Materials Science*, Eds. P. J. Grobet *et. al.*, Elsevier, Amsterdam, pp 13–27 (1988).

2. Newsam, J. M. and Vaughan, D. E. W., In *New Developments in Zeolite Science and Technology*, Eds. Y. Murakami, A. Iijima and J. W. Ward, Elsevier, Amsterdam, pp 457–464 (1986).

3. Peacor, D. R., Rouse, R. C. and Ahn, J.-H., Amer. Miner. **72**, 816 (1987).

4. Rouse, R. C., Peacor, D. R. and Merlino, S., Amer. Mineral. **74**, 1195 (1989).

5. Harvey, G. and Meier, W. M., In *Zeolites: Facts, Figures, Future*, Eds. P. A. Jacobs and R. A. van Santen, Elsevier, Amsterdam, pp 411–420 (1989).

6. Gier, T. E. and Stucky, G. D., Nature **349**, 508 (1991).

7. Nenoff, T. M., Harrison, W. T. A., Gier. T. E. and Stucky, G. D., J. Amer. Chem. Soc. **113**, 378 (1991).

8. Harrison, W. T. A., Gier, T. E., Moran, K. L., Nicol, J. M., Eckert, H. and Stucky, G. D., Chem. Mater. **3**, 27 (1991).

9. Gier, T. E., Harrison, W. T. A. and Stucky, G. D., Angew. Chem. **103**, 1191 (1991).

10. Harrison, W. T. A., Gier, T. E. and Stucky, G. D., J. Mater. Chem. **1**, 153 (1991).

11. Harrison, W. T. A., Martin, T. M., Gier, T. E. and Stucky, G. D., J. Mater. Chem. **2**, 175 (1992).

12. Shannon, R. D., Acta Crystallogr. **A32**, 751 (1976).

13. Sieber, W. and Meier, W. M., Helv. Chim. Acta **57**, 1533 (1974).

14. Baerlocher, Ch. and Schicker, P., In *Compilation of Zeolite Powder Patterns*, Ed. R. Von Ballmoos, Butterworths, Guildford, p 58 (1984).

15. Smith, J. V., Chem. Revs. **88**, 149 (1988).

16. Smolin, Yu. I., Sheplev, Yu. F., Butikova, I. K. and Kobyakov, I. B., Kristallografiya **26**. 63 (1981).

17. Kerr, I. S., Zeit. Kristallogr. **139**, 186 (1974).

18. Galli, E., In *Proc. Fifth International Conference on Zeolites*, Ed. L. V. Rees, Heyden, Philadelphia, pp 205–213 (1980).

19. Nenoff, T. M., Harrison, W. T. A., Gier, T. E., Nicol, J. M. and Stucky, G. D., Zeolites, in press (1992).

20. Watkin, D. J., Carruthers, J. R. and Betteridge, P. W., CRYSTALS USer Guide, Chemical Crystallography Laboratory, Oxford University, England (1985–1991).

21. Larson, A. C. and Von Dreele, R. B., GSAS User Guide, Los Alamos National Laboratory, Los Alamos, New Mexico, USA (1985–1990). Refinements were preformed on a μVAX-II computer.

22. Fischer, R. X., J. Appl. Cryst. **18**, 258 (1985).

23. Johnson, C. K., Oak Ridge National Laboratory Report ORNL-5138, Oak Ridge, Tennessee, USA (1976).

HYDROGEN BONDS DISTORT THE ALUMINOSILICATE FRAMEWORK OF AMMONIUM-NATROLITE

E. STUCKENSCHMIDT, D. KASSNER, W. JOSWIG and W.H. BAUR

Institut für Kristallographie und Mineralogie, Universität Frankfurt
W-6000 Frankfurt am Main, Bundesrepublik Deutschland

ABSTRACT

Cation exchange of NH_4 for Na in natrolite yields an anhydrous NH_4-natrolite, $(NH_4)_2Al_2Si_3O_{10}$. The symmetry is reduced from space group Fdd2 in the Na-form to $P2_1$ in the NH_4-form. The angles T-O-T and the T-O bonds of the framework are strongly distorted by the hydrogen bonds of the NH_4-groups. The NAT framework of natrolite is collapsible, that is all angles T-O-T (the hinges between the rigid TO_4 tetrahedra) change upon contraction or expansion in the same sense.

INTRODUCTION

Natrolite, $Na_2Al_2Si_3O_{10}.2H_2O$, is a fibrous zeolite. Pauling [1] recognized this framework as being collapsible. Exchanged natrolites have been studied recently by diffraction methods [2] and it is now known that the NAT framework responds very flexibly to cation exchange. The volume of the most open natrolite framework (the K-form) is 43 % larger than of the most collapsed natrolite (the dehydrated Na-form).

EXPERIMENTAL

A single crystal (size 0.5 x 0.25 x 0.3 mm^3) of natrolite from Salesl, CSFR, was exchanged in a melt of NH_4NO_3 at about 463 K for 16 days. Diffraction data were measured at 293 K on an Enraf Nonius CAD4 diffractometer with MoKα radiation (λ = 0.7107 Å). About 15000 F_{obs} were collected up to $\sin\theta/\lambda$ = 1.036 $Å^{-1}$. All the usual corrections were applied and after averaging 9269 F_{obs} (internal R = 2.5%) were used for the refinement. The observed systematic extinctions indicated space group $P2_1$, but for a simpler comparison with Na-natrolite, the setting $C112_1$ was used. Cell contents, Z = 8; cell parameters $\underline{a}$ = 17.899(2) Å, $\underline{b}$ = 18.390(2) Å, $\underline{c}$ = 6.529(1) Å, γ = 90°. The refinement began with the atomic coordinates of the framework atoms as known from a single crystal study of Na-natrolite from the same locality [3]. Conventional refinement of the diffraction data in space group $C112_1$ yielded a

From the Proceedings of the 9th International Zeolite Conference, Montreal 1992, Eds. R. von Ballmoos et al.,

reasonable structure and allowed a determination of the N atom positions from difference Fourier syntheses, but the R-value was too large (9 %) compared with the internal R-value (2.5 %). Analysis of the F_{obs} divided into groups of reflections showed that the crystal must be twinned by a mirror plane normal either to [100] or to [010]. After including a parameter describing the volume of the twin individuals, the R-value dropped to 5 %. The R-value was based on a comparison of Ft_{obs} and Ft_{calc}, where $Ft = (V_1F_1^2 + V_2F_2^2)^{½}$ (V_N and F_N, volumes and structure factors of twin individual N). The volumes of the two twins were found to be almost equal: 52 % and 48 %. The function minimized was $\Sigma w(Ft_o - Ft_c)^2$, weights were chosen as $w = 1/\sigma(Ft_o)^2$, where $\sigma(Ft_o)^2$ was derived from $\sigma(Ft_o^2) = 1 + \sigma_c(Ft_o^2) + 0.01 \times Ft_o^2$ with $\sigma_c(Ft_o^2)$ from counting statistics. The origin was fixed by keeping the sum of the z-coordinates of atoms Si11 and Si13 constant. In the final stage, the hydrogen atoms were localized in difference Fourier maps. Initially the NH_4 groups were refined with restraints: N-H distances at 1.00(2) Å and H-N-H angles at 109.47(2.00)° (R = 4.67 %, R_w = 2.98 % for 9269 F_{obs}). Finally the coordinates of all heavy atoms were kept constant at the values from the refinement including the high-angle reflections and the coordinates of the hydrogen atoms were refined using a smaller dataset with $\sin\theta/\lambda < 0.7$ Å^{-1} (R = 2.90 %, R_w = 2.61 % for 5419 F_{obs}).

RESULTS

In the process of exchange of NH_4 for Na in natrolite (Fdd2) the primitive unit cell is doubled in volume, the glide planes d and the twofold axes (through the central axes of the fibrous chains) are lost. What remains of the previous symmetry are the twofold rotors through the central axes of the oval pores. In NH_4-natrolite there are two crystallographically distinct tetrahedral chains and four kinds of pores, which differ in the arrangement of NH_4-ions within them (Fig. 1), while in Na-natrolite there is only one kind each of chains and pores. The total reduction in symmetry is of order four, thus one equivalent general position in Na-natrolite splits into four equivalent general positions in NH_4-natrolite. The ammonium ions exchanged for the Na-ions and for the water molecules of Na-natrolite. This is not just due to the dimensions of the NH_4-group, because that is roughly of the same size as the K-ion and K-natrolite is hydrated. The NH_4-ions

Table 1. Positional and displacement parameters of NH_4-natrolite.

Atom	x/a	y/b	z/c	U(equ)
Si11	0.49966(7)	0.49286(3)	0.0007(2)	0.0088
Si13	0.25710(3)	0.75007(7)	0.7496(2)	0.0103
Si21	0.34558(4)	0.28565(5)	0.6255(2)	0.0081
Si22	0.64876(4)	0.71517(5)	0.6245(2)	0.0077
Si23	0.40258(5)	0.53316(5)	0.3731(1)	0.0084
Si24	0.09972(5)	0.96398(5)	0.3742(2)	0.0092
Al11	0.46489(6)	0.40239(5)	0.6169(2)	0.0081
Al12	0.53541(6)	0.59234(5)	0.6170(2)	0.0082
Al13	0.28664(6)	0.65342(6)	0.3645(2)	0.0086
Al14	0.21584(6)	0.84406(6)	0.3653(2)	0.0086
O11	0.5116(1)	0.5629(1)	0.8579(4)	0.0152
O12	0.4839(1)	0.4191(1)	0.8726(4)	0.0142
O13	0.2687(1)	0.6772(1)	0.6196(4)	0.0179
O14	0.2409(1)	0.8211(1)	0.6146(4)	0.0187
O21	0.4297(1)	0.3147(1)	0.5870(4)	0.0112
O22	0.5652(1)	0.6820(1)	0.6280(4)	0.0102
O23	0.3184(1)	0.5635(1)	0.3569(5)	0.0120
O24	0.1843(1)	0.9335(1)	0.3604(5)	0.0120
O31	0.4015(1)	0.4655(1)	0.5275(5)	0.0170
O32	0.6022(1)	0.5382(1)	0.5054(5)	0.0185
O33	0.6496(1)	0.7872(1)	0.7583(5)	0.0162
O34	0.3505(1)	0.2179(1)	0.7805(4)	0.0131
O41	0.2955(1)	0.3487(1)	0.7227(4)	0.0163
O42	0.7072(1)	0.6561(1)	0.7049(4)	0.0163
O43	0.4604(1)	0.5948(1)	0.4480(4)	0.0125
O44	0.5439(1)	0.4052(2)	0.4688(4)	0.0176
O51	0.3132(1)	0.2613(2)	0.3998(4)	0.0159
O52	0.6674(1)	0.7334(1)	0.3803(3)	0.0142
O53	0.5748(2)	0.4829(1)	0.1414(4)	0.0190
O54	0.4264(1)	0.5064(1)	0.1476(4)	0.0170
N1	0.4474(1)	0.2852(1)	0.0900(3)	0.0203
H11	0.421(2)	0.292(2)	0.209(4)	0.027
H12	0.456(2)	0.332(2)	0.071(5)	0.027
H13	0.487(2)	0.268(2)	0.103(5)	0.027
H14	0.415(2)	0.262(2)	0.022(4)	0.027
N2	0.4741(1)	0.8104(1)	0.6512(4)	0.0204
H21	0.507(2)	0.838(2)	0.740(4)	0.022
H22	0.452(2)	0.833(2)	0.557(5)	0.022
H23	0.425(2)	0.791(2)	0.700(4)	0.022
H24	0.483(2)	0.772(2)	0.597(5)	0.022
N3	0.7208(2)	0.4438(2)	0.8947(5)	0.0301
H31	0.725(2)	0.411(2)	0.792(6)	0.052
H32	0.734(2)	0.424(2)	0.003(6)	0.052
H33	0.755(2)	0.485(2)	0.877(6)	0.052
H34	0.677(2)	0.459(2)	0.854(6)	0.052
N4	0.2940(2)	0.5437(2)	0.8640(7)	0.0380
H41	0.278(2)	0.576(3)	0.778(7)	0.059
H42	0.319(2)	0.524(3)	0.830(7)	0.059
H43	0.309(2)	0.568(2)	0.942(6)	0.059
H44	0.246(2)	0.514(3)	0.913(6)	0.059

occupy almost the same positions as the H_2O-molecules in Na-natrolite. The twinning is due to alternate arrangements of the NH_4-ions within the pores of the framework. The twins are related by the diagonal glide plane lost in NH_4-natrolite relative to the original natrolite. The positional and equivalent isotropic displacement parameters are given in Table 1. The locations of the individual atoms of the framework in Na-natrolite and in NH_4-natrolite are on average within 0.15 Å of each other with a maximum deviation of 0.3 Å. The equivalent isotropic displacement factors are about 45 % larger in the NH_4-form than in Na-natrolite. The T-O distances are as expected for a well ordered Si,Al distribution. The mean Si-O distances range from 1.608 Å to 1.627 Å, the individual values range from 1.588 Å to 1.663 Å. For Al-O the respective ranges are 1.727 Å to 1.741 Å, and 1.714 Å to 1.752 Å. The T-O-T angles of the oxygen atoms O2 are the smallest and they differ the least from each other in their values (Table 2). The largest T-O-T

Table 2. Angles T-O-T [°] in NH_4-natrolite and corresponding angles in Na-natrolite [*]. Hydrogen bonded oxygen atoms are marked by (h) (limit: H---O = 2.22 Å), not hydrogen bonded oxygen atoms, or those participating in bifurcated bonds by (n).

T-O-T			T-O-T		
Si11-O11-Al12	145.0(1)	n	Si21-O41-Al13	132.1(2)	n
Si11-O12-Al11	133.0(1)	h	Si22-O42-Al14	136.2(2)	n
Si13-O13-Al13	137.5(2)	h	Si23-O43-Al12	132.1(2)	h
Si13-O14-Al14	139.4(2)	h	Si24-O44-Al11	139.1(2)	n
mean	138.7		mean	134.9	
mean(h)	136.6		mean(h)	132.1	
mean(n)	145.0		mean(n)	135.8	
*Si1-O1-Al	140.9(0)		*Si2-O4-Al	135.0(0)	
Si21-O21-Al11	128.7(1)	h	Si13-O51-Si21	148.9(2)	n
Si22-O22-Al12	130.0(1)	h	Si13-O52-Si22	135.0(1)	h
Si23-O23-Al13	129.0(1)	h	Si11-O53-Si24	140.8(2)	n
Si24-O24-Al14	129.1(1)	h	Si11-O54-Si23	142.3(2)	n
mean	129.2		mean	141.8	
mean(h)	129.2		mean(h)	135.0	
			mean(n)	144.0	
*Si2-O2-Al	129.7(0)		*Si1-O5-Si2	144.5(0)	
Si23-O31-Al11	136.7(2)	n			
Si24-O32-Al12	134.2(2)	n			
Si22-O33-Al13	139.1(2)	n			
Si21-O34-Al14	132.2(2)	h			
mean	135.6				
mean(h)	132.2				
mean(n)	136.7				
*Si2-O3-Al	138.7(0)				

Table 3. Hydrogen bonds and N-O distances < 3.6 Å in NH_4-natrolite [Å],[°]. Hydrogen bond present (limit: H---O = 2.22 Å), h; bifurcated bond, b; not hydrogen bonded, n.

N-O		H---O		N-H---O		
N1-O21	2.866(3)	H13-O21	2.13(4)	N1-H13-O21	157(4)	h
N1-O12	2.917(3)	H12-O12	2.12(4)	N1-H12-O12	150(3)	h
N1-O34	2.937(3)	H14-O34	2.12(3)	N1-H14-O34	163(3)	h
N1-O51	3.171(3)	H11-O51	2.37(3)	N1-H11-O51	146(3)	b
N1-O21	3.305(3)	H11-O21	2.51(3)	N1-H11-O21	145(3)	b
N1-O21	3.344(3)					
N1-O44	3.594(4)					
N2-O43	2.859(3)	H21-O43	1.93(3)	N2-H21-O43	160(3)	h
N2-O22	2.873(3)	H24-O22	2.22(4)	N2-H24-O22	138(3)	h
N2-O11	3.027(3)	H22-O11	2.40(4)	N2-H22-O11	131(3)	n
N2-O52	3.050(3)	H23-O52	2.08(3)	N2-H23-O52	163(2)	h
N2-O22	3.195(4)					
N2-O33	3.245(4)					
N2-O22	3.490(4)					
N3-O24	2.832(4)	H33-O24	1.85(4)	N3-H33-O24	175(3)	h
N3-O14	2.928(4)	H31-O14	2.04(4)	N3-H31-O14	167(3)	h
N3-O42	3.023(4)	H32-O42	2.24(4)	N3-H32-O42	158(4)	b
N3-O24	3.116(4)	H32-O24	2.50(4)	N3-H32-O24	132(3)	b
N3-O53	3.152(4)	H34-O53	2.66(4)	N3-H34-O53	117(3)	n
N3-O32	3.267(4)					
N3-O24	3.554(4)					
N4-O23	2.816(5)	H44-O23	1.87(5)	N4-H44-O23	146(4)	h
N4-O13	2.963(5)	H41-O13	2.14(5)	N4-H41-O13	160(4)	h
N4-O54	3.084(5)					
N4-O31	3.255(5)	H42-O31	2.69(4)	N4-H42-O31	154(5)	n
N4-O23	3.269(6)	H43-O23	2.72(4)	N4-H43-O23	135(4)	n
N4-O23	3.359(6)					
N4-O41	3.459(5)					

angles (around O1 and O5) have also the largest internal variation. The hydrogen bonding around the NH_4-groups includes bifurcated and essentially nonbonded H-atoms (Table 3). Of the sixteen hydrogen atoms in the four NH_4-groups only ten are clearly involved in strong hydrogen bonds, two form rather weak bifurcated bonds and four appear not to be involved in hydrogen bonds at all, as judged by the distances H---O and the angles N-H---O. Ammonium groups 3 and 4 are less strongly hydrogen bonded than groups 1 and 2. Possibly their large displacement factors reflect static disorder due to weaker hydrogen bonding.

DISCUSSION

The distortion of the framework, when compared with Na-natrolite, is well illustrated by the T-O-T angles (Table 2). The mean values of all angles T-O-T are slightly reduced in NH_4-natrolite relative to Na-natrolite in accord with the 4 % reduction in unit cell volume in going from the Na- to the NH_4-form. The scatter of the individual angles around their mean value is remarkably large (up to 14° for Si-O5-Si). The distortion of the NH_4-framework relative to Na-natrolite is also reflected in the scatter of T-O distances within the tetrahedra. Comparison with the mean T-O distances in Na-natrolite shows that the values agree well only for the Si1-tetrahedra. The mean distances Si2-O and Al-O are distinctly larger in Na-natrolite than in NH_4-natrolite. In order to rationalize the distortion of the tetrahedral framework in NH_4-natrolite when compared with Na-natrolite, the changes in the bonds formed by the oxygen atoms to nonframework cations due to the cation exchange have to be considered. The rearrangement of the T-O bond lengths relative to the oxygen atoms (Table 4) shows that the mean O-T distances for O3 and O4 in NH_4-natrolite deviate significantly from the corresponding values in Na-natrolite. In Na-natrolite, each oxygen atom is bonded either to Na-atoms (O2, O3, O4) or by hydrogen bonding to the water molecules (O1, O5). After exchanging Na for NH_4 hydrogen bonds

Table 4. Bond lengths O-T [Å] in NH_4-natrolite (σ_4 = 2-3x10^{-3} Å) compared with those in Na-natrolite (σ = 2-4x10^{-4} Å)

	NH_4-natrolite		Na-natrolite		Δ	
	O-T1	O-T2	O-T1	O-T2	O-T1	O-T2
O11	1.604	1.717				
O12	1.618	1.731				
O13	1.600	1.752				
O14	1.602	1.740				
O1 mean	1.606	1.735	1.607	1.739	-.001	-.004
O21	1.617	1.742				
O22	1.615	1.735				
O23	1.611	1.749				
O24	1.617	1.740				
O2 mean	1.615	1.742	1.620	1.748	-.005	-.006
O31	1.602	1.724				
O32	1.611	1.718				
O33	1.588	1.725				
O34	1.607	1.737				
O3 mean	1.602	1.726	1.612	1.737	-.010	-.011
O41	1.598	1.737				
O42	1.597	1.730				
O43	1.610	1.739				
O44	1.596	1.714				
O4 mean	1.600	1.730	1.613	1.745	-.013	-.015
O51	1.609	1.646				
O52	1.626	1.663				
O53	1.639	1.621				
O54	1.644	1.610				
O5 mean	1.630	1.635	1.629	1.637	+.001	-.002

are **not** formed to **all** oxygen atoms: only one half of them are clearly hydrogen bond acceptors. Thus, the lengths of the individual T-O bonds within the various Si,Al-tetrahedra are influenced differently and the symmetry of the framework is reduced. Atoms O3 and O4, for example, are relatively strongly bonded to Na in Na-natrolite (Na-O ≈ 2.4 Å). In NH_4-natrolite, these bonds are partly replaced by much weaker hydrogen bonds, partly these O-atoms are coordinated by ammonium groups without hydrogen bonds. Therefore, O3 and O4 exhibit the largest differences of the mean O-T distances between the two forms of natrolite. On the other hand, the mean O-T distances for O1 and O5 which are hydrogen bonded in Na-natrolite as well as in NH_4-natrolite agree well in the two natrolites. Oxygen atom O2 is two-coordinated by Na in Na-natrolite but weaker bonded (Na-O ≈ 2.5 - 2.6 Å) than atoms O3 and O4. Generally it is observed that the bonds of the framework oxygen atoms to nonframework cations are weaker in NH_4-natrolite than in Na-natrolite and this results in shorter mean T-O distances in the ammonium exchanged form.

The framework of natrolite, as already realized by Pauling [1] is collapsible upon exchanges involving a reduction in volume. In collapsible frameworks all hinges distort in the same sense [4], that is they co-rotate. They collapse until either the minimum possible value of the T-O-T hinges is reached, or until they wrap themselves around the available cations [5]. This is borne out by a comparison of the NH_4- and Na-natrolites.

CONCLUSIONS

The comparison of NH_4-exchanged natrolite with the unexchanged form shows that the individual T-O distances and T-O-T angles are sensitive to different chemical bonds extended from the nonframework cations and water molecules to the oxygen atoms of the framework. The replacement of Na-O bonds in Na-natrolite by weaker H---O bonds (or not hydrogen bonded oxygen atoms) leads to shorter T-O distances in the framework and to very significant changes in T-O-T angles.

In the NAT framework all the T-O-T angles vary in the same sense when a volume change takes place. This co-rotation is an indication for collapsible zeolite structures.

We used the programs: CRYSTALS [6], STRUPLO90 [7] and SADIAN90 [8].

Acknowledgements: We thank W. Hofmeister, Mainz, for the samples of natrolite and the Bundesminister für Forschung und Technologie, Bonn, for support of this work by a grant.

REFERENCES

1. Pauling, L., Proc. Nat. Acad. Sciences, **16**, 453 (1930).
2. Baur, W.H., Kassner, D., Kim, C.-H. and Sieber, N.H.W., Eur. J. Min., **2**, 761 (1990).
3. Stuckenschmidt, E., Joswig, W. and Baur, W.H., Phys. Chem. Minerals, submitted (1992).
4. Baur, W.H., J. Solid State Chem., in press (1992).
5. Bieniok, A. and Baur, W.H., J. Solid State Chem., **90**, 173 (1991).
6. Watkin, D.J.: The CRYSTALS Primer, Oxford (1989).
7. Fischer, R.X., le Lirzin, A., Kassner, D. and Rüdinger, B. Z., Krist. Suppl. Issue, No. 3, 75 (1991).
8. Baur, W.H. and Kassner, D., Z. Krist. Suppl. Issue, No.3, 15. (1991)

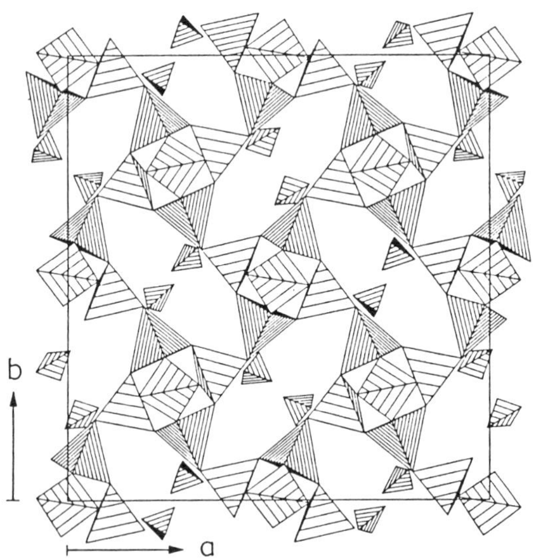

Fig. 1. View of the structure of NH_4-natrolite along the c-axis, using STRUPLO90 [7]. The SiO_4-tetrahedra are lightly shaded, the AlO_4-tetrahedra densely. The NH_4-groups are shown as small tetrahedra.

SYNTHESIS AND CHARACTERIZATION OF BRECK STRUCTURE SIX, FAUJASITE, ZSM-3, AND ZSM-20 AND THEIR STRUCTURAL RELATIONSHIPS

G.W. Skeels*, C.S. Blackwell**, K.B. Reuter**, N.K. McGuire** and C.A. Bateman**

* UOP, Tarrytown Technical Center, Tarrytown, NY 10591 USA.
**Union Carbide Chemicals and Plastics,Tarrytown Technical Center, Tarrytown, NY 10591 USA.

ABSTRACT

Breck Structure Six, (EMC-2), ZSM-20, ZSM-3 and ECR-32 have been prepared following basic synthesis methods. Characterization data for these materials are reported including X-ray powder diffraction and unit cell, thermal analysis, framework region infrared spectra, MAS-NMR, and wet chemical analysis. Electron diffraction and high-resolution electron microscopy were used to provide structural and morphological information. Structure modeling studies used in conjunction with the X-ray diffraction and electron diffraction studies provide a basis for comparison of these zeolites. The relationships between these materials within the faujasite family are discussed.

INTRODUCTION

In the structure of the natural mineral faujasite (FAU), the polyhedral cages (sodalite cages) are linked in a cubic array through double-six rings. From structures initially proposed by Moore and Smith, Breck [1] described six theoretical frameworks based on various linkages of the sodalite cages. Two of the proposed framework structures involve linkage through the double-six ring or hexagonal prism. One of the structures is that of FAU. The second proposed structure involves linkage of the sodalite cages through the double-six ring in a hexagonal rather than cubic array. Previously described as Breck Structure Six or hexagonal faujasite, this material has recently been assigned the structure-type code EMT [2]. The hexagonal polytype was once suggested as the framework structure of zeolites L [3] and ZSM-3 [4] and was possibly related to zeolite beta. However, zeolite L consists of erionite cages rather than sodalite cages linked through double-six rings [1]. The suggestion has been made that ZSM-3 consists of random intergrowths of cubic-and hexagonal-packed sodalite cages linked through the double-six ring [5]. Beta has a disordered structure unrelated to either material [6].

Recently, two separate laboratories have reported the synthesis of a material presumed to be the hexagonal polytype. Delprato [7] prepared EMC-2, and Vaughan [5] prepared ECR-30. Both materials are represented as more or less the pure hexagonal polytype. Extensive spectroscopic examination by Newsam has shown that ECR-30 is an intergrowth having approximately 85% hexagonal character and about 15% cubic character [8]. Recently, the hexagonal polytype prepared according to the procedure described by Delprato et al., was shown to contain some cubic character [9].

From the Proceedings of the 9th International Zeolite Conference, Montreal 1992, Eds. R. von Ballmoos et al.,

A number of additional synthetic zeolites are now known to be intergrowths of the cubic and hexagonal structures. For purposes of simplifying the description of an entire family of materials containing sodalite cages connected through double- six rings, the term faujasite family has been used. The term should not be taken to describe a specific structural arrangement. Additional materials that can be included in the faujasite family are ECR-4, ECR-32, CSZ-1, and ZSM-20. ECR- 4 [10] and ECR-32 [11] have a cubic morphology and the faujasite structure. CSZ-1, a cesium-containing, high-silica polymorph of faujasite is a distorted variant of FAU that crystallizes in hexagonal plates [12]. ZSM-20, synthesized by Ciric [13] was conceived as pure faujasite. However, Newsam has recently shown that the structure is an intergrowth of EMT and FAU [14]. In this paper, the properties and structural relationships of two of the more well-known intergrowths ZSM- 20 and ZSM-3 are compared with the FAU and EMT end members.

EXPERIMENTAL

Except for ZSM-3, which has been synthesized only in an inorganic system and with a SiO_2/Al_2O_3 ratio of < 3, the starting zeolites were synthesized to obtain a product SiO_2/Al_2O_3 ratio of about 7 to 8. EMC-2 (Throughout the remainder of this manuscript, the hexagonal polytype prepared by Delprato's method will be designated EMC-2), ECR-32, ZSM-20, and ZSM-3 were synthesized from gels suggested by Delprato [7], by Vaughan [5], by Dewaele, et al. [15], and by Ciric [4], respectively.

Basic characterization data (including X-ray powder diffraction and unit cell, thermal analysis, hydroxyl region and framework region infrared spectra, and wet chemical analyses) are reported for these materials. Electron diffraction and high-resolution electron microscopy were used to provide structural and morphological information. The electron microscopy technique was described earlier [9].

^{29}Si-NMR MAS can be used to determine a framework SiO_2/Al_2O_3 ratio and locate extraneous SiO_2 in the sample. Spectra obtained for this report were measured with a Bruker MSL-400 high resolution and solid state NMR operating at a field of 9.4 Tesla or 400.13 MHz for ^{1}H. The instrument was operated in the high power mode for MAS (magic angle spinning) solid state NMR. Some spectra were obtained on a Bruker CXP-200 operating at 4.7 Tesla.

RESULTS AND DISCUSSION

The product SiO_2/Al_2O_3 ratios for the as-synthesized forms of EMC-2, ECR-32, and ZSM-20 were between 7 and 8. Because the organic species remaining in the as-synthesized form of the zeolite can have an effect on standard property measurements, particularly adsorption measurements, the ECR- 32, EMC-2, and ZSM-20 were first calcined to remove the organic, and then ammonium-exchanged to establish a relative cation balance in the zeolite. The chemical composition and properties of the ammonium-exchanged zeolites, together with the as-synthesized composition and properties of the ZSM-3 are presented in Table 1. Chemical analysis of the three ammonium-exchanged samples showed no change in SiO_2/Al_2O_3 ratio as a result of calcination and ammonium exchange. The relative cation equivalents of the zeolites are near unity, indicating that most if not all of the aluminum remains in the framework

TABLE 1. COMPOSITION AND PROPERTIES OF ECR-32, EMC-2, ZSM-20 AND ZSM-3.

	ECR-32*	EMC-2*	ZSM-20*	ZSM-3
Chemical Analyses:				
Na^+/Al	0.01	0.16	0.01	0.72
M^+/Al	0.90	0.92	0.6	1.00
SiO_2/Al_2O_3	8.6	7.2	7.0	3.5
MAS NMR, SiO_2/Al_2O_3 Framework	8.7	7.0	7.1	3.4
Framework Infrared:				
Asymmetric Stretch, cm^{-1}:	1048	1032	1040	1001
Symmetric Stretch, cm^{-1}:	808	803	798	765
DTA Crystal Collapse Temp. °C	1010	986	990	839
McBain-Bakr Adsorption, wt.%:				
O_2, 100 torr, 0K	34.3	36.3	36.8	33.7
SF_6, 400 torr, 25°C	31.2	32.5	30.3	33.3
H_2O, 4.6 torr, 25°C	30.6	33.8	34.3	30.6

* Calcined 600°C to remove organic structure directing agent, then ammonium-exchanged.

following calcination and ammonium exchange. Because of its low SiO_2/Al_2O_3, ZSM-3 is inherently unstable. Since ZSM-3 contains no organic, it was not calcined and ammonium exchanged. The as-synthesized form had a cation equivalent of unity, indicating that all of the aluminum is in the framework.

The Si/Al_2 ratio and distributions in these faujasite-family materials were determined from the observed ^{29}Si-NMR MAS spectra. The assignments for the ^{29}Si lines are as follows: Si(4Al), -84 ppm; Si(3Al), -90 ppm; Si(2Al), -96 ppm; Si(1Al), -100 ppm; Si(0Al), -106 ppm; and extraneous SiO_2 at -110 ppm. These materials, whose local Si environment is closely related to faujasite, can be assigned similar to faujasite. The four crystallographic sites apparently have similar enough local environments that the ^{29}Si lines are not resolved. A calculation by the method of Ramdas and Klinowski [16], for the idealized EMT structure, predicts only a 0.2 ppm difference over 4 lines. Based on these assignments, a value for the framework Si/Al_2 was calculated from the areas of the individual lines representing each Si species. The ^{29}Si-NMR spectra for the four samples are shown in Figure 1, and the SiO_2/Al_2O_3 ratios calculated from these spectra are listed in Table 1. The calculated values are in close agreement with the bulk SiO_2/Al_2O_3 ratios obtained from chemical analysis. This agreement is further evidence that all of the aluminum is present in the framework of the respective zeolites and not in extraframework positions. The ^{27}Al-NMR results showed a single line, T_d framework Al spectrum, for all of the samples.

The X-ray powder patterns for EMC-2, ZSM-20, and ECR-32 showed the expected reflections, which were sharp and of high intensity. ZSM-3 showed the broad reflections indicative of an irregular intergrowth. The X-ray powder patterns are shown in Figure 2. McBain-Bakr adsorption values for O_2, SF_6, and water (Table 1) showed excellent adsorption capacity, which indicates that the materials were highly crystalline and had essentially unobstructed pores.

Framework-region infrared spectra for the calcined and ammonium-exchanged forms of EMC-2, ZSM-20, and ECR-32 and the as-synthesized form of ZSM-3 were similar, showing the expected range of peak positions for the asymmetric stretch and symmetric stretch bands that are indicative of the relative SiO_2/Al_2O_3 ratios (Table 1). The calcined and ammonium-exchanged EMC-2 and ZSM-20 showed the same two hydroxyl bands, at 3635 cm^{-1} and at 3550 cm^{-1}, found in FAU. The similarity in properties, particularly the MAS-NMR and infrared spectra of both framework and hydroxyl regions, suggests that the next nearest neighbor sites must be the same for these structurally different materials.

Electron Diffraction

Materials in the FAU-EMT family possess structural characteristics intermediate between the cubic and hexagonal end members. For materials that are neither purely cubic nor purely hexagonal, the structure will be described in terms of the symmetry operations that relate layers to each other. When the layers are related by inversion centers ***(i)*** through the double six rings, the cubic FAU structure is generated. If there is a mirror plane ***(m)*** across the double six rings between layers, the resulting structure is hexagonal EMT [14].

Diffraction patterns from ECR-32 can be indexed as having the FAU structure with a lattice parameter of 2.48 nm. The <110> zone axis patterns frequently show the presence of twin spots from twins on {111}, Figure 3A. Patterns with both twin spots and streaks caused by a relatively high density of closely spaced twin planes were also obtained, Figure 3B. Because each twin plane in FAU generates a unit cell of EMT it is seen that ECR-32 is not pure FAU, ie. the stacking is not pure ...***i i i*** ... There are some planes related by ***m*** operations resulting in local hexagonal symmetry.

At the other end of the family, EMC-2 is close to having the hexagonal EMT structure. <100> selected area diffraction patterns from EMC-2 are streaked indicating faults on the {001} planes of the hexagonal structure, Figure 3C. In this material the layers are nearly all related by ***m*** planes, ie. ... ***m m m*** ... stacking with some planes related by ***i*** operations.

The structures of ZSM-3 and ZSM-20 are intermediate between these two end members. Diffraction patterns from ZSM-3 were indexed on both the FAU and EMT structures. The patterns show considerable streaking due to the variability of the stacking sequence along <001> EMT ≡ <111> FAU, Figures 3D and 3E. This diffraction work is in accord with the previous studies showing that ZSM-3 contains approximately equal numbers of planes related by ***i*** operations and ***m*** planes [9]. This work also supports the idea that the hexagonal 'c' dimension is not defined in ZSM-3 because the layer stacking is so disorganized.

ZSM-20 lies structurally between ZSM-3 and EMT. A [001] EMT zone axis diffraction pattern is shown in Figure 3F. This crystal may well be faulted on planes perpendicular to the electron beam that do not contribute to the pattern. It has been reported that in ZSM-20, approximately 65 % of the stacking transitions are ***m*** planes, giving local hexagonal symmetry and 35 % are ***i*** operations resulting in cubic symmetry [14].

High Resolution TEM

High resolution [110] images of ZSM-20 show both hexagonal ***(m)*** and cubic ***(i)*** stacking sequences and faulting on more than one of the {111} FAU planes, Figure 4A. The region at the top of the micrograph contains cubic material. Also shown is an inclined fault below which there are many ***m*** related layers generating local hexagonal symmetry. Note that the hexagonal {001} planes cannot penetrate the inclined {111} cubic boundary, as reported by Newsam et al. [14].

EMC-2, when viewed along <100> EMT reveals primarily ***m*** stacking with an occasional ***i*** operation, consistent with the electron diffraction results. Figure 4B shows a typical high resolution image in the <100> orientation along the schematic of the projected structure showing the location of ***m*** and ***i*** operations. From a number of micrographs it was estimated that the material has ~ 85-92 % hexagonal symmetry. This is in accord with a recent paper by Newsam et al. [19].

CONCLUSIONS

Previously, ECR-30, which is EMT synthesized according to the method of Vaughan [5], was determined to be an intergrowth containing about 85% EMT and 15% FAU [8]. The present synthesis of EMC-2 also contains a significant amount of FAU intergrowth, as demonstrated by electron diffraction. The ratio of EMT/FAU was determined to be approximately 10:1. These data provide additional evidence that pure EMT has yet to be prepared. Structure modeling studies, coupled with X- ray diffraction and electron diffraction, indicate that ZSM-20 is an approximately 2:1 intergrowth of EMC-2 and FAU [8]. These same studies indicate that ZSM-3 is an approximately 1:1 intergrowth of EMT and FAU. The major difference between ZSM-20 and ZSM-3 is the degree of disorder in ZSM-3 along the "c" axis. This disorder is demonstrated from electron diffraction as well as by the broad reflections in the X-ray powder pattern.

REFERENCES

1. Breck, D.W., Zeolite Molecular Sieves, Wiley, NY, pp. 55- 58 (1974); Moore, P.B., and Smith, J.B., Mineral. Mag., **33**, 1008 (1964).
2. Breck Structure Six has recently been designated as EMT by the Structure Commission of the IZA.
3. Breck, D.W., and Flanigen, E.M., Molecular Sieves, Society of the Chemical Industry, London, p. 47 (1968).
4. Kokotailo, G.T., and Ciric, J., Adv. Chem. Ser. No.102, 823 (1971).

5. Vaughan, D.E.W., E.P. 0,315,461 (1988).
6. Treacy, M.M.J., and Newsam, J.M., Nature (London), **332**, 249 (1988); Newsam, J.M., Treacy, J.M.M., Koetsier, W.T., and DeGruyter, C.B., Proc. R. Soc., Ser. A. **420**, 375 (1988).
7. Delprato, F., Guth, J.L., Huve, L., Baron, J., Delmotte, L., Soulard, M., Anglerot, D., and Zivkov, C., Zeolites for the Nineties, Recent Research Reports, presented at 8th International Zeolite Conference, Amsterdam, (1989); Delprato, F., Delmotte, L., Guth, J.L., and Huve, L., Zeolites, 10, 546 (1990).
8. Treacy, M.M.J., Newsam, J.M. and Deem, M.W., Proc. R. Soc. Lond. A, **433**, 499-520 (1991).
9. Skeels, G.W., Reuter, K.B., and McGuire, N.K., Synthesis of Microporous Materials, Eds. M. Occelli, H. Robson, Van Nostrand Reinhold, New York, NY, in press.
10. Vaughan, D.E.W., U.S. 4,714,601 (1987).
11. Strohmaier, K.G., and Vaughan, D.E.W., E.P. 0,320,114 (1989).
12. Barrett, M.G., and Vaughan, D.E.W., U.S. 4,309,313 (1982).
13. Ciric, J., U.S. 3,972,983 (1976).
14. Newsam, J.M., Treacy, M.M.J., Vaughan, D.E.W., Strohmaier, K.G., and Mortier, W.J., J. Chem. Soc., Chem. Commun., 493 (1989).
15. Dewaele, N., Maistriau, L., Nagy, J.B., Gabelica, J., and Derouane, E.G., Applied Catalysis, **37**, 273-290 (1988).
16. Ramdas and Klinowski, Nature **304**, p. 521 (1984).
17. Vaughan, D.E.W., Treacy, M.M.J., Newsam, J.M., Strohmaier, K.G., and Mortier, W.J., Zeolite Synthesis, A.C.S. Symposium Series **450**, p. 544 (1983), M. Occelli, et al. editors.
18. Csencsits, R., and Gronsky, R., Mat. Res. Soc. Symp. Proc. **115,** 103-108 (1988).
19. Newsam, J.M., Treacy, M.M.J., Vaughan, D.E.W., Strohmaier, K.G. and Melchoir, M.T., Synthesis of Microporous Materials, Eds. M. Occelli, H. Robson, Van Nostrand Reinhold, New York, NY, in press.

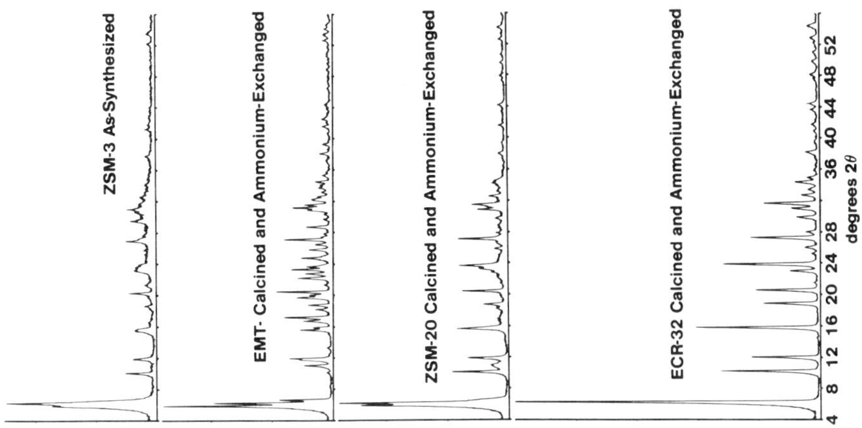

Figure 2. **X-Ray Powder Diffraction Pattern for: BSS, ECR-32, ZSM-20, and ZSM-3**

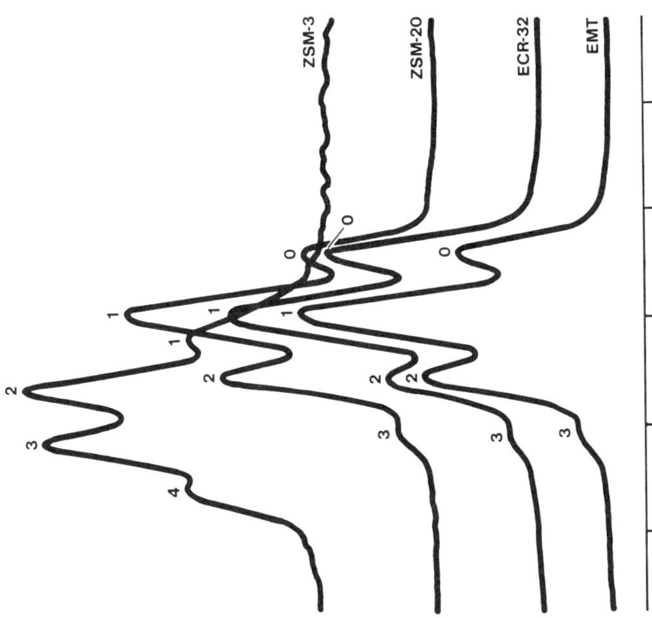

Figure 1. **^{29}Si MAS-NMR Spectra for: BSS, ECR-32, ZSM-20, and ZSM-3**

Note: The numbers above the resonance peaks refer to the number of next-nearest neighbor Al's assigned for that peak

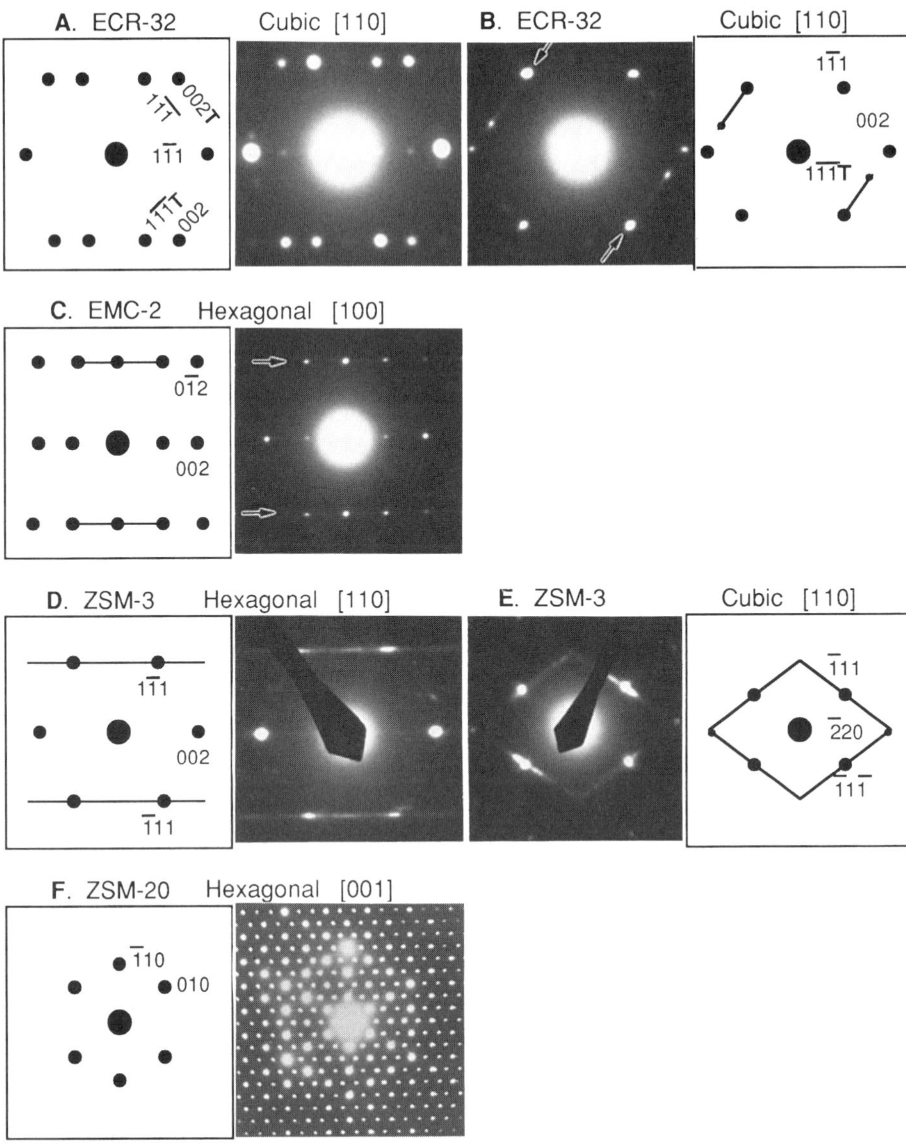

Figure 3. **Electron Diffraction Patterns from ECR-32, EMC-2, ZSM-3 and ZSM-20. Twin spots are marked "T". The ZSM-20 pattern is smaller because it was recorded at a much lower camera length.**

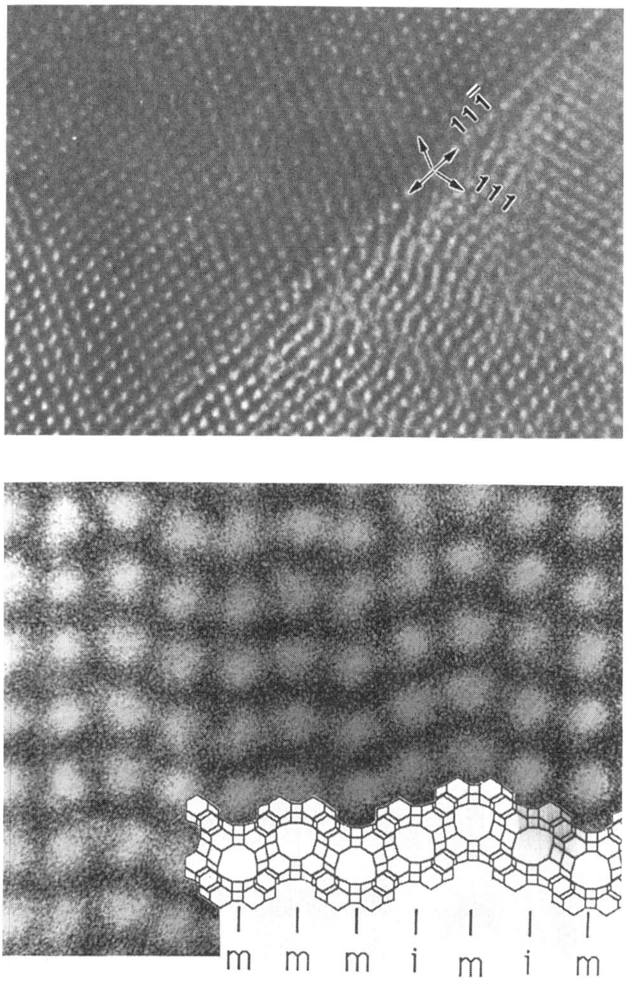

Figure 4. **High Resolution Images**
A. From ZSM-20 showing faulting on more than one of (111) FAU Planes. B. From EMC-2 showing location of "m" and "i" Operations.

SINGLE CRYSTAL STRUCTURE ANALYSIS OF ZEOLITE BETA: THE SUPERPOSITION STRUCTURE

B.Marler, R. Böhme and H.Gies

Institut für Mineralogie, Ruhr-Universität Bochum, Postbox 102148, D-4630 Bochum, Germany

ABSTRACT

A characteristic feature of the crystal structure of zeolite Beta is its onedimensionally disordered silicate framework which is built from layers as building units. Neighboring layers can be connected to each other in different ways giving rise to the observed stacking disorder. Single crystals were synthesized hydrothermally from silicious solutions using 4,4'-trimethylene dipiperidine as template. From Weißenberg and precession photographs a characteristic diffraction pattern of always two rods of diffuse reflexions between rods of sharp reflexions was obtained. The sharp reflexions represent the pattern of the superposition structure with $a_0=b_0=4.12$Å, $c_0=13.01$Å and space group $P4_2/mmc$. This structure is identical for all possible polytypes of Beta ordered as well as disordered. The superposition structure was refined from a data set of the sharp reflexions using conventional techniques (R=0.054 and $R_w=0.048$). The successful refinement confirms the Beta structure model.

INTRODUCTION

Zeolite Beta is the first synthetic large pore zeolite with a three dimensional channel system and high thermal stability. The X-ray powder diffraction pattern of zeolite Beta is characterized by a combination of broad and sharp reflexions indicating disorder of the silicate framework. NMR data, however, show high local symmetry for 9 symmetrically independent silicon atoms which could be resolved from high resolution ^{29}Si MAS NMR spectra of a silica-Beta sample [1]. Although zeolite Beta was already patented in 1967 [2] and, therefore, is one of the earliest of this class of synthetic microporous materials to be synthesized, so far, no refinement of its structure has been published.

Recently, a model of the structure was proposed based on electron diffraction and on the comparison of experimental and simulated powder XRD pattern [3,4,5,8]. The model postulates a polytypic series of structures. There are three different (hypothetic) ordered "end members" of the series each possessing a distinct stacking sequence of the **layer like building units**: Polytype A (two enantiomorphs with space groups $P4_122$ and $P4_322$), B (C2/c) and C (P2/c) (Fig. 1). The three ordered polytypes are all characterized by three sets of 12-ring channels which intersect mutually and thus form a 3-dimensional channel system. Electron diffraction experiments, however, revealed strong disorder along c* indicating either intergrowth of the different polytypes or statistical arrangement of the layer like building units. Any changes in the layer sequence do not block the channels nor change their size and dimensionality.

From the Proceedings of the 9th International Zeolite Conference, Montreal 1992, Eds. R. von Ballmoos et al.,

Although numerous syntheses routes have been reported so far only disordered Beta has been obtained in a microcrystalline form [5, p. 377]. We report here on the synthesis and structural characterization of large single crystals of Beta.

EXPERIMENTAL

Zeolite Beta has been synthesized under hydrothermal conditions from aqueous solutions containing silica (0.65 molar), boric acid (0.15 molar) and organic amines (e.g. 4,4'-trimethylene dipiperidine) as templates. The solutions were sealed in silica tubes and heated in an oven at 180°C and 200°C respectively for several months. After seperating the crystalline material from the remaining solution a few single crystals were calcined at 800°C in order to completely remove the organic material.

The boron content of the as synthesized sample was determined using a JY38 plasmaspectrograph with a ICP 1500 plasmatherm Generator.

Single crystals of Beta in the calcined and in the as synthesized form were examined by Weißenberg and precession technique.

A calcined single crystal was used to collect 890 intensities (Mo Kα, $2\vartheta_{max}$ = 65°) of the sharp reflexions using omega scan mode on a Syntex R3 diffractometer. 259 reflexions remained after merging symmetrically equivalent reflexions (R_{int} = 0.029). These were used for a conventionell refinement procedure with the XLS-system [6]. Scattering factors for neutral atoms were used as provided by the XLS-system. No absorption correction was necessary.

RESULTS and DISCUSSION

After 5 months of thermal treatment large single crystals (110x110x210μm) of Beta were obtained. No other crystalline phase was observed. Beta crystallizes in tetragonal bipyramidal morphology with an additional small basal pinacoid (Fig. 2). The clear, colorless crystals are birefringent and show a diffuse extinction under crossed nicols which is related to the disordered framework structure. These large single crystals allow a detailed x-ray structure analysis.

Analysis of single crystal X-ray photographs of zeolite Beta

Weißenberg and precession photographs show that the single crystals are disordered in one dimension.

The h0l and 0kl diffraction patterns consist of always two rods of diffuse reflexions between rods of sharp reflexions (Fig. 3). The diffuse streaks extend along c*. The spacing between the spots along the sharp columns corresponds to 13.01Å.

The hk0 diffraction pattern (Fig. 4) shows orthogonal symmetry with $a_o = b_o$ = 12.36Å. There is no streaking in this direction since the streaks seen in Figure 3 are rods running perpendicular to the hk0 plane. Many of the spots visible are points of interception between the Ewald diffraction sphere and the streaks and, therefore, do not represent truly sharp reflexions.

The presence of the sharp reflexions (in the h0l and 0kl patterns) at h = 3n and k = 3n indicates that the disorder is characterized by $1/3a_o$ and $1/3b_o$ displacements within the (001) plane. In general, the single crystal x-ray diffraction patterns of Beta are of the same type as the electron diffraction patterns obtained from small crystallites of Beta [3,4,5].

The interpretation of the single crystal x-ray photographs is consistent with the following model: The framework structure of Beta is built from identical layers as building units arranged parallel to (001). These layers possess the plane space group symmetry P4m2 and the dimensions of 12.36Åx12.36Åx6.51Å (Fig. 5). Neighboring layers are shifted by 6.51Å along the "c" direction. They are always rotated 90° against one another. The presence of the streaks in (h0l) for all h = 3n±1 and in (0kl) for all k = 3n±1 is due to the fact that all or at least most of the layers are shifted $\pm 1/3a_o$ or by $\pm 1/3b_o$ in addition to the rotation and that these shifts occur in a disordered sequence. Most of the layers will be stacked with a shift of $a_o/3$ or $b_o/3$ in the (001) plane, however, it can't be ruled out that there are pairs of layers intercalated which are connected without any displacement in the (001) plane. The latter kind of stacking the layers generates double 4-rings in the framework. Since the existence of double 4-rings is believed to be unfavourable in high silica zeolites and since there is no indication for double 4-rings in the NMR spectrum this stacking sequence is regarded to be less likely, but nevertheless has to be taken into consideration.

As described by Newsam et al. the ordered sequence of layers stacked vertically one above the other without any displacement in a and b direction produces another hypothetic polytype of Beta with space group symmetry $P4_2/mmc$ [5].

Structure refinement of the superposition structure

The sharp reflexions in the diffraction pattern determine the superposition structure which is identical for all disordered structures as well as for all ordered polytypes of Beta in question. The superposition structure is tetragonal with a unit cell of a_o = b_o = 4.12Å and c_o = 13.01Å. Systematic extinction leads to space groups P42c, $P4_2mc$ and $P4_2/mmc$. Because of the stacking disorder Beta can be described as an OD structure in the sense of the order-disorder theory of Dornberger-Schiff [7]. According to the OD theory the space group symmetry of the superposition structure is $P4_2/mmc$. The initial set of coordinates can be derived by superimposing the framework structure of any ordered polytype or disordered structures of Beta on the small subcell of the superposition structure. We used the DLS-optimized atomic coordinates of polytype A [5]. The superposition leads to 3 symmetrically independent positions for silicon atoms and 7 symmetrically independent positions for oxygen atoms. However, the two oxygen atoms O1 and O2 as well as the two oxygen atoms O3 and O4 cannot be resolved in the small unit cell of the superposition structure. Therefore, they have been treated as single scatterers (O2 and O4) in the refinement (Tab. 2). Since the chemical analysis revealed a boron content of B/(B+Si) = 0.03 it was assumed that boron partially and randomly replaces the silicon atoms in the framework. In order to make allowance for the lower refractive power of the T-positions the overall occupation factor of the silicon atoms was refined.

Best results were obtained refining the superposition structure in the space group $P4_2/mmc$. Attempts to refine the structure in one of the other possible space group symmetries yielded higher R values and less reliable results, thus confirming the theoretical treatment on the basis of the OD theory. A summary of the experimental details is given in Table 1.

a_o		4.121Å
c_o		13.010Å
Space group		$P4_2/mmc$
$2\vartheta_{max.}$	(λ = 0.7108Å)	65°
Number of reflexions		890
Number of independent reflexions		259
Number of parameters		52
R	$R = (F_{obs}-F_{calc}) / (F_{obs})$	5.4%
R_w	$R_w = ((F_{obs}-F_{calc}) * w^{1/2}) / (F_{obs} * w^{1/2})$	4.8%

Table 1: Summary of the experimental parameters of the single crystal structure refinement.

As can be seen from Table 2 the refined coordinates of the superposition structure differ only slightly from the initial coordinates confirming the proposed layer model of the Beta structure.

	Refined atomic coordinates				Initial set of coordinates			
Atom	x	y	z	K	x	y	z	K
Si1	.0000	.3827(14)	.1141(4)	.450(4)	0.0000	0.3788	0.1123	0.500
Si2	.4179(4)	.4179(4)	.2500	.450(4)	0.4217	0.4217	0.2500	0.500
Si3	.1316(5)	.3781(6)	.1123(2)	.900(7)	0.1434	0.3793	0.1144	1.000
O1					0.0000	0.0000	0.6304	0.250
O2	.0000	.0831(12)	.6287(4)	.750	0.0000	0.1489	0.6313	0.500
O3					0.0033	0.4518	-.0079	0.500
O4	.0000	.473(16)	.0000	.750	0.0000	0.4634	0.0000	0.250
O5	.5000	.503(11)	.1218(6)	.500	0.5000	0.4839	0.1215	0.500
O6	.0684(9)	.4378(12)	.1995(4)	.000	0.0786	0.4574	0.1999	1.000
O7	.3161(12)	.4455(18)	.1633(4)	.000	0.3128	0.4684	0.1633	1.000

Table 2: The refined atomic coordinates of the superposition structure and the initial coordinates.

As an effect of the superposition we find extreme short distances between some atoms, in particular for the symmetrically related atoms of the oxygen positions O4 and O5:

$d_{O4-O4'}$ = 0.22Å with O4: x=0, y=.473, z=0 and O4': x=0, y=.527, z=0;

$d_{O5-O5'}$ = 0.02Å with O5: x=1/2, y=0.503, z=0.1218 and O5': x=0.1/2, y=0.497, z=0.1218.

Consequently, high standard deviations appear for the y values of O2 and O5.

All displacement parameters (Tab. 3) have values which are expected for zeolite structures. This indicates that the model used for the refinement quite satisfactorily explains the experimental data. Since we look on a superposition structure most of the observed distances and angles do not reflect true bonds. Therefore, a detailed discussion of distances and angles is omitted here. Figure 6 shows three different projections of the superposition structure. The "bonds" shown in this drawing represent only those Si-O distances which have a value of about 1.6 Å (the expected Si-O bond length), all other Si-O distances are left out for clearity.

Atom	U11	U22	U33	U23	U13	U12
Si1	.0212(21)	.0202(14)	.0180(13)	-.0010(11)	0.0000	0.0000
Si2	.0193(8)	.0193(8)	.0182(9)	0.0011(5)	-.0011(5)	-.0004(7)
Si3	.0187(13)	.0219(9)	.0202(9)	-.0028(7)	0.0006(9)	0.0035(10)
O2	.0259(19)	.0518(45)	.0456(26)	-.0013(23)	0.0000	0.0000
O4	.0425(15)	.0376(17)	.0282(13)	0.0000	0.0000	0.0000
O5	.0174(22)	.0456(39)	.0420(36)	0.0172(22)	0.0000	0.0000
O6	.0290(32)	.0353(40)	.0302(18)	0.0077(17)	-.0031(15)	0.0032(18)
O7	.0316(21)	.0316(59)	.0398(22)	0.0008(23)	-.0130(23)	0.0047(22)

Table 3: Refined displacement parameters of the superposition structure.

CONCLUSION

The refinement of the superposition structure of zeolite beta is the first step of a complete experimental verification of the proposed structure model. It confirms the topology of the "basic layer" as the building unit. However, there is no information on the stacking arrangement or on the stacking sequence of the layers in the silicate framework of Beta. A detailed analysis of the diffuse diffraction intensities of Beta crystals synthesized in a wide range of composition is currently in progress.

REFERENCES

1. Fyfe, C.A., Strobl, H., Kokotailo, G.T. and Pasztor, C.T., Zeolites **8**, 132 (1988).
2. Wadlinger, R.L., Kerr G.T. and Rosinski, E.J., U.S. Patent No. 3, 308, 069 (1967).
3. Treacy, M.M.J. and Newsam, J.M. ,Nature **332**, 249 (1988).
4. Higgins, J.B., LaPierre, R.B., Schlenker, J.L., Rohrman, A.C., Wood, J.D., Kerr G.T. and Rohrbaugh, W.J., Zeolites **8**, 446 (1988).
5. Newsam, J.M., Treacy, M.M.J., Koetsier, W.T. and De Gruyter, C.B., Proc. R. Soc.Lond. **A 420**, 375 (1988).
6. XLS - Crystal structure refinement- 4.1, Siemens Analytical Xray Inst. Inc. (1990).
7. Dornberger-Schiff, K., Abh.Dtsch.Akad.Wiss.Berlin, Kl. Chemie Geol. Biol N.3 (1964).
8. Treacy, M.M.J., Deem M.W. Newsam, J.M., Proc. Roy. Soc. Lond. **A 433**, 499 (1991).

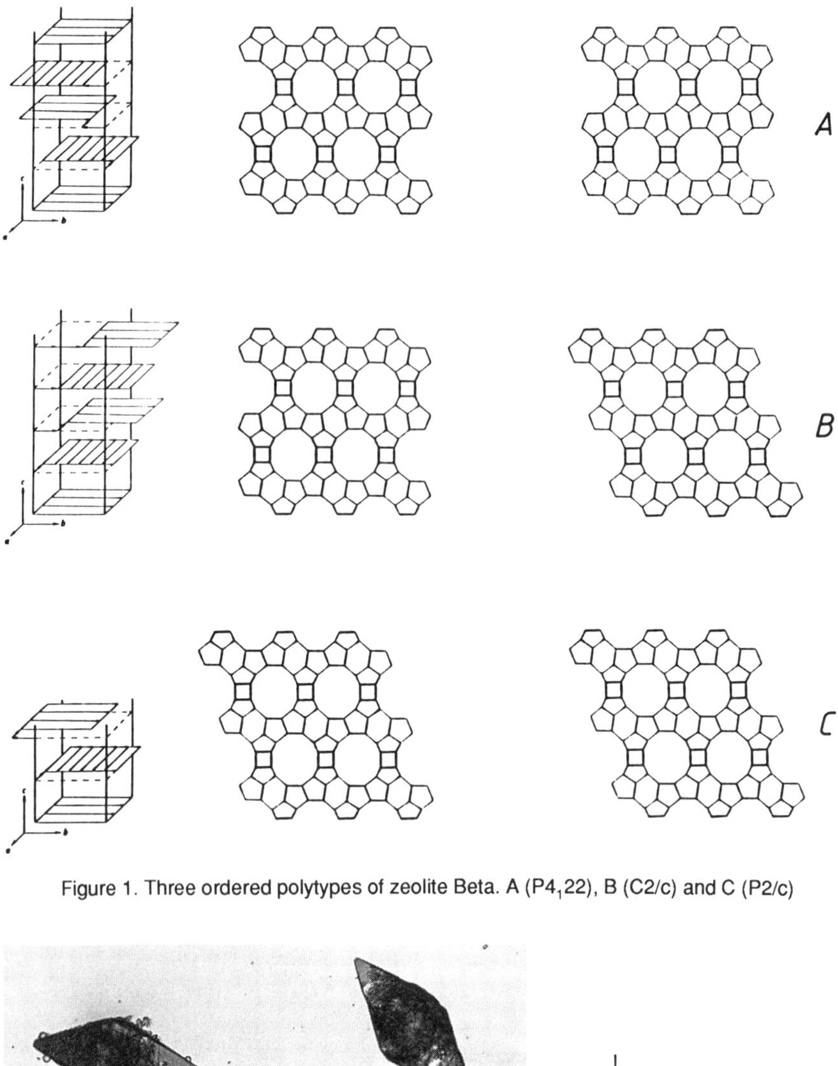

Figure 1. Three ordered polytypes of zeolite Beta. A ($P4_122$), B ($C2/c$) and C ($P2/c$)

30μm

Figure 2: The morphology of the zeolite Beta crystals showing the prevalent tetragonal bipyramid.

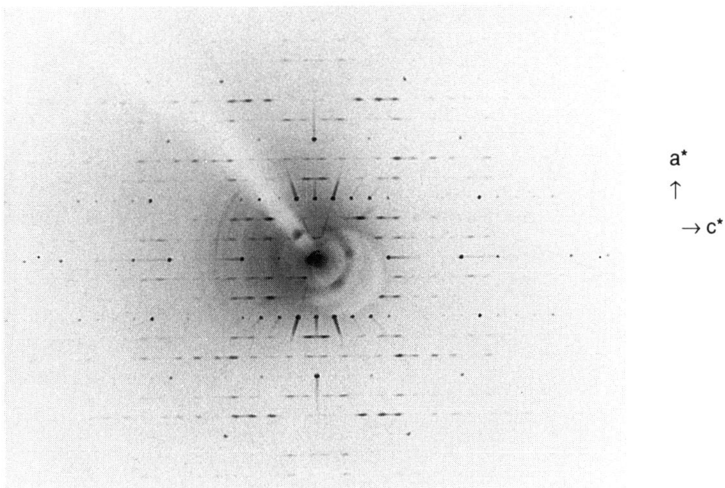

Figure 3. Precession photograph of the (h0l) plane of Beta.

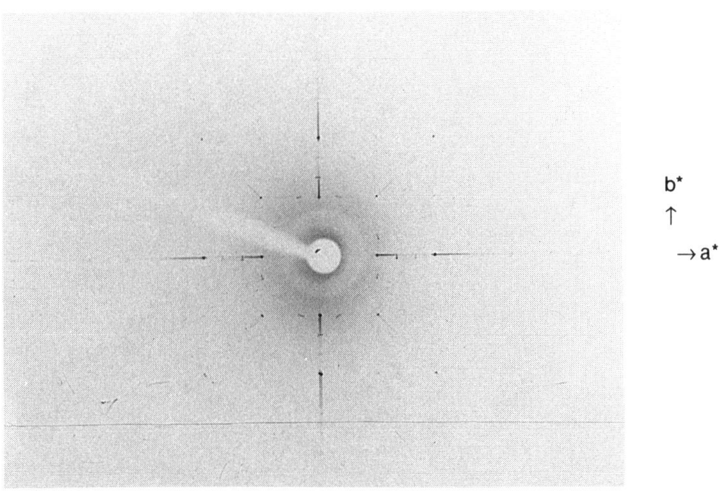

Figure 4. Precession photograph of the (hk0) plane of Beta.

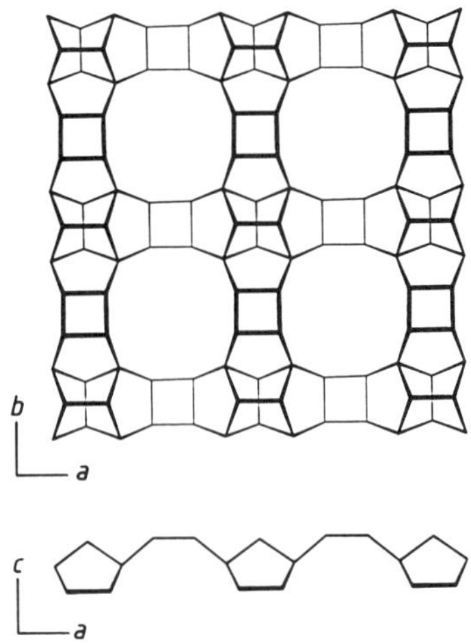

Figure 5. The layer like building unit of the silicate framework of Beta.

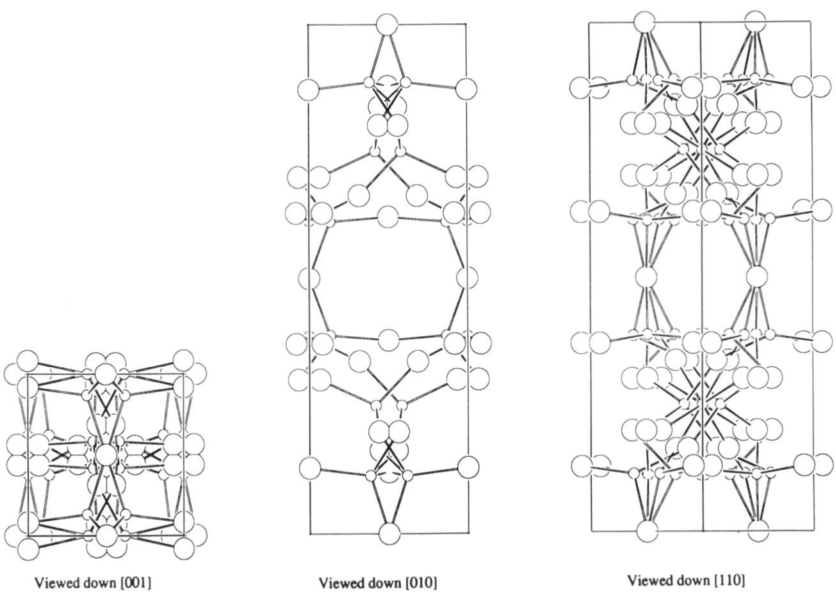

Figure 6. The superposition structure of zeolite Beta. Si = small circles, O = large circles.

RIETVELD REFINEMENT OF F^--CONTAINING $GaPO_4$-LTA

A. SIMMEN*, J. PATARIN**, and Ch. BAERLOCHER*

*Institute of Crystallography, ETH, CH-8092 Zurich, Switzerland

**E.N.S. de Chimie, 3 rue Alfred Werner, Mulhouse, France

ABSTRACT

A gallophosphate-LTA, synthesized using the fluoride method with di-n-propylamine (DPA), has been characterized with ^{13}C, ^{19}F, ^{31}P and ^{71}Ga MAS NMR and the structure refined in space group $Fm\bar{3}c$ (a = 24.035 Å) with x-ray powder data to $R_F = 0.031$ and $R_{wp} = 0.11$. The F^- ion was located in the double 4-ring making all Ga atoms pentacoordinate. The N-atom in DPA is protonated and was found near the middle of the 8-ring window where it forms a hydrogen bond to one of the framework oxygen. The propyl groups point towards the center of neighboring α-cages.

INTRODUCTION

Following the discovery of the aluminophosphate molecular sieves [1], a wide range of chemically related framework phosphates were synthesized. Many of these were counterparts of aluminosilicate zeolites. Interestingly, the LTA structure, which is easily synthesized in the aluminosilicate system, can only be synthesized in the presence of Si in the aluminophosphate system [2].

The synthesis of the LTA structure in the gallophosphate system, using di-n-propylamine (DPA) as the organic base, was successful only after HF was added to the synthesis mixture [3]. Chemical analysis showed that the F^- ions were incorporated in the structure. In a similar synthesis, but with the organic cation quinuclidine, the new large pore molecular sieve cloverite crystallized [4,5]. To investigate the role of the F^- ions and of the DPA used in the synthesis of $GaPO_4$-LTA, a structure refinement using powder diffraction data was undertaken.

EXPERIMENTAL SECTION

The X-ray data were collected on a Scintag PAD-X θ-θ diffractometer with CuKα radiation. The crystallites of the sample used were cubes with edges of 2-3 μm, so no preferred orientation was expected. A summary of the data collection parameters is given in Table 1. As expected for strictly alternating Ga- and P-atoms, strong superlattice reflections requiring a 24.035 Å unit cell were observed. Systematic absences were consistent with the space group $Fm\bar{3}c$.

From the Proceedings of the 9th International Zeolite Conference, Montreal 1992, Eds. R. von Ballmoos et al.,

Table 1. Crystallographic and experimental data for $GaPO_4$-LTA

Temperature (°C)	20
Wavelength (Å)	1.5406
Profile range in refinement (°2θ)	9.00 - 74.00
Step size (°2θ)	0.01
Standard peak (hkl,°2θ,FWHM)	664,34.99,0.083
Space group in refinement	$Fm\bar{3}c$
Cell constant a (Å)	24.035(3)
R_F	0.031
R_{wp}	0.111
R_{exp}	0.032

The unit cell chemical composition determined by different techniques [3] was found to be $[Ga_{96}P_{96}O_{384}]$ DPA_{24} $F_{24}\cdot 48\ H_2O$. The water content was measured on a sample equilibrated at $P/P_o = 0.65$ and T = 25°C.

The ^{13}C, ^{19}F, ^{31}P and ^{71}Ga MAS NMR spectra shown in Figure 1 were recorded on Bruker MSL 300 and MSL 400 (for ^{71}Ga) spectrometers. The recording conditions are given in Table 2.

Table 2. Recording conditions for the MAS NMR spectra

	^{13}C	^{19}F	^{29}Si	^{31}P	^{71}Ga
Standard	TMS	$CFCl_3$	TMS	85%H_3PO_4	$Ga(H_2O)_6^{3+}$
Frequency (MHz)	75.47	282	59.63	121.44	121.99
Recycle time(s)	5	4	10	5	0.2
Pulse width 10^{-6}(s)	6.5	3	5.5	1.8	1.0
Contact time 10^{-3}(s)	2				
Spinning rate(Hz)	3500	8000	3400	8000	9000
Number of scans	340	180	1600	16	50 000

STRUCTURE REFINEMENT

The X-ray Rietveld system XRS-82 [6] was used for the refinement process. The data were scaled, the background estimated and subtracted from the powder pattern, and an experimental peak shape function [7] calculated. A difference Fourier map calculated with idealized starting coordinates for the framework atoms ($R_F = 0.247$, $R_{wp} = 0.785$) revealed a peak inside the double 4-ring (D4R). This was interpreted as a F^- ion position. The site multiplicity of 24 agreed with the chemical analysis for the F^- ions, and the position refined with a population parameter close to one. Further difference Fourier maps showed electron density protruding from the 8-ring windows towards the centers of adjacent α-cages (Figure 2), and this was interpreted as a DPA^+ ion averaged by the symmetry of the framework. Initially, the N-atom was positioned in the center of the 8-ring but the refinement showed that with this assumption, the electron density could not be modelled satisfactorily. The N-atom was then moved away from the center but kept in the plane of the 8-ring. The C-atoms of the n-propyl groups were also positioned in such a way that they followed the observed electron density. To maintain a reasonable geometry of the DPA^+, heavily weighted restraints (common weight factor 5) were used for the C-N and C-C distances as well as for the tetrahedral angles at the N- and C-atoms. Refinement of the occupancy factor for the whole molecule converged to 1 molecule per 8-ring, so it was fixed at this value. The protons of the NH_2-, CH_2- and CH_3-groups were not explicitly included in the refinement, but were taken into account by a corresponding increase of the population parameters of the C- and N-atoms. The isotropic thermal parameters (U) of the Ga, P- and O-atoms were refined independently, but the U's of the oxygen atoms were highly correlated, so they were constrained to be equal. Since the temperature factors of the atoms of the DPA^+ ions proved to be unstable in refinement, they were fixed at a value of U = 0.02 $Å^2$. After several cycles of refinement, a further electron density map was calculated which showed only weak

Table 3. Positional, thermal and occupancy parameters for $GaPO_4$-LTA [a]

Atom	Wyckoff position	x	y	z	$U(10^{-2}Å^2)$	Occupancies [b]
Ga	96i	0	0.1896(4)	0.0817(4)	2.6(3)	96
P	96i	0	0.0981(9)	0.1808(10)	0.5(6)	96
O(1)	96i	0	0.1146(22)	0.2411(19)	1.3(9)[c]	96
O(2)	96i	0	0.1529(16)	0.1508(16)	1.3(9)[c]	96
O(3)	192j	0.0504(13)	0.0665(12)	0.1680(12)	1.3 (9)[c]	192
F	24c	1/4	0	0	4.(3)	24
NH_2[d]	96i	0.276(6)	0	0.251(17)	2.0	24
$CH_2(1)$[d]	192j	0.235(15)	0.046(9)	0.242(14)	2.0	24
$CH_2(2)$[d]	192j	0.246(20)	0.092(9)	0.285(14)	2.0	24
$CH_3(3)$[d]	192j	0.218(10)	0.146(5)	0.265(20)	2.0	24
$CH_2(4)$[d]	192j	0.251(12)	- 0.054(8)	0.231(8)	2.0	24
$CH_2(5)$[d]	192j	0.268(8)	- 0.101(4)	0.271(20)	2.0	24
$CH_3(6)$[d]	192j	0.216(12)	- 0.127(10)	0.296(10)	2.0	24
$H_2O(1)$	48e	0	0	0.120(14)	2.0	12.(4)
$H_2O(2)$	96i	0.039(13)	0.0287(18)	0	2.0	16.(3)

a Numbers in parentheses are the esd's in units of the least significant digit given. Those parameters without esd's were held fixed in least-squares refinement.

b Occupancy parameters are given as the number of atoms per unit cell. The population parameters for the atoms of the DPA^+ were constrained to be equal.

c These parameters were constraint to be equal.

d Occupancy parameters of these species were adjusted to reflect the contribution of the H atoms to the scattering.

peaks inside the sodalite cage, and these were modelled with two oxygen positions, labelled $H_2O(1)$ and $H_2O(2)$. Their occupancy factors correspond to approximately 4 water molecules per sodalite cage, and only account for 2/3 of the water found by thermogravimetric analysis. Despite their low occupancy, these positions refined well and were retained in the model.

At the beginning of the refinement, geometric restraints were also used for the distances and angles of the framework atoms. Despite this, the O-Ga-O angles were found to be significantly different from the prescribed tetrahedral angles. The F^- ion in the D4R increased the coordination number of Ga to five and the angles at Ga reflected this approach towards trigonal bipyramidal coordination. The geometric restrictions were therefore removed (except for the ones on DPA^+) and the refinement progressed to an R_{wp} of 0.111. The framework positions remained stable, the shifts were less than their corresponding esd's and a final difference Fourier did not show any significant electron density peaks.

The refinement conditions with the final R-values are summarized in Table 1, final atomic parameters are listed in Table 3 and selected bond distances and angles in Table 4. The observed and calculated powder patterns are shown in Figure 3.

Table 4. Selected interatomic distances (Å) and angles (deg)

P - O(1)	1.49(5)	O(1) - P - O(2)	104.(3)
P - O(2)	1.50(4)	O(1) - P - O(3)	109.(2)
P - O(3)	1.46(3) 2x	O(2) - P - O(3)	111.(2)
P - F	2.88(2)	O(3) - P - O(3)	111.(2)
P - $H_2O(1)$	2.8(2)		
Ga - O(1)	1.84(5)	F - Ga - O(1)	79.(2)
Ga - O(2)	1.87(4)	F - Ga - O(2)	172.(1)
Ga - O(3)	1.83(3) 2x	F - Ga - O(3)	80.(1) 2x
Ga - F	2.44(1)	O(1) - Ga - O(2)	94.(2)
Ga - $H_2O(1)$	2.6(2)	O(1) - Ga - O(3)	115.(1) 2x
		O(2) - Ga - O(3)	102.(1) 2x
		O(3) - Ga - O(3)	120.(2)
		P - O(1) - Ga	139.(4)
NH_2...O(1)	2.7(4)	P - O(2) - Ga	146.(3)
NH_2...O(2)	2.9(4) 2x	P - O(3) - Ga	138.(2)

RESULTS AND DISCUSSION

The ^{13}C MAS NMR spectrum (Figure 1a) of $GaPO_4$-LTA consists of three resonances (49.9, 20.3 and 12.1 ppm) corresponding to C_α ($N-CH_2$), C_β ($-CH_2-$) and C_γ ($-CH_3$) respectively. Comparison of the chemical shifts with previously published results [8] shows that the di-n-propylamine is protonated (DPA^+)· The ^{19}F MAS NMR spectrum (Figure 1b) has a narrow peak at

-72 ppm, with spinning side-bands marked with asterisks. Such a signal is consistent with the F^- ion located in the D4R. The broad peak at -175.5 ppm can be attributed to fluoride impurities.

The ^{31}P and ^{71}Ga MAS NMR spectra have also been reported elsewhere [3] but are included here because of their relevance to the discussion of the structure. Especially interesting is the ^{31}P spectrum (Figure 1c) which shows 2 peaks with a peak area ratio of 1:3, while only one crystallographic site for P is present in the structure. A possible explanation for this is given below. The chemical shifts of -5.3 and -11.7 ppm are markedly different from those in $AlPO_4$'s, but are comparable with the value of -9.8 ppm found by Turner et al. [9] for an anhydrous Ga-orthophosphate. There are almost no publications available on ^{71}Ga MAS NMR spectra for gallophosphates. Merrouche et al. [3] argue, that the peak at -36.8 ppm (Figure 1d) can be assigned to a penta- or hexacoordinate Ga atom.

The LTA structure is well known. Gramlich and Meier [10] showed in 1971 that the Al and Si are strictly alternating in the aluminosilicate-LTA framework, and that the cell dimensions had to be doubled. This superstructure is observed much more clearly in the gallophosphate-LTA because of the larger difference in the atomic scattering factors of Ga and P as opposed to those of Al and Si. The most striking feature of the structure, however, is the location of the F^- ion inside the D4R depicted in Figure 4. Although at first sight such a position may be surprising, it was also found in the silica polymorph octadecasil [11] (AST structure type), and, even more recently, in the gallophosphate cloverite [5]. Both were prepared using the fluoride method. Cloverite and $GaPO_4$-LTA show an interaction between the F^- ions and the Ga in the D4R, which is manifested in short F-Ga approaches of 2.44 Å in $GaPO_4$-LTA and between 2.38 - 2.48 Å for the fully connected D4R in cloverite. Each F^- ion makes four such contacts, and in $GaPO_4$-LTA all are symmetrically equivalent. The relatively short distance of the F^- ions to the Ga atoms makes the latter pentacoordinate. The F^- ion, together with the four neighboring oxygens of the framework, form a distorted trigonal bipyramid (Figure 4). As would be expected, the distance to the P position is much longer (2.9 Å), and thus precludes a stronger interaction.

Locating the DPA^+ proved to be quite difficult. Nonetheless, some useful insight into the position and orientation of this organic cation emerged. Figure 2 shows a section of a difference Fourier map through the center of the α-cage. It was calculated using the framework atoms only. Also plotted is the refined orientation of the DPA^+. This is of course only one possible orientation because the local symmetry of 4/m generates a total of 8 different orientations at this position. Since there are only 3 peaks in the ^{13}C MAS NMR spectrum, the "observed" deviation from the mirror plane in the DPA^+ may reflect only the disorder of this molecule. Therefore, it should be stressed, that the actual orientation and conformation of the DPA^+ ion may be different from the refined one. To give an idea how the DPA^+ may pack, an arrangement of six DPA^+'s is shown in Figure 5.

The refinement showed that the N-atom is shifted away from the center of the 8-ring by about 0.6 Å. This results in closer contacts to 3 of the oxygen atoms forming the 8-ring window . The distance of 2.7 Å to O(1) indicates a strong N-H...O hydrogen bridge. This could explain the somewhat surprising ^{31}P spectrum. Since there is only one DPA^+ per 8-ring window, only one hydrogen bridge per window can be present at any one time. The hydrogen bridge will pull some electron density away from the associated tetrahedron, thus, deshielding the P atom. This has the effect of moving the chemical shift of the ^{31}P atom towards less negative values. Of the four P tetrahedra in the 8-ring only one is affected in such a way, resulting in an intensity ratio of the ^{31}P signal of 1:3. This interpretation implies, that the DPA^+ is stationary and could be ordered in a certain way or does at least change its position on a much longer time scale than that of the NMR measurement.

In conclusion it can be stated, that both the F^- ion and the organic base appear to have a comparatively strong interaction with the $GaPO_4$-framework. It seems reasonable to assume that this interaction already plays a role during the crystallization process. It may be more than a coincidence that both the structure of $GaPO_4$-LTA as well as the structure of cloverite have the D4R as a secondary building unit, i.e. can be built entirely by joining only D4R units.

ACKNOWLEDGMENT

We thank Dr. L.B. McCusker for critically reading the manuscript, and for her help with the English. The work was supported in part by the Swiss National Science Foundation.

REFERENCES

1. Wilson, S.T., Lok, B.M. and Flanigen, E.M., U.S. Patent 4310440 (1982).
2. Kühl, G.H. and Schmitt, K.D., Zeolites **10**, 2-7 (1990).
3. Merrouche, A., Patarin, J., Soulard, M., Kessler, H. and Anglerot, D., in: *Zeolites and Pillared Clays Synthesis*, **1** (Ed. M. Occelli and H. Robson), van Nostrand Reinhold, New York (1992), in press.
4. Merrouche, A., Patarin, J., Kessler, H., Soulard, M., Delmotte, L., Guth, J.L. and Joly, J.F., Zeolites (1992), in press.
5. Estermann, M., McCusker, L.B., Baerlocher, Ch., Merrouche, A. and Kessler, H., Nature **352**, 320-323 (1991.
6. Baerlocher, Ch., *The X-ray Rietveld System, XRS-82*, Institute of Crystallography, ETH Zurich (1982).
7. Hepp, A. and Baerlocher, Ch., Aust. J. Phys. **41**, 229-236 (1988).

8. Patarin, J., Soulard, M., Kessler, H., Guth, J.L. and Baron, J., Zeolites **9**, 397-404 (1989).
9. Turner, G.L., Smith, K.A., Kirkpatrick, R.J. and Oldfield, E., J. Magnetic Resonance **70**, 408-415 (1986).
10. Gramlich, V. and Meier, W.M., Z. Kristallogr. **133**, 134-149 (1971).
11. Caullet, P., Guth, J.L., Hazm, J., Lamblin, J.M. and Gies, H., Eur. J. Solid St. Inorg. Chem. **28**, 345-361 (1991).

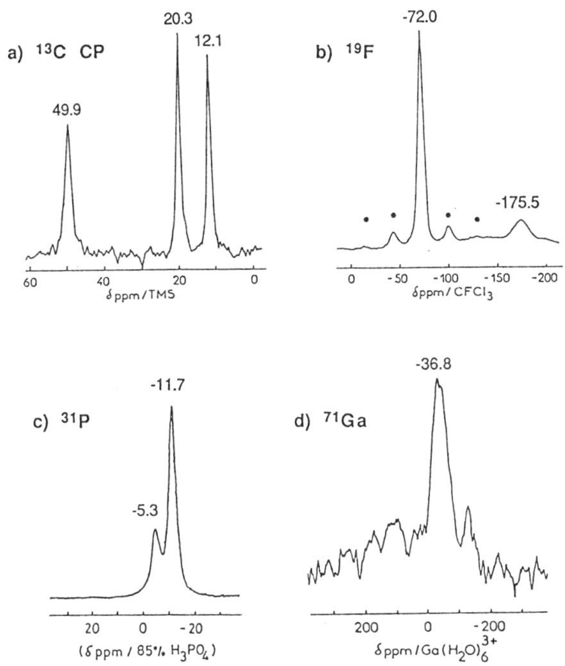

Figure 1. MAS MNR spectra of gallophosphate LTA

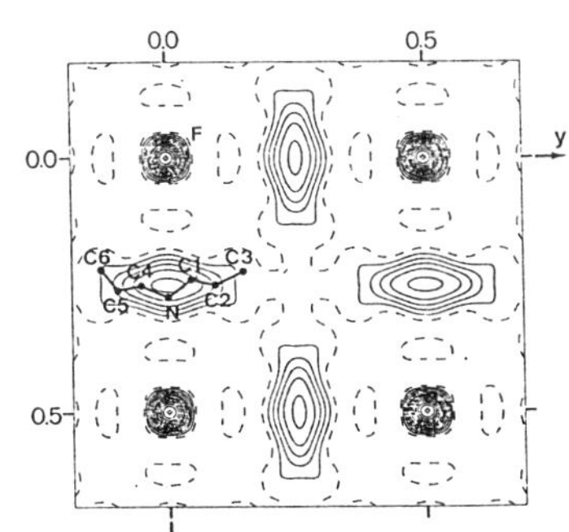

Figure 2. Difference Fourier map at z = 0 calculated using the framework atoms only. It shows the F^- ion inside the D4R and the diffuse electron density due to the DPA. The projection of the refined position of the DPA is also plotted.

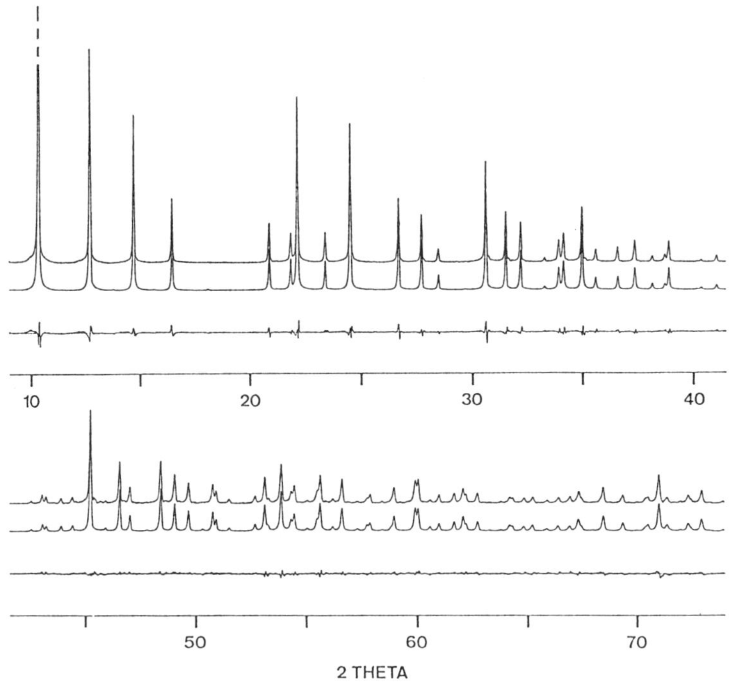

Figure 3. Observed (top), calculated (middle) and difference (bottom) patterns of $GaPO_4$-LTA. The intensity scale for the second section of the pattern has been increased by a factor of 2.

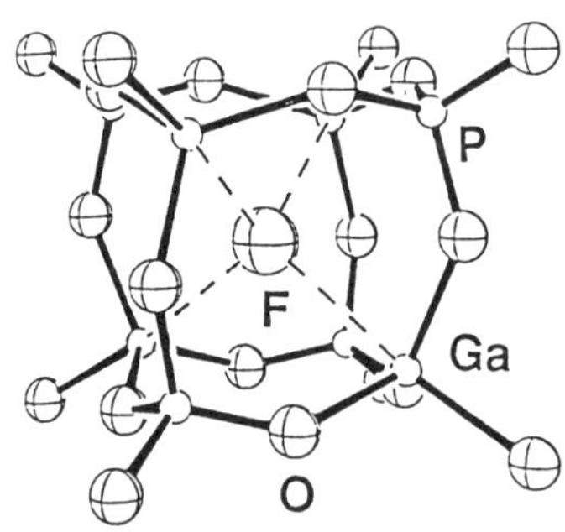

Figure 4. The double 4-ring showing the position of the fluoride ion and the 5-coordinate gallium.

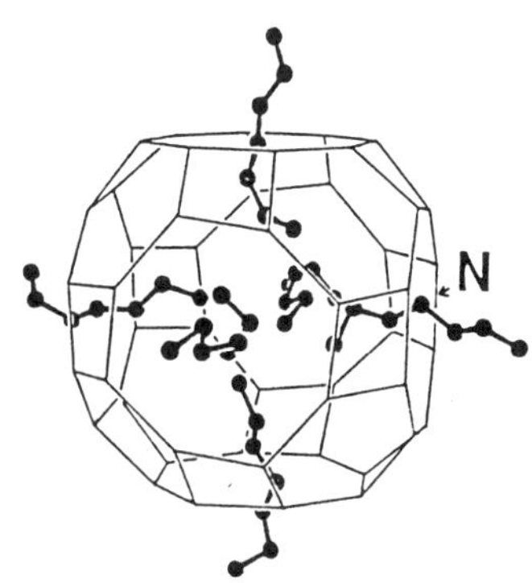

Figure 5. A possible packing of the DPA in the α-cage.

DISCUSSION - Structure

"Cation Distribution as a Structural-Chemical Probe" (J. Lievens and W. Mortier)

W. Harrison: Guest cations are highly mobile in zeolites and often occupy different sites in hydrated and dehydrated modifications of the same framework type. What is the role of water in cation siting in zeolites? Can it be modelled?

W. Mortier: The mere effect is changing the energy level of the sites. If equilibrium can be established, cations may occupy different sites in hydrated and dehydrated zeolites since the energy levels should change substantially - there should not necessarily be a relation between the energy levels of the hydrated vs. the dehydrated state.

G. Stucky: Can site occupancy changes which occur with temperature be predicted by your model?

W. Mortier/J. Lievens: The temperature is an integral part of the model. The model does indeed predict cation distribution at different temperatures, which has been confirmed by experiments. For large site energies, however, the effect of the temperature is small in the temperature range of interest.

J. Sherman: Your slide showed NaxH-GaY with X=54 having occupancies of 0.54 for site I and 0.67 for Site I', or SI + SI' = 0.54 + 0.67 = 1.21; in other words 21% of the Site I's had simultaneous occupancy of I and I'. Is this real?

W. Mortier: This should merely reflect the accuracy of the powder diffraction method and not considered as evidence for simultaneous occupancy of Sites I and I'.

"Group 2(Be/12(Zn)-15(P,As)O_4 Molecular Sieves" (W. Harrison, et al.)

R. von Ballmoos: In your experience, do the structure directing cations and organic templates used for classical zeolite syntheses result in the identical zeolite isotypes in your synthesis systems?

W. Harrison: The structure-directing effects of cations are probably distinct. For instance, Li^+ appears to be a very effective counter-cation in BePO synthesis, leading to sodium aluminosilicate isomorphs. This may be due to the good "fit" of the Li^+ cation in the BePO 6-ring window.

Question: What is the exchange capacity of the terminal OH groups found in the new $BeZnPO_4$ material from PABCD?

W. Harrison: We need to do more work in this area and plan to do so.

Characterization

TITANIUM SILICALITE-1 PEROXIDES

M.G. CLERICI, P. INGALLINA, and R. MILLINI

Eniricerche S.p.A, Via Maritano 26, 20097 S.Donato Milanese, Italy.

ABSTRACT

Stable peroxides, TS-1$(O_2^-)_n$, are produced by the reaction of titanium silicalite with hydrogen peroxide, in basic media. These have been characterized by physical chemical methods (XRD, IR, UV, TGA, TPD) and activity tests. The molar ratio $n = O_{act}/Ti$ increases with the strength and the concentration of the base ($0<n<1$). No activity is shown in oxidation reactions.

INTRODUCTION

Titanium peroxo species, formed at lattice positions, are generally believed to be active intermediates in reactions catalysed by titanium silicalite (TS-1) [1-7]. On the basis of stereochemical and kinetic studies of the oxidation of alkanes [3], epoxidation of olefins [4], and hydrolysis of C_4 epoxides [5], Ti-OOH species, forming a five-membered cyclic structure with a solvent molecule at framework positions, have been proposed. An equilibrium reaction occurs between these intermediates, free Ti sites, hydrogen peroxide and possibly with peroxides having a different configuration [5]. In the latter, according to earlier proposals, peroxo oxygen may act as bidentate ligand, forming a triangular $Ti(O_2)$ unit [6, 7]. All attempts at preparation of TS-1 peroxides have until now failed.

It has long been known that the use of a basic solution of hydrogen peroxide makes easier the synthesis and higher the stability of transition metals peroxides [8]. Applying this method, we have succeeded in the first preparation of TS-1 peroxides.

EXPERIMENTAL

Titanium silicalite was prepared and characterized by XRD, IR and elemental analysis (TiO_2 2.8 wt%, SiO_2/TiO_2 molar ratio 46) [2, 5, 9]. Unit cell parameters were determined by full-profile fitting method (Rietveld Analysis). Lattice Ti was estimated by the correlation between Ti/(Ti+Si) atomic ratio and unit cell volume [10]. $K_3[Fe(CN)_6]$ was used as internal standard for IR spectra (reference peak at 2118 cm^{-1}). TPD and TGA analyses were carried out on UTI 100C Instrument and on Mettler TG50 equipped with a microprocessor Mettler TA 3000, respectively. Ultraviolet reflectance spectra

From the Proceedings of the 9th International Zeolite Conference, Montreal 1992, Eds. R. von Ballmoos et al.,

were performed on a Perkin Elmer Lambda 9. Active oxygen of TS-1$(O_2^-)_n$ was determined by heating the peroxide (0.5-2.5 g), deareated acetic acid (20 ml), and a saturated aqueous solution of potassium iodide (2 ml), 0.5 h at 333 K. Iodine was titrated at room temperature. The oxidation of n-hexane, 1-butene, and phenol have been described [1-4].

Synthesis of TS-1$(O_2^-)_n$.

A suspension of titanium silicalite (2.5 g), a solution of the base (25 ml), and 33 wt% aqueous hydrogen peroxide (2 ml) were stirred 10 min at 273 K. The product was centrifuged (5 min), washed three times with the basic solution at 273 K (5 min, 20 ml each time) and then with deionized water (1x20 ml). It was dried at R.T. under vacuum. Li content was 1.54%, 0.71%, 0.77% for LiOH 0.5 M, 0.1 M, and 0.05 M respectively. Silicalite-1 was similarly treated. TS-1 peroxides in Table 1 were washed only with deionized water (4x20 ml). Ti released in mother liquors by TS-1 was in the range 4-6% of initial content.

TS-1 was recovered by washing TS-1$(O_2^-)_n$ with aqueous 0.1 M HCl, with water, then calcining 2 h at 823 K.

Table 1. Synthesis of TS-1$(O_2^-)_n$. Effect of bases on n.

Base (0.1M)	H_2O_2 [a] (mM)	O_{act} (mM)	O_{act}/Ti
-	0.016	0.14 [b]	0.16 [b]
NaOH	0.073	0.57	0.65
LiOH	0.083	0.55	0.63
KOH	0.067	0.70	0.80
AcONa	-	0.13	0.15
LiOH [d]	<0.005	<0.005	-

Table 2. Synthesis of TS-1$(O_2^-)_n$. Effect of [LiOH] on n.

[LiOH] (M)	H_2O_2 [a] (mM)	O_{act} (mM)	O_{act}/Ti
0.50	0.083	0.14	1.30
0.10	0.083	0.80	0.91
0.05	0.100	0.80	0.91
0.01	0.033	0.38	0.43
0.05 [c]	<0.005	0.35	0.40
0.05 [e]	-	0.39	0.74

a) mM of H_2O_2 in mother liquor of last washing.
b) mM of O_{act} before drying.
c) Washed with LiOH 0.05 M and then with water (4x20 ml).
d) Silicalite-1.
e) Solvent methanol.

RESULTS

A wide range of titanium compounds produce yellow to red orange peroxo derivatives, on reaction with basic, neutral, and acid solutions of hydrogen peroxide [8]. Titanium silicalite, suspended in aqueous or alcoholic hydrogen peroxide, exhibited a bright yellow colour. This colour gradually faded and eventually disappeared, on washing the peroxide during work up. Thus, the active oxygen content of the wet solid only accounted for 16% of the Ti sites (Table 1). It was totally lost on drying under reduced pressure.

Stable peroxides, TS-1$(O_2^-)_n$, were eventually prepared in the presence of a base (Table 1). The light yellow or whitish materials were washed and dried under vacuum without decomposition. These were stored indefinitely under nitrogen, but slowly decomposed if kept in open vessels.

The molar ratio $n = O_{act}/Ti$ was mainly a function of the strength and concentration of the base (Table 1 and 2). With alkali metal hydroxides, n approached unity at $[OH^-] \geq 0.05$ M. Lower values were shown by weak bases, by dilute lithium hydroxide or by repeatedly washing the product with water during work-up (Table 2). Depending on synthesis, TS-1 peroxides with $0.16<n<1.3$ were prepared. Silicalite-1, the pure silica parent structure, did not produce any peroxo derivative.

TGA and TPD analyses of TS-1$(O_2^-)_n$, prepared in 0.05 M NaOH, showed two different steps of weight loss (Figure 1). Water was released below 530 K. Between 530-700 K oxygen was evolved, due to the decomposition of Ti peroxo species. The ratio $O_{act}/Ti = 0.86$, as determined from TGA profile, was in good agreement with that found by iodometric titration. Faster release of water, up to 420 K, occured for TS-1 treated with neutral hydrogen peroxide. The weight decrease was also smaller. Higher adsorption by TS-1$(O_2^-)_n$ is likely the result of water coordination on alkali metal cations.

TS-1$(O_2^-)_n$ peroxides were further characterized by XRD, IR, and UV vis spectra and by catalytic activity tests. TS-1 recovered from TS-1$(O_2^-)_n$ was similarly characterized, to reveal possible damages in the crystal framework by alkalis.

Table 3. Crystallographic data and Ti content of recovered TS-1[a].

[LiOH]	a (Å)	b (Å)	c (Å)	V (Å^3)	x_i	TiO_2 (wt %)
-	20.126(2)	19.939(2)	13.412(1)	5382.1(1)	0.0202(6)	2.7
0.05	20.129(2)	19.941(2)	13.412(1)	5383.5(9)	0.0208(6)	2.8
0.10	20.128(2)	19.939(2)	13.412(1)	5382.7(9)	0.0204(6)	2.8
0.50	20.106(2)	19.932(2)	13.408(1)	5373.3(3)	0.0156(6)	2.1

a) Details in ref. [10], $x_i = Ti/(Ti + Si)$ atomic ratio.

Unit cell parameters are linearly related to lattice titanium [11]. XRD analysis showed that these were not affected by strong bases below 0.1 M concentration, while a significant decrease was caused by $[OH^-] \gtrsim 0.5$ M (Table 3). Unit cell volume of TS-1 treated with LiOH 0.5 M, corresponded to 2.1 wt% TiO_2 (25% decrease of lattice Ti). This result, compared with Ti content of mother liquors ($\lesssim$6%), suggests deposition of insoluble titanium compounds on TS-1 crystals. However, no crystalline TiO_2 phase was revealed by XRD spectra.

Table 4. IR spectra of recovered TS-1.[a]

[LiOH] (M)	550 cm^{-1}	970 cm^{-1}	TiO_2 (wt%)
-	1	1	2.8
0.05	0.96	0.95	2.7
0.05[b]	0.96	0.97	2.7
0.10	0.91	0.96	2.7
0.50	0.85	0.78	2.2

a) Relative intensities.

b) TS-1 recovered from TS-1$(O_2^-)_n$ prepared in methanol.

The 970 cm^{-1} band, originally shown in the IR spectrum of TS-1, gradually decreases and broadens, completely overlapping the strong absorption at 1100 cm^{-1}, in the IR spectra of TS-1$(O_2^-)_n$. A weak extra-peak exists at 866 cm^{-1} in spectra of peroxides prepared in 0.5 M lithium hydroxide. The intensities of the peaks at 970 and 550 cm^{-1}, in the IR spectra of recovered TS-1, decreased with increasing concentration of LiOH (Table 4). From the former absorption, due to Ti-O-Si bonds, and from the latter, assigned to skeleton vibration of TS-1, lattice Ti and crystallinity losses were estimated in the range 5-22% and 4-15%, respectively.

In the UV spectra of TS-1$(O_2^-)_n$, two broad peaks with maxima at 205 and 305 nm were observed. The spectra of fresh and recovered TS-1 were substantially identical. In both, the absorption at 205 nm was present. An extra-peak at 251 nm was shown by the peroxide prepared in methanol solvent. No absorption could be assigned to crystalline TiO_2.

Consistent with earlier studies [3, 4], anionic TS-1$(O_2^-)_n$ were not stoichiometric oxidants nor active catalysts, unless alkalinity was neutralized by acids. This was shown by the epoxidation of 1-butene and the oxidation of n-hexane, respectively (Table 5). In the hydroxylation of phenol, which is very sensitive to the presence of extra- framework titanium, the performance of recovered TS-1 was quite analogous to that of fresh TS-1, for dilute LiOH ($\leq$0.1 M). The yields were significantly reduced by using LiOH 0.5 M.

Table 5. Catalytic activity of TS-1$(O_2^-)_n$ and recovered TS-1.

LiOH (M)	Epoxidation of 1-C_4H_8 a)	Oxidation of n-C_6H_{14} b)	Hydroxylation of phenol c)
-	-	93	87
0.05	63	89	82
0.10	-	-	80
0.50	53	-	72

a) Yields of 1,2-epoxybutane (% on O_{act}). Oxidant TS-1$(O_2^-)_n$, [1-C_4H_8] 0.9 M, pH 3, 0.5 h, 283 K, solvent methanol.

b) Yields of of alcohols and ketones (% on H_2O_2). Oxidant H_2O_2, catalyst TS-1$(O_2^-)_n$ 16 g/l, [n-C_6H_{14}] 0.9 M, [H_2O_2] 0.3 M, [HCl] 0.1 M, 1 h, 328 K, solvent methanol.

c) Yields of hydroquinone and cathecol (% on H_2O_2). Recovered TS-1 3%, phenol 62%, acetone 11%, H_2O 18%, H_2O_2 6 % (all wt%), reflux.

DISCUSSION

An equilibrium reaction between titanium silicalite, hydrogen peroxide, and the solvent, shown in Scheme 1, prevents the isolation of pure TS-1 peroxides. The latter are stable as long as the solvent and excess hydrogen peroxide are both present, according to catalytic studies [3, 5], coloured reactions, and active oxygen determinations. Drying the peroxide under vacuum results into simultaneous decomposition. Basic reagents shift the equilibrium to the right, towards anionic peroxo species (III), which are isolated as stable products. These are contaminated by excess alkali metal cations, which are most likely bonded to nonbridging SiOH groups in the form of SiO^-M^+ species. Further purification causes peroxide decomposition.

In this synthesis, TS-1 crystals are attacked by basic medium. Thus, crystallinity degree and lattice titanium are both decreased. Such drawbacks are reduced to ca. 5% or lower, by minimizing the contact with strong bases, as shown by XRD, IR, and catalytic activity. Moreover, most Ti atoms released by LiOH $\leq$ 0.1 M, are loosely bound at defective sites of TS-1. In fact, even 3-4% Ti is removed by neutral hydrogen peroxide solutions.

Unity appears as the limiting value of n in TS-1$(O_2^-)_n$. Higher peroxide content, shown by LiOH > 0.1 M, is likely due to extra-lattice titanium peroxides. In the latter case, basic conditions yield $O_{act}/Ti \gtrsim 2$ [8], thus increasing overall n value beyond unity. Accordingly, for LiOH > 0.1 M, an absorption band grows in intensity at 866 cm^{-1} in the range where ν(O-O) absorptions are normally found. Coordination of 2 peroxo ligands at lattice Ti is unlikely on the grounds of steric constraints imposed by tetrahedral coordination in TS-1.

Scheme 1

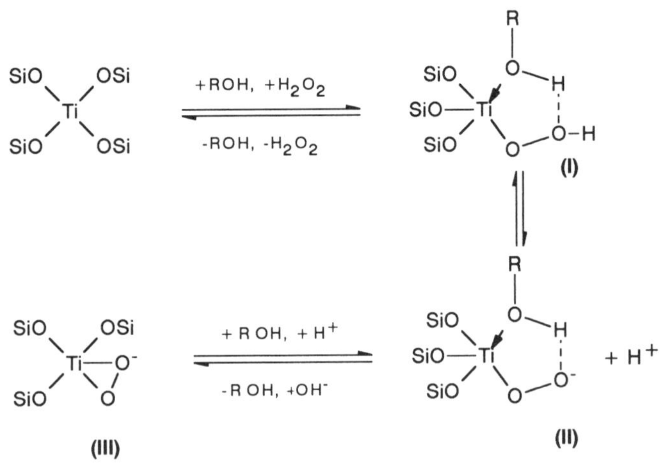

The structure of TS-1$(O_2^-)_n$, could be with peroxo-oxygen end-on or side-on coordinated at Ti center. The first proposal would imply the presence of a solvent molecule in the anion (species (II) of Scheme 1). Although (II) may help explaining the differences in the UV spectra, two points are made against it.

a) Solvent is readily released, under vacuum, by peroxide (I) prepared from neutral hydrogen peroxide. Stronger hydrogen bond will not significantly improve the stability of anionic species (II).

b) The presence of a negative charge on (II) favors, on the removal of solvent, side-on coordination of peroxide ligand against H_2O_2 release. Increased electron availability enhances the donation from filled antibonding π^*-orbitals of $O_2^=$ to empty titanium d orbitals [8].

In addition, side-on coordination of peroxo oxygen in (III) is consistent with known structures and properties of Group IV-VI transition metal peroxides, all of which show analogous bonding of peroxide ligand.

AKNOWLEDGMENTS

The authors are indebted to dr. L. Zanibelli for TPD analysis.

REFERENCES

1. Romano, U., Esposito, A., Maspero, F., Neri, C., and Clerici, M.G., Chim. Ind. (Milan) **72**, 610 (1990).
2. Clerici, M.G., Bellussi, G., and Romano, U., J. Catal. **129**, 159 (1991).
3. Clerici, M.G., Appl. Catal. **68**, 29 (1991).
4. Clerici, M.G. and Ingallina, P., submitted to J. Catal.
5. Bellussi, G., Carati, A., Clerici, M.G., Maddinelli, G., and Millini, R., J. Catal. **133**, 220 (1992).
6. Notari, B., in: *Studies in Surface Science and Catalysis*, Eds. P.J. Grobet, W.J. Mortier, E.P. Vansant, and G. Schulz-Ekloff, Elsevier, Amsterdam, Vol. **37**, pp. 413-425 (1988).
7. Huybrechts, D.R.C., De Bruycker, L., and Jacobs, P.A., Nature **345**, 240 (1990).
8. Connor, J.A. and Ebsworth, E.A.V., in: *Advances in Inorganic Chemistry and Radiochemistry*, Eds. H.J. Emeléus and A.G. Sharpe, Vol. **6**, pp. 279-381 (1964).
9. Taramasso, M., Perego, G., and Notari, B., U.S. Patent 4410501 (1983).
10. Millini, R. et al., in preparation.
11. Perego, G., Bellussi, G., Corno, C., Taramasso, M., Buonomo, F., and Esposito , A., in: *Proceedings of the 7th International Zeolite Conference,* Eds. A. Murakami, A. Ijima, and J.W.Ward, Kodansha, Tokyo, pp. 129-136 (1986).

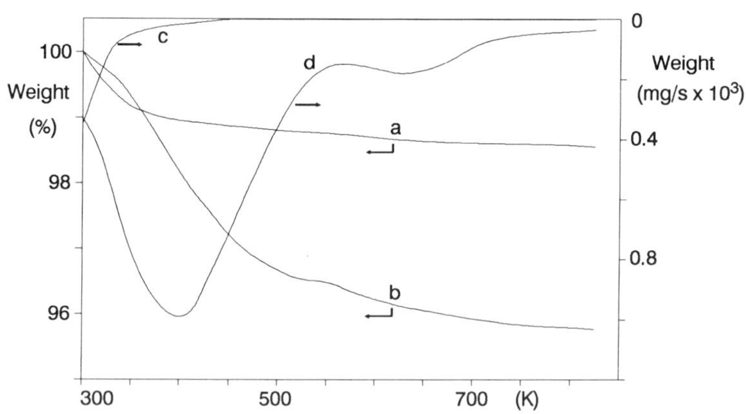

Figure 1. TGA and DTG of TS-1 treated with H_2O_2 (a, c), and of $TS(O_2^-)_n$ (b, d).

LOCAL ENVIRONMENT OF TITANIUM IN TITANIUM SILICALITE (TS-2)

A. BITTAR, D. TRONG ON, L. BONNEVIOT, S. KALIAGUINE and A. SAYARI

Department of Chemical Engineering and CERPIC, Université Laval, Québec, Canada G1K 7P4

ABSTRACT

A series of titanium silicalites with the MEL structure has been characterized by XPS, IR and catalytic activity in hydroxylation of n-hexane. Two Ti sites were revealed by XPS, a low and a high coordination species. Sodium impurities were found to decrease the intensity of 960 cm^{-1} IR band and to favor the decomposition of H_2O_2 at the expense of partial oxidation products. Based on the present data and previous XANES and EXAFS data, an overall picture of the active sites and the effect of sodium impurities is proposed.

INTRODUCTION

Titanium in pentasil zeolite framework exhibits outstanding catalytic properties in the hydroxylation and other partial oxidations of organic compounds with dilute aqueous solutions of hydrogen peroxide [1-6]. Normally Ti^{4+} has a strong preference for octahedral coordination to oxygen. According to the Pauling criterion theory, a stable tetrahedral environment is obtained for cations having ρ values in the range $0.225 < \rho < 0.414$ where $\rho = r_c/r_o$, r_c and r_o being the ionic radii of the cation and O^{-2}. Strict consideration of this theory would exclude Ti^{+4} ($\rho = 0.46$) from the list of cations capable of being isomorphously substituted in tetrahedral T positions in zeolite lattices. However, due to the flexibility of these frameworks, cations with higher ρ's have been shown to occupy such positions in zeolites (Ga^{3+} with $\rho = 0.50$ or Fe^{3+} with $\rho = 0.51$ for example) and in MeAPO's and MeAPSO's (Mg^{2+} with $\rho = 0.58$, Co^{2+} with $\rho = 0.59$, or even Mn^{2+} with $\rho = 0.66$) [7]. Thus the existence of Ti^{4+} in T positions in silicalite lattice has been postulated based on the results of a number of analytical techniques including XRD, EPR, IR, ^{29}Si MASNMR and UV-visible spectroscopy [8-11].

Recently, more powerful tools such as X-ray absorption techniques [12-15] have shown that the local environment of Ti is unique and quite complex. Behrens *et al.* [12] found in a TS-1 sample (i.e., titanium silicalite with ZSM-5 structure) that besides small amounts of Ti in tetrahedral and square pyramidal symmetries, most of the titanium was octahedrally coordinated. Lopez *et al.* [13] reported that depending on the degree of dehydration of the sample, Ti in TS-1 zeolites was either essentially tetrahedral or mainly octahedral . Our recent XANES and EXAFS studies [14, 15] of a series of TS-2 samples (i.e., titanium silicalite with ZSM-11 structure) revealed the presence of Ti

From the Proceedings of the 9th International Zeolite Conference, Montreal 1992, Eds. R. von Ballmoos et al.,

in tetrahedral, square pyramidal and octahedral coordinations. The proportions of these sites were found to be strongly dependent on Ti loading. Moreover, low coordination Ti atoms were present either as monomeric or as dimeric species. Most importantly, the local environment of tetrahedral Ti sites was found to be different from the usual environment of T atoms in zeolites. Indeed, a surprisingly short Ti---Si distance of 2.2 Å was obtained. This finding could not be explained on the ground that neighboring TiO_4 and SiO_4 units share only one corner as it is known in zeolites. The short Ti---Si distance was interpreted on the basis that these units are linked to each other via an edge not a corner.

Jacobs *et al.* [16] found that the presence of the characteristic 960 cm^{-1} IR band is a necessary but not a sufficient condition for titanium silicalites to be catalytically active. Residual acidic sites and other impurities were blamed for the poor catalytic performance of these samples. Previous work from this laboratory [17] has also shown that Na impurities have a dramatic effect on catalytic properties of titanium silicalite.

This paper deals with further characterization studies of TS-2 zeolites. X-ray photoelectron spectroscopy (XPS) data will be tentatively correlated with previous XAFS results. The effect of impurity levels of sodium on (i) the characteristic 960 cm^{-1} IR band, (ii) the distribution of tetrahedral titanium on the surface of the catalyst and (iii) the catalytic activity in the partial oxidation of n-hexane will be discussed.

EXPERIMENTAL

Titanium silicalites (TS-2) were synthesized hydrothermally using literature procedures [10, 15]. The samples were first calcined at 773 K in order to decompose the organic template. When required, removal of residual alkali metal impurities was achieved by overnight treatment with 1N HCl solution at room temperature followed by calcination at 773 K. These samples were previously characterized by IR, XANES and EXAFS [14, 15, 17]. The TS-2 samples used in this study will be designated as x-TS-2 where x refers to the atomic ratio 100 Ti/Ti+Si before acid leaching.

XPS measurements were carried out using a V.G.Scientific Escalab Mark II spectrometer with an hemispherical analyzer operated in the constant pass energy mode (20 eV). A Mg Kα X-ray source (hν=1245.6 eV) operated at 20 mA and 15 kV was used. The residual gas pressure in the spectrometer chamber during data acquisition was below 10^{-8} Torr. The measurements were performed in the following sequence: Si(2p), O(1s), Ti(2p) and Si(2p). The binding energies were referenced to the silica Si(2p) line (103.4 eV). The experimental intensity of XPS peaks were determined using linear background substraction and integration of peak areas [18, 19]. When required, XPS peaks were deconvoluted using routine procedures [18].

Infrared measurements were carried out using self-supporting wafers obtained by pressing mixtures of 2 mg of TS-2 and 100 mg of KBr. IR spectra were recorded in the range of 4000-400 cm^{-1} with 2 cm^{-1} resolution, using a Digilab *FTS-60* spectrometer.

Partial oxidation of n-hexane was performed in a pyrex reactor equipped with a reflux condenser. In a typical experiment, 0.174 mole of n-hexane (99%), 0.027 mole of H_2O_2 (in 30% aqueous solution), 1 ml of methanol and 0.3 g of catalyst were mixed and stirred at 353 K for 3 h. The reaction products were subjected to GC analysis using a 5890A Hewlett Packard gas chromatograph fitted with a DB Wax capillary column. Quantitative analysis of H_2O_2 was made by iodometric titration.

RESULTS AND DISCUSSION

Figure 1 illustrates the XPS Ti(2p) spectra before (fig. 1A) and after (fig. 1B) acid leaching of four samples with x ranging from 1.6 to 9.1. The binding energies for Ti($2p_{3/2}$) and the sodium content of these samples are given in the Table 1. Figure 1A shows that depending on Ti content, the Ti(2p) feature comprises one or two doublets. The doublet corresponding to the low binding energy (BE) may be ascribed to Ti^{4+} in octahedral coordination since its BE is close to the BE of Ti(2p) in TiO_2. The high binding energy doublet is very close to a Ti(2p) doublet reported in the literature for tetrahedral Ti in TiO_2-SiO_2 glasses [20].

It is seen from figure 1A that as the Ti content increases, the relative amount of low coordination Ti decreases. Indeed, the least Ti loaded sample (1.6 TS-2) exhibits only the high binding energy doublet corresponding to tetrahedral Ti, whereas the 9.1 TS-2 sample shows only the lines of octahedral Ti. Note that a shift of 1.2 eV in the BE of the (2p) electron of the high coordination Ti took place when the Ti/Ti+Si ratio increased from 4.2 to 9.1 % (fig. 1A). These assignments are in good agreement with the conclusions of previous XANES - EXAFS studies of the same TS-2 samples [14, 15]. In these works, it was indeed also found that sample 1.6 TS-2 contained exclusively low coordination Ti and that octahedral Ti species appeared gradually as the Ti loading increased.

Comparison of Figures 1A and 1B shows that upon acid leaching the surface tetrahedral Ti peak increases remarkably. For 4.2 TS-2, the strong line of octahedral Ti in the spectrum of the as prepared sample disappears almost entirely upon leaching. This change is due to the removal of minor amounts of a phase containing octahedral Ti (possibly sodium titanate) segregated on the external surface of the sample. The content of this acid leachable phase increases with Ti loading (Table 2).

In order to explain the shift of the high BE Ti($2p_{3/2}$) observed in figure 1A, the 4.2 TS-2 sample which is sodium free was impregnated by various amounts of $NaHCO_3$ followed by calcination at 773 K. The corresponding XPS data are summarized in Table 3. It is interesting to note that the BE of low coordination Ti(2p) line shifts downward by ca., 0.7 eV upon introduction of Na^+, indicating an increased electron density on this Ti species.

Table 1: $Ti(2p_{3/2})$ binding energy (eV) and sodium content for various titanium silicalites before and after acid leaching

Sample	As prepared			Acid leached		
	Na/Ti	$Ti(2p_{3/2})^{(a)}$	$Ti(2p_{3/2})^{(b)}$	Na/Ti	$Ti(2p_{3/2})^{(a)}$	$Ti(2p_{3/2})^{(b)}$
1.6 TS-2	0.57	-	459.8	0.012	-	460.2
4.2 TS-2	0.03	457.1	459.8	0	458.0	459.9
6.4 TS-2	0.17	457.8	460.5	0.01	458.4	460.2
9.1 TS-2	0.34	458.3	-	0	458.3	460.1
TiO_2	-	458.3	-			
TiSiG(c)	-	-	459.8			

(a) high coordination site (octahedral), (b) low coordination site (tetrahedral), (c) TiO_2 - SiO_2 glass, from ref. [20].

Table 2: Bulk and surface compositions of TS-2 before and after acid leaching

Sample	100 Ti/Ti + Si				% Ti(LC)(a)	
	Bulk		Surface			
	Before	After	Before	After	Before	After
1.6 TS-2	1.6	1.6	0.46	0.42	100	100
4.2 TS-2	4.2	4.2	2.0	1.0	29	81
6.4 TS-2	6.4	5.6	5.3	2.5	3	24
9.1 TS-2	9.1	4.2	7.2	2.8	0	19

(a) LC: low coordination Ti from XPS data.

Table 3: Effect of sodium on the BE of Ti(2p) in 4.2 TS-2.

Sample	Na/Ti	$Ti(2p_{3/2})^{(a)}$	$Ti(2p_{3/2})^{(b)}$
as prepared	0.03	457.1	459.8
acid leached	0	458.0	459.9
Na added	0.33	457.2	459.2
Na added	0.52	457.4	459.3

Sodium effect on the intensity of the 960 cm^{-1} IR band was studied using 6.4 TS-2. After acid leaching, various amounts of sodium were added to the sample as described above, and IR spectra were recorded. Data shown in figure 2 as a plot of the 960 to the 555 cm^{-1} band intensity ratio vs. the bulk Na/Ti atomic ratio indicate a clear decrease in the 960 cm^{-1} band intensity with the Na content. A similar effect was also observed upon chemisorption of pyridine [21].

The effect of sodium on the nature of accessible Ti sites was further investigated using the partial oxidation of n-hexane as a test reaction. As shown in the following scheme, several reactions occur simultaneously:

$$H_2O_2 + n\ C_6H_{14} \xrightarrow{1} \text{3-hexanol} \xrightarrow{4} \text{3-hexanone}$$

$$H_2O_2 + n\ C_6H_{14} \xrightarrow{2} \text{2-hexanol} \xrightarrow{5} \text{2-hexanone}$$

$$H_2O_2 \xrightarrow{3} H_2O + 1/2\ O_2$$

Designating by X_1, X_2 and X_3 the partial conversions of hydrogen peroxide by the three parallel routes, the total H_2O_2 conversion is $X = X_1 + X_2 + X_3$. The selectivity for oxygenates production is defined as $Y = (X_1 + X_2)/X$, where X_1 and X_2 are given by:

$$X_1 = (\ C_{3\text{-ol}} + 2C_{3\text{-one}})/C_o$$

$$X_2 = (\ C_{2\text{-ol}} + 2C_{2\text{-one}})/C_o$$

C_o being the initial concentration of H_2O_2.

Figure 3 represents the selectivity for oxygenates production (Y) along with the decomposition of H_2O_2 (X_3) over the 6.4 TS-2 sample as prepared, as well as after acid leaching and after addition of various amounts of sodium. The molar distribution of oxygenated products is reported in Table 4. Figure 3 clearly demonstrates that the acid leached sample is very selective towards the production of oxygenates and that increasing the sodium content favors the decomposition of H_2O_2. These results are in good agreement with data reported by El Hage-Al Asswed *et al.* [22] who prepared TS-1 samples by a variety of methods and found no hydroquinone (from phenol) whenever the solids contained significant amounts of Na, even though some of the inactive solids displayed an IR band at 960 cm^{-1}.

Table 4: Sodium effect on the catalytic properties of 6.4 TS-2.

Sample	Na/Ti	Y(%)	X(%)	Molar distribution of oxygenates			
				3-OL	3-ONE	2-OL	2-ONE
As prepared	0.175	8.7	94.0	51.6	12.0	13.8	22.7
Acid leached	0.010	70.0	74.8	25.0	25.0	5.4	44.6
Na added	0.086	42.0	87.3	36.9	23.1	6.3	33.8
Na added	0.171	9.7	94.3	63.9	11.2	15.0	9.9

The results obtained over the as prepared sample (Na/Ti = 0.175) were almost identical to those obtained over the same sample after acid leaching and addition of a comparable amount of sodium (Na/Ti = 0.171). It may therefore be concluded that the acid leachable Ti species present on the external surface of the as prepared catalyst do not play a significant role in the partial oxidation reactions.

CONCLUSION

The results reported in this paper on the effect of Na addition and acid leaching on XPS and IR data as well as on the partial oxidation of n-hexane on TS-2 catalysts may be given an interpretation that encompasses our previous XANES, EXAFS [14, 15], XPS [17] and IR of chemisorbed pyridine [21] results. This is conveniently addressed using scheme 1. The active site involves low coordination Ti examplified here as a tetrahedral species (species **1**) sharing edges with two neighboring SiO_4 tetrahedra from the silicalite lattice [14, 15] and bearing a δ^+ charge so that the O---Ti bond may be viewed as a Lewis acid-base pair [22]. Na^+ is located close to the charged oxygen and OH^- to the Ti ion by electrostatic interactions. This would translate into an increase of the electron density of Ti as detected by the XPS Ti(2p) binding energy shift and affect the intensity of the 960 cm^{-1} IR band which was ascribed to the stretching mode of SiO_4 bonded to a titanium ion. In the presence of NaOH (species **2**), H_2O_2 would interact primarily with the Na^+ - OH^- pair on which the decomposition to water and oxygen is favored. In the absence of NaOH, The peroxy titanate species (species **3**) which is a crucial intermediate in the partial oxidation of hydrocarbons [15] is formed.

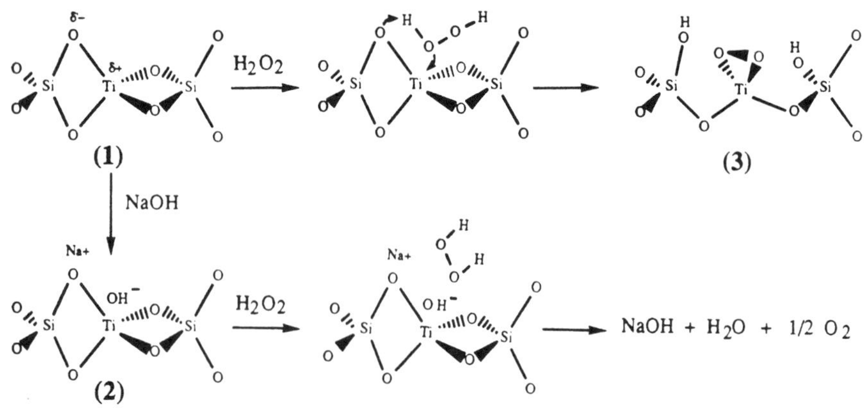

Scheme 1

REFERENCES

1. G. Perego, G. Bellussi, C. Corno, M. Taramasso, F. Buonomo and A. Esposito, Stud. Surf. Sci. Catal. **28**, 129 (1986).

2. B. Notari, Stud. Surf. Sci. Catal. **37**, 413 (1987).

3. P. Roffia, G. Leofanti, A. Cesana, M. Mantegazza, M. Padovan, G. Pertini, S. Tonti, P. Gervasutti, Stud. Surf. Sci. Catal. **55** 43 (1990).

4. M.R. Boccuti, K.M. Rao, A. Zecchina, G. Leofanti and G. Petrini, Stud. Surf. Sci. Catal. **48**, 133 (1989).

5. D.R.C. Huybrechts, L. De Bruycker and P.A. Jacobs, Nature **345**, 240 (1990).

6. M. Taramasso, G. Perego and B. Notari, US Patent 4 410 501 (1983); M. Taramasso, G. Manara, V. Fattore and B. Notari, US Patent 4 666 692 (1987).

7. (a) K.G. Ione, L.A. Vostrikova and V.M. Mastikhin, J. Mol. Catal. **31**, 355 (1985); (b) E.M. Flanigen, R.L. Patton and S.T. Wilson, Stud. Surf. Sci. Catal. **37**, 13 (1988).

8. A. Tuel, J. Diab, P. Gelin, M. Dufaux, J.F. Dutel and Y. Ben Taarit, J. Mol. Catal. **63**, 95 (1990).

9. A. Thangaraj, R. Kumar, S.P. Mirajkar and P. Ratnasamy, J. Catal. **130**, 1 (1991).

10. J.S. Reddy and R. Kumar, J. Catal. **130**, 440 (1991).

11. M.R. Boccuti, K.M. Rao, A. Zecchina, G. Leofanti and G. Petrini, Stud. Surf. Sci. Catal. **48**, 133 (1989).

12. P. Behrens, J. Felshe, S. Vetter, G. Schulz-Ekloff, N. Jaeger and W. Niemann, Chem. Soc., Chem. Commun., p. 678 (1991).

13. A. Lopez, H. Kessler, J.L. Guth, M.H. Tuilier and J.M. Popa, in: *Proc. 6th Int. Conf. X-ray Absorption Fine Structure*, York, UK, Elsevier, Amsterdam, p. 549, (1990).

14. D. Trong On, A. Bittar, A. Sayari, S. Kaliaguine and L. Bonneviot, Nature, submitted.

15. D. Trong On, L. Bonneviot, A. Bittar, A. Sayari and S. Kaliaguine, J. Mol. Catal., in press.

16. D.R.C. Huybrechts, I. Vaesen, H.X. Li and P.A. Jacobs, Catal. Lett. **8**, 237 (1991).

17. A. Bittar, A. Sayari, D. Trong On, L. Bonneviot and S. Kaliaguine, in *Proc. 6th. Brazilian Seminar Catal.*, Salvador, Brazil, p.657, (1991).

18. R. Borade, A. Sayari, A. Adnot and S. Kaliaguine, J. Phys. Chem. **94**, 5989 (1990).

19. A. Sayari, A. Adnot, S. Kaliaguine and J.R. Brown, J. Electron Spectrosc. Relat. Phenom., in press.

20. S.M. Mukhopadhyay and S.H. Garofalini, J. Non-Cryst. Solids **126**, 202 (1990).

21. A. Bittar, A. Sayari and S. Kaliaguine, Appl. Catal., submitted.

22. (a) J. El Hage Al Asswad, Ph.D. Thesis, Namur, Belgium, 1990; (b) J. El Hage Al Asswad, B. Nagy, Z. Gabelica and E.G. Derouane in "*Zeolite for the Nineties, Recent Research Reports*" 8th. Int. Conf. Zeolites, Amsterdam, J.C. Jansen, L. Moscou and M.F.M. Post, eds., Elsevier, Amsterdam, 1989, p.475.

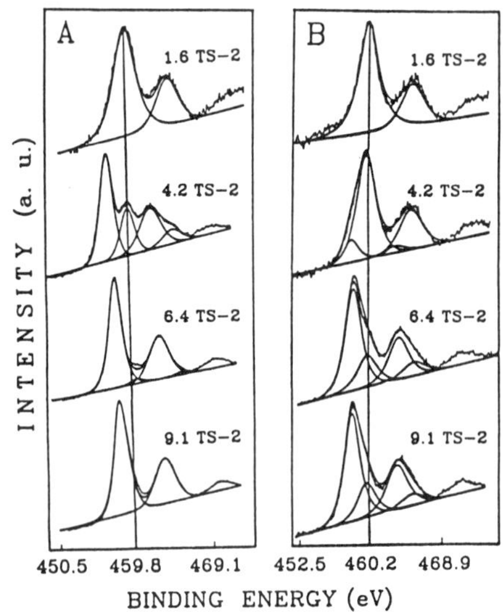

Figure 1. XPS Ti(2p) of as prepared (A) and acid leached (B) samples

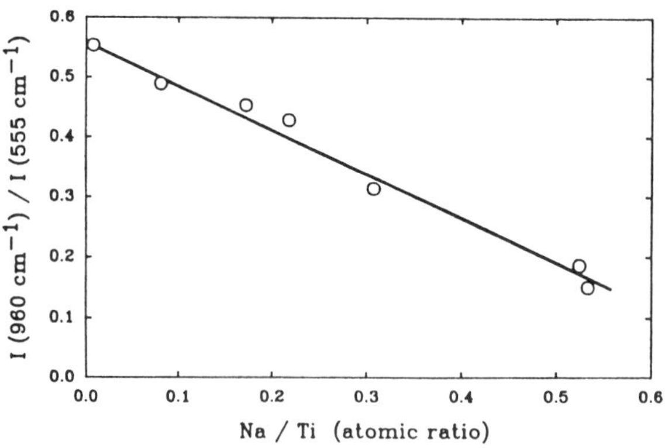

Figure 2. Effect of sodium on the 960cm^{-1} IR band intensity

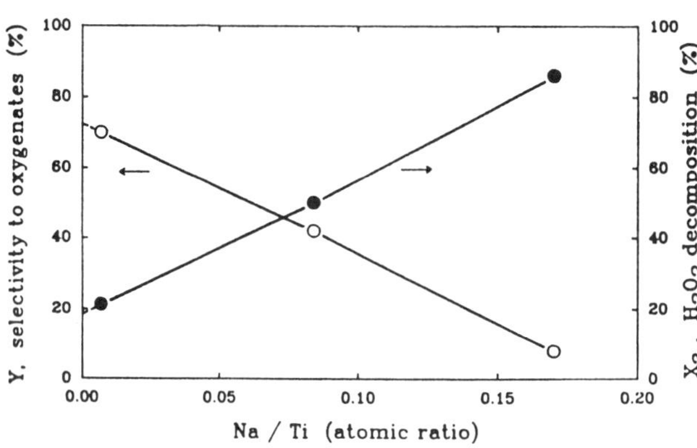

Figure 3. Effect of sodium on (o) the selectivity to oxygenated products, (●) H_2O_2 % decomposition

PROPERTIES OF INDIUM CONTAINING ZEOLITES PREPARED BY A SOLID STATE REACTION

V.KANAZIREV, Y.NEINSKA, T.TSONCHEVA and L.KOSOVA

Institute of Organic Chemistry, Bulgarian Academy of Sciences,1113 Sofia, Bulgaria

ABSTRACT

Microbalance and flow reactor TPR experiments along with TPD of ammonia, TEM/EDAX and IR measurements indicate a process of indium introduction into cationic position of the zeolite framework occurring when various zeolites: ZSM-5, Y, mordenite, offretite in a H-form are mechanically mixed with In_2O_3 and heated in the presence of hydrogen. This process results in removal of Broensted acid sites and in a dramatic decrease of the catalytic activity in both n-pentane and methanol conversion. Thus a new example of zeolite modification via a solid state reaction is demonstrated.

INTRODUCTION

The importance of gallium containing ZSM-5 zeolites for aromatization of light hydrocarbons has stimulated considerable research of proper methods for their preparation. Recently, H_2 reduction of intimate mechanical mixtures of Ga_2O_3 with HZSM-5 has been shown as leading to a rather unexpected process of gallium transfer into the zeolite [1,2]. This process differs from the known solid state ion exchange since the resulting material contains gallium ions in a lower oxidation state than gallium in the oxide source [3]. In addition, Ga-containing catalysts prepared by H_2 or hydrocarbon reduction are highly active and selective [2,4].

The purpose of the present investigation is to elucidate whether or not the H_2 reduction initiates a similar process of solid state ion introduction when another member of the IIIa group of the periodic system of elements, indium, is used as In_2O_3, as a component of the mechanically mixed catalysts.

There is a considerable lack of information in the published literature regarding the ion exchange and modification ofzeolites with indium. Zatorski has recently found that the catalytic activity of HZSM-5 zeolite in the methanol to hydrocarbons conversion was strongly suppressed in the presence of In_2O_3 as a component of the catalyst binding material [5]. It has been suggested that In_2O_3 reacts with the zeolite OH groups and In^{3+} is inserted in the cationic positions of the zeolite framework [6].

From the Proceedings of the 9th International Zeolite Conference, Montreal 1992, Eds. R. von Ballmoos et al.,

EXPERIMENTAL

The catalysts were prepared by ball milling powdered In_2O_3 (American Smelting and Refining Co) with the zeolite base in a H-form. We refer to these catalysts as **In**/zeolite catalysts with a numerical prefix indicating the indium oxide loading by weight (Table 1). All catalysts were pelletized and crushed to 0.2-0.8 mm prior to use. The TPR experiments were performed in a SETARAM TGDTA 92 microbalance. After purging the sample at 303 K in Ar flow (65 ml/min), 50 ml/min of the argon flow was replaced by an equal amount of H_2. The temperature was then raised (25 K/min) to 823 K and kept for 0.5 h. TPD of ammonia was carried out in a flow reactor as described in Ref.7. The sample was pretreated at 823 K in a He flow, then cooled to 423 K and saturated with NH_3 gas. After short purging in He flow the TPD experiment was performed by raising the temperature to 823K at 13 K/min. Finally the TPD determination of NH_3 was repeated after a TP reduction of the same sample or of a fresh one with H_2. The IR spectra were recorded on a Perkin Elmer spectrometer. A transmission electron microscope (PHILIPS EM 420) equipped with an spectral analyzer (EDAX) was used for microanalysis of some catalyst samples.
The catalytic experiments for n-pentane and methanol conversion were performed in fixed bed type reactors operated in an inert gas (He or N_2) stream at atmospheric pressure as described previously [4,8].

Table 1. Composition of the Zeolite Bases (on an Anhydrous Basis, mass%).

Zeolite	Manufacturer	NaO	K_2O	Al_2O_3	SiO_2	SiO_2/Al_2O_3(mol)	X%
Y-Type(Y)	Leuna,Germany	12.8	-	21.5	65.3	5.20	70
Mordenite(M)	Leuna,Germany	7.5	-	12.4	79.8	10.9	90
Offretite(Off)	self made *	0.3	7.4	13.4	78.2	9.95	61
ZSM-5	Union Carbide	0.03	-	3.7	94.9	43.3	-

X-Degree of exchange in a H-form. * -according to Ref.9.

RESULTS

On Figures 1,2 difference TPR spectra are depicted, registered after substracting the DTG curves of the indium free support material from the corresponding ones in the **In** containing samples. It is to be seen that up to three peaks of weight loss, denoted as low (LTP), middle (MTP) and high (HTP) temperature peaks appear on the DTG curves. All these peaks are due to the presence of In_2O_3 in the samples studied. In a separate experiment, using MS analysis, we have established that water is the sole product of the weight loss process corresponding to all three TPR peak mentioned above. We attribute the LTP to a dehydration process of the indium oxide itself since this peak is present in all cases including pure In_2O_3 in both Ar and H_2 flow. The HTP peak always appears when pure In_2O_3 or indium oxide loaded nonacidic materials like α-alumina and NaY zeolite are subjected to a temperature programmed H_2 reduction. Therefore we assume that this peak reflects the reduction process of the In_2O_3 itself to indium metal. In contrast, the presence of an acidic zeolite support is a prerequisite for MTP to appear. Moreover, the intensity of MTP increases with the In_2O_3 loading until a characteristic for the particular zeolite sample limit is attained. By further loading with In_2O_3 the MTP practically does not change and a HTP begins to grow. Some quantitative data (Table 2) from TG microbalance experiments show that

Table 2. Microbalance Estimation of the Specific Weight Loss in TPR Experiments.

Sample	Relative weight loss[a]				Selective indium loading of the zeolite support[b]		
	LTP	MTP*	HTP*	MTP+HTP*	mass ratio In/dry zeolite	atomic ratio In/Al	In/H$^+$
In_2O_3in Ar	5.8	–	–	–	–	–	
In_2O_3	5.9	–	18.9	18.9	–	–	–
25In/Al_2O_3	5.9	–	18.6	18.6	–	–	–
25In/NaY	5.7	–	18.3	18.3	–	–	–
9.2In/HZSM	5.8	18.4	–	18.4	0.091	1.08	1.10
15In/HZSM	6.0	11.5	7.6	19.1	0.097	1.15	1.18
25In/HY	5.3	15.0	6.9	21.9	0.283	0.58	0.81
35In/HY	5.3	10.4	10.1	20.5	0.310	0.66	0.93
25In/HM	5.6	13.4	5.9	19.3	0.220	0.79	0.88
25In/Off	6.0	10.2	9.4	19.6	0.171	0.58	0.95

a)weight loss(mg)/In_2O_3loaded(mg)x100;b)calculated from the weight loss corresponding to MTP;*)calculated on dry In_2O_3.

-The weight loss related to the LTP remains almost constant regardless of the sample

-In all cases of either acidic or nonacidic samples the total weight loss, calculated as a sum of MTP and HTP contributions, correlates well with the theoretical value of 17.3% corresponding to a loss of 3 atoms oxygen by the reduction of each molecule In_2O_3

-The specific zeolite loading by indium calculated on the base of the MTP weight loss seems to be limited rather more by the proton content of the zeolite than by its aluminum content. A **In**/H$^+$ ratio close to 1 is always observed despite the different zeolite bases and indium loadings used.

The TPD curves presented on Figure 3 clearly indicate both low and high temperature TPD peaks as diminishing after H_2 reduction of the indium containing HZSM samples. With a higher (about 9 wt% In_2O_3) content, the high temperature TPD peak which is usually ascribed to the presence of strong acid sites in HZSM type zeolites disappears completely. The low temperature TPD peak also decreases in size and this effect is more pronounced at a high indium loading. Moreover the decrease of the total acidity in dependence on the indium content (Figure 4) occurs in the same manner. That is the intensity of the MTP grows as registered in flow reactor TPR experiments, performed just before the acidity determination. In the same time nonreduced **In** containing samples show almost no changes in acidity (see Figure 4). Thus we observe the H_2 reduction of **In** containing HZSM as leading to strong decrease of zeolite acidity. The same effect was found also when another zeolite structure under investigation in a H-form was used as indium oxide support. This effect was further confirmed on the example of the 8.2In/HZSM sample by means of IR spectroscopy [11]. Two major features of the IR spectra recorded after H_2 reduction at about 700 K should be mentioned:

-The 3610 cm^{-1} band, ascribed to the Broensted acidic OH groups disappears completely

-The pyridine adsorbed shows no more the characteristic 1545 cm^{-1} band due to the formation of pyridin-ion on Broensted sites. In the same time a strong band at 1446cm^{-1} due to Lewis bonded pyridin has appeared.

Similarly to the approach applied to Ga containing zeolites [2], we tested the sample 5 In/HZSM before and after H_2 reduction by TEM coupled with EDAX microanalysis. Two kinds of particles belonging to the In_2O_3 phase and HZSM zeolite were visible before the reduction with H_2. After the reduction experiment the In_2O_3 phase was found to disappear completely and the microanalysis of several zeolite in particle showed that a process of indium transfer into zeolite had occurred.

Some results regarding the catalytic activity in methanol conversion of **In** containing samples before and after H_2 reduction are summarized in Table 3. At the mild reaction conditions chosen methanol dehydration to dimethyl ether and water is the main reaction route. The catalytic activity of the pure HZSM and HY samples practically does not differ from the activity of the respective indium containing samples before hydrogen treatment. In contrast,the hydrogen reduction leads to a diminishing activity of all indium loaded catalysts. The higher the indium content, the more pronounced the activity decrease until at high **In** loadings the methanol dehydration is almost fully suppressed. By further increase of the reaction temperature the selectivity of the methanol conversion is changed. Instead of dehydration the dehydrogenation of methanol, resulting in such products as H_2, CO, CO_2 and CH_4 accompanied by catalysts deactivation, appears as main reaction route now.

The conversion of n-pentane was used in the present investigation as another model reaction for catalysts evaluation. The results in Table 4 illustrate, on the example of the **In** modified HZSM, the strong decrease of the catalytic activity due to a reductive treatment of the catalyst. Unlike to the methanol conversion case, the n-pentane feed itself acts as a reducing agent. The catalytic activity of the 10 In/ZSM sample decreases strongly with the time on stream with n-pentane reagent. The subsequent H_2 reduction causes only a small additional drop in activity. From the data in Table 4 it is

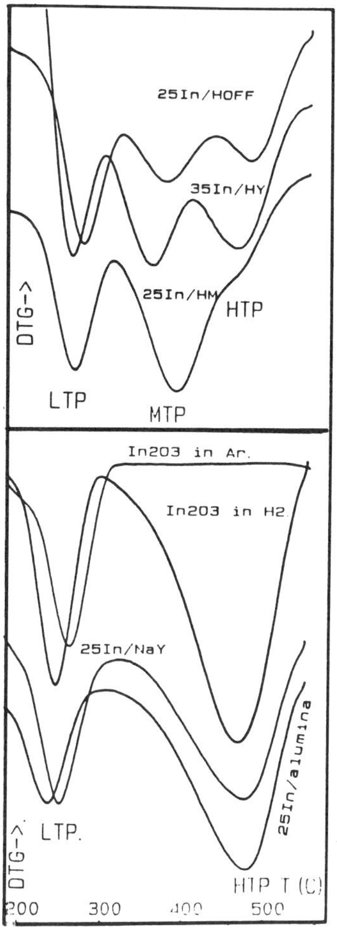

Figure 1. DTG curves of In containing samples in H_2

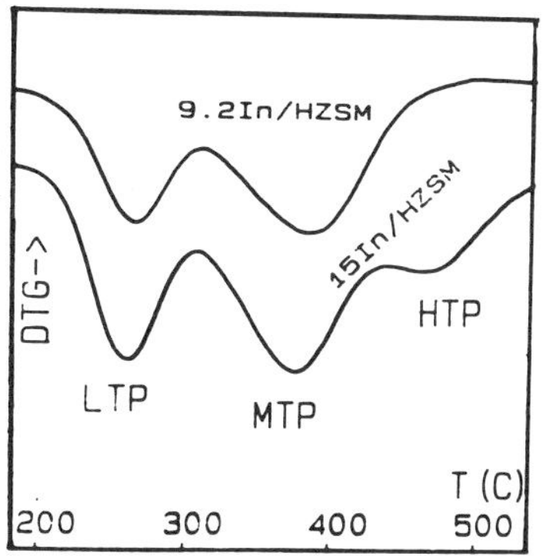

Figure 2. DTG curves in H_2.

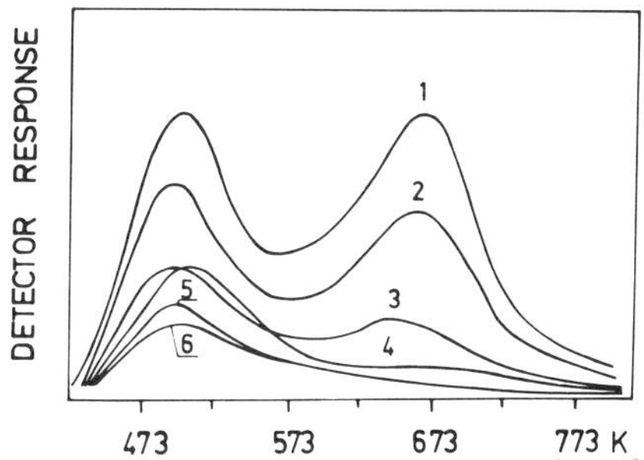

Figure 3. TPD of ammonia of H_2 reduced samples:1-HZSM;2-3In/HZSM;3-5In/HZSM 4-8.2In/HZSM;5-9.2In/HZSM;6-15In/HZSM

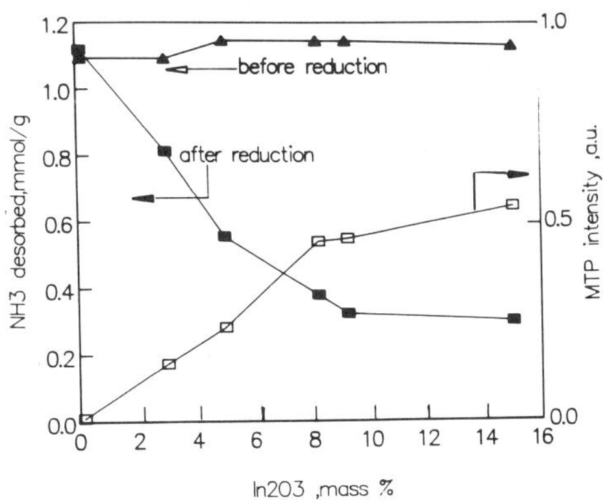

Figure 4. NH_3 desorbed and middle temperature reduction peak (MTP) depending on In_2O_3 content of HZSM.

quite clear that the degree to which the catalytic activity goes down is directly related to the amount of indium oxide loaded . A loading of 10 wt% suppresses completely the n-pentane conversion and the reduced catalysts behave like a bed of crushed glass.

Table 3.Selective Dehydration of Methanol,mol%,P_{CH3OH}=3.85kPa.

Catalyst	before reduction			after reduction		
Temperature,K	440	480	520	440	480	520
HY	17.0	44.0	75.0	-	-	-
3 In/HY	16.0	44.0	79.0	6.5	27.0	56.0
15 In/HY	17.5	47.0	79.5	-	7.0	14.5
25 In/HY	16.0	48.5	76.0	-	2.5	4.5
HZSM	56.5	82.5	86.0	-	-	-
3 In/HZSM	52.0	83.0	85.0	42.5	70.0	82.0
5 In/HZSM	53.0	81.0	86.0	35.0	58.5	79.0
8.2In/HZSM	54.0	82.0	85.5	18.5	36.5	56.5
15 In/HZSM	56.0	84.5	85.5	-	2.5	5.5

Some results in Table 4 show the differences in the catalytic action between **In** and Ga modified HZSM zeolites prepared by one and the same procedure. After H_2 reduction the 5 Ga/HZSM sample possesses catalytic activity and aromatization selectivity exceeding those of the HZSM base. The same loading with In_2O_3, however, results in manyfold decrease in activity and almost no formation of aromatics. In the case of the 3In, 4Ga/HZSM sample the possibility for simultaneous introduction of both elements through mechanical mixing and a common reduction step is exemplified. It is interesting to note that the sample above is superior in n-pentane aromatization reaction to both 5Ga/HZSM and 7Ga/HZSM catalysts.

Table 4. Pentane conversion, at 763 K, WHSV 7.5

Sample	conversion mass % * A	conversion mass % * B	Selectivity for aromatics % A	Selectivity for aromatics % B
glass	0.2	0.2	0.0	0.0
HZSM	29.9	29.7	0.8	0.5
1In/HZSM	-	21.5	-	0.6
2In/HZSM	-	19.3	-	0.6
5In/HZSM	-	5.8	-	1.0
10In/HZSM	10.3$			
10In/HZSM	3.5	0.19	0.0	0.0
15In/HZSM	-	0.18	-	0.0
5Ga/HZSM	25.3	51.3	2.0	20.0
7Ga/HZSM	28.1	49.1	3.5	13.3
3In4GaZSM	21.5	38.3	4.2	20.2

time on stream: * - 30 min; $ - 3 min; A-before, B-after reduction

Table 5. ΔG^o for reduction of Me_2O_3 to Me_2O

T,K	ΔG^o KJ/mol Ga	In
600	322.2	195.6
700	294.0	168.4
800	266.5	141.9
900	239.6	116.0
1000	213.3	90.7

DISCUSSION

Recently, the reductive treatment of mechanically mixed Ga_2O_3/HZSM catalysts has been reported as leading to introduction of Ga into the cationic positions of the zeolite framework [1-4]. Considering the position of indium in the periodic system of the elements, we expected that a similar process of ion insertion should proceed when In_2O_3, instead Ga_2O_3 is used as a catalyst component.

Generally, two main reactions of the indium oxide phase can be taken into consideration

1. $In_2O_3 + 2\ H_2(g) = In_2O(g) + 2H_2O\ (g)$
2. $In_2O_3 + 3\ H_2(g) = 2In\ (s) + 3H_2O\ (g)$ (s)-solid; (g)-gas

According to our TPR , supported by some XRD data results there is no doubt that the reduction process in the case of the pure In_2O_3 or its mixture with a nonacidic material can be rationalized in terms of reaction 2. Practically no loss due to a volatile indium product was registered under the experimental conditions used.

In the case of acidic zeolite samples we postulate the reaction 3, which is coupled with reaction 1, as playing the key role in the reductive conversion of the indium oxide component.

3. $In_2O(g) + 2\ H^+Z^-(s) = 2\ In^+Z^-(s) + H_2O(g)$ (Z)-zeolite

The estimation in Table 5 performed on basis of published thermodynamic data [10] reveals why the interaction of In_2O_3 with an acidic zeolite occurs much more easily as compared to Ga_2O_3. Since the change of standard free energy of the reaction step 1 is less positive the whole process of reductive depletion of the In_2O_3 phase is facilitated by the driving force of an acidic zeolite support to a greater extent than in the Ga_2O_3 case.

The finding that the solid state reaction above is not restricted to the pentasil-type zeolite supports is also an important result from this investigation. A number of zeolite structures like faujasite, mordenite and offretite behave similarly to the HZSM zeolite. Taking into account the accuracy of the microbalance TPR experiment along with other possible sources of experimental error, one may conclude that the **In**/H^+ ratios listed in Table 2 correlate well with a value of roughly 1. The somewhat higher values for the HZSM based samples may be due to the possible involvement of terminal silanol groups in the solid state reaction as already proposed for Ga_2O_3/HZSM [1]. Thus the proton content of a particular zeolite seems to determine its capacity for specific indium uptake. In no case have **In**/H^+ values near 0.5 or 0.33 corresponding to 2 and 3 **In** atoms respectivelly for each H^+ been observed. A value of 0.33 would be expected if the ion exchange mechanism proposed for In_2O_3/HZSM system by Zatorski [6] were true. The acidity measurements and catalytic experiments prove convincingly the specific role of the indium sites formed after hydrogen or hydrocarbon reduction. Both acidity and catalytic activity gradually decrease with the indium content until all zeolite acid sites are covered by **In**. It seems that the main difference between In and Ga/zeolite systems consists in the specific properties of the cationic species introduced by the solid state reaction under discussion. The replacement of zeolite protons by indium, existing probably as a **In**$^+$ cation results in a strong suppression of the whole acidic function of the zeolite. In contrast, the intrinsic activity of the Ga^+ cation as a Lewis acid site,through water dissociation route or by facilitating the hydrogen transfer seems to be high enough to attain a high catalytic activity in carbocationic transformations of hydrocarbons. The combination of both cations in one and the same zeolite offers also interesting possibilities for regulating the catalytic properties as it is briefly shown in the case of n-pentane conversion.

Finally, our claim is that a novel zeolite is obtained through stabilization of indium as **In**$^+$ in the interior of the zeolite crystallites. Further research is still necessary in order to elucidate whether the unusual oxidation state of the indium cation and its very high content in various zeolite structures may be utilized for appropriate catalytic transformations.

ACKNOWLEDGMENTS

V. K. thanks the Mobil Oil Austria AG and the Organizers of the 9th IZC for supporting his visit to Canada, and we thank the National Scientific Foundation of Bulgaria for financial support.

REFERENCES

1. Price, G.L. and Kanazirev, V.,J.Catal.,126,267(1990).
2. Kanazirev ,V., Price, G.L.and Dooley, K.M.,in: *Zeolite Chemistry and Catalysis,* Eds.Jacobs, P.A. et al., Elsevier Science Publishers B.V., Amsterdam,p.277(1991).
3. Kanazirev, V.,Dimitrova, R.,Price ,G.L.,Khodakov, A.Yu. ,Kustov L.M. and Kazansky V., J.Mol.Catal., 70, 111 (1991)
4. Kanazirev,V.,Mavrodinova, V.,Kozova and Price, G.L.,Catal.Lett.,9,35 (1991).
5. Zatorski, L.W., Wierzchowski, P.T. and Ciochowlas,A.A.,*Proceedings of the 6th International Symposium of Heterogeneous Catalysis,* Eds. Shopov ,D.et al.,p.205(1987).
6. Zatorski, L.W., Bull.Polish.Acad.Sci.Chem.,35,325 (1987)
7. Kanazirev, V. and Borisova, N., Zeolites, 2, 23 (1982).
8. Kanazirev, V., Tsoncheva, T. and Minchev, Ch.,Z.Phys.Chem.N.F., 149, 237 (1986).
9. Breck, D.W.,*Zeolite Molecular Sieves,* Willy, J. and Sons, New York (1974).
10. Barin, I., *Thermodinamical Data of Pure Substances,* VCH Weinheim, 1989.
11. Kanazirev,V. and Foerster, H., in preparation.

SPECTROSCOPY, ENERGETICS AND SITING OF NH_4^+ IN ZEOLITES; THEORY AND EXPERIMENT

E.H. TEUNISSEN, W.P.J.H. JACOBS, A.P.J. JANSEN and R.A. VAN SANTEN

Schuit Institute of Catalysis, Theory Group, Eindhoven University of Technology, P.O. Box 513, 5600 MB Eindhoven, The Netherlands

ABSTRACT:

The adsorption of an ammonia molecule onto an acidic zeolite and the proton transfer from the zeolite to the ammonia are studied by quantum chemical methods. The NH_4^+ is adsorbed with two or three hydrogen bonds to the zeolite. The calculated vibrational frequencies explain the experimental infra–red spectra.

INTRODUCTION

One of the most interesting applications of zeolites is their use as a catalyst [1]. Many of the catalytic properties stem from the acidic properties of the zeolites and much research has been carried out on the properties of the acidic group [2]. The interaction between the acidic hydroxyl group and the base ammonia has been studied by many techniques, like infra red spectroscopy, NMR–spectroscopy, Temperature–Programmed Desorption (TPD), X–Ray Diffraction (XRD), inelastic neutron scattering or theoretically by quantum–chemical calculations [3–40].

In this paper we will study the interaction of the zeolite acidic site and the NH_3 molecule by quantum chemical methods. Although there are methods and programs available to do calculations on periodic systems [41-43] in this paper, as in most, the zeolite is modelled by a small cluster. The cluster approach has the disadvantage that the hydroxyl–group, modelled by the cluster does not always behave the same towards adorbents or bases as in the crystalline case, as has been shown by a recent ab–initio study on the chabazite–NH_3 system [44]. The advantage of cluster calculations, however, is that a high quality basis set can be used, that geometries can be optimized, and that a normal mode analysis can be carried out. Also intensities for infra–red specta can be calculated by ab initio methods.

From the Proceedings of the 9th International Zeolite Conference, Montreal 1992, Eds. R. von Ballmoos et al.,

From ^{15}N NMR spectroscopy it is known that at low temperatures the ammonium cation is located near the aluminum tetrahedra [32]. At low temperatures the NH_4^+ is tumbling around the aluminum tetrahedrons [17,31,32]. Thus the clusters are chosen such that the siting of the ammonium tetrahedron can be described.

METHODS

The choice of the clusters

When doing quantum chemical calculations on clusters there are two choices to be made: the choice of the cluster and the choice of the basis set. In an earlier paper we have shown that the system is not described very well when a small basis set is used [39]. Thus we have to use a large basis set. The disadavantage of this is that the size of the cluster is limited.

In figure 1 the clusters that are used to model the zeolite are shown. For each coordination of the ammonium cation onto the zeolite wall a different cluster is used. In figure 1a the cluster is shown which is used if the ammonium cation is bonded to the zeolite with a single hydrogen bond. This cluster contains the characteristic acidic site, namely the Si–O**H**–Al group. In figure 1b the cluster is shown for the ammonium cation adsorbed onto the zeolitic cluster with two hydrogen bonds. In figure 1c the cluster is shown that is used if the ammonium cation is adsorbed with three hydrogen bonds. Finally in figure 1d the case is shown in which two ammonia molecules are co–adsorbed on one single site.

Quantum–chemical methods

The quantum–chemical calculations are done at the ab initio SCF–level. We use a mixed basis set with a minimal basis set on the hydrogens that are saturating the dangling bonds, a 6-31G(D) basis set on the silicon and aluminum atoms, a 6-311G(d) basis set on the nitrogen atom, a 31G(p) set on the hydrogen molecules that are attached to the nitrogen, and a 6-311+G(d) basis set on the anionic oxygens. It has been shown that the anionic oxygen needs a large basis set because of its diffuse character [39]. If a only slightly smaller basis set is used, e.g. 6-31G(d) the proton transfer energy is overestimated by about 40 kJ/mol [39]. Inclusion of electron correlation energy by means of second order Møller–Plesset perturbation theory [45] and the use of the counterpoise correction [46,47] for the binding energies increases the calculation time tenfold and it decreases the proton transfer energy by 20 kJ/mol. For all the clusters the geometries are fully optimized at the SCF–level using gradient techniques. This geometry optimization is necessary. There is experimental and theoretical evidence that the zeolite lattice is very flexible [48]. Once the geometries are optimized that infra–red spectra can be calculated using analytical second derivatives.

Experimental

The spectra of the ammonium loaded zeolites were recorded in transmission mode at a resolution of 1 cm^{-1}, with a Bruker IFS 113v spectrometer. 250 scans were coadded. Spectra were recorded using self–supporting discs of zeolite sample (4.5 mg/cm^2). The samples were obtained by exchange of the sodium form of the zeolites at 80°C. Then the samples, a 6.3:1 Si/Al mordenite and a 2.7:1 Si/Al Y zeolite, were placed in an in–situ cell which was connected to a vacuum–system and a gasdosing system. The sample was dehydrated and deammoniated at 500°C in vacuo, after this anhydrous NH_3 was adsorbed at room temperature. Before the spectra were recorded any physisorbed NH_3 was removed by evacuation at room temperature.

RESULTS AND DISCUSSION

Heats of adsorption

For the adsorption sites shown in figure 1 the adsorption energies have been calculated (Table I).

Table I. The calculated heats of adsorption (kJ/mol). * The energy refers to the adsorption of the two ammonia molecules; i.e. 30 kJ/mol each.

parameter	value
singly coordinated	−57
doubly coordinated	−110
triply coordinated	−112
coadsorbed structure	-60 *

In the case of the singly bonded structure the favorable structure is the one in which the proton is attached to the zeolitic cluster. Because the ionic structure was not a minimum the geometry could not be optimized for this case [39]. If the geometry is fixed however the difference between the hydrogen bonded form and the ionic form is 65 kJ/mol [39]. Clearly the favorable structures are those in which the ammonium cation is bonded with two or three hydrogen bonds. The calculated heats of adsorption compare very well with experimental heats of adsorption. If the heat of adsorption is measured with TPD the values for the heats of adsorption are in the range 112 ± 33 kJ/mol. [11,18,20,21,22,26,28]. With Micro-Calorimety these values are in the range of 146 ± 146 kJ/mol. [9,12,13,14,27].

The experimental spectra

Spectra have been recorded for two samples. In figure 2a the N–H stretching region of a spectrum of a NH_4^+ mordenite with a Si/Al ratio of 6.3 is shown. In figure 2b, the spectrum for a NH_4^+–Y zeolite is shown with a Si/Al ratio of 2.7. Although the spectra were recorded for different zeolites they show the same features. There are three broad overlapping bands and one sharp peak in the 2700–3500 cm $^{-1}$ region. Also spectra of other ammonium containing zeolites like zeolite beta [4], and erionite [5] shows these characteristics.

The calculated spectra

For the clusters shown in figure 1 the spectra have been calculated. They are shown in figure 3. Analytical second derivatives were used. The frequencies were scaled by 0.92. If this scaling factor is used the calculated N–H strectching frequencies of NH_3 matches the experimental ones.

The spectra of the hydrogen bonded form and of the structure where two NH_3–molecules are coadsorbed are dominated by one single peak. The spectra of the doubly bonded structure show two intense peaks and two smaller ones that are hardly visible. The structure with the triply coordinated ammonium, shows two intens peaks very close to each other and one less intense peak. From the heats of adsorption it may be expected that the latter two structures will dominate the spectrum of the ammonia loaded zeolite. The experimental spectra can be explained quite easily from the calculated ones. The two broad bands on the left side can stem from the doubly coordinated ammonium, the triply coordinated structure causes the two peaks on the right, the broad band around 3000 cm^{-1} and the relative sharp one around 3400 cm^{-1}. This sharp one is the N–H stetching of the proton pointing away from the zeolitic site.

CONCLUSIONS

The proton of the zeolite is completely transfered to the base NH_3. NH_4^+ is bonded to the zeolite wall with two or three hydrogen bonds. The calculated heats of adsorption are good agreement with the experimental ones. The calculated spectra explain the experimental ones.

References

[1] D.W. Breck, Zeolite molecular sieves, Wiley Interscience, New York (1974).

[2] 'Introduction to zeolite science and practice', H. van Bekkum, E.M. Flanigan and J.C. Jansen (eds.), Stud. in Surface Sci. and Catal. **58**, Elsevier Amsterdam (1991).

[3] W.L. Earl, P.O. Fritz, A.A.V. Gibson and J.H. Lunsford, J. Phys. Chem. **91** , 2091 (1987).

[4] S.G. Hegde, R. Kumar, R.N. Bhat and P. Ratnasamy, Zeolites **9**, 230 (1989).

[5] A. Kogelbauer, J.A. Lercher, K.-H. Steinberg, F.Roessner, A. Soellner and R.V. Dmitriev, **9**, 224 (1989).

[6] G.A. Ozin, M.D. Baker, J.Godber and C.J. Gil, J. Phys. Chem. **93**, 2899 (1989).

[7] G.A.Ozin, M.D. Baker, K. Helwig and J. Godber, J. Phys. Chem. **89**, 1846 (1985).

[8] E.A. Paukshtis, Y.D. Pankrat' ev, A.G. Pel' menschikov, E.B. Burgina, V.M. Turkov, E.N. Yurchenko and G.M. Zhidomirov, Kinetika i Kataliz **27**, 1440 (1986).

[9] R.D. Shannon, R.H. Staley, A.J. Vega, R.X. Fischer, W.H. Baur and A. Auroux, J. Phys. Chem. **93**, 2019 (1989).

[10] T. Stock, D. Dombrowski, J. Fruwert and H. Ratajczak J.C.S. Farad. Trans I,**79**, 2773 (1983).

[11] N.-Y. Topsøe, K. Pedersen and E.G. Derouane, J. Catal **70**, 41 (1990).

[12] K. Tsutumi, Y. Mitani, H. Takahashi, Bull. Chem . Soc. Jpn **56**, 1912 (1983).

[13] J.C. Vedrine, A. Auroux, V. Bolis, P. Dejaifve, C. Naccache, P. Wierzchowski, E.G. Derouane, J.B. Nagy, J.-P. Gilson, J.H.C. van Hooff, J.P. van den Berg and J. Wolthuizen, J. Catal **50**, 248 (1979).

[14] A. Auroux and J.C. Vedrine, Stud. in Surf. Sci. and Catal **20**, 311 (1985).

[15] D.R. Corbin, L. Abrams, G.A. Jones, M.M. Eddy, W.T.A. Harrison, G.D. Stucky and D.E.Cox, J. Am. Chem. Soc **112**, 4821 (1990).

[16] L.B. McCusker, Zeolites **4** , 50 (1990).

[17] D.Deininger and B. Reïmann, Z. phys. Chemie **251**, 353 (1972).

[18] E. Dima and L.V.C. Rees, Zeolites **8**, 10 (1990).

[19] B.P. Embaid, M. Hunger, M.M. Landrove and R. Martin, Z. phys. Chemie **166** (neue folge), 109 (1990).

[20] C.V. Hidalgo, H. Itoh, T. Hattori, M. Niwa and Y. Murakumi, J. Catal. **85**, 362 (1984).

[21] H.G. Karge and V. Dondur, J. Phys. Chem. **94**, 765 (1990).

[22] H.G. Karge, V. Dondur and J. Weitkamp, J. Phys. Chem **95**, 283 (1991).

[23] M.M. Mestdagh, W.E.E. Stone and J.J. Fripiat, J. Catal. **38**, 358 (1975).

[24] D. Michel, A. Germanus and H. Pfeifer, J.C.S. Farad. Trans. I, **78**, 237 (1982).

[25] L.M. Parker, D.M. Bibby and R.H. Meinhold, Zeolites **5**, 384 (1985).

[26] J.G. Post and J.H.C. van Hooff, Zeolites **8**, 9 (1984).

[27] Z.C.Shi, A. Auroux and Y.B. Taarit, Can. J. Chem **66**, 1012 (1988).

[28] W. Schirmer, H. Stach, K. Fiedler, W. Rudzinski and J. Jagiello, Zeolites **3**, 199 (1983).

[29] H. Kriegsman, A. Reklat, E. Löffler and Th. Steiger, Z. phys. Chemie **271**, 61 (1990).

[30] H. Itoh, C.V. Hidalgo, T. Hattori, M. Niwa and Y. Murakumi, J. Catal. **85**, 521 (1984).

[31] T.J. Udovic, R.R. Cavanagh, J.J. Rush, M.J. Wax, G.D. Stucky, G.A. Jones and D.R. Corbin , J. Phys. Chem **91**, 5968 (1987).

[32] A.J. Vega and Z. Luz, J. Phys. Chem **91**, 365 (1987).

[33] G.P. Babu, S.G. Hegde, S.B. Kulkarni and P. Ratnasamy, J. Catal **81**, 471 (1983).

[34] P. Dejaifve, A. Auroux, P.C. Gravelle, J.C. Vedrine, Z. Gabelica and E.G. Derouane, J. Catal **70**, 123 (1981).

[35] S.-Y. Zhang, O.Talu and D.T. Hayhurst, J. Phys. Chem **95**, 1722 (1991).

[36] M. Allavena, K. Seiti, E. Kassab, Gy. Frenczy and J.G. Ángyán, Chem. Phys. Lett. **168**, 461 (1990).

[37] A.S. Medin, V.Y. Borovkov, V.B. Kazansky, A.G. Pelmenschikov and G.M. Zhidomirov, Zeolites **10**, 668 (1990).

[38] P. Geerlings, N. Tariel, A. Botrel, R. Lissillour and W.J. Mortier, J. Phys. Chem. **88**, 5752 (1984).

[39] E.H. Teunissen, F.B. van Duijneveldt and R.A. van Santen, J. Phys. Chem. **96**, 366 (1992).

[40] E. Kassab, K. Seiti and M. Allavena, J. Phys. Chem. **95**, 9425 (1991).

[41] C. Pisani, R. Dovesi and C. Roetti, ' *Hartree–Fock Ab–initio Treatment of Crystalline Systems*, Lecture Notes in Chemistry **48**, Springer Verlag, Berlin **1988**.

[42] C. Pisani and R. Dovesi, Int. J. Quantum Chem. **17**, 501 (1980). R. Dovesi, C. Pisani, C. Roetti and V.R. Saunders, Phys. Rev. B **28**, 5781 (1983). R. Dovesi, Int. J. Quantum Chem. **29**, 1755 (1986).

[43] G. te Velde, Numerical Integration and other methodological aspects of bansdtructure calcluation, pHD–thesis VU Amsterdam (1990).

[44] E.H. Teunissen, C. Roetti, C. Pisani, A.J.M. de Man, A.P.J. Jansen, R. Orlando, R.A. van Santen and R. Dovesi, to be published.

[45] C. Møller and M.S. Plesset, Phys. Rev. **46**, 618 (1934).

[46] S.F. Boys and F. Bernardi, Mol. Phys. **19**, 553 (1970).

[47] J.H. van Lenthe, J.G.C.M. van Duijneveldt–van de Rijdt and F.B. van Duijneveldt, Adv. in Chem. Phys. **69**, 521 (1987).

[48] R.A. van Santen, A.J.M. de Man, W.P.J.H. Jacobs, E.H. Teunissen and G.J. Kramer, Catal. Lett. **9**, 273 (1991), and references therein.

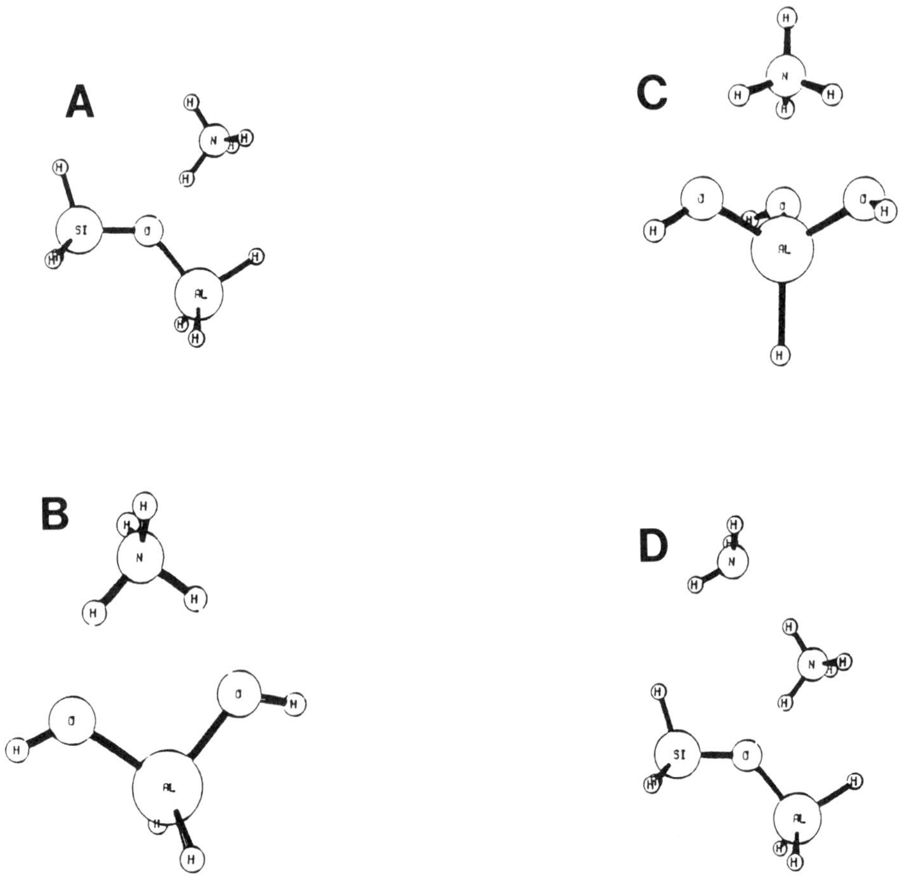

Figure 1. The clusters used to model the adsorption sites; a: the singly bonded structure, b: the doubly bonded structure, c: the triply bonded structure, d: the coadsorbed structure.

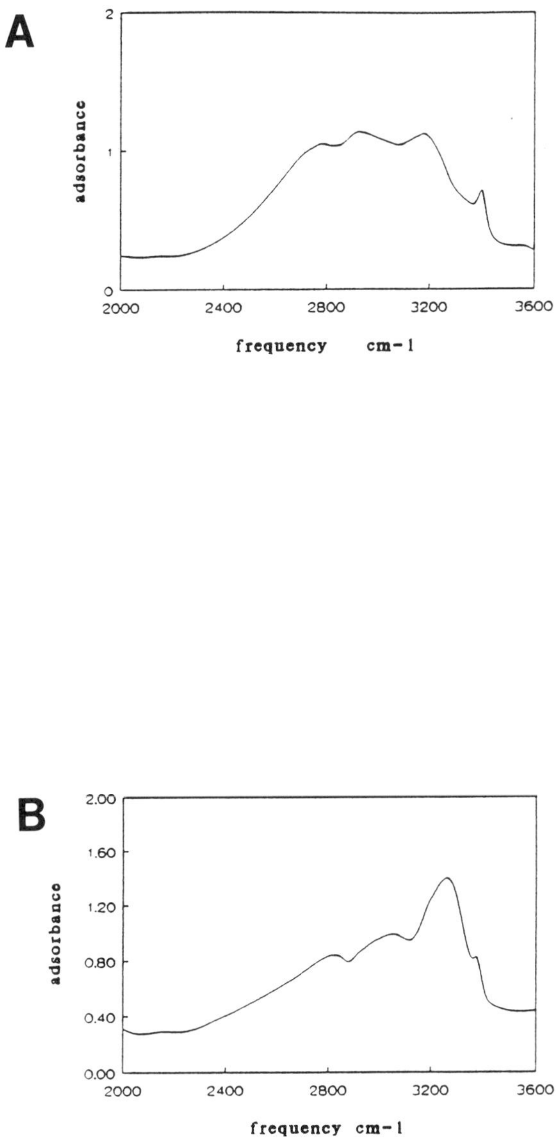

Figure 2. The experimental spectra of the ammonia reconstituted form of mordenite (a) and zeolite Y (b).

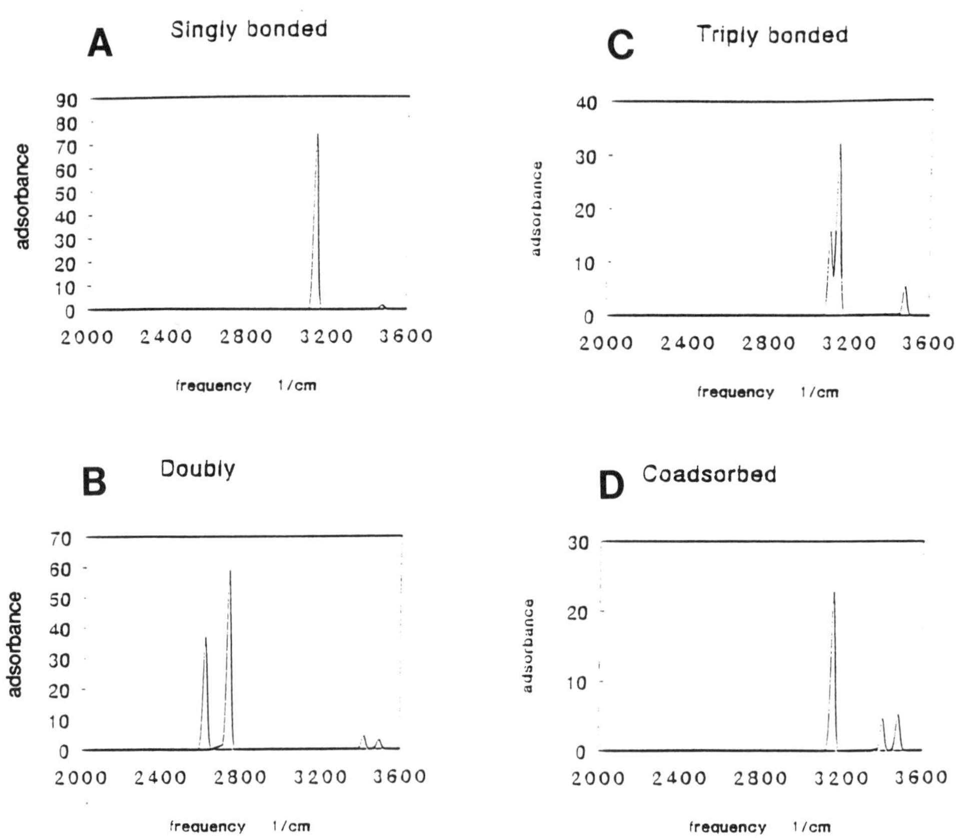

Figure 3, The calculated spectra; a: the singly bonded structure, b: the doubly bonded structure, c: the triply bonded structure, d: the coadsorbed structure.

STUDIES ON TEMPLATE MOLECULES: LOCATION, ORIENTATION, PROTONATION AND INTERACTION WITH FRAMEWORK IN HIGH SILICA ZEOLITES

YAOJUN SUN*, TAILIU WU*, LIPING WANG*, LUN FEI*, YINGCAI LONG**

* Centre of Analysis and Measurement, Fudan University, Shanghai 200433, P. R. C.

** Dept. of Chem., Fudan University, Shanghai 200433, P. R. C. Author for Correspondence.

ABSTRACT

Template molecule of TMEDA (tetramethylethylenediamine) trapped in the channel or the cage of high silica zeolites of mordenite, ZSM-5, ZSM-35, ZSM-39, and CF-3 were characterized by ^{13}C CP MAS and HPDEC MAS n.m.r. spectra, IR spectra, TG/DTA and chemical analysis. Linear relations between ^{13}C chemical shift (δ) for $-CH_3$ and $-C_2H_4-$ groups of TMEDA, and the free diameter (l) of the channel or the cage have been determined. It was found that an obvious difference in relative intensity of ^{13}C CP and HPDEC resonance, line narrowness, and great division from the δ-l linear relations can be used to deduce the position, the orientation of template molecule in the channel or the cage of the zeolites, and its interaction with framework.

INTRODUCTION

The study on the location and state of the organic template molecules trapped in the zeolite framework plays very important role in revealing the process of the crystallization of the zeolites. ^{13}C MAS n.m.r. is one of the most useful methods for such a study. Position of template [1-4], dependence of its ^{13}C chemical shift on the size of the cage and its dynamic behaviour[5,6], the protonation state[7-9], the presentation of complexes [10], and the configuration [11] have been studied.

In system of TMEDA-Na_2O-SiO_2-Al_2O_3-H_2O, mordenite (Md.), ZSM-5, ZSM-35, and clathrate zeolites ZSM-39, and CF-3 [12,13] were hydrothermally synthesized[14]. In this study, ^{13}C CP MAS and HPDEC MAS n.m.r., IR, TG/DTA, and chemical analysis have been used to investigate the position, the orientation, the occupancy percentage, and the protonation of TMEDA in the channel or the cage, and its interaction with the framework of the zeolites.

From the Proceedings of the 9th International Zeolite Conference, Montreal 1992, Eds. R. von Ballmoos et al.,

EXPERIMENTAL

Zeolite sample with high crystallinity was prepared and washed with hot distillation water, and was identified as in ref. [14]. The Al,Si and N contents were determined by chemical analysis and Na by atomic adsorption. IR spectra were recorded on FT-IR 5DX spectrometer at a scanning range of 2600 to 3200 cm^{-1} in 60 recycles. Discus of KBr and zeolites powder were used. TG/DTA measurement of 10 to 20 mg sample was carried out by using PTC-10A thermal analyzer at 10°C/min heating rate in air flow from room temperature up to 800°C. All ^{13}C MAS n.m.r. spectra were recorded at room temperature with Bruker MSL-300 spectrometer, whose ^{13}C resonance frequency was 78.468 MHz. The rotor was spun at 4.0KHz, a 2K FID was acquired and zero filled to 4K. The radio frequency field was 50KHz, corresponding $\pi/2$ pulse width 5μs, recycle time of 2s, accumulative number of 2000 to 6000. The contact time was 2ms for CP spectra. The decoupling power was 50KHz for HPDEC spectra. Adamantane was used as a second reference, its higher-field resonance line was assumed to have a chemical shift δ=28.25ppm with respect to TMS as zero.

RESULTS

The chemical composition of zeolite unit cell is summarized in Table 1. A part of framework negative charges of ZSM-5, ZSM-35 and Md. were balanced with $TMEDA^{+2}$. In these zeolites, 80 to 90% of their theoretical channel volume were occupied with TMEDA and H_2O, in which H_2O dominated more than 20%.

The DTA curves of the samples showed that the first exothermic peaks for TMEDA oxidation were at 446, 454, 477°C in Md., ZSM-5 and ZSM-35 respectively, and increased with the decreasing of channel diameter. There was no obvious thermic effect on CF-3, but there was a small endothermic peak at 587°C in ZSM-39. There were two major weight loses for TG of Md., ZSM-5 and ZSM-35, one lower than 350°C due to dehydration and the other higher's due to TMEDA escape. The weight loss for CF-3 and ZSM-39 began slowly at about 300°C and continued on when temperature was kept at 800°C. It indicated that TMEDA was difficult to be free from the cage.

IR absorption spectra of TMEDA are shown in Fig. 1.

^{13}C CP MAS and HPDEC MAS spectra collected under same experimental conditions are shown in Fig. 2, and its δ data in table 2. It was evident that ^{13}C chemical shift of each carbon group of TMEDA trapped in zeolite changes from that in D_2O, depended on the structure of zeolites. The line at 48.86ppm for ZSM-39 was the narrowest. Some obvious peak intensity changes from CP MAS to HPDEC MAS n.m.r. were visible except for CF-3. 48.86ppm line for ZSM-39 and 56.94ppm line for ZSM-35 were the most obvious.

Within the range of δ=34 to 37ppm, a new line C_X emerged in Md., ZSM-5 and ZSM-35 but not in ZSM-39 and CF-3. Peak splitting was taken place at line of $-CH_3$ for Md., ZSM-5 and ZSM-35, and of $-C_2H_4-$ for Md..

The chemical shift values of TMEDA in different zeolites were different. Fig. 3 shows the correlation between (δ) of carbon group and the average free diameter (l) of the channel or the cage. A good linear dependence was presented as: $\delta(C_X)$-l for Md., ZSM-5 and ZSM-35; $\delta(C_1)$-l for Md., ZSM-5, ZSM-35 and CF-3; $\delta(C_2)$-l for Md., ZSM-5, CF-3 and l' for ZSM-35. The split peak positions are also indicated in Fig.3. Peaks with obvious difference in CP and HPDEC spectra intensity are indicated by symbol↑. In case of ZSM-39, $\delta(C_1)$ had obvious deviation from the $\delta(C_1)$-l line, and $\delta(C_2)$ = 48.86ppm was far away.

DISCUSSION

ZSM-39 (dodecasil-3C) possesses a clathrate of $8[5^{12}]$ and $8[5^{12}6^4]$ $[SiO_4]$ hedra, the latter has structure-controlling guest molecules[15]. Since the cell composition of 136 $SiO_2\cdot$3.93 TMEDA, each $[5^{12}6^4]$ cage is occupied by 1/2 TMEDA in average. Single crystal structure determination showed that free diameter of $[5^{12}6^4]$ cage is about 6.2A [16], its free space is much smaller in size than a molecule of TMEDA (8.2X5.6A). It is reasonable that one molecule of TMEDA is shared by two neighbouring $[5^{12}6^4]$ cages in most situations. The 46.5ppm line in ZSM-39 is considerably broader. It indicates that the $-CH_3$ group can rotate around σ bond of C-N and tumbling. Its chemical environment is statistically uniform, and possesses shorter ^{13}C spin-lattice relaxation time. But the $-C_2H_4-$ group may be trapped at the center of a 6-ring oxygen window between two neighbouring $[5^{12}6^4]$ cages with the window size of 2.5A. Since the length of C-C bond and C-H bond is 1.54 and 1.10A respectively, an electronic shield effect of the window oxygen atom causes chemical shift of $-C_2H_4-$ group moving to 48.29ppm, and changes the ^{13}C spin lattice relaxation time of proton connected with the carbon atom, and thus changes the NOE effect of the carbon atom, and renders great variations on spectra intensity of CP and HPDEC. Since $-C_2H_4-$ group is fixed at centre of 6-ring window, the chemical environment around is simpler than that of $-CH_3$ group located in cage, and its motion is restricted to as a flip-flop movement around the N-C-C-N axis. This situation may cause narrowing of $-C_2H_4-$ groups resonance line. The special location and the strong interaction with the framework oxygen atoms cause stronger vibration at 2922 and 2853cm^{-1} for $-C_2H_4-$ band in IR spectra of ZSM-39, and an endothermic effect at 587°C on DTA curve as well.

There are a possibility that the free diameter of the cage will decrease to 4.8A or to 3.44A, if one or two Si-OH groups are within $[5^{12}6^{4}]$ cage. It may greatly change the chemical environment of $-CH_3$ group and thus greatly varies its ^{13}C chemical shift. As space is too small for motion, the intensity of HPDEC MAS and CP MAS varies. These might be a reason for the generation of two small lines at 57.0 and 55.5ppm in HPDEC MAS spectra.

The discussions above are in agreement with other results[17,18], Therefore, if $-CH_3$ or $-C_2H_4-$ group of TMEDA trapped in high silica zeolites is located at a special position and interacted strongly with framework, the ^{13}C MAS n.m.r. will possess following characters: an obvious difference of relative intensity for CP MAS and HPDEC MAS spectra; a distinct deviation from the δ-l linear dependence; narrowing of line.

CF-3 is a zeolite with clathrate structure of $8[5^{4}6^{4}]\cdot8[4^{1}5^{8}]\cdot8[5^{8}6^{12}]$ cage in the cell[12,13]. Only half of the cage $[5^{8}6^{12}]$ (9.7X7.9X4.9A in size) can be housed by TMEDA, which moves freely.

In cell of ZSM-35, there are two 10-ring straight channels and two 6-ring straight channels parallel [001], and two 8-ring straight channels parallel [010][19] with about two molecules of TMEDA. Four $-CH_3$ groups of each TMEDA were located in 10-ring channels, but $-C_2H_4-$ group ($\delta C_{(2)}$=56.95ppm), which strongly interacts with framework, was located in a small cage formed by intersection of 8-ring (4.8X3.4A) channel and

6-ring channel. In IR spectra, the $-C_2H_4-$ vibration at 2924 and 2853 cm^{-1} are stronger as well. As the free diameter of 10-ring channel is 4.3X5.5A and the van der Waals diameter of $-CH_3$ group is 4.0A, and as the two $-CH_3$ groups at the same end of TMEDA have to be kept in a certain bond angle, these facts may make the local chemical environment of $-CH_3$ group different and causes ^{13}C n.m.r. line splitting to 47.27 and 46.10ppm.

In cell of ZSM-5, there are two 10-ring straight channels (5.6X5.4A) parallel [010] and two 10-ring sinusoidal channels (5.5X5.1A) parallel [100] [20]. Each straight channel or sinusoidal channel is occupied by about two TMEDA. But for the TMEDA trapped in sinusoidal channel, group of $-N(CH_3)_2$ is distorted, one $-CH_3$ group ($\delta_{(C2)}$=45.95ppm) enters the 10-ring straight channel, and the motion of $-C_2H_4-$ group ($\delta_{(C2)}$=55.49ppm) and $-CH_3$ group ($\delta_{(C1)}$=44,72ppm) within sinusoidal channel are restricted by the framework.

In cell of Md., there are two 12-ring straight channels (7.0X6.7A) parallel [001] and two distorted 8-ring channels (5.7X2.5A) parallel [010][21]. In average, two TMEDA molecules lie along [010], $(CH_3)_2-N-$ group is located at a side pocket (4.3A in size) of 8-ring channel, its moving of the $-CH_3$ group ($\delta_{(C1)}$=44.77ppm) is restricted. The $-C_2H_4-N-(CH_3)_2$ group of TMEDA is located in 12-ring channel, its $-CH_3$ group ($\delta_{(C1)}$=43.65ppm) rotates freely. One $-CH_2-$ ($\delta_{(C2)}$=55,52ppm) of group $-C_2H_4-$, which is close to the side pocket, interacts with framework and its moving is restricted by 8-ring window, and lead to stronger vibration at 2927 and at 2853cm^{-1} in IR spectra. The other $-CH_2-$ ($\delta_{(C2)}$=54.00ppm) is far from the side pocket and moves freely.

New lines C_X in ^{13}C MAS n.m.r. spectra can be assigned to a protonated form of TMEDA to balance some charge due to framework Al only in Md., ZSM-5 and ZSM-35. Absorption at 2811 and 2762cm^{-1}, assigned to the vibration of $R-N(CH_3)_2$ in IR spectra, tends weaker and even disappear. For the linear correlation of $\delta(C_X)$-1, we can imagine that most of the protonated TMEDA are located at larger channel in these zeolites,and its charges at both ends of $-C_2H_4-$ group are symmetric. Since the $-CH_3$ group of protonated TMEDA may be near to Al atoms, it leads to chemical shift at higher field.

CONCLUSION

Template molecule TMEDA trapped in high silica zeolites of Md., ZSM-5, ZSM-35, ZSM-39 and CF-3 are intact and located at fixed positions. Some organic groups interact strongly with oxygen atom in the framework. Part of TMEDA in Md., ZSM-5 and ZSM-35 with greater Al content are protonated.

The δ-1 dependence, the variation in CP and HPDEC spectra relative line intensity, and the narrowing of ^{13}C n.m.r. peak can all be used to determine the location and the orientation of the template molecules trapped in the zeolites and their interaction with the framework.

Table 1. Composition of Unit Cell and Void Occupany

Zeolite	Composition of Cell									Void Occupany							
	Na	Al	Si	O	TMEDA			H_2O		Va(CC/g)**		Va/Vt %***		V_{TMEDA} %		V_{H_2O} %	
					TG*	CA*	no. of ($TMEDA^{+2}$)	TG	CA	TG	CA	TG	CA	TG	CA	TG	CA
ZSM-39			136	272		3.93											
Cf-3			88	176		1.72			4.6								
Md	2.58	3.49	44.5	96	1.19	2.23	0.46	27.9	27.1	.149	.140	90.3	84.8	20.6	40.7	79.4	59.3
ZSM-5	.09	7.14	88.9	192	8.43	8.43	3.53	28.3	22.6	.156	.146	86.6	81.1	71.0	75.8	29.0	24.2
ZSM-35	.58	2.78	33.2	72	2.76	2.22	1.11	9.9	12.4	.138	.133	86.7	83.6	69.7	60.8	30.3	39.2

*TG- determinated by TG; CA- determinated by chemical analysis.

**Va-volume occupied by H_2O and TMEDA.

***Vt-theoretical channel volume: ZSM-5-0.180 CC/g; ZSM-35-0.159 CC/g; Md-0.165 CC/g.

Table 2. ^{13}C n.m.r. Chemical Shift Data from Fig.4

Zeolite		C2($-C_2H_4-$) ppm		C1($-CH_3$) ppm		Cx ppm
Cf-3	CP	54.86		44.69		
	HPDEC	54.77		44.94		
ZSM-39	CP 57.5	48.86		46.51		
	HPDEC 57.0† *55.5†	48.29†		45.89		
Md	CP		53.60	44.91	43.62	34.05
	HPDEC	55.52†	54.00	44.77†	43.65	
ZSM-5	CP	55.19		46.15	44.86	35.63
	HPDEC	55.49†		45.95	44.72†	35.92
ZSM-35	CP	56.94		47.40	46.25	36.61
	HPDEC	56.95†		47.27	46.10†	36.48
$TMEDA/D_2O$		58.41		46.36		

*Symbol† indicates the n.m.r. line which intensity of HPDEC/MAS is different from CP/MAS.

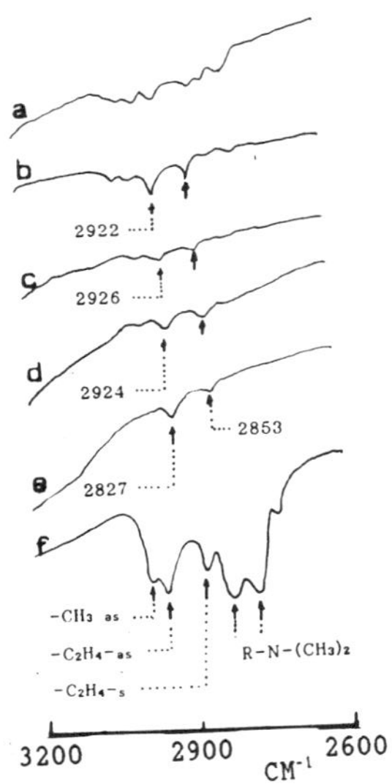

a - CF-3
b - ZSM-39
c - ZSM-5
d - ZSM-35
e - Mordenite
f - liquid TMEDA

Fig.1 IR spectra of TMEDA trapped in zeolites

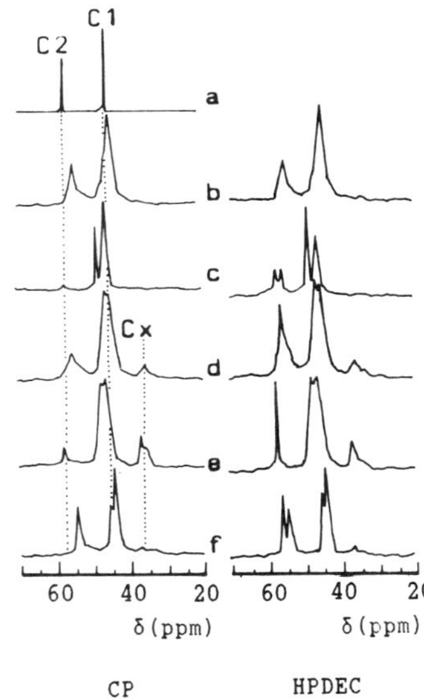

a - TMEDA in D2O
b - CF-3
c - ZSM-39
d - ZSM-5
e - ZSM-35
f - Mordenite

Fig.2 ^{13}C MAS n.m.r. spectra of TMEDA trapped in zeolites
CP-cross polarization
HPDEC-high power decoupling

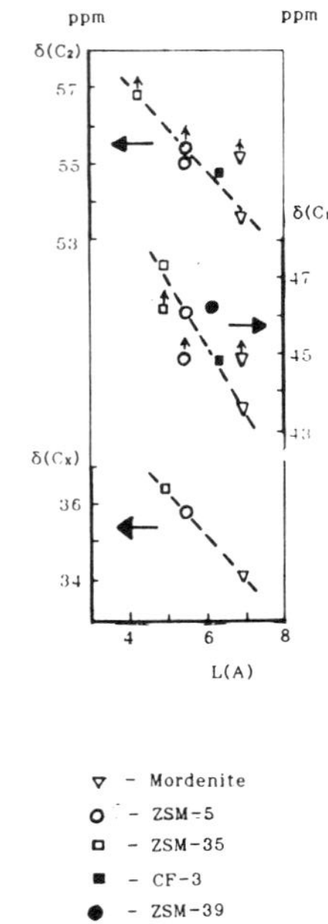

Fig.3 The ^{13}C chemical shift of TMEDA trapped in zeolites depended on the size of the channel or cage

REFERENCES

1 J.El Hage-Al Asswad, N. Dewaele, J.B.Nagy, R.A.Hubert, Z.Gabelica and E.G.Derouane;ZEOLITES,Vol 8,221(1988).

2 G.Boxhoorn, R.A.Van Santen, W.A.Van Erp,G.R.Hays,Rob Huis,and D.Clague;J.C.S.,CHEM.COMMUN.,264(1982).

3 J.B.Nagy,Z.Gabelica,and E.G.Derouane; ZEOLITES,Vol 3,43(1983).

4 R.H.Jarman,and M.T.Melchior; J.C.S.,CHEM. COMMUN.414(1984).

5 S.Hayashi,K.Suzuki and K. Hayamizu; J.C.S.,FARADAY TRANS., 1,85(9),2973 (1989).

6 S.Hayashi, K. Suzuki, S. Shin, K. Hayamizu, and O. Yamamoto; CHEMICAL PHYSICS LETTERS,Vol 113,4,368(1985).

7 N.Dewaele,Z.Gabelica,P.Bobart, J.B. Nagy, G.Giordano and E.G.Derouane; INNOVATION IN ZEOLITE MATERIALS SCIENCE,STUDIES IN SURFACE SCIENCE AND CATALYSTS, Vol 37,65(1987).

8 R.Challoner, R.K.Harris, K.J.Packer,and M.J.Taylor; ZEOLITES, Vol 10, 539(1990).

9 R.A.Hearmon, and A.Stewart; ZEOLITES,Vol 10,608(1990).

10 F.Delprato, L.Delmotte, J.L.Guth, and L.Huve;ZEOLITES, Vol 10,546(1990).

11 S.L.Zones,R.A.Van Nordstrand; ZEOLITES,Vol 8,166(1988).

12 Y. Long,W.Zhong, and X.Shen; J.INCLUSION PHENOMENA, 4,121(1986).

13 B.Marler,N.Dehnbostel, H.-H. Eulert, H.Giss and F.Liebau; J.INCLUSION PHENOMENA,4,339(1986).

14 Y.Long,H.Chun; ACTA CHIMICA SINICA,45,439(1987).

15 Y.Long,H.He,P.Zheng,G.Wu,and B.Wang;J.INCLUSION PHENOMENA,5,355(1987).

16 J.L.Schlenker,F.G.Dwyer,E.E.Jekins,W.J.Robrbaugh,G.T.Kokotailo, and W.M.Meier;NATURE,294,26,340(1982).

17 Y.Long et al.;"Studies on Crystallization Process of ZSM-39 by 13C MAS and 29Si MAS n.m.r.", to be published.

18 Y.Long et al.;"Position and Configuration of the Gust Organic Molecules Trapping in Cage of ZSM-39 by 13C MAS n.m.r.",to be published.

19 H.Gies and R.P.Gunawardane; ZEOLITES,Vol 7,442(1987).

20 H.Van Koningsveld,H.Van Bekkum,and J.C.Jansen;ACTA CRYST.,B43,127(1987).

21 W.M.Meier;Z.KRISTALLOGR.,115,439(1961).

ELECTRON SPIN ECHO STUDIES OF Mn(II) FRAMEWORK SUBSTITUTION IN MnAPO-11

GUILLAUME BROUET, XINHUA CHEN AND LARRY KEVAN

Deparment of Chemistry, University of Houston, Houston, Texas 77204-5641

ABSTRACT

Electron spin resonance (ESR) and electron spin echo modulation (ESEM) results are compared for MnAPO-11 and Mn-impregnated $AlPO_4$-11 with low Mn(II) content. Significant spectral differences are observed which indicate that MnAPO-11 has framework substituted Mn(II) while the Mn-impregnated $AlPO_4$-11 has nonframework Mn(II).

INTRODUCTION

The framework substitution of divalent ions for Al(III) in microporous aluminophosphates ($AlPO_4$-n) is of structural and catalytic interest [1]. Chemical analysis [1], Bronsted acidity [1] and powder x-ray diffraction[2] indicate that Mn(II) replaces Al(III) in MnAPO-11. ESR coupled with ESEM spectroscopy has been shown to an effective method for studying the location and local environment of transition metal ions in zeolites. [3] Direct evidence for framework Mn(II) in MnAPO-11 has recently been shown by using ESR and ESEM methods to determine the coordination geometry of water near the Mn(II). [4] Here we report additional comparative results on Mn-impregnated $AlPO_4$-11, which will be symbolized by Mn-$AlPO_4$-11, and synthesized MnAPO-11 that further demonstrate framework substitution of Mn(II) in MnAPO-11.

EXPERIMENTAL SECTION

$AlPO_4$-11 and MnAPO-11 were hydrothermally synthesized with diisopropylamine as a templating agent and were characterized by x-ray diffraction. $AlPO_4$-11 was impregnated with different amounts of Mn(II) by stirring at 90 ^oC. ESR spectra were obtained at 77 K and ESE spectra were obtained at 4 K.

From the Proceedings of the 9th International Zeolite Conference, Montreal 1992, Eds. R. von Ballmoos et al.,

RESULTS AND DISCUSSION

Synthesized MnAPO-11 with 0.1 mole % manganese shows a 35 GHz ESR spectrum with a characteristic six-line spectrum of ^{55}Mn and an isotropic hyperfine constant of 94 Gauss. The relatively clean spectrum shows that only one manganese species is present in this system. At 9 GHz the ESR spectrum also shows the characteristic six-line manganese spectrum but additionally has relatively strong forbidden transitions which make the spectrum more difficult to analyze in terms of the number of probable manganese species present. A comparison of Mn-impregnated $AlPO_4$-11 with 0.06 mole % manganese shows a similar spectrum as that observed in the synthesized MnAPO-11 material except that the resolution is significantly worse. This difference in resolution suggests that the manganese is in different sites in these two preparations.

More definitive evidence for two different sites for manganese in the synthesized MnAPO-11 versus the impregnated Mn-$AlPO_4$-11 is shown by comparing the spectrum of synthesized MnAPO-11 with 0.1 mole % manganese under two different conditions. The 9 GHz ESR spectrum for a calcined sample that has then been hydrated by allowing it to cool in air containing water shows a characteristic isotropic hyperfine coupling constant for manganese of 93.7 G. However, when the sample is calcined and then evacuated at 330 ^{o}C so that water is not present, the isotropic hyperfine constant decreases to 87.6 G. X-ray diffraction shows a significant distortion of the $AlPO_4$-11 structure on dehydration. [2] The fact that the isotropic coupling constant changes significantly on dehydration implies that the manganese is in the framework since the framework structure distorts on hydration and dehydration. One would not expect such a change in the isotropic coupling constant if the Mn were in a nonframework site in the channels of the $AlPO_4$ structure.

Further experiments with different concentrations of manganese incorporated into the synthesized MnAPO-11 materials indicate that there are two signals shown by electron spin resonance. One is a broad signal which is attributed to extra framework manganese and one is a resolved six-line signal which is attributed to framework-substituted manganese. As the small percent of manganese incorporated into the as-synthesized material increases from 0.1 to 1 mole percent the intensity of the broad signal increases monotonically. In impregnated Mn-$AlPO_4$-11 material with 0.3 mole % manganese and above only a broad signal characteristic of extra framework manganese is seen where the interaction between the manganese ions is strong enough such that the hyperfine structure is not detectable due to broadening.

Electron spin echo modulation spectrometry can give direct information about the coordination environment around a paramagnetic species. [3] In the present work coordination with deuterated water has been studied. Deuterated water is used because modulation from deuterium can be studied more quantitatively than proton modulation due to the higher nuclear spin and the lower modulation frequency and hence longer modulation period. Figure 1 shows a two-pulse electron spin echo modulation pattern of synthesized MnAPO-11 and impregnated Mn-APO_4-11 interacting with deuterated water. The normalized modulation depth is quantified by measuring the depth of the first minimum from the extrapolated unmodulated decay curve (see Figure 1) divided by the distance from the unmodulated decay curve to the baseline. It can be seen that the deuterium modulation depth is significantly greater for the synthesized compared to the impregnated sample. A greater modulation depth implies a shorter manganese to deuterium (D_2O) distance. The normalized modulation depth parameter is 0.83 in synthesized MnAPO-11 while it is only 0.53 in impregnated Mn-$AlPO_4$-11. This indicates that the manganese-to-deuterium distance is shorter in the synthesized material than in the impregnated material which is again consistent with framework substitution being present in the synthesized material and not in the impregnated material. The significant differences between the spectra in the impregnated versus synthesized materials directly indicate a difference in the location of the manganese in these two materials.

Three-pulse electron spin echo modulation can be analyzed more quantitatively than two-pulse data because any contributions from aluminum modulation are eliminated by selecting an appropriate delay time between the first two pulses so that only the deuterium modulation is detected. [3] The three-pulse electron spin echo data is very noisy for the impregnated Mn-$AlPO_4$-11 material, but an approximate simulation of the data indicates that it can be fit by a one shell model with four deuteriums at a distance of 0.28 to 0.30 nm from Mn(II). These parameters are typical for cation hydration. The three-pulse ESEM data from the synthesized MnAPO-11 shows better signal-to-noise and is best fit by a two shell model with two deuteriums at a short distance of 0.24 nm and two deuteriums at a greater distance of 0.36 nm. The short distance of 0.24 nm is not typical of cation hydration. Additional studies were made with CH_3OD and C_2H_5OD and in both cases the Mn(II) to deuterium distance was a short 0.24 nm for two deuteriums which implies two alcohol molecules interacting through their OD bonds.

Figure 2 shows a a three-pulse electron spin echo pattern of synthesized MnAPO-11 material compared with synthesized MnAPO-5 material which has a somewhat similar structure. The $AlPO_4$-11 structure involves 4, 6 and 10-rings whereas the $AlPO_4$-5 structure involves 4, 6 and 12-rings.[2] The interesting point is that there is a significant difference in the spin echo patterns between these two systems. Simulation of these patterns shows that the MnAPO-5 system is fit by a one shell model of four deuteriums at a distance of about 0.30 nm consistent with solvation of two waters around manganese cation in a nonframework position. The distance of 0.28 to 0.30 nm corresponds to the distance between manganese cation and the deuteriums on a solvent water molecule with the negative oxygen end of water oriented toward the cation. In contrast, the three pulse electron spin echo pattern of MnAPO-11 can not be fit by such a model and requires a two shell model with two deuteriums at a close distance of 0.24 nm and two at a longer distance of 0.36 nm. This pattern is similar to what has been found previously for water coordinated to small negative ions such as O_2^- and hydrated electrons in frozen solutions. [5,6] This is is consistent with two water molecules coordinated to manganese(II) in a DO bond configuration as is expected for a negatively charged site. This is what is expected if manganese(II) substitutes for Al(III) in the framework of the $AlPO_4$-11 structure. The local framework site with Mn(II) substituted for Al(III) would correspond to a negatively charged site. A proposed structure for this site is shown in Figure 3.

CONCLUSIONS

This study shows that manganese can be incorporated into the $AlPO_4$-11 framework at low manganese content (approximately 0.1 mole %). At such a low manganese content most of the manganese substitutes for aluminum in the framework. As the manganese content increases more of the manganese apparently occupies an extraframework site. It is interesting that the results indicate that manganese in the synthesis mixture forms framework-substituted manganese in MnAPO-11 but not in MnAPO-5 under the same synthesis conditions. Thus there are subtle differences between the formation of the $AlPO_4$-5 and the $AlPO_4$-11 structures in the synthesis mixture that affect the incorporation of manganese cation into a framework site. This suggests that transition metal ions may be incorporated only in specific $AlPO_4$-n materials which may serve as a route to tailored catalysts.

ACKNOWLEDGMENTS

This research was supported by the National Science Foundation, the Robert A. Welch Foundation and the Texas Advanced Research Program.

REFERENCES

1. Flanigen, E.M.; Lok, B.M.; Patton, R.L. and Wilson, S.T., in : *New Developments in Zeolite Science and Technology*, Eds. Y. Murakami, A. Iijima and J.W. Ward, Elsevier, Amsterdam, pp. 103-112 (1986).

2. Pluth, J.J.; Smith, J.V. and Richardson, W.R., Jr., J. Phys. Chem. **92,** 2734 (1988).

3. Kevan, L., Acc. Chem. Res., **20,** 1 (1987).

4. Brouet, G.; Chen, X. and Kevan, L., J. Phys. Chem. **95,** 4928 (1991).

5. Kevan, L., Acc. Chem. Res., **14** 138 (1981).

6. Narayana, P.A.; Suryanarayana, D. and Kevan, L., J. Am. Chem. Soc. **104,** 3552 (1982).

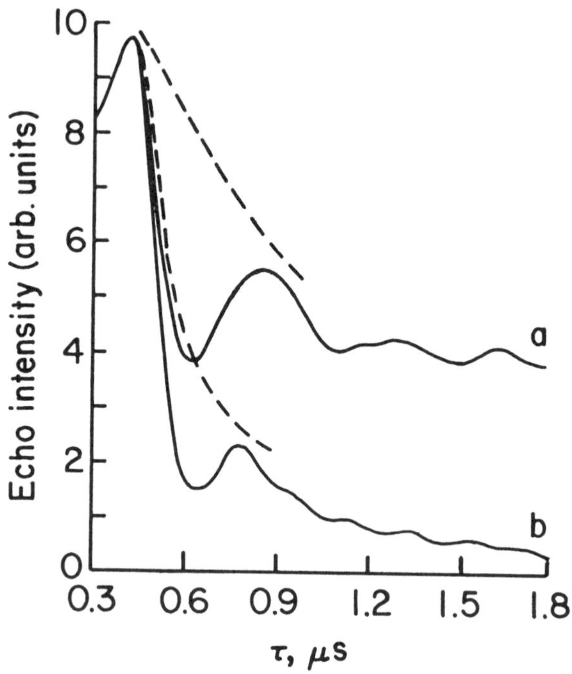

Figure 1. Two-pulse ESEM smoothed data recorded at 4 K of (a) impregnated Mn-$AlPO_4$-11 with 0.03 mol % manganese and with adsorbed D_2O and (b) synthesized MnAPO-11 with 0.05 mol % manganese and with adsorbed D_2O. The dashed lines are extrapolations of the unmodulated decay curves.

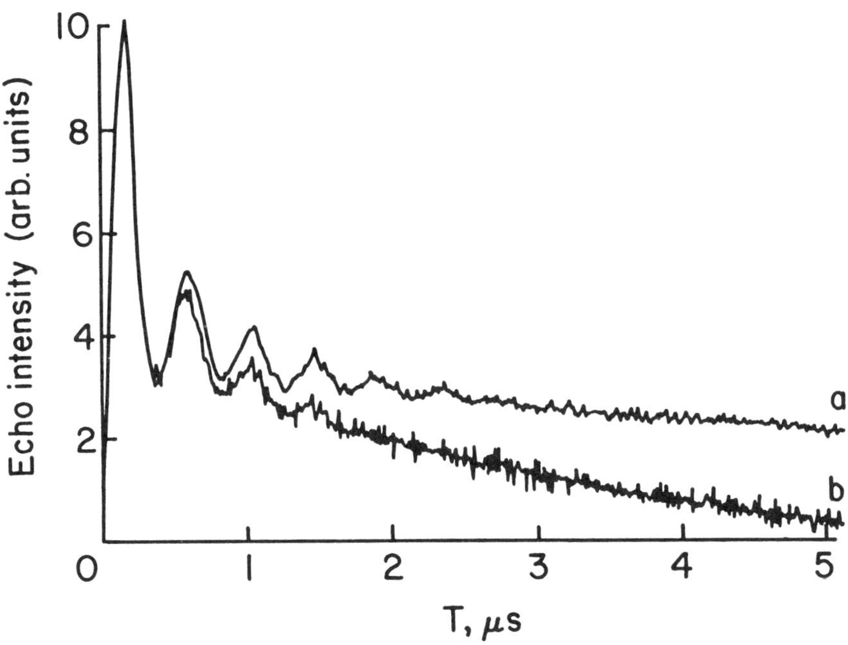

Figure 2. Three-pulse ESEM recorded at 4 K of (a) synthesized MnAPO-5 with 0.05 mol % manganese and with adsorbed D_2O and (b) synthesized MnAPO-11 with 0.05 mol % manganese and with adsorbed D_2O.

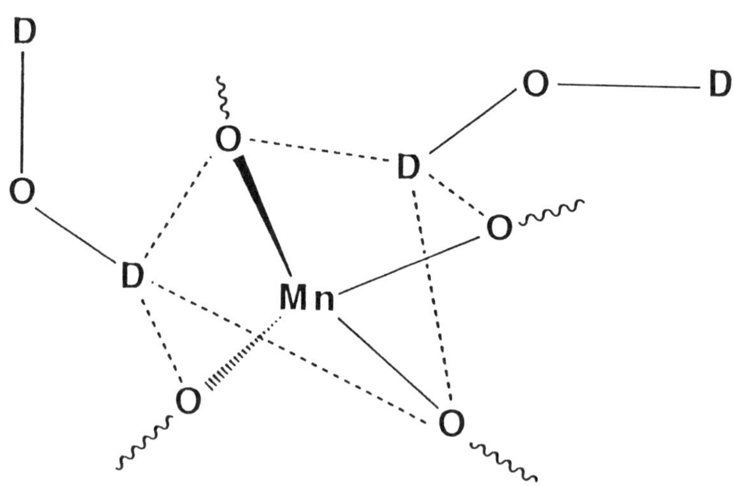

Figure 3. Proposed model of Mn coordination in synthesized MnAPO-11 hydrated with D_2O.

ELUCIDATING THE MORPHOLOGY OF CLAYS: HREM AND ESEM STUDIES

K. B. REUTER,* J. J. ALCARAZ,** J. A. TRIPHAHN,** and J. S. HOLMGREN**

*Union Carbide Chemicals and Plastics, Tarrytown, NY 10591, U.S.A.

**UOP Research Center, Des Plaines, IL 60017, U.S.A.

ABSTRACT

High resolution electron microscopy (HREM) has been used to characterize the morphology, crystallite size, and mode of aggregation of synthetic clays and pillared clays. Environmental scanning electron microscopy (ESEM) has been used to examine morphology as well as the swellability and gellation characteristics of clays. Information obtained by a combination of these techniques has been used to elucidate the impact of morphology, "swellability," and oligomer type on clay pillaring chemistry.

INTRODUCTION

Recent years have seen a revolution in the use of clays and pillared clays in catalysis. This revolution can be directly linked to the increased thermal stability that results from the pillaring of clays [1]: increased thermal stability extends the useful temperatures for clays into catalytically important regimes (>250°C). Much of the early work was confined to montmorillonite clay pillared with aluminum chlorohydrate (ACH). The potential utility of clays is currently being extended by pillaring of novel clays and developing more thermally robust pillaring oligomers.

Because of the importance of pillaring in the development of surface area and surface area stability in clays, one of the more important requirements for novel, synthetic layered materials is a propensity for pillaring. This "pillarability" appears to be directly linked to the morphology of the clay used in the pillaring procedure. However, the clay layer cannot be singly responsible for the pillaring chemistry. In fact, the oligomer used for pillaring must play an important role; after all, the physical properties of pillared clays do vary with the nature of the pillaring oligomer.

This particular work is one attempt at understanding pillaring chemistry by examining the role that clay morphology (obtained from HREM), clay aggregation (HREM), layer rigidity (HREM), clay swellability (ESEM), and the nature of the oligomer (HREM) play in the pillaring process.

EXPERIMENTAL PROCEDURES

Natural montmorillonite (HPM-20), was obtained from American Colloid; natural saponite (SapCa-1), from the Clay Minerals Society. All pillaring experiments were performed using standard procedures [2,3]. Synthetic saponite [4] and hydrotalcite [5] were prepared using published methods.

From the Proceedings of the 9th International Zeolite Conference, Montreal 1992, Eds. R. von Ballmoos et al.,

The ESEM was conducted at ElectroScan in Wilmington, Massachusetts. The samples were examined uncoated at 20 kV. The pressure in the ESEM was generally between 3 and 5 torr.

High resolution electron microscopy (HREM) was conducted on a JEOL 2000FX transmission electron microscope operating at 200 kV. The samples were ultramicrotomed, supported on a 3 mm Cu grid, and carbon coated on one side to minimize charging under the electron beam.

RESULTS AND DISCUSSION

Swellability

To prepare a pillared clay, the clay layers must be expanded enough to allow intercalation of the large inorganic oligomer that is to become the pillar. Therefore, the swellability of the clay layers is central in a pillaring experiment; the degree of swellability determines the ease of pillaring. Clay swelling can be *directly* observed with the ESEM since uncoated samples can be observed in many environments, i.e. at pressures up to 20 torr [6].

To determine the utility of this technique, the hydration of a swellable clay, montmorillonite, was compared with a non swellable clay, hydrotalcite. Three micrographs showing montmorillonite in its as-received form and at various stages during the hydration process are shown in Figure 1. The parent montmorillonite particles are sharp and discrete before hydration (Figure 1a). As hydration begins, the particle edges are lost (Figure 1b) and eventually the particles melt together to form a homogeneous mass or gel (Figure 1c). This figure illustrates the initial swelling, particle coalescence, and eventual gellation of the clay during hydration. Interestingly, this SEM progression is similar to that observed during the sintering and melting of clay particles at elevated temperatures [7].

A similar hydration experiment on the hydrotalcite sample confirms that this layered material is not swellable. The hydrotalcite retains its morphology throughout the hydration experiment.

The ESEM not only provides valuable information regarding clay swelling and gelling, but can also be used to quickly examine the morphology of clay specimens without coating the clay. Furthermore, because samples can be dried while being examined in the microscope, information that is normally lost as a result of compacting during the drying of a clay can be observed with ESEM.

Clay Morphology

To study the impact of clay morphology on pillaring chemistry, three samples were examined in their as-received and pillared forms: montmorillonite, natural saponite, and synthetic saponite. Typical transmission electron micrographs of the three as-received clays are shown in Figure 2. The relevant characteristics are presented in Table 1.

TABLE 1: MORPHOLOGY OF THE AS-RECEIVED CLAYS

Sample	Saponite			Montmorillonite
	Synthetic		Natural	
	Fine	Small		
Crystallite Size, nm	2-10	80-250	30-2000	200-1000
Aspect Ratio	4	3	3	2
Aggregate Size, nm	2	20-80	120-280	200-1000
Mode of Aggregation*	EE, EF	EE, EF	FF	FF

* EE - edge-to-edge, EF - edge-to-face, FF - face-to-face

A number of important features are immediately obvious when these specimens are examined: (1) the average crystal size of the synthetic clay sample is substantially smaller than that of the natural clays, (2) the mode of layer aggregation varies among the samples: face-to-face aggregation dominates in the natural clay samples, and (3) both natural clay samples contain a wide crystallite size distribution; the synthetic sample contains a bimodal crystallite size distribution.

Two distinct regions are observed in all three clay samples: lamellar or layered regions and flat regions. These two regions result from a difference in the orientation of the clay particles in the electron beam and can be readily distinguished by both their appearance and their electron diffraction pattern (Figure 3). These regions provide complementary information regarding the morphology of the clays. Information regarding aggregation method, layer rigidity and crystallite size was obtained from the lamellar regions. The shape and the aspect ratio were determined from the flat regions. These comparisons are tabulated in Table 1 for all three clay samples.

These data clearly show that the average crystallite size of the synthetic saponite sample is substantially smaller than that of the natural saponite and montmorillonite. These variations in crystallite size greatly affect the mode of layer aggregation for these samples. Careful examination of the micrographs shows that only a small fraction of the synthetic saponite crystals show any face-to-face aggregation. Furthermore, the face-to-face aggregation is limited to the few large crystals present in this sample. This result is consistent with this sample being delaminated: mostly edge-to-edge and edge-to-face aggregation. In contrast, both natural clays show >80% face-to-face aggregation. The degree of face-to-face aggregation has been linked to the pillarability of the clay; increased face-to-face aggregation resulting in increased pillarability [8].

From the morphology and stacking order of these three clays, one can predict that only the two natural samples should be pillarable. Pillaring experiments with all three samples confirm this assumption. This result is not only evident from HREM but is consistent with XRD and N_2 adsorption data (Table 2).

TABLE 2: PHYSICAL CHARACTERISTICS OF THE PILLARED CLAYS

Oligomer	**Rare Earth ACH**			**ACH**
Clay	**Saponite**		**Montmorillonite**	**Montmorillonite**
	Synthetic	**Natural**		
N_2 Adsorption:				
BET Surface Area, m^2/g	259	325	604	256
Micropore Volume, cc/g	–	0.15	0.26	0.11
Interlamellar Spacing, Å:				
From XRD	–	16	18	9
From HREM (Avg)	–	22	12	7
From HREM (Range)	–	13-27	6-16	3-9
Thermal Stability (°C)	550	650	>700	550

Pillaring

The clay-layer morphology appears to correlate with the pillaring chemistry for a particular clay sample. However, differences among samples pillared with different oligomers suggest that the oligomer also plays an important role. To elucidate this role, samples of montmorillonite clay pillared with ACH or rare earth ACH were compared. Examination of both PILCs revealed two obvious differences between these samples, (1) the interlayer distance in the rare earth PILC is substantially larger than that of the ACH PILC (Figure 4) and (2) the electron beam stability of the rare earth PILC is much greater. We believe that beam stability can be directly linked to the hydrothermal stability of the pillared clay.

However, further microscopic examination of these pillared clays shows that they are surprisingly similar: both clay samples have unpillared areas, and the interlamellar spacing (distance between the layers) of the pillared areas varies over a wide range (Table 2). This variability in spacing is not due to isolated regimes of clay pillared at different spacings but, rather, to variations in pillaring height within adjacent layers. In fact, the clay layers appear to buckle and bend, probably to accommodate inhomogeneities in the size of the pillaring oligomer. This ability of the clay layers to bend is clear from microscopic examination of the parent clays (Figure 2). This examination shows that the clay layers, or platelets, are not straight and flat but rather curl and buckle. The degree of curling appears to be directly related to the crystal size: as the clay crystal size increases, so does the layer curl. In recent years, the implications of this effect, referred to as "layer rigidity", have been discussed in great detail [9]. However, further experimental results are still required to clarify the relationship between layer rigidity and the stability of clays and pillared clays.

Finally, the average spacing measured from HREM images is substantially smaller than that measured from X-ray diffraction. This issue has been discussed in the literature and attributed to twisting of layers resulting in inaccurate measurements as well as dehydration of the samples in the electron microscope leading to interlayer contractions [8]. The exact reasons for this difference are still unclear and merit further investigation.

CONCLUSIONS

The development of novel, pillared materials will require a true understanding of the physiochemical aspects that control hydrothermal stability in clays. However, because clays are semicrystalline, characterization is difficult. Analyses with HREM and ESEM allow *direct* observation of important clay characteristics, such as clay morphology, mode of layer aggregation, swellability, pillaring attributes, hydrothermal stability, and layer rigidity. From this information, the propensity of a clay to pillar as well as the homogeneity and stability of the pillared products can be examined. Because of this, HREM and ESEM will become invaluable tools in future work with clays and pillared clays.

ACKNOWLEDGEMENTS

The authors wish to thank I. P. Kijak for her technical support and J. L. Methé for operating the ESEM.

REFERENCES

1. Vaughan, D. E. W., Lussier, R. J., Magee, J. S., U. S. Patent 4,176, 090.
2. McCauley, J., U. S. Patent 4,952,544.
3. Pinnavaia, T. J., Tzou, M. S., Landau, S. D., and Raythatha, R. H., J. Mol. Catal., **27**, 195 (1984).
4. Blumenthal, T., and Kaiz, H., U. S. Patent 4,749, 676.
5. Reichle, W. T., J. Catal., **94**, 547 (1985).
6. Sujate, K., and Jennings, H., MRS Bulletin, **3**, 41 (1991).
7. Occelli, M. L., Landau, S. D., and Pinnavaia, T.J., J. Catal., **104**, 331 (1987).
8. Occelli, M. L., Lynch, J., and Sanders, J. V., J. Catal., **107**, 557 (1987).
9. Lee, S., and Solin, S. A., Solid State Communications, **72**(2), 163 (1989).

A.

B.

C.

Figure 1. Secondary electron images from ESEM of montmorillonite: (a) before hydration, (b) during hydration (note the loss of particle edges), and (c) after complete hydration (note that the particles have formed a gel).

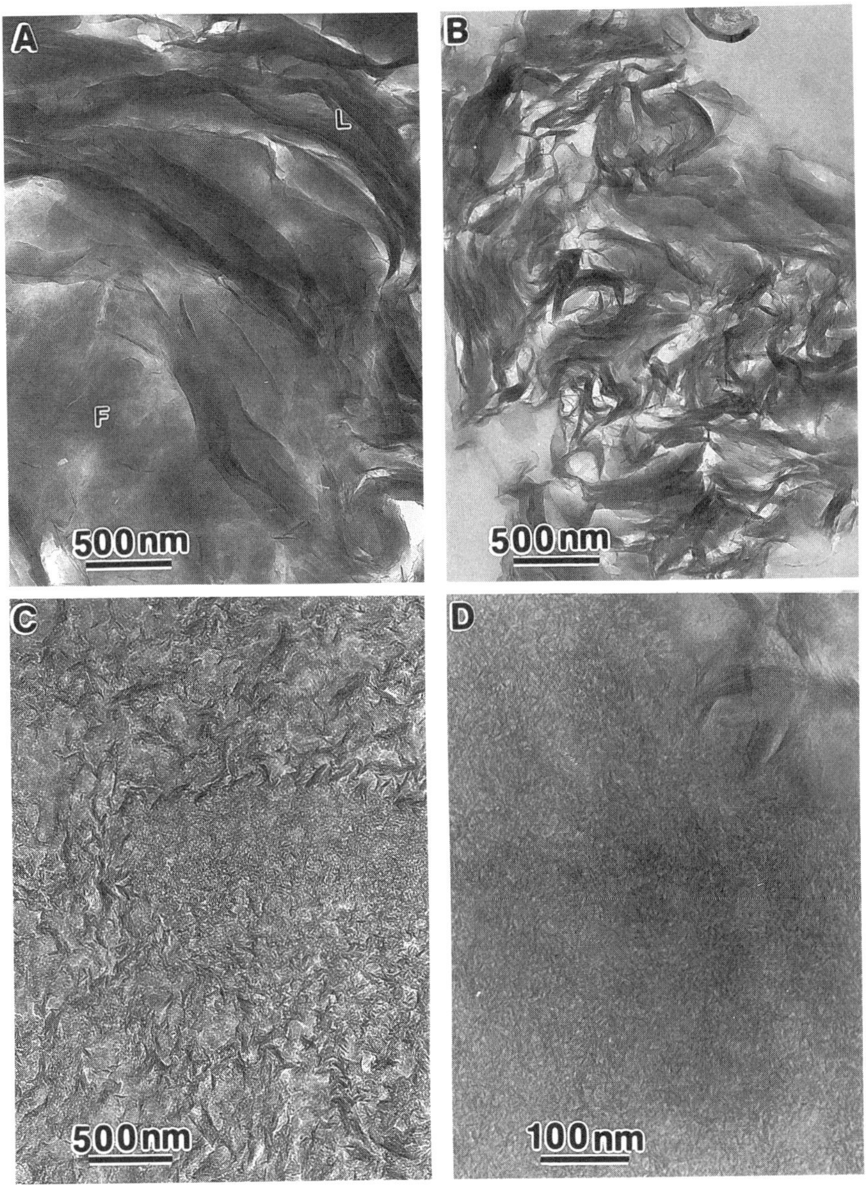

Figure 2. Typical electron micrographs of as-received clays: (a) natural saponite, (b) montmorillonite, (c) synthetic saponite, and (d) synthetic saponite at higher magnification to show small crystals. Pertinent characteristics are listed in Table 1. Note: "F" indicates flat area; "L" indicates layered area.

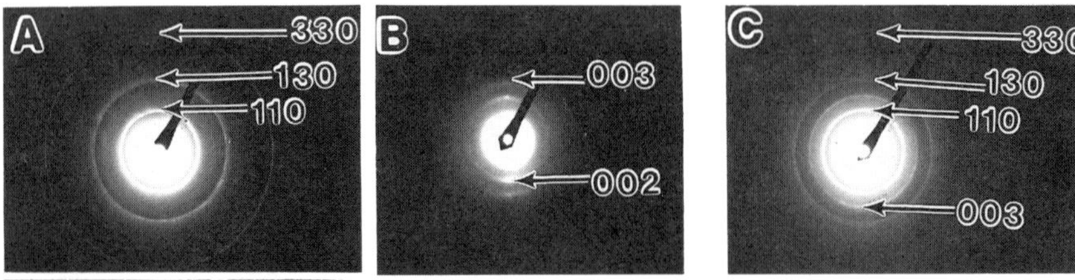

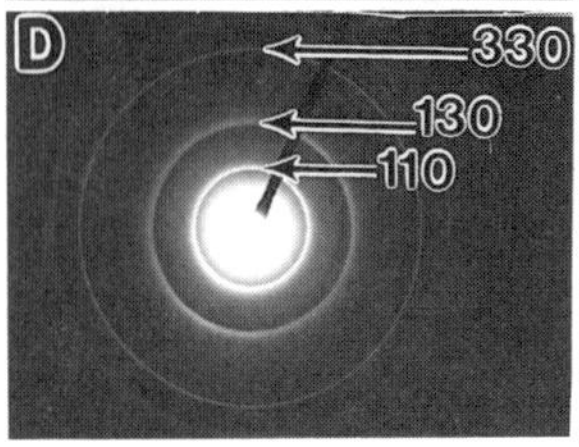

Figure 3. Electron diffraction patterns from as-received clays: natural saponite pattern (a) is obtained from flat area (Figure 2); because only hk0 rings are present, this clay area is oriented with layers perpendicular to the electron beam. Natural saponite pattern (b) is obtained from layered area (Figure 2); because only 00l rings are present, this clay area is oriented with layers parallel to the electron beam. Montmorillonite pattern (c) is obtained from layered and flat areas. In synthetic saponite, pattern (d), the rings are more complete because of the smaller size of the crystals.

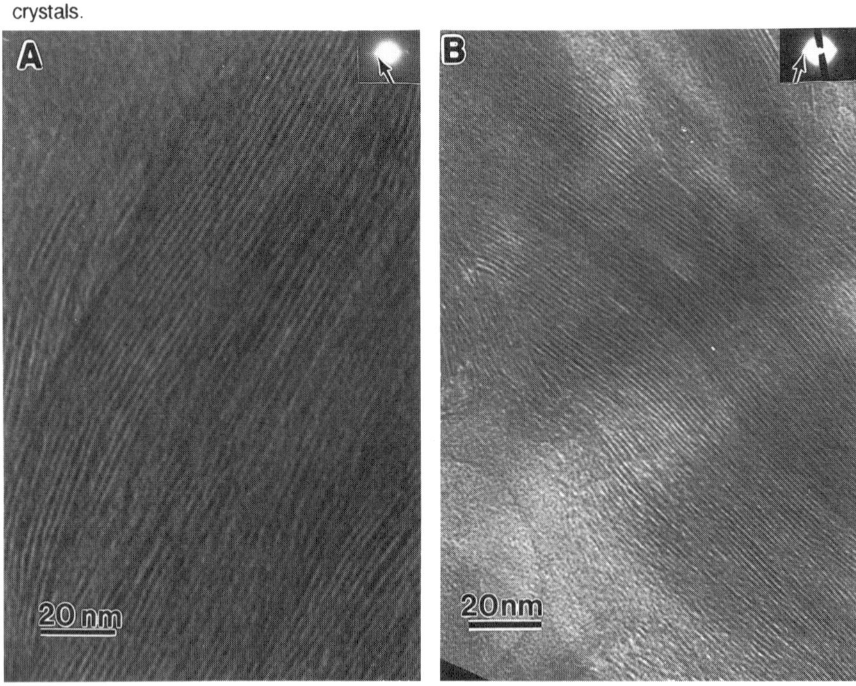

Figure 4: Typical electron micrographs of pillared clays: (a) rare earth ACH and (b) ACH. Insets show electron diffraction patterns with 00l spots indicated. The variations in interlamellar spacings for these two samples are shown in Table 2.

CHARACTERISATION AND CATALYTIC PROPERTIES OF TITANIUM INCORPORATED INTO ZEOLITES AND SILICAS FROM TITANIUM TRIFLUORIDE

P.J. KOOYMAN, J.C. JANSEN and H. van BEKKUM

Laboratory of Organic Chemistry and Catalysis, Delft University of Technology, Julianalaan 136, 2628 BL Delft, The Netherlands.

ABSTRACT

Ti was deposited c.q. chemically attached onto the external surfaces of various crystalline and amorphous silicas. Attention is focused on systems obtained with TiF_3 as the Ti-source. The direct synthesis of TS-1 in fluoride medium is discussed, as well as post-synthesis surface modification. The catalytic properties of some of the materials obtained in aromatic hydroxylation with H_2O_2 compare well with those of TS-1 synthesised in OH^- medium.

INTRODUCTION

TS-1 synthesised using OH^- as the mobilising agent is known to be a good catalyst for various selective oxidation reactions with hydrogen peroxide [1,2]. We have previously reported on the possibility to obtain good catalysts for the hydroxylation of phenol by depositing Ti from $TiCl_4$ on crystalline and amorphous silicas [3]. Now we report on systems obtained through direct synthesis of TS-1 in a fluoride medium, as well as through the deposition of Ti from TiF_3 on various silicas.

Zeolites are usually synthesised in a strongly basic environment, where OH^--ions are the mobilising agent for the anionic silicate and aluminate species that form the zeolite structure. However, Guth et al. [4,5] have synthesised zeolites and boron-containing silicates by using F^--ions as the mobilising agent. The advantage of this synthesis method is the possibility to work at moderate or even low pH. In this study a titanium containing silicalite, TS-1, was synthesised by using F^--ions as the mobilising agent and TiF_3 as the titanium source. Although titanium is mentioned by Guth et al. [4], hardly any data on the exact synthesis conditions, characterisation and catalytic properties of the materials are given. TiF_3 was chosen as the titanium source, because it is one of the few titanium compounds that do not react immediately with water to form the highly stable TiO_2, which is incorporated into the zeolite framework only with great difficulty.

This paper also reports on the preparation and catalytic properties of materials obtained by surface treatment of silicalite-1, crystobalite and amorphous silica with TiF_3.

EXPERIMENTAL

Synthesis

The synthesis of TS-1-F was performed analogous to that described by Guth et al. [4] for aluminium containing silicalite. Thus, 21.92 g of SiO_2 (Degussa Aerosil 200), 2.04 g of $NH_4F.HF$, 11.64 g of TPABr (tetrapropylammonium bromide) and 213.6 g of H_2O were stirred together until a

From the Proceedings of the 9th International Zeolite Conference, Montreal 1992, Eds. R. von Ballmoos et al.,

homogeneous mixture resulted (24 to 60 hrs). Then the TiF_3 was added, and after a further 1 to 2 hrs of stirring the mixture was autoclaved at 180°C for 23 days, yielding almost 20 g of crystals. To remove incorporated TPA, the crystals were calcined at 400°C for 3 days.
For comparison, a sample of TS-1-OH was prepared according to a procedure described in the patent literature [6].

Samples of silicalite-1 (synthesised in F^- medium), crystobalite (M006 SCR Sibelico) and 2 different amorphous silicas (Chromatografiegel, Chemische Fabrik Uetikon and silica spheres, Shell) were treated with TiF_3 in the following manner: First, a solution (solution "A") was made by dissolving 0.5 g TiF_3 in 0.2 g of 40% HF and 60 ml of water. Then, 0.5 g of the silicas was autoclaved for 90 min at 180°C with 0.25 to 5.0 ml of this solution and an amount of water to obtain a total liquid volume of 25 ml. After cooling to room temperature, the samples were filtered, washed thoroughly with water and dried overnight at 120°C. These samples will be denoted as Ti/S-1 and Ti/CRI for the silicalite and crystobalite respectively, and Ti/CFU and Ti/Sh for the amorphous silica obtained from Chemische Fabrik Uetikon and Shell, respectively.

Characterisation

Powder X-ray diffraction data were collected on a Philips PW 1840 diffractometer, using the Cu-Kα line.

Infrared spectra were recorded on a Perkin-Elmer 398 spectrophotometer, using the KBr wafer technique and the range of 4000 to 400 cm^{-1}.

SEM photographs were taken using a JEOL JSM-35 Scanning Microscope.

EDX measurements were performed using a LINK apparatus connected to a SEM.

Elemental analysis was performed using ICP-AES (inductively coupled plasm - atomic emission spectroscopy), for which the samples were dissolved in diluted H_2SO_4/HF.

BET surfaces and pore volumes were determined using N_2 isotherms.

Catalysis

Hydroxylation of phenol

Hydroxylation experiments were performed batchwise, and followed in the course of time by taking samples out of the reaction mixture. 100 mg of catalyst was used, and H_2O_2 conversion was monitored using standard iodometric titration. Products were analysed using HPLC.

10 g of phenol, 32 g of methanol (solvent) and 1 g of pentafluorobenzoic acid (internal standard) were heated to reflux (ca. 80°C) with the catalyst before 5.0 ml of 30% H_2O_2 was added dropwise in 10 min time. The main product is hydroquinone (1,4-dihydroxy benzene) and small amounts of catechol (1,2-dihydroxy benzene) and para-benzoquinone are also formed.

Oxyfunctionalisation of n-hexane

Hexane was oxidised to a mixture of 2- and 3-hexanols and 2- and 3-hexanones with aqueous H_2O_2 in batch-wise experiments in small (20 ml) stirred autoclaves. 1.0 ml of n-hexane, 10 ml of methanol (solvent), 0.80 ml of 30 % H_2O_2 and 20 mg of catalyst were heated in the autoclave at 100 °C for 24 hr. Products were analysed using GC, while H_2O_2 conversion was measured using iodometric titration.

RESULTS AND DISCUSSION

Characterisation

TS-1-OH: The X-ray diffractogram of this zeolite confirms the structure to be of the MFI-type. The IR spectrum of this zeolite shows an absorption band to be present at 960 cm^{-1}. This band is said to be indicative of Ti incorporated into the zeolite framework. Furthermore, it is said that this band is necessary to obtain catalytic activity in selective oxidation reactions with aqueous hydrogen peroxide. However, we have shown that this is not the case, and that materials containing Ti and SiO_2 can show good catalytic properties for selective oxidation reactions with aqueous hydrogen peroxide without showing the IR absorption band at 960 cm^{-1} [3].

TS-1-F: SEM micrographs (Figure 1) show the TS-1-F to consist of well-formed elongated prismatic crystals, covered with small particles. The crystals are about 35 μm long, the diameter of the particles is about 1 μm. Powder X-ray diffraction does not show any extra lines in the pattern as compared to that of ordinary silicalite. The IR spectrum does not show an absorption band at 960 cm^{-1}. EDX measurements show the small particles to contain Ti, whereas no Ti is detected on the remaining part of the external surface of the crystals. This can be seen clearly from Figure 2, which shows the EDX line-scan across the indicated line. The BET surface area of this material is 405 m^2/g and the pore volume 90 ml*NTP/g. For silicalite-1 these values are 425 and 95, respectively, indicating that the small particles on the external surface do not block the pores of the zeolite structure.

Ti/S-1: As can be seen from SEM micrographs (Figure 1) the silicalite, synthesised in F^- medium and (after calcination to remove the template) treated with TiF_3/HF, is covered with small particles, which seem to be similar to those found on TS-1-F. Neither the X-ray diffractogram nor the IR spectrum show differences before and after TiF_3/HF treatment. For these samples also, according to EDX measurements, the Ti seems to be concentrated in the small particles on the external surface. The BET surface area and the pore volume for this sample are 405 m^2/g and 91 ml*NTP/g, respectively. These values correspond with those for the untreated silicalite-1 (425 m^2/g and 90 ml*NTP/g), indicating that pore-blocking does not occur upon TiF_3/HF treatment.

Considering the formation of these particles in direct synthesis using TiF_3 as well as in post-synthesis treatment of silicalite-1 with TiF_3/HF, we think they are formed during the cooling of the autoclave before the sample is isolated, when the dissolved Ti is deposited, most probably as TiO_2. In the direct synthesis there are probably several kinds of Ti-deposition mechanisms. In one of these, Ti is incorporated into the zeolite crystals during crystal growth, possibly as T-atoms. Then, probably during the cooling period of the autoclave before the crystals are isolated, Ti that is still present in solution is deposited on the external surface of the crystals. Although we cannot detect it with EDX analysis, we cannot exclude that a TiO_2 structure similar to one or a few monolayers is formed on the silicalite crystal surface in between the TiO_2 particles. We think it is possible that these mechanisms all operate simultaneously.

In the post-synthesis treatment with TiF_3/HF, apart from the external deposition of TiO_2 as described above, it might also be possible to titanate the internal surface of the zeolite channels, e.g. by reaction with silanol groups.

Ti/CRI and Ti/CFU: Neither SEM, IR nor XRD show changes before and after treatment with TiF_3/HF for these two sets of samples.

Ti/Sh: Because of the size of the silica spheres (diameter around 2 mm) and the unknown influence of grinding, no XRD or IR data were obtained. With SEM no changes could be detected before and after TiF_3/HF treatment.

Although all the TiF_3/HF treated SiO_2 samples contain titanium (see Table 1), with SEM no particles were observed on the external surface. However, we have shown that upon treating amorphous silica (CFU) as well as silicalite-1 in the gas-phase at 450°C with $TiCl_4$, Ti-containing particles are formed on the external surface of the samples [3]. It is not clear yet why the TiF_3/HF treatment of silicas other than silicalite-1 does not result in the formation of Ti-containing particles on the external surface of these silicas.

As can be seen from Figure 3, the amount of Ti present in the samples depends linearly on the amount of solution "A" that was added to the samples.

Catalysis

Hydroxylation of phenol

The results obtained in the hydroxylation experiments are summarised in Table 1. As can be seen, all the titanated samples, as well as both direct synthesis samples, show catalytic activity for the hydroxylation of phenol with hydrogen peroxide. Most measurements show good reproducibility.

Both samples obtained through direct synthesis show good activity and efficiency. As can be seen, many of the titanated silica samples show comparable activities and efficiencies. There does not really seem to be a relation between the amount of Ti present in the samples and the catalytic properties for this reaction. For samples with Ti-containing particles on the external surface this is to be expected, because not all the Ti present in these particles is exposed to the reaction mixture. Although a general trend seems to be present in that a decreasing amount of Ti present in the samples gives rise to lower activity and efficiency, the samples with a low Ti content (a high Si/Ti ratio) show a surprisingly high activity. This indicates that only very small quantities of Ti connected to Si are required to catalyse this reaction. Blank hydroxylation experiments using the non-titanated silicas show that none of these materials has significant activity.

Oxyfunctionalisation of n-hexane

Table 2 gives some preliminary results obtained regarding the oxyfunctionalisation of n-hexane using the samples of TS-1 obtained through direct synthesis. It seems that the systems

TABLE 1: Hydroxylation of phenol.

sample	Si/Ti[1]	efficiency[2]	H_2O_2 conv[3]	sample	Si/Ti[1]	efficiency[2]	H_2O_2 conv[3]
TS-1-OH	80	55	56	Ti/CFU_A	88	53	43
TS-1-F	70	47	39	Ti/CFU_B	220	48	42
Ti/S-1_A	21	24	39	Ti/CFU_C	625	47	43
Ti/S-1_B	95	16	24	Ti/CFU_D	900	77	88
Ti/S-1_C	360	42	56	Ti/CFU_E	1500	39	82
Ti/CRI_A	37	40	27	Ti/Sh_A	84	53	78
Ti/CRI_B	145	46	14	Ti/Sh_B	280	37	26
Ti/CRI_C	850	11	27	Ti/Sh_C	900	41	76

1) bulk molar ratio as determined by ICP-AES
2) % dihydroxybenzenes formed / H_2O_2 converted
3) % H_2O_2 converted after 6 hours of reaction

TABLE 2: Oxyfunctionalisation of n-hexane.

sample	Si/Ti[1]	H_2O_2 conv[2]	2-hexanol[3]	3-hexanol[3]	2-hexanone[3]	3-hexanone[3]
TS-1-OH	80	100	12.6	6.5	14.2	2.1
TS-1-F	70	81	5.3	1.3	1.6	0.4

1) bulk molar ratio as determined by ICP-AES
2) % H_2O_2 converted after 24 hours of reaction
3) % product formed/ H_2O_2 converted; corrected by factor 2 for 2- and 3-hexanone

obtained through post-synthesis treatment with TiF_3/HF do not give products, although some H_2O_2 conversion is measured (usually 5 - 15%). As the TS-1-OH shows a much better activity and efficiency for this reaction than does the TS-1-F, which certainly contains some extra-framework Ti and probably some framework Ti, the oxyfunctionalisation of n-hexane might be a better test-reaction for these kind of systems than is the hydroxylation of phenol. Work is in progress to study other substrates, like bulky ones that cannot enter the zeolite pores and substrates with a higher intrinsic activity.

CONCLUSION

Using simple titanation methods to synthesise samples containing both SiO_2 and Ti, materials can be produced that are active catalysts in the hydroxylation of phenol with aqueous hydrogen peroxide.

ACKNOWLEDGMENTS

This research project is funded by the Dutch Organisation for Scientific Research, NWO. The authors would like to thank Mrs. C. Ferrini, ETH Zürich, Switzerland, for synthesising the TS-1-OH, and Mrs. F. van Looij, RU Utrecht, The Netherlands, for the EDX measurements.

NOTE

Recent independent work on the synthesis of TS-1 in fluoride medium [7] indicates that the particles formed on the external surface of the zeolite crystals consist of rutile (TiO_2).

REFERENCES

1) M. Taramasso, G. Perego and B. Notari, US Pat. 4,410,501 (1982).
2) G. Perego, C. Corno, M. Taramasso, F. Buonomo and A. Esposito, in Y. Murakami, A. Lijima and J.W. Ward (Eds.), Proc. 7th Int. Zeol. Conf., Tokyo, Elsevier, Amsterdam (1986) 129-136.
3) P.J. Kooyman, P. v.d. Waal, P.A.J. Verdaasdonk, J.C. Jansen and H. van Bekkum, Catal. Lett. 13(3) (1992) 229-238.
4) J.L. Guth, H. Kessler and R. Wey, in Y. Murakami, A. Lijima and J. Ward (Eds.), Proc. 7th Int. Zeol. Conf., Tokyo, Elsevier, Amsterdam (1986) 121-128.
5) J.L. Guth, H. Kessler, J.M. Higel, J.M. Lamblin, J. Patarin, A. Sieve and J.M. Chezeau, in M.L. Ocelli and H.E. Robson (Eds.), Zeolite Synthesis, ACS Symp. Ser. 398 (1989) 176-195.
6) M. Padovan, G. Leofanti and P. Roffia, Eur. Pat. 311,983 (1988).
7) B. Sulikowski and J. Klinowski, Appl. Catal. A 84 (1992) 141-153.

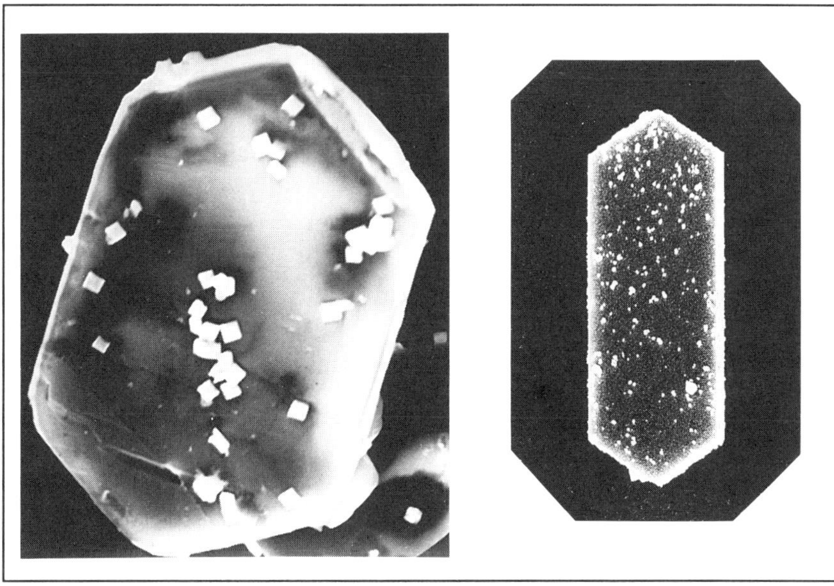

FIGURE 1: SEM micrographs of TS-1-F (left, enlargement 2400 times) and Ti/S-1, sample A, (right, enlargement 2000 times).

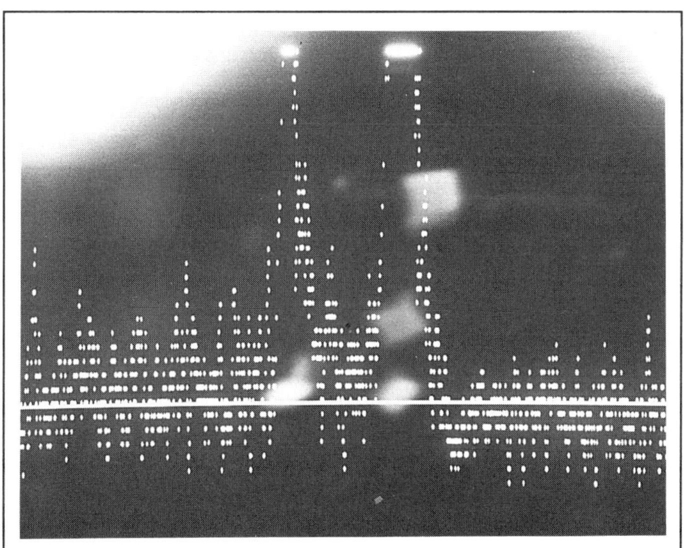

FIGURE 2: SEM micrograph of TS-1-F and EDX line-scan, Ti signal, measured along the indicated line.

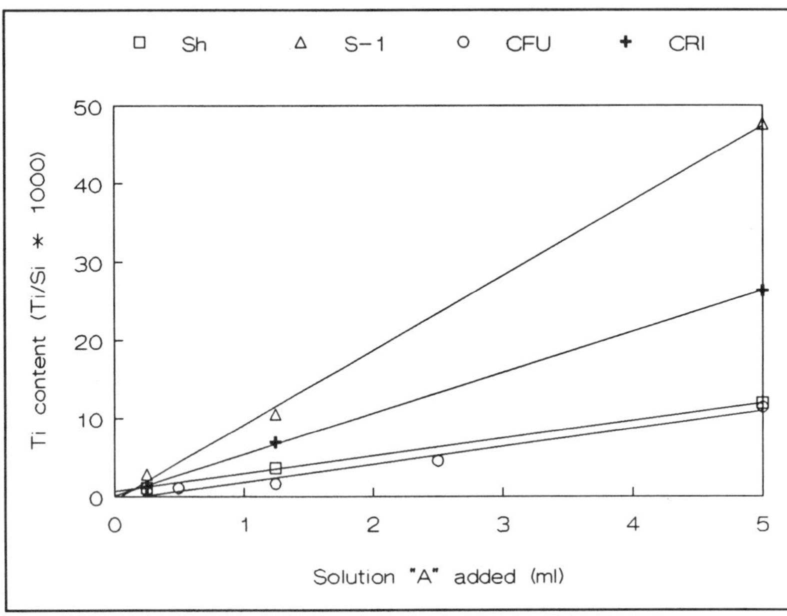

FIGURE 3: Relation between the amount of solution "A" added to the silicas and the amount of Ti found in the samples.

THE PREPARATION, CHARACTERISATION AND PROPERTIES OF ZEOLITE NU-86

J.L. CASCI*, P.A. COX*[1] and M.D. Shannon**

* ICI C&P Ltd., PO Box 90, Wilton, Middlesbrough TS6 8JE, England.

** ICI C&P Ltd., PO Box 8, Runcorn, WA7 4QD, England.

[[1] Present Address: University of Portsmouth, PO1 2DT, England]

ABSTRACT

Zeolite NU-86 is a novel high-silica zeolite with a 3-dimensional pore network containing 10, 11 and 12 ring channels. NU-86 can be prepared using long chain bis-quaternary ammonium cations to template its formation. Information is presented on its synthesis, characterisation and properties and the unique channel system described by means of pore-maps.

INTRODUCTION

New molecular sieve topologies are of great scientific interest, especially structures with multi-dimensional channel systems and those containing new ring types. Commercially, too, there is considerable interest in new medium and large pore zeolites (channels > 6Å across) because of their potential in the production and processing of hydrocarbons in oil refining and petrochemical applications. This is particularly true of 3-D channel systems since they have greater resistance to deactivation through pore-blockage by coke deposition or partial structural collapse, compared to uni-dimensional materials. This explains the considerable recent interest in "hexagonal faujasite" [1] and in trying to prepare synthetic versions of the mineral Boggsite [2].

This paper describes aspects of the synthesis, characterisation and properties of NU-86, a new high-silica zeolite [3] with an unusual 3-dimensional pore system [4]. The three intersecting channels contain 10, 11 and 12 rings: NU-86 is the first molecular sieve topology to contain an 11 ring channel.

The preparation of NU-86 was achieved using long chain bis-quaternary ammonium cations. Templates of the same general type have already had considerable success in the synthesis of zeolites and silica molecular sieves [5]. Recently there have been two reports of the synthesis of novel zeolites using related template types: ZSM-57 [6] and NU-87 [7]. Both materials have 2-D pore systems, with ZSM-57 being based on the ferrierite sheet [8] and having 10 * 8 channels while NU-87, which has a new framework topology [9], has intersecting 10- and 12-ring channels.

EXPERIMENTAL

The NU-86 samples described in this study were synthesised in 1 litre stainless steel autoclaves with continuous agitation using pitched paddle type impellers (300 rpm). Autoclave temperatures

From the Proceedings of the 9th International Zeolite Conference, Montreal 1992, Eds. R. von Ballmoos et al.,

were controlled to within ± 2°C. Samples could be withdrawn during reaction to allow the progress of the crystallisation to be monitored.

Reagents used in the synthesis will be described below. The bis-quaternary ammonium templates used were prepared by reacting the analogous dibromoalkane with trimethylamine [10].

Products were analysed by x-ray powder diffraction (Philips APD 1700 employing an automatic theta-compensating divergence slit), EM (SEM, Hitachi S650; TEM, Philips EM 301) and chemical (AES+ICP) and thermal (Stanton Redcroft STA-780) analyses. ^{13}C Cross-Polarisation Magic Angle Spinning (CPMAS) NMR spectroscopy was carried out using a Bruker MSL200 Spectrometer. Sorption measurements were made using a Micromeritics ASAP 2400 instrument (N_2) and a CI Robal microbalance for all other sorbates.

RESULTS AND DISCUSSION

NU-86: structure and pore map

NU-86 has a new framework topology enclosing a unique 3-D pore system. Two of the channels are straight while the third is sinusoidal. One of the straight channels is based on a 11 T-atom ring, while both the zig-zag and other straight channel contain alternating 10 and 12 rings. The channel system is best illustrated by means of "pore maps" (Figure 1). These are, in effect, a representation of the Van der Waals surface of the channels without showing the framework atoms themselves. The upper pore-map in the figure is a view down the 11-ring channels. At right angles to these are the straight channels containing 10- and 12-rings: these can be thought of as 12 ring channels with 10 ring constrictions, the constrictions creating 12 ring cavities. The lower pore-map in Figure 1 is a view at right angles to the upper map, and is in effect a "slice" through the structure with the 11-rings running from top to bottom. In this view both the 10 and 12-rings in the straight channels can be clearly seen in cross-section.

Synthesis

NU-86 is readily prepared [3] from the system:

$$SiO_2 - Al_2O_3 - Na_2O - RBr_2 - H_2O$$

where R is the template or structure directing agent of general formula:

$$[(CH_3)_3N\text{-}(CH_2)_m\text{-}N(CH_3)_3]^{2+}\ 2X^-$$

with m preferably 8 and 9, that is, octamethonium and nonamethonium. While NU-86 can be prepared over a range of compositions this paper will concentrate on materials prepared using octamethonium bromide, that is, octane-1,8-bis(trimethylammonium bromide). Using this template a typical synthesis mixture has a (molar) composition:

$$60\ SiO_2 - 1.846\ Al_2O_3 - 12\ Na_2O - 10\ OctaBr_2 - 3000\ H_2O$$

The reaction mixture is prepared by:

(1) Dispersing CAB-O-SIL M5 (BDH Ltd) in about 30% of the required water;

(2) Dissolving the required sodium hydroxide, and sodium aluminate solution (SOAL, Kaiser) in about 30% of the required water;

(3) Dissolving the octamethonium bromide in the remaining water.

Solution (2) is then added slowly, with stirring, to the silica dispersion (1) followed by the octamethonium bromide solution. Mixing is continued to give a smooth/homogeneous gel, which is transferred to the reactor. Time zero is taken when the autoclave reaches reaction temperature.

Such a reaction, carried out at 165°C, is illustrated in Figure 2 with plots of both pH and x-ray intensity (using the interplanar spacing at approximately 3.9Å) against reaction time. From the figure it can be seen that full crystallinity is only reached after about 425 hours (approximately 18 days) at temperature. Crystallisation is accompanied by an increase in pH, from about 11.6 to 12.1, again illustrating that this technique provides a reasonable indication of crystallisation [11] although, as predicted [12], pH lags behind crystallinity as determined by x-ray powder diffraction. The decrease in pH observed up to 350 hours is believed to be due to some (minor) template degradation.

The long crystallisation time, at such an elevated temperature, is somewhat unusual for a high-silica zeolite although it is similar to that reported for zeolite NU-87 [13]. Attempts to to reduce overall crystallisation time by simply increasing reaction temperature tend to result in the production of by-products such as mordenite.

Characterisation: XRD, Electron Microscopy, Chemical/Thermal Analysis.

Figure 3 contains powder x-ray diffraction patterns (variable slit) for NU-86 both as made and H-form, that is, after calcination and ion-exchange to remove occluded template and alkali metal ions.

NU-86 has a unique diffraction pattern reflecting the fact that it has a new molecular sieve topology. However, the pattern does have some aspects resembling that of zeolite Beta [14,]. This is largely a consequence, however, of NU-86 containing the same type of planar faulting as Beta which results in an x-ray pattern containing broad and sharp reflections. This is exacerbated by the platey morphology of the octa-NU-86 sample.

The diffraction pattern of H-NU-86 is very similar to that of the as synthesised material but there is an increase in the intensities of the low angle lines relative to those in the mid-angle region. Minor shifts in line positions indicate that there is some contraction of the unit-cell on removal of the template, hardly surprising in such an open structure.

Electron microscopy (SEM/TEM) of samples of NU-86 indicates that a range of crystal habits can be produced. Commonly crystals have a platey, or lozenge-like, morphology. Individual crystals are a few microns across but usually <500Å thick. Acicular, or rod-like, crystals can also be produced.

Based on a unit cell containing 120 T-atoms the results of chemical analysis of the octa-NU-86 produced above gives a unit cell composition of:

$$110.4\ SiO_2 - 9.6\ AlO_2 - 0.1\ Na$$

The composition illustrates that little sodium has been occluded in this particular product. Examination of a range of samples indicates a variable sodium content; believed to be due to the ability to remove sodium during the washing stage, indicating an open framework structure. Consistent with this thermal analysis gives a large template and water content. DTA suggests that the template is lost in two distinct stages since there are two major exotherms: the first of these is fairly sharp and occurs at about 425°C while the second is broader and begins at over 500°C. Overall the tga/dta trace is similar to that obtained for other zeolites prepared with similar templates [13].

Template integrity, in the as made zeolite, was confirmed by the use of ^{13}C CPMAS NMR spectroscopy. Figure 4 contains the spectra for the octa-NU-86 prepared above and (inset) for a pure sample of solid octamethonium bromide. Octa-NU-86 shows four peaks at 68.7ppm, 54.5ppm, 27.6ppm and 24.5ppm, with the peak at 27.6ppm having a large shoulder on the downfield side. The overall pattern is consistent with that for the reference octamethonium bromide. Integration of the peaks confirms that the template is intact within the zeolite, the ratios being within experimental error. The broadening observed for the cluster of peaks at 27.6ppm indicates a greater range of template conformations for molecules occluded in the zeolite compared to those in the crystalline salt.

Properties: thermal stability and sorption.

The excellent thermal stability of NU-86 is illustrated by Figure 5 which contains a plot of crystallinity against calcination temperature. The absolute intensities of lines in the mid-angle region of the powder x-ray diffraction pattern, corresponding to interplanar spacings at about 4.2, 3.9 and 3.7Å, were taken as a measure of crystallinity. In the study aliquots of calcined (ie template-free) NU-86 were calcined for 16 hours, in static air, at the temperatures indicated.

It can be seen that the structural integrity of NU-86 remains intact up to 850°C. By 950°C, however, the structure has begun to degrade and is effectively amorphous to x-rays after 16 hours at 1050°C. Thus although it has a very open framework NU-86 exhibits good thermal stability.

Figure 6 contains a sorption isotherm for nitrogen in NU-86. At the low to moderate values of p/p° illustrated, the isotherm is rectangular in shape with no evidence for hysteresis until high relative pressures. Simply taking the uptake at a p/p° of 0.4 as a measure of micropore volume gives a value of about 0.24 cm^3/g. Hydrocarbon sorption [13], not surprisingly, gives somewhat lower values with uptakes for n-heptane and cyclohexane being about 0.18 and 0.17 cm^3/g respectively. Molecular sieving with respect to m-xylene was also observed. These results are consistent with an open framework structure, access to which is controlled by windows of about 0.62nm. With its three different channels access to which is through 10, 11 and 12 ring windows the sorptive properties of

NU-86 merit a detailed study, this study is underway.

CONCLUSION

NU-86 is a novel high-silica zeolite prepared using long chain (C-8 and C-9) bis-quaternary ammonium cations. It shows excellent thermal stability. Simple measurements indicate that NU-86 has a high capacity for a variety of sorbates consistent with its 3-D channel system containing 10-,11- and 12-rings.

ACKNOWLEDGEMENTS

The authors are grateful to Dr R Higgins for helpful discussions, Dr KK Joshi for template preparation and Dr N Clayden for nmr measurements. They acknowledge Mr S Huntley, Mr TJ Hall and Mr T Muteen for technical assistance and ICI Chemicals & Polymers Ltd for permission to publish.

REFERENCES:

1 Delprato, F., Demotte, L., Guth, J.L. and Huve, L., Zeolites, **10**, 546 (1990).
2 Pluth, J.J., and Smith, J.V. Am. Mineral., **75**, 501 (1990).
3 Casci, J.L., EP-A-463,768 (1992).
4 Shannon, M.D., Proceedings of the 9th International Zeolite Conference, Eds. R. Von Ballamoos, J.B. Higgins, & M.M.J. Treacy, Butterworth-Heinemann, Stoneham, MA, (1992).
5 Casci, J.L. "Proc. of the 7th International Zeolite Conference", Kodansha, 215 (1986).
6 Valyocsik, E.W. and Page, N.M., U.S. Patent 4,873,067 (1988).
7 Casci, J.L. and Stewart, A., EP-A-377,291 (1990).
8 Schlenker, J.L., Higgins, J.B. and Valyocsik, E.W. "Zeolites for the Nineties: Recent Research Reports". Ed. J.C. Jansen, 287 (1986).
9 Shannon, M.D., Casci, J.L., Cox, P.A. and Andrews, S.J., Nature, **353**, 417 (1991).
10 Smith, P.A. and Frank, S., J. Amer. Chem. Soc., **74**, 509 (1952).
11 Casci, J.L. and Lowe, B.M., Zeolites, **3**, 186 (1983).
12 Lowe, B.M., Zeolites, **3**, 300 (1983).
13 Casci, J.L., Shannon,M.D., Andrews, S.A. and Cox, P.A., "Synthesis of Microporous Materials, Volume 1". Ed. M. Occelli and H. Robson. In Press.
14 Wadlinger, R.L., Kerr, G.T. and Rosinski, E.J., U.S. Patent 3,308,069 (1967).

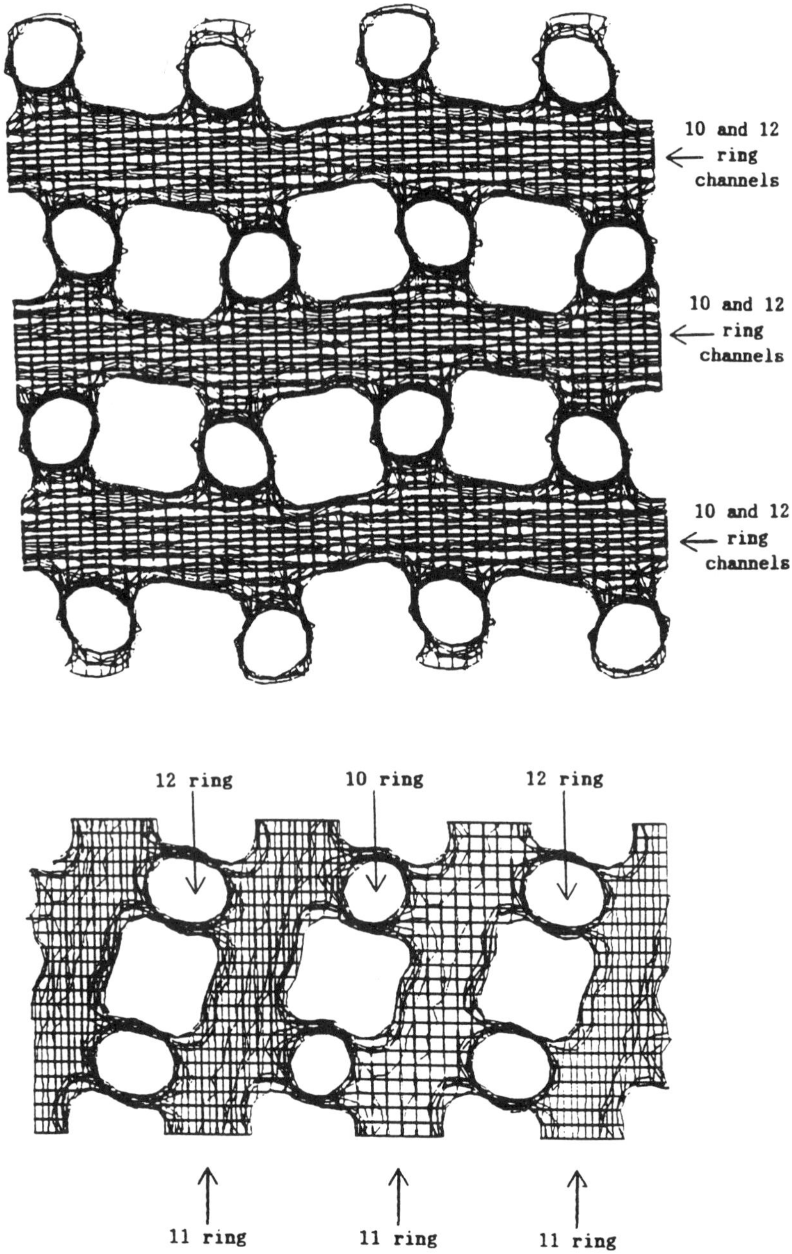

Figure 1. Pore-Maps of the NU-86 structure illustrating the 10, 11 and 12 T-atom rings.

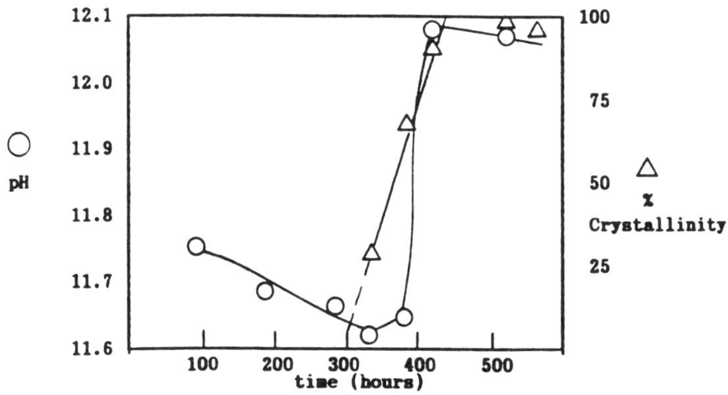

Figure 2. Crystallisation of NU-86, at 165°C, from a reaction mixture of molar composition:
60 SiO_2 - 1.846 Al_2O_3 - 12 Na_2O - 10 $OctaBr_2$ - 3000 H_2O
Plot of pH and % Crystallinity against reaction time.

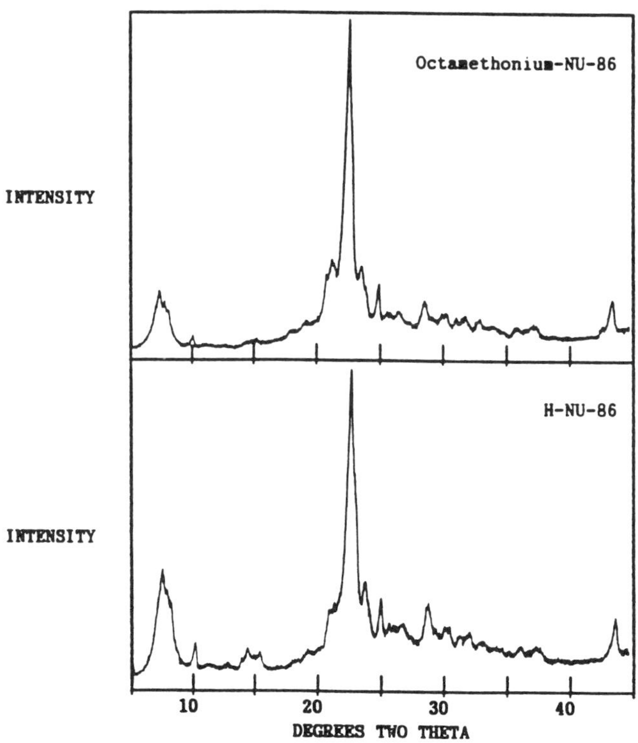

Figure 3. Powder x-ray diffraction patterns for as made and H-NU-86.

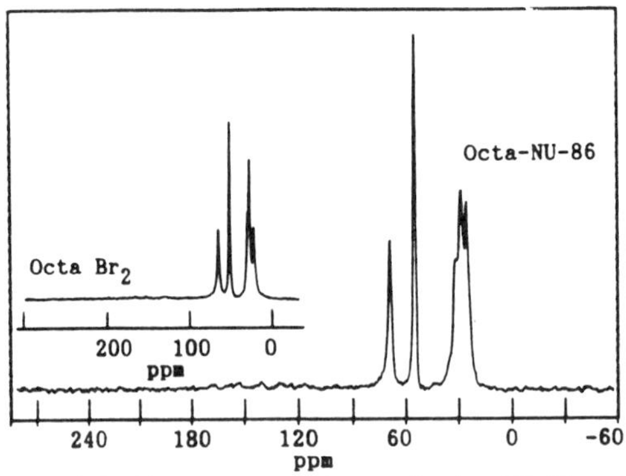

Figure 4. ^{13}C NMR (CPMAS) spectra of Octamethonium-NU-86 and (inset) octamethonium bromide.

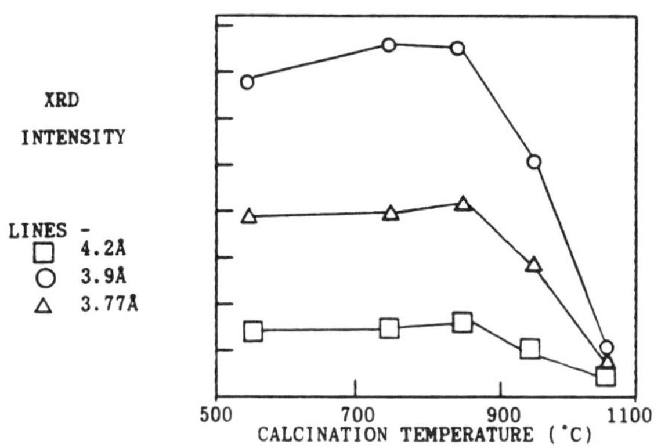

Figure 5. Thermal stability of zeolite NU-86. Plot of calcination temperature against xrd line intensity.

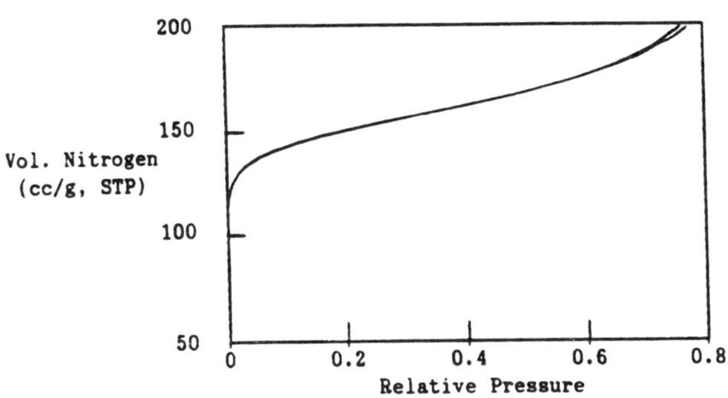

Figure 6. Nitrogen Isotherm for Zeolite NU-86.

APPLICATION OF DIFFUSE REFLECTANCE FTIR SPECTROSCOPY IN MOLECULAR SIEVE RESEARCH

E. LOEFFLER[1], CH. PEUKER[2], B. ZIBROWIUS[3], U. ZSCHERPEL[4], and K.-H. SCHNABEL[5]

[1]Adlershofer Umweltschutztechnik- und Forschungsgesellschaft mbH,
Rudower Chaussee 5, D-O-1199 Berlin, Germany
[2]Wissenschaftler-Integrationsprogramm, Geb. 19.5,
Rudower Chausse 6, D-O-1199 Berlin, Germany
[3]UMIST, Department of Chemistry, P.O. Box 88, Manchester M60 1QD, United Kingdom
[4]Bundesanstalt für Materialforschung und -prüfung, Unter den Eichen 87,
D-W-1000 Berlin 45, Germany
[5]Zentrum für Heterogene Katalyse, Rudower Chaussee 5, D-O-1199 Berlin, Germany

ABSTRACT

The capabilities of diffuse reflectance FTIR spectroscopy (DRIFT) are presented to solve various problems of molecular sieve research. For SAPO-34/$AlPO_4$-18 the results of in-situ investigations of template decomposition and hydration/dehydration processes discussed in connection with those of MAS NMR studies. If samples diluted in KBr are used, spectra of adsorbed or encaged species and framework vibrations are observed simultaneously. The overtone and combination tone bands of the lattice vibrations of undiluted ZSM-5 can be used for detecting the phase transformation caused by thermal treatment.

INTRODUCTION

While the transmission technique has been extensively applied in the IR spectroscopy to investigate molecular sieves, the diffuse reflectance has only been used more frequently in the last few years. The DRIFT technique has the great advantages of detecting weak absorption bands and of very easy sample preparation. In addition the various experimental in-situ arrangements can be realized easier by this method than in the corresponding transmission method. Kazansky and co-workers have successfully used diffuse reflectance method in measuring the OH vibration bands and spectra of adsorbed molecules in the middle and near infrared regions [1,2]. Nevertheless, only a limited number of diffuse reflectance studies have been published [3-6].

The present paper demonstrates the advantages of DRIFT measurements in various fields of molecular sieve research.

From the Proceedings of the 9th International Zeolite Conference, Montreal 1992, Eds. R. von Ballmoos et al.,

EXPERIMENTAL

The IR measurements were performed using FTIR spectrometers IRF 180 (Centre of Scientific Instrumentations, Berlin) and FTS20 (Digilab). A home-made diffuse reflectance device equipped with two off-axis elliptical mirrors was used. The beam can be focused downward to a simple heatable sample cup or to sealed MAS NMR glass tubes. The conversed mirror arrangement allows the use of a vacuum cell equipped with a special plane quartz-glass window (free of OH groups). It was shown that OH stretching vibration bands in SAPO-5 spectrum, recorded using MAS NMR glass tubes, are in a very good agreement with those obtained for the same sample activated in the vacuum DRIFT cell [7].

For DRIFT measurements, undiluted samples as well as samples mixed with KBr were used. DRIFT spectra of samples diluted in KBr are shown in Kubelka-Munk units $f(R_\infty)$[8]. Using KBr wafer technique transmission spectra were measured.

The ^{13}C, ^{27}Al, and ^{31}P MAS NMR spectra were recorded on a Bruker MSL 400 spectrometer at a resonance frequency of 100.6, 104.2 and 161.9 MHz, respectively. Spinning speeds of about 5 kHz were used. The spectra were obtained either by single pulse excitation (^{27}Al, ^{31}P) or by cross-polarization (^{13}C). High-power proton decoupling was applied.

RESULTS AND DISCUSSION

Framework vibration of SAPO-34

The framework structure and composition of molecular sieves are important factors which affect the properties of these materials. X-ray diffraction is undoubtedly the major source of structural information. However, IR spectroscopy has also provided useful information about the framework structure.

In Figure 1 the absorbance spectrum of as-synthesized SAPO-34 is compared with the diffuse relectance spectrum of the same sample. The DRIFT spectrum of SAPO-34 diluted in a KBr matrix sample (1 wt.%) is essentially identical to the absorbance spectrum of the same sample prepared as a KBr disc. Hence it follows, that DRIFT spectra are suitable to study changes in framework vibrations caused by such processes as template removal, de- and rehydration (Figure 2).

The as-synthesized SAPO-34 exhibits pronounced framework vibration bands in the absorbance spectrum (Figure 2(1)) that are absent in the case of the calcined sample (2). One might conclude that the template decomposition leads to a framework destruction. However, it must be considered that after the template removal the pore system of the molecular sieve becomes vacant, so that water can be adsorbed from the atmosphere during the preparation of the KBr wafer. Using the diffuse reflectance method, we are able to monitor the dehydration process in situ. Already for a partially dried calcined sample, the same framework vibration bands as in the template containing one are observed (3). Hence it follows, that the disappearence of IR framework vibrations is not caused by a collapse of the framework structure after template decomposition, but by a reversible distortion due to the adsorption of water.

In accordance with previous studies [9,10], pronounced changes in the ^{27}Al and ^{31}P MAS NMR spectra of SAPO-34 are observed under template removal and rehydration (Fig. 2). Whereas the ^{27}Al NMR spectra of the as-synthesized (1), and the calcined and dehydrated samples (3) are clearly dominated by lines at positions characteristic of tetrahedrally coordinated aluminium in aluminophosphates, the spectrum of the rehydrated sample (2) shows an additional line at about -13 ppm. This line position is typical of aluminium in an octahedral oxygen coordination in aluminophosphates. The change in the oxygen coordination of a part of the aluminium is found to be reversible. Also, the results of ^{31}P MAS NMR reflect severe changes in the framework structure of this silicoaluminophosphate. The position of the low-field shoulder of the broad line obtained for the calcined and rehydrated sample agrees with the position of the broad ^{31}P NMR line that is observed for reaction gels at the beginning of the aluminophosphate crystallization process. It indicates a very strong interaction with the adsorbed water molecules.

Template decomposition in AlPO$_4$-18

The template decomposition was studied by recording in-situ DRIFT spectra at increasing sample temperatures in a nitrogen stream (Figure 3). The AlPO$_4$-18 sample was synthesized with tetraethylammoniumhydroxide (TEAOH) as templating agent. At a temperature of 420 K only the CH stretching and deformation vibrations of the template are observed. The chemical shifts obtained in the ^{13}C MAS NMR spectrum of the as-synthesized AlPO$_4$-18 (52.9 ppm, 7.2 ppm) are in reasonable agreement with those observed for the tetraethylammonium ion in solution [11].

The band at 3688 cm^{-1} is due to vibration of OH ions compensating the positive charge of TEA$^+$. This band is absent in SAPO-34 [12]. At 570 K the decomposition of the template begins and after 15 min at 590 K an IR spectrum similar to that of triethylamine is observed. From these results we suggest, that at least under our conditions (in nitrogen stream) the reaction

$$[(C_2H_6)_4N]^+OH^- \rightarrow (C_2H_6)_3N + C_2H_4 + H_2O$$

is the first step of the template removal. This assumption is supported by the ^{13}C MAS NMR spectra obtained for as-synthesized AlPO$_4$-18 after vacuum treatment at 570 K and subsequent rehydration. Lines at 47.7 ppm and 8.8 ppm are observed. Taking into account the high-field shift caused by protonation of triethylamine by adsorbed water, from the chemical shift values of triethylamine in solution [11] one would expect lines at 49.1 ppm and 9.1 ppm for the α and β carbon in the triethylammonium ion, respectively. By measuring undiluted samples in the NIR region it can be shown, that the KBr matrix only influences the reaction velocity.

During the template decomposition drastical changes in the framework vibration (1300-400 cm^{-1}) are observed. The interpretation of this phenomenon needs further investigation.

Interaction of adsorbed n-hexane with hydroxyl groups of SAPO-17

The adsorption of probe molecules with different size can be used to characterize the accessibility of different types of OH groups. The interpretation of the IR spectra is often complicated by overlapping of bands of the "free" OH groups and bands due to OH groups interacting with the adsorbed molecules. Depending on the acid strength of the OH groups and the basicity of the probe molecules the stretching vibration bands (ν_{OH}) shift to lower wavenumbers, whereas the wavenumbers of bending vibration bands (δ_{OH}) increase. The linewidth increases with higher wavenumber shifts. This results in strong broadening of the combination vibration bands ($\nu_{OH}+\delta_{OH}$) which depends on the wavenumber shifts. Figure 4 shows the change of fundamental stretching vibration (ν_{OH}) as well as of combination vibration ($\nu_{OH}+\delta_{OH}$) of SAPO-17.

One type of acid OH groups is inaccessible to n-hexane and its combination vibration remains practically unchanged. The other hydroxyl groups interact with the probe molecule and the corresponding bands disappear completely due to the their strong broadening.

Remarks on DRIFT spectra of undiluted samples

In some cases diffuse reflectance spectra of undiluted samples differ due to the molecular sieve structure in the region of lattice overtone and combination tone bands (2800-1200 cm^{-1}). These bands are essentially less affected by specular reflection and dispersion than bands of the fundamental vibration region (1200-400 cm^{-1}). Thus, in principle, very simple measurements can be used to investigate molecular sieves. For instance, it can be easily shown that ZSM-5 crystals are transformed into cristobalite after thermal treatment at 1573 K. But until now the assignments of the bands are not sufficiently proved.

ACKNOWLEDGEMENTS

This work was supported by the Deutsche Forschungsgemeinschaft. The authors want to thank Dr. U. Lohse and Mrs. I. Girnus for SAPO-17 and $AlPO_4$-18/SAPO-34, respectively.

REFERENCES

1. Kustov, L.M., Borovkov, V.Yu. and Kazansky, V.B., J. Catal. **72**, 149 (1981).

2. Kazansky, V.B., Stud. Surf. Sci. Catal. **65**, 117 (1991).

3. Hanke, W. and Moeller, K., Zeolites **4**, 244 (1984).

4. Kubelkova, L., Hoser, H., Riva, A. and Trifiro, F., Zeolites **3**, 244 (1983).

5. Roessner, F., Steinberg, K.-H., Rudolf, A. and Staudte, B., Zeolites **9**, 371 (1989).

6. Johnson, S.A., Rinkus, R.-M., Diebold, T.C. and Maroni, V.A., Appl. Spectr. **42**, 1369 (1988).

7. Peuker, Ch., Finger, G., Löffler, E., Pilz, W. and Schnabel, K.-H., Z. Chem. **30**, 334 (1990).

8. Kubelka, P. and Munk, F., Z. Tech. Phys. **12**, 593 (1931).

9. Anderson, M.W., Sulikowski, B., Barrie, P.J. and Klinowski, J., J. Phys. Chem. **94**, 2730 (1990).

10. Goepper, M., Guth, F., Delmotte, L., Guth, J.L. and Kessler, H., Stud. Surf. Sci. Catal. **49**, 857 (1989).

11. Kalinowski, H.-O., Berger S. and Braun, S., ^{13}C-NMR-Spektroskopie, Georg Thieme Verlag, Stuttgart, 1984.

12. Girnus, I., Löffler, E., Lohse, U. and Neissendorfer, F., Coll. Czech. Chem. Comm., **57**, 946 (1992).

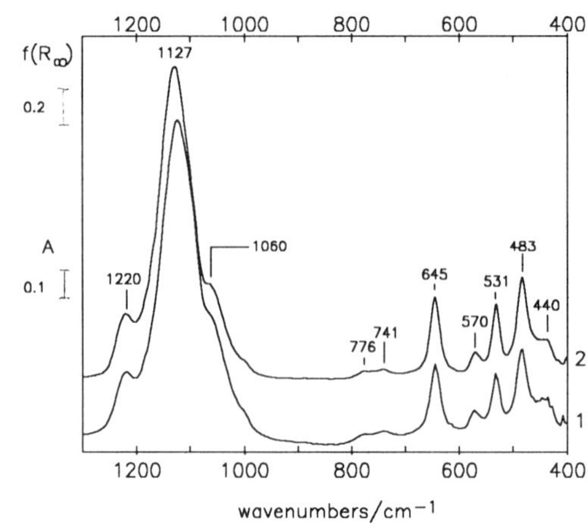

Figure 1. IR absorbance (1) and DRIFT (2) spectra of as-synthesized SAPO-34

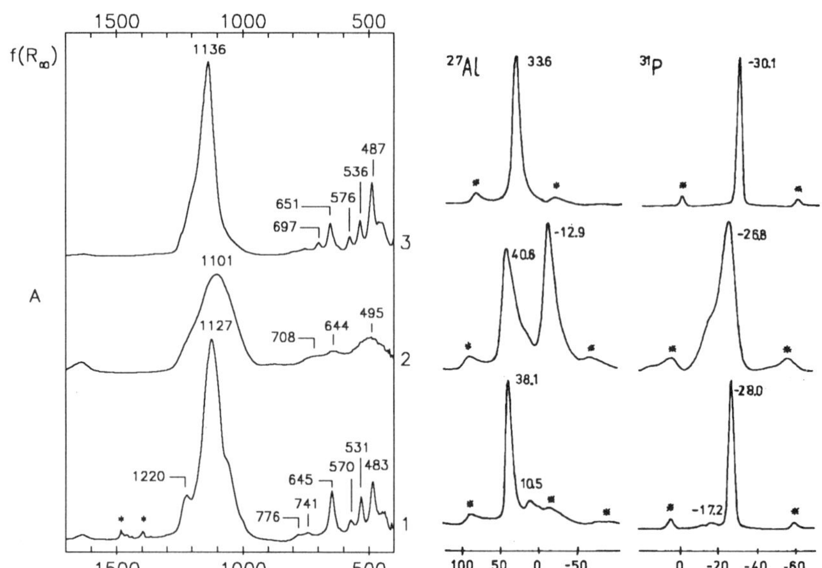

Figure 2. IR (left) and MAS NMR (right) spectra of the framework of SAPO-34: as-synthesized (1), calcined and rehydrated (2) and calcined, rehydrated and partially dehydrated (3) samples

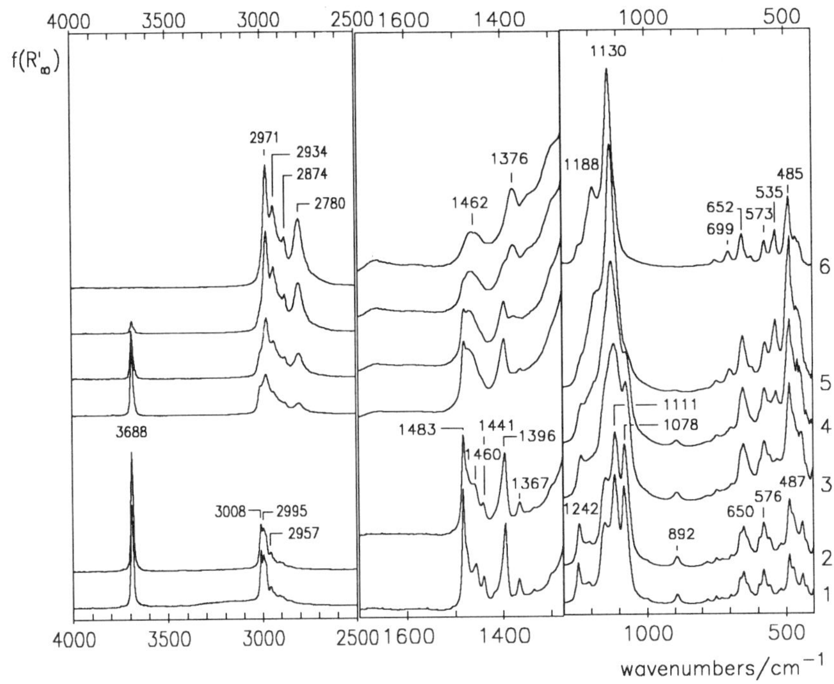

Figure 3. DRIFT spectra of as-synthesized $AlPO_4$-18 at various temperatures: 290 K (1), 30 min 420 K (2), 30 min 570 K (3), 5 min 580 K (4), 15 min 580 K (5) and 15 min 590 K (6)

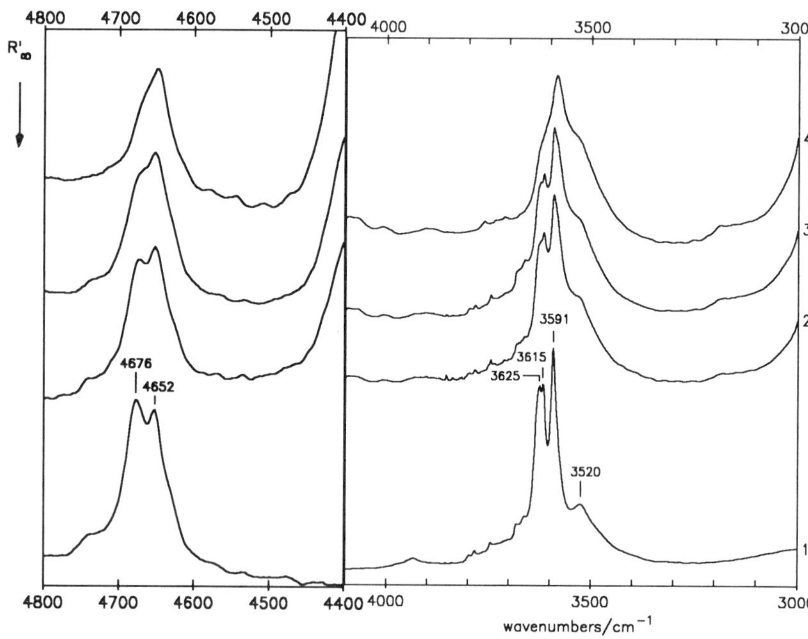

Figure 4. DRIFT spectra of OH groups of SAPO-17: 1- sample activated at 720 K; 2-4 - with an increasing amount of adsorbed n-hexane

AN INFRARED STUDY OF STRUCTURE RELAXATION IN MOLECULAR SIEVES

W.P.J.H. JACOBS, A.J.M. de MAN, J.H.M.C. van WOLPUT and R.A. van SANTEN

Schuit Institute of Catalysis, Laboratory of Inorganic Chemistry and Catalysis, Eindhoven University of Technology, P.O. Box 513, 5600 MB Eindhoven, The Netherlands.

ABSTRACT

The lattice vibrations of the isostructural analogues $AlPO_4$-5 and SSZ-24 have been studied by infrared spectroscopy. The vibrational modes have been assigned based on theoretical relaxed structures using a valence force field model to compute the lattice modes. The influence of the substitution of two silicon atoms by one phosporus and aluminium atom is discussed in relation with the experimental spectra.
The influence of protonation and exchangeable cations on lattice vibrations is demonstrated by in-situ infrared transmission studies of the reammoniation of HY and the dehydration of NiY.
The effect of substitution of the framework atoms or the extra-framework species on the lattice modes results in relaxation of the zeolitic structure, which can be monitored with infrared spectroscopy. Further in the case of protonation of the lattice coupling between the zeolitic lattice modes and the bending modes of the acidic hydroxyls is observed.

INTRODUCTION

Fourier Transform Infrared Spectroscopy (FTIR) has been applied with succes in the past in the characterisation of molecular sieves. Several groups have studied the lattice vibrations in zeolites [1,2]. According to these studies asymmetrical stretching vibrations of the T-O bonds (T represents Si or Al) are found at 950-1250 cm^{-1} (inter-tetrahedral) and at 1050-1150 cm^{-1} (intra-tetrahedral) and the corresponding symmetrical stretching vibrations are found at 650-720 cm^{-1} and at 750-820 cm^{-1} respectively; the O-T-O bending mode is located at 420-500 cm^{-1} and modes found at 500-650 cm^{-1} and at 300-420 cm^{-1} are assigned to vibrations of double rings and pore opening modes [3].
Substitution of the framework aluminium by silicon is now common practice for zeolites and infrared studies of dealuminated Y zeolites showed a linear relation between the number of aluminium atoms in the zeolitic unit cell and the position of the maxima for the T-O stretching modes [1,4]. This behaviour is explained by the increased average bond strength for the T-O bond.
After the succesful synthesis of alumino phosphate molecular sieves [5,6] it is now possible to study the changes in the infrared spectra upon replacement of Si-O-Si linkages by Al-O-P linkages. We will present here the experimental infrared spectra of the isostructrural analogues

From the Proceedings of the 9th International Zeolite Conference, Montreal 1992, Eds. R. von Ballmoos et al.,

$AlPO_4$-5 and SSZ-24, together with the spectra calculated using a valence force field model applied to energy minimized structures.

In contrast to the molecular sieves just mentioned zeolite Y contains extra-framework cations to compensate for the negativily charged framework. These cations introduce an additional parameter which can influence the lattice modes. We will demonstrate this by an in-situ FTIR study of the lattice vibrations of HY and NH_4Y obtained after ammonia adsorption. Also we will discuss the influence of dehydration on the framework modes for NiY.

EXPERIMENTAL

Infrared spectra of $AlPO_4$-5 and SSZ-24 in a KBr-matrix have been recorded in the region 250-2000 cm^{-1} using a Bruker IFS 113v FTIR spectrometer. The spectra were recorded in transmission mode at a resolution of 2 cm^{-1} and using a standard DTGS detector. At least 128 scans were coadded. In order to study the possible influence of adsorbed water a second series of measurements was performed using the molecular sieves suspended on a ZnSe support. Spectra were recorded at room temperature before and after heating for 2 hours at 773 K in an in-situ vacuum cell.

The in-situ spectra for the Y zeolites also were recorded with a support. Here we used silicon as a support and typically 500 scans were coadded at a resolution of 4 cm^{-1}. The samples used were obtained from $Na_{56}Y$ after ion exchange and had a composition given by $(NH_4)_{56}Y$ and $Ni_{17}Na_{22}Y$. The first one was converted to the proton form by heating in vacuo at 723 K for 1 hour in the in-situ cell. Then the ammonium form was obtained by subsequent adsorption of anhydrous ammonia at roomtemperature. The nickel zeolite was dehydrated by heating in vacuo at 773 K for 1 hour. Again all spectra were recorded at room temperature.

RESULTS

As can be seen from the infrared spectra, $AlPO_4$-5 (figure 1a) and SSZ-24 (figure 1b) contain some common features: the asymmetrical stretching mode at 1000-1300 cm^{-1}, a weak feature at 700-715 cm^{-1}, a band at 560-565 cm^{-1} and a sharp band at 460-470 cm^{-1}. Additional to these features the spectrum of SSZ-24 contains intense peaks at 808 cm^{-1} and 622 cm^{-1}, whereas the spectrum of $AlPO_4$-5 contains only a very weak feature at 634 cm^{-1} and additionally at 665 cm^{-1}. These features were also observed without any change after heating at 773 K using the ZnSe support. So the observed lattice vibrations are not influenced by interactions due to adsorbed water.

The spectra of HY and NH_4Y are presented in figure 3a and 3b. The ammonium form is characterised by a broad band due to asymmetrical stretching vibrations (810-1300 cm^{-1}) and a band due to symmetrical stretching vibrations (650-810 cm^{-1}). Also strong bands are centered at 575, 454 and 393 cm^{-1} and a weaker band is observed at 509 cm^{-1}. A very weak feature is present at 308 cm^{-1}. Further a broad band can be found around 1410 cm^{-1} and a weaker feature at 1670 cm^{-1} due to the two N-H bending modes of ammonium ions with a deformed tetrahedral symmetry.

In the corresponding proton form the asymmetrical and symmetrical stretching vibrations are shifted to a higher frequency by approximately 30 cm^{-1}. For the symmetrical modes also an increase in intensity is observed. The modes observed at 575 and 393 cm^{-1} for the ammonium form are absent and only some very weak modes can be found at 600 and 565 cm^{-1} and at 369 cm^{-1}. The mode found at 454 cm^{-1} for the ammonium form is now present as a broader band centered at 465 cm^{-1} with a shoulder at 500 cm^{-1} which is probably related to the feature at 509 cm^{-1} observed for the ammonium form.

The infrared spectra for the hydrated and dehydrated nickel zeolite are shown in figures 4a and 4b respectively. The spectrum of the hydrated form is comparable to that of the ammonium form. The major difference is a shift to higher frequencies for the nickel zeolite for all bands except for a band which is observed at 393 cm^{-1} in both cases. The band at 509 cm^{-1} for the ammonium zeolite is not resolved for the hydrated nickel zeolite due to the upward shift of the band located at 474 cm^{-1}. The other lattice modes are now observed as a broad band from 840 up to 1300 cm^{-1} with a maximum at 1063 cm^{-1}, as a broad band from 650 to 840 cm^{-1} with maxima at 794 and 737 cm^{-1}, and a band is centered at 585 cm^{-1}. The bending vibrations of adsorbed water are observed around 1640 cm^{-1}.

Dehydration strongly influences the infrared spectrum of the nickel zeolite. The band due to asymmetrical stretching modes is replaced by three main features: a band at 1184 cm^{-1} with a shoulder at 1155 cm^{-1}; a band with distinct maxima at 1076 and 1053 cm^{-1} and a shoulder at 975 cm^{-1}; a band at 930 cm^{-1} which is also visible as a weak shoulder in the hydrated sample. In the region were the symmetrical modes are located now a broad band with maxima at 798, 745, 703 and 647 cm^{-1} is observed. After dehydration the band at 585 cm^{-1} has almost completely vanished and only a very weak feature remains at 603 cm^{-1}. Similarly the band at 392 cm^{-1} has disappeared and a weak feature is observed at 403 cm^{-1}, together with a strong new band at 432 cm^{-1}. The band at 474 cm^{-1} is shifted to 479 cm^{-1} and due to the decreased bandwidth a shoulder at 518 cm^{-1} has evolved. Further two new bands at 323 and 305 cm^{-1} appear after dehydration.

DISCUSSION

In order to analyze the spectra for $AlPO_4$-5 and SSZ-24 infrared and raman spectra have been calculated [7]. The calculation of spectra requires a lattice energy minimization of the structure. This has been done for the two molecular sieves in a calculation using two partial charge rigid ion potential models developed by Kramer et al. [8].

The partial charge models are not suitable for calculations of vibrational spectra over the whole frequency range of interest [9]. Therefore, for the calculations of the infrared spectra, a General Valence Force Field model has been applied using the structures obtained after relaxation with the partial charge model [10]. The calculated spectra (figure 2a and 2b) compare very well with the experiment. To investigate the influence of structure and chemical composition infrared spectra were calculated using a silicon substituted $AlPO_4$-5 structure and an $AlPO_4$ substituted SSZ-24 structure [7]. It appears that only minor spectral features are due to differences in the structures. The spectra are mainly determined by the chemical composition, i.e. the force constants. For zeolites a similar situation arises when aluminium is replaced by silicon in the lattice [1].

The presented spectrum of NH_4Y is in agreement with results obtained with the standard KBr disc technique [1]. In this sample the ammonium ions are located at the cation sites $S_{I'}$, S_{II}, $S_{III'}$ and $S_{III''}$ [11]. In the proton form the proton is bonded to the Si-O-Al linkages leaving both the Si-O and Al-O bonds intact as can be concluded from the presence of both in-plane and out-of-plane O-H bending vibrations (1090 and 420 cm^{-1} respectively) [12,13]. These protons are mainly located at the O_1 and O_3 oxygen atoms. A smaller number of protons is localised near O_2 and no protons are found near O_4 [14]. Due to protonation of the lattice the zeolitic structure relaxes. The Si-O-Al angle decreases and the T-O bond strength at the acidic site decreases while the strength for the T-O bonds connected to such a site are strengthened [15,16]. Results from MAS-NMR experiments with the proton and ammonium forms of zeolite Y support this view [17]. The lattice constants for the proton and ammonium forms differ only slightly [14,18] indicating that the strengthening and weakening of the T-O bonds compensate each other. So the observed shift of the stretching modes after protonation is not related to an increase for the average bond strength of the T-O bonds due to structure relaxation. From inelastic neutron spectroscopy on Y zeolites evidence for the coupling of the protons with the lattice modes was obtained [12,13]. So this shift might be due to the coupling between the zeolitic lattice modes and the in-plane O-H bending modes [19].

Upon protonation the intensity for the modes of the double six ring vibrations almost completely disappear due to protonation of O_1, O_2 and O_3. Protonation also leads to a decrease in intensity for the pore opening modes. Finally the broadening of the band for the O-T-O bending modes indicates an increase in heterogenity for the corresponding force constants. These changes might be due to both lattice relaxation and coupling of the zeolitic lattice modes with the out-of-plane O-H bending modes.

The spectrum of hydrated NiY is quite similar to that of NH_4Y except for a shift to higher frequencies for several modes due to larger force constants. This is indicative of a weaker average interaction of the hydrated Ni^{2+} ions, that have half the concentration of the NH_4^+ ions. Maxwell et al. studied the influence of exchangeable cations for hydrated zeolites and they explained the observed downward shift in framework vibrations by an increased strength of interaction with the lattice [20].

In the hydrated form the Ni^{2+} ions are mainly located in the supercages but after dehydration they are probably situated in the hexagonal prisms at the S_I sites [21] giving rise to stronger interactions with the zeolitic lattice. A similar conclusion has been obtained from ^{29}Si-MAS-NMR studies on the dehydration of ion exchanged Y zeolites [22]. For the infrared stretching modes dehydration results in infrared bands showing more pronounced features. As in the case of protonation of the lattice the modes of the double six ring and the pore opening modes have disappeared. The band for the O-T-O bending modes in contrast has narrowed and an additional band has evolved. The changes are indicative of well-defined local distortions of the lattice.

CONCLUSION

Substitution of framework atoms and exchangeable cations results in changes in the lattice vibrational modes for molecular sieves. The vibrational changes observed by substitution of framework atoms, e.g. by substitution of Al-O-P by Si-O-Si, are determined mainly by the chemical nature of the substituting atoms. Structural differences are of minor importance.

For exchangeable cations the location and hence the interaction of these cations with the lattice determines the vibrational spectrum. For weak interactions as is the case for NH_4^+ and hydrated Ni^{2+} the spectra are much alike. Deammoniation results in a strong interaction due to the formation of O-H bonds. Dehydration of Ni^{2+} enhances the interaction with the lattice due to the migration of Ni^{2+} from the supercages into the hexagonal prisms. The oxygen atoms of the zeolite lattice now take over the stabilizing role of hydrating water molecules. For both Y zeolites studied this results in a decrease in intensity for the infrared bands of the double six ring and pore opening modes. The stretching modes are affected in a different way. Dehydration of NiY results in a splitting of the bands. Protonation provides a uniform shift to higher frequencies. In the latter case the spectra are not only determined by structural relaxation but also by the coupling of the framework modes with the in-plane and out-of-plane bending modes of the acidic hydroxyls.

The results obtained here illustrate the relaxability of the structures of zeolitic and related materials upon subsitution of the species present in the sieves.

REFERENCES

1. Flanigen, E.M., in: *Zeolite Chemistry and Catalysis*, ACS Monograph **171**, Ed. J.A. Rabo, American Chemical Society, Washington, D.C., p. 80 (1976).
2. Flanigen, E.M., Khatami H., Szymanski, H.A., in: Advances in Chemistry Series **101**, Ed. R.F. Gould, American Chemical Society, Washington, D.C., p. 201 (1971).
3. de Man, A.J.M., van Santen, R.A., Zeolites **12**, 269 (1992).
4. Sohn, J.R., Decanio J., Lunsford, J.H., O'Donnel D., Zeolites **6**, 225 (1986).
5. Wilson, S.T., Lok, B.M., Flanigen E.M., U.S. Patent 4,310,440 (1982).
6. Wilson, S.T., Lok, B.M., Messina, C.A., Cannan T.R., Flanigen, E.M., J. Am. Chem. Soc. **104**, 1146 (1982).
7. de Man, A.J.M., Jacobs, W.P.J.H., Gilson, J.P., van Santen, R.A., Zeolites, submitted.
8. Kramer G.J., Farragher N., van Beest B.W.H, van Santen R.A., Phys. Rev. B **43**, 5068 (1991).
9. de Man, A.J.M., van Beest, B.W.H., Leslie, M., van Santen, R.A., J. Phys. Chem. **94**, 2524 (1990).
10. Etchepare J., Merian, M., Smetankine, L., J. Chem. Phys. **60**, 1873 (1974).
11. Ozin, G.A., Baker, M.D., Godber, J., Gil, C.J., J. Phys. Chem. **93**, 2899 (1989).
12. Jobic, H., J. Catal. **131**, 289 (1991).
13. Jacobs, W.P.J.H., Jobic, H., van Wolput, J.H.M.C., van Santen, R.A., Zeolites **12**, 315 (1992).
14. Czjzek, M., Jobic, H., Fitch, A.N., Vogt, T., J. Phys. Chem. **96**, 1535 (1992).
15. van Santen, R.A., de Man, A.J.M., Kramer, G.J., in: *Zeolite Microporous Solids: Synthesis, Structure, and Reactivity*, Eds. E.G. Derouane et al., Kluwer Academic Publishers, p. 493 (1990).
16. Teunissen, E.H., van Duijneveldt, F.B., van Santen, R.A., J. Phys. Chem. **96**, 366 (1992).
17. Vega, J., Luz, Z., J. Phys. Chem. **91**, 365, 374 (1987).
18. Sohn, J.R., DeCanio, J., Lunsford, J.H., O'Donell, D.O., Zeolites **6**, 225 (1986).
19. Jobic, H., private communication.
20. Maxwell, I.E., Baks A., in: *Molecular Sieves*, Advances in Chemistry Series **121**, Eds. W.M. Meier, J.B. Uytterhoeven, American Chemical Society, Washington, D.C., p. 87 (1973).
21. Dooryhee, E., Greaves, G.N., Steel, A.T., Townsend, R.P., Carr, S.W., Thomas, J.M., Catlow, C.R.A., Faraday Discuss. Chem. Soc. **89**, 119 (1990).
22. Chao, K.J., Chern J.Y., J. Phys. Chem. **93**, 1401 (1989).

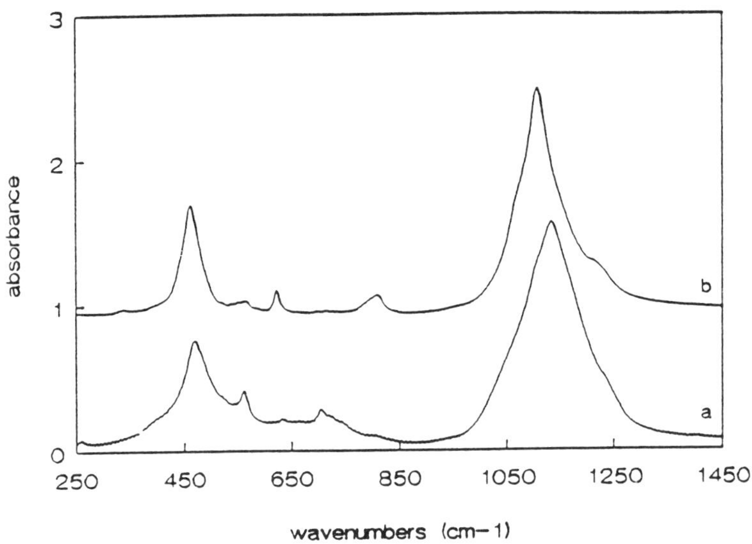

Figure 1. Experimental infrared spectra: $AlPO_4$-5 (a) and SSZ-24 (b).

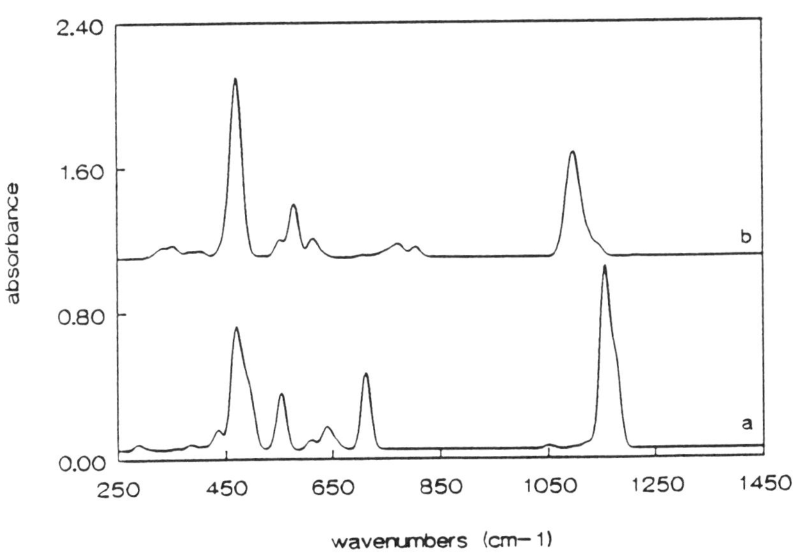

Figure 2. Calculated infrared spectra: $AlPO_4$-5 (a) and SSZ-24 (b).

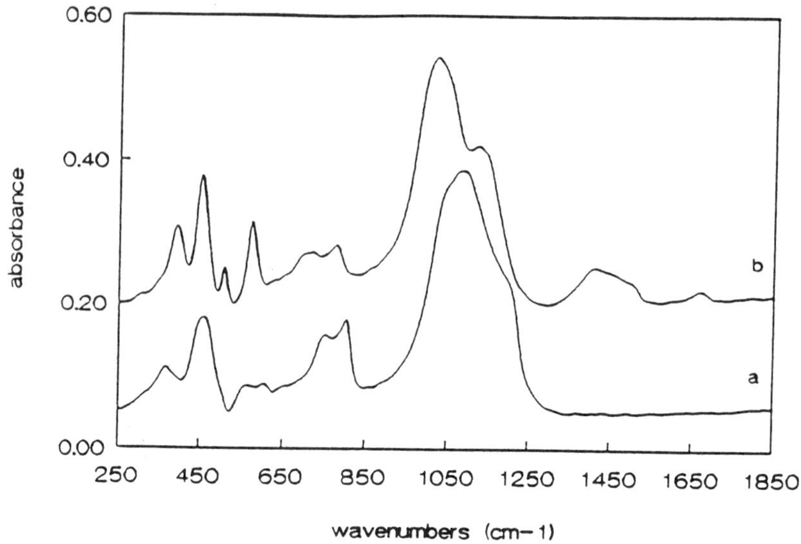

Figure 3. Experimental infrared spectra: HY (a) and NH_4Y (b).

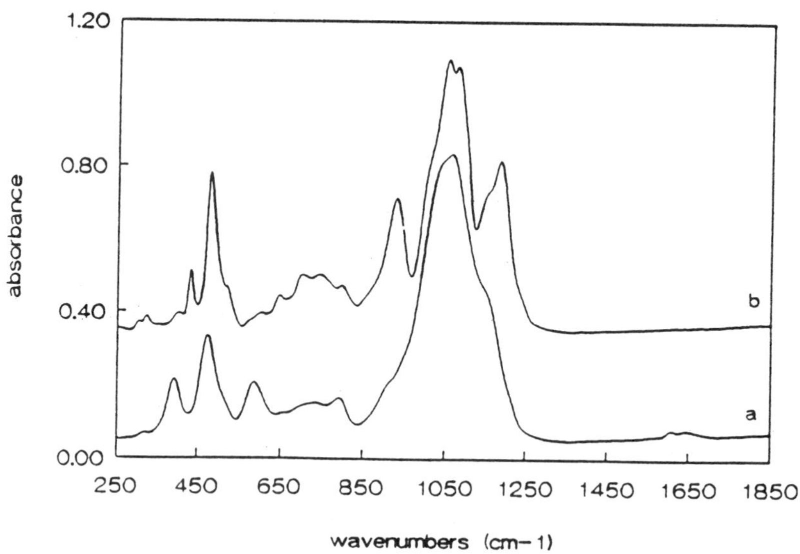

Figure 4. Experimental infrared spectra: hydrated (a) and dehydrated (b) NiY.

QUANTUM CHEMICAL INTERPRETATION OF IR DATA OF METHANOL ADSORPTION ON ZEOLITES AND SILICA

A.G.PELMENSCHIKOV*, G.MOROSI** and A.GAMBA**

*Institute of Catalysis, Lavrentieva 5, 630090 Novosibirsk, Russia
**Dipartimento di Chimica Fisica ed Elettrochimica, Universita' di Milano, Via Golgi 19, I-20133 Milano, Italy

ABSTRACT

Ab initio quantum chemical molecular models are used to interpret IR data of methanol adsorption on zeolites and silica. The calculations are carried out with 3-21G and 6-31G* basis sets, using a scaling procedure to correct the systematic error in 3-21G C-H stretching frequencies. New forms for the adsorption on the bridging and terminal OH groups are proposed, which are found both in the most agreement with the IR data and the most stable among the hypothetical structures. The calculated C-H frequencies and IR intensities are in accordance with Chung-Bennett dependence and Gutmann's rules, supporting C^{+}-H^{-} polarity of bonds in the methoxy groups.

INTRODUCTION

Successful application of quantum chemical molecular models in the interpretation of IR properties of surface OH groups of oxides ([1-3], for example) allows one to expect this approach to be no less effective in similar studies on more complex surface centers. Among these the surface methoxy groups are of great interest for catalysis, being suggested as key intermediates in various reactions of methylation and in methanol conversion by oxides [4,5]. In the present work the molecular approach is used to interpret the IR spectra [5-9] of the methoxy species formed by methanol adsorption on zeolites and silica.

METHOD AND MODELS

As shown in our recent ab initio SCF quantum chemical study [10], owing to the well known systematic character of errors [11] in a calculated vibrational frequency of a molecular fragment, in a set of closely related compounds, the simple procedure of frequency scaling

From the Proceedings of the 9th International Zeolite Conference, Montreal 1992, Eds. R. von Ballmoos et al.,

allows one to reproduce the observed O-H stretching frequencies of ≡Si-OH, ≡P-OH, ≡B-OH, ≡Al-OH surface groups within 10 cm^{-1} of the experimental data, by using $Si(OH)_4$, $OP(OH)_3$, $B(OH)_3$, $H_2O\cdot Al(OH)_3$ molecular models at the 3-21G level of treatment. In this way quantitative predictions of vibrational frequencies of some hypothetical Bronsted centers on zeolites were made [10]. In the present study the scaling procedure is used to correct the systematic error in the ab initio C-H stretching frequencies of CH_3 groups. Energies, equilibrium geometries, vibrational frequencies and infrared intensities are calculated by the ab initio SCF MO method with 3-21G basis set, using the Gaussian-88 package with the gradient optimization procedure.

ν'_{as}, ν''_{as} and ν_s stretching frequencies of the methoxy groups, defined quite accurately in the IR studies [5-9,12], are used to discriminate among the hypothetical structures. As the ≡Si-O-CH_3 group of silica, formed through esterification of the ≡Si-OH group, is practically the only surface methoxy group being explored enough to be used as a reference [5,6,8,9], the scale coefficients k'_{as} = 0.9107, k''_{as} = 0.9132 and k_s = 0.8951 are chosen to fit the calculated frequencies of the molecular model $(HO)_3SiOCH_3$ to the observed ones of the ≡Si-O-CH_3 group (see below).

RESULTS AND DISCUSSION

Interaction with Bridging OH group

The molecular models **1a** and **1b** of chemically and physically adsorbed methanol on the OH group are shown in Figure 1. Their calculated vibrational frequencies and adsorption energies, together with the experimental frequencies assigned to the chemisorbed form in Refs. [5-7], are reported in Table 1. The predicted large shifts of the frequencies of **1a** with respect to ≡Si-OCH_3 (ν'_{as}=3000, ν''_{as}=2958 and ν_s=2858 cm^{-1}) are in accordance with the empiric Chung-Bennett dependence [13], relating an electron density decrease on the CH_3 group to an increase of the C-H frequencies: the donor-acceptor interaction of the strong Lewis acid center Al≡ with the ≡Si-OCH_3 group in structure **1a** must lead to a large withdrawal of electron density from the CH_3 group and consequently to the high C-H frequencies. But such high frequencies have not been observed in experiments on methanol adsorption in zeolites [5-7]. Therefore we are inclined to think that

Table 1: Adsorption Energies (kJ/mol) and C-H Frequencies (cm^{-1})

characteristic	calcd		obsd [5-7]
	1a	1b	
adsorption energy	1	143	
ν'_{as}	3069	2986	(2977)[a]
ν''_{as}	3035	2953	
ν_s	2905	2874	2871

[a] The only observed asymmetric band is interpreted here as the result of the effective averaging of the ν'_{as} and ν''_{as} vibrational frequencies due to rapid rotation of the methyl group (see text).

the assignment [5-7] of the observed frequencies ν_{as}=2977 cm^{-1} and ν_s=2871 cm^{-1} to the methoxy group $\rangle$Si-O(CH_3)··Al$\langle$ is not correct. At the same time, the calculations show the strongly physically adsorbed form **1b** to be responsible for these bands. First, the frequency ν_s=2871 cm^{-1} practically coincides with the calculated one for **1b**. Second, considering the experimental and theoretical study by Morrow [12], the only band of asymmetric vibration at 2977-2980 cm^{-1} may be interpreted as the result of an effective averaging of two asymmetric C-H frequencies of the **1b** form, i.e. ν'_{as}=2986 and ν''_{as}=2953 cm^{-1} (Table 1), due to rapid rotation of the methyl group around the C-O bond axis. The rapid rotation of the CH_3 group of strongly adsorbed methanol in H-zeolites was found in MAS NMR investigation by Luz and Vega [14]. Third, the elementary act of **1b** formation is found to be considerably more preferable on energy grounds than that of form **1a** (Table 1). On this basis we suggest strongly physically adsorbed form **1b** as the main form for the interaction of methanol with the bridging OH group.

Physical Adsorption on Terminal OH Group

The molecular models **2a**, **2b** and **2c** of the adsorption are shown in Figure 1. Table 2 contains their calculated IR and energy characteristics together with the C-H frequencies assigned to the physically adsorbed methanol on the terminal OH group by Borello et al. [8,9]. As it follows from Table 2, form **2c** gives the best agreement to the observed bands at 2950 and 2845 cm^{-1}. Moreover, the calculations show the energy preference for this form with respect to **2a** and **2b**.

The increase of the C-H frequencies in form **2a** and the decrease in form **2b** with respect to the free methanol [5] (ν'_{as}=3005, ν''_{as}=2961 and ν_s=2847 cm^{-1}) agree with the above-mentioned Chung-Bennett dependence (cf. Figure 1 and Table 2). Considering this dependence, the experimental fact, that the C-H frequencies of the physically adsorbed methanol [9] at 3000, 2950, and 2845 cm^{-1} are very close to those of the free methanol, is a further important argument supporting **2c** as the main physically adsorbed form. The interaction of methanol with the surface is the strongest for this form, the perturbation of the C-H bonds is the smallest, because the decrease of electron density on CH_3 group, due to the $O_1 \cdot\cdot H_1$ hydrogen bond formation (Figure 1), is compensated by the increase of electron density on this group due to the simultaneous $O_2 \cdot\cdot H_2$ hydrogen bond formation. A similar phenomenon takes place also in structure **1b**.

Table 2: Adsorption Energies (kJ/mol) and C-H Frequencies (cm^{-1})

characteristic	calcd			obsd [9]
	2a	2b	2c	
adsorption energy	57	35	77	
ν'	3023	2977	2994	(3000)[a]
ν''_{as}	2977	2895	2949	2950
ν_s	2869	2816	2851	2845

[a] Only the asymmetric stretching band at 2950 cm^{-1} is discussed in Ref [9], although a shoulder at 3000 cm^{-1} is clearly seen in IR spectra of physically adsorbed methanol (see Figures 7,8,9 in Ref [9]).

As the 3-21G basis set overestimates hydrogen-bonding effects [11], the adsorption energies for structures **2a**, **2b**, **2c** are recalculated using the 6-31G* basis set. The results of these calculations (E_{ad}=30, 16 and 39 kJ/mol, respectively) confirm that form **2c** is the most stable.

Chemical Adsorption on Si-OH Group and Strained Siloxane Bridge

Besides the $\rangle$Si-O(CH_3)$\cdot\cdot$Al$\langle$ group, two other chemisorbed forms of methanol should be expected in zeolites, as being already accepted for silica [8,9]: the first is believed to take place through esterification of surface hydroxyls

$$\equiv Si\text{-}OH + CH_3OH \longrightarrow \equiv Si\text{-}OCH_3 + H_2O \qquad (1)$$

and the second is considered to be formed by methanol dissociation on the "strained" $\equiv$Si-O-Si$\equiv$ bridges

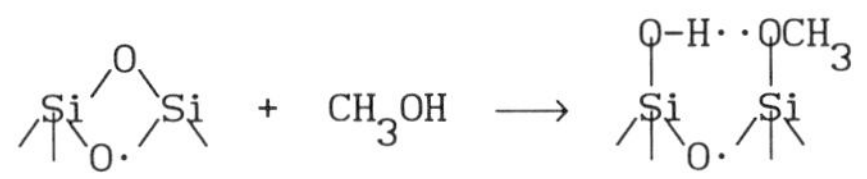

(2)

The corresponding molecular models **3a** and **3b** are presented in Figure 1. Model **3b** represents the methoxy group formed by reaction (2), with the "strained" fragment $\equiv$Si-O-Si$\equiv$ belonging to the siloxane bridge 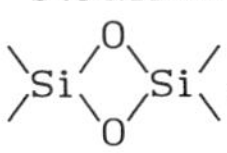. Their calculated adsorption energies and vibrational frequencies are collected in Table 3. According to the Chung-Bennett dependence, the hydrogen bonding in form **3b** increases the C-H vibrational frequencies with respect to **3a**. However, as the frequency differences are only about 10-15 cm^{-1} (Table 3) and the C-H stretching bands of methoxy groups on silica are rather broad (see for example Ref [5]), forms **3a** and **3b** are very likely not to be resolved separately if their surface densities are comparable. This may be the reason why these forms have not been separately identified by IR methods, although their existence has been proposed by many authors [8,9]. However, a change of the relative surface density of these forms might lead to shifts of the maxima of their overlapping C-H bands of about 10 cm.$^{-1}$ This hypothesis could explain the presence of the shoulder at 3010 cm^{-1} in the spectra of methanol chemisorbed on silica samples outgassed

Table 3: Adsorption Energies (kJ/mol) and C-H Frequencies (cm)

characteristic	calcd		obsd [5]
	3a[a]	3b	
adsorption energy	3	251	
ν'_{as}	3000	3010	3000
ν''_{as}	2958	2976	2958
ν_s	2858	2869	2858

[a] The calculated frequencies are equal to the observed ones, because **3a** was used as reference to define the scale coefficients.

at 650° C [8], when structures of **3b** type should be predominantly formed [8].

Frequensy/Intensity Relationship

According to the second Gutmann's rule, an intramolecular electron density shift decreasing the ionicity of a bond increases the bond strength . Therefore, the Chung-Bennett dependence as well as the results of the calculations (cf. the electron density shifts in Figure 1 with the above-mentioned shifts of the C-H frequencies) evidence the C^{+}-H^{-} polarity of C-H bonds of the surface methoxy groups. As the intensity of a molecular vibration, **I**, is proportional to the square of the derivative of the molecular dipole moment, **m**, to the normal coordinate of the vibration, **r** [15],

$$\mathbf{I} \sim (\partial \mathbf{m}/\partial \mathbf{r})^2 \qquad (3)$$

also the intensities of C-H stretching vibrations are expected to correlate with the ionicities of the C-H bonds: the larger are the ionicities, the higher must be the intensities. As shown in Figure 2, according to these ionicity/strength and ionicity/intensity relationships, the increase of the C-H stretching frequencies is accompanied by a decrease of the corresponding intensities.

The ionicities of C-H bonds being very small (due to the closeness of electronegativities of H and C atoms), a small change of charge distribution on C-H bonds can lead to a considerable relative change of the ionicities, causing a dramatic change of the relative infrared intensities. For example, the calculated intensity of the $\nu'_{as}(CH_3)$ vibration increases by 123 times (!) on going from form 1a to form 2c. This feature might complicate the experimental discovery of C-H stretching bands of surface methoxy groups.

CONCLUSION

The formation of the $\rangle Si$-$O(CH_3)\cdot\cdot Al\langle$ structure in zeolites, suggested by the assignment of the 2977 and 2871 cm^{-1} frequencies to this structure [5-7], is in considerable disagreement with the predicted vibrational frequencies. The calculations show that these frequencies belong to strongly physically adsorbed methanol, stabilized on the bridging OH group by two hydrogen bonds. Also the elementary act of such bonding is found energetically favoured by far in comparison with that of the $\rangle Si-O(CH_3)\cdot\cdot Al\langle$ group formation.

On the basis of the present calculations, the observed C-H stretching frequencies of physically adsorbed methanol on the terminal $\rangle Si-OH$ group are explained by the methanol stabilization on the group by two hydrogen bonds, with the methanol acting both as a proton acceptor and as a proton donor, simultaneously. The energy preference for this form in comparison with the other ones, is supported by the calculations with the 6-31G* basis set.

The calculations show that the C-H frequencies of the simple $\rangle Si-O-CH_3$ group are lower by only about 10-15 cm^{-1} than those of the hydrogen bonded methoxy group, formed through methanol dissociation on the siloxane bridge. Therefore it is very likely, that these chemisorbed forms cannot be resolved separately in IR spectra, but a change of their relative surface densities could lead to a shift of the maxima of the overlapping bands within about 10 cm^{-1}.

The dependences of the C-H vibrational frequencies and of their integrated intensities on the electron density of the CH_3 group are found in good agreement with Gutmann's rules, supporting a C^+-H^- polarity of the C-H bonds.

REFERENCES

1. Sauer, J. and Bleiber, A. Catalysis Today, **3**, 485(1988).
2. Beran, S. J. Mol. Catal. **26**, 31(1984).
3. Ugliengo, P., Saunders, V.R., Garrone, E. J.Phys.Chem. **94**, 2260 (1990)
4. Chang, C.D. Catal.Rev.-Sci.Eng. **25**, 1(1983).
5. Forester, T.R. and Howe, R.F. J.Amer.Chem.Soc. **109**, 5076(1987).
6. Kubelkova, L., Novakova, J., Nedomova, K. J.Catal. **124**, 441(1990).
7. Salvador, P. and Kladnig, W. J.Chem.Soc. Faraday Trans. 1, **77**, 2209 (1981).
8. Borello, E., Zecchina, A., Morterra, C. J.Phys.Chem. **71**, 2938(1967).
9. Borello, E., Zecchina, A., Morterra, C., Ghiotti, G. J.Phys.Chem. **71**, 2945 (1967).
10. Pelmenschikov, A., Morosi, G., Gamba, A. J.Phys.Chem. **95**, 10037 (1991).
11. Hehre, W.J., Radom, L., Schleyer, P.v.R., Pople, J.A.; Ab initio Molecular Orbital Theory; Wiley, New York, 1986.
12. Morrow, B.A. J.Chem.Soc.,Faraday Trans. 1, **70**, 1527 (1974).
13. Chung, J.S. and Bennett, C.O. J Catal. **92**, 173 (1985).
14. Luz, Z and Vega, A.J. J Phys.Chem. **91**, 374 (1987).
15. See, for example: Overend, J. In "Infrared Spectroscopy and Molecular Structure"; Davies, M., ed.; Elsevier: Amsterdam, 1963.

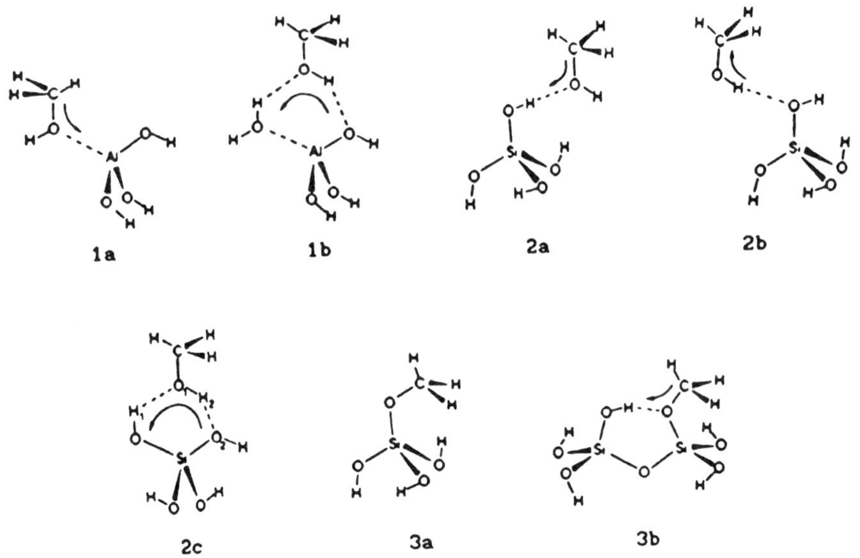

Figure 1. Molecular models of methoxy species. Arrows indicate directions of electron charge transfer to explain shifts of C-H frequencies by Chung-Bennett dependence and second Gutmann's rule.

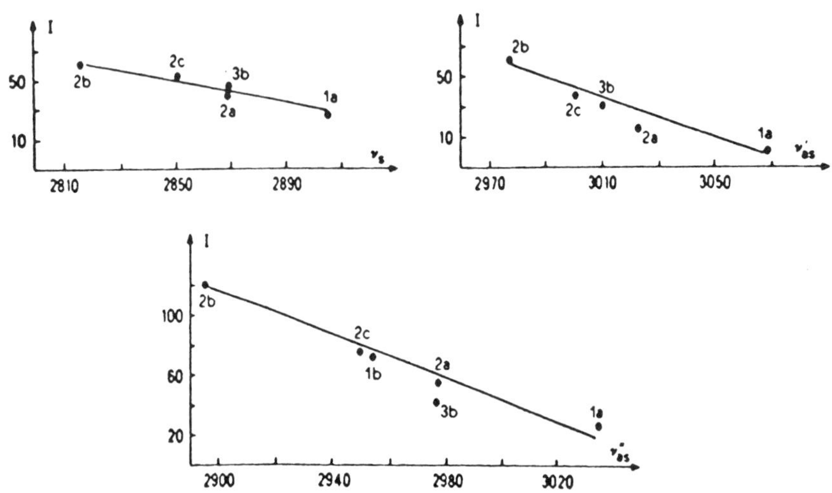

Figure 2. Infrared intensities, I (km/mol), against C-H stretching frequencies, ν'_{as}, ν''_{as} and ν_s (cm^{-1}).

THE INFLUENCE OF QUADRUPOLAR EFFECT ON HIGH-RESOLUTION ^{27}Al DOUBLE ROTATION (DOR) NMR OF ALUMINOPHOSPHATE MOLECULAR SIEVES

P.J. GROBET, H. GEERTS, J.A. MARTENS and P.A. JACOBS

Centrum voor Oppervlaktechemie en Katalyse, K.U. Leuven,
Kardinaal Mercierlaan 92, B-3001 Heverlee, Belgium.

ABSTRACT

^{27}Al DOR NMR at magnetic fields of 4.7, 8.5, 9.4 and 11.7 T is performed on a sample of calcined and rehydrated $AlPO_4$-11. The application of double-rotation (DOR) NMR in magnetic fields of medium strength proves to be particularly advantageous to enhance the dispersion of the observed shifts of ^{27}Al nuclei, located at different crystallographic sites. The spectra allow to determine umambiguously the number of crystallographic Al sites and their occupancy, and to evaluate the symmetry of the aluminium environments and the tendency to aluminium dihydrate formation. True isotropic chemical shifts for the aluminium sites in $AlPO_4$-11 are obtained by extrapolation to infinite field.

INTRODUCTION

Quadrupolar nuclei with spin systems $I > 1/2$ in solids interact not only with magnetic fields but also with electric field gradients. The combination of these interactions produces higher order, orientation-dependent NMR spectral broadening that can be reduced but not completely removed by spinning of the sample under the magic angle (MAS NMR). Spectral broadening due to anisotropic second-order quadrupolar frequency shifts has been a serious obstacle to high resolution ^{27}Al MAS NMR of aluminophospate molecular sieves [1]. With the new technique of double-rotation (DOR), involving sample spinning around two magic angles, the anisotropic line broadening can be averaged to zero [2,3]. The observed centre of mass of the ^{27}Al DOR NMR powder pattern is shifted from the zeroth-order Larmor frequency by an amount (in ppm):

$$\delta_{obs} = \delta_{iso}{}^{CS} + \delta_{iso}{}^{Q} \qquad (1)$$

where $\delta_{iso}{}^{CS}$ and $\delta_{iso}{}^{Q}$ are the isotropic chemical shift and the isotropic second-order quadrupolar shift, respectively. The experimentally observed shift δ_{obs} contains a term $\delta_{iso}{}^{Q}$ which depends on the specific quadrupolar coupling of an Al site and which is inversely proportional to the square of the magnetic field strength, B^{-2} [5]. From Eqn. 1 it follows that ^{27}Al DOR NMR at more than one field (using more than one spectrometer) allows one to determine the

From the Proceedings of the 9th International Zeolite Conference, Montreal 1992, Eds. R. von Ballmoos et al.,

true isotropic chemical shift for the Al sites [4]. The slope of the relation of δ_{obs} versus B^{-2} reflects the symmetry around a crystallographic Al site, e.g. the dispersion of Al-O-P angles [4]. Furthermore, the use of more than one field allows one to determine the number of crystallographically different Al sites unambiguously, since NMR signals that occasionally overlap at one field eventually are better resolved at another field. Such occasional signal overlapping was encountered with VPI-5 [4]. The ^{27}Al DOR NMR spectrum of VPI-5 at a field of 11.7 T shows only one tetrahedral aluminium signal. At fields of 9.4 T [4,6] and 4.7 T [4], two tetrahedral aluminium signals were nicely resolved, and based on their $\delta_{iso}{}^{Q}$ values assigned to the two tetrahedral Al sites present in the refined structure [4].

In this work we applied the ^{27}Al DOR NMR technique at four different magnetic fields to study the siting of aluminium in $AlPO_4$-11.

EXPERIMENTAL

The sample of $AlPO_4$-11 used in the present work was synthesized according to example 32 of the original patent [7]. The sample was calcined in air at a temperature of 823 K in order to remove the template and rehydrated at room temperature in a dessicator over a saturated NH_4Cl solution. The ^{27}Al NMR spectra were recorded on Bruker MSL instruments; at 4.7 and 11.7 T at Bruker Analytische Messtechnik (Reinstetten, Germany), at 8.5 T at Berkely (California, U.S.A.), at 9.4 T in our laboratory. The inner rotor of the DOR probehead was spun at 5 kHz and the outer rotor at 800 Hz; rotation-synchronized data aquisition was used. ^{27}Al MAS NMR was performed at a field of 11.4 T, using a sample spinning frequency of 3 kHz.

RESULTS AND DISCUSSION

X-ray powder diffraction studies revealed the existence of reversible phase transitions in calcined $AlPO_4$-11 and SAPO-11 upon hydration-dehydration cycles [8-10]. For the hydrated form of SAPO-11, Rietveld refinement and ab initio atomic coordinate calculations assuming a regular 4-connected framework yielded a structure belonging to the acentric $Pna2_1$ space group, with 5 P and 5 Al sites per asymmetric unit [8]. Conventional ^{27}Al MAS NMR spectra of samples of calcined and rehydrated AlPO4-11 and SAPO-11 at fields of 4.7 T and 7 T and spinning frequencies of 3 to 5 kHz consist of one major peak that is skewed upfield [1,8,11]. At 9.4 T, Jahn et al. observed a complex ^{27}Al MAS NMR spectrum with three poorly resolved maxima in the tetrahedral region, together with a signal in the octahedral region [12].

The resolution that can be achieved with ^{27}Al MAS NMR at a field of 11.4 T and a sample spinning frequency of 3 kHz is illustrated in Fig.1. The spectrum of Fig.1 shows three broad lines, at +38, +28 and -20 ppm, respectively. The lines with a positive shift are ascribed to tetrahedral Al sites, the one with the negative shift to octahedrally coordinated aluminium. This octahedral signal represents aluminium atoms coordinated to four framework oxygens and two additional oxygens of chemisorbed water molecules (aluminium dihydrate sites [13]). The improvement in spectral resolution due to the application of the DOR technique is shown in Fig.1b. The linewith of the tetrahedral signals is reduced from ca. 1500 Hz to 200 Hz in the DOR compared to MAS experiment (Fig.1).

Barrie et al. recorded ^{27}Al DOR NMR and MAS NMR spectra of a calcined and rehydrated $AlPO_4$-11 sample at a field of 11.7 T [14]. They observed three signals of tetrahedral aluminium with relative intensity ratios close to 1:3:1, in agreement with the presence of five aluminium sites in tetrahedral coordination. Besides these tetrahedral signals, a signal in the octahedral range represented one fifth of the total signal intensity. From the tetrahedral signal intensities it was concluded that the aluminiums in the five crystallographic sites were transformed to an equal extent into aluminium dihydrate sites [14].

The signals of six aluminium species are nicely resolved in a ^{27}Al DOR NMR spectrum recorded at a field of 9.4 T (Fig.2). Six lines with associated side bands are observed at +43.6, +40.1, +39.2, +37.8, +27.6 and -20.0 ppm, respectively. The presence of 5 tetrahedral Al sites in calcined-hydrated $AlPO_4$-11 is further confirmed by the ^{27}Al DOR NMR spectrum at a magnetic field of 4.7 T (Fig. 2). Even at a field of only 4.7 T, the resolution is good, except on the octahedral site (Fig. 2). Compared to the result at 9.4 T, the loss of signal dispersion at 11.7 T (Fig. 1) is remarkable. At 4.7 T, the signals are broad, but the best dispersion of the Al_3 and Al_4 signals is obtained at this field. At 11.7 T, the octahedral aluminium line at -15 ppm is ca. 200 Hz broader than the tetrahedral lines, pointing to a residual quadrupolar interaction on that site.

The deconvolution of the ^{27}Al DOR NMR spectra shown in Figs. 1 and 2 reveals the presence of 5 tetrahedral lines and one octahedral line, all of similar intensity if the spinning sidebands are taken into account. The magnetic field dependence of the observed chemical shift for the different Al sites in calcined-hydrated $AlPO_4$-11 is given in Fig.3. For the six signals, the observed chemical shift correlates linearly with B^{-2}, as expected from Eqn.1. These linear relationships support the deconvolutions made and the presence of 5 distinct tetrahedral Al sites and 1 octahedral Al site. The tetrahedral aluminium signals can be subdivised into two groups. For the Al_1, Al_2 and Al_3 signals, the observed shift varies only little with B^{-2}, indicating that these aluminiums are in a highly symmetrical environment; for the Al_4 and Al_5 signals, the observed shift is much more dependent on B^{-2}, reflecting important asymmetry and differences in the values of the Al-O-P angles around these sites. A so clear subdivision of the Al sites is not found in the available refined structure [8]. Relative signal intensities, observed chemical shifts and isotropic chemical shifts (extrapolated) are given in Table 1. At every field applied, the relative intensities of the 6 Al lines are nearly equal. This fact, together with the deconvolution requiring 5 tetrahedral sites and 1 octahedral site for all DOR spectra, implies that all five crystallographic aluminium sites have reacted with water molecules to an equal extent. Thus the octahedrally coordinated aluminium in calcined and rehydrated $AlPO_4$-11 is not located on a single crystallographic site, as was the case in the hydrated VPI-5 structure [4,13]. It is important to notice that the octahedral Al line narrows appreciably under DOR conditions and a field of 11.4 T (Fig.1); the presence of crystallographically different octahedral aluminium species does not give rise to a splitting of the octahedral signal.

CONCLUSIONS

The number and occupancy of the different crystallographic sites for aluminium in calcined, rehydrated $AlPO_4$-11 were determined with ^{27}Al DOR NMR at fields of 4.7 T, 8.5 T, 9.4 T and 11.4 T. The presence of five tetrahedral aluminium signals and one octahedral signal all of equal intensity are in agreement with the refined $AlPO_4$-11 structure [8], provided there exists no discrimination among the aluminiums at the different crystallographic sites for coordination with water molecules. From the slopes of the relations of the observed chemical shifts with B^{-2}, it is concluded that three tetrahedral Al sites have a highly symmetrical environment, while two tetrahedral Al sites are less symmetrical. Isotropic chemical shifts for the tetrahedral and octahedral aluminium sites are determined by extrapolation.

ACKNOWLEDGMENTS

PJG and JAM aknowledge the Flemish National Fund for Scientific Research for research positions as Senior Research Associate and Research Associate, respectively. This work was sponsored by the Belgian Government within the frame of an IUAP-II grant and by the Belgian Inter-University Institute for Nuclear Sciences (Contract No. 4.0010.89). The authors acknowledge the Bruker Company for the use of the Bruker 200 MSL and 500 MSL NMR facilities. The ^{27}Al DOR NMR spectrum at 8.5 T was kindly provided by Dr. Y. Wu (Berkeley, California).

REFERENCES

1. Blackwell, C.S. and Patton, R.L., J. Phys. Chem. **88**, 6135 (1984).
2. Llor, A. and Virlet, J., J. Chem. Phys. Lett. **152**, 248 (1988).
3. Samoson, A., Lippmaa, E. and Pines, A., Mol. Phys. **65**, 1013 (1988).

4. Grobet, P.J., Samoson, A., Geerts, H., Martens, J.A. and Jacobs, P.A., J. Phys. Chem. **95** , 9620 (1991).
5. Engelhardt, G. and Michel, D., High-Resolution Solid State NMR of Silicates and Zeolites, John Wiley & Sons, Chichester, p. 61 (1987).
6. Wu, Y., Chmelka, B.F., Pines, A., Davis, M.E., Grobet, P.J. and Jacobs,P.A., Nature **346**, 550 (1990).
7. Wilson, S.T., Lok, B.M., Flanigen, E.M., U.S. Patent 4,310,440 (1982).
8. Khouzami, R., Coudurier, C., Lefebvre, F., Védrine, J.C. and Mentzen, B., Zeolites **10**, 183 (1990).
9. Mentzen, B.F., Védrine, J.C., Khouzami, R., and Coudurier, C., Compt. Rend. Acad. Sci. (Paris) **305**, II, 263 (1987).
10. Goepper, M., Guth, F., Delmotte, L., Guth, J.L., and Kessler, H., in Zeolites: Facts, Figures, Future, ed. P.A. Jacobs and R.A. van Santen, Stud. Surf. Sci. Catal. No. 49B, 857 (1989).
11. Tapp, N.J., Milestone, N.B., Bowden, M.E. and Meinhold, R.H., Zeolites **10**, 105 (1990).
12. Jahn, E., Müller, D. and Becker, K., Zeolites 10, 151 (1990).
13. Grobet, P.J., Martens, J.A., Balakrishnan, I., Mertens, M. and Jacobs, P.A., Appl. Catal. **56**, L21 (1989).
14. Barrie, P.J., Smith, M.E. and Klinowski, J., Chem. Phys. Lett., **180**, 6 (1991).

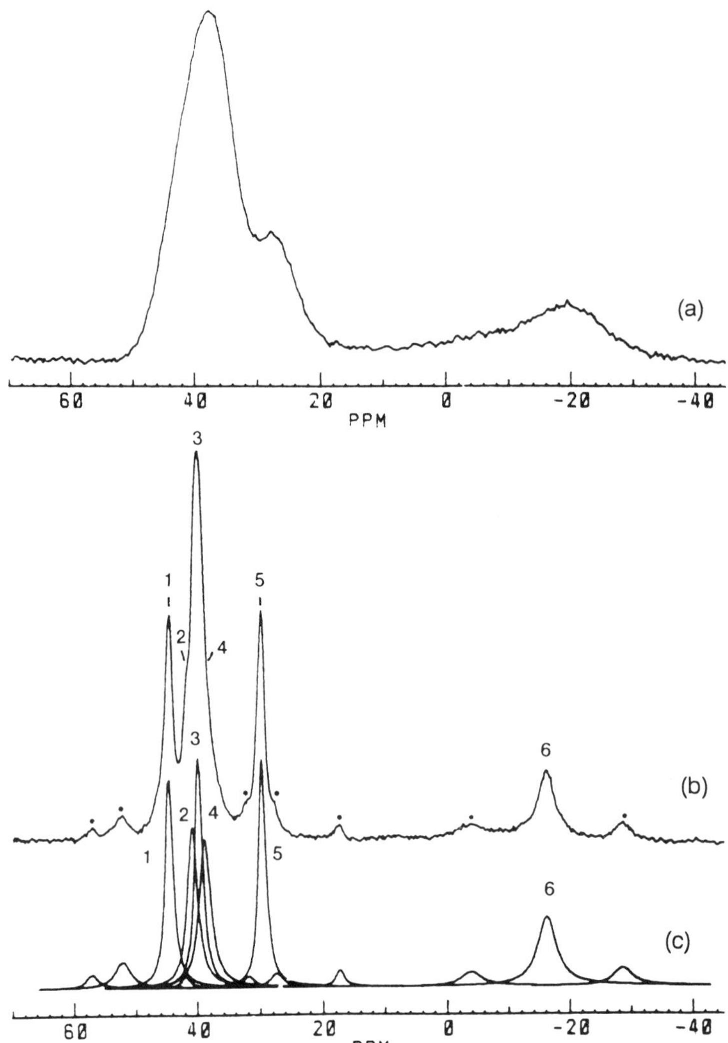

Figure 1. ^{27}Al solid state NMR of calcined, hydrated $AlPO_4$-11 in a magnetic field of 11.7 T; (a) and (b) are the MAS (using a rotation of 3 kHz) and DOR spectra, respectively; (c) is the deconvolution of (b).

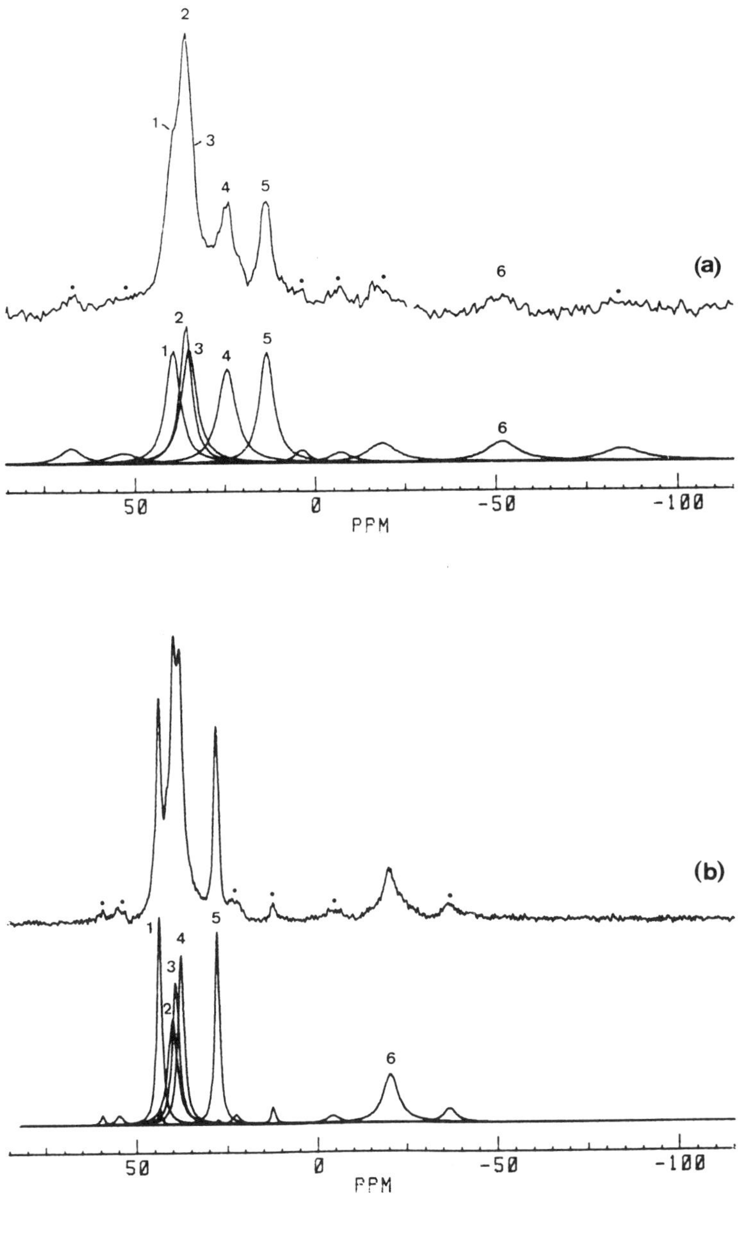

Figure 2. Experimental and deconvoluted ^{27}Al DOR NMR spectra of calcined and rehydrated $AlPO_4$-11 at magnetic fields of 4.7 T (a) and 9.4 T (b).

Table 1 : Summary of the ^{27}Al DOR NMR data of calcined-hydrated $AlPO_4$-11

Field (T)	δ_{obs}(ppm)					
	Al_1	Al_2	Al_3	Al_4	Al_5	Al_6
4.7	39.2 (16)	35.7 (16)	34.8 (17)	24.4 (17)	13.4 (16)	-51.3 (18)[a]
8.5	43.3	39.6	38.5	36.1	26.8	-22.7
9.4	43.6 (16)	40.1 (16)	39.2 (18)	37.8 (17)	27.6 (15)	-20.0 (18)
11.7	44.6 (17)	40.9 (14)	39.8 (15)	38.9 (19)	29.7 (17)	-15.8 (18)
	δ_{iso}^{CS}(ppm)					
	45.4	41.7	40.6	41.9	32.9	-8.7[c]

a, relative intensity (%) of signals (includina sidebands) in parentheses;

b, accuracy of 0.2 ppm.

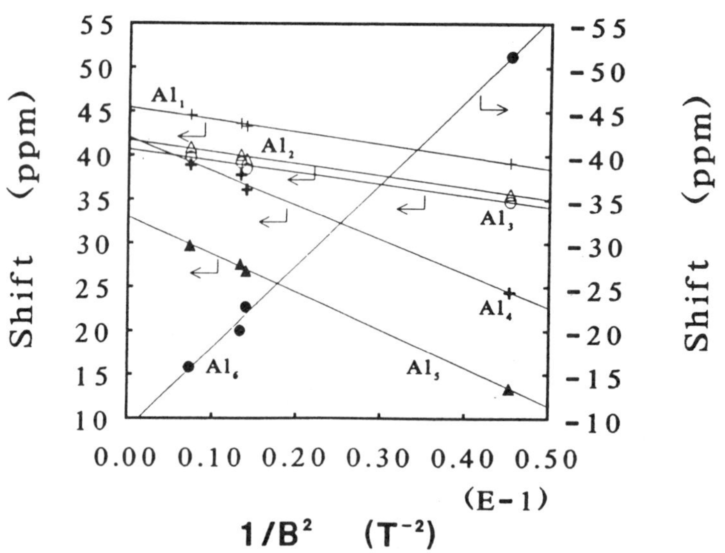

Figure 3. The magnetic field dependence of the observed shifts of the ^{27}Al DOR NMR lines of calcined-hydrated $AlPO_4$-11.

NMR STUDIES OF THE WATER MOTION IN $AlPO_4$-5, $CoAlPO_4$-5 AND VPI-5

D. GOLDFARB*, I. KUSTANOVICH* and J.O. PEREZ≠

*Chemical Physics Department, Weizmann Institute of Science, 76 100 Rehovot, Israel,
≠Department of Chemistry, Texas A&M University, College Station, TX 77843-3255

ABSTRACT

The dynamic state of the water in $AlPO_4$-5 and VPI-5 was studied by ^2H NMR of samples rehydrated with D_2O. The low temperature spectra (-155 - -30°C) of both materials are rather similar indicating the existence of two types of water molecules, bound molecules exhibiting a broad powder pattern and physisorbed molecules showing a singlet. The bound molecules are assigned to those coordinated to framework Al. Above room temperatures the two distinct signals are no longer evident and the lineshape changes suggest that exchange between the two takes place. In $AlPO_4$-5 the lineshape changes can be explained by a simple two site exchange mechanism whereas in VPI-5 this exchange is associated with an additional three-fold site jump indicating ordering of the water within the channels. The water interaction with Co(II) in $CoAlPO_4$-5 samples with different Co(II) contents was studied via proton paramagnetic shifts. For samples with Co/P > 3 atom % paramagnetic shifts of the water protons were observed, indicating that part of the water molecules binds to the Co(II).

INTRODUCTION

$AlPO_4$-5 and VPI-5 are members of the aluminophosphate molecular sieves family. Both structures exhibit large unidimensional channels. VPI-5 has the widest channels known to date with openings of 12.1Å circumscribed by 18 T atoms [1] whereas the channels of $AlPO_4$-5 are 7.3Å wide formed by 12 T atoms [2]. The aluminophosphate molecular sieves have a neutral framework and hence lack Brönsted acidity. Possible framework substitution with Si, Co, Mg and other elements [3] generates Brönsted acidity and thus enhances their catalytic potential.

An interesting property of these materials is the flexibility of their framework which allows part of the framework Al to coordinate water molecules and assume octahedral symmetry [4-5]. Although the octahedral Al have been associated with the coordination of water molecules [4-5], the dynamic state of the water molecules within the channels of these materials has not been investigated. This is particularly interesting in the case of VPI-5 where the water was found to be ordered within the channels [6]. This order should be manifested in unique dynamic properties as compared to the other members of the family. When paramagnetic cations are present, such as in the case of $CoAlPO_4$-5, the interaction of the water with the paramagnetic cations should be manifested in paramagnetic shifts of the proton resonances which, in turn, can provide information

From the Proceedings of the 9th International Zeolite Conference, Montreal 1992, Eds. R. von Ballmoos et al.,

regarding the water dynamics. In the present work we report on a multinuclear NMR study of the motional characteristics of the water within the channels of VPI-5, $AlPO_4$-5 and $CoAlPO_4$-5.

EXPERIMENTAL

$AlPO_4$-5, $CoAlPO_4$-5 and VPI-5 were synthesized according to procedures described in the literature [7-9]. Calcination of $AlPO_4$-5 and $CoAlPO_4$-5 was performed at 500-550°C and the calcined material was left in open vials for several days to adsorb water. Such samples are referred to as fully hydrated. Dehydration of VPI-5 was performed by room evacuation (24h) and calcined $AlPO_4$-5 and $CoAlPO_4$-5 were dehydrated under vacuum at room temperature and at 400°C. Dehydrated samples were exposed, at room temperature, to D_2O vapors for 2-3 hours and then sealed at 77K. The ^{31}P spectrum of VPI-5 after rehydration with D_2O was verified to ensure that no transition to $AlPO_4$-8 took place.

NMR experiments were performed on a Bruker CXP spectrometer operating at 300 MHz for ^1H, 121.44 MHz for ^{31}P, 78.17 MHz for ^{27}Al and 46.07 MHz for ^2H. The spinning rates in the MAS experiments were 3-4 KHz. Care was taken to employ 15° pulses in the Al measurements to ensure quantitative comparison of signal intensities [10]. The ^2H spectra were recorded using the quadrupole echo sequence with τ intervals of 20-30 μs. The ^{31}P T_2 relaxation rates measurements were obtained using the Carr-Purcell, Meiboom-Gill (CP-MG) echo train, with the interval between consecutive π pulses, 2τ, synchronized with the rotor frequency such that $\tau=2\pi/\omega_r$. The ^2H measurements were performed on sealed samples. Samples for the MAS experiments were kept in the sealed tubes and were transferred into tightly closed rotors in a dry nitrogen box immediately before the measurements. The ^{31}P and ^{27}Al chemical shifts are given relative to phosphoric acid and a solution of $Al(NO_3)_3$, respectively.

RESULTS AND DISCUSSION

Water dynamics in $AlPO_4$-5 and VPI-5

Representative low temperature ^2H spectra of VPI-5/D_2O and $AlPO_4$-5/D_2O are shown in Fig. 1. The spectra are rather similar and indicate the existence of two different types of signals, a singlet superimposed on a broad powder pattern with edges at ±75 MHz. In both samples the relative intensity of the powder signal increased with decreasing temperature and at −155°C the spectra approached the rigid limit powder pattern indicating the freezing of all types of water molecules. We assign the singlet to physisorbed water molecules reorienting rather isotropically within the channels and the powder pattern to bound water molecules, most probably those coordinated to framework Al. Raising the temperature above room temperature revealed interesting lineshape changes in both samples as shown in Fig. 2. The powder signal disappeared, the singlet broadened and eventually developed into a uniaxial powder pattern. The splitting between the perpendicular singularities of the powder pattern, ω_Q, is 6.1 KHz for VPI-5 at 64°C whereas in $AlPO_4$-5 it is 4.3 KHz at 80°C.

In principle one could attribute the high temperature limit spectrum in both materials to anisotropy in the motion of the "free" molecules exerted by a uniaxial potential representing the axial shape of the channels. If this were, however the case, one would expect more averaging in the case of VPI-5, yielding a final spectrum with a smaller ω_Q than in $AlPO_4$-5, since in VPI-5 channels are wider. In both materials, the appearance of the powder characteristics at the high temperature range, namely the development of parallel and perpendicular singularities, is associated with the disappearance of the broad powder pattern assigned to the bound molecules at lower temperatures. Accordingly, we explain the lineshape evolution in terms of the following simplified two sites exchange model:

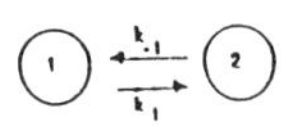

Site 2 corresponds to "free" water molecules, in which the quadrupole coupling constant (e^2qQ/h) is averaged to 0, and site 1 to bound water molecules with $e^2qQ/h>0$. Simulations based on this model are shown in Fig. 2b [11]. In this particular case the bound molecules were taken as completely rigid, namely $e^2qQ/h=210$ KHz. However, a temperature dependence of the relative population, given by P, had to be included in order to reproduce the experimental lineshapes. The exchange rate, $1/\tau=k_1P$, for each temperature is indicated on the figure. The quadrupole splitting of the fast limit spectrum (80°C) is given by:

$$\omega_Q = 3/4 \cdot (e^2qQ/h) \cdot P .$$

Hence, it is possible to obtain similar calculated spectra with an averaged e^2qQ/h due to fast local motion of the bound molecules and increase P accordingly. The latter is correct only if the local motion is such that it results in a reduced axial tensor at this temperature range.

In general, the lineshape changes observed in VPI-5/D_2O (Fig. 2c) look similar to those of $AlPO_4$-5, a close observation, however, discloses significant differences manifested mainly in the evolution of the parallel singularities at ±6 KHz. These lineshape changes are very similar to those exhibited by a three-fold site jump with a jump angle close to the tetrahedral angle. The jump angle is the angle between the direction of the principal axis of the quadrupole tensor of the deuterium nucleus and the C_3 rotation axis. To simulate the VPI-5 spectra, the following exchange mechanism which combines a three-fold jump and exchange between bound and free molecules is proposed:

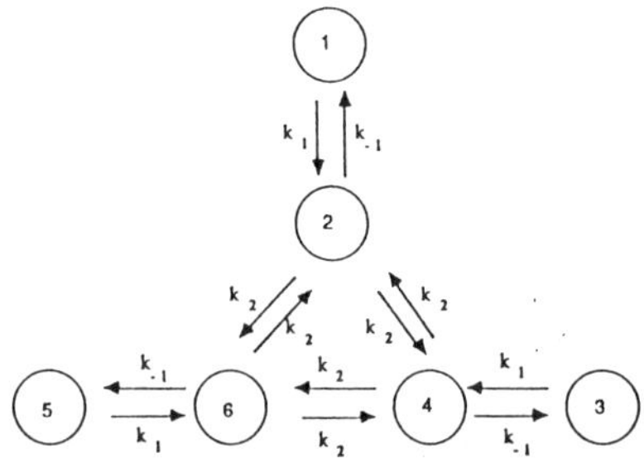

In this case sites 1,3,5 represent bound molecules and sites 2,4,6 free molecules. The populations of each of the bound sites and of the free sites are P and (1-3P)/3, respectively, the exchange rates are given by: $1/\tau_1=k_1P$ and $1/\tau_2=k_2$. The simulated spectra are shown in Fig. 2d. Note, that unlike the case of $AlPO_4$-5, the population was taken as temperature independent within the range –29 - 64°C. The three bound sites were correlated by a jump angle of 71°. The bound water molecules were assumed to be completely rigid with e^2qQ/h=210 KHz. As in the case of $AlPO_4$-5, it is possible to obtain a similar lineshape evolution taking a larger population of the bound sites and a reduced averaged quadrupole coupling constant due to a fast local motion.

Comparison of the relative amounts of the bound and free water molecules obtained from the simulations with the relative amount of the octahedral Al and the total water capacity suggests that, indeed, in both materials a local motion of the bound molecule exists yielding a reduced averaged quadrupole coupling constant. Unfortunately in both cases the lineshape evolution throughout the low temperature range investigated is not sufficient to unambiguously determine the type of the local motion of the bound molecules. An additional complication is the gradual freezing of the "free molecules" at very low temperatures. The sharp edges at ±75 at low temperatures may imply a π flip about the Al-O axis [12]. This, however, has to change at higher temperature to produce a uniaxial averaged tensor which is necessary to reproduce the high temperature lineshapes observed. One possibility will be a free rotation about the Al-0 axis as opposed to a hindered π flip at low temperatures.

The exchange mechanism found in VPI-5 indicates that the water molecules do form some structure within the channel which does not exist in $AlPO_4$-5. This mechanism has features which appear to be consistent with the water structure proposed by McCusker et al. [6]. The bound molecules correspond to those coordinated to framework Al and the "free" molecules are the hydrogen bonded molecules forming the triple helix structure down the channel. Exchange within the pairs corresponds to exchange between the two types of molecules whereas the three-fold jump is attributed to jumps between the three chains forming the helix. It should, however, be noted that although the exchange mechanism obtained by the NMR results is consistent with the above model it does not restrict the location of either the bound or the free water molecules to the large channel only, the free water molecules may include water from all voids undergoing fast exchange. The water ordering observed in VPI-5 is also in agreement with adsorption of water vapor studies which suggest the presence of a highly ordered hydrogen bonded water network [13]. The lack of water ordering in $AlPO_4$-5 as compared to VPI-5 may suggest that while in VPI-5 the octahedral Al are ordered, namely, they occupy specific framework sites which are related by a three fold symmetry axis [6] no such correlation exists between the octahedral Al in $AlPO_4$-5.

Recent results obtained from variable temperature ^{31}P MAS NMR VPI-5 are also in agreement with our findings [14]. In this study the temperature dependence of the ^{31}P spectra was interpreted in terms of the existence of a hydrogen bonded structure which leads to framework distortions below 60°C whereas above 60°C the distortions are relaxed due to increased water mobility yielding a spectrum with only two peaks.

In the above discussion the water motion characteristics were interpreted in terms of exchange of the whole water molecules as opposed to D-D exchange which may lead to similar effects. We assume that in the case of the aluminophosphate studied, the D-D exchange should be much slower than the water mobility. This is justified since the binding of the water to the Al is rather weak, thus facilitating water exchange. Framework Al in $AlPO_4$-5 undergoes an easy and reversible reaction with water to form octahedral Al [15] and prolonged room temperature evacuation removes all octahedral Al in VPI-5 and $AlPO_4$-5 [12]. Moreover the materials do not contain the necessary acidic or basic environment to catalyze D-D exchange. The D-D exchange contribution can be experimentally tested by comparison with ^{17}O NMR of D_2O labeled water molecule.

Water Interaction with Co(II) in $CoAlPO_4$-5

Several samples of $CoAlPO_4$-5 with different Co contents were studied, their ^{31}P and ^{1}H spectra are shown in Fig. 3. The ^{31}P MAS NMR spectra show the same sideband characteristics as reported for $MnAlPO_4$-5 [16] but with a considerably narrower linewidth attributed to the much shorter longitudinal spin relaxation time of Co(II). The linewidth is comparable to ^{31}P in $AlPO_4$-5 and undergoes the same variation upon hydration and rehydration [12]. The position of the centerband is similar to that of $AlPO_4$-5 indicating no paramagnetic shift. The numerous sidebands are due to the dipolar interaction between the Co(II) and the ^{31}P nuclei [16]. The corresponding ^{1}H spectra show at the lower Co(II) contents three resolved peaks, assigned to water (~4 ppm) and to the amine methylene and methyl groups (2-3 ppm). Note that the sidebands are centered about the amine signals just as in the case of $AlPO_4$-5 [16]. When the amount of Co(II) in the sample increases the relative intensity of the water center band decreases, resolution is lost and a broad signal at ~ -110 ppm appears. Moreover, the intensity of the sidebands and their breadth increase since more ^{31}P nuclei are in the close vicinity of Co(II). Note that in the spectrum of 5% Co, the water centerband is hardly detectable. We assign the broad shifted peak to water protons with an isotropic paramagnetic shift indicating that at least part of the water molecules binds to the Co(II).

After calcination and rehydration, the ^{31}P spectrum shows only some linewidth increase whereas the ^{1}H spectrum undergoes significant changes (Fig. 4). The ^{1}H spectrum of $CoAlPO_4$-5 (0.5%) is essentially similar to that of $AlPO_4$-5 [16], showing the two types of water, free and bound, the latter responsible for the numerous sidebands [16]. In the 5% sample, the relatively intense sideband pattern observed in the as-synthesized sample disappeared, confirming its assignment to the template protons, and indicating that the amine is less mobile than the water. A total of three different signals can be identified: a narrow non shifted centerband with a wide sideband pattern, a broad line shifted from the center by 30 KHz and a broad non shifted centerband. The broad shifted line is assigned to water protons with a paramagnetic shift as in the as-synthesized samples, the peak however is broader and the shift somewhat larger. The narrow line, associated with the sideband pattern, is assigned to bound water molecules. Comparison with the ^{1}H spectrum of $CoAlPO_4$-5 (0.5%) suggests that it is the "free" water molecules which are mostly affected by the interaction with the paramagnetic Co(II), namely, part of them bind to the Co(II), yielding the broad shifted line. Slow exchange between these molecules and the other physisorbed molecules takes place causing broadening of the "free" water peak. Unfortunately, the ^{2}H spectrum of a similar sample after rehydration with D_2O was too noisy and reliable powder pattern lineshapes could not be obtained. It should be noted that at a Co(II) content of 5% the majority of the Co(II) is not incorporated into the framework [8], hence, what we observed is mainly water binding to Co(II) located either within the channels or on the external surface.

^{31}P T_2 and linewidth measurements of $CoAlPO_4$-5 (0.5%) gave results very close to those of $AlPO_4$-5 [16], and both the relaxation rates and the linewidths showed the following trend: hydrated; $\Delta\nu$ = 550 Hz, $1/T_2$ = 98 sec^{-1}, $\stackrel{\sim}{>}$ as-synthesized; $\Delta\nu$ = $1/T_2$ = 560 Hz, $1/T_2$ = 97 sec^{-1} > dehydrated; $\Delta\nu$ = 325 Hz, $1/T_2$ = 30 sec^{-1}. The same line narrowing upon dehydration was observed in $CoAlPO_4$-5 (5%) indicating, that unlike in $MnAlPO_4$-5 [16] the narrowing is due to the removal of the water, as in $AlPO_4$-5 and not due to an increase in the Co(II) spin lattice relaxation rates. The ^{31}P longitudinal relaxation time, T_1, however, decreased significantly with the increase of the Co(II) concentration.

CONCLUSIONS

The ^{2}H spectra of D_2O in $AlPO_4$-5 and VPI-5 reflect the special dynamic properties of the water molecules within the sieves' channels which were not observed in any of the closely related aluminosilicate zeolites. The unique dynamic behavior arises from exchange between water molecules bound to framework Al and water molecules reorienting isotropically within the channels. In VPI-5 a water structure with a three fold symmetry has been found. An analogous structure was not found in $AlPO_4$-5. In both as-synthesized and calcined $CoAlPO_4$-5 with Co/P > 1.5 atom % the Co(II) binds water molecules.

ACKNOWLEDGEMENT

This study was made possible by funds granted to D.G. through a fellowship program sponsored by the Charles H. Revson Foundation. We are most grateful to Prof. M.E. Davis for giving us samples of VPI-5 and for very industrious discussions. We also thank Prof. A. Clearfield for his cooperation regarding the $CoAlPO_4$-5 studies.

REFERENCES

1. Davis, M.E., Saldarriaga, C., Montes, C., Garces, J. and Crowder, C., Nature **331**, 698 (1988).
2. Bennet, J.M., Cohen, J.P., Flanigen, E.M., Pluth, J.J. and Smith, J.V., Am. Chem. Soc. Sym. Ser. **218**, 109 (1983).
3. Flanigen, E.M., Lok, B.M., Patton, R.L. and Wilson, S.T., in: **New Developments In Zeolite Science Technology**, Eds. Y. Murakami, A. Ijima and J.W. Ward, Kodansha, Tokyo, p. 103 (1986).
4. Meinhold, R.H. and Tapp, N.J., J. Chem. Soc. Chem. Commun. p. 219 (1990).
5. Blackwell, C.S. and Patton, R.L., J. Phys. Chem. **88**, 6135 (1984).
6. McCusker, L.B., Baerlocher, Ch., Jahn, E. and Bulow, M., Zeolites **11**, 308 (1991).
7. Lok, B.M., Messina, C.A., Patton, R.L., Gajek, R.T., Cannan, T.R. and Flanigen, E.M., U.S. Patent 4,440,871 (1984).
8. Shiralkar, V.P., Saldrriaga, C.H., Perez, J.O., Charfield, A., Chen. M., Anthony R.G. and Donohue, J.A., Zeolites **9**, 474 (1989).
9. VPI-5 was kindly given by M.E. Davis, synthesized as described in: Davis, M.E. and Young, D., Stud. Sur. Sci. Catal. **60**, 53 (1991).
10. Luz, Z. and Vega, A.J., J. Phys. Chem. **90**, 4903 (1986).
11. The ^{2}D lineshape simulations were done using well established theories described in: Mehring, M. in: **NMR-Basic Principles and Progress**, Eds. P. Diehl, E. Fluck, and R. Kosteld, Spring-Verlag, Berlin 1976; R.G. Griffin, Methods in Enzmol. **72**, 108 (1981).
12. Kustanovich, I. and Goldfarb, D., J. Phys. Chem. **95**, 8818 (1991).
13. Kenny, M.B., Sing, K.S.W. and Theocharis, C.R., J. Chem. Soc. Chem. Commun. 974 (1991).
14. van Braam Houckgeest, J.P., Kraushaar-Czarnetzki, B., Dogterom, R.J. and de Groot, A., J. Chem. Soc. Chem. Commun. 666 (1991).
15. Meinhold, R.H. and Tapp, N.J., Zeolites **11**, 401 (1991).
16. Goldfarb, D., Zeolites **9**, 509 (1989).

Figure 1. ^2H NMR spectra of $AlPO_4$-5 and VPI-5 adsorbed with D_2O recorded at -55°C and -78°C respectively.

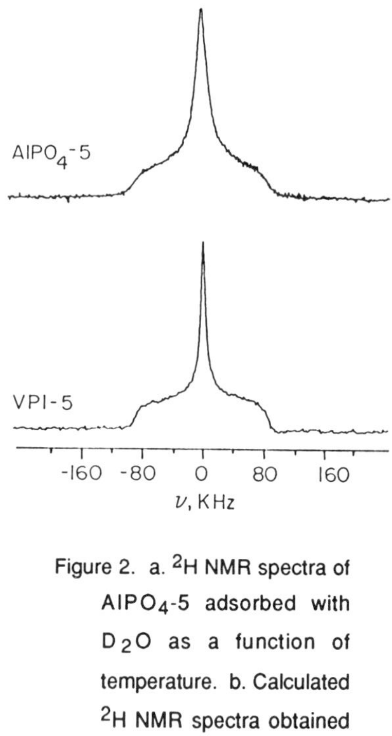

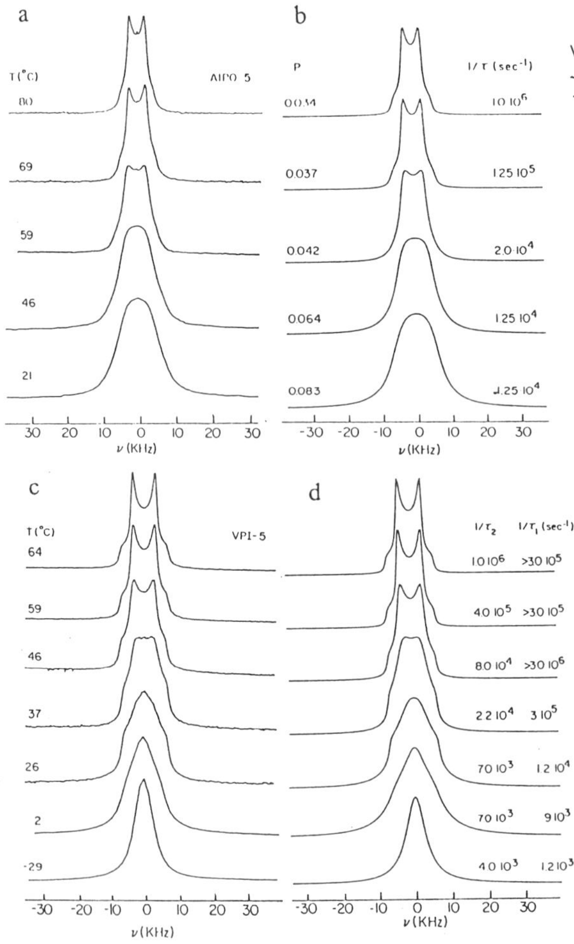

Figure 2. a. ^2H NMR spectra of $AlPO_4$-5 adsorbed with D_2O as a function of temperature. b. Calculated ^2H NMR spectra obtained for a model of two site exchange with $^2qQ/h$=210 KHz, T_{2b}=4 msec, T_{2f}=2 msec, P and $1/\tau$ are indicated on the figure. c. ^2H NMR spectra of VPI-5 adsorbed with D_2O as a function of temperature. d. Calculated ^2H NMR spectra for a 6 site exchange model. P=0.04, T_{2b}=4 msec, T_{2f}=2.8 msec and θ=71°. The subscripts b and f stand for bound and free.

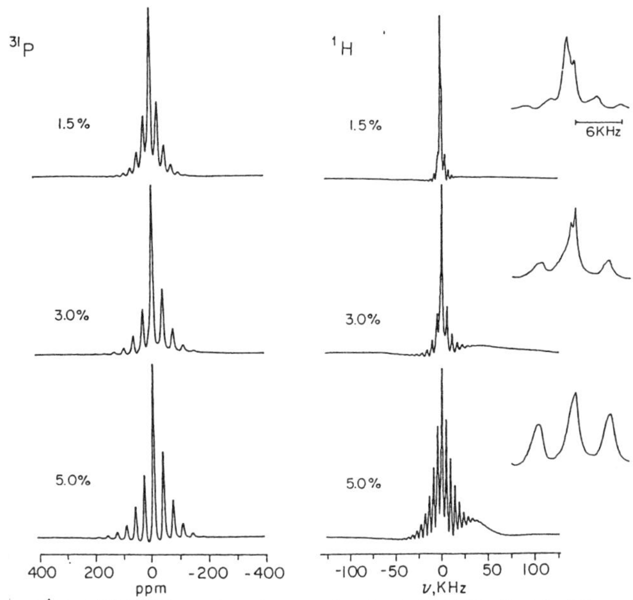

Figure 3. ^{31}P and ^{1}H MAS NMR spectra of as-synthesized CoAlPO$_4$-5 with different amounts of Co(II) (in terms of Co/P). The insert is a blow up of the centerband and the next two sidebands.

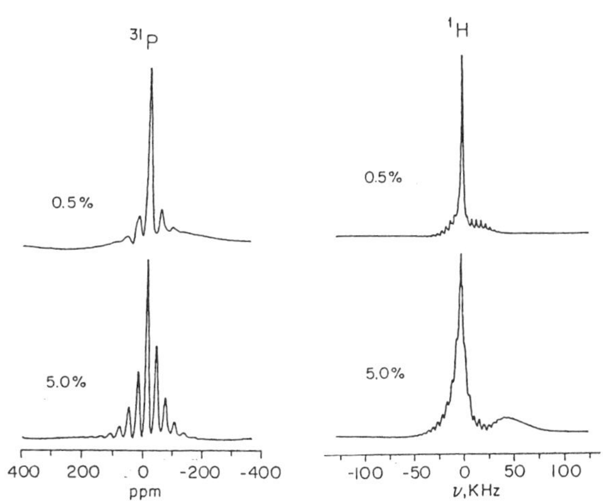

Figure 4. ^{31}P and ^{1}H MAS NMR spectra of calcined, rehydrated CoAlPO$_4$-5 with 0.5 and 5% Co(II).

SOLID STATE MAS NMR CHARACTERISATION OF THE TRANSFORMATION OF VPI-5 TO $ALPO_4$-8: PRESERVATION OF THE OCTAHEDRAL ALUMINIUM SITES?

D.E. Akporiaye and M. Stöcker

Centre for Industrial Research, P.O.Box 124 Blindern, N-0314 Oslo 3, Norway.

ABSTRACT

Parallel ^{31}P and ^{27}Al *in situ* VT MAS NMR studies have been carried out on hydrated-VPI-5 and hydrated-$AlPO_4$-8. The ^{31}P *in situ* MAS NMR results confirm the same behaviour of the low-field signal at -23.3 and -24.0 ppm respectively. This would appear to indicate the same octahedral aluminium environment for the 4/4 site assigned to the signal in the $AlPO_4$-8 structure as has been concluded for VPI-5. Further changes in the ^{31}P and ^{27}Al spectra of $AlPO_4$-8 at higher VT temperatures indicate a more complex situation for hydrated-$AlPO_4$-8, needing a reaccessment of the specific assignments to individual crystallographic sites and consideration of the number of octahedral aluminium sites.

INTRODUCTION

The original interest in novel applications of this new molecular sieve VPI-5 [1] has been somewhat mitigated by its tendancy to transform to another phase under particular conditions of thermal treatment [2-4]. The new phase, $AlPO_4$-8, reported in the original Union Carbide patent [5], has now been elucidated [6] and shows an obvious relationship to the VPI-5 structure, Figure 1. The specific conditions for the transformation process and the possibilities for stabilisation of the VPI-5 phase are still under investigation. Of particular importance in these studies is the use of solid state MAS NMR spectroscopy, which is proving to be even more sensitive to the subtleties of the transformation than powder X-ray diffraction. In our study we present further characterisation of the hydrated-$AlPO_4$-8 material with respect to the effect of *in situ* variable temperature (VT) treatment on the ^{31}P and ^{27}Al MAS NMR spectra.

EXPERIMENTAL

Synthesis of VPI-5:

From the Proceedings of the 9th International Zeolite Conference, Montreal 1992, Eds. R. von Ballmoos et al.,

VPI-5 was synthesised using dipropylamine [7] (DPA). The composition of the gel was: Al_2O_3 : P_2O_5 : DPA : 40 H_2O and the preparation procedure as follows. Alumina (Catapal B) was dispersed in most of the water, to which was added a solution of phosphoric acid. After mixing the gel thoroughly, the amine was added and the mixture blended. Thereafter the gel was transferred to a teflon lined autoclav and heated to 137°C for 24 hr. After recovery of the solid material, all samples were thoroughly washed with distilled water and dried under room temperature conditions. Samples were confirmed by powder X-ray diffraction (XRD) to be crystalline VPI-5.

Preparation of $AlPO_4$-8:

$AlPO_4$-8 was prepared from the synthesised VPI-5 material by calcination for 12 hr at 400°C in air. The absence of the former VPI-5 phase and the appearance of the new $AlPO_4$-8 phase was confirmed by XRD. Hydration of the calcined $AlPO_4$-8 material was achieved by washing thoroughly with deionized water and drying under room temperature conditions.

Solid-state NMR:

The solid-state NMR spectra were recorded on a VARIAN VXR 300 S WB spectrometer, equipped with both Jakobsen- and Doty-high spinning speed MAS probes using 7mm zirconia rotors. The registration conditions were as follows:

^{27}Al: Frequency: 78.157 MHz, sweep width: 50000 Hz, pulse width: 3.0 μ sec (90 degree pulse: 10.5 μ sec), repetition time: 1.0 sec, aquisition time: 0.05 sec, number of scans: 1000, MAS spinning speed: 5-6 kHz. The lines were referenced to 1M Al-nitrate solution = 0 ppm.

^{31}P: Frequency: 121.421 MHz, sweep width: 15500 Hz, pulse width: 8.0 μ sec (90 degree pulse: 12.0 μ sec), repetition time: 20 sec, aquisition time: 0.5 sec, number of scans: 16, MAS spinning speed: 4.6 - 5 kHz. The lines were referenced to 85% phosphoric acid solution. The experimental error for the chemical shifts was less than 0.1 ppm.

RESULTS AND DISCUSSIONS

The ^{31}P MAS NMR spectra of the original VPI-5 material and the effect of calcination and rehydration are shown in Figure 2a,c,d. As has been recently reported by a number of studies [8,9], the three characteristic signals of hydrated VPI-5 at -23.3, -27.5 and -33.3 ppm are replaced by a broad single peak at -29.8 ppm and a small peak at -12.1 ppm after the high temperature calcination in open conditions. This material now is confirmed to be $AlPO_4$-8 formed by the transformation of VPI-5. Following rehydration, this $AlPO_4$-8 material regains three signals at -24.0, -28.1 and -32.4 ppm at a significantly different intensity ratio (1:2:6) to that observed for the original VPI-5 (1:1:1). This latter spectra is now known to be characteristic of hydrated-$AlPO_4$-8.

It has been shown that the three signals observed in the ^{31}P MAS NMR of VPI-5 are associated with the individual crystallographic sites [8,9] of the new $P6_3$ space group structural determination [10]. Detailed studies of the transformation of VPI-5 to $AlPO_4$-8 [8] have indicated that the specific assignments of the VPI-5 signals presented in Table 1 are consistant with the observed changes in the relative intensities during the progressive transformation and allow specific assignments of the hydrated-$AlPO_4$-8 signals, also presented in Table 1.

Table 1. Comparison of the chemical shift and possible crystallographic assignments for ^{31}P MAS NMR spectra of hydrated VPI-5 and hydrated-$AlPO_4$-8

VPI-5		
Chemical Shift / ppm	Relative No. of atoms	Assignment of Site [8,9]
-33.3		
-27.5	6	6/4
-23.3	3	4/4

$AlPO_4$-8		
Assignment of Site [8]	Relative No. of atoms	Chemical Shift / ppm
6/6	2	-28.1
6/6/4	2	
6/4	2	-32.4
6*/4	2	
4/4	1	-24.0

Our *in situ* variable temperature study of hydrated-VPI-5, Figure 2b, and hydrated-AlPO-8, Figure 3, give further evidence for the general validity of the relation of the VPI-5 and $AlPO_4$-8 crystallographic assignments before and after the transformation. *In situ* VT experiments on VPI-5 show the shift of the low-field signal to fuse with the adjacent high-field peak, resulting in two signals, shown in Figure 2b, with the ratio of 2:1. Similar *in situ* VT experiments on hydrated $AlPO_4$-8 also show the same behaviour, with the low field signal at -24.0 ppm again shifting up-field to fuse with the adjacent peak, as shown in Figure 3a at 110°C, resulting in the approximate intensity ratio of 1:2. It has been concluded [8,9] that the low-field signal of hydrated-VPI-5 should be assigned to the phosphorus in the 4/4 site (see Figure 1) and consequently the low-field signal of hydrated-$AlPO_4$-8 was similarly assigned to the 4/4 sites in the $AlPO_4$-8 structure. It was also indicated that the VT behaviour of the low-field signal of hydrated VPI-5 was related to the presence of some degree of interaction with water associated with the structure. The same behaviour observed for $AlPO_4$-8 gives strong indication that the same environment is preserved in the hydrated-$AlPO_4$-8 structure.

Continuation of the VT experiments to higher temperatures has also revealed further important changes in the ^{31}P and ^{27}Al spectra of hydrated-$AlPO_4$-8. As shown in Figure 3a, the 1:2 ratio of the two signals of the ^{31}P spectra at -27.1 and -32.5 ppm at 110°C is approximately reversed giving two signals at -29.3 and -33.2 ppm at 120°C and above. Upon cooling to 105°C an apparent intermediate spectra between those obtained at 110°C and 120°C is observed, the original spectrum being regained when the sample is cooled to RT. These changes are not observed for VPI-5, which remains essentially unchanged in the temperature range of 60-150°C.

In the parallel *in situ* VT ^{27}Al study presented in Figure 3b, with the temperature range 20 to 110°C, very few changes are observed in the two signals at -13.4 and 37.2 ppm. However, at 130°C, the strong signal at -13.2 ppm, generally associated with octahedral aluminium, is observed to disappear in correspondence to the observed intensity changes in the ^{31}P VT experiment. This signal is regained to a large extent when the sample is cooled back to RT.

Table 2 Distribution of sites in the ^{31}P VT MAS NMR spectra of hydrated-$AlPO_4$-8, (1=4/4 site and 2=6/6, 6/6/4, 6/4, or 6*/4 sites).

Chemical shift */ ppm	Distribution of sites at various VT temperatures* 20°C	110°C	130°C	105°C
-32.4	2+2+2	2+2+2	2	2+2
-28.1	2	1+2	1+2+2+2	1+2+2
-24.0	1	-	-	-

* Equivalent peak to the initial RT spectra

Two important implications of the VT results presented here relate to the assignments of the ^{31}P signals of hydrated-$AlPO_4$-8 to specific crystallographic sites, and the apparent number and location of octahedral aluminium sites. As indicated in Table 1, according to Maistriau et al.[8], the two types of sites in the $AlPO_4$-8 structure, defined by us as 6/4 and 6*/4, Figure 1, are considered to be NMR equivalent and are assigned to the signal at -32.4 ppm. However, the changes in the hydrated spectra in the temperature range of 110 to 130°C indicate that these two sites cannot be considered as equivalent as is illustrated in Table 2. This shows the probable distribution of the sites assigned to the ^{31}P NMR signals at the various VT temperatures in terms of the relative number of atoms of the five crystallographic sites of $AlPO_4$-8, indicating that all five sites must be considered as NMR non-equivalent and that specific assignments must be reconsidered.

More importantly, the approximate 2:1 ratio observed between the tetrahedral and octahedral signals in the RT ^{27}Al NMR spectra of Figure 3 are not consistant with the proposition that, like VPI-5, the 4/4 site is the only location at which octahedrally coordinated aluminium is sited. If this were the case, then the expected intensitiy ratio would be 8:1. It might have been feasible to consider that a significant part of this octahedral signal was related to some form of extra-framework aluminium material, however the apparent correspondence between the changes observed in the ^{31}P and the ^{27}Al spectra seem to indicate otherwise. One possible explaination is the retention of the siting of the octahedral aluminium in the transformation. This could mean that, according to Table 1, the 6*/4 sites formed from two-thirds of the former 4/4 sites in the VPI-5 structure still retain their association with water in hydrated-$AlPO_4$-8.

ACKNOWLEDGEMENTS

The authors are indebted to Phillips Petroleum Company, Bartlesville, Oklahoma (USA), for financial support of this study.

REFERENCES

1. Davis, M.E., Saldarriaga, C., Montes, C., Garces, J. and Crowder, C., Nature 331, 698 (1988).
2. Vogt, E.T.C., Richardson, J.W., J. Solid State Chem 87, 469 (1990).
3. Stöcker, M., Akporiaye, D.E., Lillerud, K.-P., Applied Catal. 69, L7 (1991).
4. Maistriau, L., Gabelica, Z. and Derouane, E.G., Appl. Catal. 67, L11 (1991).
5. Wilson, S.T., Lok, B.M. and Flanigen, E.M., U.S. Patent 4 310 440 (1982).
6. Dessau, R.M., Schlenker, J.L., Higgins, J.B., Zeolites 10, 522 (1990).
7. Davis, M.E., Montes, C., Hathaway, P.E. and Garces, J.M., in: Zeolites: Facts, Figures, Future, Studies in Surface Science and Catalysis, Eds. P.A. Jacobs and R.A. van Santen, Elsevier, Amsterdam, Vol. 49A, p. 199 (1989).
8. Maistriau, L., Gabelica, Z., Derouane, E.G., Vogt, E.T.C., and van Oene, J. Zeolites 11, 583 (1991).
9. Akporiaye, D.E. and Stöcker, Zeolites (in press).
10. McCusker, L.B., Baerlocher, Ch., Jahn, E. and Bülow, M. Zeolites 11, 308 (1991).

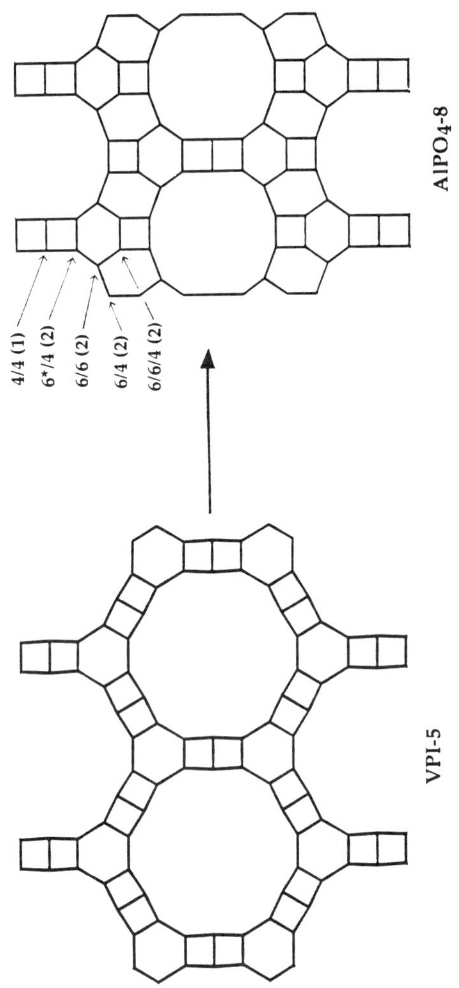

Figure 1 Projection of the framework topology of VPI-5 and $AlPO_4$-8. The relative number of atoms per site are given in brackets.

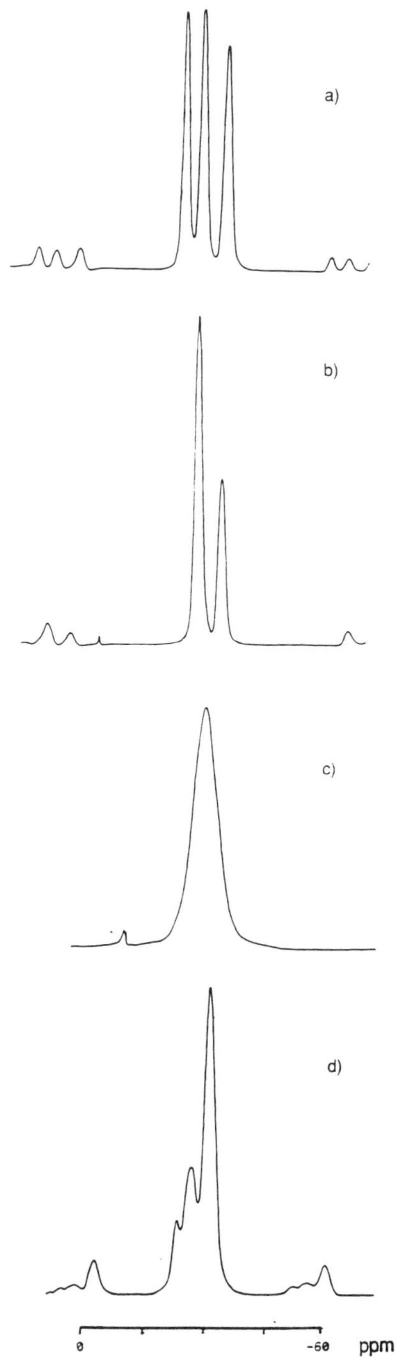

Figure 2 Effect of temperature and rehydration treatments on ^{31}P MAS NMR of VPI-5: a) RT, b) in situ VT 100°C, c) calcined 400°C / 12 hr, d) rehydrated after calcination at 400°C.

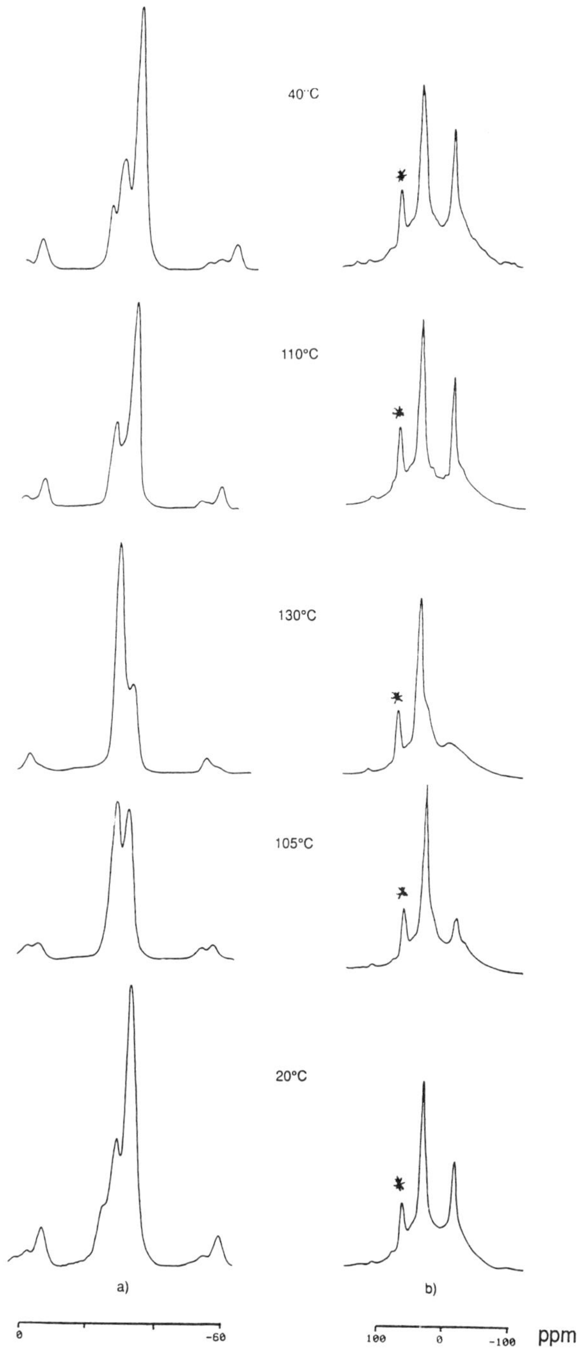

Figure 3 Effect of in situ VT on a) ^{31}P and b) ^{27}Al MAS NMR spectra of hydrated $AlPO_4$-8 (* denotes background signal from rotor cap).

^{129}Xe NMR AS A PROBE OF STRUCTURE AND DYNAMICS IN MICROPOROUS SOLIDS *

J.A.RIPMEESTER and C.I.RATCLIFFE

Steacie Institute for Molecular Sciences, National Research Council of Canada, Ottawa, Ontario K1A 0R6, Canada

ABSTRACT

Anisotropic chemical shifts were measured for ^{129}Xe sorbed in ALPO-11 and NMe_4 montmorillonite. The loading dependence in ALPO-11 was explained in terms of dynamic averaging of a statistical distribution of three distinguishable xenon types. Discrete populations of xenon atoms in AgA zeolite give information on Ag siting, and diffusion processes. Evidence is presented that low temperature NMR properties of xenon in NaY often are artifacts produced by poor thermal conduction.

INTRODUCTION

^{129}Xe NMR spectroscopy as a technique to probe void space in the solid state has been in use for some 10 years [1,2]. Much valuable information has been derived on various aspects of zeolite chemistry [3], although the emphasis has been on the use of empirical relationships. Some promising correlations have been suggested, but there still is no detailed understanding of the fundamental parameters such as chemical shifts, longitudinal and transverse relaxation times. The work presented here is aimed at exploring some of the fundamental relationships, at identifying the parameters which are important in obtaining reliable data, and at finding novel applications of ^{129}Xe NMR spectroscopy.

EXPERIMENTAL

Zeolites NaA and NaY were commercial samples (Linde), ALPO-11 was a gift of UOP, Tarreytown and Na Montmorillonite (Wyoming) was obtained from the Clay Minerals Society. AgA zeolite and NMe_4 montmorillonite were prepared by two cycles of ion exchange with excess 1N solutions containing the appropriate cation. Samples were dehydrated at appropriate temperatures before loading with measured quantities of xenon gas. Samples were sealed under a low pressure of He gas to make them suitable for low temperature work. ^{129}Xe NMR measurements were made at 49.8 or 83.0 MHz on Bruker CXP-180 or MSL-300 NMR spectrometers employing variable temperature solenoid probes. Pulse experiments carried out employed the phase-cycled single pulse, phase-cycled Hahn echo, non-selective and selective inversion recovery pulse sequences.

RESULTS AND DISCUSSION

From the Proceedings of the 9th International Zeolite Conference, Montreal 1992, Eds. R. von Ballmoos et al.,

Anisotropic Chemical shifts

It is clear that a simple explanation of the isotropic chemical shift in terms of structure would be of great and practical benefit. Several correlations have been observed or suggested [4,5,6,7,8]. We note, however, the obvious limitations of interpreting isotropic chemical shifts in terms of three dimensional pore structure which often is quite anisotropic. Also, given that the interaction of a xenon atom with a surface must result in at best an anisotropic chemical shift of axial symmetry and at worst a completely asymmetric tensor, the isotropic shifts observed must be dynamic averages. Therefore it would be quite attractive to correlate individual shielding tensor components with structure and external variables such as temperature and loading.

The short range and anisotropic nature of xenon-surface interactions is easy to substantiate from a few preliminary ab-initio calculations [J.S.Tse, private communication]. ^{129}Xe shielding tensor components were calculated for a xenon atom approaching the O atom of a water molecule along the C_2 axis (zz component along internuclear axis, xx component defined by the plane of the molecule) with respect to a reference Xe atom at infinity. Shielding tensor components δ_{zz}, δ_{yy} and δ_{xx} were -0.68, 53.84, 48.24 ppm and -0.55, 4.21 and 2.59 ppm for internuclear distances of 3 and 4 Å.

Anisotropic chemical shifts have been observed for xenon trapped in a number of clathrates and molecular sieves such as ALPO-11 [9] and DD3-R [10]. Below we report new results for ALPO-11 and a pillared montmorillonite.

ALPO-11

ALPO-11 is a convenient material for study, as the pore space can be thought of as an elliptical cylinder [11] with loading seen as a one-dimensional filling of the cylinder. Springuel-Huet and Fraissard [9] previously reported the isotropic chemical shifts as a function of loading, and also observed that the anisotropy changed sign on going from low to high loading. We have studied these changes in more detail and have developed a statistical dynamic averaging model which adequately accounts for the observed changes. Lineshapes are shown in fig.1. At low coverages the lineshapes are approximately axially symmetric. At higher coverages three tensor components are visible until again two components coincide and an axial lineshape with opposite asymmetry arises. Finally, the lineshape becomes non-axial again. A plot of the shielding tensor components versus coverage (fig.2) shows that one component is always insensitive to degree of loading, whereas the other two vary linearly with the degree of loading. A simple analysis of the problem, in terms of mean free path, suggests that only the path length along the channel should vary with loading, with the two at right angles to the channel remaining constant. Obviously, the data cannot be explained this way, as only one shift component is loading independent. We have noted similar difficulties in dealing with the anisotropies of xenon trapped in clathrate cages [7]. As mentioned above, the ALPO-11 pore space superficially resembles a one-dimensional elliptical cylinder.

Closer examination of van der Waals' contacts of xenon atoms with the channel walls shows in fact that the cylinder is a chain of single, more or less well defined xenon sites, with four such sites per crystallographic unit cell. When xenon is loaded into the channel three distinguishable situations arise: a) xenon with two empty neighbouring sites (OXO), b) xenon with one empty and one full neighbouring site ((XXO) or (OXX)), c) xenon with both neighbouring sites filled (XXX). Each of the central xenon atoms in these three situations can be considered to have different shielding tensors with the tensor principal axes oriented along the three axes of the cylinder. The observed tensor at each loading is then a dynamic average of the three tensors weighted by their statistical populations, as movement to empty sites appears to be rapid on an NMR timescale. This model predicts a linear dependence of the tensor components on loading.

Tetramethylammonium Montmorillonite

Clays form another class of highly anisotropic materials suitable for study by ^{129}Xe NMR. A recent study [12] of a swelling clay pillared with Al Keggin ions did not show obvious signs of chemical shift anisotropy of the ^{129}Xe resonances, although the extremely broad lines might well have an anisotropic component rendered indistinct due to dynamic processes or distributions.

Montmorillonite exchanged with tetramethylammonium cations (TMA-M),with an interlayer spacing of 4.2 Å, gives layers pillared almost perfectly to match the xenon van der Waals' diameter at 4.4 Å [13]. The lineshape of ^{129}Xe trapped in TMA-M was difficult to define due to its great width and relatively poorly defined axial edge. We found that much of the uncertainty in defining the lineshape and eliminating baseline roll could be accomplished by using a phase-cycled Hahn echo [14] rather than simple free induction decays. The line shown in fig.3a is best defined as an axial powder pattern with an isotropic chemical shift of 175 ppm and an anisotropy of 155±5ppm. The latter is one of the largest values yet observed for trapped Xe atoms and reflects a highly anisotropic environment with the axial clay layers carring a negative charge , and the equatorial TMA pillars positively charged. The trends seen for chemical shift anisotropy with cage symmetry [7] suggest that the trapping site can be thought of as a prolate spheroid due to the negative anisotropy, although this approach neglects the presence of charges. The unique, long axis of the spheroid must then pass at right angles through the clay layers. This assignment can be checked for oriented samples prepared by pressing flat, platelike clay particles in a die to form a pellet before dehydration and loading with xenon. Xe lineshapes are shown in fig.3b,c for two orientations of a pellet, and prove the previous assignment. One cautionary statement is in order, as it is relatively easy to get orientation effects on powders of platelet-like particles, even when these are not wanted. In this case the spectrum can be changed simply by tapping the sample tube with the tube held either vertically or horizontally.

The weak dependence of spectral parameters on loading or temperature suggests the siting of xenon atoms in relatively isolated sites [15].

Chemical shifts and Xe transport in AgA zeolite

Several laboratories have now reported spectra of discrete populations of xenon atoms of up to 8 atoms/cage in isolated cages of NaA zeolite [8,16,17]. The populations have been explained in terms of binomial and hypergeometric distributions at low and higher loading, respectively [16], although exceptions have been noted favouring formation of Xe_6 clusters [17]. One important result is that the Xe chemical shift dependence on loading for the exchange-averaged populations in CaA zeolite can be explained in terms of the discrete population distributions observed for NaA zeolite [17].

We have also identified discrete populations of up to 8 Xe atoms/cage in activated Ag exchanged A zeolite (fig.4). The signal for 8 Xe atoms/cage shows a distinct chemical shift anisotropy, and suggests that the Xe_8 cluster has a fixed, definite geometry. Simple packing considerations suggest a square antiprism which completely fills the supercage. There also are implications for the siting of Ag ions and charged Ag atom clusters over which there still is considerable controversy [18,19,20]. The observation of non-exchanging xenon populations implies that the 8-rings which connect supercages are blocked by silver ions. On the other hand, the charged silver atom clusters must be present in the sodalite cages, as the ^{129}Xe chemical shifts are not too different from those observed for NaA zeolite. When xenon atoms are in direct contact with the silver clusters, eg. as in AgX zeolite, a negative shift is observed [21].

The presence of discrete populations also offers a unique opportunity to study the loading of the zeolite as a function of time. The integrated intensity of the xenon populations, for an AgA sample with a quantity of xenon sealed into the tube, as a function of time shows that the uptake process is finished after about 6hrs. (fig.5,top). However, the xenon populations continue to readjust for another 4 days (10hrs. shown in fig.5,bottom) before internal equilibrium is achieved. The actual cage-to-cage exchange process can be observed as line-broadening and coalescence of the individual lines with increasing temperature. From the greater line broadening it appears that cages with larger populations have shorter lifetimes at any one temperature. These observations offer a unique opportunity to study diffusion processes at a molecular level.

Effects of bulk properties on ^{129}Xe NMR in NaY zeolite

Despite the fact that this was one of the first materials probed by ^{129}Xe NMR [1], and that it has received a lot of attention subsequently [3], all the factors which determine the chemical shifts, linewidths and relaxation times in these materials are still not defined. There is now evidence from many sources [22,23] that a xenon atom samples a number of particles on the NMR timescale as a result of the large pore openings. The pore system probed by Xe NMR therefore includes both internal and interparticle pore space (i.e. between particles). Since many finely powdered samples have very low bulk densities (0.3 g/cm^3) , and therefore a large interparticle volume which can be 70-80% of the total sample volume, the extraparticle xenon plays an extremely important role. For spectra obtained at room temperature, the effect can be illustrated by loading Xe into samples containing samples compressed to different bulk densities [24].

Other evidence for the importance of bulk property effects comes from temperature dependent studies. The description of fluids in small pores is of course of great fundamental interest, and the behaviour of many Xe-zeolite systems at 144K has been investigated by Cheung[25], who described a gas-liquid transition in a number of Y zeolites. We also have investigated the NaY zeolite system as a function of loading and temperature, with significantly different results from those reported previously. During the course of the experiments we found it essential to include a small pressure of He gas for efficient heat transfer, as the low temperature behaviour was dominated completely by the effect of thermal gradients. This is illustrated in fig.6, which shows a set of data for loosely powdered samples with and without He gas added. At cooling rates of 2°C/min. the sample without He gas developed very broad lines below 230K. By observing the spectrum of a sample cooled at 2°C/min. to 143K as a function of time, it can be seen that equilibrium will not be attained for a considerable length of time. Xenon appears to behave as a viscous liquid at this temperature, with Xe transport taking place only at points of interparticle contact. Helium-free samples stored at 77K continue to show spectral changes 10 months after quench-cooling the samples.

REFERENCES

* Published as NRCC no. 32948.

1) Ito, T. and Fraissard, J. , in Proceedings of the 5th Conference on Zeolites, Heyden, London, 1981; p.510; J. Chem. Phys. **76**, 5225 (1982)

2) Ripmeester, J. A. , J. Am. Chem. Soc. **104**, 209 (1982); J. Magn. Reson. **56**, 247 (1984)

3) Dybowski, C. and Bansal, N. in Ann. Rev. Phys. Chem. **42**, 433 (1991)

4) Demarquay, J. and Fraissard, J. , Chem. Phys. Lett. **136**, 314 (1987)

5) Derouane, E. G. and Nagy, J. B. , Chem. Phys. Lett. **137**, 341 (1987)

6) Johnson, D. W. and Griffiths, L. , Zeolites **7**, 484 (1987)

7) Ripmeester, J. A. , Ratcliffe, C. I. and Tse, J. S. , J. Chem. Soc. Farad. Trans. I **84**, 3731 (1988)

8) Ripmeester, J. A. and Ratcliffe, C. I. J. Phys. Chem. **94**, 7652 (1990)

9) Springuel-Huet, M. A. and Fraissard, J., Chem. Phys. Lett. **154**, 299 (1989)

10) Ripmeester, J. A. and Ratcliffe, C. I., 8th International Zeolite Conference, Amsterdam, 1990

11) Bennett, J. M. , Richardson Jr. , J. W. , Pluth, J. J. and Smith, J. V. , Zeolites **7**, 160 (1987)

12) Barrie,P. J. , McCann, G. F. , Gameson, I. , Rayment, T. and Klinowski, J. , J. Phys Chem. **95**, 9416 (1991)

13) Barrer, R.M. and Reay, J.S.S., Trans. Farad. Soc. **53**, 1253 (1957)

14) Rance, M. and Byrd, R. A., J. Magnetic Reson. **52**, 221 (1983)

15) Graves-Smith, H., Detellier, C. , Ratcliffe, C. I. and Ripmeester, J. A. , to be published.

16) Chmelka, B. F. , Raftery, D. , McCormick, A. V. , de Menorval, L. C. , Levine, R. D. and Pines, A., Phys. Rev. Lett. **66**, 580 (1990); **67**, 931 (1991)

17) Jameson, C. J. , Jameson, A. K. , Gerald II , R. and de Dios, A. C. , in press

18) Kim, Y. and Seff, K. , J. Am. Chem. Soc. **100**, 6989 (1978)

19) Gellens, L. R. , Mortier, W. J., Schoonheydt, R. A. and Uytterhoeven, J. B. , J. Phys. Chem. **85**, 2783 (1981)

20) Texter, J. , Kellerman, R. , Gonsiorowsky, T. , J. Phys. Chem. **90**, 2118 (1986)

21) Gedeon, A. , Burmeister, R. , Grosse, R. , Boddenberg, B. and Fraissard, J. , Chem. Phys. Lett. **179**, 191 (1991)

22) Shoemaker, R. , Apple, T., J. Phys. Chem. **91**, 4024 (1987)

23) Ryoo, R. , Cho, S. J. , Pak, C. ,Kim, J. , Ihm, S. , and Lee, J. Y. , J. Am. Chem. Soc. **114**, 76 (1992)

24) Ripmeester, J. A. and Ratcliffe, C. I. , Proc. Materials Research Society Meeting, Anaheim, Spring 1991. p.281

25) Cheung, T. T. P., Fu, C. M. and Wharry, S., J. Phys. Chem., **92**, 5170 (1988)

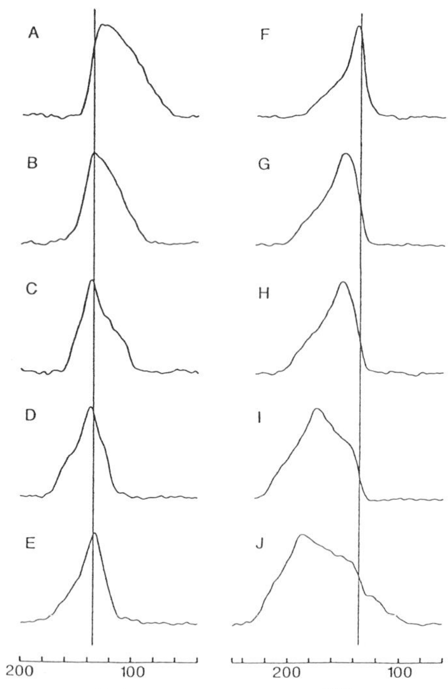

Figure 1. ^{129}Xe NMR lineshapes for xenon sorbed in ALPO-11. Increasing loading levels A-J correspond to points plotted in figure 2.

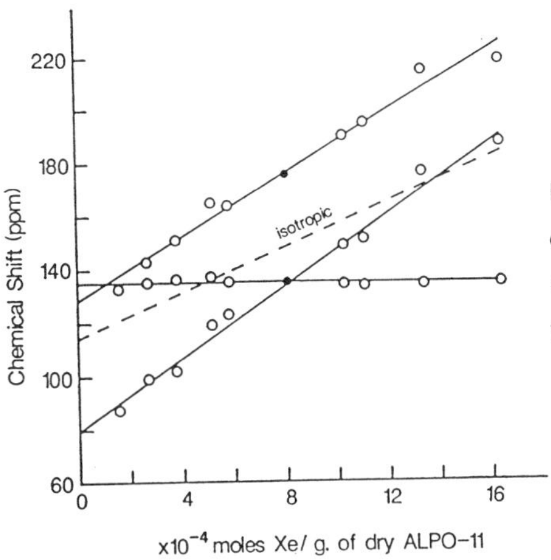

Figure 2. ^{129}Xe chemical shift tensor components as a function of loading for xenon sorbed in ALPO-11. Solid lines represent the fit to the model.

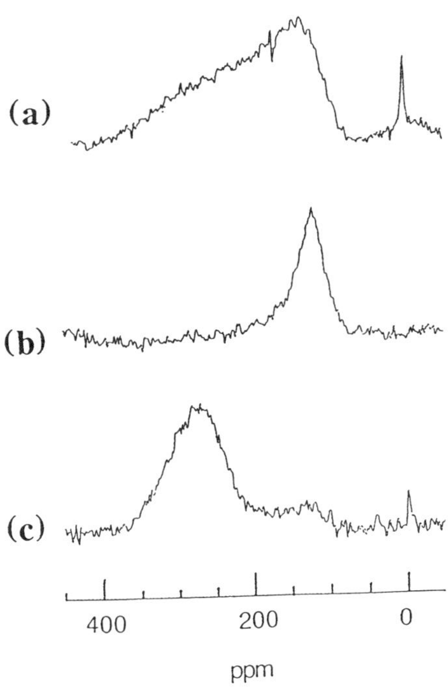

Figure 3. ^{129}Xe NMR lineshapes for xenon sorbed in NMe_4 montmorillonite (0.54 mmol xenon/gram clay) a) powdered sample b) pressed cylindrical pellet with long axis perpendicular to field c) pressed cylindrical pellet with long axis parallel to field.

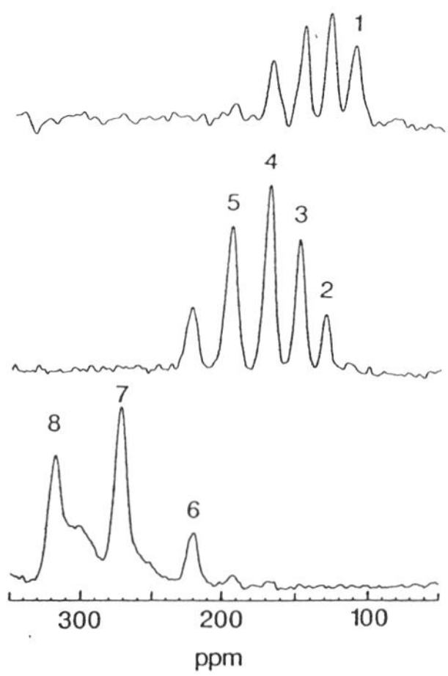

Figure 4. ^{129}Xe NMR lineshapes for xenon sorbed in AgA zeolite at different loading levels. Cage populations from 1-8 xenon atoms per cage are evident.

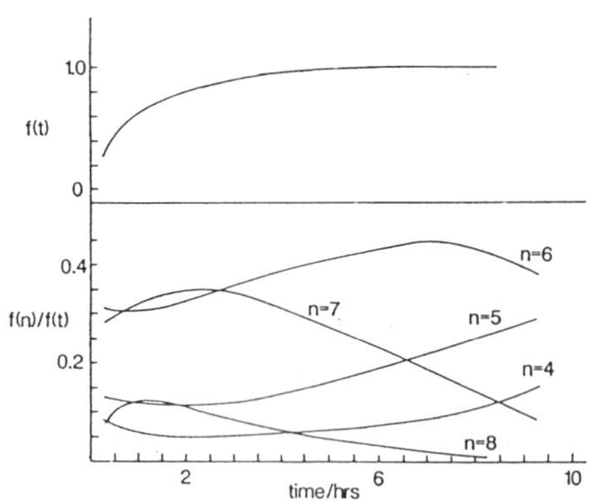

Figure 5.(top) Fraction of xenon f(t) sorbed into AgA zeolite as a function of time; (bottom) normalized fraction f(n) of cages with xenon population n as a function of time.

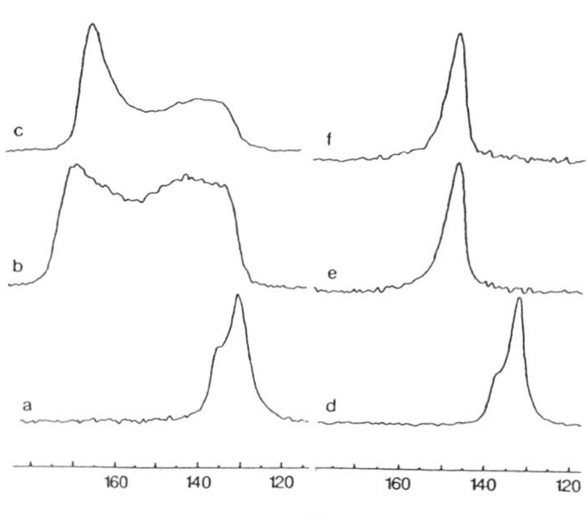

Figure 6.Effect of He gas on ^{129}Xe NMR lineshapes for xenon in NaY zeolite (7.4 Xe atoms/cage) a,b,c) sample without He gas; d,e,f) sample with He gas ; a,d) room temperature; b,e) lineshapes after cooling at 2K/min. to 143K ; c,f) lineshapes after waiting 5 min. at 143K.

ZEOLITE INTRINSIC FRAMEWORK REACTIVITY AND SENSITIVITY ANALYSIS OF FRAMEWORK PERTURBATIONS

B. G. BAEKELANDT, J. L. LIEVENS, W. J. MORTIER, R. A. SCHOONHEYDT

Centrum voor Oppervlaktescheikunde en Katalyse, Materials Research Center, K.U.Leuven, Kardinaal Mercierlaan 92, B-3001 Heverlee, BELGIUM

ABSTRACT

The intrinsic electronic properties of 30 zeolite structure types with hypothetical SiO_2 composition are estimated by the Electronegativity Equalization Method (EEM). The structure-dependent average (global) softness and the connectivity dependent local softnesses (Fukui functions) for all crystallographically different atom types are calculated. It is shown how these quantities depend on the structural features of the crystal and some general rules are formulated. Also some framework perturbations (by cations) are studied, the results of which are rationalized in terms of the rules derived. Finally, the electron donor capacity of the structural oxygens, and its relation with the hardness/softness concepts, is commented upon.

INTRODUCTION

A way of growing importance to model crystalline solids is the empirical treatment by force fields. This so-called 'molecular mechanics' technique has been successfully applied to study and predict mechanical, diffusional, thermodynamical and vibrational properties of crystalline solids and their interaction with molecules and ions. However, electronic effects cannot be considered by the very fact that the charges (which arise in the electrostatic terms) are fixed and treated as constant parameters.

A complementary method, known as the Electronegativity Equalization Method (EEM), has been developed to study charge distributions and charge reorganizations in molecules and solids. EEM is a semi-empirical approach to Density Functional Theory, developed by Mortier *et al.* [1]. With the structure as input the original EEM formalism enables the direct calculation of atomic charges (a finite difference approximation to the electron density distribution function) and the system's electronic chemical potential (minus the electronegativity), and is particularly suitable for solid state applications [2].

It has recently been extended [3] to calculate several other 'sensitivity coefficients' (partial derivatives of the energy functional E[N,v]) such as the Fukui function, the local and global hardness and softness.

METHOD

We restrict ourselves to define the relevant quantities of the sensitivity analysis and to briefly comment upon their physical significance. The method is explained in detail elsewhere [3,4].

From the Proceedings of the 9th International Zeolite Conference, Montreal 1992, Eds. R. von Ballmoos et al.,

'Sensitivity' is defined as the response of a property P to a perturbation in N or v, *i.e.* the derivative of P with respect to N or v: $(dP/dN)_v$ and $(dP/dv)_N$. N and v are the total number of electrons and the external potential respectively, quantities completely defining the system. Various fundamental properties of a system can in this way be defined, such as:

- the charge	q_a	=	$(dE/dv_a)_N$		
- the electronegativity	X	=	$-(dE/dN)_v$	=	$(dE/dq)_v$
- the global hardness	H	=	$(d^2E/dN^2)_v$	=	$-(dX/dN)_v$
- the Fukui function	f_a	=	$-(dq_a/dN)_v$	=	$(dX/dv_a)_N$

with E the total molecular electronic energy and 'a' denotes an atom.

Electronegativity is commonly used as an intuitive concept to rationalize electronic effects: *electronegativity* differences drive the electron transfer and determine the *direction of the flow*. Hardness generally refers to resistance to change or deformation. *Chemical hardness* is the resistance to electron displacements or fluctuations, and thus governs the *amount of charge transfer*. It is associated with polarizability and is proportional to the energy gap between the frontier orbitals (HOMO-LUMO gap) [5]. The Fukui function is a local property measuring the charge shift on an atom upon changing the total number of electrons in the system. Electrodynamically it can be considered as a charge capacitance and in terms of MO theory, it is the electron density of the frontier orbitals [6]. Formally it is a relative measure (since it is normalized and integrates to one) of local softness. Absolute local softnesses (comparable from molecule to molecule) can be obtained by dividing f_a by the global hardness: $s_a = f_a/H$.

Within the sensitivity analysis, an understanding of chemical reactivity requires the knowledge of density relaxation effects, *i.e.* the adaptation of the electronic charge distribution of a reactant to the perturbing electrostatic potential produced by the other reactant. Especially in the case of inorganic solids (such as zeolites) where the ionic contributions to the total energy may be dominant, the charge shifts will be significant and the charge sensitivities determine both activity and selectivity.

RESULTS AND DISCUSSION

Intrinsic framework properties

The intrinsic electronic properties of 30 zeolite structure types with hypothetical SiO_2 composition have been estimated by the Electronegativity Equalization Method (EEM). Full structural information and parametrization is given in reference [7] (all structures were optimized by using DLS). The structure-dependent average (global) softness ($S = 1/H$) and the connectivity dependent local softnesses (Fukui functions) for all crystallographically different atom types were calculated. The results for the global softness are compiled in Table 1 and the local softness data are summarized in Figure 1, giving the local softness s_a as a function of the local topological density d_a. The latter

ABW	0.22636	GIS	0.19510	MEL	0.17884
ANA	0.24682	GME	0.14811	MER	0.18436
BIK	0.25924	HEU	0.18935	MFI	0.17375
BRE	0.20186	KFI	0.14730	MOR	0.16770
CAN	0.19304	LAU	0.18497	OFF	0.15938
CHA	0.15537	LEV	0.17384	PHI	0.18875
EAB	0.17289	LOS	0.05188	RHO	0.13867
EPI	0.19508	LTA	0.13127	SOD	0.22158
ERI	0.16167	LTL	0.13245	STI	0.16582
FAU	0.10062	MAZ	0.16427	THO	0.18705

Table 1: Global softness S (in eV/TO_2) for various zeolite structures with hypothetical SiO_2 composition.

is defined as $\Sigma 1/R_{ab}$ and measures the number of atoms at their respective distances from a central point or atom. It allows to formally distinguish between (atoms in) dense and open parts of a structure.

One of the most important rules which can be derived from the theory is the existence of an inverse relation between the softness of an atom in a molecule and the softness of its environment [4]. With this rule, we are in a position to rationalize the trend observed in Figure 1, namely the inverse relation between the local topological density of an atom and its local softness: since in all cases the local softness is positive, and since local softness is additive [4], each atom in the environment of a central atom will contribute to generate a positive softness potential at that atom; because of the inverse relation between the softness of an atom in a molecule and its environment, we can expect an increased local softness upon decreasing the local density. Therefore, atoms in open parts of the crystal are predicted to be the softest because they have the hardest environment. Thus, in zeolites, the softest oxygens are located in the walls of the large channels and cavities; these are the most basic. Also the Si/Al ratio comes into play. Al atoms (since they are hard), soften the neighbouring oxygens [8]. This is confirmed by several experimental data, showing an increased basicity upon increasing the Al-content (catalysis, XPS O(1s) binding energies; see reference [8]).

The data also reveal that intrinsically, T-atoms are softer than oxygens. This means that T-atoms are the electron reservoirs in such systems, which agrees with the electrodynamical and statistical mechanical interpretation for local softness (charge capacitance and low energy density fluctuations respectively) and confirms previous work concerning the charge delocalization in zeolites [9].

Framework perturbations

The intrinsic framework properties serve as a tool to rationalize and predict the effect of a perturbation. Two examples are discussed, namely the effect of a cation on the framework charges (dv perturbation) and the relation between the oxygen softness and the charge transfer to the cation (dN).

Perturbation sphere of a cation: charge shifts versus distance from the perturbation

Figure 2 represents the charge shifts (dq) on the framework atoms of a Faujasite type structure, caused by an incoming cation, as a function of the distance (R) from the locus of the perturbation. A different sensitivity of the system has been obtained, according to its global hardness (not shown): the radius of the perturbation sphere is larger for soft structures (high Si/Al ratio) than for harder ones. When looking at the perturbation sphere in detail, some special characteristics appear:

(i) the perturbation is restricted to a sphere of radius 7 Å. This perturbation is compensated by polarization effects or charge redistributions over the lattice atoms. A perturbation potential is hereby created which contributes significantly to the final potential: cations generate their own site energy and this is easier in soft structures because of the increased ability to perturb the framework. In soft structures, all cations can easily perturb the lattice at every possible site so as to create a negative potential. In harder structures, the interaction energy is more dependent on the existing charge distribution [8]. Therefore, an increase in the energy differences between the different cation positions is predicted with increasing Al-content (increasing hardness), and has indeed been observed experimentally [10].

(ii) returning to the sensitivity analysis, it can be seen that the charge shifts for the different atom types follow the order $dq_T > dq_O$, in full agreement with the order of sensitivities or local softnesses: $s_T > s_O$: T-atom charges are most flexible while the oxygen electron populations are rather rigid.

Perturbation potential versus softness potential

Intuitively the softness potential S_a has been identified with the ability to perturb the environment [4]. S_a indeed contains the Fukui function which is a measure of the *charge shifts* on the atoms upon perturbation: $S_a = \Sigma\, dq_b/R_{ab}$. The electrostatic potential reflects the *charges* in the surroundings: $v_a = \Sigma\, q_b/R_{ab}$. To clarify the physical meaning the framework perturbations from the previous section were rationalized in terms of the charge sensitivities: charge shifts caused by the cations should correlate with the softness potential.

In Faujasite type zeolites, the most probable cation sites (I, I' and II) are located on the threefold axis. A Na^+ cation was displaced along this axis (whereby each time the framework charges were recalculated) and at regular distances the electrostatic potential was calculated (v_2). The same was done without any cation present (v_1). The difference v_1-v_2 (the perturbation potential, dv) is due to the charge shifts on the lattice, caused by the cation. On the same axis the softness potential was calculated. The results are represented in Figure 3a and 3b. It is seen that the perturbation potential (dv) and the softness potential exhibit the same maxima and minima. For a maximum in the softness potential the framework relaxes most easily, *i.e.* low energy electron fluctuations take place.

Local softness versus local charge transfer properties

Global charge transfer properties in zeolites and oxides have been rationalized in terms of the (optical) electronegativity concept [11]: the 'electron attractive power' of the oxygens governs the ligand to metal charge transfer. The frequency of the band maximum in the electronic spectra has indeed been shown to be related to the average electronegativity [12]. However, in this way structural effects cannot be accounted for: different structure types with the same chemical composition cannot be distinguished and within one structure one cannot separate the different components of the band envelope, due to different locations of the TMI in the structure.

Detailed theoretical analyses of the spectroscopic data of Cu^{2+} ions in several zeolites, allowed to assign all signals and to distinguish between different sites within one structure [13,14]. The difference in the spectroscopic parameters so obtained has a physical significance for the coordinative bond and reflects the local charge transfer properties of the TMI complexes. As is obvious by now the local softnesses also reflect local charge transfer properties. We thus expect the spectroscopic and softness parameters to correlate. This is illustrated by the following example. For Cu^{2+} on different sites in some zeolites, the EPR parameters are plotted against the local softnesses of the oxygens of the first coordination sphere in Figure 4 (when there is more than one oxygen type around Cu^{2+} in the first coordination shell the arithmetic mean is taken). The general trend is that $g_{//}$ decreases and $A_{//}$ increases with increasing softness of the oxygens. This is expected for an increasing covalency of the Cu-O bond. The correlation is not perfect and should be improved by taking into account the T-atoms as well, since these were found to be the electron reservoirs. Also Al-containing models should be investigated, since Al enhances the oxygen softness considerably [8]. Calculations which take into account the whole environment (*i.e.* the softness potential instead of s_O only) are in progress.

Thus increasing the local softness of the oxygens also means increasing the covalent character of the coordinative bond, a clear manifestation of the HSAB principle (since, generally, TMI's are known to be soft). An increased oxygen softness, which can *e.g.* be induced by a nearby hard Al-atom, reflects an increased electron donor capacity (basicity), *i.e.* the easiness with which dN-perturbations take place and the amount of charge transfer during the perturbation.

CONCLUSIONS AND PROSPECTS

When dealing with inorganic solids, the knowledge of the charge (re)distribution is crucial to understand and rationalize the trends observed in chemical reactivity, since the ionic contribution to the total energy is significant. Therefore the charge sensitivity analysis is a relevant and attractive tool for the study of materials such as zeolites. Beside the quantitative information, it provides a conceptual framework for the experimental chemist interested in guidelines which allow the qualitative prediction of properties.

The main purpose of the present paper was to illustrate the way of thinking and to give some specific applications. The next step is to investigate the activation of adsorbed molecules. It is *e.g.* of interest to analyze the preferential polarization channels, *i.e.* independent patterns of electron displacements through which the system can relax its charge distribution (formally equal to the eigenvalues of the hardness matrix [4]). A perturbation of the system by the field of a nearby active surface, is strong enough to open ('unlock') many polarization channels through which an adsorbed substrate can adjust its charge density [15]. Studies in this respect are promising and have shown to be useful to understand and explain reaction and sorption mechanisms [16].

REFERENCES

1. Mortier, W.J., Ghosh, S.K., Shankar, S., J. Am. Chem. Soc.,**108**, 4315 (1986).
2. Mortier, W.J., Structure and Bonding, **66**, 125 (1987).
3. Baekelandt, B.G., Mortier, W.J., Lievens, J.L., Schoonheydt, R.A., J. Am. Chem. Soc., **113**, 6730 (1991).
4. Baekelandt, B.G., Mortier, W.J., Schoonheydt, R.A., Structure and Bonding: Chemical Hardness, Ed. K.D. Sen, in press (1992).
5. Pearson, R.G., Proc. Natl. Acad. Sci. USA, **83**, 8440 (1986).
6. Parr, R.G., Yang, W., J. Am. Chem. Soc., **106**, 4049 (1984).
7. Van Genechten, K., Mortier, W.J., Zeolites, **8**, 273 (1988).
8. Baekelandt, B.G., Mortier, W.J., Schoonheydt, R.A., in *"Modeling of Structure and Reactivity in Zeolites"*, Ed. R.A. Catlow, in press (1992).
9. Mortier, W.J., Geerlings, P., Van Alsenoy, C., Figeys, H.P., J. Phys. Chem., **83**, 855 (1979).
10. Van Dun, J.J., Mortier, W.J., J. Phys. Chem., **92**, 6749 (1988).
11. Duffy, J.A., J. Solid State Chem., **61**, 3426 (1986).
12. Packet, D., Ph.D. Thesis no. 151, K.U. Leuven, Fac. Agronomy (Feb. 1987).
13. Packet, D., Schoonheydt, R.A., ACS Symp. Ser., **368**, 203 (1988).
14. De Tavernier, S., Schoonheydt, R.A., Zeolites, **11**, 155, (1991).
15. Nalewajski, R.F., Koninski, M., Z. Naturforsch., **42a**, 451 (1987).
16. Nalewajski, R.F., Structure and Bonding: Chemical Hardness, Ed. K.D. Sen, in press (1992).

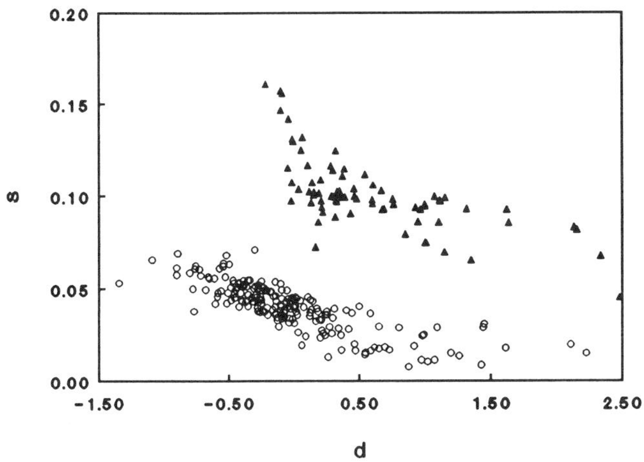

- Figure 1 : Local softness versus local topological density for O- and T-atoms in 30 zeolite structure types.(O:○; T:▲)

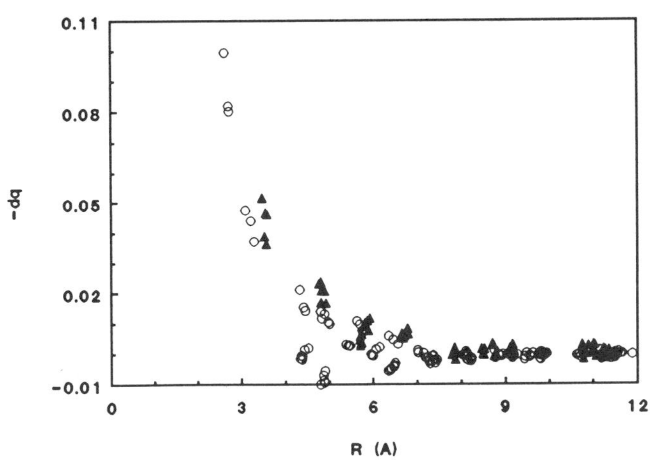

- Figure 2 : Perturbation sphere of a cation (Na^+ at site I' in Faujasite; Si/Al=3): charge shifts on all framework atoms versus distance (in Å) from the cation. (O:○; T:▲)

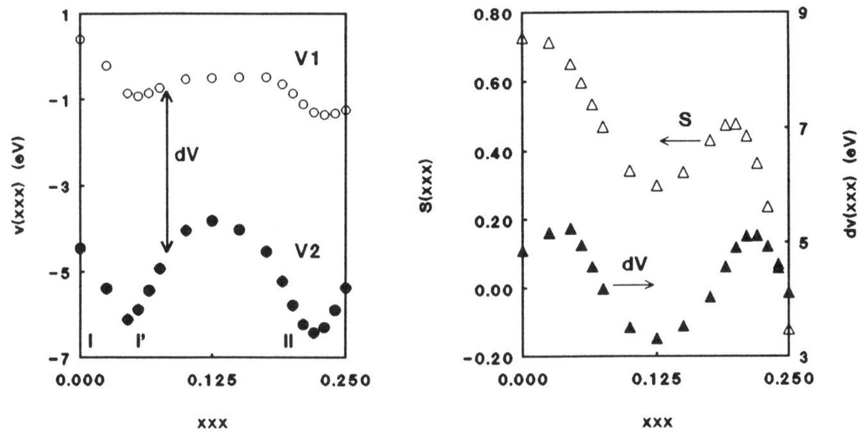

- Figure 3a: Correlation between perturbation potential of a cation ($dv = v_1 - v_2$ (eV); ○ :v_1 = potential without any cation; ● : v_2 = potential with cation present) and the softness potential S in Faujasite. 3b: both dv (▲) and S (△) along the trigonal axis (xxx) show the same maxima.

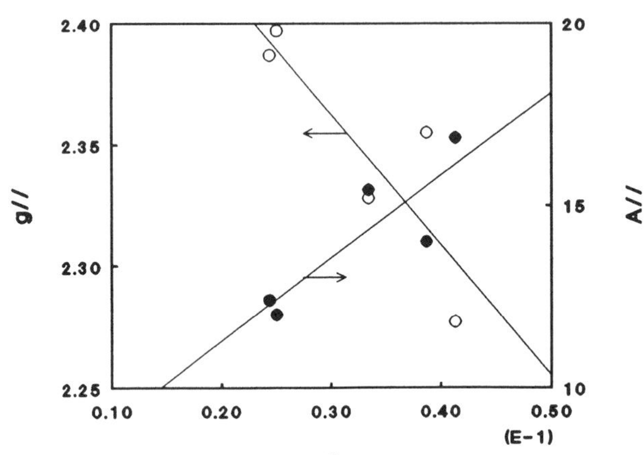

- Figure 4 : Correlation between the EPR parameters ($g_{//}$ (○), $A_{//}$ (●)) of Cu^{2+} at different sites in various zeolites, and the softness of the oxygens in the first coordination sphere.

DEPENDENCE OF CHARGE DENSITY AND ELECTRIC FIELD WITHIN THE CAGES OF X AND Y ZEOLITES ON THE CATION: PHOTOPHYSICAL PROBES

V. RAMAMURTHY* AND D. F. EATON
Central Research and Development,
Experimental Station,The Du Pont Company,
Wilmington, DE 19880-0328, USA.
Coontribution Number: 6129

ABSTRACT

Pyrene, pyrenealdehyde and *para*-dimethylaminobenzonitrile have been used as photophysical probes to monitor the polarity of zeolite interior. As per pyrenealdehyde and *para*-dimethylaminobenzonitrile the cage is highly polar or possess high electric field when the cation is Li or Na. Also the polarity/electric field of the cage decreases with the cation size (Li>Na>K>Rb>Cs). Furthermore, the polarity of the supercages is higher in the case of X than in Y zeolites. Although the results with pyrene are generally consistent with those of the other two probes, pyrene fails to sense the difference between Li, Na and other cations. The photophysical technique described here is established to be general and has been utilized to estimate the polarity of other members of zeolite family.

INTRODUCTION

The use of heterogeneous systems such as micelles, solid supports, polymers, and membranes for controlling photochemical and photophysical properties of organic molecules is currently an area of considerable activity [1]. In this context we and others have been investigating the use of zeolites [2]. In industrial applications, acidic and basic activities and molecular sieving effects of zeolites have attracted considerable attention. However, zeolites as a simple medium for photoreactions calls for understanding zeolites as a solvent for organic guests rather than as a catalyst. In the case of solutions, the effect of a solvent can be readily estimated by taking into account the properties of the molecules involved such as polarity, polarizability or dielectric constant.

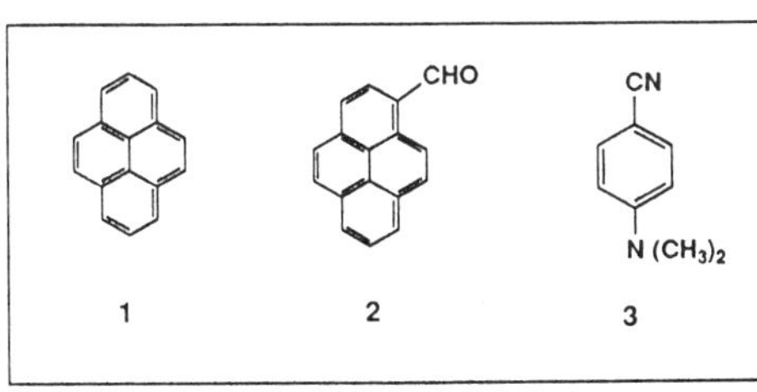

The characterization of zeolites as a solvent is much less tractable in a quantitative sense. In spite of the growing general knowledge in zeolite science [3] some of the basic questions remain unanswered. One of the basic properties of zeolites namely micro-polarity is being addressed, in a quantitative manner, only recently [4]. The *micropolarity* of the zeolite interior is a reflection of the electric field provided by cations that are present within the cages and channels. The notion of polarity within zeolite cages and channels does not have precisely the same meaning as in solution. We present in this article a photophysical technique utilizing pyrene (**1**), pyrenealdehyde (**2**) and *para*-dimethylaminobenzonitrile (**3**) as probes to estimate the micropolarity of the zeolite interior. This simple and easy-to-use technique has the potential of becoming a common tool in zeolite chemistry.

RESULTS AND DISCUSSION

Background

A brief background on the ability of pyrene, pyrenealdehyde and *para*-dimethylaminobenzonitrile to report the microenvironmental polarity is in order [5]. Condensed aromatic molecules such as pyrene and anthracene show fairly well-resolved fluorescence spectra in solution at ambient temperature. Variation in the nature of the solvent leads to large variations in the intensities of the vibronic bands without much spectral shift and the phenomenon is known as the *Ham Effect*. For pyrene, due to the Ham effect, band 1 (0 – 0 transition) located at ~ 372 nm shows significant intensity enhancement in polar solvents [6]. The peak ratio (I_1/I_n where n = 3, 4 or 5) can serve as a sensitive guide of the environmental micropolarity. Detailed interpretation of the magnitude of the peak ratio can be complex due to the fact that the exact mechanism of how polar solvents enhance the intensity of symmetry-forbidden vibronic transitions by means of a reduction in local symmetry is not fully understood.

Occurrence of low-lying closely spaced n,π* and π,π* levels is a common feature in the photophysics of organic heterocycles. Since π,π* excited states are considerably more sensitive to solvent polarity than n,π* states, the emission maxima for a number of molecules that have the π,π* state as the lowest state are solvent dependent. There exists a qualitative relationship between the polarity of the medium and the emission maximum. Fluorescence probes such as pyrenealdehyde show this type of behavior [7].

Numerous organic molecules with electron donor and acceptor groups linked by a single bond react in the excited state with the formation of a highly polar state with mutually perpendicular conformation of the D^+ and A^- subunits [8]. A classical example of this kind is *para*-dimethylaminobenzonitrile (DMABN, **3**). A unique feature of this molecule is the existence of a single broad fluorescence in nonpolar solvents centered around 350 nm and dual fluorescence in polar organic solvents (one at ~350 nm and the other with a solvent dependent maximum between 420 and 600 nm) [9]. A linear correlation is observed in plots of emission energy of the longer wavelength band *vs.* solvent polarity parameters. The two emissions are generally believed to occur from two conformationally distinct singlet excited states—the short wavelength emission from a planar state and the long-wavelength emission from a twisted internal charge transfer state (TICT).

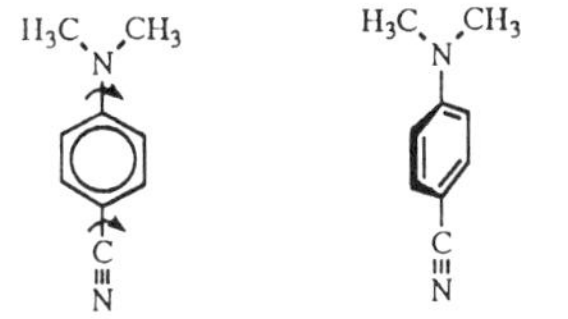

Planar TICT

X and Y Zeolites Probed

While the phenomenon monitored and the mechanism responsible for the medium dependent change is different in each one of the above three probes (**1-3**), they are capable of reporting, at least in a qualitative manner, the micropolarity of zeolite supercages when they are housed in them. We have used all three molecules as polarity probes for X and Y zeolite cavities. Photophysical measurements, in general, were made on solvent free dry complexes under aerated conditions. Complexes were prepared as described in the experimental section. Inclusion of guests within zeolites was confirmed by diffuse reflectance spectra of the complexes. Also, gc analysis of the supernatant solvent from the complexation process revealed that no guest remained in the solvent. Results of the photophysical measurements with **1-3** as probes are summarized in **Table 1**. Also listed in the Table, for reference, are the measurements made for each probe in four solvents of varying polarity. In Figures 1-3 the spectra of the three probes in several zeolites are provided.

TABLE 1: Micropolarity of zeolite cages as probed by three conventional photochemical probes

Zeolite	Pyrene[a] I_5/I_1	DMABN[b] emission λ_{max} in nm	ε [c] in D	$E_T(30)$[c] kcal/mole	Pyrenealdehyde[d] emission λ_{max} in nm
LiX	0.58	337 and 535	68	59	481
NaX	0.60	338 and 528	56	57	473
KX	0.49	336 and 510	42	55	460
RbX	0.53	339 and 477	16	49	454
CsX	0.60	342 and 471	11	48	450
LiY	0.81	334 and 523	55	57	467
NaY	0.73	334 and 512	44	55	463
KY	0.56	335 and 500	37	54	450
RbY	0.60	335 and 463	4	46	443
CsY	0.71	335 and 440[e]	–	–	440
Cyclohexane[f]	1.25	340 only	2.0	30.9	392
Acetonitrile	0.60	337 and 485	35.9	46.7	415
Methanol	0.69	352 and 500	32.6	55.4	450
Water	–	347 and 553	78.3	63.1	480

a. Intensity of the monomer fluorescence at 392 nm *vs* 372 nm; 0.001 mg of pyrene in 250 mg of zeolite.
b. λ_{max} of the normal emission and emission from TICT state; 0.01 mg of DMABN in 250 mg of zeolite.
c. Solvent parameters, dipole moment (ε) and $E_T(30)$ were obtained by utilizing the plot provided in ref. 9c.
d. λ_{max} of the monomer emission; 0.01 mg of pyrenealdehyde in 250 mg of zeolite.
e. emission corresponds to phosphorescence.
f. All solvent numbers correlate well with literature reports: see refs. 6-9. Solvent parameters taken from ref. 5b.

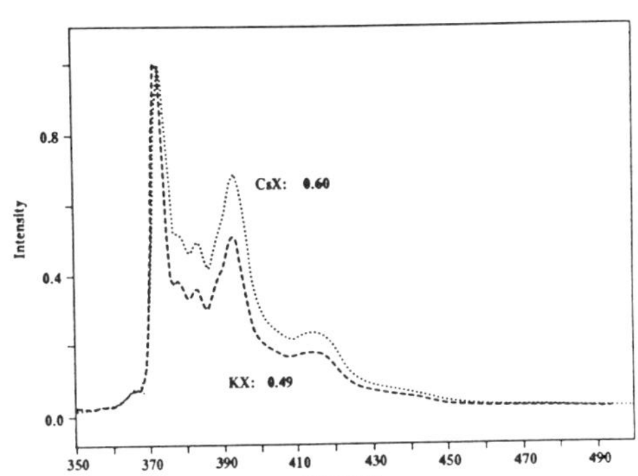

Figure 1: Emission spectra of pyrene in K X and Cs X. Note the change in the intensity ratio of the bands

Analysis of **Table 1**, on the basis of the background of each probe described above, reveals the following: (a) supercages of cation exchanged X and Y zeolites are relatively polar; (b) the supercage is very polar or possesses higher electric field when the cation is Li or Na; (c) the polarity/electric field of the cage decreases with the cation size (Li>Na>K>Rb>Cs); and (d) the cage is more polar in the case of X zeolites than in Y zeolites. The above observations are in line with the calculations of Ward who showed that the electric field as well as the electrostatic potential within the cage decreases with the increase in the size of the cation [10]. Also the fields and the potential depend on the number of cations. An increase in the micropolarity within the supercages of X relative to that in Y zeolites is consistent with the difference in the number of cations in these two zeolites: X zeolite has more Type III cations than does Y. Thus it is apparent that the nature and the number of cations determine the micropolarity of the cages.

The cations are indeed responsible for the micropolarity of the cage is further supported by the measurements made in the case of X and Y zeolites containing probe, *para*-dimethylaminobenzonitrile, co-included with water or hexane. Quartz cells containing *para*-dimethylaminobenzonitrile included M^+ X zeolites were filled with water and the spectra recorded. Similarly spectra for zeolite–hexane slurries prepared as above using hexane instead of water were recorded. Results are summarized in **Table 2**. Two emissions corresponding to the normal and the TICT states are seen in all three cases–dry, and water and hexane included zeolites. A comparison between the dry and the wet zeolites reveals that in the latter the polarity of the supercage is independent of the cation. Also the polarity of the supercage co-included with water is much lower than the dry ones.

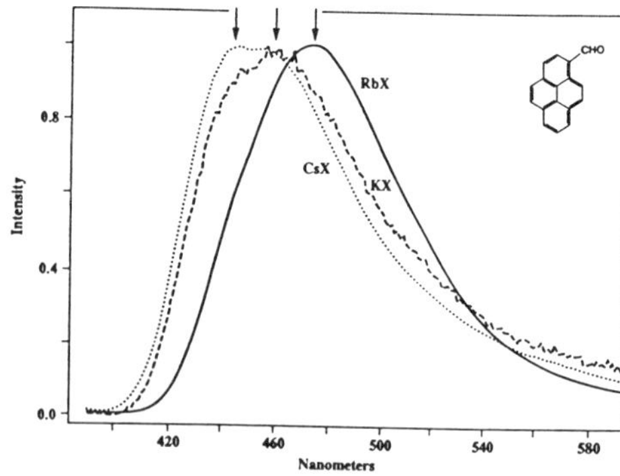

Figure 2: Emission spectra of pyrenealdehyde. Note the change in emission maxima.

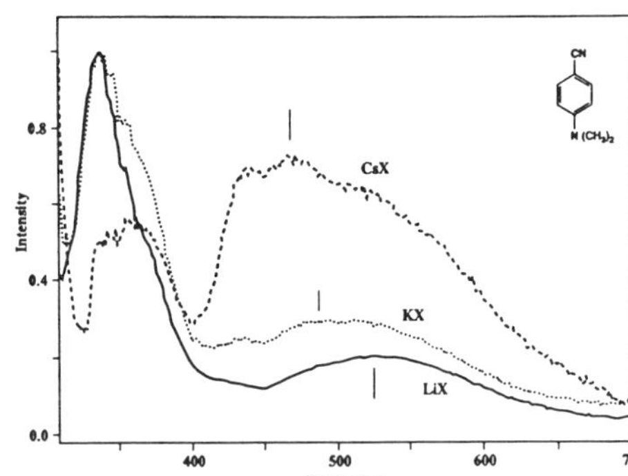

Figure 3: Emission spectra of *para*-dimethylamino-benzonitrile. Note the change in the long wavelength emission maxima and its intensity.

(This inference assumes that the probe's location is not disturbed by the co-included water. Both absorption and emission spectra of the aqueous portion of the slurry revealed that the probe is still within the zeolite and none is present in the aqueous layer.) We believe that the reduction in polarity and its cation independence are the result of water co-ordinating to the cation. Such a co-ordination forbids the probe's access to the cation. Therefore, the micropolarity sensed is that of the ordered water present within the cages. It is of interest to note that within the cages of X zeolite the polarity of the water is considerably reduced relative to the bulk solvent water probably owing to highly organized structure. While water had a significant influence on the micropolarity of the supercage, hexane did not. This is consistent with the known incapacity of hexane to co-ordinate to the cation. Apparently even in the presence of hexane the probe is still close to the cation and the micropolarity measured is that, reduced slightly by the surrounding hexane molecules, of the cation.

Table 2. Micropolarity of zeolite cages co-included with hexane or water as probed by *para*-dimethylaminobenzonitrile

Zeolite	λ_{max} of the normal and TICT emissions		
	Dry	Hexane Slurry	Water Slurry
LiX	337 and 535	342 and 522	337 and 475
NaX	338 and 528	344 and 520	336 and 478
KX	336 and 510	337 and 480	340 and 473
RbX	339 and 477	338 and 477	342 and 471
CsX	342 and 471	338 and 466	343 and 472

Where and what do these probes sense?

To interpret the information provided by the probes about their surroundings one should know their exact location within these structures. We assume in this study, based on literature precedence [11], that at low loading levels the included probe molecule resides within a supercage, probably close to the cation. While both pyrenealdehyde and *para*-dimethylaminobenzonitrile are expected to be close to the cation due to their strong acid-base type of interaction with it (lone pair with M^+), pyrene, because of its size and shape, may not be able to reach very close to the location of the cation. This is expected to be more so in the case of Li and Na as cations which are expected to be buried in Type III sites within the oxygen rings. Both pyrenealdehyde and *para*-dimethylaminobenzonitrile have a small handle (CHO and $N(CH_3)_2$/CN respectively) which can penetrate into narrow regions. The above logical expectation is indeed realized when one compares the polarities reported by the three probes. Closer perusal of **Table 1** reveals that pyrene does not sense the difference between Li^+, Na^+ and other cations. This is evident from the values of I_5/I_1 which are much higher in Li^+ and Na^+ than in K^+ exchanged X and Y zeolites. However, pyrene's effectiveness as a probe is obvious when one notices that it is able to sense the difference in polarity between the cages containing K, Rb and Cs as cations and that between X and Y zeolites.

Solvent polarity is being represented by more than 30 different polarity scales [5b]. Of these, scales based on dielectric constant, E_T (30), Z, Y, and (α, β and π^*) are popular. Since the E_T (30) scale takes into account both non-specific solute-solvent interactions as well as Lewis acid-base type interactions, this scale is particularly appealing for investigating guests within zeolites. Lewis acid-base type interaction is expected between the cation and the probes described in this study. By utilizing the reported correlations of of solvent parameters in the case of *para*-dimethylaminobenzonitrile [9c] we were able to estimate roughly the polarity within the supercages. The dielecric constant and E_T (30) for Li, Na, K, Rb and Cs–X and –Y are provided in Table 1. Supercages of Li, Na and K X have polarities similar to that of methanol-water mixture while those of Rb and Cs X have close to that of propanol. There have been recent attempts to monitor the micropolarity of the supercages of Na-X and Na-Y. The earliest such study is due to Baretz and Turro and this utilized pyrenealdehyde as the probe [4a]. Although this investigation was complicated by excimer formation due to high loading levels, supercages were inferred to be polar. Yoon and Kochi have concluded recently that supercages of methyl viologen (MV^{2+}) exchanged Na-Y zeolite possess polarity comparable to that of 50% aqueous acetonitrile [4b]. Dutta and Turbeville have suggested on the basis of salicylidenes as indicators that supercages of Na-X and Na-Y zeolites exhibit polarity similar to that of polar hydroxylic solvents [4c]. Thomas and co-workers have also concluded on the basis of pyrene as the probe that the supercages of X and Y zeolites are polar [4d]. The present study in addition to confirming the above observations, provides a wealth of additional information about zeolite interior.

Limitations

The effectiveness of the probes **1-3** to monitor the polarity of a supercage is dependant on the loading level (number of probe molecules to the number of cages, also termed as occupancy number and represented as <S>) and on the temperature. The variation in I_5/I_1 values in various X zeolites with respect to loading level of pyrene is presented in **Table 3**. Results suggest that as the loading level increases the supercage is less polar. This, we believe, is an artifact of the variation in the supercage occupancy. At higher levels of loading there is a distribution of supercages with different levels of occupancies–single and double. Therefore, the polarity estimated based on I_5/I_1 values of the monomer emission is an average of the polarities of singly and doubly occupied cages. The supercages containing two pyrene molecules would be less polar than those containing single pyrene molecules. When there is a large percentage of doubly occupied cages, which is the case at higher loading levels, the micropolarity reported by pyrene should be small and is indeed the case. Pyrenealdehyde also showed a similar behavior and excimer formation and double occupancies at higher loading levels complicated the micropolarity measurements of the supercage. At higher loading levels, *para*-dimethylaminobenzonitrile tended to aggregate and this gave rise to new emissions complicating the emission from the twisted state [12]. Therefore, all three probes described in this report are effective only at low loading levels (<S> less than 0.01).

Table 3: Pyrene as a Polarity Probe: Polarity Dependence on the Loading Level

Zeolite	Intensity of 392 nm band/Intensity of 372 nm band				
<S>[a]	0.00005	0.0005	0.005	0.05[b]	0.3[b]
Li X	0.57	0.58	0.61	0.70 (e)	0.71 (e)
Na X	0.58	0.60	0.60	0.66 (e)	0.76 (e)
K X	0.49	0.49	0.54	0.65 (e)	0.68 (e)
Rb X	0.55	0.53	0.53	0.56 (e)	0.64 (e)
Cs X	0.63	0.60	0.60	0.68	0.64

a. <S> refers to loading level.
b. 'e' in parentheses refers to the presence of excimer emission.

While both pyrene and pyrenealdehyde could be used as probes at any temperatures in the range 77 to 300 K, *para*-dimethylaminobenzonitrile could be used only at temperatures above ~200 K. At low temperatures the thermally activated rotation of the phenyl–nitrogen bond, required for the formation of the TICT state, is unable to compete with the intersystem crossing from S_1 to T_1 state of the planar form. Since the rate of intersystem crossing is dependent on the cation, due to the variation in the spin orbit coupling parameter [13], the exact temperature at which the probe is useful varies with the cation (200–250 K).

Generality—Other zeolites

We have explored, in addition to X and Y, zeolites such as mordenite, omega, L and ZSM-5 with *para*-dimethylaminobenzonitrile as the probe. Pyrene and pyrenealdehyde could not be used as they did not fit into the narrow channels of some of the above zeolites. In all cases except ZSM-5 the above probe was able to sense the micropolarity of the region where it is residing. Emission due to the TICT state was seen and its maximum varied with the cation and with the zeolite. Na mordenite, Na L and Na omega possess polarities close to that of acetonitrile (TICT emission maxima were 485, 480 and 478 respectively).

However, in the channels of ZSM-5 (Si/Al ratio 27, 41, 71, 120 and 500) *para*-dimethylaminobenzonitrile failed to show any emission due to the TICT state. One would have expected the channels of ZSM-5 with high aluminum content (Si/Al ratio 22) to be slightly polar. Absence of TICT emission, could be due to either the narrow channel restricts the rotation of the phenyl–nitrogen bond [14] or the probe is not located near the aluminum site. It is suggested that the aluminum is located at the intersections [15]. The probe being small (~10Å) may be sitting between intersections (distance between intersections ~18Å) and thus polarity difference between ZSM-5 with various Si/Al ratio is not be sensed.

CONCLUSIONS

This study utilizing organic photophysical probes indicates that the supercages of X and Y zeolites have polarities somewhere between that of water and methanol. The exact micropolarity of the cage is dependent on the nature and number of the cations. Slight variations in the polarity as sensed by the three probes is suggested to result from the differences in the exact location of these probes within the supercages. The technique described has been applied to other zeolites as well. Electrostatic fields within supercages are proposed to play an important role in polarizing guest reactant molecules [16]. It is clear from our work that a careful choice of the cation is necessary to achieve the maximum polarizing effect.

EXPERIMENTAL

Zeolites 13X (NaX), LZ-Y52 (NaY), mordenite, omega and L were obtained from Linde. The cation of interest was exchanged into these powders by contacting the material with the appropriate nitrate solution at 90° C. For each gram of zeolite, 10 mL of a 10% nitrate solution was used. This was repeated a number of times. The samples were then thoroughly washed with water and dried. Exchange loadings were typically between 70 and 85%. Prior to use these samples were heated in a furnace at 500° C in air for about 10 h. Activated zeolites were used immediately. Pyrene, pyrenealdehyde and *para*-dimethylaminobenzonitrile were obtained from Aldrich and recrystallized three times from ethanol.

Known amounts of **1**, **2** or **3** and the activated zeolites were stirred together in 20 mL of hexane for about 10 h. In a typical preparation 250 mg of zeolite and 0.05 mg of the probe was stirred in 20 mL of hexane. White powder collected by filtration of the solvent was washed with dry hexane several times and dried under nitrogen. Samples were taken in Pyrex cells fitted with Teflon stopcocks and degassed thoroughly (10^{-5} mm) and sealed.

Diffuse reflectance spectra of the zeolite solid samples were measured in 2 mm path length quartz cells using a Varian 2400 spectrometer equipped with an integrating sphere (Varian) and using barium sulphate (Kodak, White Reflectance Standard) as the reference. For comparison, spectra of the empty zeolites were also recorded. Data were recorded digitally and appropriate background corrections were carried out using the computer program, SpectraCalc (Galactic Industries). Emission spectra were recorded in supracil quartz EPR tubes under degassed conditions with a Spex Fluorolog 212 spectrofluorometer.

ACKNOWLEDGEMENT

It is indeed a pleasure to thank D. Sanderson for technical assistance and D. Corbin and P. Hollins for some of the zeolite samples.

REFERENCES

1. Kalyanasundaram, K. *Photochemistry in Microheterogeneous Systems*, Academic Press, New York, 1987.
   Ramamurthy, V. Ed., *Photochemistry in Organized and Constrained Media*, VCH, New York, 1991.
2. For a summary see: Ramamurthy, V., in: *Photochemistry in Organized and Constrained Media*, Ed., Ramamurthy, V., VCH, New York, 1991, p. 429.
   Turro, N. J. and Garcia-Garibay, M., in: *Photochemistry in Organized and Constrained Media*, Ed., Ramamurthy, V., VCH, New York, 1991, p. 1.
3. Breck, D. W., *Zeolite Molecular Sieves: Structure, Chemistry, and Use*, John Wiley and Sons, New York, 1974.
   Dyer, A., *An Introduction to Zeolite Molecular Sieves*, John Wiley and Sons, Bath, 1988.
   Introduction to Zeolite Science and Practice, Eds., van Bekkum, H.; Flanigen, E. M.; Jansen, J. C. Elsevier, Amsterdam, 1991.
4. a. Baretz, B. and Turro, N. J., *J. Photochem.*, **1984**, *24*, 201.
   b. Yoon, K. B. and Kochi, J. K., *J. Phys. Chem.*, **1991**, *95*, 1348.
   c. Dutta, P. K. and Tubeville, W., *J. Phys. Chem.*, **1991**, *95*, 4087.
   d. Iu, K. K. and Thomas, J. K., *Langmuir*, **1990**, *6*, 471.
5. a. Kalyanasundaram, K., in: *Photochemistry in Organized and Constrained Media*, Ed., Ramamurthy, V., VCH Publishers, New York, 1991, p. 39.
   b. Reichardt, C., *Solvents and Solvent effects in Organic Chemistry*, VCH Publishers, New York, 1988.
6. Kalyanasundaram, K. and Thomas, J. K., *J. Am. Chem. Soc.*, **1977**, *99*, 2039.
   Dong, D. C. and Winnik, M. A., *Photochem. Photobiol.*, **1982**, *35*, 17.
7. Kalyanasundaram, K. and Thomas, J. K., *J. Phys. Chem.*, **1977**, *81*, 2176.
   Dederen, J. C., Coosemans, L., De Schryver, F. C. and Van Dormael, A., *Photochem. Photobiol.*, **1979**, *30*, 447.
8. Rettig, W., *Angew. Chem. Int. Ed. Engl.*, **1986**, *25*, 971.
9. a. Wang, Y. and Eisenthal, K. B., *J. Chem. Phys.*, **1982**, *77*, 6076.
   b. Auweraer, M. V., Grabowski, Z. and Rettig, W., *J. Phys. Chem.*, **1991**, *95*, 2083.
   c. Cox, G. S., Hauptman, P. J. and Turro, N. J., Photochem. Photobiol., **1984**, *39*, 597.
10. Ward, J. W., *J. Catalysis*, **1968**, *10*, 34.
    Ward, J. W., *Appl. Ind. Catal.*, **1984**, *3*, 271.
11. Hepp, M., Ramamurthy, V., Corbin, D. R. and Dybowski, C., *J. Phys. Chem.*, **1992**, *96*, 2629.
12. Rotkiewicz, K., Leismann, H. and Rettig, W., J. *Photochem. Photobiol. A: Chemistry*, **1989**, *49*, 347.
    Meech, S. R. and Philips, D., *J. Chem. Soc.*, **1987**, *83*, 1941.
    Weisenborn, P. C. M., Huizer, A. H. and Varma, C. A. G. O., *Chem. Phys.*, **1989**, *133*, 437.
13. Ramamurthy, V., Caspar, J. V., Corbin, D. R. and Eaton, D. F., *J. Photochem. Photobiol., A: Chemistry*, **1989**, *50*, 157.
    Ramamurthy, V., Caspar, J. V. and Corbin, D. R., *Tetrahedron Letters*, **1990**, *31*, 1097.
    Caspar, J. V., Ramamurthy, V. and Corbin, D. R., *Coordination Chem. Rev.*, **1990**, *97*, 225.
    Caspar, J. V., Ramamurthy, V., Corbin, D. R., Schlyer, B. D., and Maki, A. H., *J. Phys. Chem.*, **1990**, *94*, 3391.
14. Ramamurthy, V., Caspar, J. V., Corbin, D. R., Eaton, D. F., Kauffman, J. and Dybowski, C., *J. Photochem. Photobiol, A: Chemistry*, *51*, 259, **1990**.
    Ramamurthy, V., in: *Inclusion Phenomena and Molecular Recognition*, Ed., Atwood, J. A., Plenum Press, New York, 1990, p. 351.
15. Huddersman, K. D. and Rees, L. V. C., *Zeolite*, **1991**, *11*, 270 and references therein.
16. Wada, Y., Yasunori, Y. and Morikawa, A., *J. Chem. Soc., Chem. Commun.*, **1990**, 319.
    Ozin, G. A., Ozkar, S., Patore, H. O., Poe, A. J. and Vichi, E. J. S., *J. Chem. Soc., Chem. Commun.*, **1991**, 141.
    Ramamurthy, V., Corbin, D. R., Turro, N. J. and Sato, Y., *Tetrahedron Letters*, **1989**, *30*, 5829.

LOCATION OF IRON IN FRAMEWORK-SUBSTITUTED ZEOLITE L USING ANOMALOUS POWDER X-RAY DIFFRACTION AND X-RAY ABSORPTION FINE STRUCTURE

I. J. PICKERING‡, D. E. W. VAUGHAN, K. G. STROHMAIER, G. N. GEORGE‡ and G. H. VIA

Exxon Research and Engineering Company, Route 22 East, Annandale, NJ 08801, USA

ABSTRACT

Iron substituted zeolite LTL has been synthesized with one third of the framework aluminium replaced by iron. X-ray absorption fine structure at the iron K absorption edge reveals that the iron is substituted into the framework, with a first coordination shell of oxygen with 4 Fe-O distances of 1.85Å and an outer shell of silicon. Anomalous dispersion powder diffraction combined with Rietveld profile analysis has been used to locate the iron with respect to the two crystallographically distinct framework sites of the LTL structure, showing it to be essentially randomly distributed between them. Anomalous dispersion powder diffraction is also used to verify that the extra-framework cation sites contain negligible iron.

INTRODUCTION

There has been much interest in the synthesis of high-silica zeolites with heteroatoms such as transition metals substituted for silicon or aluminum in the framework T-sites [1,2], but there is little convincing evidence that transition metals can be incorporated in large amounts into conventional zeolite frameworks. Such modified zeolites may have subtly different properties than their unmodified analogues, particularly in having different acidities and providing novel metal-support interactions for noble metal catalysts.

A major problem is that transition metal incorporation during zeolite synthesis is not necessarily only into the framework. The transition metal ion may additionally be incorporated as extra-framework cations (balancing the framework charge-deficit), or as an interstitial phase of finely-divided particles, or metal cluster complexes, growing either in the zeolite cavities or intimately associated with the crystal matrix. Many reports in the literature of supposed transition metal framework substitution may in reality be such an interstitial phase having crystallites too small to be identified by X-ray diffraction. The problem is further complicated in that the level of framework substitution is frequently low. Recent work on ferrialuminosilicate sodalite shows that at framework occupancies of about 5%, Fe^{3+} begins to partition between the sodalite framework and nanocrystal matrix haematite [3].

The matter of determining whether a transition metal has actually substituted into the zeolite framework is far from trivial. Various techniques for structural characterization [4] have been applied to the problem, but many of these leave ambiguities. Techniques which are direct structural probes of the location and environment of the heteroatom ought to provide more definitive evidence of framework substitution, if present. In this paper we use a combination of X-ray absorption fine structure (XAFS) and anomalous dispersion powder X-ray diffraction (ADPD); the former to determine whether framework substitution has been achieved, and the latter to probe the precise crystallographic location of the substituted heteroatoms.

---

‡ Present address: Stanford Synchrotron Radiation Laboratory, P.O. Box 4349, Bin 69, Stanford, CA 94309, U.S.A.

From the Proceedings of the 9th International Zeolite Conference, Montreal 1992, Eds. R. von Ballmoos et al.,

The technique of XAFS [5] can be used to probe the local coordination of a specific element in a sample, and can yield estimates of the number and type of atoms surrounding the central absorbing atom, together with an accurate estimate of the interatomic distance. Whereas XAFS is sensitive to the local atomic environment, diffraction probes the long-range structure of the solid. In this paper the technique of ADPD [6] is used, in which powder diffraction data is measured at more than one energy, one of which is close to the absorption edge of the heteroatom (in this case the K absorption edge of iron). In the proximity of the absorption edge the otherwise energy-independent atomic scattering factor f_o is moderated by real and imaginary components of the anomalous dispersion correction, so that the atomic scattering factor may be expressed as $f = f_o + f' + if''$. Since the scattering terms of the other atoms are invariant in this energy range, the change in scattering of the iron alone can serve to highlight its position in the unit cell. It can also be used to verify the shared site occupancy of an unfilled cation site in the zeolite.

The structure of zeolite L [7] was solved initially by Barrer and Villiger [8]. The structure is built from cancrinite cages, which consist of tetrahedrally coordinated silicon, aluminium or in this case iron atoms linked through oxygen atoms. The cages are stacked via double 6-rings of T-atoms to form columns, which are further linked together to form a large channel parallel to the columns which has a 12 T-atom window (see Figure 1). There are two crystallographically distinct T-atom sites: T1 is on the 12-ring and T2 is on the 6-ring. Extra-framework cations are incorporated into the structure in order to balance the deficit of charge caused by the replacement of Si^{4+} by Al^{3+} or Fe^{3+} (see Figure 1). In the hydrated form water molecules also occupy space in the channel.

EXPERIMENTAL

Iron substituted zeolite L was prepared from a ferric chloride modified sodium aluminosilicate gel [9] of overall composition:

$$5.0K_2O : 11.8SiO_2 : (.67Al, .33Fe)_2O_3 : 176H_2O : (KCl)$$

A 50 wt% solution of ferric chloride hexahydrate dissolved in water was added as the last component during the gelation procedure, in a manner analogous to the use of acidic aluminum salts in conventional aluminosilicate syntheses. The preparation was heated at 100°C for 5 days (1 atm. pressure) yielding a light tan colored product. Powder diffraction analysis using CuK_α radiation revealed the product to be single phase zeolite L. Chemical analysis using ICP-AES (Leeman Laboratories, Plasma-Spec 2.5) gave the following percentages by weight: Fe - 6.42, K - 13.55, Si - 26.77, Al - 6.29. The corresponding atomic ratios of $\frac{K^+}{Al^{3+}}$ = 1.49 and $\frac{K^+}{(Al^{3+} + Fe^{3+})}$ = 0.996 suggest that Fe is present in the framework. If it is assumed that all the iron is indeed in the framework, then the formula of the 36 T-atom unit cell can be written as:

$$K_{9.59}(Si_{26.37}Al_{6.45}Fe_{3.18}O_{72}).nH_2O.$$

XAFS spectra were collected at NSLS beamline X10C and SSRL beamline SB07-3. In the latter case the data were collected at close to 4K. Energy calibration was carried out by simultaneously recording the spectrum of iron metal, the lowest energy K-edge inflection of which is assumed to be 7111.3eV. The

EXAFS (extended X-ray absorption fine structure) oscillations were quantitatively analyzed by a curve-fitting procedure [10] to the approximate expression:

$$\chi(k) = \sum_{i=1}^{n} \frac{N_i A_i(k,R_i)}{kR^2} \exp\left(\frac{-2R_i}{\lambda(k,R_i)}\right) \exp(-2\sigma_i^2 k_i^2) \sin[2kR_i + \phi(k,R_i)]$$

Here **k** is the photoelectron wave number, N_i is the number of i-type atoms at a mean distance R_i from the absorber atom (in this case iron). The summation is over all sets of equivalent atoms, n. The functions A(k,R), λ(k,R) and ϕ(k,R) are the curved-wave total EXAFS amplitude, photoelectron mean free path, and total EXAFS phase functions, respectively. These were calculated by using the program ***feff*** of Rehr and co-workers [11,12].

Synchrotron X-ray powder diffraction patterns of hydrated Fe LTL were collected in flat-plate geometry on NSLS beamline X10B using a wavelength of 0.9706Å, and on NSLS beamline X10C with a wavelength of 1.7397Å (close to the K absorption edge of iron), both with a stepsize of 0.01° 2θ. Rietveld profile refinements [13] of the substituted LTL structure were performed using the GSAS (Generalized Structure Analysis System) suite of programs [14]. The space-group P6/mmm (no. 191 [15]) was assumed. The starting atomic parameters for the refinement were taken from the powder diffraction study of K^+-LTL by Barrer and Villiger [8]. Background functions were subtracted from the profiles using a point by point interpolation. The profile peak shape function used was a multi-term Simpson's rule integration [16] of a pseudo-Voigt [17] with five coefficients. Scattering factors were assumed for Si^{4+}, Fe^{3+}, K^+ and O^-. Three temperature factors B were refined, one for all the T-site atoms, another for the framework oxygens and the third for the extra-framework cations, with appropriate constraints applied. The temperature factors for the oxygens of the water molecules were fixed at 15$Å^2$ to model the disorder.

RESULTS

X-ray absorption fine structure

Figure 2 (inset) shows the iron K absorption edge spectrum of iron-substituted zeolite L. Edge spectra are very sensitive to local electronic structures, but are quite difficult to interpret in a quantitative manner. Nevertheless, the spectrum in Figure 2 shows features characteristic of the tetrahedral geometry expected for an iron substituted into the zeolite framework. In particular, the intensity of the dipole-forbidden 1s-3d transition at 7113 eV indicates the presence of significant *p-d* mixing which is expected for a tetrahedral site.

EXAFS (Extended X-ray Absorption Fine Structure) spectra, on the other hand, are rather simpler to interpret in a quantitative manner. Figure 2 also shows the Fe K-edge EXAFS Fourier transform of Fe-LTL, together with the Fourier transform of the results of a curve-fitting analysis. The analysis indicates the presence of four iron-oxygen scatters at 1.85Å, and inclusion of iron-silicon interactions at 3.13 and 3.47Å significantly improved the fit. Taken together these results strongly suggest that the iron is present at a site in the zeolite framework.

Multiple wavelength powder X-ray diffraction

Two independent data sets were collected at X-ray wavelengths of 0.9706Å and 1.7397Å. The latter wavelength (1.7397Å, 7126.7eV) is close to the K absorption edge of iron (see inset, Figure 2), where the X-ray atomic scattering factor is modified by the so-called anomalous scattering, which has real and imaginary components f' and f''. The diffraction patterns clearly show that the sample contains zeolite L as the single crystalline phase, with no evidence for impurity crystalline phases

Initially, refinements used the data set collected at 0.9706Å in the range $5 \leq 2\theta \leq 50°$, neglecting the iron and assuming only Si^{4+} occupancy of the T-sites. Structural parameters were varied starting with the tetrahedral atom positions and then adding in successively the framework oxygen positions and the fractional occupancies and positions of the extra-framework cations. The initial water oxygen positions of Barrer and Villiger [8] were then refined, and those with an occupancy of less than 10% were removed from the refinement. Atomic parameters were converged, and this refinement was then used as a starting model for locating the Fe^{3+} within the framework.

Three different models for substitution of Fe^{3+} were tested, firstly with all the iron located on the T1 (12-ring) site, secondly with all the iron on the T2 (6-ring) site, and thirdly with the iron being distributed over both sites. The limitations of space prevent us from describing the analyses in detail here, and a full description will be presented in a future publication. It was found that the third model, involving distribution of Fe^{3+} between the two T-sites gave the best fit (weighted R-factor = 4.64, compared with 5.06 for all Fe on T1 and 4.88 for all on T2).

In order to better define the location of the iron atoms, the second data set collected at the "anomalous" wavelength of 1.7397Å was used together with the first in a simultaneous refinement of the model using a total of 595 diffraction peaks. The value of f', the real part of the anomalous dispersion correction, was refined for the 1.7397Å data set to give a value of -10.4(10) (the value of f' at the original wavelength is considered to be negligible). In the final refinement the distribution of iron between the two sites was allowed to vary, the overall level of iron being constrained equal to that derived from chemical analysis.

A further refinement was carried out to test for the presence of Fe^{3+} in the extra framework cation positions, in which occupancy of iron in these sites was allowed to vary without constraint. The resulting occupancies obtained were negligible, indicating that little or no iron is present in these sites.

The final observed, calculated, and difference diffraction pattern for the 1.7397Å data set is shown in Figure 3, demonstrating the excellent fit obtained. A similar quality of fit was obtained for the 0.9706Å wavelength data. A summary of the refinement results is given in Table 1.

DISCUSSION

The XAFS results show conclusively that the majority of the iron in the Fe LTL sample is indeed substituted into the framework. The Fe-O distance and coordination of four is exactly as would be expected for such an environment. The Fe-Si outer shell is also strong evidence for framework substitution. Attempts to fit an outer shell of iron, which might indicate the presence of a dense oxide or hydroxide amorphous to powder diffraction, were unsuccessful. The EXAFS results are therefore consistent with iron being present only as the framework substituted phase within the limit of the technique.

The data presented here show that the iron is essentially randomly distributed between the two crystallographically distinct T-sites in the zeolite L framework. The ordering of substituted atoms with respect to these sites has been commented on by different authors. In the gallium form, both Wright *et al.* [18] and Newsam [19] are in agreement that the gallium is randomly distributed between the two T-sites. In the case of aluminium, some variability is suggested since earlier X-ray diffraction studies [8,20] show a preference of aluminium for the T2 site, whereas powder neutron diffraction data [21] show instead a preference for the T1 site using both site occupancy and T-O bond length arguments. The extent to which these differences may be caused by different synthesis techniques (raw materials sources; crystallization conditions, etc.) has not yet been investigated. In iron L, the refined T-site occupancies show that the iron is essentially randomly distributed between the two crystallographically distinct T-sites. The use of bond length data to verify this is much more difficult in the case of iron substituted L since we are dealing with the substitution of a mixture of Fe^{3+} and Al^{3+} rather than the single species as in Al- or Ga-substituted L. The average T-O bond lengths for T1 (1.68(4)Å) are slightly longer than for the T2 site (1.63(2)Å), possibly indicating that there is an enrichment of aluminium on this site. The mean T-O distance for all the sites of 1.65Å is in agreement with the weighted sum of bond distances from Si-O, Al-O and Fe-O in the proportions from the chemical analysis.

The anomalous dispersion powder diffraction data allows a full description of the location of the extra-framework cations in the hydrated zeolite. Site A, in the double 6-ring, has negligible scattering density. Site B, at the center of the cancrinite cage (see figure 1), is fully occupied with potassium ions and is coordinated by a total of 12 oxygens from the 6-rings (6 O(3) at 2.83Å and 6 O(5) at 3.33Å). Site C, located between the cancrinite cages, is also filled with potassium and is surrounded by 4 O(5) at 3.02Å and 8 O(4) at 3.29Å. By contrast site D, in the wall of the 12-ring channel, is only about 78% occupied by potassium. The cation is coordinated to framework species at 2.13Å (2) and 3.21Å (4), but is most closely bound by water molecules $O_w(7)$ located in the channel at a distance of 2.86Å. There is no evidence for any occupancy by iron of this or other cation sites, as deduced from the ADPD studies. The total cation count, assuming full occupation of B and C, negligible occupancy of A and the refined value of occupancy for D gives 9.75 K^+ per unit cell, in excellent agreement with the value of 9.59 from chemical analysis.

CONCLUSION

Iron substituted zeolite LTL has been synthesized with Fe^{3+} replacing one third of the Al^{3+}. XAFS spectroscopy has been used to verify that the iron is indeed substituted into the framework, showing a first coordination shell of oxygen and second shell of silicon with bond lengths and coordination numbers commensurate with tetrahedral framework substitution. ADPD has been used to locate the iron within the L framework. Site occupancies show that it is essentially randomly distributed between the two distinct crystallographic sites, with negligible iron on any of the extra-framework cation sites. Attempts to substitute higher levels of Fe^{3+} result in coprecipitation of ferric hydroxide.

ACKNOWLEDGEMENTS

We thank the staff of both NSLS and SSRL for their assistance in making this work possible. The National Synchrotron Light Source is funded by the Division of Materials Sciences, U.S. Department of Energy, under contract DE–AC02–76CH–0016. Stanford Synchrotron Radiation Laboratory is funded by the Department of Energy under contract DE–AC03–82ER–13000, Office of Basic Energy Sciences, Division of Chemical Sciences, and NIH, Biotechnology Resource Program, Division of Research Resources.

REFERENCES

1. Barrer, R. M. *Hydrothermal Chemistry of Zeolites,* Academic Press, London, 294 (1982).
2. Ratnasamy, P. and Kumar, R., *Catalysis Today* **9**, 329 (1991).
3. Pickering, I. J., Prince, R. C., Bare, R., Vaughan, D. E. W., Strohmaier, K. S., and George, G. N., *J. Am. Chem. Soc.,* manuscript in preparation.
4. Thomas, J. M. and Vaughan, D. E. W. *J. Phys. Chem. Solids,* **50**, 449 (1989).
5. Koningsberger, D. C. and Prins, R. (eds) *X-ray Absorption: Principles, Applications, Techniques of EXAFS, SEXAFS and XANES* Wiley Interscience (1988).
6. James, R.W. *The Optical Principles of the Diffraction of X-rays,* Ox Bow Press, CT, pp. 135-192 (1982).
7. Breck, D. W. and Acara, N. A., U.S. Patent 711,565 (1958).
8. Barrer, R.M. and Villiger, H., *Z. Kristallogr. Kristallgeom., Kristallphys., Kristallchem.* **128**, 352 (1969).
9. Vaughan, D. E. W., U.S. Patent 4,657,749 (1987).
10. George, G. N., Kipke, C. A., Prince, R. C., Sunde, R. A., Enemark, J. H. and Cramer, S. P. *Biochemistry* **28**, 5075 (1989).
11. Rehr, J.J., Mustre de Leon, J., Zabinsky, S.I. and Albers, R.C., *J. Am. Chem. Soc.* **113**, 5135 (1991).
12. Mustre de Leon, J., Rehr, J.J. and Zabinsky, S.I., *Phys. Rev. B* **44**, 4146 (1991).
13. Rietveld, H. M., *J. Appl. Cryst.* **2**, 65 (1969).
14. Larson, A.C. and Von Dreele, R.B., "General Structure Analysis System", 1985-1991, LANSCE, MS-H805, Los Alamos National Laboratory, Los Alamos, NM 78545.
15. Hahn, Th., Ed. *Intl. Tables for X-ray Crystallography,* Riedel, Dordrecht, Holland, Vol.A (1983).
16. Howard, C.J., J, Appl. Cryst. 15, 615 (1982).
17. Thompson, P. Cox, D.E. and Hastings, J.B., *J. Appl. Cryst.* **20**, 79 (1987).
18. Wright, P. A., Thomas, J. M., Cheetham, A. K. and Nowak, A. K., *Nature (London),* **318**, 611 (1985).
19. Newsam, J. M., *Mat. Res. Bull.,* **21**, 661 (1986).
20. Baerlocher, C. and Barrer, R.M., *Z. Kristallogr. Kristallgeom., Kristallphys., Kristallchem.* **136**, 245 (1972).
21. Newsam, J. M., *J. Phys. Chem.,* **93**, 7689 (1989).

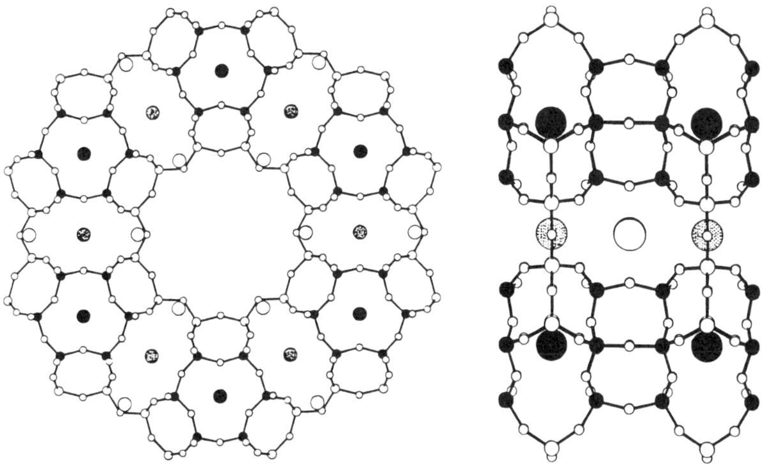

Figure 1. The structure of zeolite L. The medium circles represent the T-sites: T1-open (on the 12-ring) T2-filled (on the 6-ring). The small circles are the framework oxygens. The large circles are the populated cation sites, as follows: B-solid (center of the cancrinite cage) ; C-stippled (between the columns, adjacent to the cancrinite cage) D-open (in the wall of the channel). (Not shown are A (in the double 6-ring) and E (adjacent to the double 6-ring)). a) Projection down the c-axis showing the12-ring channel. b) Projection down the a-axis showing a side view of 4 cancrinite cages.

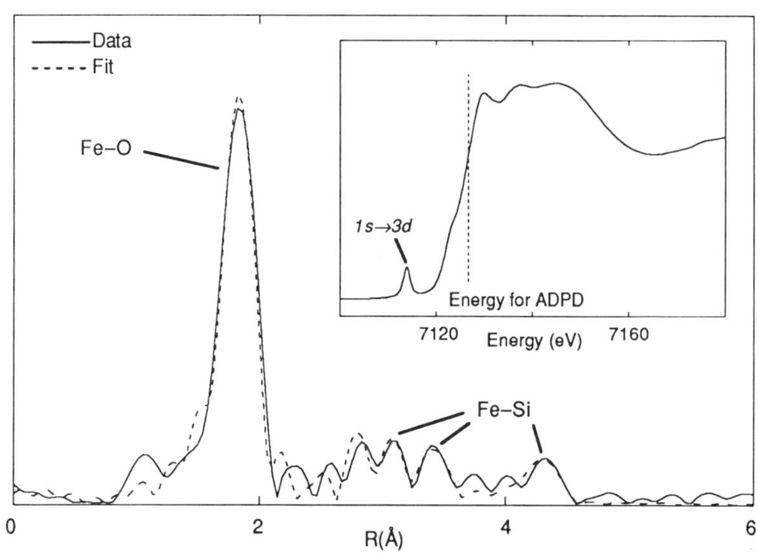

Figure 2. Fe K absorption edge spectrum (inset) and EXAFS Fourier transforms of iron-substituted LTL.

Table 1. Summary of final atomic parameters for the structure refinement of Fe LTL using powder X-ray diffraction data sets collected at 0.9706Å and 1.7397Å. Reduced $\chi^2 = M/(N_{obs} - N_{var}) = 1.107$ and $R_{wp} = \sqrt{(M/\Sigma\, wI_o^2)} = 4.94$. Quantity minimized, $M = \Sigma\, w(I_o - I_c)$, I_o, I_c are observed and calculated intensities and w a weight related to the error. N_{obs}, total number of observations is 9898 and N_{var}, number of variables is 53. Refined lattice parameters a_o, c_o are 18.4797(5), 7.5456(2)Å, respectively. *Occupancies of the T-sites are specified as the fractional site occupancy by iron; the balance of the site is refined as Si^{4+}.

Atom	Occupancy	x	y	z
T(1)	0.098(3)*	0.09118(18)	0.3551(2)	0.5
T(2)	0.083(2)*	0.1701(2)	0.50038(19)	0.2117(3)
O(1)	1	0	0.2670(5)	0.5
O(2)	1	0.16731(19)	2x	0.5
O(3)	1	0.26632(14)	2x	0.2560(8)
O(4)	1	0.1054(3)	0.4182(3)	0.3212(5)
O(5)	1	0.42240(16)	2x	0.2733(7)
O(6)	1	0.1478(3)	0.4828(3)	0
K(A)	0.058(6)	0.333333	0.666667	0
K(B)	1.059(7)	0.333333	0.666667	0.5
K(C)	0.959(5)	0	0.5	0.5
K(D)	0.791(4)	0	0.2981(2)	0
$O_w(1)$	0.755(9)	0.0697(3)	2x	0.5
$O_w(3)$	0.29(3)	0	0	0.5
$O_w(4)$	0.752(8)	0	0.1311(4)	0.2081(14)
$O_w(6)$	0.556(14)	0	0	0.129(3)
$O_w(7)$	0.930(8)	0.1246(3)	2x	0

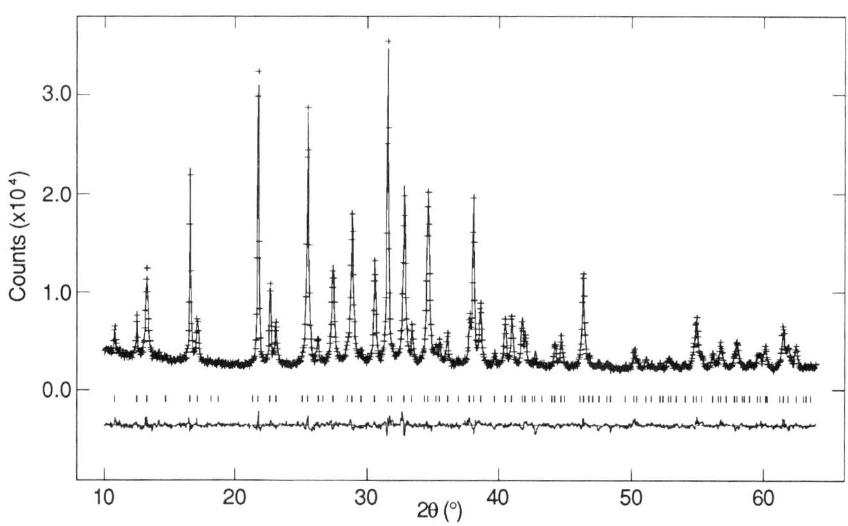

Figure 3. ADPD pattern of Fe LTL at 1.7397Å. In the upper trace the points show experimental data and the line the calculated best fit. The lower trace shows the fit residual, and the ticks mark the peak positions.

ELECTRON-ACCEPTING STRENGTH OF ALUMINUM SPECIES IN H-ZEOLITES

F. R. CHEN and J. J. FRIPIAT

Department of Chemistry and Laboratory for Surface Studies, University of Wisconsin-Milwaukee, P.O. Box 413, Milwaukee, Wisconsin 53201 USA

ABSTRACT

The strength of Lewis acid sites in H-mordenite, HL, HY zeolites and transition alumina was studied by EPR with aniline and N,N-dimethylaniline (DMA) as molecular probes. The following order was established : HM > HL > HY > alumina. Non-framework Al species in zeolites was shown to be an important source of Lewis acidity. Introduction of oxygen after the adsorption of organic reagent resulted in the formation of O_2, most probably located on the non-framework Al species. The measurement of the hyperfine splitting constant, A_{zz}, of O_2 confirms the above order of acid strength. The effect of zeolite structure on the acid strength of both framework and non-framework Al is discussed.

INTRODUCTION

Knowledge of zeolite acidity has been much advanced in recent years. However, owing to the complexity of zeolites (difference in structure, texture, Si/Al ratio, ...) some very basic problems remain unsolved. Numerous results obtained by different physicochemical methods have suggested that the acid strength is influenced by structural features [1-3], but some authors believe that the acid strength of different types of zeolites without non-framework Al or other defects should be the same [4].

EPR has been shown to be a powerful method for studying the acid strength of zeolites, either by investigating their ability to ionize organic molecules [5] or by measuring the variation of hyperfine splitting constants of the formed radicals [6,7].

In the present work, the electron-accepting strength (or Lewis acid strength) of H-mordenite (HM), HL, HY and transition alumina was investigated in combining EPR with MAS NMR technique.

EXPERIMENTAL

Linde H-mordenite (HM-1) and Linde HY (steamed) were further exchanged several times with ammonium chloride to insure a high exchange rate. HM-2 is a Zeolon H-mordenite dealuminated by acid leaching (6N HCl, reflux). HL is obtained by exchanging KL (TSZ-520 KOA, Toya Soda Co. Ltd.) 4 times at 80°C with ammonium chloride. The zeolites were then washed, dried at 120°C and calcined (except HM-2) at 550 °C in air for 4 h. The transition

From the Proceedings of the 9th International Zeolite Conference, Montreal 1992, Eds. R. von Ballmoos et al.,

alumina was derived from calcining a mechanically ground Boehmite.

100 mg samples were treated in an EPR tube under 760 Torr oxygen at 550°C for 2 h and then outgassed at the same temperature overnight. The adsorption of aniline and DMA was carried out at room temperature at the equilibrium vapor pressure. The ionization potentials of aniline and DMA are 7.7 and 7.12 eV, respectively. EPR measurements were performed with a Varian E-15 spectrometer (X-band) at room temperature. The quantitative estimate of radical concentration, expressed as spins per gram of sample, was obtained by double integration and comparison with DPPH standard.

The characterization of the samples was achieved by ^{29}Si and ^{27}Al high resolution solid state NMR with a GN-500 spectrometer. The experimental conditions and calculation procedure were described elsewhere [8].

Table 1. Properties of the prepared samples

Samples	HM-1	HM-2	HL	HY	alumina
Si/Al^a	12.3	14.2	3.1	4.7	0
Si/Al^b	35	17	4.9	8.2	0
S (m^2/g)	480	495	380	540	76
D (Å)^c	6.7x7.0	6.7x7.0	7.1	~15^d	> 20
Al_{NF}/Al_F^e	1.7	0.18	0.44	0.61	-

a: chemical analysis, b: ^{29}Si MAS NMR, c: dimension of the cavities or pores (alumina), d: dimension of the supercages and e: the ratio of non-framework Al to framework Al.

RESULTS

Formation of organic radicals

The adsorption of aniline at room temperature on the two HM and HL generated immediately radicals with well defined hyperfine structure (Fig. 1). The formed radical was identified to have the structure of $C_6H_7N^{+\bullet}$ [8]. Although the two HM have similar Al content by chemical analysis, the radical concentration is quite different: $2.3x10^{17}$ spins/g for HM-2 and $16x10^{17}$ spins/g for HM-1. This will be discussed later. The hyperfine splitting constants of the radical on the two HM are very similar. The overall width of the spectra a and b, (the distance between the peaks of the two outer lines as indicated in the figure) is about 45 Gauss. However, the overall width of the spectrum observed on HL is larger, being about 48

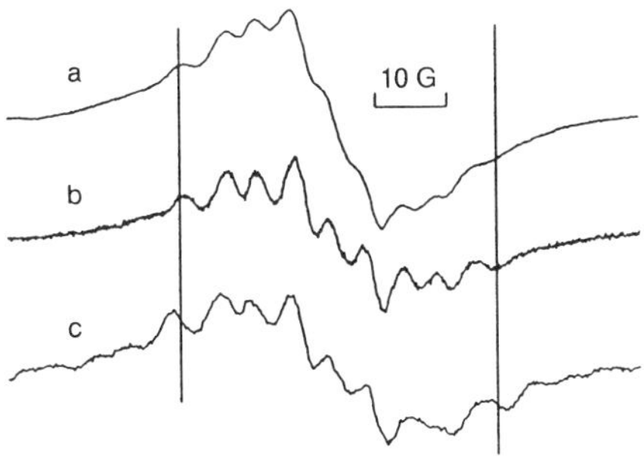

Fig. 1 EPR spectra of aniline radical cation at room temperature (a): HM-1, (b): HM-2 and (c): HL.

Gauss. The radical concentration on HL is 3×10^{17} spins/g, close to the value obtained for HM-2.

The value of the hyperfine splitting constants is proportional to the electron density due to the unpaired electron on the interacting nuclei [9]. A smaller value means that less electron density remains on the interacting nuclei or that more electron density is shifted from the radical to an electron-withdrawing center. Al Lewis acid sites were considered as the most probable electron-accepting centers [8]. Therefore, the value of the hyperfine splitting constants reflects the strength of the corresponding Lewis acid center which is stronger on HM than on HL.

Aniline adsorption on HY yields radicals only in the presence of oxygen without well defined hyperfine structure, while on the alumina no radical was observed even in the presence of oxygen. Oxygen increases the electron-accepting strength of Lewis acid sites [8]. Organic radical formation was detected on the alumina only in the adsorption of DMA [10]. These results demonstrate that the Lewis acid strength of the above materials scales as HM > HL > HY > alumina.

Formation of superoxide ion O_2^- and its interaction with Al species

In a previous paper [8], it was reported that introduction of oxygen into the aniline-HM-1 and HY systems resulted in the formation of superoxide ion identified in the superimposed spectral contribution of O_2^- and of the aniline radical cation formed prior (HM-1) or simultaneously (HY) to oxygen exposure. The interaction of the superoxide ion with an Al center ($I = 5/2$) was clearly observed, as evidenced by the sextet superhyperfine structure of the g_{zz} component of O_2^- (Fig. 2a and c). Since the g_{xx} and g_{yy} components are superimposed with the absorption of aniline radical cation, figure 2 shows only the expanded

g_{zz} components to have a better view of the superhyperfine splitting constant, A_{zz}. Such a superimposed spectrum was also obtained when oxygen was introduced into aniline-HL and DMA-alumina systems and the g_{zz} components are shown in Fig. 2b and d.

The value of A_{zz} varies sensibly with the nature of solid acids. It measures 6.3, 5.8 and 5.4 Gauss for HM-1, HL and HY, respectively. Even though the g_{zz} component on alumina is obviously smaller than the three others above, the A_{zz} value can not be measured directly.

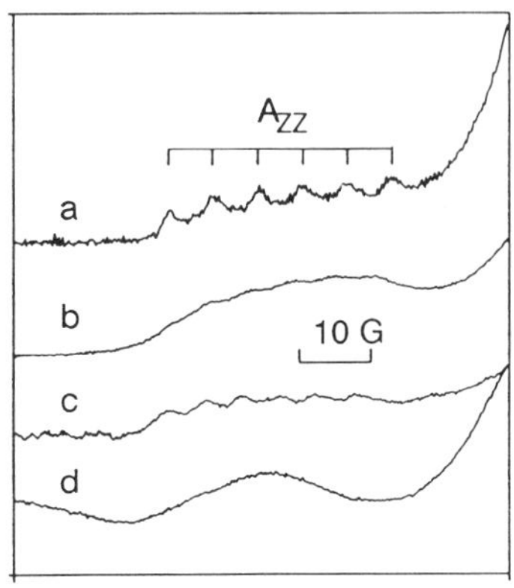

Fig. 2 EPR spectra of the g_{zz} components of O_2^- on (a): HM-1, (b): HL, (c) HY and (d): alumina.

However, a theoretical simulation with a Gaussian line shape and the following parameters: g_{zz} = 2.040, g_{yy} = 2.008, g_{xx} = 2.004, and A_{zz} = 4.9, A_{yy} = 3.8, A_{xx} = 3.7 Gauss gives a reasonable fit of the g_{zz} component. Therefore, the value of A_{zz} = 4.9 Gauss can be a good estimate for the alumina. The superhyperfine splitting observed in the present case is generally considered to arise from the delocalization of the unpaired electron of O_2^- onto an Al nucleus [11]. A smaller value of A_{zz} means less spin density on the Al nucleus resulting from a weaker electron-withdrawing ability or a weaker Lewis acid center. The order of the strength of Lewis acid center for the studied samples is thus: HM-1 > HL > HY > alumina, consistent with the order observed in organic radical formation.

The g_{zz} values for HM-1, HL, HY and alumina are 2.040, 2.042, 2.045 and 2.040, respectively, which show also some variation from sample to sample. The variation may arise from the difference in the effective crystal field at the adsorption sites [12] on different samples.

Introduction of oxygen into the aniline-HM-2 system did not result in any detectable formation of O_2^-. The remarkable difference between HM-1 and HM-2 appeared once again. It seems unlikely that the absence of O_2^- on HM-2 is due to the lower concentration in aniline radical cation since a strong signal of superoxide ion was observed on HL, on which the concentration in aniline radical before introduction of oxygen is similar to that on HM-2.

^{27}Al MAS NMR investigation

In order to understand the particular behavior of HM-2 the ^{27}Al MAS NMR characterization was carried out on samples recovered after EPR measurement. The results are shown in Fig. 3. The assignments of the ^{27}Al lines have been well documented [10, 13-15]. The line at about 60 ppm arises from the framework and non-framework Al in tetrahedral coordination, and the amount of each can be separately calculated [8]. The line at about 0 ppm (~9 ppm for alumina) is due to the octahedral non-framework Al, and the line at about 33 ppm is attributed to non-framework penta-coordinated Al. From the spectra in Fig. 3 the distinctive feature of HM-2 is clearly seen: no signal in between 56 ppm and 0 ppm, and the line at 56 ppm is narrower than the similar lines in other samples. Calculation indicated that the non-framework Al species in HM-2 is almost exclusively in octahedral coordination, while in other samples they exist in all three possible coordinations. The results show the involvement of tetra- and/or penta- coordinated non-framework Al species in the formation of organic radical and superoxide ion.

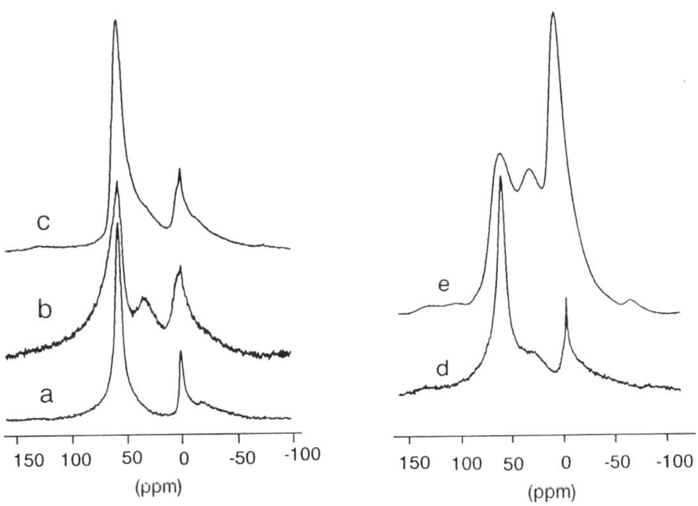

Fig. 3 ^{27}Al MAS NMR spectra (a): HM-2, (b): HM-1, (c): HL (d): HY and (e) alumina.

In order to investigate further the participation of tetra- and/or penta- coordinated non-framework Al in the formation of radicals, HM-2 was subjected to a dealumination treatment. The dealumination was achieved by calcination at 640 and 700°C for 2 h and then the samples were outgassed and cooled to room temperature for the adsorption of aniline. After EPR measurement the samples were recovered and investigated by ^{27}Al MAS NMR.

EPR measurement showed a progressive increase of the concentration in aniline radical cation with the calcination temperature, from the previously observed 2.3×10^{17} spins/g for

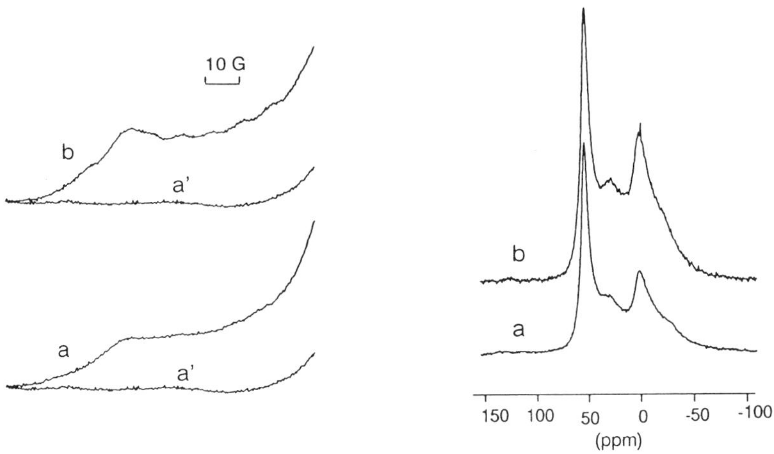

Fig. 4 EPR spectra of the g_{zz} components of O_2^- (left, a' from samples calcined at 540°C) and ^{27}Al MAS NMR spectra (right) of HM-2 dealuminated at (a): 640°C and (b) 700°C.

the sample calcined at 540°C to 6.4 and 14.9×10^{17} spins/g for the samples calcined at 640 and 700°C, respectively. After the introduction of oxygen the formation of superoxide ion was observed on the two last samples and the intensity of the g_{zz} component is higher on the sample calcined at 700°C than at 640°C (Fig. 4, left). The ^{27}Al MAS NMR spectra of the two samples showed an increasing intensity of the penta-coordinated Al as the calcination temperature increases (Fig. 4, right). Calculation demonstrated that the additional non-framework Al contains Al in both tetra- and penta- coordination. Two important points can be deduced from these results: 1) tetra- and/or penta- coordinated non-framework Al species are electron-accepting centers and 2) the superoxide ions are most probably located on them. The Al in octahedral coordination is unlikely to play any role in forming radicals probably because of its complete coordination shell. When Al species are in zeolite framework, only those in the large cavities are accessible to organic molecular probes. Dealumination forms non-framework Al species, exposed to adsorbates, and thus increases the amount of formed radicals.

DISCUSSIONS

From the above results it is clear that the acid strength of both framework and non-framework Al species is very different for the studied zeolite samples. However, the factors affecting zeolite acid strength are quite complex, for which Barthomeuf has provided a good description [2,3]. It seems that, for an H-type zeolite, the major factors having effect on the acid strength of framework Al are: 1) framework Si/Al ratio which determine the closeness of acid sites from each other and a high value of which, to certain extent, increases the strength of the remaining acid sites, 2) the amount of non-framework Al which can give inductive effects, and 3) structure type of the lattice which changes the topology of TO_4 tetrahedra (T: Si or Al) and the geometry of the cages (shape and size). While for non-framework Al another important factor should be added: its dispersion on zeolite surface.

Among the three types of zeolites studied, H-mordenite has strongest Lewis acid sites measured either by organic radical formation or by the A_{zz} value of superoxide ion. However, it is difficult to compare H-mordenite with the other two types of zeolites because its Si/Al ratio and amount of non-framework Al are quite different from those of the other two (Table 1). In order to see how zeolite structure affects its acid strength let us take the HY and HL for comparison. The amount of non-framework Al of the two samples is quite similar, and the Si/Al ratio of HY is higher than that of HL. Higher Si/Al ratio would predict that the acid sites on HY are stronger than on HL if structure effect is not taken into account. In fact, the results obtained both by the formation of aniline radical cation and by the interaction of superoxide ion with Al center have clearly shown that HY has Lewis acid sites less strong than HL. This indicates that zeolite structure has an important effect on its acid strength.

How does zeolite structure influence its acid strength? Zeolite structure defines its Al topology density and cage shape. The limit Si/Al ratio beyond which no AlO_4 has another AlO_4 tetrahedra in a next nearest neighbor position, determined by the topology of TO_4, is 5.8 and 7.3 for HY and HL, respectively [3]. The HY (Si/Al = 8.2) used in the present work reaches its maximum acid strength, while the HL (Si/Al = 4.9) does not. Therefore, the cage shape should play a determining role in this case. It has been shown that oxygen can not enter into the β-cages of HY [16]. The radicals should be located in the supercages whose dimension is ~15 Å, while HL has unidimensional channel with a 12-ring opening (7.1 Å) [17] accessible to adsorbates. Thus, the confinement effect of the two locations is very different, which may probably account for the difference in acid strength of these two types of zeolites.

The transition alumina may be considered as an extreme case of zeolites with Si/Al ratio equal to zero, and its pore dimension is larger than zeolite micropores. This may explain the much lower acid strength of the material.

CONCLUSIONS

Formation of organic radicals and superoxide ion can be used to probe the acid strength of solid acids from which the following order was observed: HM > HL > HY > alumina. Non-framework Al species are important sources of zeolite Lewis acidity. Introduction of oxygen into the systems composed of solid acids and organic radicals results in the formation of superoxide ion, which is most probably bound to non-framework Al species in zeolites. The zeolite structure affects the acid strength of both framework and non-framework Al species.

REFERENCES:

1. Naccache, C., Chen, F.R. and Coudurier, G., in: *Zeolites: Facts, Figures, Future,* Eds. P.A. Jacobs and R.A. van Santen, Elsevier, Amsterdam, pp. 661-668 (1989).
2. Barthomeuf, D., in: *Catalysis,* Ed. J.W. Ward, Elsevier, Amsterdam, pp. 177-186 (1987).
3. Barthomeuf, D., Materials Chemistry and Physics **17** 49-71 (1987).
4. Lombardo, E.A. and Hall, W.K., J. Catal. **112** 565 (1988).
5. Jacobs, P.A., *Carboniogenic Activity of Zeolites,* Elsevier, p.42 (1977).
6. Chen, F.R. and Guo, X.X., J. Chem. Soc., Chem. Commun., 1682 (1989).
7. Chen, F.R., Sheng, S.S., Zhang, W.M., Guo, X.X. and Fripiat, J.J., J. Chem. Soc., Faraday Trans, **88** 887-890 (1992).
8. Chen, F.R. and Fripiat, J.J., J. Phys. Chem., **96** 819-823 (1992).
9. Ayscough, P.B., *Electron Spin Resonance in Chemistry,* London, p. 299 (1967).
10. Chen, F.R., Davis, J.G. and Fripiat, J.J., J. Catal., **133** 263-278 (1992).
11. Miller, D.J. and Haneman, D., Phys. Rev., **3B** 2918 (1971).
12. Che, M. and Tench, A.J., *Advances in Catalysis,* **32** (1983).
13. Sanz, J., Fornes, V. and Corma, A., J. Chem. Soc., Faraday Trans. I **84** 3113 (1988).
14. Ray, G.J., Meyers, B.L. and Marshall, C.L., Zeolites **7** 307 (1987).
15. Rocha, J. and Klinowski, J., J. Chem. Soc., Chem. Commun., 1121 (1991).
16. Imai, T and Habgood, H.W., *J. Phys. Chem.,* **77** 927 (1973).
17. Breck, D.W., *Zeolite Molecular Sieves,* R.E. Krieger Publishing Company, (1984).

ACKNOWLEDGEMENT

DOE grant DE-FG02-90 ER1430 is gratefully acknowledged.

F^- : A MULTIFUNCTIONAL TOOL FOR MICROPOROUS SOLIDS b) ^{19}F NMR OF AS-SYNTHESIZED AND POST-SYNTHESIS TREATED MATERIALS

E. KLOCK, L. DELMOTTE, M. SOULARD and J.L. GUTH

Laboratoire de Matériaux Minéraux (URA-CNRS 428) ENSCMu-UHA,
3, rue Alfred Werner, 68093 MULHOUSE CEDEX (FRANCE)

ABSTRACT

^{19}F MAS NMR technique was used to study crystalline microporous solids (MFI, TON, MTT,AST, FAU, CLO, LTA, SOD, AFI, AEL and a CHA-type structures) in which F was introduced during synthesis. The correlations between the observed chemical shift values and the local structural and chemical environment of F allows the use of this element as a probe. This new method can be extended to other materials after post-synthesis incorporation of F^- anions. Thus, the different chemical shifts observed in MFI-type zeolites are related to the Si/Al ratio and to the presence of different species in the channels. This can be used to distinguish samples prepared in OH^- medium from these prepared in F^- medium. Similar conclusions can be drawn from the ^{19}F spectra of F^- treated FAU-type zeolites with different Si/Al ratios obtained by dealumination.

INTRODUCTION

F^- is a mineralizer which may replace advantageously OH^- in the hydrothermal synthesis of zeolites and other microporous solids [1-7]. But it is also often a structure directing [5,7] and templating [2-4,6,7] species. In this last case its incorporation into the structure can be monitored by MAS NMR spectroscopy which is a powerful tool for the study of the local environment of the atoms [8]. Moreover, ^{19}F is generally considered to be an excellent nucleus for NMR study : nuclear spin of 1/2, high sensitivity and large chemical shift range. The results of a preliminary study [9] of as-synthesized zeolites prepared in fluoride medium demonstrated the interest of this method. But the limited number of samples available at that moment, the lack of structural information on the exact localization of the fluoride and more generally the lack of published data on the ^{19}F chemical shifts in reference inorganic compounds,did not allow a complete and exact assigning of the observed chemical shift values. In addition this first study showed the possibility to introduce F^- in a MFI-type zeolite after the removing of the template. Thus, the two aims of this paper are :

- to bring new results on chemically and structurally well characterized samples in order to show the correlations between the chemical shift values and the different localizations of F
- to test the post-synthesis introduction method of F^- for two different classes of zeolites (MFI and FAU-type) in order to get chemical and structural information from the ^{19}F chemical shift values.

From the Proceedings of the 9th International Zeolite Conference, Montreal 1992, Eds. R. von Ballmoos et al.,

EXPERIMENTAL PART

Synthesis

- Synthesis of the as-synthesized F^- containing samples (Table 1) : Details on synthesis conditions are available in reference [7].

Table 1. Chemical shifts of ^{19}F in the as-synthesized microporous solids with various framework compositions and topologies.

Struct. ref.	Framework elements	Unit-cell or sub-cell formula	^{19}F chem.shift (ppm/$CFCl_3$)
AST	Si	$[Si_{20}O_{40}]$ • 2Quinuclidinium fluoride	-38.2
MFI	Si	$[Si_{96}O_{192}]$ • $4Pr_4NF$	-64.0
TON	Si	$[Si_{24}O_{48}]$ • $Pent_2NH_2F$	-74.2
MTT	Si	$[Si_{24}O_{48}]$ • Pr_2NH_2F	-74.8
SOD	Si	$[Si_{12}O_{24}]$ • yEG•xNaF[(a)]	-123
SOD	Si,Al	$Na_6[Si_6Al_6O_{24}]$•2NaF•xH_2O[(b)]	-179
NU1	Si,Al	$[Si_xAl_yO_{2(x+y)}]$•zMe_4NF[(b)] (Si/Al > 20)	-114
AST	Al,P	$[Al_{20}P_{20}O_{80}]$ • x Quinuclidinium fluoride	-87.6;-143.2;-145.5
AFI	Al,P	$[Al_{12}P_{12}O_{48}]$•Pr_4NF	-119.4
AEL	Al,P	$[Al_{20}P_{20}O_{80}]$ • $2Pr_2NH_2F$ • $6H_2O$	-119.5
CHA-like	Al,P	$[Al_6P_6O_{24}]$•2 Morpholinium fluoride	-127
CHA	Al,P(Si)	$(Et_4N)_2[Si_4Al_{17}P_{15}O_{72}]$ •Et_4NF•$9H_2O$	-122.8
FAU	Al,P(Si)	$(Pr_4N)_{14}(Me_4N)_6[Si_{24}Al_{94}P_{74}O_{384}]$•$2Pr_4NF$	-122.9
LTA	Ga,P	$[Ga_{12}P_{12}O_{48}]$ • $3Pr_2NH_2F$ • xH_2O	-69.7
CLO	Ga,P	$[Ga_{96}P_{96}O_{372}]$ • 24Quinuclidinium fluoride • xH_2O	-68.5

(a) EG = ethyleneglycol, x and y not determined, prepared according to ref.[10] in presence of NaF
(b) Prepared in alkaline media according to a not yet published method.

- Synthesis and modification of the samples for the post-synthesis introduction of F^- : A first series of MFI-type samples was prepared in our laboratory with the fluoride method in presence of tetrapropylammonium bromide [1] and calcined at 550°C in air during 4 hours. The second series corresponds to calcined and H^+-exchanged samples synthesized by Conteka in OH^- media. Three additional silicalite samples were synthesized according to the conventional method in presence of Pr_4NOH. The FAU-type samples, referred to as (n)FAU were all in the H^+-form. Sample (2.7)FAU was obtained from a NaY zeolite by NH_4^+ exchange followed by shallow bed calcination at 550°C. All the other samples correspond to different dealumination products, obtained by steaming followed by acid leaching of sample (2.7)FAU.

Post-synthesis introduction of F^-

- MFI-type samples : about 500 mg of the calcined sample were impregnated with 5 ml of aqueous solution of Pr_2NH_2F (2.5 M, pH ~ 7) during 6 hours at room temperature, then washed and filtered twice with 5 ml of water and dried at 150°C for 2 hours.
- FAU-type samples : about 250 mg were blended with 1 or 2 ml of aqueous solution of CH_3NH_3F (0.25 M) and with CH_3NH_2 necessary to neutralize the H^+ present in each sample. After 1 hour impregnation at room temperature the mixture was evaporated and dried at 80°C for 2 hours.

^{19}F mas n m r

A Bruker MSL 300 N.M.R. spectrometer with a high speed Doty probe was used. The measurements were performed at 282 MHz, with a spectral width of 125 kHz, sample spinning rate from 7 to 10 kHz, 5μs pulse duration, 5 s recycle time and180 scans (δ/$CFCl_3$).

RESULTS AND DISCUSSION

^{19}F NMR chemical shifts in as-synthesized materials

In the 15 materials of Table 1, F^- is the charge compensating anion of the organic cations used as template. In a first hypothesis, based on ^{13}C MAS NMR results obtained with Pr_4NF in siliceous MFI-type zeolite, it has been assumed that F^- was localized in the channels close to the nitrogen of Pr_4N^+ [11]. But a recent ^{19}F MAS NMR study [9] showed that the nature of the associated organic cation had only a weak influence of the ^{19}F chemical shift. Its value was chiefly determined by the nature of the elements in the framework and the structure of the latter. This led us to the hypothesis that F^- has stuck into interstices of the framework which was confirmed by the structure determination and refinement of several solids of Table 1. Thus, in the MFI-type structure, F^- is located between two five-membered rings [12] and in the AST-, LTA- and CLO-type structures, F^- is trapped in the middle of the double-four ring cube [2,13,14]. At last, in the AFI-type structure, F^- is also between two four-membered rings which are however linked together by only two oxygens, thus F^- has only 10 oxygen neighbourgs instead of 12 in the preceeding cube [3]. In all these structures F^- is not strongly coordinated to a framework element according to the MAS NMR spectra of these elements. The situation is somewhat different for the CHA-like phase [15,16] where two fluoride anions bridge a four-membered ring leading to two six-fold coordinated aluminium with, as a consequence, two different chemical shifts for this element.

It may be noticed that several other cases are known in which the anion (e.g., OH^-, HPO_4^-) of the templating ionic pair is in the framework and not beside the cation [17,18]. But these anions are generally coordinated to a framework element (e.g. Al or Ga) as for the CHA-like phase.

These MAS NMR results allow for the first correlations between the chemical shift of ^{19}F and the structural and chemical environment of F^- anions.

If we assume that in the samples of Table 2, F^- is not strongly coordinated to a framework element, one can see that three main parameters seem to govern its chemical shift : increase in the interstices size, number of nearly oxygens and electropositivity of the framework elements lead to a more negative chemical shift. The shape of the interstice has probably a minor effect. These correlations were used to localise F- in the microporous solids marked with an* in Table 2.

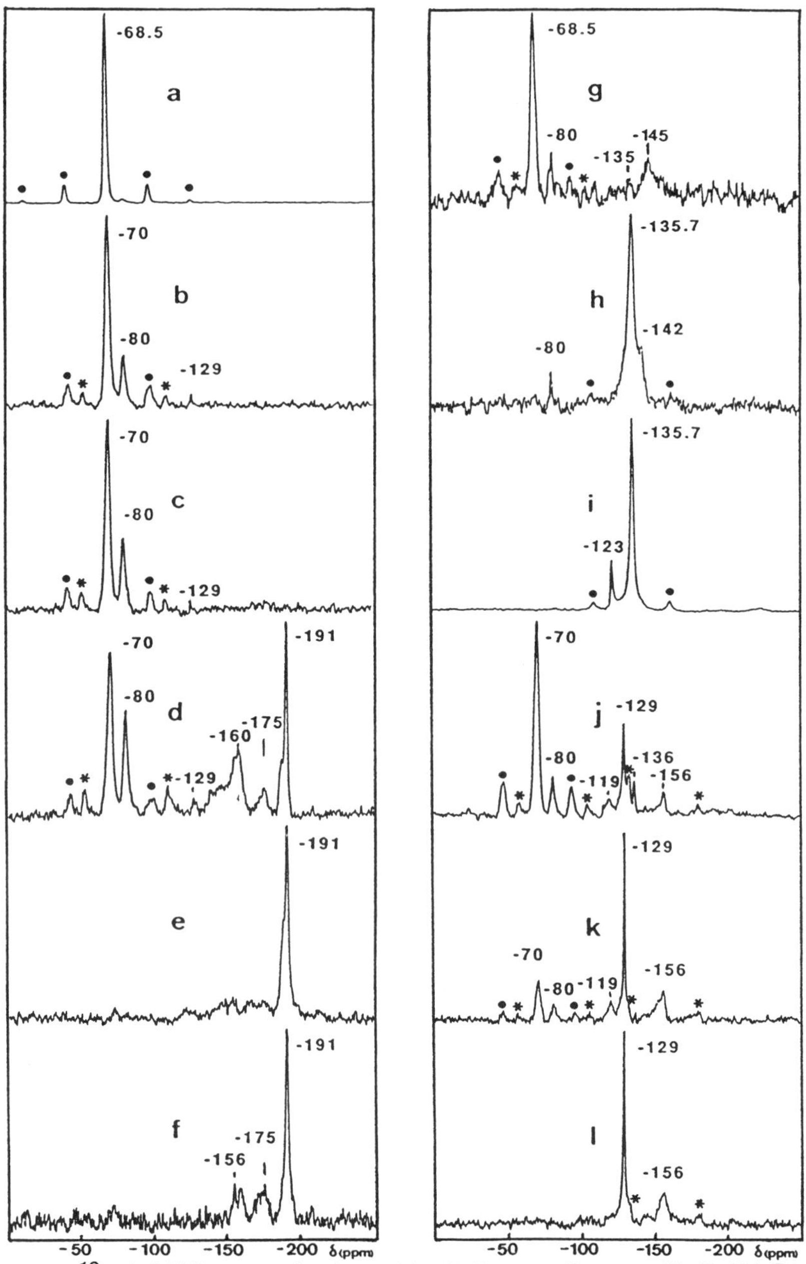

Figure 1 : ^{19}F MAS NMR spectra of various calcined MFI-type zeolites treated by Pr_2NH_2F: a, b, c, d, e and f samples synthesized with F^- ; g, h, i, j, k and l samples synthesized with OH^- ; Si/Al = ∞(a), 60(b), 45(c), 30(d), 22(e), 12(f), ∞(g,h,i), 75(j), 30(k), 17(l) ; • or * = spinning side bands.

Table 2. Chemical shift values in function of the structural and chemical environment of F^-.

Struct. units	Number of O near F^-	Increasing chemical shift values for different framework compositions
2 times bound D4R	10	Al,P-AFI or AEL* -119,4 ; -119,5
4 times bound D4R	12	Si-AST -38.2; Ga,P-CLO and LTA -68.5 ; -69.7; Al,P-AST* -87.6
2 times bound D5R	12	Si-MFI -64.0; Si-TON* and MTT* -74.2 ; 74.8
6 times boundD6R	18	Al,P(Si)-CHA*and FAU* -122.8 ; -122.9
Sodalite cage	36	Si-SOD* -123; Al,Si-SOD* -179.0
troncated rhomb.	48	Si(Al)NU1* -114; Al,P-AST* 143.2 ; 145.5

* assumed assignment, not controlled by structure determination.

^{19}F NMR chemical shifts observed on post-synthesis treated MFI-type samples

- samples synthesized with F^-: The ^{19}F spectrum of a highly siliceous sample,characterized by a thin single peak centred at -68.5 ppm with spinning side-bands (Figure 1a), is similar to that obtained for an as-synthesized sample (Table 1). For the Al-containing samples with a Si/Al ratio ≥ 40 , the previous signal becomes broader and is slightly shifted to -70 ppm (Figure 1b, c). Besides, an additional line appears near -80 ppm [9].The intensity of the peak at -70 ppm significantly decreases in the spectra of samples with higher Al content (figure 1d-f) and a new line at -191 ppm, which may be ascribed to Al-F bonds belonging to aluminum oxide clusters in the channels, appears. Other weak extra peaks, poorly defined, ranging from -156 to -175 pppm, may be observed in the spectra of Al-containing samples. They may be attributed to fluoroaluminates species less condensed than the previous ones.

- samples synthesized with OH^-: As for the sample prepared in F medium, a major peak at -68.5 ppm is observed (Figure 1g) with a silicalite having a low defect content [19] (preparation at pH ~10 in the absence of alkali cations). The spectrum relevant to a silicalite obtained by increasing the basicity (pH ~11.5, absence of alkali cations) becomes completely different (Figure 1h) : the peak assigned to F^- occurring as an ionic pair, disappears entirely and is replaced by a broader peak centred at -135.7 ppm with two spinning side-bands, and also by another one, poorly resolved around -142 ppm. For a silicalite containing about 1 Na per unit cell (preparation at pH ~13.5 in presence of NaOH), another one appears at -123 ppm (Figure 1i) in addition to the previous peak.Although the F signal at -135.7 ppm is connected with the lattice defects of the silicalite, it has not been hitherto assigned to determined species.

Three samples characterized by a Si/Al ratio of 75, 29 and 17, respectively, were also examined. The spectrum recorded for the former one consists of numerous lines, among which two signals at -70 and -80 ppm are ascribed to F^- of an ionic pair and that at -129 ppm is assigned to

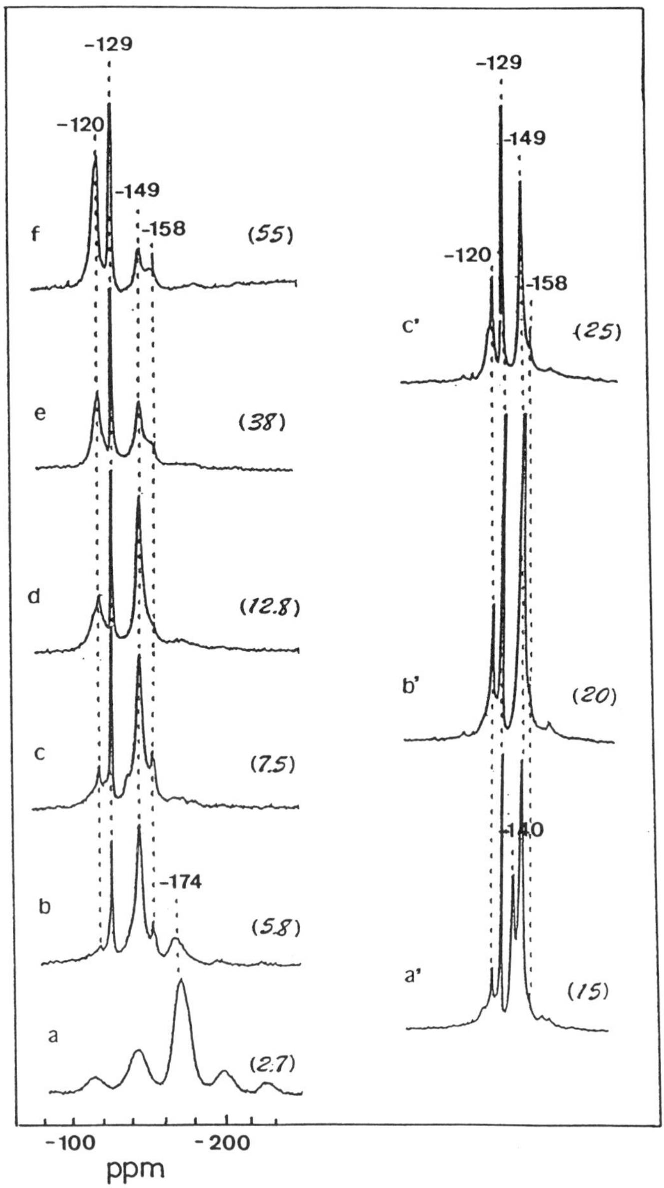

Figure 2 : ^{19}F MAS NMR spectra of various dealuminated FAU-type zeolites treated by CH_3NH_3F a, b, c, d, e and f samples treated with 1 ml ; a', b' and c' samples treated with 2 ml. The peak at -129 ppm, in the spectra c-f and a'-c', is troncated. Number between brackets represent Si/Al ratio.

SiF_6^{2-} (Figure 1j) [20]. Finally, a peak around -136 ppm is also present. For higher aluminium contents the intensity of the peaks corresponding to paired F^- ions, decreases (Figure 1k) or is cancelled (Figure 1l), whereas a thin line at -129 ppm and a broader one centred at -156 ppm with two spinning side-bands, appear. Thus, for the aluminosilicate samples obtained in OH^- medium there are no Al species in the channels and the F introduced would attack preferentially the siliceous species and defects,with the resultant formation of SiF_6^{2-}or Si-F bonds in the framework. In summary, these measurements provide indeed, a means of discrimination between the zeolites synthesized with F^- as mobilizing agent and those obtained with OH^-.

^{19}F NMR chemical shifts observed for post-synthesis treated FAU-type samples.

Figure 2 (a-f, a'-c') shows the ^{19}F spectra of the different treated faujasites. Except for the two spectra corresponding to the lowest Si/Al ratio, the peak at -129 ppm reveals to be the most intense. It is assigned to fluoride in SiF_6^{2-}[20] which may form from siliceous "debris" remaining in the steamed and acid washed faujasites. Moreover its intensity decreases with Si/Al (decreasing of the amount of "debris"). On the other hand the ratio of the intensity of the peak at -120 ppm on the intensity of the peak at -149 ppm increases with the Si/Al ratio (e.g., from 0.1 to 2 for the samples with Si/Al ratios from 5.8 to 55). Taking into account the chemical shift values observed for ^{19}F in as-synthesized sodalites (Table 1) which are -123 ppm for Si/Al ratio of ∞ and -179 ppm for a Si/Al ratio of 1, the assumption may be made that the two former chemical shifts values correspond to fluoride anions present in sodalite-type cavities of the faujasites.If the peak at -120 ppm may thus be assigned to the presence of F^- in a siliceous cavity, it is more difficult to estimate the Si/Al ratio of the cavity with F^- responsible for the peak at -149 ppm. A value close to 5 or 6 may be possible. Peaks characteristic of intermediate ratios cannot be excluded. For instance, a part of the intensity of the peak at -129 ppm could be attributed to F^- in a silica rich cavity. In the same manner peaks, at -140, -158 and -174 ppm may also correspond to F^- in sodalite type cavities,which are richer and richer in aluminum. But it cannot be excluded that these peaks arise also from the presence of alumineous "debris" remaining after acid washing or formed during the post-synthesis treatment with F^- (e.g.,(2.7)-FAU and (5.8)-FAU))[21].

CONCLUSION

Like ^{129}Xe, which is a probe employed in a MAS NMR method sensitive to the geometry (shape and size) of the cavities and channels in microporous solids, ^{19}F can be used as a chemical and structural probe in as-synthesized or post-synthesis treated samples. It is sensitive to the structural and chemical composition of the framework and species present in the micropores.

REFERENCES

1. Guth, J.L., Kessler, H., HIGEL, J.M., Lamblin, J.M., Patarin, J., Seive, A., Chézeau, J.M. and Wey, R., in *Zeolite Synthesis*, Eds. M.L. Occelli and H.E. Robson, ACS Symp. Series 398, ACS, Washington D.C. pp. 176-195 (1989).

2. Caullet, P., Guth, J.L., Hazm, J., Lamblin, J.M. and Gies, H., Eur. J. Solid State Inorg. Chem. **28**, 345 (1991).

3. Qui, S., Pang, N., Kessler, H. and Guth, J.L., Zeolites **9**(5), 440 (1989).

4. Merrouche, A., Patarin, J., Kessler, H. and Anglerot, D., French Patent Appl. 91/01106 (1991).

5. Caullet, P., Hazm, J. and Guth, J.L., Zeolites (accepted).

6. Merrouche, A., Patarin, J., Kessler, H., Soulard, M., Delmotte, L. and Guth, J.L., Zeolites (accepted).

7. Kessler, H., Guth, J.L., Caullet, P., Hazm, J., Merrouche, A. and Patarin, J. in *Proceeding of the 9th International Zeolite Conference*, Eds. R. von Ballmoos, J.B., Higgins and M.M.J., Treacy, Butterworth-Heinemann, Stoneham, M.A. (1992).

8. Klinowski, J., Chem. Rev. **91**, 1459 (1991).

9. Delmotte, L., Soulard, M., Guth, F., Seive, A., Lopez, A. and Guth, J.L., Zeolites **10**(8), 778 (1990).

10. Bibby, D.M., Bakter, N.I., Grant-Taylor, D. and Parker, L.M., in *Zeolite Synthesis*, Eds. M.L. Occelli and H.E. Robson, ACS Symp. Series 398, ACS, Washington D.C., pp. 209-220 (1989).

11. Chézeau, J.M., Delmotte, L., Guth, J.L. and Soulard, M., Zeolites **9**(1), 78 (1989).

12. Mentzen, B.F., Sacerdote-Peronnet, M, Guth, J.L and Kessler, H., C.R. Acad. Sci. Paris **313** (II), 177 (1991).

13. Simmen, A., Patarin, J. and Berlocher, C., in *Proceeding of the 9th International Zeolite Conference*, Eds. R. von Ballmoos, J.B., Higgins and M.M.J., Treacy, Butterworth-Heinemann, Stoneham, M.A. (1992).

14. Estermann, M. McCusker, L.B., Baerlocher, C., Merrouche, A. and Kessler, H., Nature **352**, 320 (1992).

15. Guth, F., Thesis Université de Haute Alsace (France) (1989).

16. Simmen, A.,Thesis ETH Zurich (Switzerland) (1992).

17. Richardson, J.W., Pluth, J.J. and Smith, J.V., Naturwissenschaften, **76**, 469 (1989).

18. Parise, J.B., Acta. Cryst., **C42**, 670 (1986).

19. Chézeau, J.M., Delmotte, L., Guth, J.L. and Gabelica, Z., Zeolites **11**(6), 598 (1991).

20. Duke, C.V.A., Miller, J.M., Clark, J.H. and Kybett, A.P., Spetrochim. Acta **46A**, 1381 (1990).

21. Duke, C.V.A., Miller, J.M. Clark, J.H. and Kybett, A.P., Mol. Catal., **62**, 233 (1990).

SYNCHROTRON BASED DIFFRACTION STUDIES OF SORPTION IN ZEOLITE Y AT ROOM TEMPERATURE

Z.A. KASZKUR *, J.W. COUVES*, R.H. JONES*, D. WALLER*, C. R.A. CATLOW * and J.M.THOMAS *

* Davy Faraday Research Laboratory, The Royal Institution of Great Britain,21, Albemarle St., London. W1X 4BS, UK.

ABSTRACT

Examples are presented of successful interpretation of room temperature, high resolution X-ray powder diffraction patterns of several halogenated hydrocarbons sorbed into Na-zeolite Y. The methods applied vary from the extensive use of geometrical constraints and molecular graphics to Molecular Dynamics, which are combined with Rietveld profile refinement.

INTRODUCTION

The mechanism of the selectivity of zeolitic solids can be understood only when the local environment of sorbed molecules is fully characterized. Powder diffraction is commonly used in such studies since sufficiently large single crystals are often unavailable. Several previous studies [1-6] have reported positions and site-occupancies of complex neutral molecules in zeolites, mainly based on neutron diffraction data. More recently the use of synchrotron radiation enabled high resolution X-ray powder diffraction studies of this problem [4,7]. Analysis of the data can provide atomic positions for the molecules containing atoms of atomic number at least comparable to that of the silicon atom. Refinements of this type are greatly assisted by geometrical constraints and molecular graphics with estimation of the energetics of the sorbate-zeolite complex. For large, flexible molecules, molecular dynamics simulations can give a detailed understanding of the dynamics of the adsorbate and of likely conformations of the molecule to which the X-ray diffraction data are not very sensitive. In this paper we present examples of recent results illustrating applications of these approaches. We show how, by a combination of synchrotron based diffraction with molecular modelling techniques, we are able to characterize in detail the sorption sites of molecules within zeolite pores.

EXPERIMENTAL

All experiments were based on samples of Na-zeolite Y (Laporte Industries Ltd, SiO_2/Al_2O_3=4.74) dehydrated at 300°C for 16 hours in a vacuum of at least 8x10^{-5} T, held in

From the Proceedings of the 9th International Zeolite Conference, Montreal 1992, Eds. R. von Ballmoos et al.,

quartz capillaries (0.3 mm). The adsorbates were purified by repeated freeze-thaw cycles and then sorbed at room temperature into the zeolite samples. The adsorbates used were:(a) chloroform (2T); (b) chloroform (200T); (c) meta-dichlorobenzene (5.6T); (d) 1,4 dibromobutane (0.2T). The samples were left to equilibrate for several hours prior to flame sealing of the capillaries.

High resolution powder diffraction patterns were collected at room temperature (R.T.) on station 9.1 of the SRS Daresbury Laboratory using X-rays of wavelength 0.9959(6)Å (a,c), 1.0099(8)Å (b), and 1.00010(4)Å (d). To obtain a deeper insight into the dynamics of the adsorbate in sample (d), two data sets: at R.T. and at 40K were collected. The typical range of the measurements was 3-60 deg (2θ) with a step of 0.01 deg. The data range covers thus about 400 Bragg reflections. The details of the data handling and refinement is given elsewhere [4,8].

RESULTS AND DISCUSSION

First, regarding the framework and extra-framework cation positions, we note that the interaction of the Na cations exposed to the supercage with different adsorbate molecules influences the occupancy factor for the S_{II} cations which was determined for the samples to be 32 (a and d), 26.7(8) (b) and 27.4(6) (total occupancy for c) (Table 1). Also the population of the $S_{I'}$ and S_I cations depends significantly on the kind of adsorbate.

Profile refinement of the data for (a)-(c) provides final positions of the adsorbate molecules (Table 1). Detailed description of the refinement for the samples (a)-(c) has already been reported [4]. The final sites are characterized by low profile agreement indices and low energies of the system for sample (a). For (b) and (c) the interaction with the framework atoms is much stronger influencing the population of the S_{II} cations and causing static displacements of the framework atoms, as is detectable by the appearance of the apparent short Cl-O(framework) distances: 2.73Å(O1), 2.98Å(O4) (sample b), and 2.69Å (Cl1-O4), 2.83Å (Cl1-O1) (sample c). Estimation of the energy of the system in these cases is much more difficult and was not undertaken.

One of the positions of the chloroform molecule for sample (b) is close to those found for sample (a) with Cl atoms in the vicinity of the hexagonal prism joining the sodalite cages. This position may be responsible for the interaction with S_I and $S_{I'}$ cations. Two sites (sample b) with occupancies less than 2/supercage and 3/supercage respectively, together form a close packed molecular system of 5 molecules within one supercage. This enables the determination of the orientations of the molecules because there is no room for the H atoms of the chloroform to point towards the framework.

For sample (c) the excessively short C-Na(S_{II}) distance (2.42Å and 2.50Å) for two carbon atoms in the benzene ring was explained by splitting the density of the S_{II} peak (slightly washed out in <111> direction) into two equal peaks. One of these peaks gave a distance ~2.6Å which is in good agreement with the sum of the ionic Na^+ and Van der Waals C radii, the second was assigned to the cation position when no neighbouring dichlorobenzene molecule is present. The

TABLE 1

The atomic parameters resulting from the Rietveld refinement for the samples (a)-(d); x,y,z-fractional coordinates, B- Deby'e-Waller factor [Å^2], O- occupancy .

		(a)	(b)	(c)	(d) RT	(d) 40K
T=Si,Al (192i)	x	0.1236(2)	0.1231(2)	0.1253(1)	0.12496(21)	0.1242(19)
	y	0.9454(2)	0.9449(2)	0.9465(1)	0.94518(18)	0.9452(19)
	z	0.0349(2)	0.0365(2)	0.0361(1)	0.03604(19)	0.0364(20)
	B	2.98(6)	4.94(8)	2.13(3)	1.2(2)	2.3(3)
O(1) (96h)	x=-y	0.1092(3)	0.1086(3)	0.1076(2)	0.1063(3)	0.107(4)
	z	0.0	0.0	0.0	0.0	0.0
	B	3.0(3)	5.2(5)	3.2(2)	1.9(2)	3.3(3)
O(2) (96g)	x=y	0.2519(3)	0.2545(3)	0.2516(2)	0.2506(3)	0.250(4)
	z	0.1410(4)	0.1450(5)	0.1436(2)	0.1395(4)	0.145(5)
	B	3.6(3)	6.9(5)	3.1(2)	1.9(2)	3.3(3)
O(3) (96g)	x=y	0.1782(3)	0.1804(4)	0.1761(2)	0.1760(3)	0.176(4)
	z	0.9680(5)	0.9667(7)	0.9670(3)	0.9671(5)	0.970(6)
	B	4.1(4)	5.0(7)	2.2(2)	1.9(2)	3.3(3)
O(4) (96g)	x=y	0.1758(3)	0.1732(5)	0.1785(2)	0.1784(3)	0.180(4)
	z	0.3198(4)	0.3175(5)	0.3233(3)	0.3204(5)	0.320(6)
	B	3.8(4)	4.8(6)	3.3(2)	1.9(2)	3.3(3)
Na(I) (16c)	O	7.9(3)	12.1(5)	3.2(2)		
	B	4.9(4)	4.1(4)	4.9(3)		
Na(I') (32e)	x=y=z	0.0534(5)	0.0558(5)	0.0624(3)	0.0493(4)	0.052(5)
	O	15.8(6)	10.0(8)	21.1(5)	22.6(7)	21.0(8)
	B	4.9(4)	4.1(4)	+	2.4(3)	2.1(3)
Na(II) (32e)	x=y=z	0.2356(17)	0.2373(19)	0.2345(2) 0.2430(2)	0.23467(30)	0.2341(29)
	O	32	26.7(8)	O1=O2=13.7(3)	32	32
	B	4.9(4)	3.6(5)	B1=B2=3.8(2)	2.4(3)	2.1(3)
(II') (32e)	x=y=z	0.148(3),0.172(5)	0.168(6)	0.1581(4)		
	O	3.7(6),2.2(6)	5.1(6)	8.5(3)		
	B	4.9(4)	3.6(5)	4.2(3)		
Br1	x=y				0.273(8)	0.268(30)
	z				0.362(13)	0.35(4)
	O				0.7(3)	1.2(5)
	B				1.6*	1.6*
Br2,3	x=y				- 0.33(3)	0.32(3)
	z				0.57(4)	0.28(4)
	O				1.7(4)	2.0(5)
	B				1.6*	1.6*
R_I		5.81	7.41	4.99	4.72	4.46
R_{wp}		13.69	14.41	10.70	7.04	6.73
R_e		13.64	12.36	6.39	3.68	4.67
a_0 (nm)		2.47885(3)	2.48030(3)	2.47384(2)	2.48882(5)	2.4889(4)

+ - refined as anisotropic with $B_{11}=B_{22}=B_{33}=0.0031(2)$ and $B_{12}=B_{13}=B_{23}=0.0019(2)$

* - not refined.

necessity of such a procedure for the self consistency of the model suggests that the adsorption of the dichlorobenzene molecule displaces the Na cation towards the sodalite cage. The refined position for the meta-dichlorobenzene is shifted from the <111> symmetry axis (Figure 1) and the plane of the molecule is not strictly perpendicular to this axis. This R.T. result is thus in disagreement with the low temperature studies of benzene [2] as well as of o-,m-, and p-xylene [9] where the aromatic ring was found to be normal to the <111> axis and close to an S_{II} cation. This discrepancy suggests that interactions involving the chlorine atoms are dominant in this case.

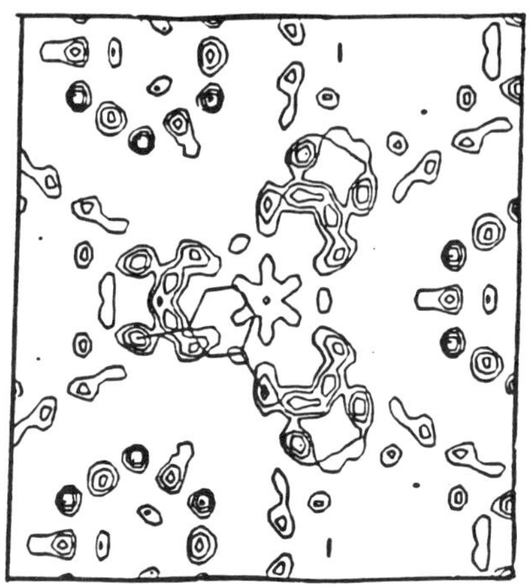

FIG.1.

Difference Fourier map perpendicular to [1,1,1]; x is paralel to [-2,1,1] and y paralel to [0,-1,1]. The section is about (0.29,0.29,0.29) showing positions of Cl atoms corresponding to schematically drawn molecule of meta dichloro-benzene.

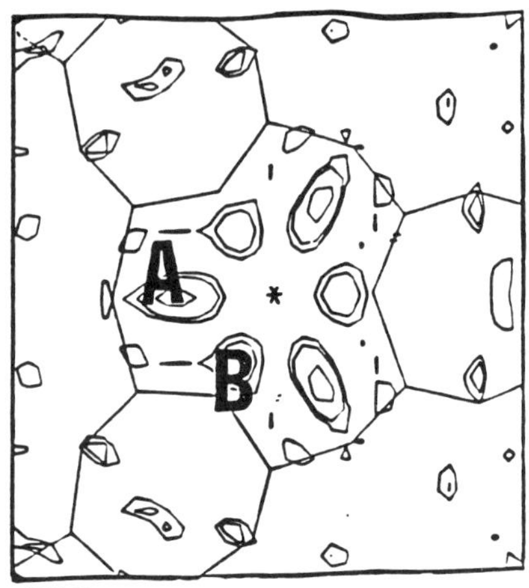

FIG.2.

The observed Fourier map for 1,4 dibromobutane adsorbed in Na-zeolite Y measured at 40K. The section displayed is perpendicular to the [111] axis at (0.3,0.3,0.3). Shown are both adsorption sites: Br1- shifted from the centre toward a neighbour supercage, marked as A, and Br3 - shifted towards a neighbouring sodalite cage, marked as B.

Analysis of the diffraction pattern of the sample (d) enabled location of the Br atoms (Table 1). Carbon atoms have an X-ray scattering factor 0.17 times that of Br atoms and in this case cannot be located. The R.T. data reveal two electron density peaks: one (Br1- Table1) close to the S_{II} cation, and the second close to the centre of symmetry in the middle of the 12-ring window between two supercages. To explain significance of these sites, the low temperature (40K) data were analysed, revealing the same Br1 electron density peak of the Fourier map, and two further maxima Br2,3 (Table 1). The section of the Fourier map Fobs perpendicular to the <111> axis at

(0.3,0.3,0.3) (Figure2) displays two peaks assigned to Br1,Br3 and denoted as A and B respectively.

To gain an insight into behaviour of the whole molecule we have carried out a complementary set of molecular dynamics (MD) simulations. The system studied is well suited for simulation because of the full occupancy of the S_{II} cation sites which therefore define strictly the environment for the molecule in the supercage. In setting up the force field for calculations the local charges on every atom of the adsorbate molecule were estimated using semi-empirical calculations (MNDO)[10] which gave the charges listed in the Table 2. The local charge on the Na cation was assumed to be +1 and neutralised equally by six neighbouring framework oxygen atoms. All the potentials were taken from the parameterisation available in the DISCOVER code of BIOSYM Technologies Inc.[11]. The latter code was used in all our simulations which employed no periodic boundary conditions and both, the framework and the cations on the S_{II} sites were fixed during the MD run. The positions of the Na cations were taken from the experimental results. Each MD run compassed 10000 steps of 1fs time-interval. When all the adsorbate atoms were allowed to move, the initial MD simulation at a high temperature (1200K), starting from the energy minimized position, shows both bromines exhibiting extensive freedom but moving within a short distance of the sodium cations (3-5Å). On 'cooling' the system to 400K (Figure 3), one of the bromine atoms locates itself approx. 3Å away from the sodium cation. But it remains, however, in a small well defined area at the site of symmetry which is close to that found experimentally. The

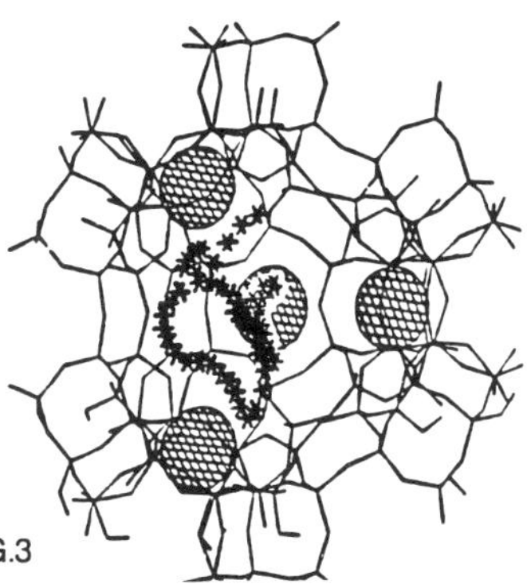

FIG.3
Projection of the model based on one supercage on the plane perpendicular to the <111> axis. Cations in front of the hexagonal entries to the sodalite cages are marked as hatched spheres, the traces of the moving Br atoms from both ends of the molecule in the MD run at 400K are marked as light and heavy stars respectively.

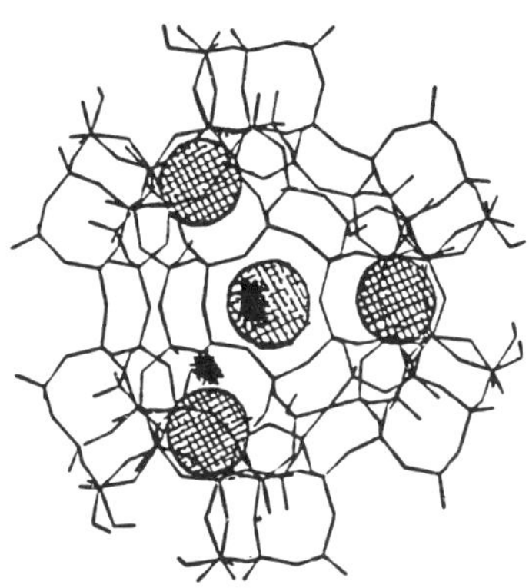

FIG.4.
The MD run of the model based on one supercage at 40K; view projected on the plane perpendicular to the <111> axis; trace of the moving Br atoms of the molecule of dibromobutane- stars; cations- hatched spheres.

second bromine atom at the other end of the chain molecule is much more mobile conserving the distance of ~3-5Å from one of the cations but increasingly limited in its motion by the localisation of the first Br atom. At 40K the simulation shows both Br atoms located close to the <111> axis, on similar sites, each at ~3Å away from a cation on the S_{II} site (Figure 4).

TABLE 2

Atomic charges used in MD simulations.

Br1=Br2	-0.1358
C1	0.0402
C2	-0.0210
H11=H12	0.0347
H21=H22	0.0236

The above simulations qualitatively explain the distribution of the Br atoms' density within the supercage - deduced from the R.T. data. The real adsorption of the Br terminal atoms of the 1,4 dibromobutane takes place in the vicinity of the S_{II} cations. The occurrence of the two sites A and B (Figure 2) is a more subtle effect, which is discussed further below. The R.T. electron density peak close to the centre of symmetry is formed upon heating the system and does not occur in the 40K data. We ascribe this peak to the effect of increased electron density in the area of the 12-ring due to the thermal motion of the Br atom of the molecule, which is constrained only by the localization of the other terminal Br atom of this molecule (i.e. by the Br-Br distance which for the free molecule is ~7.0Å). This peak in the electron density is detected on the <111> axis because of the high symmetry of this axis. The likelihood of this explanation is reinforced by our observation of a similar peak in the Fourier map from an unpublished X-ray diffraction data of 1,6 dibromohexane in Na-zeolite Y. In that case however, the peak is shifted to the point of higher symmetry (0.5,0.5,0.5) due to the greater length of the molecule.

In order to explain the occurrence of the trapped Br atom at the two sites A and B (Figure 2) at low temperature, we have proposed two complementary models: (1) involving two supercages (8 cations) with the molecule spanning the two and the Br atoms attracted each to the S_{II} cation of the neighbouring supercages; and (2) using only one supercage with some cations added on the $S_{I'}$ sites (as guided by the refinement). As can be seen from the trace of the Br atoms during the respective MD run (Figure 5,6), model (1) can explain site A whereas model (2) tends to shift the most probable Br position towards B. The latter model illustrates the significance of the $S_{I'}$ cations for which distribution over the available sites was chosen arbitrarily. However, sorption at site A should be the stronger, as this remains in the R.T. data.

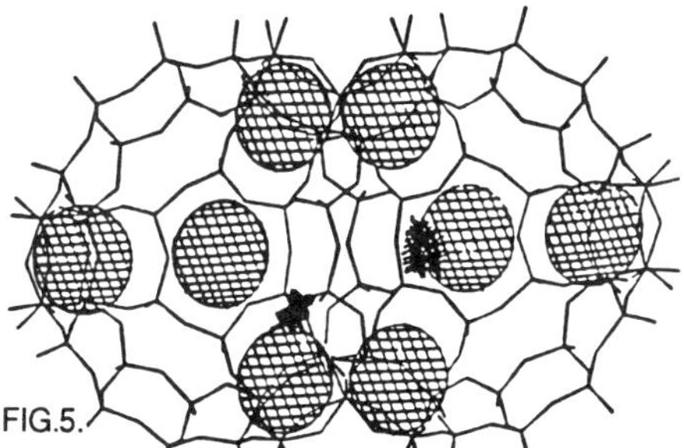

FIG.5.
The model of two supercages with 8 S_{II} cations (hatched spheres). The view is projected on the plane parallel to the 6-member window of the sodalite cage to which the S_{II} cation and the attracted Br atom (group of stars on the right) is attached. The position of the group of stars (trace of the MD run at 40K) corresponds to the site A (compare Fig.2). Position of the other end of the undisplayed molecule (group of stars on the left) has similar symmetry when seen projected on the 6-member window of the closest S_{II} cation.

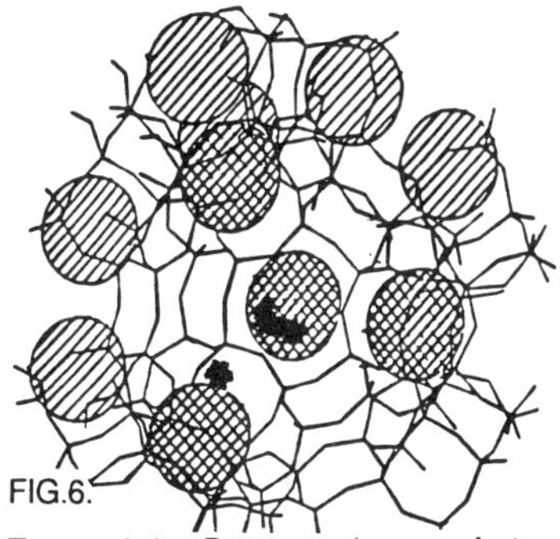

FIG.6.
Trace of the Br atoms (group of stars) from both ends of the dibromobutane molecule during MD run (40K) in the model of one supercage including cations on the S_{II} (hatched) and $S_{I'}$ site (shaded spheres). The projections shown are on the planes containing 6-member window of the S_{II} cations attracted to respective Br atoms. The model assumes arbitrary occupancy of the $S_{I'}$ sites available and shows shift towards site B of the end of the molecule (compare Fig.2).

CONCLUSIONS

The work presented in this paper describes a series of successful interpretations of room temperature, synchrotron based high resolution X-ray powder diffraction patterns of a series of halogenated hydrocarbons sorbed in Na-zeolite Y. Understanding of such data is of importance for catalysis as real catalytical processes occur at higher temperatures. Fourier maps derived from these R.T. data sets are difficult to interpret as many features are washed out by the dynamic and static disorder. We have however shown that: (1) for rigid molecules these can be interpreted in terms of the average position of the molecule which best fits the electron density distribution while at the same time producing minimum profile agreement indices; (2) for flexible molecules, the X-ray data can be interpreted with use of the molecular dynamics simulations of the sorbate in the zeolite. In both cases the presence of the 'heavy' atoms in the molecules of the sorbate is of primary importance when X-ray techniques are employed, but of course is no longer necessary when neutron methods are used.

The results of our study also yield valuable insight into the nature and consequences of sorption in zeolites. In the case of chloroform, high density of packing is clearly possible, promoted by strong Van der Waals interactions between the sorbed molecules. We find evidence for significant framework distortions following sorption - a result which is of general significance. In the case of 1,4 dibromobutane, we find an interesting partial detrapping of the molecule on raising the temperature. Further studies using our combination of theoretical and experimental techniques are in progress.

ACKNOWLEDGEMENTS

The authors wish to thank Dr J. Gale for making calculations of the charge distribution within dibromobutane used in our MD studies. We are also grateful to Dr C. M. Freeman for his cooperation in the initial stages of our MD calculations and to Dr R. Cernik for his advice and assistance. The Science and Engineering Research Council is thanked for provision of a supporting grant, and BIOSYM Technologies for assistance with the graphical representation of our simulations.

REFERENCES

1 Wright P.A.; Thomas J.M.; Cheetham A.K.; Nowak A.K., Nature, **1985**, 318, 611.

2. Fitch A.N.; Jobic H.; Renouprez A., J.Phys.Chem., **1986**, 90,1311.

3. Wright P.A.; Thomas J.M.; Cheetham A.K.; Ramdas S., J.Chem.Soc. Chem.Comm., **1984**, 1338.

4. Kaszkur Z.A.; Jones R.H.; Couves J.W.; Waller D.; Catlow C.R.A.; Thomas J.M., J.Phys.Chem.Solids, **1991**, 52, 1219.

5. Newsam J.M., J.Phys.Chem., **1989**, 93, 7689.

6 Goyal R.; Fitch A.N.; Jobic H., J.Chem.Soc., Chem.Commun., **1990**, 1152.

7. Dooryhee E.; Greaves G.N.; Steel A.T.; Townsend R.P.; Carr S.W.; Thomas J.M.; Catlow C.R.A., Faraday Discussion Chem.Soc., **1990**, 89, 119.

8. Kaszkur Z.A.; Jones R.H.; Waller D.; Catlow C.R.A.; Thomas J.M., Combined Rietveld-Molecular Dynamics Powder Diffraction Approach to the Location of Molecules in Porous Solids: Application to Dibromobutane in Zeolite Y, submitted to J.Phys.Chem..

9. Czjzek M.; Fuess H.; Vogt T., J.Phys.Chem., **1991**, 95, 5255.

10. Dewar M.J.S. and Thiel W., J.Am.Chem.Soc., **1977**, 99, 4899.

11. Molecular simulation program Discover, version 2.7.0, March 1991, Biosym Technologies, 10065 Barnes Canyon Road, San Diego, CA 92121, USA.

COMBINED *IN-SITU* X-RAY POWDER DIFFRACTION AND X-RAY ABSORPTION SPECTROSCOPIC STUDIES OF TRANSITION-METAL-INCORPORATED ALPO AND SAPO CATALYSTS

J. W. COUVES*, G. SANKAR*, J.M. THOMAS*, J. CHEN*, C.R.A. CATLOW* , R. XU ** and G.N. GREAVES***

* The Davy Faraday Research Laboratory, The Royal institution of GB, 21 Albemarle street, London W1X 4BS, UK.

** Department of Chemistry, Jilin university, Chungchun, Peoples republic of China.

*** The SERC Daresbury Laboratory, Daresbury, Warrington, Cheshire, WA4 4AD, UK.

ABSTRACT

Using a position sensitive detector and a photo-diode array detector combined *in-situ* X-ray absorption spectroscopy and X-ray powder diffraction measurements were performed on a variety of Co and Ni substituted AlPO and SAPO catalysts. It was shown that the Co site acted as a redox centre, during the catalytic conversion of methanol to dimethyl ether. The nickel-substituted materials had a greater selectivity towards olefins in the conversion of methanol and were also shown to be considerably more stable than those of the cobalt analogues.

INTRODUCTION

Synchrotron radiation continues to offer further scope for characterizing micro-crystalline solids of wide structural types. An increasing trend is to deploy a combination of two or more parallel techniques. One such combination that is currently of interest as a means of monitoring the early stages of phase separation is that which entails X-ray absorption near edge structure (XANES) and anomalous small-angle X-ray scattering (ASAXS). Another which is the combination highlighted here, is that of X-ray diffraction(XRD) and X-ray absorption (XRA). This combination is used to chracterize a small sub-set of the important class of microporous microcrystalline alumino phosphates (AlPO-n) discovered in the early 1980's[1]. These materials possess three-dimensional pore structures and exhibit adsorptive and catalytic properties similar to the aluminosilicate zeolites. Incorporation of transition metal ions into AlPO's and the silico modified form produce an interesting class of uniform heterogenous catalysts[2,3], where essentially all the atoms of the bulk of the solid are accessible to, and may participate in, the catalytic process of the overall reaction. Recent studies[4,5] have shown that in particular Ni-SAPO-34, which possess a chabazite-like structure[6] is a highly selective catalyst for converting methanol into ethene.

From the Proceedings of the 9th International Zeolite Conference, Montreal 1992, Eds. R. von Ballmoos et al.,

X-ray absorption spectroscopy and X-ray powder diffraction provide powerful analytical tools for the characterization of such uniform heterogenous catalysts. XRA probes the local environment(<6Å) of a specific absorbing atom; whereas XRD probes the long range order of all atoms simultaneously. Independently these techniques have been extensively used for probing zeolitic catalysts[7,8]. Recent studies using both these techniques to probe zeolites have provided new insights in to such systems[9]. The combination of these complementary techniques in a single experiment provides a powerful, experimental means for elucidating both the atomic and electronic structure of these substituted alumino phosphate molecular sieves.

EXPERIMENTAL

Synthesis of the cobalt and nickel AlPO and SAPO's was achieved using a hydrothermal crystallization technique[10,5]. This technique allowed the synthesis of CoAlPO-5, CoSAPO-11, CoSAPO-34 and the nickel equivalents. Chemical analyses were consistent with the transition metal ion substitution for Al in the mircorporous structure. X-ray diffraction patterns recorded in a Siemens D500 were consistent with the structure types reported.

Conventional X-ray absorption spectra were recorded on station 7.1 at the SERC Daresbury Synchrotron radiation source using a Si[111] double crystal monochromator. Details of the experimental arrangement[11] and the data processing procedures[12] have been given elsewhere.

Combined XRA/XRD experiments were carried out on station 7.4 at the Daresbury laboratory. The experimental set up is shown schematically in Fig.1. X-ray absorption spectra

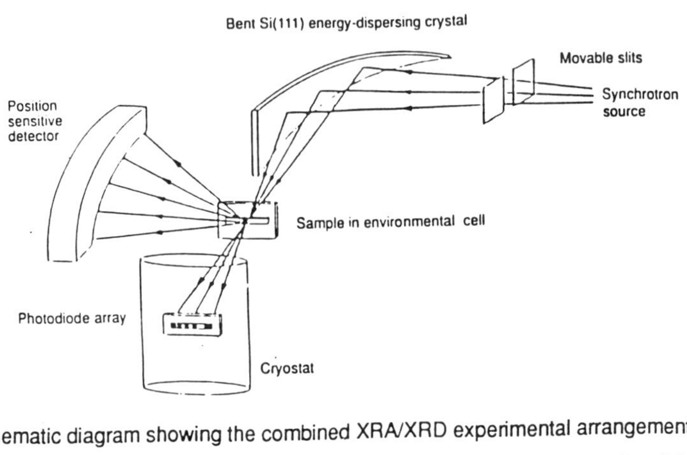

Fig.1 Schematic diagram showing the combined XRA/XRD experimental arrangement

were collected in the transmission geometry using an energy dispersive mode. A bent, tapered Si[111] single crystal monochromator produced an energy band of approximately 600eV (which includes the pre-edge, edge and extended x-ray absorption fine structure (EXAFS) regions of the element of interest). A 120° INEL, position sensitive detector was used to intersect the Debye-Scherrer diffraction rings. Using this experimental setup complete diffraction data were collected in 5 minutes and using a photo-diode array detector, complete XRA data were collected in 500 ms. The details of the experimental arrangements have been discussed in detail recently[13-15]. The changes occurring at the specific metal cation site and to the bulk structure are monitored during the calcination and catalysis. a self supporting wafer of the sample was mounted in a specailly constructed cell (designed and constructed at the Davy Faraday Laboratory) for combined XRA/XRD, which has the facility to place the sample under reactive conditions and monitor the gases with on-line GC and mass spectrometric techniques[16].

RESULTS AND DISCUSSION

Co-substituted molecular sieves

Conventional XRA data of the as-prepared Co-AlPO and SAPO molecular sieves indicated that the Co^{2+} cation occupies a tetrahedral site in the macroanionic framework. (The presence of a pre-edge 1s-->3d transition, on the Co-K-edge (see Fig.2) is a strong

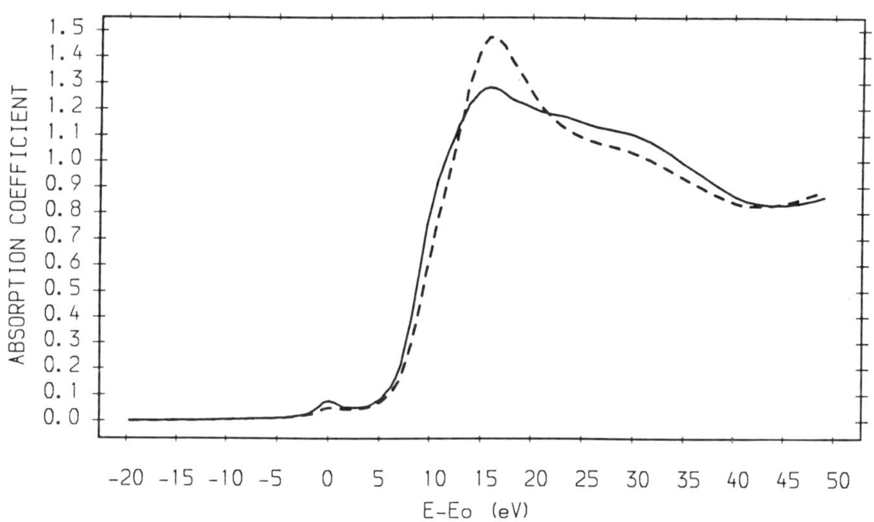

Fig.2 Typical Co K-edge X-ray absorption near edge structure (XANES) of as-prepared Co-ALPO-5 (solid line) and calcined Co-ALPO-5 (dashed line) molecular sieves.

qualitative indication that the metal ion is in a tetrahedral geometry). Analysis of the EXAFS data confirms that the Co is coordinated to four oxygens at a distance of 1.90 to 1.99Å (depending on the structure type, see tabel 1). A 4-coordinated phosphorous shell is also observed between 3.0 and 3.2Å.

Table 1: The refined EXAFS data of the Ni and Co incorporated ALPO-5 and SAPO-34 molecular sieves.

	Atom Pair	N	R (Å)
Ni-ALPO-5 as-prepared	Ni-O	4.6	1.99
	Ni-P	3.6	3.16
Ni-ALPO-5 calcined	Ni-O	4.5	2.00
	Ni-P	3.6	3.14
Ni-SAPO-34 as-prepared	Ni-O	4.6	1.99
	Ni-P	4.3	3.18
Ni-SAPO-34 calcined	Ni-O	4.0	2.00
	Ni-P	4.3	3.16
Co-ALPO-5 as-prepared	Co-O	4.1	1.92
	Co-P	3.7	3.08
Co-ALPO-5 calcined	Co-O	2.6	1.90
	Co-O	2.9	2.06
	Co-P	2.6	3.00
	Co-P	2.7	3.18
Co-SAPO-34 as-prepared	Co-O	4.5	1.92
	Co-P	4.1	3.11
Co-SAPO-34 calcined [a]	Co-O	5.0	1.99
	Co-P	3.7	3.07

[a] Co-SAPO-34 calcined data was analyzed based on an average coordination and bond distance of the distorted Co environment

Calcination of these materials results in loss of the temeplate material accompanied by loss in crystallinity (see XRD pattern in Fig.3) and change in oxidation state of Co^{2+} to Co^{3+} as seen by the shift in the XRA edge data to higher energy. However, we see clearly the pre-edge peak(reduced in intensity as compared to the as-prepared) similar to the as-prepared material indicating that the cobalt is still predominantly retained in the tetrahedral coordination. The

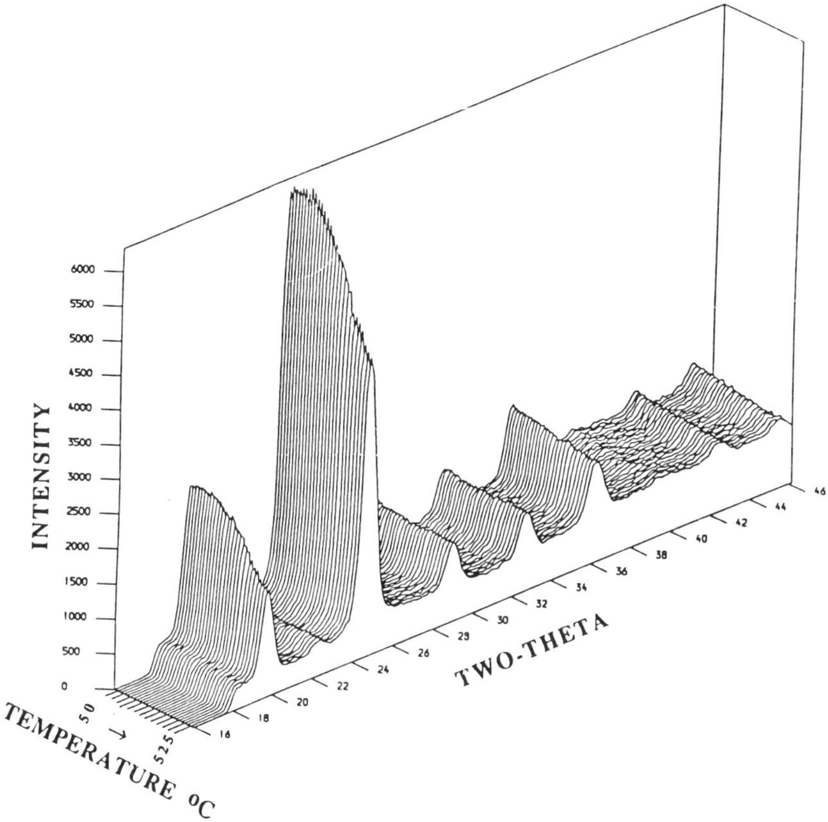

Fig.3 Typical XRD pattern of Co-SAPO-34 catalyst recorded during the course of calcination.

analysis of the EXAFS data shows that there are atleast two distinct Co sites having different Co-O distances. The XRA results corroborate those from UV/VIS and ESR spectroscopies[17-19] which indicate that the majority of the Co^{3+} remains in a tetrahedral geometry. This is rather unusual for a d^6 ion[20]. The distortion in the Co-O interatom distance is a result of the highly unfavourable crystal-field-stabilization of a tetrahedral d^6 ion.

Absorption of methanol causes reduction of the Co^{3+} to Co^{2+}, as seen by the XRA edge shift to a lower energy. Temperature programmed reaction spectroscopy(TPRS) recorded during the *in-situ* XRA/XRD measurement (see Fig.4) shows that the production of dimethyl ether under reaction conditions is predominant.

Ni-substituted molecular sieves

The X-ray absorption studies of the Ni-AlPO and SAPO molecular sieves indicate that the coordination geometry, state of oxidation of the Ni remains unaffected by calcination. EXAFS being best described by four oxygen atoms at 1.97-1.99Å and 4 phosporous atoms at 3.1-3.2Å (see Table.1). The Pre-edge, 1s-->3d, feature on the Ni-K-edge is largely unaffected by calcination. The presence of Ni^{2+} in the AlPO structure is shown to cause a large local lattice distortion, with the T-O-T bond angle decreasing by as-much-as 20° in the proximity of the nickel ion. Only small changes of lattice parameter are observed on calcination from the *in-situ* XRD data.

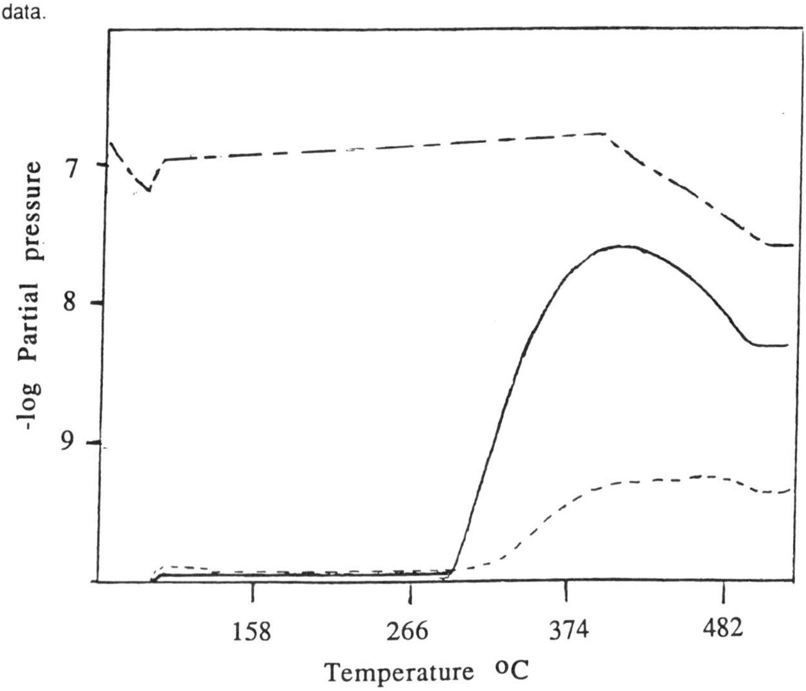

Fig.4 *In-situ* TPRS of Co-SAPO recorded during the combined XRA/XRDexperiment, under the reaction condition (methanol/N2; temperature rate 10°C/min); — - —m=32 (methanol); ---------- m=42 (C3H6); ———— m=46 (dimethyl ether)

The catalytic performance of the Ni-SAPO and AlPO's recorded using TPRS, was such that they had the expected high selectivity towards olefins (mainly ethene) whereas only dimethyl ether (predominant product) was formed with Co-AlPO with methanol as reactanct .

Comparison of Co and Ni Molecular sieves

The XRA recorded during the course of catalysis shows no evidence for any direct interaction between the transition metal and adsorbed species, but at present the quality of data attainable by energy dispersive EXAFS is not of a suitable quality to yield reliable quantitative evidence of such dynamic interactions.

The high catalytic activity and selectivity of the Ni-SAPO-34 has previously been shown to be a result of modification of framework Bronsted acidity by Ni^{2+} ions incorporation[5]. For the Cobalt system, as well as changes in Bronsted acid site density, arising through Co^{2+} isomorphous substitution, the Lewis acidity is increased and more significantly a redox centre, capable of methane activation[17] is produced.

CONCLUSION

Combined XRA/XRD studies of Ni and Co Alumino phosphate molecular sieves have yielded useful insights into the correlation between structure and catalytic performance. In particular they have shown that the poor activity of Co-AlPO's in the methanol to olefin conversion is a result of the presence of a redox centre at the transition metal. Ni-SAPO-34 shows particularly good activity, an effect that appears to be correlated with modification of the Bronsted acidity by the framework-nickel. Further studies of the nature and density of the acid sites in these material are called for.

ACKNOWLEDGEMENTS

The authors wish to thank Dr's A. Dent, G.E. Derbyshire, D. Waller and S. Natarajan for technical help with the combined XRA/XRD experiments. We acknowledge with gratitude the support of the Science and Engineering Research Council, the Royal Society, British Petroleum and Unilever plc.

REFERENCES

1. Wilson, S.T., Lok, B.M., Messina, C.A., Camana, T.R. and Flanigan, E.M., J.Am.Cem.Soc., **104**, 1146, (1982).
2. Thomas, J.M., Angew.Chem.Int.Ed. Engl. **27**, 1673 (1988).
3. Thomas, J.M., Philos.Trans.R.Soc. London, A **333**, 173 (1990).
4. Inui, T., Phatanasri, S., Mutsuda, H., J.Chem.Soc.,Chem.Commun. 205, (1990).
5. Thomas, J.M., Xu, Y., Catlow, C.R.A. and Couves, J.W., Chem.Materials **3**, 667 (1991).
6. Ito, M., Shimoyama, Y., Saito, Y., Yasushi, Y.,and Otake, M., Acta Crystallogr. C **41**, 1698, (1985).

7. Couves, J.W. , Jones, R.H. and Thomas, J.M., Advanced Materials, **2,** 181 (1990)

8. Sano.M., Maruo. T, Yanatera.H., Suzuki.M., Saito.Y, J. Am. Chem. Soc. **109**, 52 (1987)

9. Doorhyee E., Catlow.C.R.A., Couves.J.W., Maddox.P.J., Thomas.J.M , Greaves.G.N., Steel.A.T., and Townsend. R.P., J.Phys.Chem **95,** 4514 (1991)

10. Xu, Y., Maddox, P.J., and Couves, J.W., J.Chem.Soc., Faraday Tans. **86**. 425, (1990)

11. Doorhyee, E., Greaves, G.N., Steel, A.T., Townsend, R.P., Carr, S.W., Thomas, J.M., and Catlow, C.R.A., Faraday Discuss. Chem.Soc., **89**, 119 (1990).

12. Binstead, N., Gurman, S.J., and Ross, I.J., J.Phys. C, **17**, 143, (1984); SERC Daresbury Laboratory Update 1990.

13. Couves, J.W., Thomas, J.M., Catlow, C.R.A., Greaves, G.N., and Dent, A.J., J.Phys.Chem, **94**, 6517, (1990).

14. Dent A.J., Derbyshire. G.E., Greaves. G.N., Ramsdale.C.A., Couves.J.W., Jones.R.H., Catlow.C.R.A., and Thomas. J.M., in X-rays in Materials analysis: Novel applications and recent developments.

15. Couves J.W., Thomas.J.M., Waller.D., Dent.A.J., Derbyshire.G.E., and Greaves.G.N. Nature **354**, 465, (1991)

16. Couves J.W. ,Sankar. G, Waller. D, Madill. D, Sheehy. M, Jones. R.H. and Thomas. J.M., (in preparation.)

17. Montes C., Davis,M.E., Murray.B., and Narayana.M., J.Phys.Chem., **94**, 6425, (1990)

18. Iton .L.E., Choi.I., Desjarlins, J.A., and Maroni.V.A., Zeolites **9**, 535 (1989)

19. Kraushur-Czametzki , H., Hoogervorst. W.G.M., Andrea. R.R., Emieis, C.A., and Stork. W.H.J., J.Chem.Soc., Faraday Trans. **87**, 891 (1991)

20. Chen.J., Sankar.G., Thomas.J.M. and Couves.J.W. (in preparation)

CORRELATIONS BETWEEN ACID STRENGTH AND Si ENVIRONMENT IN FAUJASITE TYPE MOLECULAR SIEVES

M. Briend, D. Barthomeuf

Laboratoire de Réactivité de Surface et Structure, URA 1106, CNRS, Université P. et M. Curie, 4 place Jussieu, 75252 Paris Cédex 05, France, (FAX (1) 44-27-60-33)

ABSTRACT

A series of faujasite type zeolites (NaHX, dealuminated Y, LZY-82, SAPO-37) are compared in the TPD of ammonia. The acid strength is related to the environment of Si atoms in the Si-OH-Al species. In NaHX the weaker sites correspond to Si(4 Al) with 9 Si as second neighbours. The presence of P as second neighbours generate a stronger acidity in SAPO-37 for Si(4 Al) sites. SAPO-37 has the most homogeneous acid strength distribution of all the samples. The acid strength of the four samples is connected to the ^{29}Si NMR chemical shifts.

INTRODUCTION

It has been known for a long time that the strength of the protonic acidity of zeolites (i.e. also the charge on the hydrogen and oxygen atoms) depends on the environment of the SiO_4 and AlO_4 tetrahedra involved in the Si-OH-Al species [1-3]. For a zeolite in which all the T sites are equivalent this effect depends only on the number of next nearest neighbours (NNN) sites occupied by Al or Si [3] . The Si-Al faujasite belongs to this category and the strength of protons may be explained by the Al content and the topology of the Al location i.e. the number of Al in NNN location [3-5]. It has been shown independently that the NMR chemical shift of ^{29}Si depends not only on the number of Al or Si in the first shell but also on that in the second one (NNN sites) particularly at low Al content and for Si(0 Al) and Si(1 Al) centers [6]. This evidences the importance on the local properties of the nature of the atom located in a NNN position with regard to a central Si atom. In SAPO's, by contrast with Si-Al zeolites, the acidity is created by the presence of Si atoms. The first and second neighbours of Si should then govern the acid strength.

Several methods can be used to characterize the strength of protons in zeolites. Comparison of infrared and TPD studies of ammonia adsorption showed that in mordenites (7,8) or faujasites (9) the ammonia desorbed at low or high temperatures corresponds to the Bronsted and Lewis acidity respectively. A large number of studies used the TPD method to characterize, either the nature of acid sites [7-10], their strength [8,10-15], the influence of acid sites on the diffusion of molecules [13-15] or the influence on acid strength of the Si, Al distribution in mordenite [16] or dealuminated Y [17]. The approach consisting of pretreatment of the solid step by step at increasing temperatures tells about the acid centers still existing after each treatment [11,13]. It is a time consuming way but which has the advantage of giving very valuable information. It will be used

From the Proceedings of the 9th International Zeolite Conference, Montreal 1992, Eds. R. von Ballmoos et al.,

here. The present paper compares four samples with the faujasite structure (i.e. same topology) but with different environments for the protons i.e. Si, Al or P atoms in the first or second layer.

EXPERIMENTAL

Materials

A sample of NaX has been partly exchanged to give $Na_{70}H_{16}X$ (NaHX). A Y zeolite dealuminated with $(NH_4)_2SiF_6$ gives HYD with a framework Si/Al of 7 while it is 4.5 for LZY-82 (ultrastable Y). SAPO-37 prepared as in (18) gives the sample $Si_{0.13}Al_{0.49}P_{0.38}O_2$. The materials (50 mg) are pretreated at 673 K for NaHX and 873 K for the three others under flowing oxygen for 6 hr, then under vacuum for 6 hr.

TPD of ammonia

In the first step i) ammonia (100 Torr) is adsorbed at room temperature for 6 hr, ii) the excess of ammonia is evacuated at a temperature $T_1 \geqslant 423$ K for 6 hr, T_1 is varied from 423 K to 773 K from one experiment to the other, iii) the temperature is decreased from T_1 to RT and the TPD of ammonia is performed at a 300 K/hr heating rate up to 873 K (673 K for X sample). The gases evolved are analyzed with a Leybold Heraeus quadrupole mass spectrometer Quadruvac PGA 100. Different experiments after pretreatment at increasing T_1 values are conducted with the same sample. After the complete evolution of ammonia in vacuum for 6 hr at 873 K the process starts again in i). The evacuation temperature T_1 in ii) increases at each step by 50 K from 423 K to 773 K. In this way each curve in figure 1 is a TPD experiment conducted on a newly pretreated sample. The amount of ammonia evolved is expressed as A_{873} for the base removed up to 873 K at a heating rate of 300 K/hr or as A_T for the total ammonia release obtained in maintaining the temperature constant at 873 K until no NH_3 is detected by the mass spectrometer.

RESULTS

Acid strength and ^{29}Si NMR

The protons in Si-OH-Al species arise from the presence of Al in Si-Al zeolites and of Si in SAPO's. Information is known from NMR on the environment of Si and of Al atoms (population of Si(n Al) species and chemical shifts). Attempts can be made to correlate the acidity to the neighbours of the Si atoms.This approach is particularly useful to compare Si-Al zeolites with SAPO's where a large number of Al atoms are connected to P and do not generate acidity (Al(4 P)). In both of these zeolitic materials the Si atoms may have only Al or Si as first neighbours.

In Si-Al zeolites it is well accepted that the acid strength increases as the number of Al atoms in NNN location decreases (3,4). The chemical environment of the Al atoms is strongly correlated to that of the Si atoms. An "isolated Al" (i.e. with no Al in NNN location) which generates a strong protonic site on the corresponding Si-OH-Al species [3-5] implies four Si in the first shell of tetrahedra and nine in the second. This may be written as Al(4 Si) (9 Si). This corresponds to Si atoms surrounded at the most by 1 Al in the first layer and 9 Si or less in the second, depending on the overall Al content. This means that Si atoms around the isolated Al are of the type Si (1 Al) (x Si, y Al) with $x + y = 9$ in faujasites. They are connected, when $x = 9$, to the strongest acidity referred to above. Increasing the Al content decreases the number of isolated Al and that of Si surrounded by only one Al and increases the number of Al in NNN location, together with the number of Si(3 Al) and Si(4 Al) species. It has been stated that the ^{29}Si NMR chemical shift of the various Si(n Al) sites can be explained by a weighted linear combination of the deshielding effects of Al atoms in the first and second shell [6]. The acid strength, decreasing as the Al content increases would give the following order for the protonic acid strength associated to Si(n Al) in the Si-OH-Al species : Si(4 Al) < Si(3 Al) < Si(2 A) < Si(1 Al). This follows the rank of chemical shifts δ in ^{29}Si NMR which decreases from -84.7 ppm for Si(4 Al) to -107.8 ppm for Si(0 Al) in faujasites [6].

^{29}Si NMR gives also information on the population of the various Si(n Al) (9(Al+Si)) species. In X zeolites most of the Si atoms are of the type Si(4 Al) (9 Si). For a Si/Al ratio of 1.19 as in the present sample the population of the various species is in % Si(4 Al) : 64, Si(3 Al) : 26.5, Si(2 Al) : 6.2, Si(1 Al) : 1.4 and Si(0 Al) : 1.9 [19]. In SAPO-37 around 90-95 % of the Si are described by Si(4 Al) (9 P), the rest being mainly Si(3 Al) and Si(0 Al) [18]. In HYD a variety of environments exists which may be written as Si(n Al) (x Si, y Al) with $n < 3$, $x + y = 9$ and y depending on n. The distribution of Si atoms in Si(3 Al), Si(2 Al), Si(1 Al) and Si(0 Al) species is around 0, 10, 40, 50 % respectively [21]. For LZY-82 the percentages are 7, 14, 38 and 41 respectively [20]. In addition extraframework Al is present in this last sample. It may be noted that only NaHX and SAPO-37 have Si(4 Al) species.

From these results, it is expected that the four samples with their different Si(n Al) population will generate specific acid strengths, seen in TPD.

NH_3 TPD

The figure 1A gives for SAPO-37 the ammonia TPD results up to 873 K. The full desorption obtained by keeping the temperature constant at 873 K up to the point of no NH_3 evolution is reported in figure 1B. Each curve referred to by the temperature T_1 in figure 1 gives access to the number (area under the curve) and to the mean strength (temperature T_m of the maximum of the peak) of the acidic sites which exist after evacuation of ammonia at the temperature T_1. The amount of acid sites retaining NH_3 between two temperatures T_1 is easily obtained by difference. It decreases as the temperature T_1 increases. The temperature T_m of the maximum of the peak expresses the contribution of the larger fraction of remaining sites which have a same strength after the evacuation of NH_3 at T_1. It increases simultaneously with T_1.

Figures similar to figure 1 are obtained for the four samples with differences in amount desorbed and temperatures T_m. The results are collected in figures 2 and 3. The figure 2 gives for the four samples the decrease in amount remaining adsorbed as a function of T_1, for the ammonia evolved up to 873 K (673 K for NaHX) with the heating rate of 5 K/min. (A_{873}). After the full evacuation of ammonia at 873 (A_T) the points are displaced upwards by around 5 %. The temperature T_m of the maximum of the desorption peak reported in figure 2 shows for each amount desorbed the mean acid strength of the remaining sites after evacuation at T_1. The figure 3 gives the amount of ammonia desorbed between two successive temperatures T_1 and (T_1 + 50) as a function of T_1. It gives access to the number of acid sites which strength can be characterized by the temperature range T_1-(T_1 + 50), i.e. to the distribution of acid strengths.

DISCUSSION

One should first keep in mind that TPD technique has intrinsic limitations since for instance T_m depends on the total amount of acidic sites. This might arise from a non statiscal distribution of acid sites or from diffusion and readsorption phenoma. In the present case the theoretical number of protons is 16 for NaHX, 24 for HYD, 25 for H-SAPO-37 and 35 for LZY-82. The values are very similar for HYD and H-SAPO-37. In addition both samples have the same structure. It follows that diffusion or/and readsorption phenoma should be comparable.

Comparing first the theoretical number of protons to the results described it may be observed at first that in NaHX the sites are known to be weak and it may occur that the first evacuation at 423 K removes not only the physically adsorbed NH_3 but also some base adsorbed on weak acid sites. For the three other samples the ammonia remaining after this evacuation is, in arbitrary units, 390, 430 and 520 for HYD, H-SAPO-37 and LZY-82 respectively. This follows rather well the theoretical number of protons.

Figure 2 shows that NaHX has a small number of acid sites desorbing mainly at 423 K. The temperature of the maximum of the peak at 530 K for this T_1 value increases after the pretreatments at 473 and 523 K but after the two last treatments only very small amounts of NH_3 are desorbed. The homogeneity in the acid strength is also seen in figure 3 where a sharp distribution of strengths is shown. The sites observed should correspond to the Si(4 Al) (9 Si) species with a large number of Al in a NNN configuration. The few Si(3 Al) (x Si, y Al) sites existing in X zeolites [19] would be responsible for the very small number of stronger sites for T_1 equals 473 and 523 K. For H-SAPO-37 the major part of ammonia is desorbed after evacuation at 473 K (fig. 2), the larger amount being evolved between 423 and 473 K (fig. 3). The very sharp peak in the distribution of acid strengths (fig. 3) should correspond to the sites the more frequent in H-SAPO-37 i.e. the Si(4 Al) (9 P) species which represent 90-95 % of the acid centers [18]. The value of T_m (peak maximum) after the 423 K evacuation is 548 K i.e. significantly higher than for NaHX (530 K). This suggests than the weakest sites in H-SAPO-37 corresponding to Si(4 Al) (9 P) environments are stronger that the weak sites in NaHX (Si(4 Al) (9 Si)). This means that P as second nearest neighbour of a Si atom in Si-OH-Al species increases the acid strength of the proton. This is also in line with a more negative value of the chemical shift in the ^{29}Si NMR studies for Si(4 Al) which gives -84.7 for NaHX [19] and -89.2 for SAPO-37 [18] (table 1).

In SAPO-37 the sites retaining ammonia between 473 and 523 K could belong to the Si(3 Al) (6 P, 3 Si) or Si(3 Al) (5 P, 4 Si) centers shown to exist preferentially when Si islands are present [18]. Besides these sites, the environments Si(n Al) with n < 3 [18] generate a small number of strong sites as strong as in LZY-82 or HYD which retain ammonia up to high temperatures (fig. 2 and 3).

A main difference between NaHX/SAPO-37 on the one side and LZY-82/HYD on the other side is the inhomogeneity of the strength distribution in the two last samples as seen in figure 3. A large number of acid strengths should exist in these materials due to the variety of possible second shell environments for Si(n Al) with n < 3. In addition LZY-82 is known to have a large number of Lewis acid sites which also adsorb NH_3 and contribute to the results described in figures 2 and 3. HYD might also contain some of them after the pretreatment at 873 K. This renders a detailed comparison of these samples with NaHX and SAPO-37 rather difficult. Some main features are nevertheless observable. SAPO-37 and HYD have similar theoretical number of protonic acid sites (25 and 24 respectively) which is rather well described by the peak areas in arbitrary units (430 and 390 respectively). The NMR results tell that HYD has neither Si(4 Al) nor Si(3 Al) environments [21]. The weakest acid sites should be of the type Si(2 Al) (x Si,y Al). The figure 2 shows that the temperature at the maximum of the peak for these sites is 620 K. This is quite higher that for the weakest sites in SAPO-37 of the type Si(4 Al) (9 P) and for which T_m is 548 K. This observation suggests to compare for all the samples the T_m values for the weakest sites which are the easiest to assign since they loose NH_3 after the treatment at 423 K. Figure 2 gives a T_m value of 568 K for LZY-82. Table 1 shows an order for increasing temperature for T_m : NaHX < H-SAPO-37 < LZY-82 < HYD. This may be correlated to the chemical shifts of ^{29}Si NMR for the peak in each sample corresponding to the highest existing Al environment. This order should correspond to the weakest acid sites (3,4). A similar trend is seen in table 1 between T_m and δ.

As to the strongest sites, figure 2 shows after the pretreatment at 773 K the following order : HYD ~LZY-82 > H-SAPO-37. Other studies indicated the presence of a very small number of very strong protonic sites in H-SAPO-37, even stronger than in HYD or LZY-82 [18,22]. This apparent discrepancy may be due to the presence of very strong Lewis sites in LZY-82 and possibly in HYD after its pretreatment at 873 K. They retain ammonia at high temperature, possibly even higher than the strong protons of SAPO-37.

In conclusion the acidity of materials with the faujasite structure shows that P as second neighbours of a central Si atom increase the acid strength in the Si(4 Al) (9 P) centers compared to the Si(4 Al) (9 Si) ones present in X. This difference can be correlated with the ^{29}Si NMR chemical shift. A general trend is proposed between this NMR parameter and the acid strength for four different faujasite-type materials. It may be expected that further correlations will be found since both protonic strength [3-5] and ^{29}Si NMR chemical shift [6] are depending upon structural relationship between a central Si atom and its non silicic neighbours in the first and second shells of tetrahedra.

ACKNOWLEDGEMENTS

We thank D. Delafosse, M.J. Peltre and P. Man for very interesting discussions and J. Maquet for NMR experiments on HYD.

REFERENCES

1. Barthomeuf D. and Beaumont R., J. Catal. **30**, 288 (1973).
2. Gibbs G.V., Meagher E.P., Smith J.V. and Pluth J.J., in *Molecular sieves II*, Ed. J.R. Katzer, ACS Symp. Series **40**, 19 (1977).
3. Wachter W.A., in *Proceed. 6th Intern. Zeolite Conf.*, Eds D. Olson and A. Bisio, Butterworths, Guildford, UK, pp. 141-150 (1984).
4. Barthomeuf D., Materials Chemistry and Physics, **17**, 49 (1987).
5. Barthomeuf D., in *Catalysis 1987*, Ed. J.W. Ward, Elsevier, Stud. Surf. Sci. Catal., **38**, 177 (1988).
6. Melchior M.T. and Newsam J.M., in *Zeolites : Facts, Figures, Future*, Eds. P.A. Jacobs and R.A. Van Santen, Elsevier, Stud. Surf. Sci. Catal., **49B**, 805 (1989).
7. Mirodatos C., Ha B.H., Otsuka K. and Barthomeuf D., in *Proceed. 5th Int. Conf. Zeol.*, Heyden, London, pp. 382-393 (1980).
8. Karge H.G. and Dondur V., J. Phys. Chem., **94**, 765 (1990).
9. Karge H., Dondur V. and Weitkamp J., J. Phys. Chem., **95**, 283 (1991).
10. Reschetilowski W., Unger B. and Wendlandt K.P., J. Chem. Soc., Faraday Trans., **85**, 2941 (1989).
11. Hashimoto K., Masuda T., Mori T., in *Proceed 7th Intern. Zeol. Conf.*, Eds. Y. Murakami, A. Iijima, J.W. Ward, New Developments in Zeolite Science and Technology, Jodanska-Elsevier, Tokyo, pp. 503-510 (1986).
12. Hunger B. and Hoffmann J., J. Thermal Analysis, **33**, 933 (1988).
13. Forni L., Peratello S., Ortoleva E., Solinas V., Monaci R. and Ferino I., Gazz. Chim. Italiana, **120**, 535 (1990).
14. Sawa M., Niwa M. and Murakami Y., Zeolites, **10**, 307 (1990).
15. Sawa M., Niwa M. and Murakami Y., Zeolites, **11**, 93 (1991).
16. Sawa M., Niwa M. and Murakami Y., Zeolites, **10**, 532 (1990).
17. Corma A., Fornes V., Melo F.V. and Herrero J., Zeolites, **7**, 559 (1987).
18. Man P.P., Briend M., Peltre M.J., Lamy A., Beaunier P., Barthomeuf D., Zeolites, **11**, 563 (1991).
19. Klinowski J., Progr. in NMR Spectroscopy, **16**, 237 (1984).
20. Aouali L., Jeanjean J., Dereigne A., Tougne P., Delafosse D., Zeolites, **8**, 517 (1988).
21. Barthomeuf D. and Baizoumi Zoua, unpublished results.
22. Su Bao Lian, Lamy A., Dzwigaj S., Briend M., Barthomeuf D., App. Catal., **75**, 311 (1991).

Table 1

T_{max} of the NH_3 desorption peak after pretreatment at 423 K[(a)] and [(b)] chemical shift of $^{29}Si(n\ Al)$ with highest existing n value for each sample

Sample	NaHX	SAPO-37	LZY-82	HYD
T_m[(a)] (K)	530	548	568	620
δ[(b)]	- 84.7[(c)]	- 89.2[(d)]	- 91[(e)]	- 96[(f)]

(a) Pretreatment at 423 K
(b) Chemical shift in ppm
(c) Si(4 Al) (9 Si) [6]
(d) Si(4 Al) (9 P) [18]
e) Si(3 Al) (x Si, y Al) [20]
(f) Si(2 Al) (x' Si,y' Al) [21]

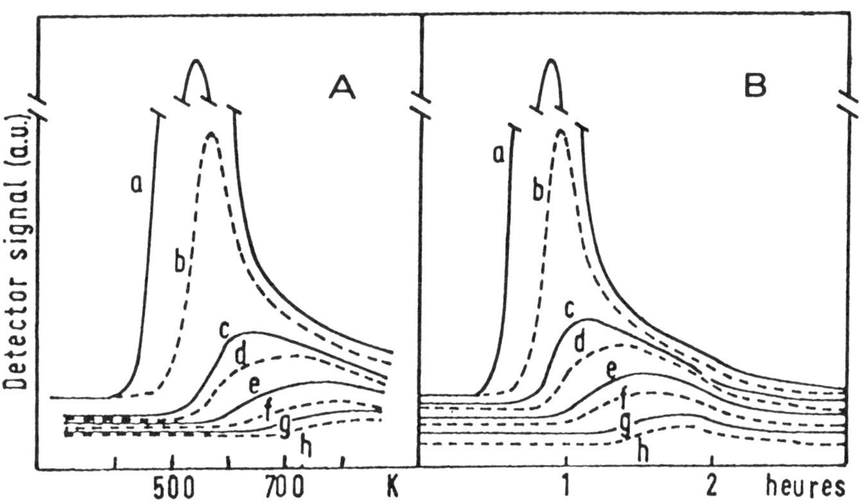

Figure 1. Ammonia TPD curves for the sample H-SAPO-37 for the desorption as a function of temperature up to 873 K (A) or as a function of time up to the complete desorption of ammonia at 873 K (B). Prior pretreatment temperature T_1 (K) : a (423), b (473), c (523), d (573), e (623), f (673), g (723), h (773) (ordinates displaced).

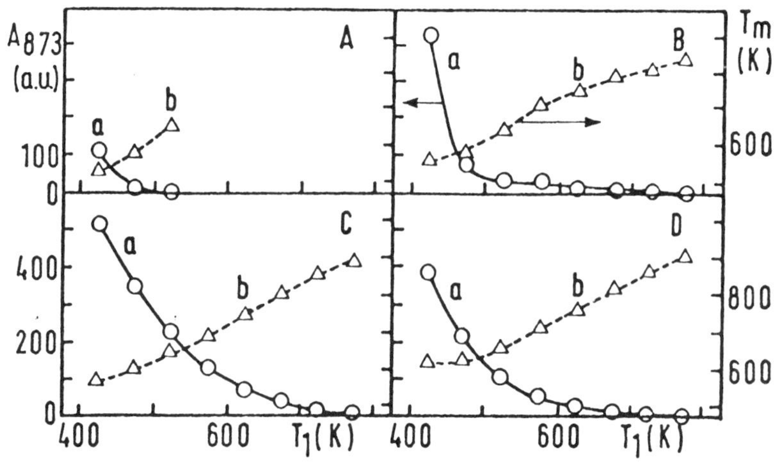

Figure 2. Changes as a function of the pretreatment temperature T_1 of the amount of ammonia desorbed (a) and of the temperature T_m of the maximum of the peak (b) in ammonia TPD. Sample NaHX (A), H-SAPO-37 (B), LZY-82 (C), HYD (D).

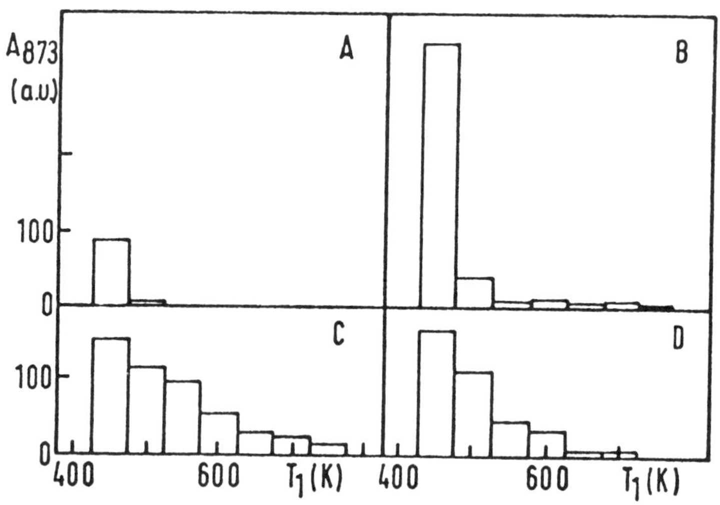

Figure 3. Changes as a function of the pretreatment temperature T_1 of the amount of ammonia desorbed between T_1 and $T_1 + 50$. Samples NaHX (A), H-SAPO-37 (B), LZY-82 (C), HYD (D).

TEMPLATE-FREE AND ALKALI-FREE SYNTHESIZED SILICATES WITH MFI-STRUCTURE -A COMPARATIVE STUDY

M. WALLAU*, F. SCHÜTH*, A. BRENNER*, S.MELSON*, R. SPICHTINGER*,K. UNGER*, A. TISSLER** AND B. ZIBROWIUS***

* Institut für Anorganische Chemie und Analytische Chemie, Johannes Gutenberg- Universität Mainz, J.J. Becherweg 24, W- 6500 Mainz

** Vereinigte Aluminiumwerke AG, Georg-von Boeselager-Straße 25, W-5300 Bonn

*** UMIST, Department of Chemistry, POB 88, Manchester M 601 QD

ABSTRACT

Zeolites and gallosilicates with an MFI-structure which were crystallized from inorganic reaction systems and from alkali-free, tetrapropylammoniumbromide (TPABr) containing systems are compared.
The strength, nature and distribution of the acidic sites were determined by FTIR-spectroscopy, TPD, electron microprobing, ^{29}Si, ^{27}Al and ^{1}H MAS NMR spectroscopy.
The influences of the different trivalent elements as well as the strength, type and distribution of the acidic sites on the catalytic performance of the different MFI-silicates are discussed.

INTRODUCTION

The crystallization of MFI-silicates from an alkali-free reaction gel which contains templating agents such as TPABr is preferred for scientific studies as it gives pure MFI-silicates with a well defined crystal morphology [1]. However, in their industrial production the use of organic templates causes environmental problems.

The wish to avoid the expensive purification of the waste water and waste air led to the development of a template-free synthesis [2-4].

The aim of this work was to investigate the influence of the different synthesis methods on the kind, distribution and strength of the acidic sites of the MFI-silicates. These results were used to explain the different catalytic behavior observed.

From the Proceedings of the 9th International Zeolite Conference, Montreal 1992, Eds. R. von Ballmoos et al.,

EXPERIMENTAL

Synthesis

The template-free synthesized (Si,Al)-MFI was supplied by VAW AG (Bonn, FRG). It was synthesized in a VAW-patented continuous process based on a synthesis procedure developped at the University of Mainz which is described elsewhere [3].

The synthesis of (Si,Al)MFI with an "organic template" (TPABr without alkali cations), was carried out in an alkali-free system containing dissolved ammonia in static, teflon-lined autoclaves at 453 K. A detailed description of the synthesis is given elsewhere [1].

The (Si,Ga)MFI was also synthesized in an alkali-free system. In this case, solid hexamethylenetetramine was used to generate the required ammonia "in situ" in the autoclave.

To achieve pure (Si,Ga)MFI from wholly inorganic gels it was necessary to study the crystallization parameters by varying the molar ratios and the specific surface area of the silica source. These experiments were carried out in static autoclaves and finally scaled-up to stirred autoclaves.

IR-spectrocopy and TPD-mesurements

The IR-spectra were recorded with a Nicolet 5 SXB spectrometer linked to an Ultra High Vacuum (UHV)-cell. The sample preparation and the UHV-cell will be described elsewhere [5].

The UHV-cell which was equipped with a mass-spectrometer (Balzers QME 125) allowed the temperature programmed desorption of pyridine. By employing simultaneous FTIR-spectroscopy during the pyridine desorption it was possible to relate the TPD-peaks to the different acidic sites of the silicates.

MAS NMR spectroscopy

The MAS NMR measurements were conducted at room temperature using a Bruker MSL 400 spectrometer at a resonanance frequency of 79.5 MHz for ^{29}Si NMR and 104.3 MHz for ^{27}Al NMR spectroscopy. Magic angle spinning was achieved at a frequency between 3 and 4 kHz on uncalcined samples.

The TPD of pyridine monitored by simultaneous FTIR-measurements allowed the follwowing order of the acidic sites with decreasing strength(desorption maxima are given in parentheses).

Al_{Lewis} (773 K) > Ga_{Lewis} (743 K) > $Al_{Broenstedt}$ (673 K) > $Ga_{Broenstedt}$ (593 K)

Propane aromatisation

The aromatisation of propane was carried out in a fixed-bed reactor at a temperature of 823 K (WHSV = 1 h^{-1}). For the catalytic test the silicate powder samples were pressed into pellets and than subsequently crushed and sieved. The resulting split, with diameters between 0.5 to 1 mm, was used as the catalyst. The product stream was analyzed on-line with a gas chromatograph.

Electron microprobing

For the electron microprobing crystals were prepared by embedding them in an epoxy resin and subsequently polishing with a diamond paste to remove their outer surface. The X-ray spectrum which was induced spot wise along one crystal axis by a focussed electron beam was analyzed.

RESULTS AND DISCUSSION

Synthesis

The (Si,Al)MFI was the only phase to crystallize in the TPABr containing, alkali-free reaction system. The crystallization of (Si,Al)MFI from "template-free systems" (hydrated Na^+ without organic cations) is already described [3].

The template-free synthesis of (Si,Ga)MFI in static teflon-lined autoclaves had a chemical composition of following molar ratios: H_2O/SiO_2 = 80;100; OH^-/SiO_2 = 0.2-0.6; SiO_2/Ga_2O_3 = 20-140; Na_2O/SiO_2 = 0.2-1.0 with a crystallization duration of three days. - However dense silica phases were found. The crystallinity of the (Si,Ga)MFI-phase amounted 50% of the XRD- crystallinity of the(Si,Ga)MFI sample which were synthesized with TPABr at most. The amount of dense silica could subsequently be reduced by decreasing the surface area of the silica source from 380 to 22 m^2 /g. A molar ratio of SiO_2/Ga_2O_3 = 119-140 gave the largest amount of (Si,Ga)MFI (70%).

The scale-up of the optimized reaction procedure to the stirred autoclaves indicated that there is a strong influence of gel-aging on the yield of the (SI,Ga)MFI. It was seen that stepwise aging of the reaction mixture for several hours at 413 K, 433 K and 453 K was necessary to obtain (Si,Ga)MFI.

Further investigations into the influence of gel-aging showed that the formation of dense silica was totally suppressed when a seeding gel (a gallo-silica gel which was aged for several days at a temperature between 293 and 363 K) was used in the synthesis. NaOH, amorphous silica, $GaCl_3$ -solution and water was added to this still X-ray amorphous gel. The resulting reaction gels were heated to 453 K over 12 h. After 25 h at this temperature pure (Si,Ga)MFI was obtained.

Electron Microprobing

As it was reported elsewhere [3] the template-free synthesized MFI-zeolites exhibited a homogeneous Al distribution. Also the template-free synthesized (Si,Ga)MFI showed a constant Si/Ga-ratio along the length of the crystal.

Figure 1 indicates, as reported elsewhere [1,3], that the aluminium is concentrated at the surface of the (Si,Al)MFI crystals when they were synthesized with TPABr. As shown in Figure 2 the gallium is enriched in the crystal core when it was synthesized with TPABr.; this is contrary to the aluminium distribution. These distributions are observed independent on the Si/T^{III} ratios.

NMR- spectroscopy

The ^{27}Al MAS NMR-spectra of the uncalcined (Si,Al)MFI materials studied showed only a resonance at 54 ppm, which characterizes tetrahedrally coordinated aluminum. This allowed the conclusion that in the template-free and the alkali-free synthezised (Si,Al)MFI crystals the Al atoms are incorporated in the framework.

The ^{29}Si-MAS-NMR spectra of the (Si,Ga)MFI showed a broad resonance at -104 ppm which is due to silicon that is coordinated to gallium. This indicated that at least part of the gallium was incorporated in the MFI-framework.

The concentration of Broensted acidic protons in the (Si,Ga)MFI calculated from the ^{1}H-MAS-NMR is fairly exact to the concentration of gallium atoms in the (Si,Ga)MFI crystals determined by elemental analysis.

Concentration of gallium: $43 \cdot 10^{19}$ Ga/g (template-free); $21 \cdot 10^{19}$ Ga/g (organic template)
Concentration of Broensted acidic protons: $(40 \pm 2) \cdot 10^{19}$ H/g (template-free); $(19 \pm 2) \cdot 10^{19}$ H/g (organic template)

These results allowed the conclusion that also in the (Si,Ga)MFI no extra-framework gallium species exist.

FTIR spectroscopy and TPD measurements

The spectra of pyridine adsorbed on the MFI-type zeolites are given in Figure 3. They show that the MFI-type zeolites synthesized from template containing gels possess almost exclusively Broensted-sites while in the template-free synthesized MFI-type zeolites a relatively large number of Lewis sites was observed. The NMR-data given above indicated that these Lewis sites cannot be due to extra- framework species. The nature of these sites is not yet known. It could be that they were formed during the activation of the samples.

Propane aromatisation

The different MFI-type zeolites showed decreasing activity with respect to their propane conversion (T_R = 823K, WHSV = $1h^{-1}$). This is easily explained by the decreasing number and strength of the acidic sites of these MFI-type zeolites as can be seen in table 1, below;

Template	Sample	SiO_2/Ga_2O_3	SiO_2/Al_2O_3	Propane	S(BTX)
	ratio	ratio	conversion	%	
none	A: (Al,Si)MFI	-	24	63%	40
none	B: (Ga,Si)MFI	26	-	40%	50
Organic	C: (Al,Si)MFI	-	60	7%	8.3
Organic	D: (Ga,Si)MFI	60	-	3%	79

T_R = 823K; WHSV = $1h^{-1}$

In comparison to the (Al,Si)MFI-zeolites the (Ga,Si)MFI-zeolites produce BTX-aromatics with a higher selectivity. This is due to the lower acidity of the (Ga,Si)MFI which also leads to a reduction in the formation of cracking products.

Another explanation could be that gallium can, in contrast to the aluminium, be reduced to its monovalent state by dehydrogenating the alkanes to olefines. This process which is discussed in the literature [6] could also reduce the formation of lower alkanes.

The very low yield of BTX-aromatics from the organic template synthesized MFI-zeolite might be due to the nearly exclusive presence of Broensted sites which could catalyse the cracking of propane. The homologous gallosilicate also possess a large proportion of Broensted sites such that one would also expect a selectivity to lower alkanes. However, in contrast to this the gallosilicate showed the highest selectivity to the described aromatics.

It might be possible from the present results to explain the BTX-selectivities with the different distribution of the acidic sites in the investigated silicates.

To obtain BTX-aromatics, the dehydrogenated olefin, which is mainly propene, must di- or trimerise to cyclic- six membered hydrocarbons, which would then be dehydrogenated to the aromatics. It might be that the probability of the occurence of this oligomerisation step is higher in the channels than at the crystal surface. In the organic template (TPA^+ without alkali cations) synthesized zeolite where the catalytically active sites are concentrated at the crystal surface the propene formed by the dehydrogenation of propane cannot oligomerize because the lower olefins will have left the reaction zone before oligomerization can take place. This would explain the high amount of lower hydrocarbons in the product stream of this catalyst.

The other three catalysts also had active sites in the crystal core, so that the propene molecules that formed at active sites in the pores could not leave the reaction zone as fast as the olefines formed at the crystal surface. Thus the probability for their oligomerisation was much higher. This would specifically explain the very high BTX-selectivity of the alkali-free synthesized (Ga,Si)MFI where the gallium is concentrated in the center of the crystal (see Figure 2).

To verify this hypothesis catalytic tests with MFI-silicates coated with an aluminium- and gallium-free surface or catalysts the surface of which was inactivated by a suitable method are currently under investigation.

CONCLUSION

It could be shown in this work that it is possible to produce (Ga,Si)MFI-type zeolites from a wholly inorganic reaction mixture to avoid environmental problems arising from the synthesis of this catalyst.1

The investigation of the MFI-zeolites with different NMR-techniques pointed out that the trivalent elements (aluminium, gallium) were incorporated into the lattice during the synthesis.

The catalytic results showed that the activity of the MFI-zeolites is increased with the number and strength of the acidic sites while the selectivity of BTX-aromatics increases when (Ga,Si)MFI were used as catalysts. From the present results it seems that the selectivity of the MFI-zeolites for BTX-aromatics in the aromatization of propane could also be influenced by the distribution of the active sites.

ACKNOWLEGDEMENTS

The authors gratefully acknowledge the assistance of G. Engelhardt (Universität Konstanz) and P.J. Grobet (Katholieke Universiteit, Leuven) for carrying out the ^{29}Si and the ^{27}Al MAS NMR spectroscopy of the zeolites. Also acknowledged is the assistance of M. Hunger (Universität Leipzig) for calculating the concentration of Broensted acidic sites in the (Ga,Si)MFI by ^{1}H MAS NMR spectroscopy. This project was supported by the German BMFT (Project No.: 03 C 259 1).

REFERENCES

1. Müller, U., Brenner A., Reich A., Unger,K.in:Zeolite synthesis in ACS Symp. Ser.,H.E.Robson,M.L.Occelli,Los Angeles, **398**, 346 (1989)
2. Grose, R. and Flanigen,E., US Patent 4 257 886 (1981)
3. Tißler, A., Polanek, P., Girrbach, U., Müller, U., Unger, K. in: Stud. Surf. Sci. Catal.,H.G.Karge,J.Weitkamp,Zeolites as catalysts,sorbents and detergent builders,Elsevier, **46**, 399 (1989)
4. Bergk, K., Fürtig, H., Schwieger, W., Hädicke, U. in: Stud. Surf. Sci. Catal.,G.Öhlmann,H.Pfeifer,R.Fricke,Catalysis and adsorption by zeolites,Elsevier, **65**, 185 (1991)
5. Schüth, F., J. Catal. submitted
6. Petit, L., Bouronvill,J., Raatz, F. in: Stud.Surf. Sci. Catal.,P.A.Jacobs,R.A.van
2 Santen,Zeolites: Facts,Figures,Future,Elsevier, **49**, 1163 (1989)

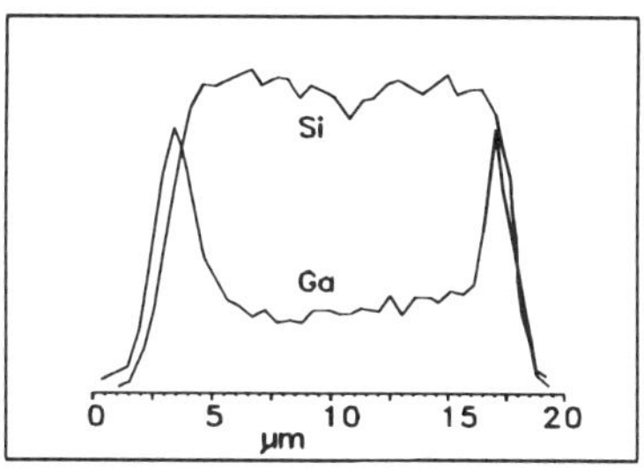

Figure 1: Al-distribution in (Si,Al)MFI synthesized with TPABr

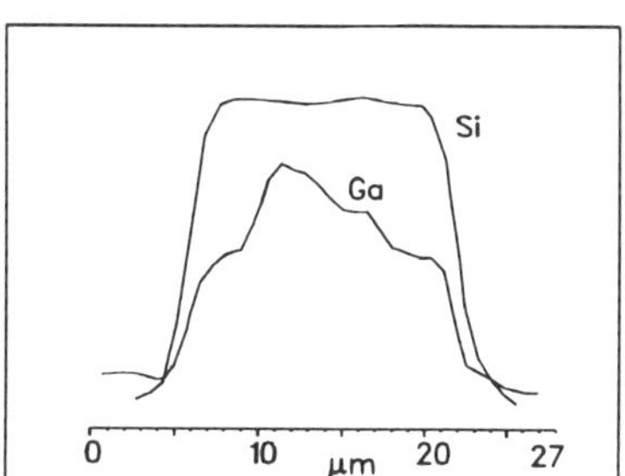

Figure 2: Ga-distribution in (Si,Ga)MFI sample D synthesized with TPABr

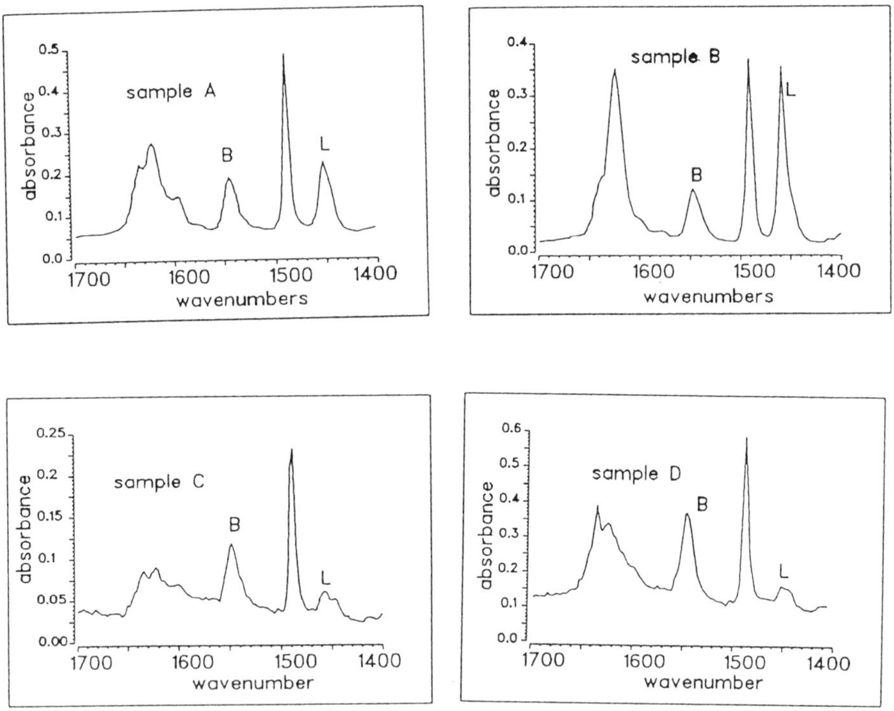

Figure 3: FTIR-spectra of pyridine:

The samples were activated under high vacuum of $5x10^{-8}$ Torr at 623 K for 12h. Pyridin was adsorbed at room temperature at $5x10^{-3}$ Torr until saturation was reached. Information about the Si/T^{III} ratio is given in table 1.

Sample A: (Si,Al)MFI (template-free); Sample B: (Si,Ga)MFI (template-free)

Sample C: (Si,Al)MFI (alkali-free); Sample D: (Si,Ga)MFI (alkali-free)

(B= Broensted sites; L= Lewis sites)

SPECTROSCOPIC INVESTIGATION OF THE REDOX PROPERTIES OF CoAPO MOLECULAR SIEVES

M.P.J. Peeters[*], J.H.C. van Hooff[*], R.A. Sheldon[**], V.L Zholobenko[***], L.M. Kustov[***] and V.B. Kazansky[***]

[*] ---Schuit Institute of Catalysis, Eindhoven University of Technology, P.O. Box 513, 5600 MB Eindhoven, The Netherlands,

[**] -Delft University of Technology, Organic chemistry, 2628 BL Delft,

[***] N.D. Zelinsky Institute of Organic Chemistry, Academy of Sciences of the USSR, Moscow, Leninsky prosp. 47, 117334, USSR.

ABSTRACT

Acid sites in CoAPO-5 and CoAPO-11 and redox properties of these Co-substituted aluminophosphates have been investigated using UV- and IR-spectroscopy. The data obtained show that CoAPO exhibits quite unusual acidic and redox properties. The absence of the bridged OH-group seems to be a common feature of Me^{2+} containing MeAPOs. The behaviour of Co^{2+}/Co^{3+} couple in redox reactions turns out to be very different from that characteristic for supported cobalt ions.

INTRODUCTION

Crystalline aluminophosphates, silicoaluminophosphates and isomorphously substituted aluminophosphates represent a new class of molecular sieves which are considered as potential catalysts and adsorbents [1-3]. It is commonly accepted that aluminophosphates containing bivalent cations (MeAPOs) possess negatively charged framework fragments where each Me^{2+}-ion can give rise to the formation of acidic groups. However, the corresponding bridged hydroxyl groups have not yet been directly observed for different MeAPOs.

In the case of Co-substituted aluminophosphates one has a system which may combine both redox and acidic properties [2-3]. According to [2] reduction of Co^{3+} to Co^{2+} in CoAPOs leads to the increase of the concentration of acidic sites. However the real nature and properties of these acidic sites in CoAPOs and other MeAPOs are still not clear. The introduction of transition metal ions creates redox properties which makes MeAPOs perspective catalysts for redox type reactions.

From the Proceedings of the 9th International Zeolite Conference, Montreal 1992, Eds. R. von Ballmoos et al.,

In the present work acidic sites in CoAPO-5 and CoAPO-11 and redox properties of these Co substituted aluminophosphates have been investigated using UV- and IR-spectroscopies.

EXPERIMENTAL

CoAPO-5 and CoAPO-11 were prepared from gels with the following composition by high shear mixing:

$$P_2O_5 : (1\text{-}x/2)\ Al_2O_3 : x\ CoO : \sim 40\ H_2O : y\ \text{Template}$$

With: $0 < x < 0.05$
CoAPO-5 : $y = 1.5$, Template = TriEthylAmine (TEA)
CoAPO-11: $y = 1.0$, Template = Di-iso-PropylAmine (DiPA).

Pseudoboehmiet (Condea) and aluminium-iso-propoxide was used for the synthesis of CoAPO-5 respectively CoAPO-11. Crystallization of the gels was performed in teflon lined static autoclaves at 463 K for two days. Calcination was performed by heating the samples in dry oxygen at 393 K for 1 hour and 823 K for 5 hours. The heating rate was 5 K min^{-1}. Calcined samples were then heated in vacuum at 770 K for 5 hours. Reduction in H_2 or CO and oxidation in O_2 or N_2O were performed at 570-770 K and pressure of 50 Torr followed by further evacuation. UV- and IR-diffuse reflectance spectra were recorded using "Hitachi-340" and "Perkin Elmer 580 B" spectrometers supplied with diffuse reflectance units [4].

To identify Brønsted and Lewis acid sites adsorbed pyridine, CD_3CN and CO were used as IR-molecular probes. Ethene oligomerization was used to identify the active acid sites.

RESULTS AND DISCUSSION

The UV-VIS spectra of the as-synthesized CoAPOs (containing template) and after calcination in O_2 at 823 K as well as the spectra of the calcined sample reduced in CO and H_2 at various temperatures are shown in figure 1. The appearance of the absorption bands at 520, 580 and 630 nm in the UV-VIS spectra of as-made CoAPOs is responsible for the intense blue colour of these catalysts. This is considered as proof of tetrahedral coordination of cobalt [2-3] and isomorphous substitution of Al^{3+} for Co^{2+}. These three bands are characteristic of *d-d* electron transition of tetrahedrally coordinated Co-ions with d^7 configuration with distortion caused by the Jahn-Teller effect. Upon calcination in O_2 CoAPO-5 and CoAPO-11 show a clear colour change from blue to yellow-green. The intensity of the absorption bands at 500-650 nm strongly decreases and new absorption bands at 320 and 400 nm appear. The latter were assigned to ligand-to-metal charge transfer transition in Co^{3+} ions [3] which were formed by oxidation of Co^{2+}.

It should however be noted that even after prolonged treatment of CoAPO-5 & CoAPO-11 in O_2 or N_2O at 720-820 K remaining absorption bands at 500-650 nm are still present in the UV-VIS spectra. This implies therefore that there is a part of Co^{2+} in CoAPO which cannot be oxidized. It is known that under such conditions non-framework Co^{2+} ions introduced by ion-exchange in zeolites or supported on other catalysts are hardly oxidized into Co^{3+} [5].

Therefore, it is possible to assume that CoAPO contains in addition to isomorphously substituted cobalt also non-stoichiometric species like CoO or spinel. At the same time the existence of exchangeable cations or even framework Co^{2+} species which are resistant toward oxidation cannot be excluded.

Reduction of CoAPO in CO or H_2 at a temperature higher than 570 K leads to step-by-step transformation of Co^{3+} to Co^{2+} and to a corresponding decrease of the intensities of the bands at 320-400 nm and to a growth of the bands at 500-650 nm. Thus the transformation of Co^{3+} to Co^{2+} is reversible.

Figure 2 shows the IR-spectra of CO adsorbed at 300 K on two CoAPO-11 samples (different Co content) calcined in O_2 and those reduced in H_2. The band at 2185 cm^{-1} observed in these spectra could be assigned to CO complexes with coordinatively unsaturated Co^{2+} ions. The intensity of the bands does not change after reduction of the sample. Therefore, this band may be attributed to those Co^{2+} species which cannot be oxidized. The intensity of this band however increases with the amount of cobalt present in the samples.

It is generally accepted that the substitution of Al^{3+} ions for Me^{3+} ions in AlPOs does not change the electroneutrality of the framework and therefore it should not result in any Brønsted acidity. On the contrary introduction of Me^{2+} ions into aluminophosphate frameworks is considered to result in formation of Brønsted acidic sites and thereby increases the catalytic activity [1]. Thus, according to this point the reduction of CoAPOs in H_2 which converts Co^{3+} to Co^{2+} might be accompanied by generation of either bridged ≡P-(OH)-Al≡ groups or terminal P-OH groups if the interaction of bridged OH-groups and Co^{2+} is weak:

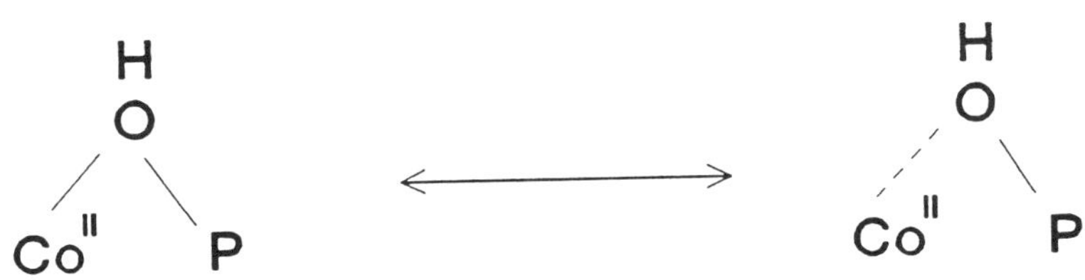

On the contrary, the treatment of CoAPOs with CO should result in anionic vacancies (Lewis acid sites) appearing due to oxygen removal from the framework:

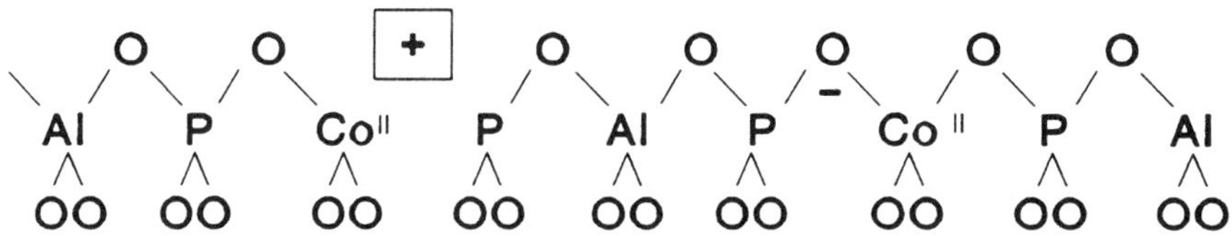

Figure 3 represents the IR spectra of calcined CoAPO-11 before (1) and after reduction in H_2 (2) and CO (3) at 770 K. The bands at 3800 and 3680 and a low-intensity band at 3740 cm^{-1} correspond to the terminal AlOH- and POH-groups located at the outer surface of AlPOs [6]. A broad band near 3400 cm^{-1} could be ascribed to H-bonded OH-groups. It should be noted that the IR spectra of both oxidized and reduced samples are very similar to each other. Thus a conclusion may be drawn that reduction of CoAPO-11 in CO or H_2 does not lead to appearance of neither bridged OH-groups nor additional POH groups.

It is possible to propose that during reduction of CoAPO-11 in H_2 as in the case of CO oxygen atoms are removed from the framework with formation of H_2O rather than the formation of structural OH-groups. Water may be easily desorbed from CoAPO surfaces at the reduction temperature (higher than 570 K). It was mentioned above that reduction of Co^{3+} in CoAPO-11 by H_2 or CO may be observed at the temperature higher than 570-670 K. On the other hand, it is known that this reaction takes place at room temperature in the presence of methanol [2]. We found also that unlike H_2 or CO , any organic molecule adsorbed on Co^{3+}APOs, for instance CH_3CN and C_2H_4 are able to reduce Co^{3+} to Co^{2+} in very mild conditions (at T ~ 300-350 K). Moreover water was found to act similar to methanol, i.e. it reduced cobalt ions in the framework. Figure 4 shows the UV-VIS spectra of CoAPO-5 measured before and after water adsorption at 330 K. Water adsorption leads to a fast colour change from yellow-green to blue. Simultaneously the intensities of the bands at 500-650 nm increase and those at 320-400 nm decrease. Thus, the interaction of water at 330 K leads to the partial reduction of Co^{3+} to Co^{2+}. Heating CoAPO-5 in water vapours at 370-720 K results in further reduction of Co^{3+}. As reported in [3] reduction of CoAPO-11 by methanol occurred with the formation of CH_2O. When water is used as reducing agent one expects the formation of O_2 or H_2O_2. However our attempts to find these products by means of mass spectrometry failed.

Thus CoAPOs exhibit unique redox properties, different from supported cobalt catalysts; this makes these kind of molecular sieves particularly interesting for oxidation catalysis. Soluble cobalt salts are widely used as homogeneous oxidation catalysts in industrial processes [7], e.g. the catalytic oxidation of p-xylene to terephtalic acid.

Ethylene oligomerization is known [8] to proceed on both Brønsted and Lewis acid sites of zeolites at room temperature and may be used as a test for the reactivity of active sites. Previously it was shown [8] that H-D exchange takes place during C_2D_4 oligomerization and is a necessary step on Brønsted acid sites. Therefore the observation of such an exchange after C_2D_4 adsorption on the samples of CoAPO-11 oxidized or reduced in H_2 or CO, may be considered as an indication of the presence of acidic OH-groups. If adsorption of the olefin will result only in oligomerization without simultaneous H-D exchange it may be a proof of the presence and reactivity of Lewis acid sites. Figure 5 shows the IR spectra of CoAPO-11 measured after exposure of the samples to C_2H_4 or C_2D_4 for 6 hours. Prior to the adsorption the samples were oxidized with O_2 or reduced with H_2 or CO. Adsorption of C_2D_4 on the oxidized sample leads to a very slow oligomerization with formation in a low concentration of branched species characterized by the adsorption bands at $\upsilon^{as}{}_{(CH2)}$ = 2790 cm^{-1}, $\upsilon^{s}{}_{(CH3)}$ = 2870 cm^{-1}, $\upsilon^{as}{}_{(CH3)}$ = 2930 cm^{-1} and $\upsilon^{s}{}_{(CH2)}$ = 2860 cm^{-1}. Unlike the oxidized sample, CoAPO-11 reduced in H_2 or CO exhibits higher activity in oligomerization which is however still much lower compared with HZSM-5 zeolites for example. This reaction results in formation of oligomers with a low degree of branching. Moreover, the spectra of CoAPO-11 reduced in H_2 or CO are similar to each other. No H-D exchange was found for oligomerization of C_2D_4 neither for the oxidized nor for reduced samples, which might result in the appearance of OD-groups (υ_{OD} = 2750-2500 cm^{-1}) or C-H species (υ_{CH} = 1300-2800 cm^{-1}). These data confirm the absence of Brønsted acid sites capable of olefin oligomerization and show that mainly Lewis acid sites due to the presence of Co^{2+} species formed after reduction in H_2 or CO are responsible for the activity of CoAPO-11 in C_2H_4 conversion.

CONCLUSIONS

The data obtained show that CoAPO-5 and CoAPO-11 exhibit quite unusual acidic and redox properties. The absence of the bridged hydroxyl groups probably seems to be a common feature of Me containing MeAPOs; previously we have found that the bridged OH-groups were not formed in different MeAPOs (Me = Mg, Zn). The behaviour of Co^{2+}/Co^{3+} couple in redox reactions turns out to be very different from that one characteristic for supported cobalt ions. It may be explained in terms of stabilizing effects of the ionic lattice of aluminophosphates which are more important for Co^{3+} species.

REFERENCES

1 E.M. Flanigen, B.M. Lok, R.L. Patton, S.T. Wilson, New developments in Zeolite Science an Technology, 1986, Tokio p.103-112.

2 B. Kraushaar-Czarnetzki, W.G.M. Hoogervorst, R.R. Andrea, C.A. Emeis, W.H.J.Stork, J. Chem. Soc. Faraday Trans., 1991, 87(6), 891-895

3 L.E. Iton, I. Choi, J.A. Desjardins, V.A. Maroni, Zeolites, 1989, 9, 535-538.

4 Kazansky V.B., Borovkov V. Yu., Kustov L.M., Proc 8 Intern. Congr. on Catalysis, 1984, West Berlin, v.3, p.3-14.

5 J.C. Kim, S.I. Woo, Appl. Catal., 1988, 39, 107.

6 S.G. Hedge, P. Ratnasamy, L.M. Kustov, V.B. Kazansky, Zeolites, 1988, 8,137.

7 R.A. Sheldon and J. Kochi, "Metal catalyzed oxidations of organic compounds", Academic Press, New York, 1981.

8 L. Kubelková, J. Nováková, B. Wichterlová and P. Jíru, Collection Czechoslov. Chem. Commun., vol. 45, 1980, p.2290-2299

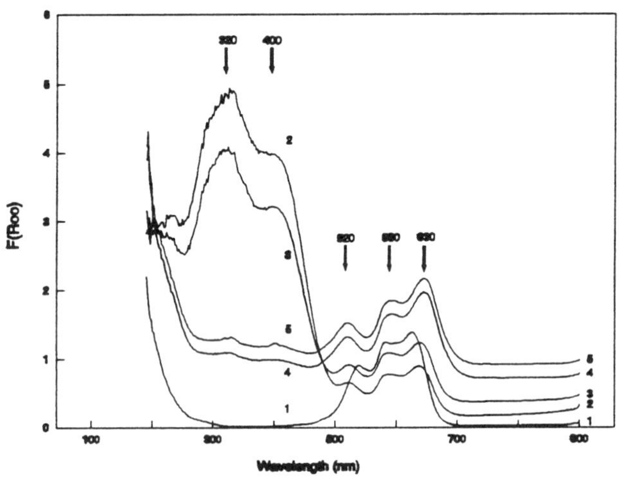

Figure 1: UV-VIS spectra of (1) As-synthesized CoAPO-5, (2) Calcined in O_2, (3,4) reduced in H_2 (525 resp. 575 K) and (5) CO (770 K).

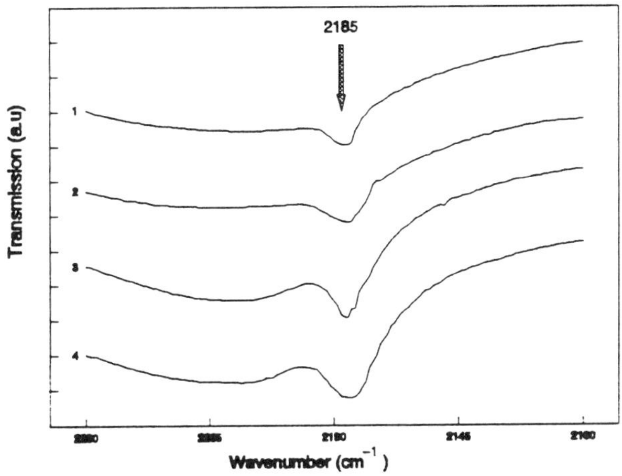

Figure 2: IR spectra of CO adsorbed at 300K on CoAPO-11 samples calcined in O_2 (1,3) and reduced in H_2 (2,4); (1,2) x = 0.01, (3,4) x = 0.005

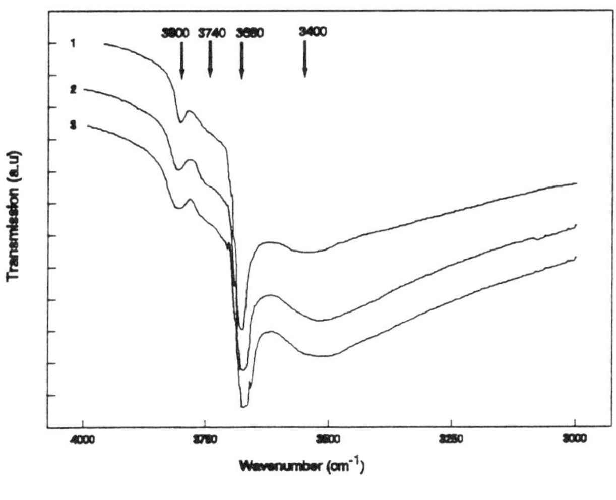

Figure 3 : IR spectra of calcined CoAPO-11 before (1) and after reduction in H_2 (2) and CO (3) at 770 K.

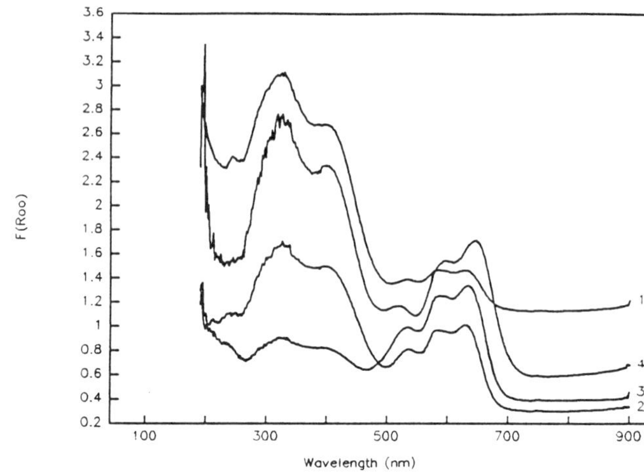

Figure 4: UV-VIS spectra of CoAPO-5 before (1), after (2,3) water adsorption at 340 K and reoxidized sample (4).

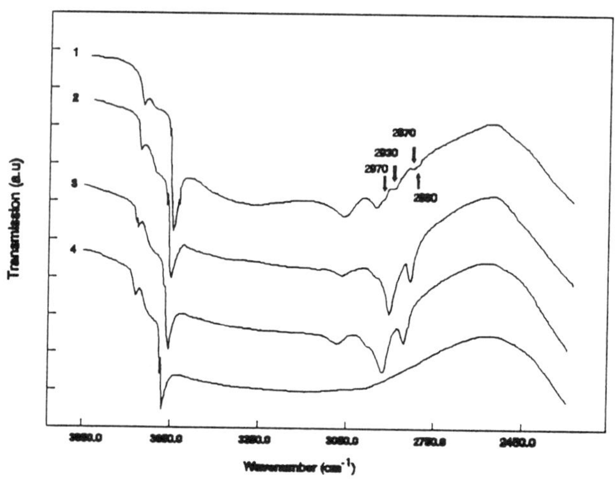

Figure 5 : IR spectra of CoAPO-11 oxidized in O_2 (1) and reduced in H_2 (2,4) or CO (3) after adsorption of C_2H_4 (1-3) or C_2D_4 (4).

EFFECT OF DEHYDRATION AND CATIONIC EXCHANGE ON THE ^{23}Na AND ^{7}Li MAS NMR SPECTRA OF X AND Y ZEOLITES

J.M. CHEZEAU, A. SAADA, L. DELMOTTE and J.L. GUTH

Laboratoire de Matériaux Minéraux (URA-CNRS 428) ENSCMu-UHA,
3, rue Alfred Werner, 68093 MULHOUSE CEDEX (FRANCE)

ABSTRACT

High resolution MAS NMR was used to investigate the progressive dehydration of NaX and NaY zeolites and to study the cation distribution in a series of dehydrated partially exchanged (Li, Ca)NaX. The ^{23}Na spectrum of a totally dehydrated NaX exhibits a line at -40ppm which differs from that observed in NaY, demonstrating an influence of the Si/Al ratio. In partially exchanged (Li,Ca)NaX, Li^{+} and Ca^{2+} seem to act in opposite ways. Li^{+} has a tendancy to substitute the Na^{+} in the more constrained sites, leading to narrow ^{23}Na lines, whereas Ca^{2+} seems to force the remaining Na^{+} into highly asymmetric environments where it gives broad NMR signals which can only be partially observed. The ^{7}Li NMR signal is quantitatively detected and at least two clearly separated lines are observed in dehydrated samples.

INTRODUCTION

The nature and location of cations are known to influence sorption and diffusion of molecules in zeolites. The aim of the present study is to observe ^{23}Na and ^{7}Li MAS NMR spectra to monitor the substitution of Ca^{2+} for Li^{+}, at an approximatively constant Na^{+} content, in a FAU-type zeolite Although NMR is expected to be very sensitive to site occupancy, hydration and motion of the compensating cations, there exists relatively few such studies compared to the number of those dealing with the framework atoms [1]. This is probably due to the quadrupolar nature of the cationic nuclei and to the difficulty to spin sealed dehydrated samples as well. According to Mortier [2], cations in the FAU framework are mainly located at the center of the hexagonal prisms (site I), in the sodalite cages in the vicinity of the double six-ring (site I') and in the supercages in the vicinity of the single six-ring (site II).An analysis of the cations positions in an Y zeolite in which Na^{+} was progressively substituted by NH_4^{+}, Ca^{2+} or La^{3+} has been conducted by Welsh et al. by ^{23}Na MAS NMR in both hydrated and dehydrated samples [3,4]. Nakata et al. report on the effect of exchanging Na^{+} by H^{+} and K^{+} [5]. However, even in the case of non-exchanged NaY zeolites, discrepancies exist between the results of previously reported ^{23}Na MAS NMR studies of both hydrated and dehydrated samples [3-8].So, prior to the study of cationic exchange, we examined dehydration of a NaX zeolite and a possible influence of the Si/Al ratio on the NMR spectra.

From the Proceedings of the 9th International Zeolite Conference, Montreal 1992, Eds. R. von Ballmoos et al.,

EXPERIMENTAL

Hydrated samples of NaX zeolite (13X Union Carbide, Si/Al = 1.2) were left for 24 hours in an oven at temperatures ranging from 70°C to 350°C. Their water content, determined by thermogravimetry, is given in Table 1.

Table 1 : Water content of NaX after dehydration in air at different temperatures.

Temperature (°C°	Wt% H_2O	Temperature (°C°	Wt% H_2O
20	25.9	170	3.6
70	13.4	350	~0
115	7.2		

4g of the same NaX were stirred at 60°C for 2h with 30 ml of an aqueous solution containing various concentrations of Ca^{2+} and Li^+ ions. The samples were washed with distilled water, dried at 60°C and finally equilibrated with water vapour at room temperature. For the highest Li^+ exchange rates, the process had to be repeated two or even three times. The cationic content of the samples, as determined by chemical analysis, appears in Table 2.

Among the many proposals which have been published in the literature in order to spin air-tight NMR samples, those which do not need any modification of the probe were chosen. A 5 mm DOTY probe can receive spinners fitted with caps having Viton O-rings. A spinning rate of 8 kHz was easily achieved but the sample volume was limited to 45 µl [9]. The spinners of a 7 mm Bruker probe can be fitted with a tightly machined insert [10]. The sample volume was 200 µl but the spinning rate appeared to be limited to 4 kHz. In both cases, the spinners were filled with dehydrated samples in a glove box in a dry nitrogen atmosphere and could be kept for several weeks without showing any weight increase which would indicate rehydration.

High resolution MAS NMR spectra were obtained on a Bruker MSL 300 spectrometer. The resonance frequencies were 79.39 MHz and 116.6 MHz for ^{23}Na and ^7Li respectively. An ordinary one-pulse cycle, with a 5s repetition time, was used without proton decoupling. A very short pulse-length (0.7 µsec) was chosen to be less or equal to 1/12 of a ninety degree pulse in solution. Spectra are referenced to aqueous molar solutions of NaCl and LiCl respectively. The number of scans and spinning rates appear in the figures. The spectra were processed by an appropriate broadening.

RESULTS

Dehydration study

The ^{23}Na MAS NMR spectra of the NaX of various water contents are shown in figure 1. The progressive change of the spectra demonstrates that the problem of the air-tightness of the

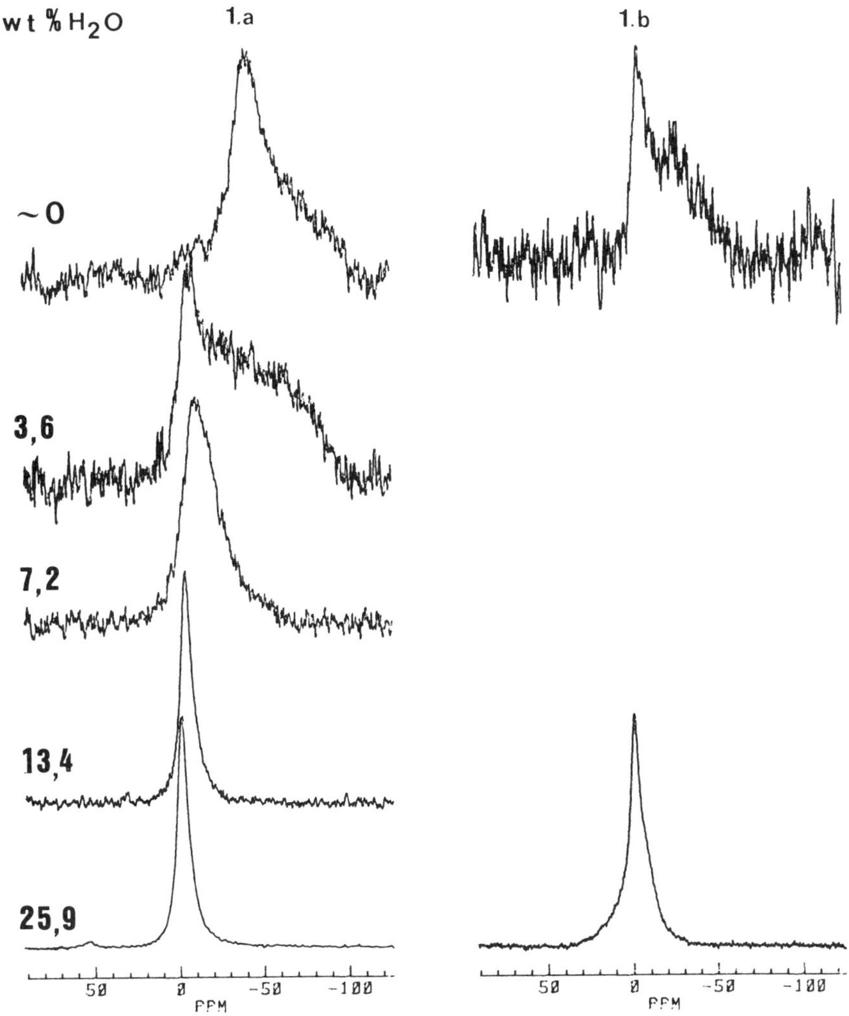

Figure 1 : ^{23}Na MAS NMR spectra (Spinning rate 8 kHz, DOTY probe, scan numbers between 200 and 600).

a) NaX (Si/Al = 1.2) of variable water content

b) hydrated and dehydrated NaY (Si/Al = 1,98)

spinner has been overcome. The hydrated sample exhibits, at approximatively -2 ppm, a fairly symmetrical line of 650 Hz width.The sample dehydrated at 350°C shows a broader line (≈ 1800 Hz) whose maximum is largely shifted upfield, appearing in the vicinity of -40 ppm. Intermediate dehydration states lead to composite lines.

In order to test the generality of such an evolution, spectra were run on both hydrated and dehydrated (350°C) samples of NaY zeolites of same structure but of higher Si/Al ratio, (Si/Al = 2.1 and 1,98 respectively). As already reported by Kundla [6], the spectra of the hydrated samples do not differ widely. The maximum remains in the vicinity of -2 ppm, but, as Si/Al ratio increases, the spectrum becomes asymmetric, with a second line at a higher field, like that observed by Welsh [3]. On the contrary, the spectra of dehydrated NaY samples strongly differ from that of the NaX. A line is observed ad ≈ -5 ppm, whose signal to noise ratio is greatly reduced, and there is no evidence of any signal at - 40 ppm.

Cationic exchange

The substitution of part of the Na^+ content by either Li^+, Ca^{2+} or variable mixtures of Li^+ and Ca^{2+} does not greatly alter the ^{23}Na MAS spectra of the approximately 20 % remaining Na^+ in the hydrated zeolite. Line width and signal to noise ratio are much alike, whereas a small upfield shift is observed on substituting Na^+, which is greater for Ca^{2+} than for Li^+ subsitution (Table 2).

Table 2 : Chemical shifts δ(ppm) and linewidth $\Delta\nu_{1/2}$ (Hz) of ^{23}Na MAS spectra in five X samples with various (Li,Ca) contents

Sample		1	2	3	4	5
Cationic charges (%)	Na	100	30	25	18	14
	Li	0	70	35	5	0
	Ca	0	0	40	77	86
Hydrated	δ(ppm	-1.8	-3.8	-4.3	-4.6	-5.2
	$\Delta\nu_{1/2}$(Hz)	660	900	830	750	740
dehydrated	δ(ppm)	-40.2	-26.3	-24	-23,-130	-22,-130
	$\Delta\nu_{1/2}$ (Hz)	1900	660	1100	2 broad peaks of weak intensity	

On the contrary, dramatic changes appear on the ^{23}Na spectra of the dehydrated samples. When Li^+ alone substitutes for Na+, the line appears remarkably narrow (≈ 650 Hz) in the vicinity of -26.3 ppm. When Ca^{2+} alone substitutes for Na^+,only a very poor signal can be observed, with two broad lines at approximatively -22 and -130 ppm. Mixtures of Li^+ and Ca^{2+} give intermediate spectra, as shown in figure 2.

^7Li MAS spectra of a series of hydrated samples exhibit a single narrow line, ≈ 80 Hz at -0.2 ppm, as shown in figure 5a. A graph of the signal area as a function of the amount of Li in the sample appears to be linear on figure 3. So, we may reasonably infer that all the Li is detected. For the corresponding dehydrated samples the spectra (figure 5b) obviously exhibit at least two narrow lines appearing respectively at 0.3 ppm and -0.7 ppm. Again, the overall signal area is proportional to the Li content (figure3) but the relative intensities of the two lines can be determined and appear on figure 4.

DISCUSSION

As shown by the experimental results described above, the ^{23}Na spectra are very sensitive to several factors such as dehydration rate, presence of cations other than Na^+ and Si/Al ratio. Moreover, NMR parameters such as pulse width and intensity of the static magnetic field may interfere in the case of quadrupolar nuclei. So, comparison with results already published must proceed with caution, as already pointed out by Welsh [3]. The chemical shifts of the spectra here obtained for the hydrated Na-samples are coherent with those of Kundla [6] and Welsh [3]. The asymmetry noted by Welsh seems related to a high Si/Al ratio of the sample. However, different chemical shifts were found by both Hayashi [7] and Nakata [5] that cannot be explained at the moment. According to our thermogravimetric measurements, activation at 150°C for 24 hours as conducted by Welsh et al [4] should lead to incomplete dehydration. The spectrum of their so-called dehydrated NaY sample is effectively intermediate between those obtained for NaX activated at 115°C and 170°C despite different Si/Al ratios. For fully dehydrated samples, the difference observed between the NaX and the NaY samples will need more detailed investigation before a conclusion can be drawn concerning respective location of the Na^+ cations. A careful quantitative determination of both observed and unobserved sodium seems also necessary, as very contradicting observations have been made so far [4,7,11].

Since the ^{23}Na spectra of the hydrated zeolites are not greatly modified by partial cationic exchange, we may infer that highly mobile hydrated Na^+ cations are observed, whereas the situation is different in the case of the dehydrated samples. Substitution of Na^+ by Li^+ leads to narrow ^{23}Na lines appearing at a lower field than for the dehydrated NaX sample. This seems to indicate that Li^+ has a tendancy to substitute the Na^+ which are located in the more constrained sites, leading to an homogeneous population of Na^+ situated on open sites. On the contrary, substitution by Ca^{2+} seems to force the remaining Na^+ into highly asymmetric strongly constrained sites, leading to partially observed broad NMR lines at a higer field.

Complementary information is given by ^{7}Li NMR. As expected in hydrated samples, the single narrow line is coherent with mobile hydrated Li^+ cations, whereas the observation of two or even three lines in dehydrated samples indicates located, non exchanging Li^+ cations. Splitting of the ^{7}Li MAS line in a (Li,Na)A zeolite was interpreted by Melchior et al. [12] as being due to multiple

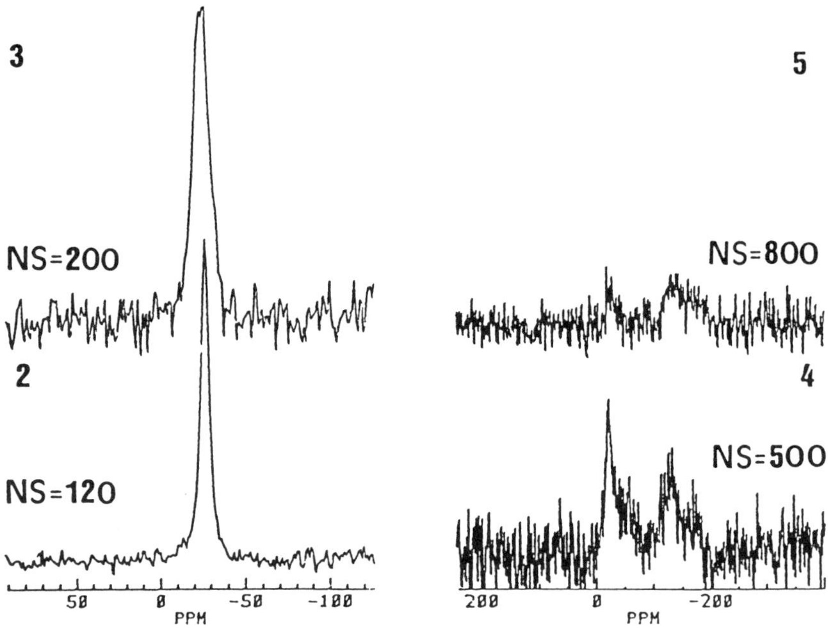

Figure 2 : ^{23}Na MAS NMR spectra of the dehydrated (Li,Ca)NaX zeolite samples numbered as in Table 2. (Spinning rate 8kHz, DOTY probe, number of scans on the spectra).

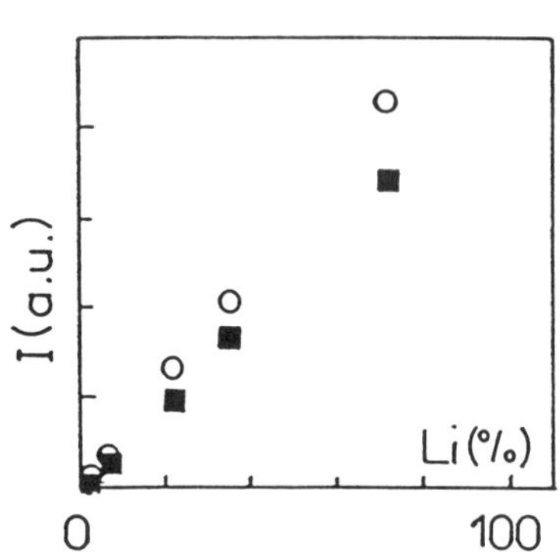

Figure 3 : ^{7}Li signal area as a function of Li amount in the sample. ■ hydrated
O dehydrated (the two sets of data have independant arbitratry units)

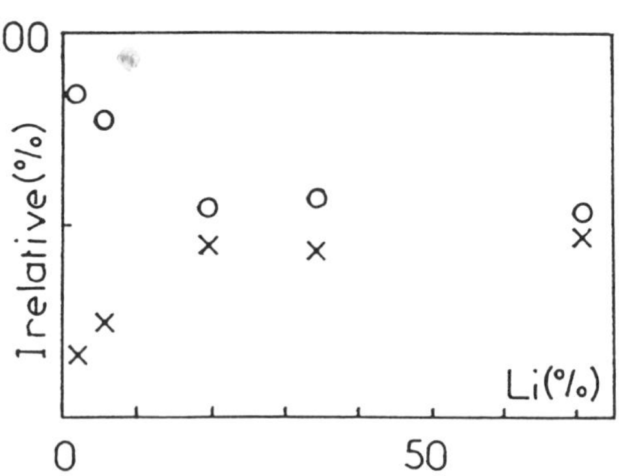

Figure 4 : Respective intensities of the two ^{7}Li signals observed in dehydrated (Li,Ca)NaX

O upper field signal δ ~-0.7 ppm

x lower field signal δ ~0.3 ppm

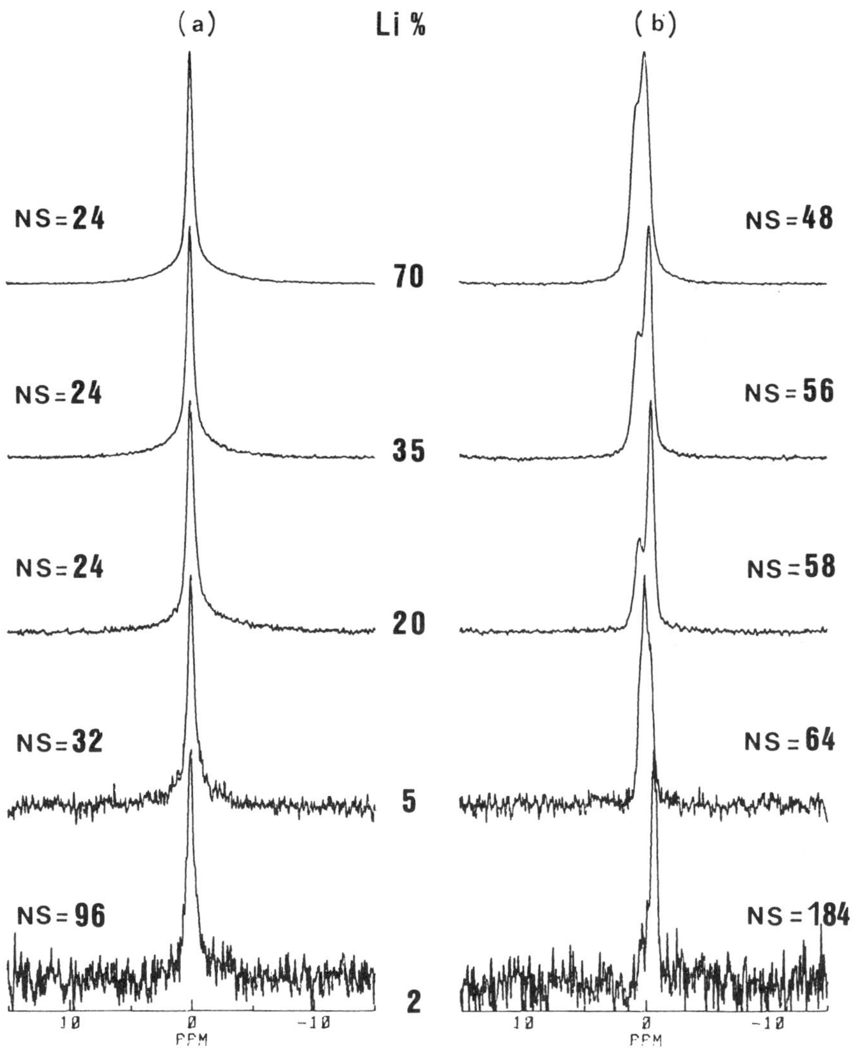

Figure 5 : ^7Li MAS NMR spectra of

a) hydrated

b) dehydrated (Li,Ca)NaX zeolites

(Spining rate 8kHz, Doty probe, scan numbers on the spectra)

Samples containing respectively 70, 35, 5 % Li are those appearing in Table 2.

Samples with 20 and 2 % Li were prepared by the same way.

environments of Li^+ on a single site, but occupancy of different sites by Li^+ seems more probable in the present case.

CONCLUSION

The present study demonstrates the inverse behaviour of Li^+ and Ca^{2+} when substituting Na^+. Obviously, more experiments are still needed to fully understand the detailed repartition of these ions in the available sites of the FAU structure. Due to the smaller quadrupolar constant of lithium, MAS NMR allows a quantitative observation of this nucleus, even in dehydrated samples but the small chemical shift range might be a limiting factor.

REFERENCES

1. Engelhardt, G. and Michel, D., High-Resolution Solid-State NMR of Silicates and Zeolites, J. Wiley, Chichester,(1987).
2. Mortier, W.J., Compilation of Extra Framework Sites in Zeolites, Butterworth, Guilford,(1982).
3. Welsh, L.B. and Lambert, S.L., in *Perspective in Molecular Sieve Science*, ACS Symposium Series, **368**, Eds W.H. Flank and T.E. Whyte Jr., American Chemical Society, Washington D.C., pp. 33-47, (1988).
4. Welsh, L.B. and Lambert, S.L., in *Characterization and Catalyst Development*, ACS Symposium Series, **411**, American Chemical Society, Washington D.C., pp. 262-272, (1989).
5. Nakata , S., Asaoka, S. and Takahashi, H., Nippon Kagaku Kaishi, 393, (1989)
6. Kundla, E., Samoson, A. and Lippmaa, E., Chem. Phys. Lett., **83**, 229, (1981)
7. Hayashi, S., Hayamizu, K. and Yamamoto, O., Bull. Chem. Soc., Jpn., **60**, 105, (1987).
8. Challoner, R. and Harris, R.K, Zeolites, **11**, 265, (1991).
9. Eckman, R. and Doty, F.D., J. Magn. Reson., **69**, 527, (1986).
10. Merwin, L.H., Sebald, A., Espidel, J.E. and Harris, R.K., J. Magn. Reson., **84**, 367, (1989)
11. Basler, W.D., Colloids and Surfaces, **12**, 59 (1984).
12. Melchior, M.T., Vaughan, D.E.W., Jacobson, A.J. and Pictroski, C.F., Proceedings of the 6th International Zeolite Conference (Reno, 1983), p. 684, Butterworth Guilford, (1985).

DISCUSSION - Characterization

"Titanium Silicalite-1 Peroxides" (M. Clerici & Roberto Millini)

P. Ratnasamy: Does your mechanism account for the significant effect of the dielectric constant of the solvent on the para/ortho ratio in phenol hydroxylation?

M. Clerici: The effect of the solvent cannot be explained in the usual way, because the oxidations catalyzed by TS-1 occur inside pores having 5.5 nm average diameter. Its effect might be the result of different actions: cleaning of pore mouth, protection in one of the elementary steps of the reaction (see ref. 4,5), and others. The data available today does not allow a reliable correlation between dielectric constant and selectivity.

P. A. Jacobs: You were discussing the effect of the treatment on product yields for alkane oxidation and phenol hydroxylation in the reactions mentioned?

M. Clerici: The effect of TS-1 pretreatment on selectivity is dependent on the concentration of LiOH solution and the contact time with it. Product distribution in the oxidation of n-hexane or phenol catalyzed by recovered TS-1, does not change significantly, provided [LiOH]>0.1M, the O/P ratio of dehydroxybenzenes increases from 0.7 to higher values.

"Local Environment of Titanium In Titanium Silicalite (TS-2) (A. Bittar, et al.)

R. Prins: A Ti-Si contribution was found at 2.20Å and a Ti-Ti contribution at 2.56Å. If true, these contributions in the EXAFS spectrum would provide important information about the Ti site geometry. But if we apply Brillioun's theorem,Δk. ΔR. $2/\pi$=N (the number of parameters which can be determined in a Δ k observation range and a range ΔR of interest) we see that 10. 0.5 $2/\pi$ $\simeq$ 3. Since we need 4 parameters per EXAFS contribution (N,R, E_o, $\Delta\sigma^2$), it is not allowed to analyze your spectrum in the range 2.1 - 2.6Å with more than one shell of neighbours. Thus, it is impossible to say if both Si at 2.2 <u>and</u> Ti at 2.56 is present in your sample.

L. Bonneviot: Including the baseline fit, the number of independent parameters was 18 according to the formula you refer to. Five (5) parameters were necessary for the baseline and four for each O, Si and Ti shells (Γ was kept constant and taken to one). The calculation was performed with these shells and 17 parameters leaving 1 degree of freedom. We were allowed to introduce Ti. Nevertheless, the number of Ti pairs is not known with precision.

G. Bellussi: How can you be sure that titanium species you observe are due to framework and not to extra framework silicate impurities? Did you observe any lattice variation in samples used for EXAFS investigation, using XRD analysis?

L. Bonneviot: We have not mesured the lattice expansion but all the other characteristics (XPS, IR, reactivity) are consistant with previous findings on TS-1 where Ti was claimed to be a framework species. Furthermore, in all the Ti silicates or (Ti, Na) silicates, titanium ions occupy an octahedral site unlike our silicalite of ZSM-11 type of structure.

"Properties of Indium Containing Zeolites Prepared by a Solid State Reaction" (V. Kanazirev, et al.)

J. Fraissard: What is the temperature of reduction of the system zeolite and indium oxide? Did you detect any decrease of the zeolite crystallinity after such treatment?

V. Kanazirev: 1. The TPR peak at about 640K corresponds to the process leading to In introduction into cationic positions of the zeolite. Under isothermal conditions the reduction proceeds easily at a temperature of 570K or even lower depending on the duration and H_2 partial pressure. 2. High crystallinity is attained even at a rather high In loadings as detected by XRD and IR measurements.

W. Hoelderich: Are you not afraid that In will be washed out in reactions you mentioned such as the dehydration of methanol, i.e., in presence of water? The toxicity will cause problems for mechanical application.

V. Kanazirev: 1. We did not observe any appreciable loss of In by means of microbalance experiments despite the much more severe treatment conditions used than the conditions of the methanol to dimethyl ether conversion. 2. Unfortunately there are no data available regarding the toxicity of the In-zeolite system. We do believe that there will be no serious toxicity problems taking into account the stabilizing impact of the rather large zeolite anions.

J. Kornatowski: 1. What is the influence of a calcination process or better to say of a thermal treatment on your indium containing materials? 2. Did you notice an interaction between the aluminum and indium ions?

V. Kanazirev: 1. The pretreatment in an inert gas, oxygen or vacuum does not lead to any significant difference in the temperature programmed reduction step. 2. We have convincing evidence that In is transferred into cationic positions of the zeolite, probably as In^+. The intensity of the 3610 cm^{-7} OH band strongly decreases after H_2 reduction and a band (IR) at 1446 cm^{-7} due to the adsorbed pyridine on Lewis sites increases.

"Spectroscopy, Energetics, and Siting of NH^-_4 in Zeolites: Theory and Experiment" (E. Teunissen, et al.)

J. Sauer: In the first part of your talk you interpreted the IR spectra of NH_4^+ as due to a super position of a twofold and a threefold coordinated structure. In the calculations you used different models with two and three oxygen binding sites. If both structures were real, you should find both structures as two different local minima in the calculations for the larger H-Al$(OH)_3^-$ model.

Answer: We have also optimized the twofold structure on the H-Al$(OH)_3^-$ structure and it appeared a local minimum.

D. Koningsberger: Could you please comment on how close theory predicts the position of IR bands?

Answer: The protons are hydrogen bonded, thus the potential is anharmonic because of this they are off by 100 cm^{-1}. However, they can lead to a good qualitative understanding of the experimental spectrum.

J. Sauer: First, I wuld like to congratualte you for your significant comparison of periodic and cluster calculations. We should be aware of the fact that the Al-content of your chabasite model is rather high. Therefore, for high-silica zeolite catalysts, we may expect smaller effects of the modeling field. You have already stressed the significant effect of the geometry relaxation. How did you choose the position of the NH_3/NH_4^+ species within the series of calculations performed?

Answer: The chabasite used was a 3:1 Si/Al chabasite, the natural form has a 2:1 Si/Al ratio. The position of the NH_3 and NH_4^+ was only relaxed in one direction.

"Electron Spin ECHO Studies of Mn(II) Framework Substitution In MnAPO-11" (G. Brouet, et al.)

R. Gorte: In your model with D_2O coordination around positive and negative ions, you don't include the D^+ which should balance the framework change for a framework species (D_2O will cause rapid H-D exchange with the Bronsted site). Would this change your model or conclusions?

L. Kevan: By ESEM we are sensitive to distance out to 0.5 to 0.6 nm from the paramagnetic species. The D^+ of a Bronsted site should be within this distance. It would certainly be worthwhile to incorporate this into the model and see if it should be distinguishable in the ESEM data.

R. Prins: Your results indicate that there is a problem in how much Mn (and Co) can be put into the $AlPO_4$-11 and $AlPO_4$-5 lattices. The low solubility found in ESR and ESE is based on the assumption that clustered Mn is extra framework Mn. Maybe this is the reason for the discrepancy.

L. Kevan: This is a possibility, but we have looked at several low concentrations like 0.05, 0.1, 0.15, etc. atom % Mn and we see evidence for two signals rather than one signal that gradually broadens.

G. N. George: Have you observed couplings from ^{27}Al and ^{31}P in your experiments?

L. Kevan: Yes, but the ^{27}Al can't be analyzed very quantitatively with respect to distance because it has a large quadrupole interaction which adds more parameters. The analysis of ^{31}P interaction seems more promising although it is weak since it has $I=1/2$ and the modulation depth goes as $I(I+1)$. The difficulty here is that Mn(II) is actually a $S=5/2$ system and there are other transitions that needed to also be evaluated. To date this hasn't been worked out.

"Elucidating the Morphology of Clays: HREM and ESEM Studies" (K. B. Reuter and J. S. Holmgren)

C. S. Cundy: You have drawn attention to the importance of sample preparation in the study of these materials. Have you compared the procedures you have specified with those of critical point drying.

J. Holmgren: We have not had the opportunity to try critical point drying.

L. Kevan: Interstratification is often used as an explanation of variations in the interlayer spacing of clays. Is this presently applicable to your samples?

J. Holmgren: I do not believe that interstratification of the clay is responsible for the variations in the interlayer spacing. The HREM shows that the variation is in the interlamellar space layer itself.

Question: You stated that clays are not thermally stable. Could you describe the methods you used in order to avoid radiation damage?

J. S. Holmgren: First, our accelerating voltage is 200kv which causes less electron beam/sample interaction than lower accelerating voltages. Secondly, an image intensifier is attached to the microscope, allowing very low beam currents to be used during most of the examination, minimizing the damage.

M. Prakash: Have you studied pillars other than alumina, e.g., TiO_2, SiO_2-TiO_2, etc.? The stability may well depend on the nature of pillars too!

J. S. Holmgren: We have not studied pillars other than alumina by HREM. I agree that the pillar composition plays an important role in the hydrothermal stability of the clay. We have prepared ZrO_2, TiO_2 and SiO_2/Al_2O_3 pillared clays and have DTA and adsorption data which shows that the pillar composition affects the stability. We intend to follow up with HREM.

A. Moini: What is the dependence of pillar size variation on the charge density of the clay layers?

J. Holmgren: I believe that the charge density affects the oligomer/solution chemistry. Furthermore, the layer charge density affects the pillar density within the gallery. Although we have nont done a systematic

study to address this question, a number of other workers, including Tom Pinnavaia, are working in this area.

R. Schoonheydt: Solution chemistry is very important for pillaring. How did you perform the exchange of the pillars?

J. Holmgren: Solution chemistry is certainly a key in pillaring. I believe that the hydrothermal stability of the pillared clays can be improved by "tightening" the distribution of oligomers in the sol used for pillaring.

"Characterization and Catalytic Properties of Titanium Incorporated Into Zeolites and Silicas From TiF_3 (P. J. Kooyman, et al.)

J. H. C. van Hooff: 1.) Do you treat the samples before the use as catalyst to transfer them from fluorides into oxides? 2.) What are the reasons for the low efficiencies in the n-hexane oxidation? Is the rest of the H_2O_2 decomposed to O_2 or are there other oxidation products formed?

Answer: 1.) No, the samples are only dried at 120°C, but the TiF_3 most probably hydrolyses upon deposition. 2.) The rest of the H_2O_2 is decomposed.

R. Prins: You showed that there are small Ti-containing particles on the zeolite and silica. It is well known that anatase and rutile do not show the typical oxidation catalysis demonstrated by TS-1. So, what form of Ti is giving the catalysis that you observed?

Answer: We don't have data to prove it yet, but we think it is either a monolayer-like structure in between the particles or the interface between the particles and the silica that gives the catalytic activity.

A. Ojo: You reported the absence of the infrared band at 960cm^{-1}. Do you have any reasons for this as it has been well documented in the literature that this band proves the existence of titanium in the zeolite framework?

Answer: It is still a matter of discussion whether the 960cm^{-1} IR band is due to titanium incorporated in the framework or to -OH bands accompanying the Ti. A very similar band has been reported for B and V containing silicalite.

M. Clerici: Which is the ratio between the products of the 2 and 3 position in the oxidation of n-hexane with Euro-TS-1 and TS-1-F?

Answer: This question is answered in table 2 of the full paper.

S. Nicholas: Have you studied the stability of the small particles of titanium deposit on the large crystals?

Answer: The particles cannot be removed by ultrasonic treatment and after the catalytic experiments they were still there.

"Application of Diffuse Reflectance FTIR Spectroscopy In Molecular Sieve Research" (Ch. Peuker, et al.)

E. Cohen DeLara: Is it possible to perform quantitative measurements by dilute reflectance IR spectroscopy? How is it possible to determine the number of vibrators responsible of the adsorption band?

E. Loeffler: In principle, quantitative measurements are possible if the conditions of the Kubelka-Mink theory are considered. This means, e.g., weak bands are measured in a spectral region without any other adsorptions. For the given species the number of vibrators has to be determined, firstly, by another method (e.g., by H MAS NMR in the case of the vibrations of OH groups).

"An Infrared Study of Structure Relaxation in Molecular Sieves" (W. P. J. H. Jacobs, et al.)

E. Vogt: In your theoretical study of $AlPO_4$-5 and SSZ-24, you find a shift of layers that is not experimentally observed. Also, in recent force field work by E. de Vos Ruerhart of Delft University this shift is not found. Do you think the effect is realistic and, if not, what feature of the force field is responsible for the effect?

W. Jacobs: The shift of layers observed in the calculation is not realistic. It is an artifact caused by an underestimation of the charges used in the partial charge model.

R. Schoonheydt: 1.) Do the calculations provide confirmation of earlier interpretations of lattice vibrations in terms of ring vibrations and of other structural subunits? 2.) Is it allowed to decouple exchangeable cation vibrations from lattice vibrations in the calculations?

W. Jacobs: 1.) According to calculations performed by DeMan, A. J. M, Van Santen, R. A. (Zeolites, accepted) no correlations between the presence of specific rings or other structural subunits could be found. 2.) Exchangeable cation vibrations are observed in the fir region below 250 cm^{-1}, so coupling with the lattice modes which are observed at much higher frequencies is not likely to occur. This holds especially for the stretching modes.

B. Sulikowski: Do you have any evidence of the possible change of the Si to Al ratio in the framework of faujasite after the decomposition of its ammonium form? The shift of the band at 1000 cm^{-1} may be partly ascribed to the zeolite dealumination.

W. Jacobs: Dealumination has not occurred because readsorption of ammonia results in an IR spectrum very similar to the one observed for the starting ammonium zeolite Y. Also, we observed a shift to lower wave numbers for a band near 510 cm^{-1} for samples with increasing degree of dealumination. This band is observed to increase in frequency.

"Quantum Chemical Interpretation of IR Data of Methanol Adsorption on Zeolites and Silica" (A. G. Pelmenschikov, G. Morosi, A. Gamba)

J. A. Lercher: 1.) On the basis of the similarity between adsorption of methanol on silica and on silicalite, could one deduce that SiOH groups are most important adsorption sites in silicalite? 2.) You proposed dissociative adsorption of methanol on silicalite. Our own experiments suggest that (for kinetic reasons) dissociation does not take place to a significant extent below 373K.

A. Gamba: The IR spectra of methanol adsorbed on ZSM5-like silicalite dehydroxilated at 773 K (see the applied Figure) show two forms of the interactions. The one appearing at low coverage (spectra 2-3) manifests itself by bands at ν'_{as}=3000, ν''_{as}=2960 and ν_s=2860 cm^{-1} in the CH fundamental region. Its formation is accompanied by a parallel growth of a broad O-H band at 3420 cm^{-1} ($\Delta\nu_{1/2}\simeq$200 cm^{-1}) indicating the formation of a hydrogen bond of moderate strength. These C-H bands were found by Borello et al. (8,9) also for the first stages of methanol adsorption on dehydroxilated silica and were assigned to chemisorption on the strained siloxane bridges according to the scheme

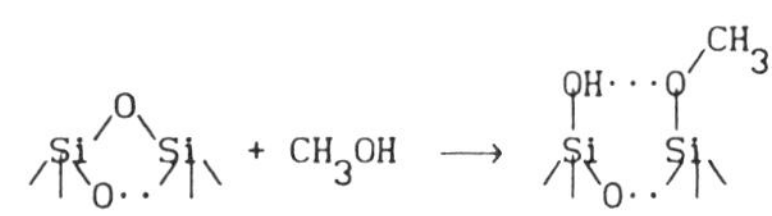

1 2

By increasing the coverage (spectra 4-6) the maxima of the 3000 and 2960 cm^{-1} bands shift gradually to 2994 and 2950 cm^{-1}, and a new band appears at 2845 cm^{-1} very close to the 2860 cm^{-1} band. This is due to appearance of a new form with ν'_{as}, ν''_{as} and ν_s shifted to lower frequencies, with respect to those of structure 2, by about 6, 10 and 15 cm^{-1}, respectively. The 2950 and 2845 cm^{-1} bands were detected by Borello et al. (8,9) in the methanol adsorption of dehydroxilated silica (after the stage of chemisorption) and ascribed to the physical adsorption on the siloxane OH groups. Our ab-initio calculations of the hypothetical adsorbed structures show that the best agreement between these experimental and the calculated C-H frequencies is presented by structure 5, which is also the most energetically stable. The appearance of the new form is accompanied by the parallel growth of a band at 3632 cm^{-1}, this band was observed also in the adsorption on silica and assigned to the O-H stretching frequency of the physically adsorbed methanol. Such small shift of about 50 cm^{-1} of the methanol O-H frequency under the influence of hydrogen bonding in structure 5, with respect to the gas-phase methanol O-H frequency of 3681 cm^{-1} [6], can be explained by the strong distortion of the O-H··O angle from the optimum value about 180° in this structure. So, one can deduce that the strained siloxane bridges are the most important chemisorption sites while the therminal SiOH groups are responsible for the physical adsorption in methanol interaction with severely dehydroxilated silicalite. 2.) Your proposed dissociative adsorption of methanol on silicalite. Our own experiments suggest that (for kinetic reasons) dissociation does not take place to a significant extent below 373 K. As follows from the above, the dissociation of methanol takes place on the strained siloxane bridges. If these sites have been formed during the preliminary dehydroxilation of silicalite samples at high temperature, the dissociative adsorption takes place without any kinetic difficulties at room temperature (similarly to that for dehydroxilated silica [8,9,12].

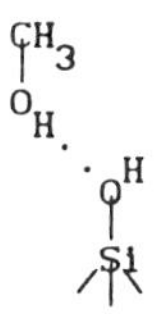

3

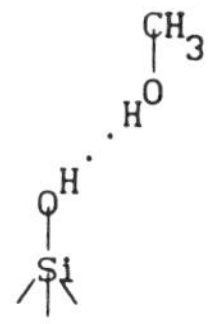

4

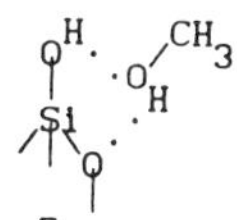

5

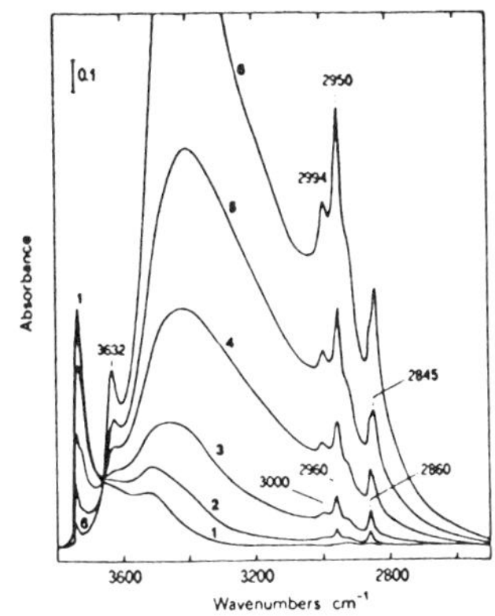

V. B. Kazansky: Among the different possibilities you considered interaction of methanol with strained oxygen bridges resulting in their opening. Does this concern silica or also silicalite?

A. Gamba: As shown by Borello et al. [8,9] the 3000, 2960 and 2860 cm^{-1} bands appearing at the first stages of methanol adsorption on severely dehydroxilated silica belong to the methanol chemisorbed on
the strained siloxane bridges. The same frequencies are found by Zecchina et al. (see Figure) for the first stages of methanol adsorption on dehydroxilated ZSM-5-like silicalite.

"The Influence of Quadrupolar Effect on High-Resolution ^{27}Al Double Rotation (DOR) NMR of Aluminophosphate Molecular Sieves" (P. J. Grobet, et al.)

F. R. Chen: It is quite nice to see that the DOR method can identify Al in different crystallographic sites. I wonder if this technique allows us to distinguish tetra-coordinated, extra-framework Al from framework Al, since these two kinds of Al species have very similar chemical shift. If this technique is not successful in this matter, caution may have to be taken when you assign the tetra-coordinated Al only to unidentical crystallographic sites, that in some of these tetra-coordinated Al can be on the surface.

P. J. Grobet: Till now, we have not applied the DOR technique on this matter.

A. Cheetham: 1.) Are you able to perform high power decoupling and cross-polarisation experiments with your DOR probe? 2.) Do you think that the inability to proton decouple may be responsible for the problems that you mention with aluminosilicates?

P. J. Grobet: 1.) With our actual DOR probehead, these techniques are not possible. 2.) No, I don't think so.

D. Barthomeuf: You described results for AlPO's. In the case of SAPO's, is the technique able to distinguish between Al atoms surrounded only by P atoms and those which have Si atoms as neighbors?

P. J. Grobet: In the case of SAPO's, we do not get sufficient line-narrowing of the ^{27}Al signals. The reason for that is not yet clear.

M. P. J. Peeters: DOR spectra of hydrated AlPO-11 were fitted with 5 signals in the tetrahedral region. What was the intensity of these signals? Is simulation with 4 peaks impossible. Our DOR exp. at 3.0T could be satisfactorily fitted with 4 signals with equal areas, suggesting the preferential hydration of one of the Al-sites, as opposed to the a-selective hydration of 20% of the 5 different Al-sites.

P. J. Grobet: We found the best fit with 5 tetrahedral signals at four different magnetic fields. To our opinion, there is indeed an a-selective hydration in this system.

G. Haller: Since VPI-5 has only 2 crystalline Al sites, but three Al sites are seen in the hydrated sample by DOR-NMR; 1.) can the three sites be absolutely assigned to the three dispersion versus field curves seen by DOR-NMR, and 2.) can one understand why only one site becomes octahedral by hydration?

P. J. Grobet: 1.) The slopes of the three dispersion vesus field curves seen by DOR NMR are proportional to the square of the quadrupolar coupling constant. This coupling constant can be correlated to the structural features (Al-O band lengths and O-Al-O band angles) of the different Al-sites. (See Ghose et al., American Mineralogist 58,748 (1973). 2.) The fused four-ring Al-site has to be stabilized by dehydrates (see L. B. McCusker et al., Zeolites 11,308 (1991)).

"NMR Studies of the Water Motion in $AlPO_4$-5, $CoAlPO_4$-5 and VPI-5" (D. Goldfarb, et al.)

D. King: How do you define the fully hydrated samples. How much bound water is there relative to the number of Al-sites? Similarly with ammonia. How much NH_3 is present relative to the number of Al-sites? How did you prepare the samples? What happens at lower H_2O loadings. Do you get both bound and free water in the same proportions?

D. Goldfarb: 1.) Sample preparation: VPI-5 and $AlPO_4$-5 were evacuated at room temperature (24 h) to a residual pressure of $\sim 10^{-5}$ torr. They were then exposed for 2-3 hours to water vapors and sealed at 77K. ND_3 adsorption was done by exposure to 600 torr ammonia for 2-3 hours. Prior to sealing at 77K the pressure was reduced to 37 torr. Samples were kept in sealed ampules of 5 mm diameter and or 2 cm length in which the ^2H measurements were performed. The total amount of adsorbed ND_3 was 23 wt% as determined by TGA measurement. 2.) The water content of the VPI-5/D_2O sample has not been determined but from the ^{31}P spectrum that showed 3 peaks of equal intensity at the same positions as previously reported and from the appearance of the intensity of the octrahedral Al peak we estimate the water content to be close to full hydration (24 wt%) and definitely higher than 15% (see J. O. Perez et al., J. Phys. Chem., 1991, 95, 994. 3.) We did not conduct a systematic study of VPI-5 samples with lower water loadings.

Question: On silica and clay minerals there is also NMR evidence for bound and free water. Can you compare AlPO's with silicas and clays?

D. Goldfarb: I am no so familiar with the NMR results in silicas and clay minerals. However, I know that in zeolites, at higher hydration levels as those we looked at, the ^2H NMR spectra of D_2O do now show well defined averaged powder pattern but feature less broad signals, which I believe is due to the multitude of adsorption sites between which exchange takes place. The uniqueness of the AlPO's is probably due to their well defined coordination sites of the water molecules which even under fast exchange result in well defined powder patterns with axial symmetry.

"Solid State MAS NMR Characterisation of the Transformation of VPI-5 to $AlPO_4$-8: Preservation of the Octahedral Aluminum Sites?" (D. Akporiaye & M. Stoecker)

R. Jelinek: Comment-- TEM results showed high degree of stacking disorder in AlPO-8 which might explain the "disorder" of adsorbed water in the hydrated sample.

M. Stoecker: Thanks for the comment. This might be the case and could be in line with our findings.

E. Vogt: Your data suggests there are several sites in AlPO-8 that are hydrated. One of these could be the 4-4 position, which leaves two more possible sitings to be explained. Do you think that the water structure in AlPO-8 is as ordered as in VPI-5, and if so, where would you propose the other two octahedral aluminums could be?

M. Stoecker: We are not sure about the ordering of water in the hydrated $AlPO_4$-8, but we believe that, besides the 4/4 site, water molecules could be coordinated to some of the 6*/4 or 6*/4 sites. Those sites are formed from 2/3 of the former 4/4 sites in the VPI-5. DOR NMR could give more details about this topic.

"Zeolite Intrinsic Framework Reactivity and Sensitivity Analysis of Framework Perturbations: The Density Functional Viewpoint" (B. Baekelandt, et al.)

D. Barthomeuf: 1.) The "topological density" you consider is related to the inverse of a distance. It seems to be different to the "topological density" reported in the areas of zeolite structure (related to the number of close TO_4 tetrahedra). 2.) In the curve giving O--->Cu^{2+} charge transfer is the global softness calculated taking into account both the structure and the Al content?

R. Schoonheydt: 1.) Indeed, our definition of topological density is :

$$\sum_{\beta \neq \alpha} R_{\alpha\beta}^{-1}$$

We should use another name to avoid confusion. 2.) Global softness for SiO_2 compositions were used; the difference induced by IR are small.

"Location of Iron in Framework-Substituted Zeolite L Using Anomalous Powder X-ray Diffraction and X-ray Absorption Fine Structure" (I. Pickering, et al.)

E. Coker: How was the iron introduced into the synthesis mixture and were iron oxide/hydroxide precipitates observed?

I. Pickering: The iron was added as ferric chloride as the last component during the gelation procedure. The colour of the reaction mixture was initially brown, but the final product was light tan; in agreement with our finding that most of the iron was incorporated into the zeolite framework. We observed no iron oxide or hydroxide precipitator other than the small amount of interstitial phase observed in the final product.

W. Harrison: What are the restrictions on element-type for the EXAFS and anomalous-diffraction techniques? Are the restrictions "intrinsic" or instrumental?

I. Pickering: XAFS is much less restricted than anomalous-diffraction. Practically speaking, the most useful energy range is from the PK-edge (2.1KeV) upwards. We would like to be able to do SiK and AlK-edge EXAFS but there are instrumental difficulties with these energies. There are no real instrumental problems at other energies. Anomalous diffraction is limited by choice of an appropriate wavelength (energy) to allow one to collect enough data to refine a structure. A practical limit is 4.5 KeV - 17.5 KeV and a great many elements fall within this range.

Modelling

PROTON TRANSFER FROM ACIDIC SITES TO WATER, METHANOL AND AMMONIA. A COMPARATIVE AB INITIO STUDY

Joachim Sauer[a,b], Christoph Kölmel[a], Frank Haase[b,c] and Reinhart Ahlrichs[c]

[a]BIOSYM Technologies Inc., 9685 Scranton Road, San Diego, CA 92121, USA

[b]Quantum Chemistry Group, Max Planck Society, Rudower Chausse 5, D-O-1199 Berlin, Germany

[c]Institute of Physical Chemistry and Electrochemistry, University of Karlsruhe, Kaiserstrasse 12, D-W-7500 Karlsruhe, Germany

ABSTRACT

An ultimate goal of computational studies of zeolite catalysis is to model reaction pathways. Illustrative of substantial progress in this direction we describe here calculations of geometries and relative stabilities of the initial physisorption and the final ion-pair structures of complexes of zeolitic cluster models with water, methanol, and ammonia molecules. The final estimates of the proton transfer energies (-5, -12 and -17 kJ/mol) indicate formation of hydroxonium, methoxonium and ammonium cations on acidic sites. For the enthalpy of adsorption the following estimates are obtained: -62 (water), -73 (methanol) and -87 (ammonia). The uncertainty of the calculations is ±20 kJ/mol.

INTRODUCTION

Up to now, kinetic and spectroscopic experiments have provided, and possibly can only provide, limited information about the elementary steps of reactions catalyzed by acidic zeolites. Theoretical tools, notably quantum chemical ab initio calculations, add significantly to the picture emerging from experiments. They can predict atomic structures , geometries, and relative energies of reactants, reaction intermediates and transition structures. More advanced approaches strive for a full description of the energy change and the change of harmonic nuclear motions along the complete reaction path.

We address here the initial step of a catalytic reaction, the proton transfer from the acidic site ZO-H onto a substrate molecule S (which subsequently undergoes chemical transformations),

$$\text{ZO-H} + \text{S} \longleftrightarrow \text{ZO}^{-}\cdots\text{HS}^{+}. \quad (1)$$

In particular, we are interested in the relative stability of the physisorption complex, ZO-H···S, which may be initially formed,

$$\text{ZO-H} + \text{S} \longleftrightarrow \text{ZO-H}\cdots\text{S}, \quad (2)$$

and of the final ion-pair structure of the surface complex, ZO^{-}···HS^{+},

From the Proceedings of the 9th International Zeolite Conference, Montreal 1992, Eds. R. von Ballmoos et al.,

$$ZO\text{-}H\cdots S \longleftrightarrow ZO^{-}\cdots HS^{+}. \quad (3)$$

We study first the relatively simple molecules H_2O, NH_3 and CH_3OH. Methanol is a feed in the real MTG catalytic process. The significance of NH_3 is its use for characterising the acid strength of surface sites in TPD experiments. Moreover, the ammonium form of zeolites, $ZO^{-}NH_4^{+}$, is an important intermediate in the preparation of the acidic form ZO-H. In the case of ammonia there is no doubt that the ion-pair structure is the more stable one. This is also inferred from IR data [1]. For water and methanol, conclusions drawn from IR and NMR experiments [2-4] are not fully conclusive, however.

The results of the calculations for water are taken from reference [5], a preliminary report of some of the results for methanol and ammonia has been given before [6]. Calculations for water [7] and ammmonia [8,9] have been also reported by others.

CALCULATIONS

The calculations involve the following steps:

-complete geometry optimizations of both initial and final structures (SCF approximation)

-estimation of the electron correlation energy at the SCF geometries (single point MP2 calculation)

-extraction of the zero point energy (ZPE) from vibrational frequencies.

To make the calculations feasible, the zeolitic Brønsted site is replaced by a finite model, $H_3SiOH{\cdot}Al(OH)_2OSiH_3$, which represents essentially three tetrahedra (see ref. 5 and the references therein for details). This model which is the standard model of the present study involves a complete AlO_4 tetrahedron and two Al-O-Si sites one of which is protonated. The bordering oxygen atoms of the two silicon atoms are replaced by hydrogen atoms thus forming SiH_3 groups. The other two oxygen atoms of the central AlO_4 tetrahedron are saturated by hydrogen atoms which represent silicon atoms of the lattice. The fortunate fact that in zeolite models hydrogen atoms can replace both oxygen and silicon atoms of the lattice has been discussed in ref. 10. While the latter model essentially represents three tetrahedra, in the case of ammonia, larger models, $H_3SiOH{\cdot}Al(OSiH_3)_2OH$ and $H_3SiOH{\cdot}Al(OSiH_3)_3$ consisting of three and four tetrahedra, respectively, are also considered.

Calculations of vibrational frequencies from analytical second derivatives and geometry optimizations including electron correlation effects from analytical MP2 gradients are the computationally most demanding part of this study. In these calculations, a smaller model, $HOH{\cdot}Al(OH)_3$, is adopted in which all four silicon atoms bonded to the central AlO_4 tetrahedron are replaced by hydrogen atoms.

The SCF geometry optimizations and MP2 energy calculations were performed with the TURBOMOLE code [13,14] on workstations. For the MP2 geometry optimizations the MPGRAD code [15] was employed. A double-zeta plus polarization basis set was adopted for all elements except for oxygen. For the valence part of the latter a triple-zeta description was selected. The relatively large number of functions on O relative to Al and Si takes account of the fact that O is the electronegative component in the molecules studied. It also guarantees that basis set superposition errors are small. Details of the basis sets are given in ref. [5].

RESULTS AND DISCUSSION

Decomposition of the reaction energies

Hypothetically, the proton transfer which converts the physisorption structure into the ion-pair structure can be decomposed into four steps:

(1) deprotonation of the acidic Brønsted site:

$$ZO\text{-}H \longrightarrow ZO^- + H^+, \tag{4.1}$$

(2) protonation of the substrate molecule:

$$S + H^+ \longrightarrow SH^+, \tag{4.2}$$

(3) dissociation of the physisorption complex:

$$ZO\text{-}H...S \longrightarrow ZO\text{-}H + S, \tag{4.3}$$

(4) formation of the ion-pair complex:

$$ZO^- + HS^+ \longrightarrow ZO^-HS^+. \tag{4.4}$$

The reaction energy of step (1) is the deprotonation energy, the corresponding Gibbs free energy is the acidity. The reaction heat of step (2) is defined as proton affinity of the molecule S. In the gas phase, the proton transfer energy depends only on the acid strength of the active site (which is the same for different substrates if the catalytic site is the same) and on the proton affinity of the substrate molecule onto which a proton is transferred:

$$\Delta E_{PT}(\text{gas phase}) = \Delta E(4.1) + \Delta E(4.2), \tag{5}$$

For a heterogeneous reaction, the proton transfer energy, ΔE_{PT} is the sum of the hypothetical gas phase energy, ΔE_{PT}(gas phase), and of the change of the binding energies of the substrate onto the surface on protonation, ΔE(surface):

$$\Delta E_{PT} = \Delta E_{PT}(\text{gas phase}) + \Delta E(\text{surface}), \tag{6}$$

with

$$\Delta E(\text{surface}) = \Delta E(4.3) + \Delta E(4.4). \tag{7}$$

This decomposition will be used to analyse the results and to estimate errors since some reference data are available for the individual steps, but not for the overall reaction.

Structures of surface complexes

Figure 1 shows the predicted geometries of the surface complexes considered. All of the complexes, both physisorption and ion-pair structures, are stabilized by two hydrogen bonds. For the surface complexes of NH_3 with the HOH·Al(OH)$_3$ model geometry optimizations were performed including electron correlation effects (MP2 approximation). No significant changes compared with the SCF results are observed. As expected, the hydrogen bonds become shorter.

Reaction energies

Table 1 shows that the calculated proton affinities of water, methanol and ammonia which reproduce the observed values within a few kJ/mol. Zero point energy corrections, ZPE, are more important than correlation energy contributions.

Table 1. Proton affinities, ΔH (kJ/mol).

Substrate	Energy, ΔE(4.2)		ZPE	ΔH_{298}(4.2)		
	SCF	MP2		Calc.	Obsd.[11]	Obsd.[12]
H_2O	-736	5	36	-694	-697	-724
CH_3OH	-802	12	11	-778		-774
NH_3	-907	6	42	-858	-854	-858

Table 2 shows the proton transfer energies for the three molecules studied. The MP2 results indicate that electron correlation makes significant contributions to these energies. They have been obtained at geometries optimized within the SCF approximation. Optimizing the structures at the MP2 level changes the total reaction energy for NH_3 from 0 (Table 2) to +3 kJ/mol and the SCF and MP2 contributions from +13 to +9 and from -13 to -6 kJ/mol, respectively. These results show that the calculation of MP2 energies at geometries optimized at the SCF level is a valid approximation.

The total results for the three molecules follow the trend of their proton affinities, but the differences are smaller. The reason is an opposite trend for the stabilization of the ion-pairs (ΔE(4.4) in Table 3). The proton transfer energy is a small difference between the hypothetical gas phase reaction energy and the surface contribution. Differences of the former for different molecules come solely from proton affinity differences (the acidity contribution is constant), while differences for the latter are dominated by differences between the stabilization energies of the ion-pair structures. From the data of Table 3 it is also obvious that it is only the interaction with the surface that makes the proton transfer possible.

Table 2. Reaction energies (kJ/mol) of proton transfer for two different models of the acidic site[a]

Substrate	$H_3SiOH{\cdot}Al(OH)_2OSiH_3$			$HOH{\cdot}Al(OH)_3$			
	SCF	MP2	total	SCF	MP2	total	ZPE
H_2O	45	-43	2	44	-42	2	-7
				(48)	(-38)	(10)	
CH_3OH	31	-38	-7	37	-38	-1	-5
	(37)			(40)	(-32)	(+8)	
NH_3	-10	-11	-21	13	-13	0	+5

[a] Numbers in parenthesis are values corrected for the basis set superposition error

Table 3. Energies of individual steps (kJ/mol).

Substrate	gas phase	interaction	ΔE(3)		ΔE(4)	
	ΔE(1,2)	ΔE(3,4)	SCF	MP2	SCF	MP2
H_2O	585	-583	47	20	-585	-64
CH_3OH	526	-533	49	22	-534	-69
NH_3	415	-436	57	23	-479	-38

For ammonia, Table 4 shows the dependence of the reaction energies on the size of the model (SCF energies only). It is monoton for the individual components, but not for the total reaction energy. We conclude that the uncertainty of our standard model results ($H_3SiOH{\cdot}Al(OH)_2OSiH_3$) is about 20 kJ/mol. Other sources of uncertainty are the quantum chemical approximations used. Results for the gas phase reaction,

$$H_3SiOH + H_2O \longrightarrow H_3SiO^- + H_3O^+, \qquad (8)$$

obtained within the same approximation as used in this study deviate by +5.5 kJ/mol from the best available energies [5]. This fits in to the general observation that the uncertainty of MP2 energy differences may amount to ±10 kJ/mol. The basis set superposition errors (Table 2) are also within these limits.

The zero point energies given in Table 2 allow final estimates for the proton transfer energies of -5, -12, and -17 kJ/mol which indicate formation of H_3O^+, $CH_3OH_2^+$, and NH_4^+ species on acidic sites. The absolute uncertainty limit of these energies is ±20 kJ/mol reflecting both model and methodological incompleteness. We expect hopwever that the uncertainty limits for relative proton transfer energies are significantly smaller since the proton affinities are reproduced within a few kJ/mol.

Table 4. Dependence of the reaction energy and of the energies for individual steps on the size of the model for ammonia as proton acceptor (SCF approximation, units: kJ/mol).

Model	total	deprotonation	physisorption	ion-pair
	ΔE_{PT}	$\Delta E(4.1)$	$\Delta E(4.3)$	$\Delta E(4.4)$
$HOH \cdot Al(OH)_3$	+13	1398	22	-500
$H_3SiOH \cdot Al(OH)_2OSiH_3$	-10	1319	57	-479
$H_3SiOH \cdot Al(OSiH_3)_2OH$	+14	1300	60	-440
$H_3SiOH \cdot Al(OSiH_3)_3$	+7	1274	63	-424

The heat of adsorption which is observed in, e.g. calorimetric, experiments compares with $\Delta H_{ads}(1) = -\Delta H(4.3) + \Delta H_{PT}$.

In order to obtain DH values from the energies of Tables 2 and 3 estimates of nuclear motion corrections are necessary. For step (3), formation of hydrogen bonded surface complexes, an estimate of +10 kJ/mol is available [16], for the proton transfer zero point energy correction are given in Table 2 and thermal corrections up to room temperature are negligible. Hence, the following estimates for DH_{ads} are obtained: -62 (H_2O), -73 (CH_3OH) and -87 kJ/mol (NH_3). The uncertainty of the calculated results is ±20 kJ/mol (vide supra). The experimental data available are -51±4 kJ/mol for H_2O [17] and about -100 to -150 kJ/mol for NH_3 (-120...-140 H-Y [18], -100...-120 H-ZSM-5 [19], about -150 H-ZSM-5 [20]). While the calculated and observed data for H_2O more or less agree within uncertainty limits, the predicted heat of adsorption for ammonia seems to be too small. A possible explanation is that long range electrostatic effects in zeolites stabilize the ion pair structure more than the initial physisorption structure. A recent comparison of periodic calculations [22] with finite cluster calculations based on rigid geometries and on the SCF approximation employing the STO-3G basis set indicates that this may indeed be the case for the protonation of ammonia [21]. The extra stabilization found in the periodic calculation was about 45-55 kJ/mol.

REFERENCES

1. Earl, W.L.; Fritz, P.O.; Gibson, A.A.V. and Lunsford, J.H., J. Phys. Chem. **91**, 209 (1987).
2. Jentys, A.; Warecka, G.; Derewinski, M. and Lercher, J.A., J. Phys. Chem. **93**, 8437 (1989).
3. Mirth, G.; Lercher, J.A.; Anderson, M.W. and Klinowski, J., J. Chem. Soc., Faraday Trans. **86**, 3039 (1990).

4. Anderson, M.W.; Barrie, P.J. and Klinowski, J., J. Phys. Chem. **95**, 235 (1991).
5. Sauer, J.; Horn, H.; Häser, M. and Ahlrichs, R., Chem. Phys. Lett. **173**, 26 (1990).
6. Sauer, J.; Haase, F. and Ahlrichs, R., *7th International Congress on Quantum Chemistry, Menton/France, 2-5 July, 1991,* Poster.
7. Gorb, L.G.; Il'chenko, N.N.; Senchenya, I.N.; Goncharuk, V.V.; Kazansky, V.B., Kinet. and Catal., in press.
8. Allavena, M.; Seiti, K.; Kassab, E.; Ferency, G. and Angyan, J.G., Chem. Phys. Lett. **168,** 461 (1990); Kassab, E.; Seiti, K. and Allavena, M., J. Phys. Chem., **95**, 9425 (1991).
9. Teunissen, E.H.; van Duijneveldt, F.B. and van Santen, R.A., J. Phys. Chem., **96**, 366 (1992).
10. Sauer, J., Chem Rev. **89**, 199 (1989).
11. Lias, S.G.; Liebmann, J.F. and Levin, R.D., J. Phys. Chem. Ref. Data **13**, 695 (1984).
12. Aue, D.H. and Bowers, M.T. in: "Gas Phase Ion Chemistry", Vol. 2, M.T. Bowers (Ed.), Academic Press, new York, 1979, pp. 1-51.
13. Ahlrichs, R.; Bär, M.; Häser, M.; Horn, H. and Kölmel, C., Chem. Phys. Lett. **162**, 165 (1989).
14. Horn, H.; Weiß, H.; Häser, M.; Ehrig, M. and Ahlrichs, R., J. Comput. Chem. **12**, 1058 (1991).
15. Haase, F. and Ahlrichs, R., to be published.
16. Sauer, J. In: *Modeling of Structure and Reactivity in Zeolites*, C.R.A. Catlow (ed.). Academic Press, in press, Table 8.9.
17. Ison, A.; Gorte, R.J., J. Catal. **89**, 150 (1984).
18. Aurox, A.; Vedrine, J.C., Stud. Surf. Sci. and Catal., **20**, 311 (1985).
19. Lercher, J., personal communication 1991.
20. Karge, H. and Jozefowicz, L., personal communication 1991.
21. Sauer, J.; Kölmel, C.; Teunissen, E.; Roetti, C. and Ugliengo, P. *CECAM Workshop "Periodic-versus Cluster-ab initio Treatment of Condensed Systems"* Orsay (Paris), Oct. 10-18, 1991.
22. Teunissen, E.; van Santen, R.; Dovesi, R. and Roetti, C., Lecture presented at the CECAM workshop *"Periodic-versus Cluster-ab initio Treatment of Condensed Systems"* Orsay (Paris), Oct. 10-18, 1991.

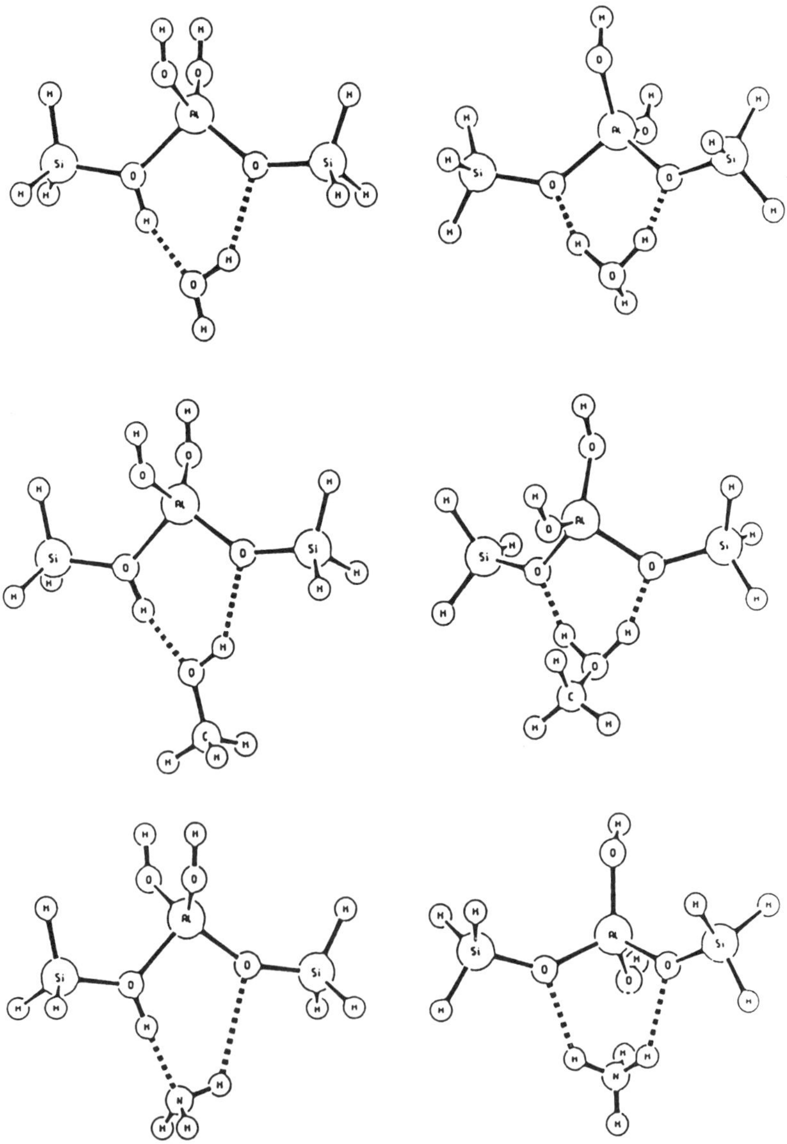

Figure 1. SCF optimized physisorption (left) and ion-pair structures (right) of the surface complexes with water (top), methanol (middle) and ammonia (bottom)

SITING OF AL AND BRIDGING HYDROXYL GROUPS IN ZEOLITE CATALYSTS. COMPUTER SIMULATIONS OF THEIR STRUCTURE, VIBRATIONAL SPECTRA AND ACIDITY

K.-P. SCHRÖDER and J. SAUER

Central Institute of Physical Chemistry, O-1199 Berlin, Germany, and BIOSYM Technologies, Inc., San Diego, CA 92121-2777, USA

ABSTRACT

Recently proposed potentials for lattice energy minimizations of acidic catalysts are applied for the silicon-rich zeolites FAU, MFI and MOR with isolated bridging OH groups. From energetical reasons it is concluded that Al atoms and bridging hydroxyl groups are distributed over several framework sites in MFI and MOR. In FAU the protonation of O1 and O3 is preferred. O1H and O3H stretching vibrations give rise to the HF and LF bands observed in infrared spectra of faujasite.

INTRODUCTION

The catalytic activity of zeolites is strongly connected with their Brønsted acidity, which arises from bridging hydroxyl groups. The local geometry of these active sites as well as their location inside the channel system of the zeolite is of high interest. Experimental techniques like X-ray diffraction are not well suited to answer questions such as distinguishing Si from Al and locating protons. Because of the low concentration of hydroxyl sites in the most interesting silicon-rich zeolite catalysts even the more sensitive neutron diffraction techniques fail. The active sites that are not ordered into cells are probed only in an averaged way by diffraction experiments. The structural model emerging from spectroscopic techniques is still incomplete [1,2]. Quantum chemical structure predictions by ab initio methods [3,4] are limited to relatively small molecular models which are not specific for individual frameworks and, thus, neglect environmental effects.

Lattice energy minimization methods using analytical potential functions have proved successful in predicting structures and properties for a variety of silicates and zeolites [5-8]. Parameters for hydroxyl groups have been recently derived [9] allowing the treatment of protonated zeolites.

Here we report on lattice energy minimization and defect calculations for H-forms of dealuminated zeolite Y (FAU), ZSM-5 (MFI) and mordenite (MOR), which all are important catalysts with a high Si/Al ratio. An attempt is made to determine the structure and preferred location of bridging hydroxyl sites (if there are any) and to understand their IR spectra. Preliminary results for

From the Proceedings of the 9th International Zeolite Conference, Montreal 1992, Eds. R. von Ballmoos et al.,

faujasite and ZSM-5 have been recently presented [9, 10].

CALCULATIONS

Two different computational approaches are available for modelling active sites in solids on the basis of classical potential functions: (i) lattice energy minimizations assuming lattice periodicity and (ii) defect energy calculations which use a generalized Mott-Littleton procedure. In the former case the hydroxyl site is part of the unit cell, in the latter the site is completely isolated within a SiO_2 lattice. Both methods have been used for the present study.

The calculations have been performed within the shell model. This potential function, in contrast to the rigid ion model, allows for the polarization of oxygen atoms. For faujasite and mordenite there was also tested a potential [8] recently derived from quantum chemical data.

RESULTS AND DISCUSSION

Faujasite

Lattice energy minimizations were performed for an isolated bridging hydroxyl group in the smallest possible unit cell of faujasite with $a=b=c\approx 17.2$ Å and $\alpha=\beta=\gamma=60°$. Thus the Si/Al ratio was 47. Since all tetrahedral atom sites in faujasite are equivalent, there are only four possible bridging hydroxyl groups corresponding to protonation of the O1, O2, O3 and O4 oxygens as shown in Figure 1.

The calculated relative lattice energies show that O1 and O3 are the preferred sites for protonation, with O3H being slightly more stable. However, the energy difference of about 5 $kJ \cdot mol^{-1}$ between O3H and O1H is small, so that we expect both hydroxyl groups to be present in H-faujasite. Since the structure after deprotonation and relaxation is the same for the four hydroxyl groups considered it follows that differences in the calculated energies give directly the differences in deprotonation energies (but have opposite signs). Since the deprotonation energy is a measure of acid strength [11], we conclude that the O1H group is slightly more acidic than the O3H group. The proton affinities follow the sequence O3>O1>O2>O4 as previously predicted by CNDO/2 calculations [12].

Table 2 shows the bond distances and angles for all four bridging hydroxyl groups. They are in close agreement with results of quantum chemical model calculations [4]. In particular, the surprisingly large Al-O(H) distance predicted by the quantum chemical study is confirmed by our calculations. This structural parameter and the flexible Si-O(H)-Al angle show the largest variation for the four possible bridging hydroxyl groups. The energetically favoured hydroxyl groups, O1H and O3H, have geometries which are closest to the geometry predicted for completely free hydroxyl groups by quantum chemical calculations. Further support for the reliability of the description of hydroxyl group structures by the lattice energy minimization calculations is provided by the fact that the average Al-O distance of 174 pm for the whole AlO_4 tetrahedron is close to the value of 173 pm observed for zeolite A. Structural information for bridging hydroxyl groups is sparse in H-faujasite. It is known that the Al-H distance, found by Stevenson [1] using ^1H NMR and later by Freude et al.[2] combining ^1H, ^{27}Al and ^{29}Si solid state NMR, is 238±4 pm. Very recently Al-H distances of 248±4 and 237±4 pm were deduced from the ^1H MAS NMR sideband patterns of bridging hydroxyl groups O1H and O3H, respectively, in zeolite H-Y [13], which compare with calculated distances of 239 and 233 pm (Table 2).

Table 1. Bridging OH groups in faujasite. Results of lattice energy minimization

	E_{rel} (kJ·mol^{-1})	ν_{OH} (cm^{-1})		δ_{OH} (cm^{-1})	
	calc.[a]	calc.	exp.	calc.[b]	exp.
O1H	5.3 (35.2)	3622	3623	937	1015
O2H	19.8 (65.5)	3552	-	963	-
O3H	0.0 (0.0)	3586	3550	951	1055
O4H	23.7 (83.2)	3601	-	944	-

[a] energies in parentheses for the quantum chemical potential [8], [b] harmonic

Table 2. Geometries (distances in pm, angles in degrees) of the four bridging hydroxyls in silicon-rich H-faujasite

	r_{SiO}	r_{AlO}	r_{AlH}	∠AlOSi	∠SiOH	$<r_{SiO}>$[a]	$<r_{AlO}>$[a]
O1H	169.4	191.0	238.6	131.1	123.0	160.3	173.8
O2H	168.7	190.3	229.6	142.4	117.6	160.7	174.3
O3H	169.7	193.0	233.2	138.7	120.7	160.7	174.2
O4H	168.8	190.6	236.3	134.0	121.4	160.3	174.3
q. ch.[b]	170.0	194.5	246.1	131.5	117.5	-	-

[a] average distance for the whole TO_4 tetrahedron. [b] predicted structure on the basis of quantum chemical ab initio calculations on small models [4]

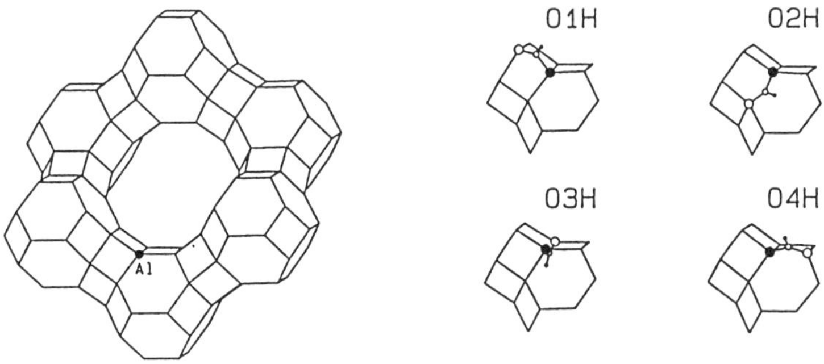

Figure 1. The four different types of bridging OH groups in silicon-rich faujasite. Left: Location of the aluminum atom within the framework. Right: Orientation of the OH groups found by adding a proton to the four crystallographically different oxygen atoms.

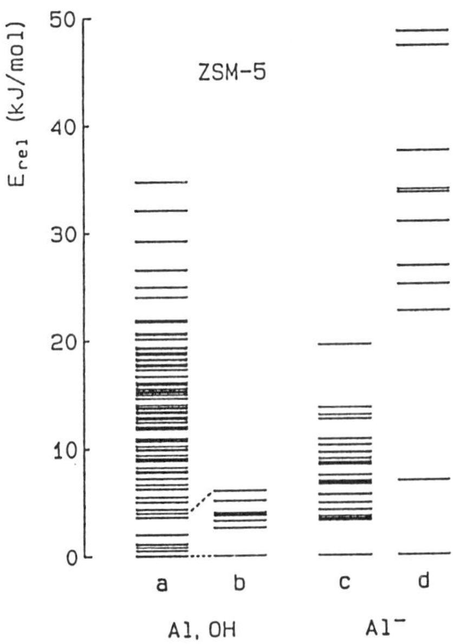

Figure 2. Relative defect energies for (a) the 96 bridging OH groups (lower accuracy); (b) the nine bridging OH groups with lowest defect energies (higher accuracy); (c) Al insertion on the 24 T sites in monoclinic ZSM-5; and (d) Al insertion on the 12 T sites in orthorhombic ZSM-5 calculated for rigid clusters [20]

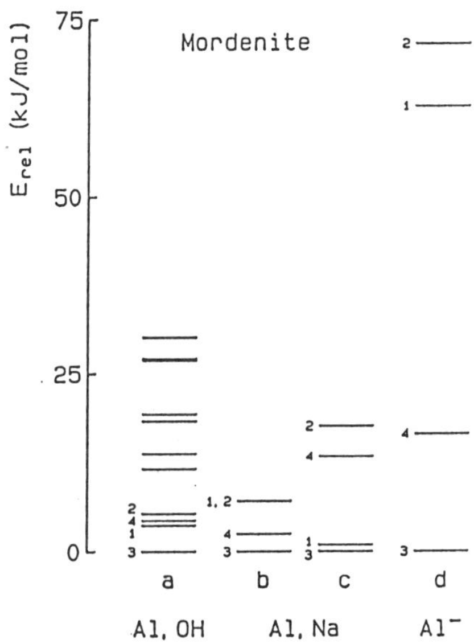

Figure 3. Relative energies for (a) the 14 bridging OH groups; (b) and (c) Al insertion on the 4 T sites with charge compensating Na^+ cation using the two parameter sets of ref. [5], resp. (see text); and (d) Al insertion on the 4 T sites calculated for rigid clusters [25]

Table 1 also shows the calculated OH stretching and SiOH bending frequencies calculated from the harmonic frequencies by

$$\nu_{OH} = \omega_{OH} - 2\,x_{OH}\ , \qquad (1)$$

where the anharmonicity constant x_{OH} is known to be about 75 cm^{-1} for bridging OH groups in zeolites [14,15]. The higher stretching frequency of the O1H bridging hydroxyl group compared with that of the O3H group indicates that the well-known HF band of the infrared spectrum should be assigned to O1H, whereas the O3H group is the source of the LF band. The HF band is observed between 3610 [16] and 3623 [17] cm^{-1} for Si/Al ratios higher than 7, whereas the LF band is found at about 3550 cm^{-1}. The calculated frequencies fall into the observed range although the difference between the HF and LF frequencies is somewhat underestimated. Since the anharmonicity constant for the SiOH bend is known to be small [15, 18] the calculated harmonic frequencies can be directly compared with observed fundamentals (Table 1). They are somewhat underestimated but the simulation agrees with the experiment in predicting the higher value for the O3H hydroxyl group.

There is no clear relation between the relative energy (and the deprotonation energy) of the bridging hydroxyl group and its OH stretching frequency. This confirms the conclusion from theoretical considerations [11] that the OH band in infrared spectra is not a proper measure of acid strength.

The tested quantum chemical potential [8] provides the same ordering in relative energies like the shell model does. However, the energy differences between the different OH groups are much greater. A clear preference of the O3H group should be the consequence, which is not observed. Unrealistic Al-O distances of 205-219 pm are obtained.

ZSM-5

Active sites in ZSM-5 were treated as defects within a perfect lattice of silicalite (aluminum-free ZSM-5). Defect energy calculations were performed for two kinds of defects: (i) substitution of one Si ion by Al with a charge of -1 and (ii) substitution of Si by Al with charge compensation by protonation of one of the adjacent oxygen atoms, thus forming a bridging hydroxyl group. All 96 possible hydroxyl groups of monoclinic ZSM-5 were treated with full relaxation of about 50-60 ions. The energetically most favoured hydroxyl groups and all Al substitutionals were treated in a second step with higher accuracy allowing about 230 ions to relax.

The defect energies of all bridging hydroxyl groups were found within a range of about 35 $kJ{\cdot}mol^{-1}$. The 11 lowest energies differ by less than 5 $kJ{\cdot}mol^{-1}$, and the 27 lowest energies by less than 10 $kJ{\cdot}mol^{-1}$. This result is confirmed by the more accurate calculations for the nine most stable hydroxyls. As seen in Figure 2, the range of the energy differences increases from 4.4 to 6.1 $kJ{\cdot}mol^{-1}$, but the narrow distribution of energy levels persists.

A similar shape of the energy distribution is found for insertion of Al into one of the 24 different T sites (Figure 2). The defect energies fall into a range of 20 kJ·mol^{-1}. The 18 lowest energies differ by less than 10 kJ·mol^{-1}. From the small energy differences, we conclude that bridging hydroxyl groups in ZSM-5 will be distributed over a broad range of framework sites even at low temperature provided that this is controlled by normal statistical mechanical factors.

This clearly contradicts results of quantum chemical calculations [19-21] on small clusters representing the Al atom at the different lattice positions (Figure 2). The cluster geometries were either fixed at the observed structures [19, 20] or relaxation was limited onto the Al atom and the four adjacent O atoms [21]. Since an Al-O bond is longer than a Si-O bond, the average T-O distances will be slightly longer for sites where Al occupation is preferred, and it was hoped that this effect would show up in lower cluster energies. The differences to our results may be caused by two problems which are connected with such an approach. The first, which is relatively minor, is that the X-ray structure refinement may not be good enough to render small differences between bond lengths meaningful. The second, and more severe problem, however, is that quantum chemical energy differences are only reliable if they correspond to a fully relaxed structure.

The question arises whether a broad distribution of bridging hydroxyl groups is consistent with the fact that only a single infrared band at 3609 cm^{-1} is reported for ZSM-5 in the hydroxyl stretching region. However, this band is broad with a half-width of 39 cm^{-1} [22]. Indeed Datka [23] suggested that the 3609 cm^{-1} OH group in ZSM-5 is heterogeneous and of various acid strengths. We find that using eq.(1) the harmonic frequencies of the nine bridging hydroxyl groups with the lowest defect energies are found between 3582 and 3619 cm^{-1}. It is concluded that the observation of a single, but broad peak at 3609 cm^{-1} is not in conflict with the predicted distribution of bridging hydroxyl groups over a large number of lattice sites.

Mordenite

The distribution of Al atoms and bridging OH groups within the mordenite lattice was studied employing the method of lattice energy minimization. Protons as well as Na^+ cations were used for charge compensation. In the latter case both Na···O potential parameter sets of ref. [5] were adopted.

The framework position T3 in the 4-membered ring is found as the energetically preferred one for Al insertion (Figure 3). Such a prediction is consistent with suggestions that come both from experiment [24] and from ab initio calculations [25]. However, the obtained differences for Al insertion on the different lattice sites are small indicating an Al distribution over all lattice sites.

Again the calculated energy differences are much smaller than those from ab initio calculations on rigid clusters [25] although they agree with our results in assigning the deepest and highest energy to T3 and T2 position, respectively.

In analogy to ZSM-5 there is not a single preferred bridging OH group in mordenite. The energies of five of them are within a range of 5.5 $kJ \cdot mol^{-1}$. At least these hydroxyls should exist at relatively low temperatures. Their OH stretching frequencies range from 3594 to 3640 cm^{-1}. The observed value is 3610 cm^{-1} [14].

The rigid ion potential [8] failed sometimes in modeling the structure of H-mordenite. Frameworks containing highly strained two-membered rings are predicted with lower lattice energies than the perfect mordenite structure.

CONCLUSIONS

The structure of silicon-rich zeolites containing active sites has been simulated within the classical shell model with a refined potential for bridging hydroxyl groups. The differences in structure and properties of the four possible bridging OH groups in dealuminated faujasite could be studied in detail. It has been shown that O1H and O3H hydroxyls give rise to the infrared HF and LF bands, respectively. Their difference in acid strength is minor.

The calculations further indicate that in ZSM-5 and mordenite there is little energetic preference for bridging hydroxyl groups at specific framework positions. Hence, a distribution over a range of framework sites can be expected provided that energy differences govern this distribution. Calculated OH stretching frequencies are in agreement with observed bands.

A recent quantum chemical rigid ion potential [8] for zeolites that enables an implicit treatment of hydroxyl groups showed some weakness in describing framework structures.

ACKNOWLEDGEMENT

Financial support of the "Volkswagen Stiftung" and of the "Fonds der chemischen Industrie" is gratefully acknowledged.

REFERENCES

1. Stevenson, R.L., J. Catal. **21,** 113 (1971).
2. Freude, D, Klinowski, J., and Hamdan, H., Chem. Phys. Letters **149**, 355 (1988).
3. Sauer, J., J. Phys. Chem. **91,** 2315 (1987).
4. Sauer, J., Kölmel, C.M., Hill, J.-R., and Ahlrichs, R., Chem. Phys. Letters **164**, 193 (1989).
5. Jackson, R.A. and Catlow, C.R.A., Molec. Simul. **1**, 207 (1988).
6. de Man, A.J.M., van Beest, B.W.H., Leslie, M., and van Santen, R.A., J. Phys. Chem., **94,** 2524 (1990).
7. Winkler, B., Dove, M.T., and Leslie, M., Amer. Mineral. **76**, 313 (1991).
8. Kramer, G.J., de Man, A.J.M., and van Santen, R.A., J. Am. Chem. Soc. **113**, 6435 (1991).

9. Schröder, K.-P., Sauer, J., Leslie, M., Catlow, C.R.A., and Thomas, J.M., Chem. Phys. Letters **188**, 320 (1992).
10. Schröder, K.-P., Sauer, J., Leslie, M., and Catlow, C.R.A., Zeolites **12**, 20 (1992).
11. Mortier, W.J., Sauer, J., Lercher, J.A., and Noller, H., J. Phys. Chem. **88**, 905 (1984).
12. Dubsky, J., Beran, S., and Bosacek, V., J. Molec. Catal. **6**, 321 (1979).
13. Fenzke, D., Hunger, M., and Pfeifer, H., J. Magn. Res., in press.
14. Kustov, L.M., Borovkov, V.Yu., and Kazansky, V.B., J. Catal. **72**, 149 (1981).
15. Mix, H., Sauer, J., Schröder, K.-P., and Merkel, A., Coll. Czechoslov. Chem. Commun. **53**, 2191 (1988).
16. Jacobs, P.A., Martens, J.A., Weitkamp, J., and Beyer, H.K., Faraday Disc. Chem. Soc. **72**, 353 (1981).
17. Anderson, M.W. and Klinowski, J., Zeolites **6**, 455 (1986).
18. Pfeifer, H., Coll. Surf. **36**, 169 (1989).
19. Beran, S., Z. Phys. Chem. Neue Folge **137**, 89 (1983).
20. Derouane, E.G. and Fripiat, J.G., Zeolites **5,** 165 (1985).
21. Lonsinger, S.R., Chakraborty, A.K., Theodorou, D.N., and Bell, A.T., Catal. Letters **11**, 209 (1991).
22. Jacobs, P.A. and von Ballmoos, R., J. Phys. Chem. **86**, 3050 (1982).
23. Datka, J., Wiss. Z. Friedrich-Schiller-Univ. Jena, Naturwiss. R. **37**, 809 (1988).
    Datka, J., Boczar, M., and Rymarowicz, P., J. Catal. **114**, 368 (1988).
24. Meier, W.M., Meier, R., and Gramlich, V., Z. Kristallogr. **147**, 329 (1978).
25. Derouane, E.G. and Fripiat, E.G., in *Proceedings of the 6th International Zeolite Conference*, Eds. D.H. Olson and A. Bisio, pp. 717-726 (1983).

STRUCTURAL AND DYNAMICAL PROPERTIES OF HYDROCARBONS IN SILICALITE

E. HERNANDEZ*, M. KAWANO*, A.A. SHUBIN**, C.M. FREEMAN***, C.R.A. CATLOW*, J.M. THOMAS* and K.I. ZAMARAEV**

*The Royal Institution, 21 Albemarle Street, London W1X 4BS, UK
**Institute of Catalysis, Novosibirisk, 630090, RUSSIA
***BIOSYM Technologies Inc., 10065 Barnes Canyon Road, San Diego, CA 92121, USA

ABSTRACT

Simulation studies employing energy minimisation, Monte Carlo (MC) and molecular dynamics (MD) techniques are now powerful techniques for the study of the docking and migration of sorbed molecules in zeolites. In this paper we present two recent and related applications of the techniques to contemporary problems in zeolite science.

INTRODUCTION

Computer simulation methods have been extensively used in recent years to investigate the structures, energetics and dynamics of sorbed molecules in zeolites. Energy minimisation (EM) methods have successfully located the sites at which sorbed molecules 'dock' [1], and have yielded calculated sorption energies that accord well with experiment [2]. Monte Carlo (MC) techniques [3] have been used to model the temperature dependence of the distribution of molecules over sorption sites within zeolites while, most recently, molecular dynamics (MD) simulations [4-7] have investigated explicitly the dynamics of both the sorbed molecules and the framework structures in sorbed zeolite systems. In this paper we show how the scope of the methods may be substantially extended by development in techniques relating firstly to the reliable locations of low energy minima for sorbed molecules, and secondly to the full and detailed treatment of framework dynamics in MD simulations. The systems investigated include the topical ones firstly of butene in silicalite and of butanols in silicalite and H-ZMS 5, and secondly of the dynamics of hydrocarbon diffusion in silicalite. The results we obtain amplify and extend the information available from experimental studies of the systems.

TECHNIQUES

We employ two main types of simulation methodology: energy minimisation supplemented by a MC/MD algorithm to assist location of global minima; and MD with full treatment of framework dynamics. As these techniques are described in detail elsewhere [6-8], our account here is brief.

From the Proceedings of the 9th International Zeolite Conference, Montreal 1992, Eds. R. von Ballmoos et al.,

Combined EM/MC/MD Algorithms for Locating Low Energy Minima

The application of EM methods may be severely restricted by the local minimum problem, *i.e.* the profusion of energy minima whose number increases with complexity of the system investigated. There are two strategies available for dealing with this difficulty: the first involves the use of dynamics enabling the system to jump over the barriers separating different minima and thereby (hopefully) enabling the system to find the lowest energy configuration; the second entails generating starting points for minimisations in a systematic but economical manner; following minimisation from all these configurations the global minimum is identified. In this study, the second approach is used. As described elsewhere(8), our method comprises the following steps:

(i) MD simulations are undertaken on the isolated sorbant molecule at a high temperature for a few hundred time steps in order to generate a 'library' of conformational states of the molecule.

(ii) All the resulting conformational states of each molecule are introduced at random into the zeolite; the resulting configuration is retained if its energy is less than a specified threshold value. By repeating the procedure many times, we are able, using this crude MC method, to build up an ensemble of configurations with reasonable energies; those with excessively high energies due, for example, to high repulsive interactions between the molecule and the zeolite framework will have been rejected. An alternative procedure is to select a certain number of the lowest energy configurations. Thus in our work on butanols, the 100 lowest energy configurations were used.

(iii) Energy minimisation is applied to all members of the ensemble generated in (ii) above, to generate a second ensemble of energy minimised configurations. The lowest energy minimum and its sorption energy may then be identified; in addition, the average energy of the minimised ensemble may be calculated.

In our earlier study [8], the minimisation procedure in step (iii) used a rigid framework for reasons of computational economy. The method is extended in this study by taking the lowest energy configurations generated by this procedure and undertaking a full minimisation in which both the framework and the molecule are allowed to relax. The relaxation energies so obtained are appreciable.

Steps (i)-(iii) of the above procedure may be implemented with commercial software; and in our study the DISCOVER [9] package of BIOSYM Technologies Inc. was employed; the INSIGHT code of the same supplier was used for graphical display of the results of the simulations. For the work on butanols in silicalite and H-ZSM-5 a new optimised MC docking algorithm was developed. In the studies of the same system, the final stage in which framework relaxation is included employed the CASCADE code [10]. The interatomic potentials employed for *intra*-molecular, *intra*-framework and molecule-framework interactions are taken from references [2] and [11-15]; they have all been successfully employed previously in simulations of sorption in zeolites.

MD Simulations of Hydrocarbon Diffusion in Zeolites

MD is based upon the numerical integration of the classical equations of motion of an ensemble of particles which represent the system simulated; a time step (typically 10^{-15} s) is specified following which positions and velocities of all particles are updated employing forces calculated using specified interatomic potentials. Following an initial equilibration period of a few ps (several thousand time steps) the simulation is run typically for 50-100 ps in contemporary studies (although longer runs may be needed for simulation of sorbed molecules in zeolites).

Analysis of the results yields a wealth of dynamical detail on the system. Using graphical techniques, molecular trajectories may be analysed and migration mechanisms deduced; and by studying the variation of the mean square displacement (MSD) of particles with time, diffusion coefficients may be deduced using the relationship

$<R_\alpha^2(t)> = 6D_\alpha t + B_\alpha$ where $<R_\alpha^2>$ is the MSD of the particle type,

D_α is the diffusion coefficient, B_α is the quantity related to the vibrational amplitude of the mobile species

In the present study, standard microcanonical ensemble (NVE) MD techniques were employed, with a simulation box comprising 576 atoms (2 unit cells) of the zeolite framework. Concentrations of 1 and 2 molecules per unit cell were considered for each type of adsorbate (methane, ethane, propane, *n*-butane and *n*-hexane).Periodic boundary conditions were applied to generate an infinite system. The time step used was 1 fs, and the target temperature was 300K.

Initial velocities were generated at random according to the Maxwell-Boltzmann distribution defined by the target temperature, with the restriction of zero total momentum, to avoid translation of the centre of mass. Each system was allowed to equilibrate for a period of 5 ps (5000 time steps) prior to the production run. During this time, velocity scaling was used to drive the system towards the desired temperature. Once equilibrium was achieved, production runs of 100 ps were carried out. The FUNGUS simulation code was used with the potential parameters given in reference (7) which also contains greater details of the techniques.

RESULTS

Butene Isomers in Silicalite

The incentive for studying this system is provided by the importance of the closely related H ZSM-5 as a shape selective catalyst for the conversion of but-1-ene into 2-methyl propene (*iso*-butene) which is used in the synthesis of the MTBE (methyl-*tertiary*-butyl ether) fuel additive. We have therefore undertaken, as an initial step, a comprehensive survey of the energetics of sorption of the four isomers (but-1-ene, *cis*-butene, *trans*-butene and *iso*-butene) in silicalite using the combined EM/MC/MD approach outlined in the section on Techniques; calculations on H-ZSM-5 will be reported subsequently. The results are summarised in Table 1 which reports the sorption energies both for the global minimum for each isomer and for the average of the ensemble of accepted, minimised configurations. We recall that framework relaxation is <u>not</u> included in the minimisations; its inclusion would be expected to lower the energies of the sorbed molecules, *i.e.* increase their sorption energies.

Table 1: Butene isomer binding energies in silicalite, calculated and relative

Isomer	Minimum energy (Calc.) (kJ/mol)	Average energy (Calc.) (kJ/mol)	Minimum energy (Rel.) (kJ/mol)	Average energy (Rel.) (kJ/mol)
2-methyl propene	-38.58	-28.09	0.00	0.00
But-1-ene	-54.30	-44.41	-15.72	-16.32
cis- but-2-ene	-45.45	-31.64	-6.87	-3.55
trans-but-2-ene	-48.76	-38.44	-10.18	-10.35

It is interesting to note that *iso*-butene has the lowest sorption energy of the four isomers — a result which we consider to be consistent with its dominance in the product distribution of the isomerisation reaction. Examination of the configuration of the lowest energy minima for *iso*-butene, shown in Figure 1, gives a clue as to the origin of the weaker sorption for *iso*-butene, as even in this most favoured configuration the carbon atoms of the hydrocarbon do not achieve very satisfactory registry with the oxygen of the framework, thereby reducing the magnitude of the van der Waals interaction between molecules and the framework. In essence, *iso*-butene has the wrong shape to bind effectively to the walls of the silicalite pores.

Butanols in Silicalite and H-ZSM-5

Interest in this system is once more stimulated by recent catalytic studies [17-20]. We again use the techniques outlined above, but on this occasion we consider more polar molecules *i.e.* the isomer of butanol in both silicalite and ZSM-5; the latter is modelled by including one Al (located at a variety of T sites) per unit cell. Charge neutrality is conserved by replacing a neighbouring oxygen atom by an OH group. We also note, that unlike the calculations summarised in the previous section, in the present case we added a final stage of energy minimisation in which framework relaxation was included.

The results obtained for silicalite and H ZSM-5 are summarised in Tables 2a and 2b respectively. Configurations of two energy minima are displayed in Figures 2a and 2b.

Table 2a: Low energy sorption sites and energies for butanol isomers in silicalite*. (All sorption energies are with respect to the zeolite and molecules at infinity and are given in kJ/mole.)

Butanol-1		Butanol-2		*tert*-butanol		*iso*-butanol	
-132.25	(z)	-117.47	(z)	-114.18	(z)	-121.31	(s)
-117.44	(z)	-111.69	(s)	-113.85	(s)	-119.87	(z)
-117.36	(s)	-102.67	(z)	-111.56	(z)	-119.79	(si)
-116.47	(si)	-102.27	(zi)	-109.54	(si)	-118.11	(si)
		-100.58	(zi)	-108.93	(zi)	-108.04	(zi)

* Channels: s - straight ; si - straight near intersection; z - zig-zag; zi - zig-zag near intersection

Table 2b: Low energy sorption sites and energies for butanol isomers on H-ZSM-5 (kJ/mol)

(x *marks positions for which CASCADE run was unsuccessful due to significant distortions in the framework structure.*)

Crystallographic Position of OH [16]							
Al	O	Si		Butanol-1	Butanol-2	*tert*-butanol	*iso*-butanol
2	13	8	(s)	-115.09	-112.93	-116.67	x
7	17	16	(z)	x	x	-129.06	-119.10
20	33	19	(si)	-119.52	-126.50	-124.32	-119.88
7	7	8	(si)	x	x	x	x
19	43	4	(z)	-138.42	-126.83	-120.37	-114.57
16	17	7	(z)	-120.42	-109.18	-105.09	x
14	32	18	(z)	-127.36	x	-110.98	-117.20
1	1	2	(s)	x	-113.08	x	x

As is seen from Table 2b, some of the CASCADE calculations were unsuccessful. In all these cases there were substantial distortions of the framework structure around the alcohol molecule. The exact origin of these difficulties is currently under investigation. It seems that they reflect the fact that the molecular dimensions of the various isomeric butyl alcohols are close to those of the zeolite channels. In this case, lattice relaxation is more dependent on the choice of potentials for the alcohol-framework interaction. There are currently two methods to avoid such difficulties: the first is to reduce the Coulomb interaction between the alcohol and framework atoms; the second is to include some (even a very small) non-bonded interaction with a repulsive term between the alcohol oxygen and the framework silicon. We note also that the alcohol-framework potentials used in this and the majority of previous studies were taken from [14] where they were obtained using the assumption of a rigid guest molecule. More detailed discussion of these points will follow in subsequent publications.

We note also that the absolute values of the energies obtained for different choices of the framework-alcohol interaction show appreciable differences. Nevertheless the energy differences and alcohol positions inside the framework structure are similar. The general conclusion we draw for butanol adsorption from the above results are firstly that for all four butanol isomers the predicted adsorption energies are only slightly different. Secondly, the total adsorption energy increases only by 6-15 kJ/mol for adsorption on H-ZSM-5 compared with silicalite for all butanols except *iso*-butanol for which the energies obtained are almost the same. The difference of 6-15 kJ/mol is less than the expected value of 25-30 kJ/mol value. This may be due to pore-confinement effects in adsorption of C4 alcohols in H-ZSM-5. It seems that it is difficult to find any conformation for the C4 alcohol in the vicinity of the bridging OH-group in H-ZSM-5 without increasing the internal strain of the alcohol molecule and of the zeolite lattice. For *iso*-butanol, we can explain the similar adsorption energies in both silicalite and H-ZSM-5 as this is a bulky isomer while for both butanol-1 and butanol-2 and also for *t*-butanol it is easier to optimise their positions within each one of the zeolite channels.

We note, finally, that, as a rule, for adsorption on both silicalite and H-ZSM-5, the energy values obtained for positions in straight and zig-zag channels and channel intersections are only slightly different. We cannot therefore offer definite predictions as to the dominant sorption site.

Diffusion of Hydrocarbons in Silicalite

As described in the section on Techniques,we have used MD techniques to study the diffusion of methane, ethane, propane, *n*-butane and *n*-hexane in silicalite at 300K and loadings of 1 and 2 molecules per unit cell. The mean square displacements were calculated using the expression

$$\langle X^2(t) \rangle = \frac{1}{N_m N_{t_0}} \sum_{t_0=1}^{N_{t_0}} \sum_{i=1}^{N_m} \left[X_t(t+t_0) - X_i(t_0) \right]^2$$

where N_m is the number of diffusing molecules, N_{t_0} is the number of time origins used in calculating the average and X_i is the coordinate of the centre of mass of molecule i.

Equivalent expressions could be written for the *y* and *z* coordinates and from these the total mean square displacements can easily be obtained. In Figure 3 we plot the calculated MSDs for one ethane molecule per unit cell.

From the slope of the line fitted to the MSD's the diffusion coefficients can then be calculated. Our results are given in Table 3. Our calculated diffusion coefficients for ethane and propane are in good agreement with experimental measurements obtained with the single-step frequency response (SSFR) [21] and pulsed-field gradient (PFG) NMR techniques [21-22]. As noted by Van Den Begin *et al.* [21], values obtained with thePGF NMR technique are approximately five times bigger than those obtained with the SSFR method. Our values agree well with both experimental measurements. The results for butane compare very well with those reported by Shen *et al.* [23], measured with the SSFR technique, but those for hexane are higher than the reported experimental values [21].

Analysis of the molecular trajectories shows that the molecules diffuse by means of 'jumps' whose distance is approximately equal to 0.5 lattice units (~10Å). For paraffins the channel intersections are energetically less favourable [24] so the molecules tend to reside in either the straight or the sinusoidal channels, and diffusion along a given channel will consist of a series of jumps across the channel intersections. However, a molecule that was originally diffusing along a straight channel can leave it through the channel intersection and commence diffusing through the sinusoidal channel. Examples of this kind of motion have been identified in our trajectory plots.

Table 3: Calculated Diffusion Coefficients (cm^2/s).
Numbers in parenthesis indicate concentration in molecules per unit cell.

Molecule type	$Dx10^7$	Molecule type	$Dx10^7$	Molecule type	$Dx10^7$	Molecule type	$Dx10^7$
Methane (1)	410	Methane (2)	1340	Ethane (1)	692	Ethane (2)	620
Propane (1)	56.3	Propane (2)	68	Butane (1)	7	Butane (2)	14
Hexane (1)	14.4	Hexane (2)	9.6				

A decrease in the diffusivity with increasing concentration of adsorbed molecules has been observed experimentally [21-22] and predicted by simulation [25-26]. As pointed out by Van-Den-Begin *et al.* [21], when the diffusion takes place by means of jumps, there are two possible mechanisms to explain this behaviour: (a) a reduction in the mean molecular jump length and/or (b) a decrease in the frequency of jumps. In our simulations we see no decrease in the mean jump length although we cannot discount this type of mechanism at higher loadings.

DISCUSSION AND CONCLUSIONS

The results described in this paper show that static and dynamic simulation techniques may now be used in relatively routine manner to investigate the detailed behaviour of sorbed molecules in zeolites. The study of butene isomers in silicalite show how sorption is subtly influenced by molecular shape and how these effects can be accurately modelled by our simulation techniques. The work on butanol sorption illustrates how modification of the zeolite (in this case introduction of framework Al) modifies the sorption properties. Finally our MD studies demonstrate the capacity of simulation to yield accurate diffusion coefficients and to probe mechanistic aspects of molecular diffusion within zeolites. Further studies of other catalytically relevant systems will shortly be reported.

ACKNOWLEDGEMENTS

We are grateful to Dr R.A. van Santen for useful discussions. We are grateful to Mr K. Bains for help with the earlier stages of the calculations on butane diffusion, and Professor L. Rees for useful discussions. Dr A. Shubin wishes to thank The Royal Society for a visiting fellowship. This work has also in part been supported by a grant from the Commission of European Communities.

REFERENCES

1. Wright, P.A., Thomas, J.M., Cheetham, A.K. and Nowak, A.K.,Nature **318**, 611 (1985).
2. Titiloye, J.O., Parker,S.C., Stone, F.S and Catlow, C.R.A., J.Phys.Chem **95**, 4038 (1991).
3. Yashonath, S., Thomas, J.M., Nowak, A.K. and Cheetham, A.K., Nature **331**, 601 (1988).
4. Pickett, S.D., Nowak, A.K., Cheetham, A.K. and Thomas, J.M., Mol.Simul. **2**,353 (1989).
5. Yashonath, S., Demontis, P. and Klein, M., Chem.Phys.Lett. **153**, 551 (1988).
6. Demontis, P., Suffritti, G.B., Quartieri, S., Fois, E.S. and Gamba, A., J.Phys.Chem. **92**, 867 (1988).
7. Catlow, C.R.A., Freeman, C.M., Vessal, B., Tomlinson, S.M.and Leslie, M.,J.Chem.Soc., Faraday Trans. **87**,1947 (1991).
8. Freeman, C.M., Catlow, C.R.A., Thomas, J.M. and Brode, S., Chem.Phys.Lett., **186**, 137 (1991).
9. Hagler, A.T., Lifson, S. and Dauber, P., J.Am.Chem.Soc., **101**, 5122 (1979).
10. Leslie, M., in: *Technical Memorandum of SERC Daresbury Laboratory,* No. DL/SCI/TM 31T (1982).
11. Sanders, M.J., Leslie, M. and Catlow, C.R.A., J.Chem.Soc., Chem. Commun. **19**, 1271-1273 (1984).
12. Schroder, K-P., Sauer, J., Leslie, M.,Catlow, C.R.A. and Thomas, J.M. Chem.Phys.Lett., **188**, 320-325 (1992).
13. Dauber-Osguthorpe, P., Roberts, V.A. , Osguthorpe, D.J., Wolff, J., Genest, M. and Hagler, A.T., Proteins: Structure, Function and Genetics, **4**, 31-47 (1988).
14. Bezus, A.G., Kocirik, M., Kiselev, A.V., Lopatkin, A.A. and Vasilyeva, E.A., Zeolites **6**, 1001 (1986).
15. Vetrivel, R., Catlow, C.R.A. and Colbourn, E.A., J.Chem.Soc. Faraday Trans.2 **85**(5), 497-503 (1989).
16. Koningsveld, H., Jansen J.C. and Bekkum, H. Zeolites, **10**, 235 (1990).
17. Makarova, M.A., Williams, C., Thomas, J.M. and Zamaraev, K.I., Catalysis Letters **4**, 261-264 (1991).
18. Makarova, M.A., Williams, C., Romannikov, V.N., Zamaraev, K.I. and Thomas, J.M., J. Chem. Soc. Faraday Trans. **86(3)**, 581-584 (1990).
19. Williams, C., Makarova, M.A., Malysheva, L.V., Paukshtis, Zamaraev, K.I. and Thomas, J.M., J. Chem.Soc., Faraday Trans. **86**(20), 3473-3485 (1990).
20. Williams, C., Makarova, M.A., Malysheva, L.V., Paukshtis, E.A., Talsi, E.P., Thomas, J.M. and Zamaraev, K.I., Journal of Catalysis **127**, 377-392 (1991).
21. Van-Den-Begin, N., Rees, L.V.C., Caro, J. and Bülow, M., Zeolites **9**, 287-291 (1989).
22. Caro, J., Bülow, M., Schirmer, W., Kärger, J., Heink, W., Pfeifer, H. and Zdanov, S.P., J. Chem. Soc., Faraday Trans. 1 **81**, 2541-2550 (1985).
23. Shen, D.M. and Rees, L.V.C., Zeolites **11**, 684-690 (1991).
24. June, R.L., Bell, A.T. and Theodorou, D.N., J.Phys.Chem., **94**, 4329-4334 (1990).
25. June, R.L., Bell, A.T. and Theodorou, D.N., J.Phys.Chem., **94**, 8232-8240 (1990).
26. Goodbody, S.J., Watanabe, K., MacGowan, D., Walton, J.P.R.B. and Quirke, N., J.Chem.Soc., Faraday Trans. **87**(13), 1951-1958 (1991)

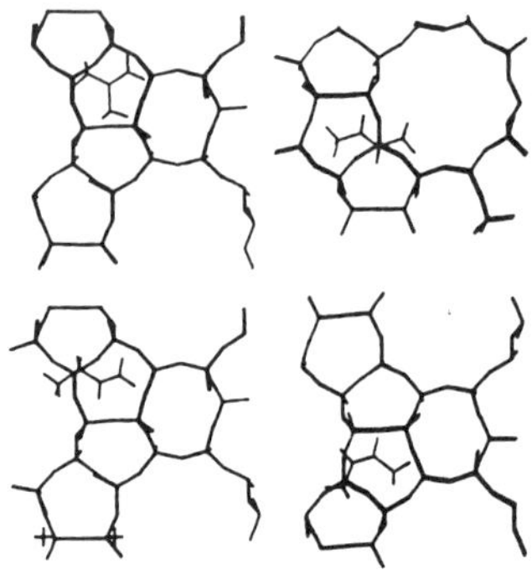

Figure 1: Typical minimum energy conformation for four isomers in silicalite.

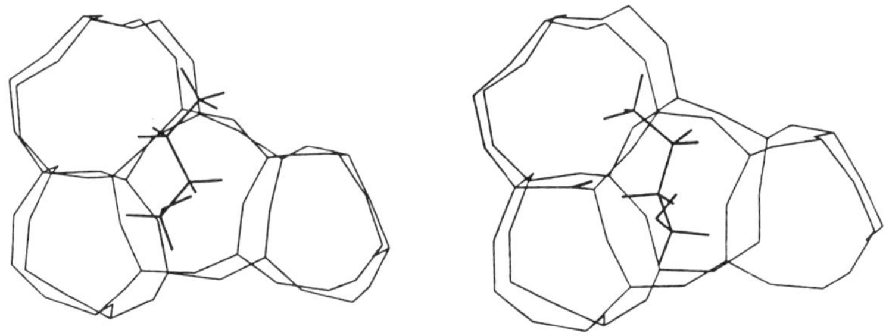

Figure 2: An example of final butanol-1 positions in relaxed framework structure.
a) butanol-1 in zig-zag channel of silicalite and b) butanol-1 and bridging hydroxyl group (shown in bold lines) at position (19,43,4) in H-ZSM-5 [10].

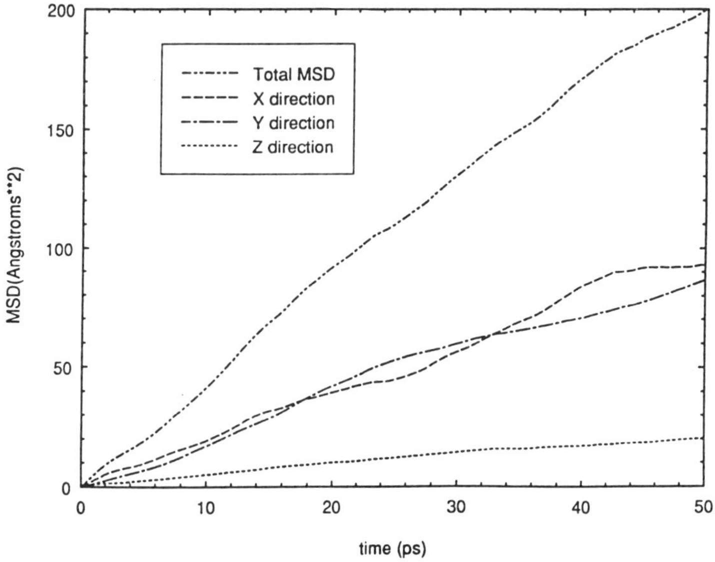

Figure 3: MSDs for ethane (1 molecule per unit cell) at 300K.

COMPUTER MODELLING OF MICROPOROUS SOLIDS

R.G. BELL [1], R.A. JACKSON [2*] and C.R.A. CATLOW[1]

(1) Davy Faraday Research Laboratory, Royal Institution, 21 Albemarle Street, London W1X 4BS, U.K.

(2) Department of Chemistry, University of Keele, Staffs ST5 5BG, U.K.

ABSTRACT

This paper is concerned with recent developments in the computer modelling of the structural properties of zeolites and related germanate and ALPO materials. A detailed study of the cation distribution in mordenite is presented, and the current status of structural modelling of germanates and ALPOs described, together with examples of recent calculations.

INTRODUCTION

Computer modelling, based on lattice energy minimisation, is now a routinely applied method in the study of the structural and defect properties of ionic and semi-ionic solids, including zeolites and related materials [1]. Recent applications to zeolites have included a study of the polymorphs of zeolite beta [2], the prediction of the monoclinic phase transition in silicalite [3] and a study of Löwenstein's rule in zeolite A [4].

This paper first describes the methods to be used, followed by applications to the cation distribution in mordenite, and structural modelling of germanates and ALPOs.

METHODOLOGY

The methods employed in this paper have been extensively described elsewhere (see e.g. [1]) and will be summarised here. Lattice energy minimisation involves taking a given starting structure, calculating its lattice energy using specified interatomic potentials, and adjusting the structure until the lattice energy is at a minimum either with respect to atomic positions (constant volume minimisation) or, additionally, with respect to unit cell parameters (constant pressure minimisation). It is also possible to carry out the minimisation in stages by selectively 'freezing' parts of the lattice.

Attention must be paid to the interatomic potentials used as these are crucial to the reliability of any computer simulation. The potentials used in this study were mainly obtained by fitting to the structure and lattice properties of alpha-quartz [1].

From the Proceedings of the 9th International Zeolite Conference, Montreal 1992, Eds. R. von Ballmoos et al.,

CATION POSITIONS AND ION EXCHANGE IN MORDENITE

The zeolite mordenite has a number of important catalytic applications in its Ni-exchanged form. It has been observed however that at low aluminium content, it is not possible to exchange divalent cations fully up to the maximum ratio of $M^{2+}/Al = 0.5$, whereas monovalent cations can always be fully exchanged up to $M^{+}/Al = 1$. A previous study [5] examined the question of location of Ni^{2+} ions in mordenite, and concluded that they are preferentially coordinated to Al-O-$(Si\text{-}O)_n$-Al sequences in the lattice, where n=1 or 2, and also that 8-ring or side-pocket sites are preferred over sites in the 12-ring main channel.

In this work, the effects of site preference and Ni^{2+} ion concentration are studied in a combined manner by calculating the energy of inclusion of 1 or 2 Ni^{2+} ions at various sites in the dehydrated lattice. For comparison with the results for monovalent cations, similar calculations are carried out for 2 and 4 Na^{+} ions. The aluminium atoms are included explicitly, but distributed randomly, subject to a Löwensteinian constraint. Three different distributions are used in each calculation, and the calculated energies averaged. It should be noted that for each calculation a full energy minimisation is carried out , and that substitution energies are calculated by taking the difference between the minimised energy and the lattice energy for purely siliceous mordenite.

Table 1 gives the relative energies of substitution of Ni^{2+} and Na^{+} ions in mordenite, using the site nomenclature due to Mortier [6]. Where there are gaps in the table, no energy minima could be found. The sites are illustrated in figure 1.

The results reported in table 1 may be interpreted in terms of both (i) the observed site preferences for the individual ions, and (ii) the energetics of Na^{+} - Ni^{2+} exchange:

(i) it is evident that, for Ni^{2+}, there is a significant preference for the more closely-coordinated sites: the sites A (8-ring) and B(side-pocket) are favoured by some 6 eV over the 12-ring sites F and F'. For Na^{+} the same energy trend is observed, although less marked. This trend is consistent with that found in [5]. It should be noted, however, that in this case, even if the aluminiums are located in the 12-ring, this does not affect the preferential occupation of the 8-ring site by Ni^{2+}. It is also found that the minimum energy positions for cations in the 12-ring channel are at the sides, rather than at the centre, and that the energy surface is very flat with various local minimum energy positions. This is to be expected since the 12-ring positions are not normally occupied in the dehydrated lattice. By contrast the minimum energy positions in the 8-ring channel and side-pockets are very close to the ideal crystallographic sites.

(ii) comparing the figures for Na^{+} with those of Ni^{2+} it can be seen that the energy difference on going from one Ni^{2+} to two Ni^{2+} ions is slightly less than that found between two Na^{+} and four Na^{+} ions. This result is consistent with the experimental observation that Ni^{2+} is more easily exchanged at higher framework aluminium content.

Further calculations are in progress, on higher and lower ion concentrations. It should also be noted that a full comparison between Ni^{2+} and Na^{+} exchange will require the calculation of solution energies. This will be the subject of a further publication.

MODELLING OF GERMANATES

Many zeolitic materials exist in which germanium is substituted for silicon at tetrahedral framework sites. These materials may be modelled using the techniques discussed in this paper provided a suitable interatomic potential is available to describe the Ge-O interaction. Such a potential has been obtained by empirical fitting to the alpha-quartz form of GeO_2. The potential obtained must be regarded as preliminary, having been fitted to structure and dielectric constants only. No experimentally determined elastic constants were available for inclusion in the fitting process.

This potential has been used in modelling the sodalite structured aluminogermanate $Na_8(AlGeO_4)_6Cl_2$, and a comparison of the simulated and experimental structure is given in table 2. It can be seen that, in general, good agreement is obtained between the simulated and experimental structures. It may be noted however that the Ge-O bond distance is calculated to be slightly smaller than the experimental value, leading to a slightly larger Ge-O-Al angle. Refinements to the potential are in progress, but very recent calculations [8] show that it also predicts stability for the GeO_2 forms of the faujasite, mordenite and silicalite structures, which is an additional confirmation of validity.

MODELLING OF ALPOs

As with the modelling of germanates, a necessary prerequisite for modelling ALPOs is an interatomic potential, this time to describe the P-O interaction. A potential is available in the literature [7], but this employs partial charges for all ions, and is inconsistent with other potentials used in this work, although it has been applied to a number of wide pore ALPOs, including VPI-5 [9]. At the time of writing this paper, a suitable consistent potential is under development [10], and full details will be reported in due course.

CONCLUSIONS

This paper has summarised some recent developments in the computer modelling of zeolites and related materials. Emphasis has been placed on the importance of reliable interatomic potentials, and comparison made with experiment where possible.

ACKNOWLEDGEMENTS

The authors are grateful to Shell Research BV, and to Biosym Technologies Inc. for financial support, and to the SERC for provision of computer time.

REFERENCES

1. Jackson, R.A. and Catlow, C.R.A. *Mol. Sim.* , **1**, 207 (1988).
2. Tomlinson, S.M., Jackson, R.A. and Catlow, C.R.A., *J. Chem. Soc. Chem. Commun.*, 813 (1990).
3. Bell, R.G., Jackson, R.A. and Catlow, C.R.A., *J. Chem. Soc. Chem. Commun.*, 782 (1990).
4. Bell, R.G., Jackson, R.A. and Catlow, C.R.A., *Zeolites*, accepted for publication.
5. den Ouden, C.J.J., Jackson, R.A., Catlow, C.R.A. and Post, M.F.M.,*J. Phys. Chem.*, **94**, 5286 (1990).
6. Mortier, W.J., *Compilation of Extra Framework Sites in Zeolites*, Butterworth, London (1982).
7. van Beest, B.W.H., Kramer, G.J. and van Santen, R. A.,*Phys. Rev. Lett.*, **16**, 1955 (1990).
8. George, A.R., Catlow, C.R.A. and Thomas, J.M., to be published.
9. de Man, A.J.M., PhD thesis, Technical University of Eindhoven (1991).
10. J. D. Gale and N. J. Henson, in preparation.

Table 1. Relative energies of substitution of Ni^{2+} and Na^+ ions in mordenite (energies in eV/48 T-atom unit cell)

site	Ni^{2+}	Na^+
A	59.09	66.32
B	59.47	66.37
D	61.99	67.90
F	65.94	-
F'	65.66	68.90
	$2Ni^{2+}$	$4Na^+$
A	116.62	134.20
B	117.19	-
D	118.59	-
F	-	-
F'	119.51	140.74

Table 2. Comparison of experimental and simulated structure for $Na_8(AlGeO_4)_6Cl_2$ (lattice parameters and bond distances in Ångströms)

	simulated	experimental
a	9.030	9.044
Ge-O	1.69	1.73
Al-O	1.73	1.74
Na-Cl	2.82	2.71
Na-Na	4.85	4.93
Cl-Cl	7.82	7.83
Na-O	2.37	2.40
Cl-O	4.37	4.45
Ge-O-Al	137.40°	133.40°

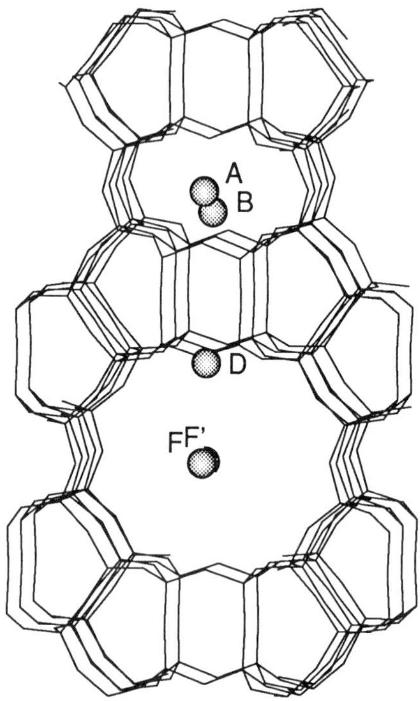

Figure 1. Diagram of the structure of mordenite showing the extra-framework cation positions

DISCUSSION - Modelling

"Proton Transfer From Acidic Sites to Water, Methanol and Ammonia. A Comparative ab initio Study" (J. Sauer, et al.)

S. Lee: What difference would you expect as the calculations were performed with an without considering the periodicity of the zeolite lattice?

J. Sauer: For high-silica zeolite catalysts, I expect the difference to be rather small. The extra-stabilization found in the periodic calculations by Teunissen et al. of about 50 kJ/mol is certainly an upper limit. The reasons are both in the zeolite considered (high Al content) and in the details of the calculation (small basis set, rigid structures).

J. Lercher: There has been a recent paper by the group of Prof. van Santen on adsorption of ammonia on acidic clusters. The paper describes a threefold coordinated ammonium ion to be the most stable structure. In your presentation you stressed that ammonia occupies a twofold adsorption structure. Can you outline the differences in the two approaches?

J. Sauer: The study you are referring to (ref. 9) adopts the $Al(OH)_4^-$ model for the zeolite surface site. For this model we also find a threefold coordination. Our calculations for larger models (Table 4), in particular for the $Al(OSiH_3)_4^-$ model however show that this threefold coordination is an artifact of the small model. The failure of the $Al(OH)_4^-$ model is easy to understand. Oxygen is more basic in Al-O-H groups than it is in Al-O-Si groups and therefore enforces an unrealistically strong binding (Table 4).

"Siting of Al and Bridging Hydroxyl Groups in Zeolite Catalysts. Computer Simulations of Their Structure, Vibrational Spectra and Acidity" (K. P. Schroder and J. Sauer)

D. Barthomeuf: You said that in faujasite Al are more "comfortable" when they are next nearest neighbors. It is generally accepted from experimental results that in faujasites the Al atoms which are easily extracted are the ones which belong to a pair of next nearest neighbors Al. How does this correlate with your results?

J. Sauer: Our simulation is done for a very low aluminum content, Al/(Al+Si) = 2/48. The observations you describe have been made for much higher aluminum contents. Therefore there is no conflict between both results.

R. Schoonheydt: Preferential siting of protons is on oxygens of large channels and supercages, giving OH groups pointing into the large cages. Is this in agreement with your calculations?

J. Sauer: The high frequency band has been shown to be primarily affected by pyridine adsorption. We assign this band to OH [or O1H, ed.] hydroxyl groups which point into the large cages. Hence, there is agreement between these findings and our calculations.

B. Nagy: Do you think that Al-O-Si-O-Al pairs are also found in ZSM-5 and MOR, as you have shown for faujasite? These results would be very important to understand acid properties as well as catalytic activities.

J. Sauer: We did not perform such simulations for H-ZSM-5 and H- Mordenite so far.

Author Index

I and II refer to volumes I and II

Subject Index

I and II refer to volumes I and II